## Conversion Factors



**Mass**

$1\ g = 10^{-3}\ kg$
$1\ kg = 10^{3}\ g$
$1\ u = 1.66 \times 10^{-24}\ g = 1.66 \times 10^{-27}\ kg$
1 metric ton = 1000 kg

**Length**

$1\ nm = 10^{-9}\ m$
$1\ cm = 10^{-2}\ m = 0.394\ in.$
$1\ m = 10^{-3}\ km = 3.28\ ft = 39.4\ in.$
$1\ km = 10^{3}\ m = 0.621\ mi$
$1\ in. = 2.54\ cm = 2.54 \times 10^{-2}\ m$
1 ft = 0.305 m = 30.5 cm
1 mi = 5280 ft = 1609 m = 1.609 km

**Area**

$1\ cm^2 = 10^{-4}\ m^2 = 0.1550\ in^2 = 1.08 \times 10^{-3}\ ft^2$
$1\ m^2 = 10^{4}\ cm^2 = 10.76\ ft^2 = 1550\ in^2$
$1\ in^2 = 6.94 \times 10^{-3}\ ft^2 = 6.45\ cm^2 = 6.45 \times 10^{-4}\ m^2$
$1\ ft^2 = 144\ in^2 = 9.29 \times 10^{-2}\ m^2 = 929\ cm^2$

**Volume**

$1\ cm^3 = 10^{-6}\ m^3 = 3.53 \times 10^{-5}\ ft^3 = 6.10 \times 10^{-2}\ in^3$
$1\ m^3 = 10^{6}\ cm^3 = 10^{3}\ L = 35.3\ ft^3 = 6.10 \times 10^{4}\ in^3 = 264\ gal$
$1\ liter = 10^{3}\ cm^3 = 10^{-3}\ m^3 = 1.056\ qt = 0.264\ gal = 0.0353\ ft^3$
$1\ in^3 = 5.79 \times 10^{-4}\ ft^3 = 16.4\ cm^3 = 1.64 \times 10^{-5}\ m^3$
$1\ ft^3 = 1728\ in^3 = 7.48\ gal = 0.0283\ m^3 = 28.3\ L$
$1\ qt = 2\ pt = 946\ cm^3 = 0.946\ L$
$1\ gal = 4\ qt = 231\ in^3 = 0.134\ ft^3 = 3.785\ L$

**Time**

1 h = 60 min = 3600 s
$1\ day = 24\ h = 1440\ min = 8.64 \times 10^{4}\ s$
$1\ y = 365\ days = 8.76 \times 10^{3}\ h = 5.26 \times 10^{5}\ min = 3.16 \times 10^{7}\ s$

**Angle**

$1\ rad = 57.3°$

| | |
|---|---|
| $1° = 0.0175\ rad$ | $60° = \pi/3\ rad$ |
| $15° = \pi/12\ rad$ | $90° = \pi/2\ rad$ |
| $30° = \pi/6\ rad$ | $180° = \pi\ rad$ |
| $45° = \pi/4\ rad$ | $360° = 2\pi\ rad$ |

$1\ rev/min = (\pi/30)\ rad/s = 0.1047\ rad/s$

**Speed**

1 m/s = 3.60 km/h = 3.28 ft/s = 2.24 mi/h
1 km/h = 0.278 m/s = 0.621 mi/h = 0.911 ft/s
1 ft/s = 0.682 mi/h = 0.305 m/s = 1.10 km/h
1 mi/h = 1.467 ft/s = 1.609 km/h = 0.447 m/s
60 mi/h = 88 ft/s

**Force**

1 N = 0.225 lb
1 lb = 4.45 N
Equivalent weight of a mass of 1 kg on Earth's surface = 2.2 lb = 9.8 N

**Pressure**

$1\ Pa\ (N/m^2) = 1.45 \times 10^{-4}\ lb/in^2 = 7.5 \times 10^{-3}\ torr\ (mm\ Hg)$
$1\ torr\ (mm\ Hg) = 133\ Pa\ (N/m^2) = 0.02\ lb/in^2$
$1\ atm = 14.7\ lb/in^2 = 1.013 \times 10^{5}\ N/m^2 = 30\ in.\ Hg = 76\ cm\ Hg$
$1\ lb/in^2 = 6.90 \times 10^{3}\ Pa\ (N/m^2)$
$1\ bar = 10^{5}\ Pa$
$1\ millibar = 10^{2}\ Pa$

**Energy**

$1\ J = 0.738\ ft \cdot lb = 0.239\ cal = 9.48 \times 10^{-4}\ Btu = 6.24 \times 10^{18}\ eV$
1 kcal = 4186 J = 3.968 Btu
$1\ Btu = 1055\ J = 778\ ft \cdot lb = 0.252\ kcal$
$1\ cal = 4.186\ J = 3.97 \times 10^{-3}\ Btu = 3.09\ ft \cdot lb$
$1\ ft \cdot lb = 1.36\ J = 1.29 \times 10^{-3}\ Btu$
$1\ eV = 1.60 \times 10^{-19}\ J$
$1\ kWh = 3.6 \times 10^{6}\ J$

**Power**

$1\ W = 0.738\ ft \cdot lb/s = 1.34 \times 10^{-3}\ hp = 3.41\ Btu/h$
$1\ ft \cdot lb/s = 1.36\ W = 1.82 \times 10^{-3}\ hp$
$1\ hp = 550\ ft \cdot lb/s = 745.7\ W = 2545\ Btu/h$

**Mass–Energy Equivalents**

$1\ u = 1.66 \times 10^{-27}\ kg \leftrightarrow 931.5\ MeV$
$1\ electron\ mass = 9.11 \times 10^{-31}\ kg = 5.49 \times 10^{-4}\ u \leftrightarrow 0.511\ MeV$
$1\ proton\ mass = 1.67262 \times 10^{-27}\ kg = 1.007276\ u \leftrightarrow 938.27\ MeV$
$1\ neutron\ mass = 1.67493 \times 10^{-27}\ kg = 1.008665\ u \leftrightarrow 939.57\ MeV$

**Temperature**

$T_F = \frac{9}{5} T_C + 32$
$T_C = \frac{5}{9} (T_F - 32)$
$T = T_C + 273$

**cgs Force**

$1\ dyne = 10^{-5}\ N = 2.25 \times 10^{-6}\ lb$

**cgs Energy**

$1\ erg = 10^{-7}\ J = 7.38 \times 10^{-6}\ ft \cdot lb$

# College Physics Essentials

## Electricity and Magnetism
## Optics
## Modern Physics

### Eighth Edition
### (Volume Two)

# College Physics Essentials

## Electricity and Magnetism
## Optics
## Modern Physics

### Eighth Edition
### (Volume Two)

Jerry D. Wilson
Anthony J. Buffa
Bo Lou

CRC Press is an imprint of the
Taylor & Francis Group, an informa business

CRC Press
Taylor & Francis Group
6000 Broken Sound Parkway NW, Suite 300
Boca Raton, FL 33487-2742

First issued in paperback 2022



ISBN-13: 978-1-138-47608-0 (hbk)
ISBN-13: 978-1-03-233727-2 (pbk)
DOI: 10.1201/9780429323379



Publisher's Note

The publisher has gone to great lengths to ensure the quality of this reprint but points out that some imperfections in the original copies may be apparent.

**Visit the Taylor & Francis Web site at**
**http://www.taylorandfrancis.com**

**and the CRC Press Web site at**
**http://www.crcpress.com**

# Contents

# Authors

**Jerry D. Wilson** is a professor emeritus of physics and former chair of the Division of Biological and Physical Sciences at Lander University in Greenwood, South Carolina. He received his BS degree from Ohio University, MS degree from Union College, and, in 1970, a PhD from Ohio University. He earned his MS degree while employed as a materials behavior physicist by the General Electric Corporation. As a doctoral graduate student, Professor Wilson held the faculty rank of instructor and began teaching physical science courses. During this time, he coauthored a physical science text. In conjunction with his teaching career, Professor Wilson has continued his writing and has authored or coauthored six titles.

**Anthony J. Buffa** is an emeritus professor of physics at California Polytechnic State University, San Luis Obispo, where he was also a research associate at the Radioanalytical Facility. During his career, he taught courses ranging from introductory physical science to quantum mechanics, while he developed and revised many laboratory experiments. He received his BS in physics from Rensselaer Polytechnic Institute and both his MS and PhD in physics from the University of Illinois, Urbana-Champaign. In retirement, Professor Buffa continues to be involved in teaching and writing. Combining physics with his interests in art and architecture, Dr. Buffa develops his own artwork and sketches, which he uses to increase his effectiveness in teaching. In his spare time, he enjoys his children and grandsons, gardening, walking the dog, and traveling with his wife.

**Bo Lou** is professor in the Department of Physical Sciences at Ferris State University, and earned his PhD in physics from Emory University. He teaches a variety of undergraduate physics courses and is the coauthor of several physics textbooks. He emphasizes the importance of conceptual understanding of the basic laws and principles of physics and their practical applications to the real world.

# 15

# Electric Charge, Forces, and Fields

**Spectacular lightning strikes like this are caused by the electric force – the focus of this chapter.**

Few natural processes deliver such an enormous amount of energy in a fraction of a second as a lightning bolt (see chapter opening photo). Yet most people have never experienced its power at close range; luckily, only a few hundred people per year are struck by lightning in the United States.

It might surprise you to realize that you have almost certainly had a similar experience, at least in a physics context. Have you ever walked across a carpeted room and received a shock when reaching for a metallic doorknob? Although the scale is dramatically different, the physical process involved (static electricity discharge) is analogous to being struck by lightning – mini-lightning, in this case.

Clearly electricity can be dangerous, but it can, of course, be "domesticated." In the home or office, its usefulness is taken for granted. Indeed, our dependence on electric energy becomes evident only when the power goes off unexpectedly, providing a dramatic reminder of the role it plays in our lives. Yet in the early part of the twentieth century, there were none of the electrical applications that are all around us today.

Physicists now know that the electricity is related to magnetism, which will be studied later (Chapter 20). In this chapter, our study begins with the electric charge.

## 15.1 Electric Charge

Fundamentally, electricity involves the study of the interaction between *electrically charged* objects. To demonstrate this, our study begins with the simplest situation, *electrostatics*, when the electrical charges on objects are static or at rest.

Like mass, **electric charge** is a fundamental property of matter. Electric charge is associated with particles that make up the atom: the electron and the proton. The simplistic solar system model of the atom, as illustrated in ▼**Figure 15.1**, likens its structure to that of planets orbiting a star. The electrons are viewed as orbiting a nucleus – a core containing most of the atom's mass in the form of protons and electrically neutral particles called neutrons. Instead of the force of gravity between the planets and the star, the force that keeps electrons in orbit is the electrical force. There are important distinctions between gravitational and electrical forces.

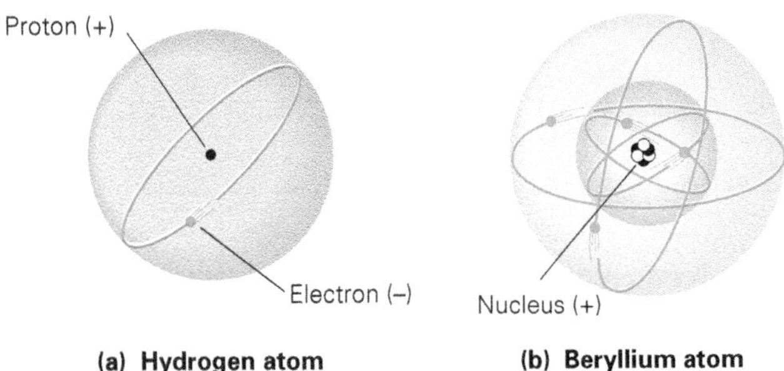


▲FIGURE 15.1 **Simplistic model of atoms** The so-called solar system model of **(a)** a hydrogen atom and **(b)** a beryllium atom views the electrons (negatively charged) as orbiting the nucleus (positively charged), analogously to the planets orbiting a star. The electronic structure of atoms is actually much more complicated than this.

One difference is that there is only one type of mass, and gravitational forces are always attractive. Electric charge, however, comes in two types, distinguished by the labels positive (+) and negative (−). Protons are designated as having a positive charge, and electrons as having a negative charge. Different combinations of the two types of charge can produce *either* attractive *or* repulsive net electrical forces.

The directions of the electric forces on isolated charged particles are given by the following principle, called the **charge-force law**:

Like charges repel, unlike charges attract.

That is, two negatively charged particles or two positively charged particles repel each other, whereas particles with opposite charges attract each other (▶**Figure 15.2**). The repulsive and attractive forces are equal and opposite, and act on different objects, in keeping with Newton's third law (Section 4.3).

Note that the electron and proton have exactly equal but opposite charge. The magnitude of the electron's charge is abbreviated as *e* and is the fundamental unit of charge, because it is the smallest charge observed in nature.*

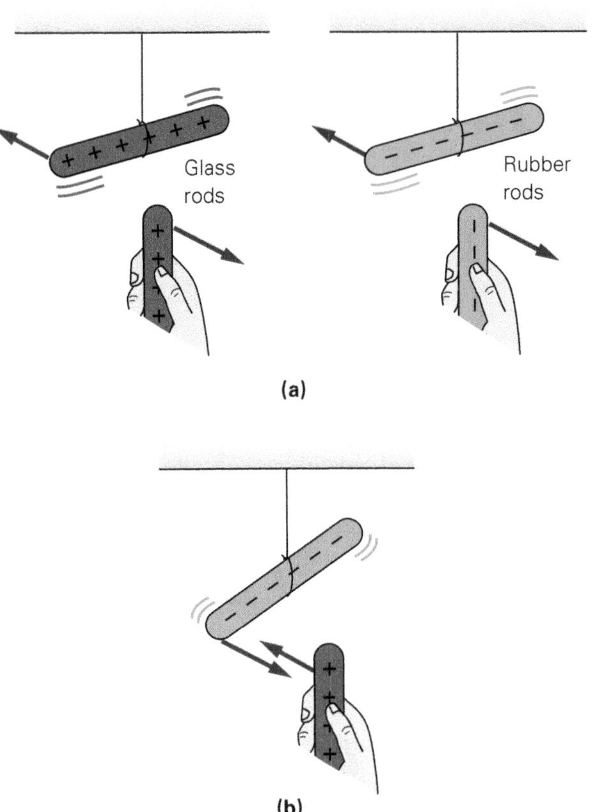


▲FIGURE 15.2 **The charge-force law, or law of charges (a)** Like charges repel. **(b)** Unlike charges attract.

The SI unit of charge is the **coulomb (C)**, named for the French physicist/engineer Charles A. de Coulomb (1736–1806), who discovered the relationship between electric force and charge (Section 15.3). The charges and masses of the electron, proton, and neutron are given in ▼**Table 15.1**. From the charge values, $e = 1.602 \times 10^{-19}$ C. Our general symbol for charge will be $q$ or $Q$. Thus the charge on the electron can be written as $q_e = -e = -1.602 \times 10^{-19}$ C, and that on the proton as $q_p = +e = +1.602 \times 10^{-19}$ C.

**TABLE 15.1** Properties of Subatomic Particles

| Particle | Electric Charge | Mass |
|---|---|---|
| Electron | $-1.602 \times 10^{-19}$ C | $9.109 \times 10^{-31}$ kg |
| Proton | $+1.602 \times 10^{-19}$ C | $1.673 \times 10^{-27}$ kg |
| Neutron | 0 | $1.675 \times 10^{-27}$ kg |

Other terms are frequently used when discussing charged objects. Saying that an object has a **net charge** means that it has an excess charge due to either an excess of positive or of negative charges. (It is common, however, to talk about the "charge" of an object when we really mean net charge.) As will be seen in Section 15.2, excess charge on everyday materials is most commonly produced by a transfer of electrons, not protons. (Protons are bound in the nucleus and in most situations, stay fixed.)

* Protons, as well as neutrons and other subatomic particles, are now known to consist of more fundamental particles called *quarks*, which carry charges of $\pm\frac{1}{3}$ and $\pm\frac{2}{3}$ of *e*.

As an example, suppose an object has a (net) charge of $+1.6 \times 10^{-18}$ C. This means that it has had ten electrons removed from it because $10 \times 1.6 \times 10^{-19}\ \text{C} = 1.6 \times 10^{-18}$ C. That is, the total negative charge of the electrons in the object no longer completely cancels the total positive charge of all its protons – resulting in a net positive charge. On an atomic level, this situation means that some of the atoms that compose the object are now deficient in electrons. Such positively charged atoms are termed *positive ions*. Atoms with an excess of electrons are *negative ions*.

An object having a net charge on the order of one coulomb is rarely seen in everyday situations. Therefore, it is common to express amounts of charge using *microcoulombs* ($\mu$C, $10^{-6}$ C), *nanocoulombs* (nC, $10^{-9}$ C), and *picocoulombs* (pC, $10^{-12}$ C).

In our study the electric charge on an object will usually be caused by either a deficiency or an excess of electrons. In this case, the net charge must be an integer multiple of the charge on one electron. In other words, the net charge on any object is "quantized"; that is, it can occur only in integral multiples of the charge on the electron (with the appropriate sign). Thus, the (net) charge of an object, can be written as

$$q = \pm ne \quad \text{SI unit of charge: coulomb (C)} \quad \text{where } n = 1, 2, 3, \ldots \tag{15.1}$$

In dealing with any electrical phenomena, another important principle is **conservation of charge**:

The net charge of an isolated system remains constant.

For example, suppose that a system consists initially of two electrically neutral objects, and some electrons are transferred from one object to the other. The object with the added electrons will then have a net negative charge, and the object with the reduced number of electrons will have a net positive charge of equal magnitude. But the net charge of the system, defined as both objects, remains constant (here zero). It is important to realize that this principle applies to an isolated system even if it has a non-zero net charge; that is, the net charge stays the same, but it need not be zero. (See Example 15.1.)

Note that charge conservation does not prohibit the creation or destruction of charged particles. In fact, physicists have known for a long time that charged particles can be created and destroyed on the atomic and nuclear levels. Considering the Universe as a whole, conservation of charge means that the net charge of the Universe is constant. Thus these charged particles must be created or destroyed only in pairs of equal and opposite charges.

**INTEGRATED EXAMPLE 15.1: ON THE CARPET – CONSERVATION OF QUANTIZED CHARGE**

When you shuffle across a carpeted floor on a dry day, the carpet acquires a net positive charge. (a) Will you have (1) a deficiency or (2) an excess of electrons? (b) If the charge the carpet acquired has a magnitude of 2.15 nC, how many electrons were transferred?

**(a) CONCEPTUAL REASONING.** (a) Since the carpet has a net positive charge, it must have lost electrons and you must have gained them. Thus, your charge is negative, indicating an excess of electrons, and the correct answer is (2).

**(b) QUANTITATIVE REASONING AND SOLUTION.** Because the charge of one electron is known, the excess of electrons can be quantified. Express the charge in coulombs, and state what is to be found.

***Given***:

$$q_{\text{carpet}} = +(2.15\ \text{nC})\left(\frac{10^{-9}\ \text{C}}{1\ \text{nC}}\right) = +2.15 \times 10^{-9}\ \text{C}$$

$$q_e = -1.60 \times 10^{-19}\ \text{C (from Table 15.1)}$$

***Find:*** $n$ (number of electrons transferred to you)

Considering [you + carpet] as the system, charge conservation requires that net charge on you is

$$q_{\text{you}} = -q_{\text{carpet}} = -2.15 \times 10^{-9}\ \text{C}$$

Thus

$$n = \frac{q_{\text{you}}}{q_e} = \frac{-2.15 \times 10^{-9}\ \text{C}}{-1.60 \times 10^{-19}\ \text{C/electron}} = 1.34 \times 10^{10}\ \text{electrons}$$

As can be seen, net charges, even in everyday situations, can involve huge numbers of electrons (here, more than 13 billion), because the charge of any one electron is very small.

**FOLLOW-UP EXERCISE.** In this Example, if your mass is 80 kg, by what percentage has your mass increased due to the excess electrons? *(Answers to all Follow-Up Exercises are given in Appendix V at the back of the book.)*

## 15.2 Electrostatic Charging

The existence of two types of electric charge along with the attractive and repulsive electrical forces is easily demonstrated. Before learning how this is done, let's distinguish between electrical conductors and insulators. What distinguishes these two groups of substances is their ability to conduct, or transmit, electric charge. Some materials, particularly metals, are good **conductors** of electric charge. Others, such as glass, rubber, and most plastics, are **insulators**, or poor electrical conductors.

In conductors, the outermost electrons (called *valence electrons*) of the atoms are loosely bound. As a result, they can be relatively easily removed from the atom and moved about in the bulk of the conductor material, or even leave it altogether. That is, valence electrons are not permanently bound to a particular atom. In insulators, however, even the loosest bound electrons are not easily removed from their atoms. Thus, charge is not available to move through or be removed from an insulator.

There is a "middle" class of materials called **semiconductors**. Their ability to conduct charge is intermediate between insulators and conductors. Their electrical property description is beyond the scope of this book. However, it is interesting to note that beginning in the 1940s, scientists began to create applications with these materials. Semiconductors first were used to create transistors, then solid-state circuits, and eventually, modern computer microchips – which are responsible for today's high-speed computer technology.

Now knowing the electrical difference between conductors and insulators, let's investigate a way of determining the sign of the charge on an object. The *electroscope* is one of the simplest devices used to determine electric charge (▼ **Figure 15.3**). In its most basic form, it consists of a metal rod with a metallic bulb at one end. The rod is attached to a solid, piece of metal that has an attached foil "leaf," usually made of gold.

When a charged object is brought close to the bulb, electrons in the bulb/rod/leaf assembly are either attracted to or repelled by the charges on that object. Thus if a negatively charged rod is brought near the bulb, electrons in the bulb are repelled, and the bulb is left with a positive charge. The electrons are repelled (conducted) down to the foil leaf, which then will swing away, because of the like (negative) charges on the metal and leaf (Figure 15.3b). Similarly, if a positively charged rod is brought near the bulb, the leaf also will swing out. (Can you explain why?)

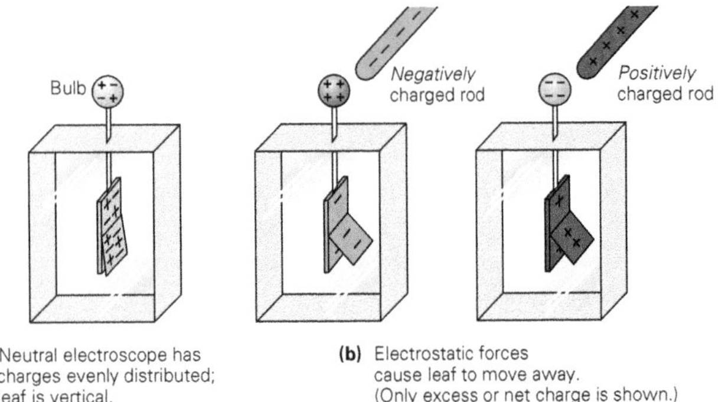
Bulb
Negatively charged rod
Positively charged rod
(a) Neutral electroscope has charges evenly distributed; leaf is vertical.
(b) Electrostatic forces cause leaf to move away. (Only excess or net charge is shown.)

▲ **FIGURE 15.3 The electroscope** An electroscope can be used to determine whether an object is electrically charged. When a charged object is brought near the bulb, the leaf moves away from the metal piece. See text for description.

Notice that the net charge on the electroscope remains zero in these instances. Because the device is isolated, only the *distribution* of charge is altered. However, it is possible to give an electroscope (and other objects) a net (non-zero) charge by different methods, all of which are said to involve **electrostatic charging**. Consider how this can be done using the following types of charging processes.

### 15.2.1 Charging by Friction

In the frictional charging process, certain insulator materials are rubbed, typically with cloth or fur, and they become electrically charged by a transfer of charge. For example, if a hard rubber rod is rubbed with fur, the rubber will acquire a net negative charge; rubbing a glass rod with silk will give the glass a net positive charge. This process is called **charging by friction**. The transfer of charge is due to the frictional contact between the materials, and the amount and sign of charge transferred depends, as you might expect, on the nature of those materials.

Example 15.1 was an example of frictional charging in which you picked up a net negative charge. Then, when reaching for a metal object such as a doorknob, you might have been "zapped" by a spark. As your hand approaches the knob, because you have a net negative charge, the close end of the knob becomes positively charged, thus attracting electrons from your hand. As the electrons travel, they collide with, and excite, the atoms of the air, which give off light as they de-excite (lose energy). This light is seen as the spark of "mini-lightning" between your hand and the knob.

### 15.2.2 Charging by Conduction (Contact)

Bringing a charged rod close to an electroscope will reveal that the rod is charged, but as we have seen it does *not* tell you the sign of the charge on the rod because the leaf will swing out regardless of the sign of charge on the rod. The sign can be determined, however, if the electroscope is first given a known type of (net) charge. For example, electrons can be transferred to the electroscope from a negatively charged object, as illustrated in ▶ **Figure 15.4a**. The electrons in the rod repel one another, and thus some will transfer onto the electroscope. Now the leaf will remain permanently in the "swung out" position. This electroscope has been **charged by contact** or by **conduction** (Figure 15.4b). "Conduction" refers to the flow of charge during the short period of time the electrons are transferred. This negatively charged electroscope can now be used to determine the unknown sign of charged objects.

For example, if a negatively charged rod is brought close to this negatively charged electroscope, the leaf will diverge even further as more electrons are repelled down from the bulb (Figure 15.4c). However, a positively charged rod will cause the leaf to collapse by attracting electrons up to the bulb and away from the leaf area (Figure 15.4d).

### 15.2.3 Charging by Induction

It is possible to create an electroscope that is positively charged using a negatively charged rubber rod (already charged by friction). This can be accomplished by a process called **induction**. Starting with an uncharged electroscope, you first touch the bulb with a finger, which *grounds* the electroscope – that is, provides a path by which electrons can escape from the bulb to the ground (▶ **Figure 15.5**). Then, when a negatively charged rod is brought

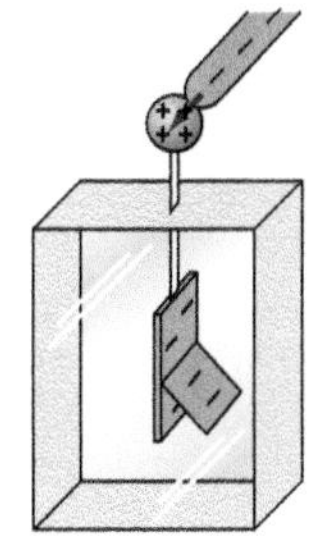

(a) Neutral electroscope is touched with negatively charged rod.

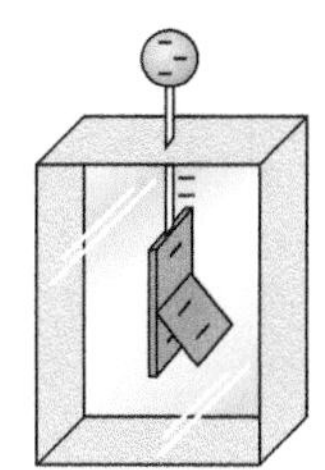

(b) Charges are transferred to bulb; electroscope has net negative charge; leaf moves out.

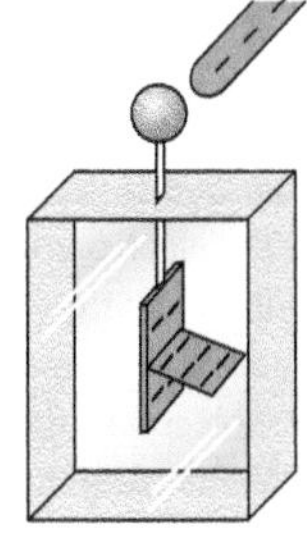

(c) Negatively charged rod repels electrons; leaf moves further out.

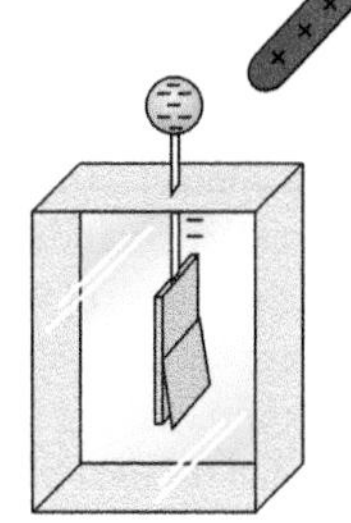

(d) Positively charged rod attracts electrons; leaf collapses.

▲ FIGURE 15.4 **Charging by conduction** (a) The electroscope is initially neutral (but the charges are separated), as a charged rod touches the bulb. (b) Charge is transferred to the electroscope. (c) When a rod of the same charge is brought near the bulb, the leaf moves farther apart. (d) When an oppositely charged rod is brought nearby, the leaf collapses.

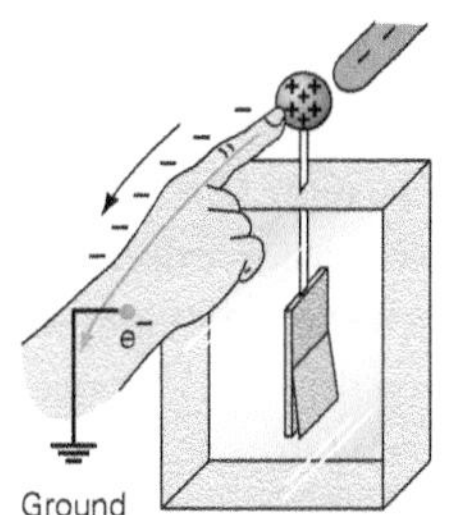


(a) Repelled by the nearby negatively charged rod, electrons are transferred to ground through hand.

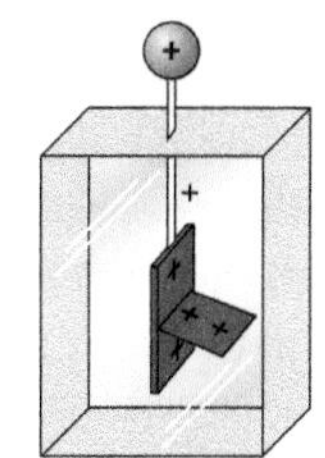

(b) After removing the finger first, then later the rod, the electroscope is left with a net positive charge.

▲ FIGURE 15.5 **Charging by induction** (a) Touching the bulb with a finger provides a path to ground for charge transfer. The symbol $e^-$ stands for "electron." (b) When the finger is removed, the electroscope is left with a net positive charge, opposite that of the rod.

close to (but not touching) the bulb, the rod repels electrons from the bulb through your finger and body and down into the Earth (we say that the electroscope has been *grounded*). Removing your finger while the charged rod is kept nearby leaves the electroscope with a net positive charge. This is because when the rod is removed, the electrons that moved to the Earth have no return path.

### 15.2.4 Charge Separation by Polarization

Charging by contact and charging by induction create a net charge on an object by moving charge to or from that object. It is possible, however, that charge can be moved within the object while keeping its net charge zero. For example, the induction process described previously initially causes **polarization**, or separation of positive and negative charge. If the object is not grounded, it will remain electrically neutral, but have regions of equal and opposite charge. In this situation, it is said that the object has become *polarized*. On the molecular level, molecules polarized like this (called **electric dipoles** – two poles, one of each sign) can be permanent; that is, they don't need a nearby charged object to retain charge separation. A good example of this is the water molecule.

Examples of both permanent and nonpermanent electric dipoles and forces that can act on them are shown in ▶ **Figure 15.6**. Now you can understand why, when you rub a balloon on your sweater, it can stick to the wall. The balloon is charged by friction, and bringing it near the wall polarizes the wall. The opposite sign charge on the wall's nearest surface creates an attractive force on the balloon.

The electrostatic force can be annoying, as when it causes clothes to stick together, or even dangerous, as when electrostatic spark discharges start a fire or cause an explosion in the presence of a flammable gas. To discharge electric charge, many large trucks have dangling metal chains in contact with the ground. However, electrostatic forces can also be beneficial. For example, the air we breathe is cleaner because of electrostatic precipitators used in smokestacks. In these devices, electrical discharges cause the particles (by-products of fuel combustion) to acquire a net charge. The charged particles can then be removed from the flue gases by attracting them to electrically charged surfaces. On a smaller scale, electrostatic air cleaners are available for use in the home.

## 15.3 Electric Force

The directions of electric forces on interacting charges are given by the charge-force law. However, what about their *magnitudes*? This was investigated by Coulomb, who found that the magnitude of the electric force between two "point" (very small) charges $q_1$ and $q_2$ depended directly on the product of the magnitude of the charges and inversely on the square of the distance $r$ between them. That is, $F_e \propto |q_1||q_2|/r^2$. (*Note:* $|q|$ denotes the magnitude of $q$.) This relationship is mathematically similar to that for the force of gravity between two point masses ($F_g \propto m_1m_2/r^2$), see Section 7.5.

Coulomb's measurements provided an experimental value for a constant of proportionality, $k$, so the electric force could be written as an equation known as **Coulomb's law**:

$$F_e = \frac{kq_1q_2}{r^2} \quad \text{(point charges only, } q \text{ means magnitude)} \tag{15.2}$$

where $k = 8.988 \times 10^9\ \text{N·m}^2/\text{C}^2 \approx 9.00 \times 10^9\ \text{N·m}^2/\text{C}^2$.

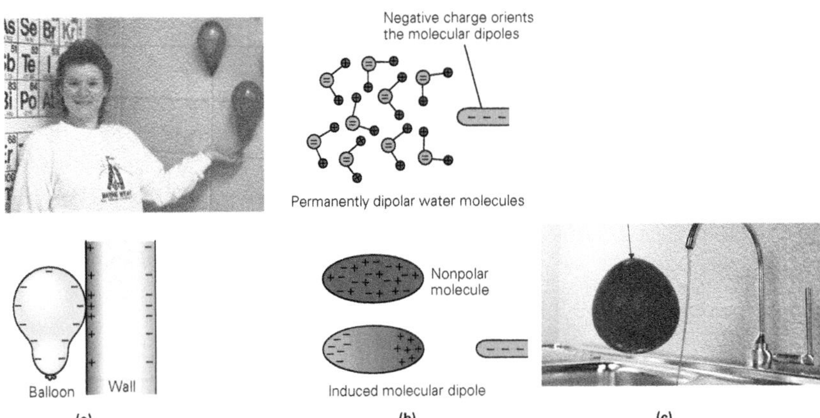


▲ **FIGURE 15.6 Polarization (a)** When the balloons are charged by friction and placed in contact with the wall, the wall is polarized. That is, an opposite charge is induced on the wall's surface, to which the balloons then stick by the force of electrostatic attraction. The electrons on the balloon do not leave the balloon because its material (rubber) is a poor conductor. **(b)** Some molecules, such as those of water, are polar by nature; that is, they have permanently separated regions of positive and negative charge. But even some molecules that are not normally dipolar can be polarized temporarily by the presence of a nearby charged object. The electric force induces a separation of charge and, consequently, temporary molecular dipoles. **(c)** A stream of water bends toward a charged balloon. The negatively charged balloon attracts the positive ends of the water molecules, causing the stream to bend.

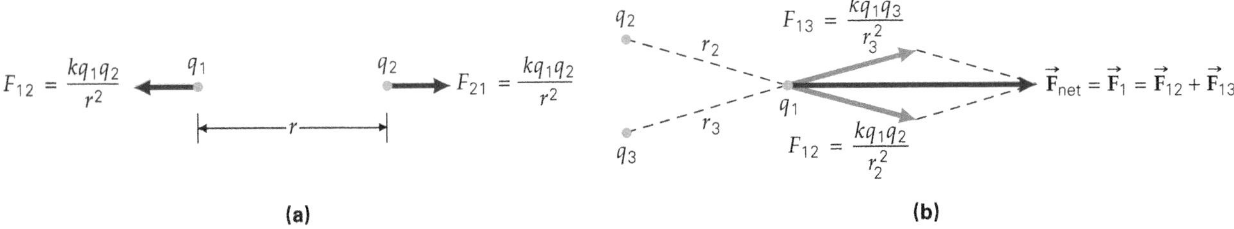


▲ **FIGURE 15.7 Coulomb's law (a)** The mutual electrostatic forces between two point charges are equal and opposite. **(b)** For a configuration of two- or more point charges, the force on a particular charge is the vector sum of the forces on it due to all the other charges. (*Note*: In each of these situations, all of the charges are of the same sign. How can we tell that this is true? Can you tell their *sign?* What is the direction of the force on $q_2$ due to $q_3$?)

Equation 15.2 gives the force between any two charged particles as shown in ▲ **Figure 15.7a**. In general, for several charges, the mathematics involves vector addition, as shown in Figure 15.7b. The following Example provides some insight into the qualitative use of Coulomb's law.

### CONCEPTUAL EXAMPLE 15.2: FREE OF CHARGE - ELECTRIC FORCES

A rubber comb combed through dry hair can acquire a net negative charge. That comb will then attract small pieces of *uncharged* paper. This would seem to violate Coulomb's force law. Since the paper has no net charge, you might expect there to be no electric force on it. Which charging mechanism, (a) conduction, (b) friction, or (c) polarization, explains this phenomenon, and how does it explain it?

**REASONING AND ANSWER.** Because the comb doesn't touch the paper, the paper cannot be charged by either conduction or friction, because both of these require contact. Thus, the answer must be (c). When the charged comb is near the paper, the paper becomes polarized (▶ **Figure 15.8**). The key here is that the charged ends of the paper are not the same distance from the comb. The positive end is closer than the negative end. Since the electric force decreases with distance, the attractive force ($\vec{F}_1$) between the comb and the positive end of the paper is greater than the repulsive ($\vec{F}_2$) between the comb and the negative end. Adding these two forces vectorially, the net force on the paper is seen to point toward the comb. If the paper has a small mass, it will accelerate in that direction.

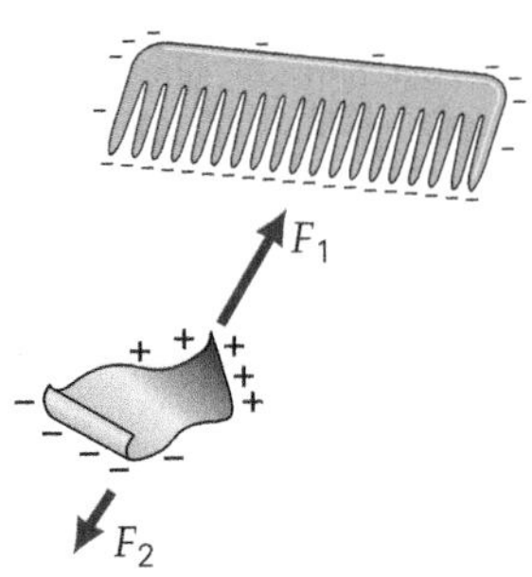


▲ **FIGURE 15.8 Comb and paper** A neutral object can feel an electric force.

**FOLLOW-UP EXERCISE.** Does the phenomenon described in this Example tell you the sign of the charge on the comb? Why or why not?

The next Example illustrates the direct application of Coulomb's law for two charges as well as the need for vector addition when more than two charges are involved.

### EXAMPLE 15.3: COULOMB'S LAW - VECTOR ADDITION INVOLVING TRIGONOMETRY

(a) Two point charges of −1.0 nC and +2.0 nC are separated by a distance of 0.30 m (▼ **Figure 15.9a**). What is the electric force on each particle? (b) A configuration of three charges is shown in Figure 15.9b. What is the net electric force on $q_3$?

**THINKING IT THROUGH.** Adding electric forces is no different from adding any other type of force. The only difference here is that Coulomb's law must first be used to calculate the force magnitudes. Then it is just a matter of computing components. (a) For the two point charges, Coulomb's law is used, noting that the forces are attractive. (Why?) (b) Here components must be used to vectorially add the two forces acting on $q_3$ and the angle $\theta$ is necessary to calculate these components. $\theta$ can be found from the given distances. (See the Problem-Solving Hint immediately following this Example.)

**SOLUTION**

Listing the data and converting nanocoulombs to coulombs,

***Given:***

**(a)** $$q_1 = -(1.0\,\text{nC})\left(\frac{10^{-9}\,\text{C}}{\text{nC}}\right) = -1.0\times10^{-9}\,\text{C}$$

$$q_2 = +(2.0\,\text{nC})\left(\frac{10^{-9}\,\text{C}}{\text{nC}}\right) = +2.0\times10^{-9}\,\text{C}$$

**(b)** Data given in Figure 15.10b. Convert charges to coulombs as in part (a).

***Find:***

**(a)** $\vec{F}_{12}$ and $\vec{F}_{21}$

**(b)** $\vec{F}_3$

**(a)** Coulomb's law gives the magnitude of the force on each charge. Using the given values:

$$\begin{aligned} F_{12} = F_{21} &= \frac{kq_1q_2}{r^2} \\ &= \frac{(9.00\times10^9\,\text{N}\cdot\text{m}^2/\text{C}^2)(1.0\times10^{-9}\,\text{C})(2.0\times10^{-9}\,\text{C})}{(0.30\,\text{m})^2} \\ &= 0.20\times10^{-6}\,\text{N} = 0.20\,\mu\text{N} \end{aligned}$$

Coulomb's law gives only the force's magnitude not direction, but because the charges are of opposite sign, the forces are mutually attractive, as shown in Figure 15.9a.

**(b)** Like all forces, $\vec{F}_{31}$ and $\vec{F}_{32}$ must be added vectorially to find the net force. Since all the charges are positive, the forces are repulsive, as shown in the vector diagram in Figure 15.9b. Since $q_1 = q_2$ and the charges are equidistant from $q_3$, it follows that $\vec{F}_{31}$ and $\vec{F}_{32}$ have the same magnitude. Since $r_{31} = r_{32} = 0.50$ m. (How do you know this?) Coulomb's law gives the magnitude of the force of 2 on 3:

$$\begin{aligned} F_{32} &= \frac{kq_2q_3}{r_{32}^2} \\ &= \frac{(9.00\times10^9\,\text{N}\cdot\text{m}^2/\text{C}^2)(2.5\times10^{-9}\,\text{C})(3.0\times10^{-9}\,\text{C})}{(0.50\text{ m})^2} \\ &= 0.27\times10^{-6}\,\text{N} = 0.27\,\mu\text{N} \end{aligned}$$

Taking into account the directions of $\vec{F}_{31}$ and $\vec{F}_{32}$ by symmetry their *y*-components cancel to produce zero net vertical force. Thus, $\vec{F}_3$ (the net force on $q_3$) acts horizontally along the positive *x*-axis and has a magnitude of $F_3 = F_{31x} + F_{32x} = 2F_{31x}$ because $F_{31} = F_{32}$. The angle $\theta$ can be computed directly from the distances: $\theta = \tan^{-1}(0.30\text{ m}/0.40\text{ m}) = 37°$. Thus $\vec{F}_3$ has a magnitude of

$$\begin{aligned} F_3 &= 2F_{32}\cos\theta \\ &= 2(0.27\,\mu\text{N})\cos 37° = 0.43\,\mu\text{N} \end{aligned}$$

and acts in the positive *x*-direction (to the right).

**FOLLOW-UP EXERCISE.** In part (b) of this Example, calculate the net force on $q_1$.

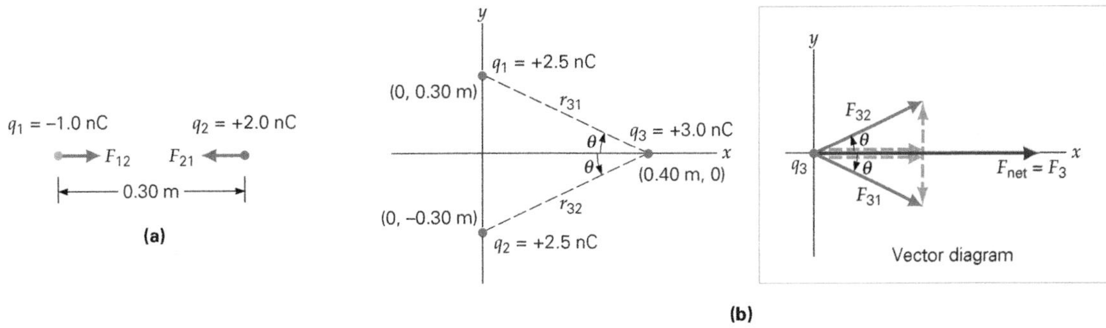


▲ **FIGURE 15.9 Electric forces** These charges and charge configurations are used in Example 15.3 which emphasizes the use of vectors to find net electric forces.

### 15.3.1 Problem-Solving Hint

When forces are two-dimensional, such as in part (b) of the preceding Example, this generally requires determining components. Equation 15.2 should be used to calculate the magnitude of the force, using only the magnitude of the charges (as in the Example). Then the charge-force law determines the direction of the force between each pair of charges. (Draw a sketch and put in the angles.) Last, use trigonometry to calculate each force's components and then combine them appropriately. This approach is strongly recommended and the one that will be used in this text.

The magnitudes of the charges in Example 15.3 are typical of the magnitudes of static charges produced by frictional rubbing; that is, they are tiny. Thus, the forces involved are very small by everyday standards, much smaller than any force we have studied so far. However, on the atomic scale, electric forces can produce large forces and accelerations, because the particles (such as electrons and protons) are very close together and have extremely small masses. Compare the magnitude of the answers in Example 15.4 with those in Example 15.3.

---

**EXAMPLE 15.4: INSIDE THE NUCLEUS – REPULSIVE ELECTROSTATIC FORCES**

(a) What is the magnitude of the repulsive electrostatic force between two protons in a nucleus? Treat them as point charges and take the distance between them to be $3.00 \times 10^{-15}$ m (which is a typical value). (b) If the protons were released from rest, how would the magnitude of their initial acceleration compare with that of the acceleration they might experience due to gravity near the Earth's surface, $g$?

**THINKING IT THROUGH.** (a) Coulomb's law must be applied to find the repulsive force. (b) To find the initial acceleration, Newton's second law ($\vec{\mathbf{F}}_{\text{net}} = m\vec{\mathbf{a}}$) can be used.

**SOLUTION**

Listing the known quantities:

***Given:***

$r = 3.00 \times 10^{-15}$ m
$q_1 = q_2 = +1.60 \times 10^{-19}$ C (from Table 15.1)
$m_p = 1.67 \times 10^{-27}$ kg (from Table 15.1)

***Find:***

**(a)** $F_e$ (magnitude of force)
**(b)** $\frac{a}{g}$ (ratio of acceleration to $g$)

**(a)** Using Coulomb's law (Equation 15.2),

$$F_e = \frac{kq_1q_2}{r^2} = \frac{(9.00\times10^9\,\text{N}\cdot\text{m}^2/\text{C}^2)(1.60\times10^{-19}\,\text{C})(1.60\times10^{-19}\,\text{C})}{(3.00\times10^{-15}\,\text{m})^2} = 25.6\,\text{N}$$

This force is much larger than that in Example 15.3 and is equivalent to the gravitational force (weight) acting on an object with a mass of about 2.5 kg. With its small mass, the proton should experience a huge acceleration.

**(b)** If this force acted alone on a proton, it would produce an acceleration of

$$a = \frac{F_e}{m_p} = \frac{25.6\,\text{N}}{1.67\times10^{-27}\,\text{kg}} = 1.53\times10^{28}\,\text{m/s}^2$$

or, expressed instead as a multiple of $g$:

$$\frac{a}{g} = \frac{1.53\times10^{28}\,\text{m/s}^2}{9.8\,\text{m/s}^2} = 1.56\times10^{27}$$

Most nuclei contain more than two protons. With these enormous repulsive forces, you might therefore expect nuclei to fly apart. Because this doesn't generally happen, there must be a stronger attractive force holding the nucleus together. This is the *nuclear* (or *strong*) force, and will be discussed in Chapters 29 and 30.

**FOLLOW-UP EXERCISE.** Suppose you had a fixed isolated proton and wished to delicately place a second one directly above the first so that the second proton was balanced in static equilibrium. How far apart would the protons be?

---

Although there is a striking similarity between the mathematical form of the expressions for the electric and gravitational forces, there is a huge difference in the relative strengths of the two forces, as is shown in Example 15.5.

---

**EXAMPLE 15.5: INSIDE THE ATOM – ELECTRIC FORCE VERSUS GRAVITATIONAL FORCE**

How many times larger is the electric force than the gravitational force between a proton and an electron at any given distance?

**THINKING IT THROUGH.** The distance between the proton and electron is not given. Notice that asking "how many times larger" implies that the answer should be in the form of ratio. Since both the electrical force and the gravitational force vary as the inverse square of the distance, any distance dependence should cancel out in a ratio. Using Coulomb's law and Newton's law of gravitation (Section 7.5), the ratio can be determined using the known values for charges, masses, and electric and gravitational constants.

**SOLUTION**

All the needed values are in Table 15.1 and are listed below.

***Given:***

$q_e = -1.60 \times 10^{-19}$ C
$q_p = +1.60 \times 10^{-19}$ C
$m_e = 9.11 \times 10^{-31}$ kg
$m_p = 1.67 \times 10^{-27}$ kg

***Find***: $F_e/F_g$ (ratio of forces)

The expressions for the force magnitudes are $F_e = (kq_eq_p)/r^2$ and $F_g = (Gm_em_p)/r^2$. Forming a ratio:

$$\frac{F_e}{F_g} = \frac{kq_eq_p}{Gm_em_p}$$
$$= \frac{(9.00\times10^9\ \text{N}\cdot\text{m}^2/\text{C}^2)(1.60\times10^{-19}\ \text{C})^2}{(6.67\times10^{-11}\ \text{N}\cdot\text{m}^2/\text{kg}^2)(9.11\times10^{-31}\ \text{kg})(1.67\times10^{-27}\ \text{kg})}$$
$$= 2.27\times10^{39}$$

The electrostatic force between a proton and an electron is more than $10^{39}$ times that of the gravitational force! With this enormous ratio, it should be clear that the gravitational force between charged particles can generally be neglected in electrostatics.

**FOLLOW-UP EXERCISE.** With respect to the results of this Example, show that the force of gravity between two electrons is even more negligible compared with the electric force between them. Explain why this is so.

## 15.4 Electric Field

The electric force, like the gravitational force, is an "action-at-a-distance" force. Since the range of the electric force between point charges is infinite ($F_e \propto 1/r^2$ and approaches zero only as $r$ approaches infinity), it stands to reason that a configuration or assembly of charged particles can have an effect on other charges located nearby.

The idea of a force acting across empty space was difficult for early investigators to accept, and the modern concept of a *force field*, or simply a field, was introduced. They envisioned an **electric field** as surrounding every arrangement of charges. To them, this field would represent the electrical effect of that arrangement on the nearby space. The field represents what is different about this nearby space because those charges are present. This field concept envisions charges as interacting with the electric field created by other charges, not directly with the charges "at a distance."

The electric field concept is as follows: A configuration of charges creates its electric field in the space nearby. This field describes the charge configuration's "electric region of influence" and if another charge is placed in this region, it is the field that exerts the electric force on that charge. In other words:

> Charge configurations can create electric fields, and it is these fields that, in turn, exert electric forces on other charges in that field.

An electric field is a vector field (symbol $\vec{\mathbf{E}}$) since it has direction as well as magnitude. Once known, it determines the force (including direction) exerted on a charge placed anywhere in the field region. *Note carefully however, the electric field is not that force.*

The magnitude ($E$, sometimes called "field strength") is defined as the magnitude of the electric force exerted per unit charge. Determining the field's strength may be pictured using the following "thought" procedure: imagine placing a very small charge (called a *test charge*) at the location of interest near an assembly of charges. Now, determine the force acting on that test charge and divide by its charge; that is, find the force exerted at that location *per coulomb*. Then imagine removing the test charge. The force at that location disappears (why?), but the field remains. This is because the field is created by the nearby charge assembly which remains. Repeating this process in many locations results in a "map" of the field strength $E$ but we have no direction as yet. Thus at this point the "mapping" is incomplete.

The field direction is specified by the direction of the force on the test charge. Thus it depends on whether the test charge is positive or negative. The convention is that a positive test charge ($q_+$) is used for determining electric field direction (see ▼ **Figure 15.10**). That is:

> the direction of $\vec{\mathbf{E}}$ at any location is in the direction of the force on a small *positive* test charge at that location.

Once the electric field's magnitude and direction due to a charge configuration are known, you can ignore the "source" charge configuration and talk in terms of the field it produces. This way of visualizing electric interactions between charges often facilitates calculations.

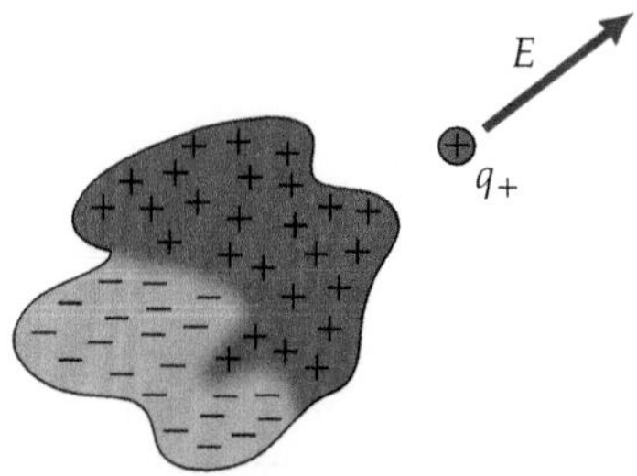


▲ **FIGURE 15.10** **Electric field direction** By convention, the direction of the electric field $\vec{\mathbf{E}}$ is the same direction as the force experienced by a positive test charge. Thus the direction is that in which the test charge would accelerate if released. Here the "system of charges" produces a (net) electric field upward and to the right at the location of the test charge. Can you explain this direction by noting the signs and locations of charges in this specific system?

To summarize mathematically, the **electric field** $\vec{\mathbf{E}}$ at any location is defined as:

$$\vec{\mathbf{E}} = \frac{\vec{\mathbf{F}}_{q+}}{q_+} \tag{15.3}$$

SI unit of electric field: newton/coulomb (N/C)

Once $\vec{\mathbf{E}}$ is known, the force on any charge $q$ can be determined by solving Equation 15.3: $\vec{\mathbf{F}} = q\vec{\mathbf{E}}$ or $F = qE$ if just the force magnitude is desired.

For the special case of the field due to a single point charge, Coulomb's law can be used. To determine the magnitude of the electric field due to a point charge at a distance $r$ from that point charge, use Equation 15.3 to express the electric force. The result is:

$$E = \frac{F_{q+}}{q_+} = \frac{(kqq_+/r^2)}{q_+} = \frac{kq}{r^2}$$

Notice that $q_+$, canceled out as it must, because the field is produced by other charges, not the test charge $q_+$. Thus a point charge's electric field is given by

$$E = \frac{kq}{r^2} \quad \text{(magnitude of electric field due to point charge } q) \qquad (15.4)$$

Some electric field vectors in the vicinity of a positive charge are shown in ▼**Figure 15.11a**. Note that their directions are away from the positive charge because a positive test charge would feel a force in that direction. The magnitude of the field (proportional to the arrow length) decreases with increasing $r$ as expressed in Equation 15.4.

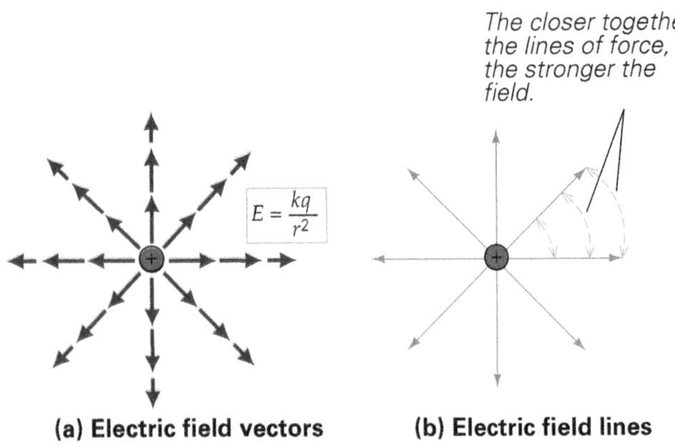


▲FIGURE 15.11 **Electric field** **(a)** The electric field points away from a positive point charge, in the direction a force would be exerted on a small positive test charge. The field's magnitude is proportional to the vector arrow length and decreases with distance from the charge. **(b)** In this simple case, the vectors are easily connected to give the electric field line pattern due to a positive point charge.

If there is more than one charge creating an electric field, then the total, or net, electric field at any point is found using the **superposition principle for electric fields**:

> For any configuration of charges, the total, or net, electric field at any point is the vector sum of the electric fields due to the individual charges of the configuration.

The application of this principle is demonstrated in the next two Examples.

### EXAMPLE 15.6: ELECTRIC FIELDS IN ONE DIMENSION - ZERO FIELD BY SUPERPOSITION

Two point charges (values shown) are placed on the $x$-axis as in ▶**Figure 15.12**. Find all locations on the axis where the electric field is zero.

**THINKING IT THROUGH.** Each point charge produces its own field. By the superposition principle, the electric field is the vector sum of the two fields. Thus the question is asking for locations where these fields are equal and opposite, so they will cancel and give no net electric field. Since both charges are positive, their fields point to the right at all locations to the right of $q_2$. Therefore, the fields cannot cancel in that region. Similarly, to the left of $q_1$, both fields point to the left and cannot cancel. The only possibility of cancellation is in the region between the charges. In that region, the two fields will cancel if their magnitudes are equal, because they are oppositely directed.

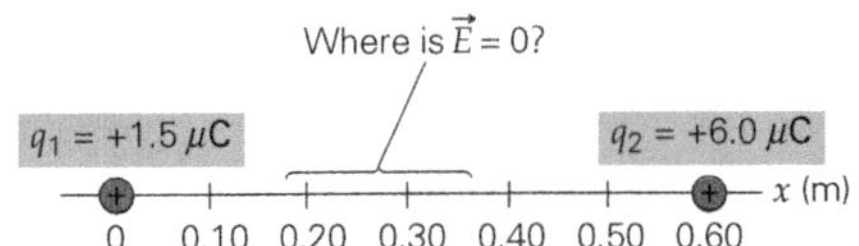


▲FIGURE 15.12 **Electric field in one dimension**

**SOLUTION**

Let us specify the location as a distance $x$ from $q_1$ (located at $x = 0$) and convert charges from microcoulombs to coulombs as usual.

***Given:***

$d = 0.60$ m (distance between charges)
$q_1 = +1.5\ \mu\text{C} = +1.5 \times 10^{-6}$ C
$q_2 = +6.0\ \mu\text{C} = +6.0 \times 10^{-6}$ C

***Find:*** $x$ [the location(s) of zero $E$]

Setting the magnitudes of the individual fields equal and solving for $x$:

$$E_1 = E_2 \quad \text{or} \quad \frac{kq_1}{x^2} = \frac{kq_2}{(d-x)^2}$$

Rearranging the last expression by canceling $k$, putting in the charge ratio $q_2/q_1 = 4$ and then taking the square root of both sides, we have:

$$\frac{1}{x^2} = \frac{(q_2/q_1)}{(d-x)^2} \quad \text{or} \quad \frac{1}{x} = \frac{2}{d-x}$$

Solving, $x = d/3 = 0.20$ m. (Why not use the negative square root? Try it and see.) The result being closer to $q_1$ makes sense physically. This is because $q_2$ is the larger charge, thus for the fields to be equal in magnitude, the location for zero field must be closer to $q_1$.

**FOLLOW-UP EXERCISE.** Repeat this Example, changing only the sign of $q_2$.

### INTEGRATED EXAMPLE 15.7: $\vec{E}$ FIELDS IN TWO DIMENSIONS - USING VECTOR COMPONENTS WITH THE SUPERPOSITION PRINCIPLE

▶**Figure 15.13a** shows a configuration of three point charges. (a) In what quadrant is the electric field direction at the origin: (1) first quadrant, (2) second quadrant, or (3) third quadrant? Explain your reasoning, using the superposition principle. (b) Calculate the magnitude and direction of the electric field at the origin due to these charges.

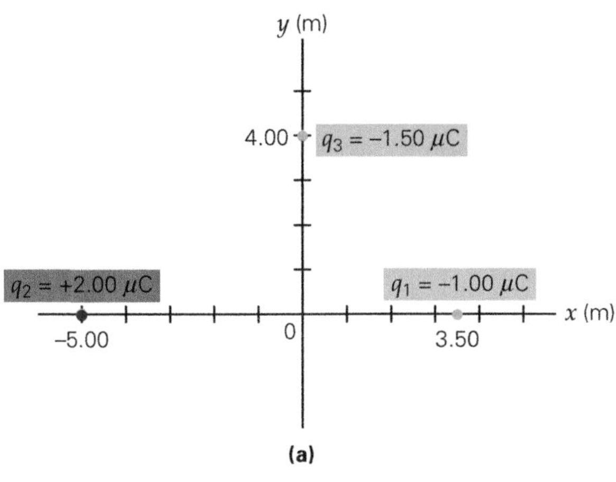


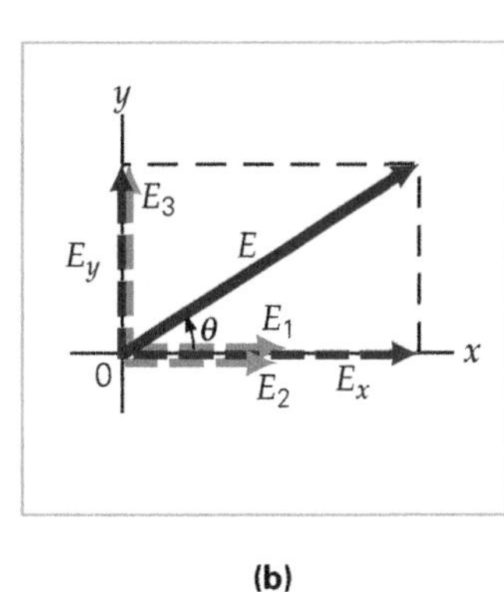


▲ FIGURE 15.13 **Finding the electric field** Vector addition requires the use of components.

(a) **CONCEPTUAL REASONING.** In general, $\vec{\mathbf{E}}$ due to a *single* point charge points toward negative charges and away from positive ones. Therefore, $\vec{\mathbf{E}}_1$ and $\vec{\mathbf{E}}_2$ are in the positive $x$-direction and $\vec{\mathbf{E}}_3$ is in the positive $y$-direction. The (net) electric field at the origin, which is the vector sum of these three fields, thus has both positive components. Therefore, $\vec{\mathbf{E}}$ is in the first quadrant (Figure 15.13b) and the correct answer is (1).

(b) **QUANTITATIVE REASONING AND SOLUTION.** The directions of the individual $\vec{\mathbf{E}}$ fields are shown in Figure 15.13b. According to the superposition principle, these fields are added vectorially to find the net electric field $(\vec{\mathbf{E}} = \vec{\mathbf{E}}_1 + \vec{\mathbf{E}}_2 + \vec{\mathbf{E}}_3)$.

Listing the data given and converting the charges to coulombs:

***Given*:**

$q_1 = -1.00\ \mu C = -1.00 \times 10^{-6}$ C
$q_2 = +2.00\ \mu C = +2.00 \times 10^{-6}$ C
$q_3 = -1.50\ \mu C = -1.50 \times 10^{-6}$ C
$r_1 = 3.50$ m
$r_2 = 5.00$ m
$r_3 = 4.00$ m

***Find*:** $\vec{\mathbf{E}}$ (electric field at origin)

From the sketch it is seen that the electric field a the origin has a component $E_y$ entirely due to $\vec{\mathbf{E}}_3$, and $E_x$ is the sum of the magnitudes of $\vec{\mathbf{E}}_1$ and $\vec{\mathbf{E}}_2$. The magnitudes of the three fields are determined from Coulomb's law.

$$E_1 = \frac{kq_1}{r_1^2} = \frac{(9.00\times10^9\ \text{N}\cdot\text{m}^2/\text{C}^2)(1.00\times10^{-6}\ \text{C})}{(3.50\,\text{m})^2}$$
$$= 7.35\times10^2\ \text{N/C}$$

$$E_2 = \frac{kq_2}{r_2^2} = \frac{(9.00\times10^9\ \text{N}\cdot\text{m}^2/\text{C}^2)(2.00\times10^{-6}\ \text{C})}{(5.00\,\text{m})^2}$$
$$= 7.20\times10^2\ \text{N/C}$$

$$E_3 = \frac{kq_3}{r_3^3} = \frac{(9.00\times10^9\ \text{N}\cdot\text{m}^2/\text{C}^2)(1.50\times10^{-6}\ \text{C})}{(4.00\,\text{m})^2}$$
$$= 8.44\times10^2\ \text{N/C}$$

Thus the components of the field are

$$E_x = E_1 + E_2 = +7.35\times10^2\ \text{N/C} + 7.20\times10^2\ \text{N/C}$$
$$= +1.46\times10^3\ \text{N/C}$$

and

$$E_y = E_3 = +8.44\times10^2\ \text{N/C}$$

or in component form

$$\vec{\mathbf{E}} = E_x\hat{\mathbf{x}}\ + E_y\hat{\mathbf{y}} = (1.46\times10^3\ \text{N/C})\hat{\mathbf{x}} + (8.44\times10^2\ \text{N/C})\hat{\mathbf{y}}$$

You should be able to show that in magnitude-angle form an alternative expression for $\vec{\mathbf{E}}$ is

$$E = 1.69\times10^3\ \text{N/C at } \theta = +30.0^\circ$$
$$(\theta \text{ ccw from the } +x\text{-axis})$$

**FOLLOW-UP EXERCISE.** In this Example, suppose $q_1$ was moved to the origin. Find the electric field at its former location.

## 15.4.1 Electric Lines of Force (Electric Field Lines)

A convenient way of graphically representing the electric field is by use of *electric lines of force*, or since they represent a field and not a force, more correctly, **electric field lines**. To start, consider the electric field vectors near a positive point charge, as in Figure 15.11a. In Figure 15.11b these vectors have been "connected." This method constructs the *electric field line pattern* due to a positive point charge. Notice that the field lines are closer together (their spacing decreases) nearer the charge, because the field increases in strength. Also note that at any location on a field line, the electric field direction is tangent to the line. (The lines usually have arrows attached to them that indicate the general field direction.) It should be clear that electric field lines can't cross. If they did, it would mean that at the crossing spot there would be two directions for the force on a charge placed there – a physically unreasonable result.

The general rules for sketching and interpreting electric field lines are:

1. The closer together the field lines, the stronger the electric field.
2. At any location, the direction of the electric field is tangent to the field line.
3. Electric field lines start at positive charges and end at negative charges.
4. The number of lines leaving or entering a charge is proportional to the magnitude of that charge. (See ▼**Figure 15.14**.)
5. Electric field lines can never cross.

To illustrate these rules, see the following Conceptual Example.

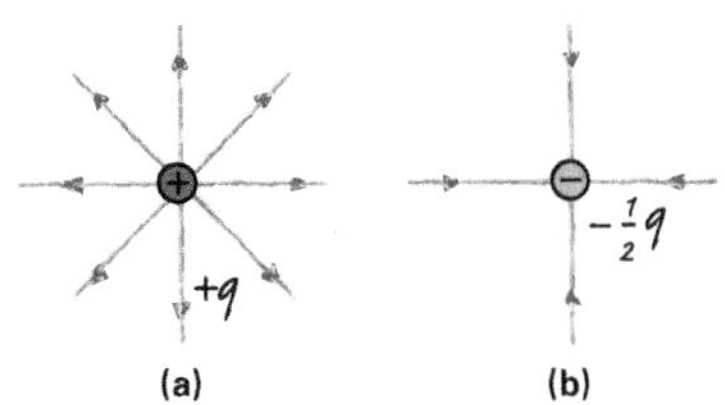


▲FIGURE 15.14 **Sketching electric field lines** see Conceptual Example 15.8 for details.

### CONCEPTUAL EXAMPLE 15.8: SKETCHING ELECTRIC LINES OF FORCE FOR VARIOUS POINT CHARGES

Based on the field due to a positive point charge shown in Figure 15.14a, sketch the pattern if, instead, the charge was half as large and negative?

**REASONING AND ANSWER.** Since the charge is now negative, the electric field direction should be reversed to point toward the charge. In addition, since the charge is only half as large (in magnitude) the field should be half as large at any given distance. Thus the spacing of the lines should be doubled. As shown in Figure 15.14b, this is accomplished by reducing the number of lines that start on the charge from 8 to 4, since the amount of charge is halved.

**FOLLOW-UP EXERCISE.** Sketch the electric field line pattern if the charge were $-1.5q$ instead.

In principle, the superposition principle should enable us to "map" the electric field line pattern for any configuration of charges. The resulting map is an equivalent alternative (to the previous vector arrow picture) and provides a visual representation of the electric field in a region. As an example, consider the pattern due to an electric dipole in Example 15.9. An **electric dipole** consists of two equal, but opposite, electric charges (or "poles"). Even though the net charge on the dipole is zero, it does create an electric field because the charges are separated. If the charges were at the same location, their fields would cancel everywhere.

### EXAMPLE 15.9: CONSTRUCTING THE ELECTRIC FIELD PATTERN DUE TO A DIPOLE

Use the superposition principle and the electric field line rules to construct a typical electric field line due to an electric dipole.

**THINKING IT THROUGH.** The construction involves vector addition of the individual electric fields from the two opposite ends of the dipole.

**SOLUTION**

***Given:***
An electric dipole of two equal and opposite charges ($q_+$ and $q_-$) separated by a distance $d$

***Find:*** a typical electric field line

An electric dipole is shown in ▼**Figure 15.15a**. To keep track of the two fields, let's designate them using the same subscripts: $\vec{\mathbf{E}}_+$ and $\vec{\mathbf{E}}_-$.

Let's begin at location A, near charge $q_+$. We know that $\vec{\mathbf{E}}_+$ will *always* point away from $q_+$ and $\vec{\mathbf{E}}_-$ will *always* point toward $q_-$. Because A is closer to $q_+$, it follows that $E_+ > E_-$. Putting these two facts together enables us to qualitatively draw the two fields and the parallelogram method determines their vector sum called $\vec{\mathbf{E}}_A$.

Since we are trying to follow field lines, the direction of $\vec{\mathbf{E}}_A$ will point us to our next location, B. At B, there is a reduced magnitude (why?) and slight directional change for both $\vec{\mathbf{E}}_+$ and $\vec{\mathbf{E}}_-$. You should now be able continue on to see how the fields at C and D are determined. Location D is special because it is on the perpendicular bisector of the dipole axis (the line that connects the two charges). $\vec{\mathbf{E}}$ points downward anywhere on this line. You should then be able to continue the construction at points E, F, and G.

Last, to construct the electric field *line*, start at the positive end of the dipole, because the field lines start at positive charges. Create the electric field line so the vectors are tangent to it and follow from one location to the next. (Other lines can be drawn starting at different locations than A, and thus complete the dipole field pattern shown in Figure 15.15b.)

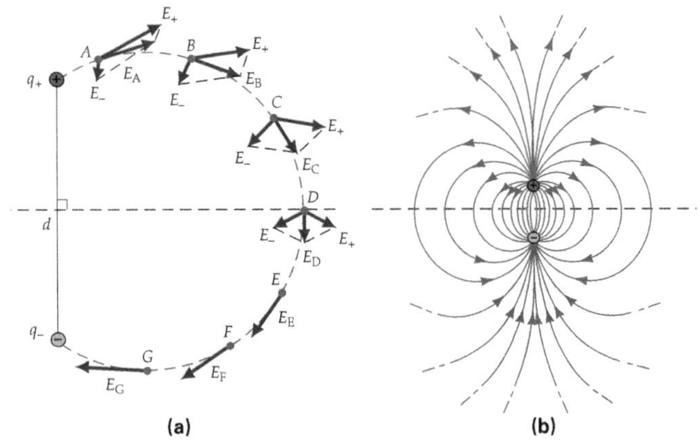


▲FIGURE 15.15 **Mapping the electric field due to a dipole**

**FOLLOW-UP EXERCISE.** Using the techniques in this Example, construct the field lines that are (a) directly above the positive charge, (b) directly below the negative charge, and (c) directly below the positive charge.

In addition to providing a guide to electric field sketching, dipoles are important in themselves, because they often occur in nature. For example, electric dipoles can serve as a model for important polarized molecules, such as the water molecule. (See Figure 15.7.)

▼ **Figure 15.16a** shows the use of the superposition principle to construct the electric field line pattern due to a large uniformly charged positive plate. Notice that the field points perpendicularly away from the plate on both sides. Figure 15.16b shows the result if the plate is negatively charged, the only difference being the field direction. Putting these two together, the field *between* two closely spaced and oppositely charged plates is found. The result is the pattern in Figure 15.16c. Due to the cancellation of the horizontal field components (as long as we stay away from the plate edges), the electric field is uniform and points from the positive to negative. (Think of the direction of the force acting on a positive test charge placed between the plates.)

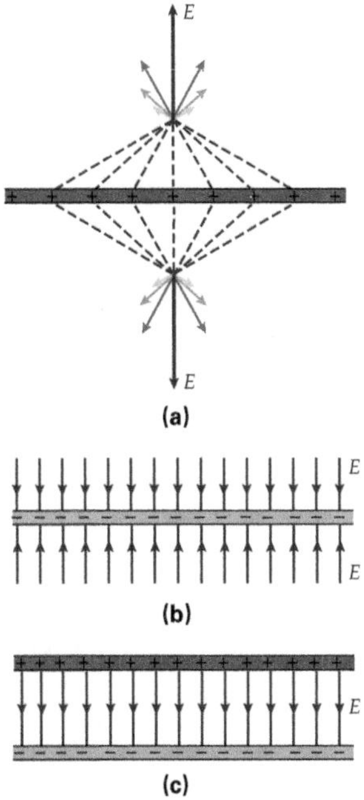


▲ FIGURE 15.16 **Electric field due to very large parallel plates** **(a)** Above a positively charged plate, the net electric field points upward. Here, the horizontal components of the electric fields from various locations on the plate cancel. Below the plate, $\vec{E}$ points downward. **(b)** For a negatively charged plate, the electric field directions (shown on both sides of the plate) are reversed. **(c)** For two oppositely charged, closely spaced plates, there is field cancellation outside the plates resulting in almost no field there. Between the plates the fields from the two plates add, resulting in (approximately) a uniform field in that region. (The field at the edges of the plates is not shown.)

The derivation of the expression for the electric field magnitude between two closely spaced plates is beyond the scope of this text. However, the result is

$$E = \frac{4\pi kQ}{A} \quad \text{(electric field between parallel plates, not near edges)} \tag{15.5}$$

where $Q$ is the magnitude of the total charge on one plate and $A$ is the area of one plate. Parallel plates are common in electronic applications. For example, in Chapter 16 an important circuit element called a *capacitor* will be studied more closely. In its simplest form a capacitor is a pair of parallel plates used to store electric charge and energy. Cloud-to-ground lightning can be approximated by closely spaced parallel plates as in the next Example.

**EXAMPLE 15.10: PARALLEL PLATES - ESTIMATING THE CHARGE ON STORM CLOUDS**

The electric field (magnitude) $E$ required to ionize moist air (that is, to remove electrons from the air molecules) is about $1.0 \times 10^6$ N/C. When the field reaches this value, the least bound electrons are are removed from the molecules, leading to a lightning stroke. Let us make a crude model of this atmospheric situation by picturing the bottoms of the cloud layer and the ground as the two "plates" of a parallel plate capacitor. Typically, the cloud is negatively charged and the ground is positive during lightning storms. Take the area involved (ground and clouds) to be squares 10 miles on each side. Estimate the magnitude of the total negative charge on the lower cloud surface when the field reaches the value required for a lightning strike.

**THINKING IT THROUGH.** The electric field is given, so Equation 15.5 can be used to estimate Q. The cloud area $A$ (one of the "plates") must be expressed in square meters.

**SOLUTION**

***Given:***

$E = 1.0 \times 10^6$ N/C
$d = 10 \text{ mi} \approx 1.6 \times 10^4$ m

***Find:*** $Q$ (the magnitude of the charge on the lower cloud surface)

Using $A = d^2$ for the area of a square and solving Equation 15.5 for the *magnitude* of the charge on each:

$$Q = \frac{EA}{4\pi k} = \frac{(1.0\times10^6\ \text{N/C})(1.6\times10^4\ \text{m})^2}{4\pi(9.0\times10^9\ \text{N}\cdot\text{m}^2/\text{C}^2)} = 2300\ \text{C}$$

This expression is justified if the distance between the clouds and the ground is much less than their size, so we can neglect the "edges" - an assumption that Equation 15.5 makes. This charge is huge compared with the frictional static charges that usually develop when, say, someone shuffles across a carpet.

**FOLLOW-UP EXERCISE.** In this Example, (a) what is the direction of the electric field between the cloud and the Earth? (b) What would the electric field between the cloud and ground be if the charge on each were only 1000 C?

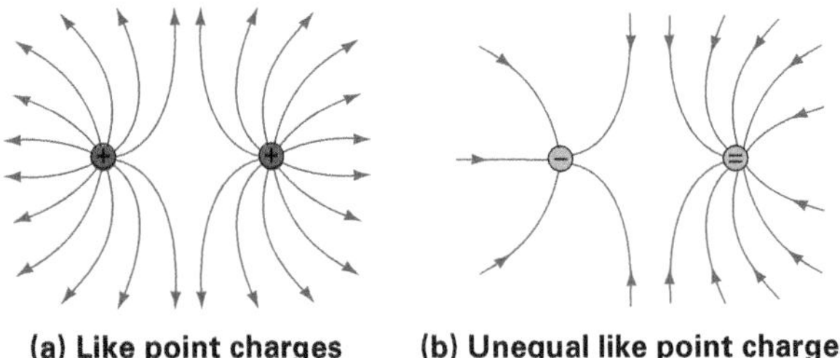


▲FIGURE 15.17 **Electric fields** Electric fields for **(a)** like point charges and **(b)** unequal like point charges.

The electric field patterns for some other point charge configurations are shown in ▲**Figure 15.17**. You should be able to see how they are sketched qualitatively. Note that the electric field lines always begin on positive charges and end on negative ones (or at infinity when there is no nearby negative charge). Choose the number of lines emanating from or ending at a charge in proportion to the magnitude of that charge. (See Conceptual Example 15.8.)

## 15.5 Conductors and Electric Fields

The electric fields associated with charged conductors have several interesting properties. By definition in electrostatics, the charges are at rest. Because conductors possess some electrons that are free to move but don't under these conditions, the electrons must experience no electric force and thus no electric field. Hence we conclude that:

> The electric field is zero *inside* a charged conductor.

Excess charges on a conductor tend to get as far away from each other as possible, because they are highly mobile. Then:

> Any *excess* charge on an isolated conductor resides entirely on the surface of the conductor.

Another property of static electric fields and conductors is that there cannot be a tangential component of the field at the surface of the conductor. If this were not true, charges would move *along* the surface, contrary to our assumption of a static situation. Thus:

> The electric field at the surface of a charged conductor is perpendicular to the surface.

Last, the excess charge on a conductor of irregular shape is most closely packed where the surface is highly curved (at the sharpest points). Since the charge is densest there, the electric field will also be the largest at these locations. That is:

> Excess charge tends to accumulate at sharp points, or locations of highest curvature, on charged conductors. As a result, the electric field is greatest at such locations.

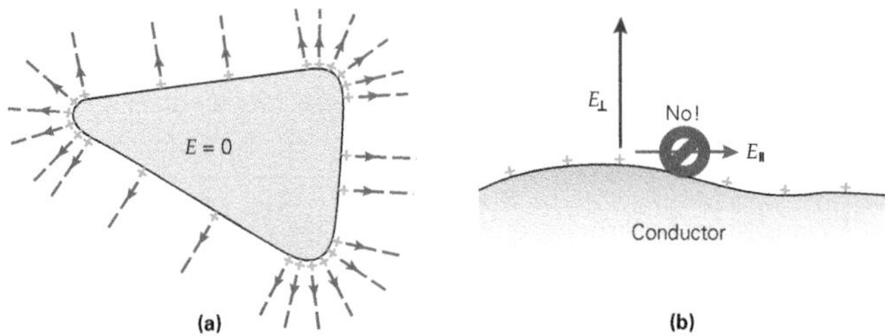


▲FIGURE 15.18 **Electric fields and conductors (a)** Under static conditions, the electric field is zero inside a conductor. Any excess charge resides on the conductor's surface. For an irregularly shaped conductor, the excess charge accumulates in the regions of highest curvature (the sharpest points), as shown. The electric field near the surface is perpendicular to that surface and strongest where the charge is densest. **(b)** Under static conditions, the electric field cannot have a component tangential to the conductor's surface.

These properties are summarized in ▲**Figure 15.18**. Note that they are true only for conductors under static conditions. Electric fields can, in fact, exist inside non-conducting materials and inside conductors when conditions vary with time.

To understand *why* most of the charge accumulates in the highly curved surface regions, consider the forces acting *between* charges on the surface of the conductor. (See ▼**Figure 15.19a.**) Where the surface is fairly flat, these forces will be directed nearly parallel to the surface. The charges will spread out to the sharper ends where the forces between charges are directed more nearly perpendicular to the surface. Here there will be little tendency for the charges to move parallel to the surface. Thus the more highly curved regions of the surface accumulate the highest concentration of charge.

An interesting situation occurs when there is a large concentration of charge on a conductor with a sharp point (Figure 15.19b). The electric field above the point may be high enough to ionize air molecules (to pull or push electrons off the

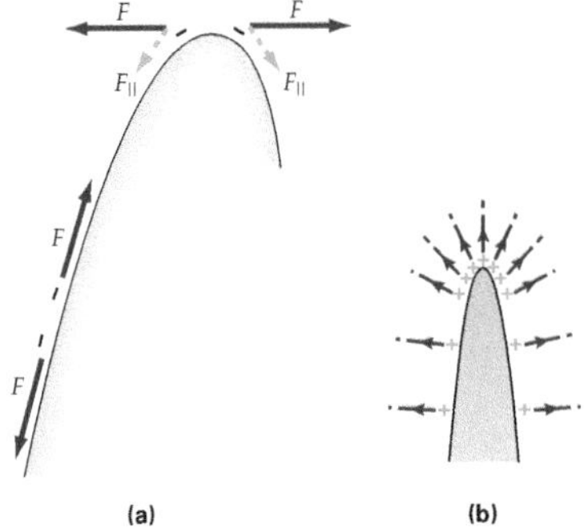


▲FIGURE 15.19 **Concentration of charge on a curved surface (a)** On a flat surface, the repulsive forces between excess charges are parallel to the surface and tend to push the charges apart. On a sharply curved surface, in contrast, these forces are directed at an angle to the surface. Their components parallel to the surface are smaller, allowing charge to concentrate in such areas. **(b)** Taken to the extreme, a sharply pointed metallic needle has a dense concentration of charge at the tip. This produces a large electric field in the region above the tip, which is the principle of the lightning rod.

molecules). The freed electrons are then further accelerated by the field and can cause secondary ionizations by striking other molecules. This results in an "avalanche" of electrons, visible as a spark discharge – for example, from your finger to a doorknob when you accumulate a net charge by shuffling along a carpet.

As an illustration of an early experiment done on conductors with excess charge, consider the following Example.

---

CONCEPTUAL EXAMPLE 15.11: THE CLASSIC ICE PAIL EXPERIMENT

A positively charged rod is held inside an isolated metal container that has uncharged electroscopes conductively attached to its inside surface and to its outside surface (▼ **Figure 15.20**). What will happen to the leaves of the electroscopes? (a) Neither electroscope's leaf will show a deflection; (b) only the outside-connected electroscope's leaf will show a deflection; (c) only the inside-connected electroscope's leaf will show a deflection; or (d) the leaves of both electroscopes will show deflections? (Justify your answer.)

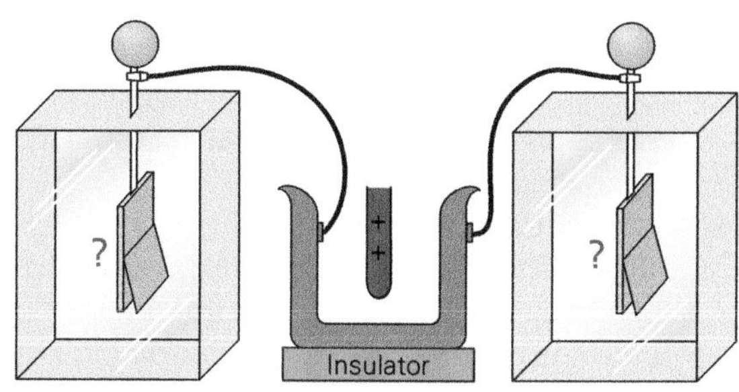


▲ FIGURE 15.20 **An ice pail experiment**

**REASONING AND ANSWER.** The positively charged rod will attract negative charges, causing the inside of the metal container to become negatively charged. The outside electroscope will thus acquire a positive charge. Hence, both electroscopes will be charged (though with opposite signs) and show deflections, so the answer is (d). This experiment was performed by the nineteenth-century English physicist Michael Faraday using ice pails; hence this is often called *Faraday's ice pail experiment.*

**FOLLOW-UP EXERCISE.** Suppose in this Example that the positively charged rod actually *touched* the metal container. How would the electroscopes react now?

---

## Chapter 15 Review

- The **law of charges**, or **charge-force law**: Like charges repel and opposite charges attract.
- **Charge conservation** means that the net charge of an isolated system remains constant.
- **Conductors** easily gain or lose electric charge. They conduct electric charge readily.
- **Insulators** are poor conductors of electric charge and do not easily gain or lose electric charge.
- **Electrostatic charging** involves using friction, contact (conduction), and induction charging to enable an object to gain a net charge.
- **Electric polarization** of an object involves creating a separation of equal amounts of positive and negative charge on that object.
- **Coulomb's law** expresses the magnitude of the force between two point charges:

$$F_e = \frac{kq_1q_2}{r^2} \quad \text{(point charges only, } q \text{ means magnitude)} \qquad (15.2)$$

where $k \approx 9.00 \times 10^9\ \text{N·m}^2/\text{C}^2$.

- The **electric field** is a vector field that describes how an assembly of electric charges modify the space around them. At any location it is defined as the electric force per unit positive charge, and has a direction of the force that a positive charge at that location would experience:

$$\vec{\mathbf{E}} = \frac{\vec{\mathbf{F}}_{q+}}{q_+} \qquad (15.3)$$

- The **superposition principle** for electric fields states that the (net) electric field at any location due to a nearby assembly of charges is the vector sum of the individual electric fields from individual charges of that assembly.
- $\vec{\mathbf{E}}$ **between two closely spaced parallel plates** is uniform and points from positive to negative plate. Its magnitude is given by

$$E = \frac{4\pi kQ}{A} \quad \text{(electric field between parallel plates, not near edges)} \qquad (15.5)$$

- **Electric field lines** provide a visualization of the electric field. The line spacing at any location is inversely related to the field strength, and tangents to the lines give the field direction. Field lines start at positive charges and end at negative charges.
- Under electrostatic conditions, $\vec{\mathbf{E}}$ **fields associated with charged conductors** have the following properties: In the conductor's interior, the field is zero. Excess charge on such a conductor resides entirely on its surface. The electric field near its surface is perpendicular to that surface. Excess charge is most dense at locations of highest surface curvature and thus the surface electric field is greatest there also.

## End of Chapter Questions and Exercises

### Multiple Choice Questions

**15.1 Electric Charge**

1. A combination of two electrons and three protons would have a net charge of (a) +1, (b) − 1, (c) $+1.6\times10^{-19}$ C, (d) $-1.6\times10^{-19}$ C.
2. An electron is just above a fixed proton. The direction of the force on the proton is (a) up, (b) down, (c) zero.
3. In Question 2, which one feels the greater magnitude force: (a) the electron, (b) the proton, or (c) both feel the same?

### 15.2 Electrostatic Charging

4. A rubber rod is rubbed with fur. The fur is then quickly brought near the bulb of an uncharged electroscope. The sign of the charge on the leaves of the electroscope is (a) positive, (b) negative, (c) zero.
5. A stream of water is deflected toward a nearby electrically charged object that is brought close to it. The sign of the charge on the object (a) is positive, (b) is negative, (c) is zero, (d) can't be determined by the data given.
6. A balloon is charged and then clings to a wall. The sign of the charge on the balloon (a) is positive, (b) is negative, (c) is zero, (d) can't be determined by the data given.

### 15.3 Electric Force

7. How does the magnitude of the electric force between two point charges change as the distance between them is increased? The force (a) decreases, (b) increases, (c) stays the same.
8. Compared with the electric force, the gravitational force between two protons is (a) about the same, (b) somewhat larger, (c) very much larger, (d) very much smaller.
9. If the distance between two charged particles is tripled, what happens to the magnitude of the electric force each exerts on the other: (a) it stays the same, (b) it is reduced to one-third its original value, or (c) it is reduced to one-ninth its original value?
10. In Question 9, if you wanted to change the amount of charge on each of the particles by the same amount so that the force between them went back to its original value, what would you do: (a) increase each charge by three times, (b) increase each charge by nine times, or (c) decrease each charge to one-third its original value?

### 15.4 Electric Field

11. How is the magnitude of the electric field due to a point charge reduced when the distance from that charge is tripled? (a) It stays the same, (b) it is reduced to one-third of its original value, (c) it is reduced to one-ninth of its original value, or (d) it is reduced to one-twenty-seventh of its original value.
12. The SI units of electric field are (a) C, (b) N/C, (c) N, (d) J.
13. At a point in space, an electric force acts vertically upward on an electron. The direction of the electric field at that point is (a) down, (b) up, (c) zero, (d) undetermined.
14. Two electrons are placed on the (vertical) $y$-axis, one at $y = +20$ cm and the other at $y = -20$ cm. What is the direction of the electric field at the location $y = 0$ cm, $x = +40$ cm: (a) right, (b) left, (c) up, or (d) down?
15. In the previous question, what is the field direction at the location $y = +40$ cm, $x = 0$ cm: (a) right, (b) left, (c) up, or (d) down?

### 15.5 Conductors and Electric Fields

16. In electrostatic equilibrium, is the electric field just below the surface of a charged sphere of radius $R$ (a) the same value as the field just above the surface, (b) zero, (c) dependent on the amount of charge on the conductor, or (d) given by $kq/R^2$?
17. An uncharged thin metal slab is oriented perpendicularly to an external electric field that points horizontally to the left. What is the direction of the electric field inside the slab? (a) left, (b) right, or (c) it has none because it is zero.
18. The direction of the electric field at the surface of a charged conductor under electrostatic conditions (a) is parallel to the surface, (b) is perpendicular to the surface, (c) is at a 45° angle to the surface, or (d) depends on the charge on the conductor.

## Conceptual Questions

### 15.1 Electric Charge

1. If the charge on the electron were designated as positive and that of the proton negative, what do you think would be the overall effect on the physical universe as we know it?
2. An electrically neutral object can be given a net charge by several means. Does this violate the law of conservation of charge? Explain.
3. If a neutral piece of metal becomes negatively charged, does its mass increase or decrease? What if it becomes positively charged?

### 15.2 Electrostatic Charging

4. Fuel trucks often have metal chains reaching from their frames to the ground. Why is this important?
5. Is there an overall gain or loss of electrons by an object when it is electrically polarized? Explain.
6. Explain carefully the steps you would use to create an electroscope that is positively charged by induction. After you are done, suppose a charged object was brought near the top of the electroscope and the leaves collapsed. What is the sign of the charge on this object?
7. Suppose the negatively charged balloon in Figure 15.6c was replaced by a positively charged glass rod. Which way would the water stream bend now, if at all? Explain your reasoning.

### 15.3 Electric Force

8. Two point charges initially exert an electric force of magnitude $F$ on one another. Suppose the charge of one was halved and that of the other was quadrupled. Write the new force be in terms of $F$ and explain your reasoning.
9. Two nearby electrons would fly apart if released. How could you prevent this by placing a single charge in their neighborhood? Explain clearly what the sign of the charge and its location would have to be.
10. Two point charges are initially separated by a distance $d$. Suppose the charge of one is increased by twenty-seven times while the charge of the other is reduced to one-third its initial value. What would their separation distance have to be, expressed in terms of $d$, so as to keep the force between them the same? Explain your reasoning.

11. A small charged object is placed and held just above the positive end of an electric dipole. The dipole starts to accelerate downward when released. (a) What is the sign of the charge on the object? (b) What would happen to the dipole if released when same charged object was held just below the negative end of the dipole?

**15.4 Electric Field**

12. How is the relative magnitude of an electric field at different locations determined from an electric field vector diagram?
13. How can the relative magnitudes of an electric field at different locations be determined from an electric field line pattern?
14. Explain clearly why electric field lines can never cross.
15. A positive charge is placed inside at the center an isolated thick metal spherical shell. Describe the electric field in the following three regions: between the charge and the inside surface of the shell, inside the shell, and outside the outer shell surface. What is the sign of the charge on the two shell surfaces? How would your answers change if the charge at the center were negative?
16. At a certain location, the electric field near the Earth's surface points downward. What is the sign of the charge on the Earth's surface at that location? Explain.
17. (a) Could the electric field due to two identical negative charges ever be zero at some location(s) nearby? Explain. If your answer is yes, describe and sketch the situation. (b) How would your answer change if the charges were equal but oppositely charged? Explain.
18. A large square (finite size) flat plate is uniformly positively charged and in the horizontal plane. Determine the direction of the electric field at the following locations: (a) just above the center of the plate, (b) just off any edge of the plate and in the same horizontal plane as the plate, and (c) a vertical distance above any edge of the plate on the same order of magnitude as the length of one side of the plate.
19. At a distance extremely far from *any* positively charged object (regardless of its shape), what pattern would its electric field line map approximate? Explain. [*Hint:* When any object is viewed from a very large distance, it will appear geometrically as what type of object?]

**15.5 Conductors and Electric Field**

20. Is it safe to stay inside a car during a lightning storm? Explain.
21. Under electrostatic conditions, it is found that the excess charge on a conductor is uniformly spread over its surface. What is the shape of the surface?
22. Tall buildings have lightning rods to protect them from lightning strikes. Explain why the rods are pointed in shape and taller than the buildings.
23. Sketch the electric field line pattern that results when a metal slab is placed between a pair of closely spaced, equal (but oppositely charged) parallel plates. (Assume the slab has the same area as the plates and is oriented parallel to them but not touching.)
24. Repeat the previous question, but this time insert a solid small metal sphere into the middle of the plate region.

## Exercises*

*Integrated Exercises (IEs) are two-part exercises. The first part typically requires a conceptual answer choice based on physical thinking and basic principles. The following part is quantitative calculations associated with the conceptual choice made in the first part of the exercise.*

**15.1 Electric Charge**

1. • What is the net charge of an object that has 1.0 million excess electrons?
2. • In walking across a carpet, you acquire a net negative charge of 50 $\mu$C. How many excess electrons do you have?
3. •• An alpha particle is the nucleus of a helium atom. (a) What would be the charge on two alpha particles? (b) How many electrons would need to be added to make an alpha particle into a neutral helium atom?
4. IE •• A glass rod rubbed with silk acquires a charge of $+8.0 \times 10^{-10}$ C. (a) Is the charge on the silk (1) positive, (2) zero, or (3) negative? Why? (b) What is the charge on the silk, and how many electrons have been transferred to the silk? (c) How much mass has the glass rod gained or lost?
5. IE •• A rubber rod rubbed with fur acquires a charge of $-4.8 \times 10^{-9}$ C. (a) Is the charge on the fur (1) positive, (2) zero, or (3) negative? Why? (b) What is the charge on the fur, and how much mass is transferred to or from the rod? (c) How much mass has the rubber rod lost or gained?

**15.2 Electrostatic Charging**

6. • An initially uncharged electroscope is polarized by bringing a negatively charged rubber rod near the bulb. If the bulb end of the electroscope acquires a net charge of +2.50 pC, how many excess electrons are on the leaf end?
7. • An initially uncharged electroscope is charged by induction by bringing a positively charged object near. If $3.22 \times 10^{8}$ electrons flow through the ground wire to Earth and the ground wire is then removed, what is the net charge on the electroscope?

* The bullets denote the degree of difficulty of the exercises: •, simple; ••, medium; and •••, more difficult.

### 15.3 Electric Force

8. **IE** • An electron at a certain distance from a proton is acted on by the electrical force. (a) If the electron were moved to twice that distance from the proton, would the electrical force be (1) 2, (2) 1/2, (3) 4, or (4) 1/4 times the original force? Explain. (b) If the initial electric force is called *F*, and the electron were moved to one-third the original distance toward the proton, what would be the new electrical force in terms of *F*?
9. • Two identical point charges are a fixed distance apart. By what factor would the magnitude of the electric force between them change if (a) one of their charges were doubled and the other were halved, (b) both their charges were halved, and (c) one charge were halved and the other were left unchanged?
10. • In a certain organic molecule, the nuclei of two carbon atoms are separated by a distance of 0.25 nm. What is the magnitude of the electric force between them?
11. • An isolated system consists of just an electron and proton separated by 2.0 nm. (a) What is the magnitude of the force on the electron? (b) What is the net force on the system?
12. **IE** • Two charges originally separated by a certain distance are moved farther apart until the force between them has decreased by a factor of 10. (a) Is the new distance (1) less than 10, (2) equal to 10, or (3) greater than 10 times the original distance? Why? (b) If the original distance was 30 cm, how far apart are the charges?
13. • Two charges are brought together until they are 100 cm apart, causing the electric force between them to increase by a factor of exactly 5. What was their initial separation distance?
14. • The distance between neighboring singly charged sodium and chlorine ions in crystals of table salt (NaCl) is $2.82 \times 10^{-10}$ m. What is the attractive electric force between the ions?
15. •• Two charges, $q_1$ and $q_2$, are located at the origin and at (0.50 m, 0), respectively. Where on the *x*-axis must a third charge, $q_3$, of arbitrary sign be placed to be in electrostatic equilibrium if (a) $q_1$ and $q_2$ are like charges of equal magnitude, (b) $q_1$ and $q_2$ are unlike charges of equal magnitude, and (c) $q_1 = +3.0\ \mu C$ and $q_2 = -7.0\ \mu C$?
16. •• Two negative point charges are separated by 10.0 cm and feel a mutual repulsive force of 3.15 $\mu$N. The charge of one is three times that of the other. (a) How much charge does each have? (b) What would be the force if the total charge were instead distributed equally between the charges?
17. •• An electron is placed on a line connecting two fixed point charges of equal charge but opposite sign. The distance between the charges is 30.0 cm and the charge of each is 4.50 pC. (a) Compute the force on the electron at 5.0-cm intervals starting 5.0 cm from the leftmost charge and ending 5.0 cm from the rightmost charge. (b) Plot the force on the electron-versus-electron location using your computed values. From the plot, can you make an educated guess as to where the electron feels the least force?
18. ••• Three charges are located at the corners of an equilateral triangle, as depicted in ▼**Figure 15.21**. What are the magnitude and the direction of the force on $q_1$?

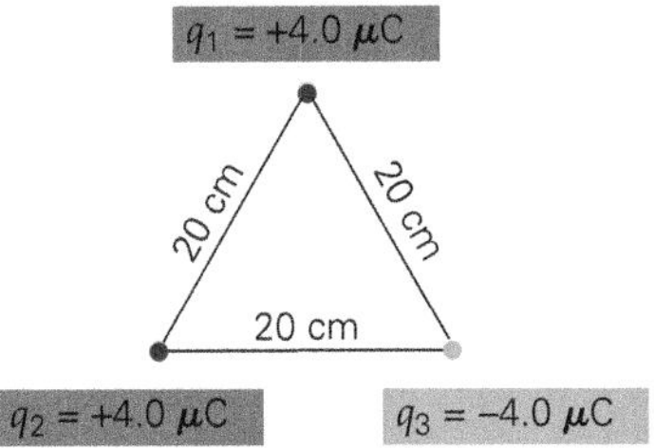


▲**FIGURE 15.21 Charge triangle** See Exercises 18, 31, and 32.

19. ••• Four charges are located at the corners of a square, as illustrated in ▼**Figure 15.22**. What are the magnitude and the direction of the force (a) on charge $q_2$ and (b) on charge $q_4$?

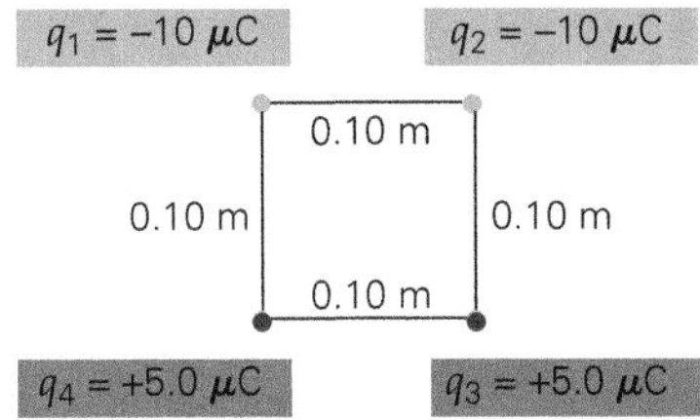


▲**FIGURE 15.22 Charge square** See Exercises 19, 33, and 37.

20. ••• Two 0.10-g cork balls are suspended from a point by insulating threads each 30 cm long. When the balls are given equal charges, they come to rest 18 cm apart, as shown in ▼**Figure 15.23**. What is the magnitude of the charge on each ball? (Neglect the mass of the thread.)

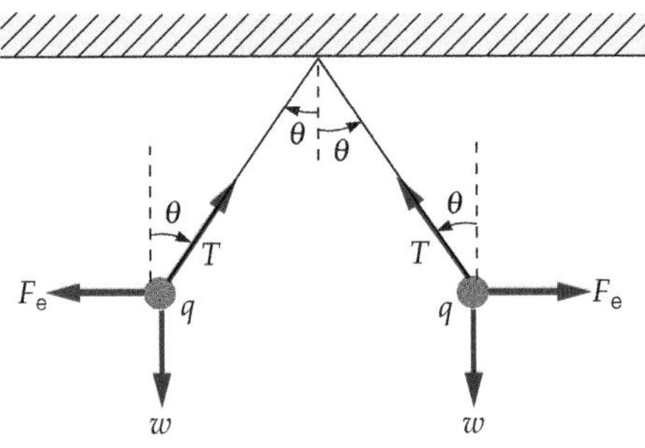


▲**FIGURE 15.23 Repelling pith balls** See Exercise 20.

### 15.4 Electric Field

21. IE • (a) If the distance from a charge is doubled, is the magnitude of the electric field (1) increased, (2) decreased, or (3) the same compared to the initial value? (b) If the original electric field due to a charge is $1.0 \times 10^{-4}$ N/C, what is the magnitude of the new electric field at twice the distance from the charge?
22. • An electron is acted on by an electric force of $3.2 \times 10^{-14}$ N. What is the magnitude of the electric field at the electron's location?
23. • An electron is acted on by two electric forces, one of $2.7 \times 10^{-14}$ N acting upward and a second of $3.8 \times 10^{-14}$ N acting to the right. What is the magnitude of the electric field at the electron's location?
24. • What are the magnitude and direction of the electric field at a point 0.75 cm away from a point charge of +2.0 pC?
25. • At what distance from a proton is the magnitude of its electric field $1.0 \times 10^5$ N/C?
26. IE •• Two fixed charges, −4.0 $\mu$C and −5.0 $\mu$C, are separated by a certain distance. (a) Is the net electric field at a location halfway between the two charges (1) directed toward the −4.0 $\mu$C charge, (2) zero, or (3) directed toward the −5.0 $\mu$C charge? Why? (b) If the charges are separated by 20 cm, calculate the magnitude of the net electric field halfway between the charges.
27. •• What would be the magnitude and the direction of an electric field that would just support the weight of a proton near the surface of the Earth? What about an electron?
28. IE •• Two charges, −3.0 $\mu$C and −4.0 $\mu$C, are located at (−0.50 m, 0) and (0.50 m, 0), respectively. There is a point on the $x$-axis between the two charges where the electric field is zero. (a) Is that point (1) left of the origin, (2) at the origin, or (3) right of the origin? (b) Find the location of the point where the electric field is zero.
29. •• Three charges, +2.5 $\mu$C, −4.8 $\mu$C, and −6.3 $\mu$C, are located at (−0.20 m, 0.15 m), (0.50 m, −0.35 m), and (−0.42 m, −0.32 m), respectively. What is the electric field at the origin?
30. •• Two charges of +4.0 $\mu$C and +9.0 $\mu$C are 30 cm apart. Where on the line joining the charges is the electric field zero?
31. •• What is the electric field at the center of the triangle in Figure 15.21?
32. •• Compute the electric field at a point midway between charges $q_1$ and $q_2$ in Figure 15.21.
33. •• What is the electric field at the center of the square in Figure 15.22?
34. •• A particle with a mass of $2.0 \times 10^{-5}$ kg and a charge of +2.0 $\mu$C is released in a (parallel plate) uniform horizontal electric field of 12 N/C. (a) How far horizontally does the particle travel in 0.50 s? (b) What is the horizontal component of its velocity at that point? (c) If the plates are 5.0 cm on each side, how much charge is on each?
35. •• Two very large parallel plates are oppositely and uniformly charged. If the field between the plates is $1.7 \times 10^6$ N/C, (a) how dense is the charge on each plate (in $\mu$C/m$^2$)? (b) How much total charge is on each plate if they are 15.0 cm on a side?
36. ••• Two square, oppositely charged conducting plates measure 20 cm on each side. The plates are close together and parallel to each other. They each have a total charge of +4.0 nC and −4.0 nC, respectively. (a) What is the electric field between the plates? (b) What force is exerted on an electron located between the plates? (c) What would be the electron's acceleration if it were released from rest?
37. ••• Compute the electric field at a point 4.0 cm from $q_2$ along a line running toward $q_3$ in Figure 15.22.
38. ••• Two equal and opposite point charges form a dipole, as shown in ▼ **Figure 15.24**. (a) Add the electric fields due to each end at point P, thus graphically determining the direction of the field there. (b) Derive a symbolic expression for the magnitude of the electric field at point P, in terms of $k$, $q$, $d$, and $x$. (c) If point P is very far away, use the exact result to show that $E \approx kqd/x^3$. (d) Why is it an inverse-*cube* falloff instead of inverse-*square*? Explain why it is not an inverse square falloff.

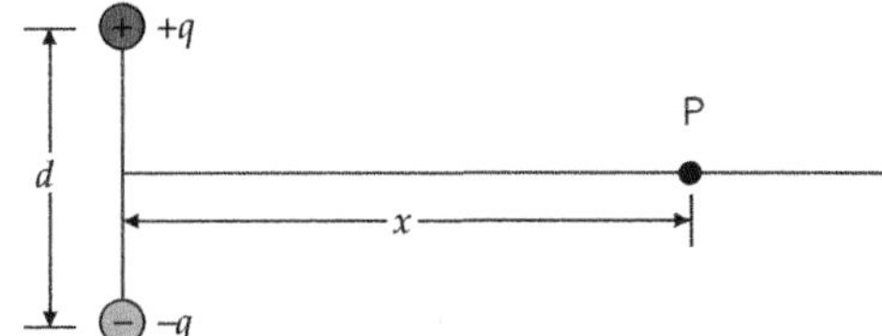


▲ FIGURE 15.24 **Electric dipole field** See Exercise 38.

### 15.5 Conductors and Electric Fields

39. IE • A solid conducting sphere is surrounded by a thick, spherical conducting shell. Assume that a total charge +$Q$ is placed at the center of the sphere and released. (a) After equilibrium is reached, the inner surface of the shell will have (1) negative, (2) zero, (3) positive charge. (b) In terms of $Q$, how much charge is on the interior of the sphere? (c) The surface of the sphere? (d) The inner surface of the shell? (e) The outer surface of the shell?
40. • In Exercise 39, what is the electric field direction (a) in the interior of the solid sphere, (b) between the sphere and the shell, (c) inside the shell, and (d) outside the shell?
41. •• In Exercise 39, write expressions for the electric field magnitude (a) in the interior of the solid sphere, (b) between the sphere and the shell, (c) inside the shell, and (d) outside the shell. Your answer should be in terms of $Q$, $r$ (the distance from the center of the sphere), and $k$.

42. •• A flat, triangular piece of metal with rounded corners has a net positive charge on it. Sketch the charge distribution on the surface and the electric field lines in two dimensions near the surface of the metal (including their direction).
43. •• Approximate a metal needle as a long cylinder with a very pointed, but slightly rounded, end. Sketch the charge distribution and outside electric field lines if the needle has an excess of electrons on it.
44. •• An electrically neutral thin, square metal slab, measuring 5.00 cm on a side, is placed in a uniform external field that is perpendicular to its square area. (a) If the top of the slab becomes negatively charged, what is the direction of the external field? (b) If the external field strength is 1250 N/C, what are the direction and strength of the field that is generated by the charges induced on the slab? (c) What is the total charge on the negative side of the slab?
45. •• An pair of identical and initially neutral thin, square metal slabs, measuring 5.00 cm on a side have a 2.00 mm spacing between them. With the slabs oriented horizontally, together they are placed in a vertical uniform external field of field strength 500 N/C. After a few seconds, it is noted that the top surface of the top slab is positively charged. (a) What is the direction of the external field, up or down? (b) What is the sign of the charge on the other three slab surfaces? (c) What is the direction and strength of the field between the plates? (d) What is the charge (magnitude) on each surface of the slabs?
46. •• The early twentieth century model of the hydrogen atom pictured the electron as in a circular orbit (radius $5.3 \times 10^{-11}$ m) with the proton at rest at the center. (a) What is the electric force on the electron? (b) What is the electron's orbital speed? (c) What is the electron's centripetal acceleration in units of $g$?
47. ••• A small negatively charged cork ball (mass $6.00 \times 10^{-3}$ g, charge $-1.50$ nC) is suspended vertically from a light non-conducting string of length 15.5 cm. This apparatus is then placed in a horizontal uniform electric field. After being released, the ball comes to a stable position at an angle of 12.3° to the left of the vertical. (a) What is the direction of the external electric field? (b) Determine the magnitude of the electric field.
48. ••• A positively charged particle with a charge of 9.35 pC is suspended in equilibrium in the electric field between two oppositely charged horizontal parallel plates. The square plates each have a charge of $5.50 \times 10^{-5}$ C, are separated by 6.25 mm, and have an edge length of 11.0 cm. (a) Which plate must be positively charged? (b) Determine the mass of the particle.
49. ••• An electron starts from one plate of a charged closely spaced (vertical) parallel plate arrangement with a velocity of $1.63 \times 10^4$ m/s to the right. Its speed on reaching the other plate, 2.10 cm away, is $4.15 \times 10^4$ m/s. (a) What type of charge is on each plate? (b) What is the direction of the electric field between the plates? (c) If the plates are square with an edge length of 25.4 cm, determine the charge on each.
50. ••• A proton is fired into a uniform electric field, opposite to the direction of the field. The proton's speed upon entering the field is $3.15 \times 10^5$ m/s, and it comes to rest 5.25 cm after entering the field. (a) What is the electric field strength? (b) What is the proton's velocity when it is only 3.50 cm into the field? [*Hint:* There is more than one answer. Why?]

# 16

# Electric Potential, Energy, and Capacitance

**This important warning includes an important electrical concept called voltage that is explored in this chapter.**

The sign in this photo warns us about the transformers behind the locked fence at this energy station. It certainly looks menacing and clearly something that should be avoided, but just what is "voltage?" Is it always harmful? What is meant by "high" voltage? You probably know that our everyday household circuits (in the United States) operate at 120 volts and can give a dangerous shock, so shouldn't they have such a label? You will find answers to these questions, along with explanations of other electrical phenomena, in this and following chapters. Here we will start with basics – a discussion of electric potential and potential difference (voltage) as well as several other important electrical quantities.

Also in this chapter are discussions involving practical applications. For example, your dentist's X-ray machine works by using high voltage to accelerate electrons. Heart defibrillators use capacitors to store the electrical energy required to stimulate a heart into correct rhythm. Here, and in following chapters, practical uses of electricity, such as appliances, medical instruments, electrical energy distribution systems, and household wiring, will be presented.

## 16.1 Electric Potential Energy and Electric Potential Difference

In Chapter 15, electrical effects were analyzed in terms of the electric field – a vector quantity. Recall that our initial study of mechanics used a vector approach – Newton's laws with forces (vectors). A scalar approach to mechanics led us to an energy (conservation) approach, employing such concepts as kinetic and potential energy. In many cases, the energy approach enabled a simpler path to solve problems in comparison to the vector (force) approach. In the same spirit, the principles in this chapter show how to extend energy methods to the study of electric fields.

### 16.1.1 Electric Potential Energy

To investigate electric potential energy, let's start with a simple electric field pattern: the field between two large, oppositely charged parallel plates. As was seen in Chapter 15, near the center of the plates the field is uniform in magnitude and direction (▼ **Figure 16.1a**). As shown, suppose a small positive charge $q_+$ is moved at constant velocity directly against the electric field from the negative (A) to the positive plate (B). To do this, an external force $F_{ext}$ must be applied to the charge. This force must have the same magnitude as the electric force on the charge, and therefore $F_{ext} = q_+E$. Because this force and the charge displacement are in the same direction, the work done by this external force is $W_{ext} = F_{ext}(\cos 0°)d = q_+Ed$.

Suppose the positive charge is now released from plate B. It will accelerate toward plate A, gaining kinetic energy. This kinetic energy gain is a result of the work done on the charge. Since the initial energy at plate B was *not* kinetic, it must have been some form of potential energy. Thus in moving the charge from A to B, the charge's **electric potential energy,** $U_e$, has increased; that is, $U_B > U_A$. This change is equal to the work done by the external force, or $\Delta U = U_B - U_A = q_+Ed$.

The gravitational analogy to this electric field is the uniform gravitational field near the Earth's surface. When an object is raised a vertical distance $h$, the change in its *gravitational* potential energy is positive ($U_B > U_A$) and equal to the work done by the external (lifting) force. At constant velocity, the external force must balance the object's weight, or $F_{ext} = w = mg$ (Figure 16.1b). Thus the increase in gravitational potential energy is $\Delta U_g = U_B - U_A = F_{ext}h = mgh$.

### 16.1.2 Electric Potential Difference

At any location, the electric field is defined as the electric force per unit positive test charge. Calculating the field $\vec{E}$ using this definition eliminates any dependence on the test charge and enables us to find the force on *any* charge $q$ placed at that location from $\vec{F}_e = q\vec{E}$. In a similar (but scalar) fashion, the **electric potential difference** ($\Delta V$) between any two points is defined as the change in potential energy per unit positive test charge moved:

$$\Delta V = \frac{\Delta U_e}{q_+} \quad \text{(electric potential difference)} \qquad (16.1)$$

The SI unit of electric potential difference is the joule per coulomb (J/C). This unit is named the **volt (V)** in honor of Alessandro Volta (1745–1827), an Italian scientist who constructed the first battery, thus 1 V = 1 J/C. Potential difference is commonly called **voltage**.

Notice a crucial point: Electric potential difference, although based on electric potential *energy* difference, is not the same. Electric potential difference is electric potential energy difference per unit charge, and therefore, like its vector counterpart $\vec{E}$, does not depend on the charge $q_+$ moved. Like electric field rather than electric force, electric potential difference is more useful than electric potential energy difference. This is because, once $\Delta V$ is known, $\Delta U_e$ can then be determined for any charge $q$ moved.

To illustrate this, let's calculate the electric potential difference associated with the uniform field (magnitude $E$) between two parallel plates:

$$\Delta V = \frac{\Delta U_e}{q_+} = \frac{q_+Ed}{q_+} = Ed \quad \begin{array}{c}\text{potential difference}\\ \text{(for parallel plates)}\end{array} \qquad (16.2)$$

Notice that the charge moved, $q_+$, cancels out. Thus the potential difference $\Delta V$ depends on only the plates – their field ($E$) and separation ($d$). This result can be described in electric potential language as follows: For a pair of oppositely charged parallel plates, the positively charged plate is at a higher electric potential than the negatively charged plate by an amount equal to $\Delta V$.

Notice that electric potential *difference* is defined without defining electric potential ($V$) itself. Although this may seem strange, there is good reason for it. Of the two, electric

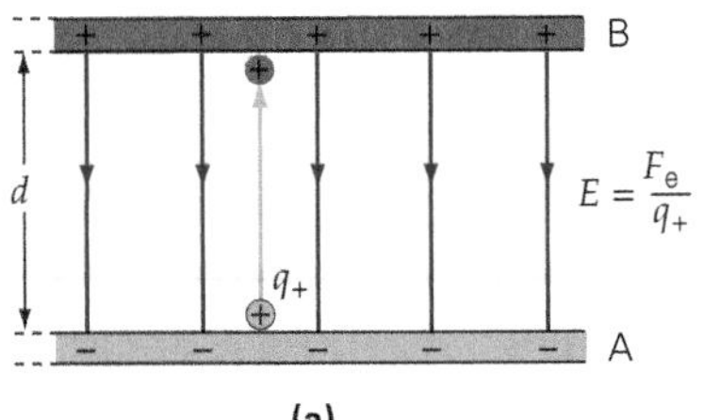


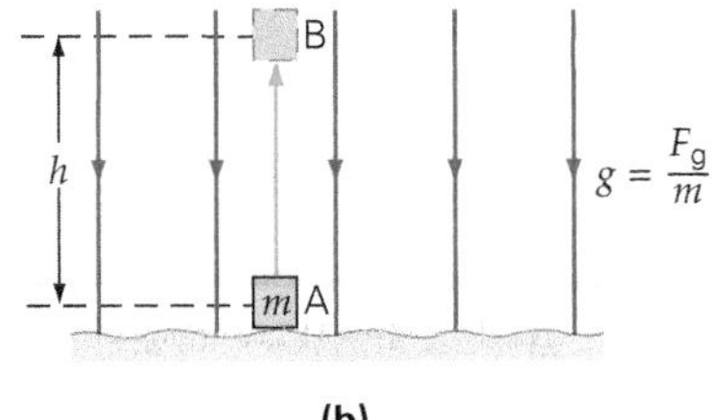


▲ FIGURE 16.1 **Changes in potential energy in uniform electric and gravitational fields (a)** Moving a positive charge $q_+$ against the electric field requires positive work and involves an increase in electric potential energy. **(b)** Moving a mass $m$ against the gravitational field requires positive work and involves an increase in gravitational potential energy.

potential difference (or voltage) is the only physically meaningful quantity – the quantity actually measured. The electric potential $V$, in contrast, isn't definable in an absolute way – it depends entirely on the choice of a reference point. This means that an arbitrary constant can be added to, or subtracted from, potential values ($V$), changing them, while producing no change to the physically meaningful quantity of potential difference.

This concept was encountered during the study of mechanical forms of potential energy where only *changes* in potential energies ($\Delta U$) were important. Specific values of potential energy ($U$) could be determined, but only if a zero reference point was chosen. For example, in the case of gravity, gravitational potential energy is sometimes chosen to be zero at the Earth's surface. However, it is just as correct, and sometimes more convenient, to define the zero point at an infinite distance from Earth (see Section 7.5).

The same concepts hold for electric potential energy and electric potential. The electric potential may be chosen as zero at the negative plate of a pair of parallel plates. However, it is sometimes convenient to locate the zero value at infinity. Either way, electric potential *differences* are unaffected. For a visualization of this, refer to ▼**Figure 16.2**. For one choice of zero electric potential, A is at a potential of +100 V and B is at +300 V. With a different zero reference choice, the potential at A might be +1100 V, in which case the potential at B would be +1300 V. Regardless of the location of zero potential chosen, B will always be 200 V higher in potential than A. That is the only physically meaningful quantity.

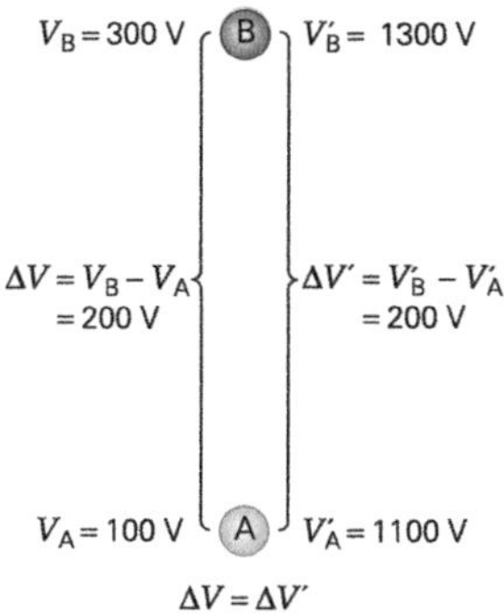


▲FIGURE 16.2 **Electric potential difference** The electric potential difference $\Delta V$ between any two locations A and B is independent of the choice of reference point for electric potential $V$.

### EXAMPLE 16.1: ENERGY METHODS IN MOVING A PROTON – ELECTRIC POTENTIAL ENERGY VERSUS ELECTRIC POTENTIAL

Imagine moving a proton from the negative plate to the positive plate of a parallel plate arrangement (▶**Figure 16.3a**). Assume plates are 1.50 cm apart, and the field is uniform with a magnitude of 1500 N/C. (a) What is the change in the proton's electric potential energy? (b) What is the electric potential difference (voltage) between the plates? (c) If the proton is released from rest at the positive plate (Figure 16.3b), what speed will it have just before it hits the negative plate?

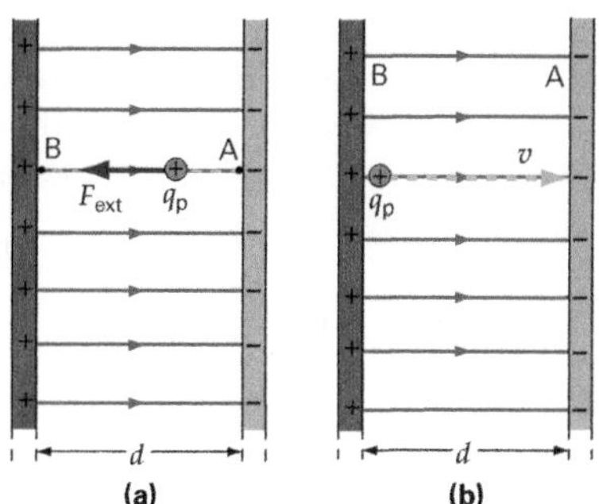


▲FIGURE 16.3 **Accelerating a charge (a)** Moving a proton from the negative to the positive plate increases the proton's potential energy. (See Example 16.1.) **(b)** When it is released from the positive plate, the proton accelerates toward the negative plate, gaining kinetic energy at the expense of electric potential energy.

**THINKING IT THROUGH.** (a) The change in potential energy of the proton is equal to the work done to move the proton. (b) The electric potential difference between the plates is found by dividing this potential energy change by the charge moved. (c) When the proton is released, its electric potential energy change (loss) is converted into kinetic energy. Since the proton's mass is known, its speed can be calculated.

**SOLUTION**

The magnitude of the electric field, $E$, is given. Because a proton is involved, its mass and charge can be found in Table 15.1.

***Given:***

$$E = 1500\,\text{N/C}$$
$$q_p = +1.60 \times 10^{-19}\,\text{C}$$
$$m_p = 1.67 \times 10^{-27}\,\text{kg}$$
$$d = 1.50\,\text{cm} = 1.50 \times 10^{-2}\,\text{m}$$

***Find:***

**(a)** $\Delta U_e$ (potential energy difference of proton)
**(b)** $\Delta V$ (potential difference between plates)
**(c)** $v$ (speed of proton just before it reaches negative plate)

**(a)** The electric potential energy increases as positive work is done to move the proton against the field, toward the positive plate:

$$\Delta U_e = q_p E d = (+1.60 \times 10^{-19}\,\text{C})(1500\,\text{N/C})(1.50 \times 10^{-2}\,\text{m}) = +3.60 \times 10^{-18}\,\text{J}$$

**(b)** The potential difference, or voltage, is the potential energy change per unit charge Equation 16.1:

$$\Delta V = \frac{\Delta U_e}{q_p} = \frac{+3.60 \times 10^{-18}\,\text{J}}{+1.60 \times 10^{-19}\,\text{C}} = +22.5\,\text{V}$$

Thus one would describe this by saying the positive plate is 22.5 V higher in electric potential than the negative one.

(c) The total energy of the proton is constant; therefore, $\Delta K + \Delta U_e = 0$. The proton has no initial kinetic energy ($K_o = 0$). Hence, $\Delta K = K - K_o = K$. Thus the change (gain) in kinetic energy is just the opposite of the change (loss) of electric potential energy or

$$\Delta K = K = -\Delta U_e$$

hence

$$\frac{1}{2}m_p v^2 = -\Delta U_e$$

When it reaches the negative plate, the proton's potential energy change is negative thus $\Delta U_e = -3.60 \times 10^{-18}$ J and its speed is given by

$$v = \sqrt{\frac{2(-\Delta U_e)}{m_p}} = \sqrt{\frac{2[-(-3.60\times10^{-18}\text{ J})]}{1.67\times10^{-27}\text{ kg}}} = 6.57\times10^4 \text{ m/s}$$

Notice that even though the kinetic energy gained is very small, the proton acquires a high speed because its mass is extremely small.

**FOLLOW-UP EXERCISE.** In this Example, what would be your answers if an alpha particle were moved instead of a proton? (An alpha particle is the nucleus of a helium atom and has a charge of $+2e$ and a mass approximately four times that of a proton.)

In Example 16.1 it might be asked whether the result depends on the path. As depicted in ▼**Figure 16.4**, the work done in moving the proton from plate A to plate B is the same, regardless of the route. The alternative wiggly paths from A to B (as well as A to B′) require the same work as do the straight-line paths, because no work is done by a force acting at right angles to the movement of the charge. So whether the charge ends at B or B′ does not matter, and the path to get there also does not matter. Hence it can be concluded that B and B′ (and all points on plate B) have the same value electric potential.

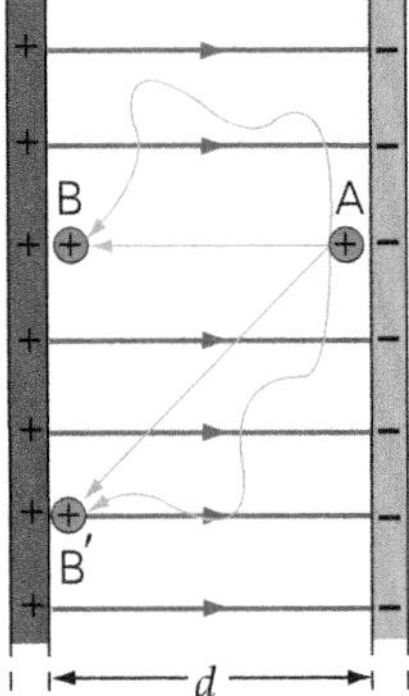


▲**FIGURE 16.4 Electric potential energy and electric potential differences are path independent** The work done to move a charge between any two points in an electric field, such as A to B, or A to B′, is independent of the path. Here B and B′ are at the same potential. Although shown for a uniform field, the result is true for any shape of electric field.

The gravitational analogy to Example 16.1 is that of raising an object in a uniform gravitational field, thus doing positive work on it and increasing its gravitational potential energy. However, the analogy is not perfect. This is because, unlike gravity, which is only attractive and has just one type of mass, with electricity there are two types of charge, and the forces can be repulsive or attractive.

To understand this difference, consider how Example 16.1 would differ if an electron instead of a proton were moved. Because an electron is negative, it is attracted to plate B, and thus the external force would be opposite the electron's displacement. Hence to move an electron requires negative work to be done on it, thereby decreasing the electric potential energy. Unlike the proton, the electron is attracted to the positive plate (at a higher electric potential). Thus, if released, it would move toward the plate with higher potential. The proton would have started moving toward the region of lower potential. Regardless, both the proton and electron ended up losing electric potential energy and gaining kinetic energy.

The behavior of charged particles in electric fields can be summarized in "potential language" as follows:

> Positive charges, when released in an electric field, accelerate toward regions of lower electric potential, losing electric potential energy and gaining kinetic energy.
>
> Negative charges, when released in an electric field, accelerate toward regions of higher electric potential, also losing electric potential energy and gaining kinetic energy.

Keep these in mind as you study Example 16.2 which illustrates a medical application – the creation of X-rays from fast-moving electrons, accelerated by large electric potential differences (voltages).

**EXAMPLE 16.2: CREATING X-RAYS – ACCELERATING ELECTRONS**

Modern dental offices use X-ray machines for diagnosing hidden dental problems (▶**Figure 16.5a**). Electrons are typically accelerated from rest through electric potential differences (voltages) of 25 000 V. When they hit the positive plate, they are stopped and their kinetic energy is converted into high-energy particles of light called *X-ray photons* (Figure 16.3b). Suppose a single electron's kinetic energy is distributed equally among five X-ray photons. How much energy would each photon have?

**THINKING IT THROUGH.** Since electrons start from rest, they have no initial kinetic energy. Thus energy conservation, the kinetic energy of one electron is equal in magnitude to the electric potential energy it loses. From the kinetic energy lost by one electron, the energy of one X-ray photon can be calculated.

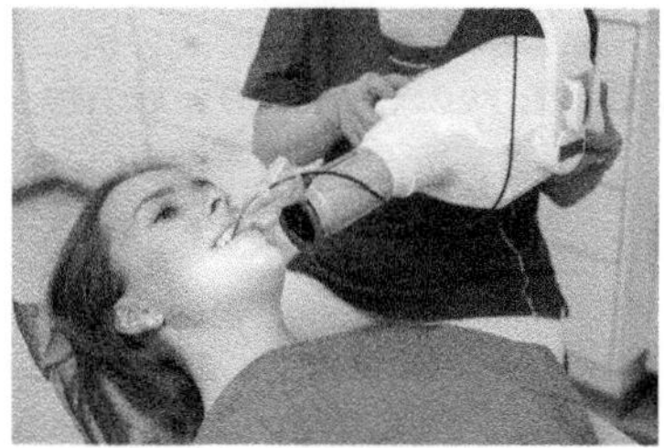
(a)

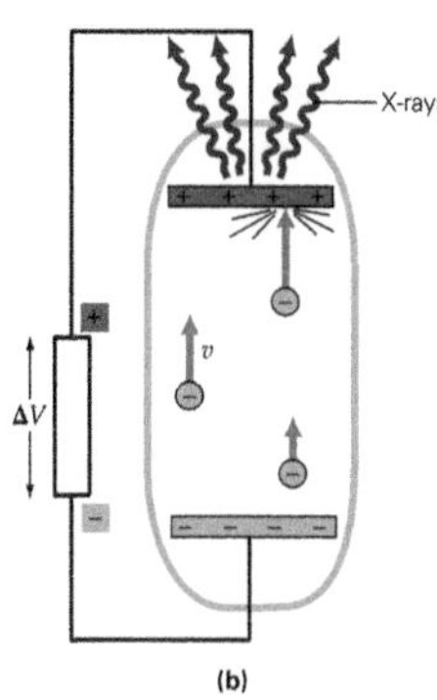

(b)

▲ **FIGURE 16.5 X-ray production** **(a)** A dental X-ray machine in operation. **(b)** A schematic diagram of the X-ray production using accelerated electrons. See Example 16.2.

**SOLUTION**

The charge of an electron is known (Table 15.1), and the accelerating voltage is given.

***Given:***

$$q = -1.60 \times 10^{-19}\,\text{C}$$
$$\Delta V = 2.5 \times 10^{4}\,\text{V}$$

***Find:*** $E_x$ (energy of one X-ray photon)

The electron leaves the negatively charged plate and moves toward the positive plate, which is the region of highest electric potential (i.e., it accelerates "uphill" in terms of potential). Thus, the change in its electric potential energy is

$$\Delta U_e = q\Delta V = (-1.60 \times 10^{-19}\,\text{C})(+2.5 \times 10^{4}\,\text{V}) = -4.0 \times 10^{-15}\,\text{J}$$

The gain in kinetic energy comes from this loss in electric potential energy. Because the electrons have no appreciable kinetic energy at the start, their final kinetic energy is

$$K = |\Delta U_e| = 4.0 \times 10^{-15}\,\text{J}$$

Since this is equally shared, each of the five photons will have an energy of

$$E_x = \frac{K}{5} = 8.0 \times 10^{-16}\,\text{J}$$

**FOLLOW-UP EXERCISE.** In this Example, use energy methods to determine the speed of one electron when it is halfway to the positive plate.

### 16.1.3 Electric Potential Difference Due to a Point Charge

In nonuniform electric fields, the potential difference between two points is, in principle, determined by applying the definition given in Equation 16.1. However, because the field (and thus work done) varies along the path, the calculation is beyond the scope of this text. In our study, the only nonuniform field considered in any detail will be that due to a point charge (▼ **Figure 16.6**). The potential difference (voltage) between two locations at distances $r_A$ and $r_B$ from a point charge $q$ is given by:

$$\Delta V = \frac{kq}{r_B} - \frac{kq}{r_A} \quad \text{(point charges only)} \qquad (16.3)$$

In Figure 16.6, the point charge is positive. Since point B is closer to the charge than A, the potential difference is positive; that is, $V_B - V_A > 0$, or $V_B > V_A$. Remember that changes in potential are determined by visualizing the movement of a positive test charge. Here it takes positive work to move such a charge from A to B, thus $V_B > V_A$. In general, then, electric potential increases as one moves nearer to a positive charge. Just as between the parallel plates, the work done during path II is the same as that for path I.

But what would the result be if the central point charge were negative? In this case, B would be at a lower potential than A because the work required to move a positive test charge closer would be negative (why?).

In summary, changes in electric potential follow these rules:

> Electric potential increases when moving nearer to positive charges or farther from negative charges
>
> and
>
> Electric potential decreases when moving farther from positive charges or nearer to negative charges.

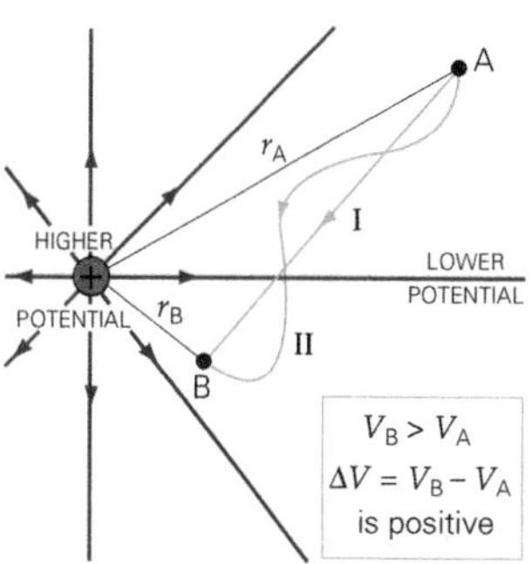


▲ **FIGURE 16.6 Electric field and potential due to a point charge** Using a positive test charge shows that electric potential increases as you move closer to a positive charge. Thus, B is at a higher potential than A. In addition, the difference is independent of the path taken.

The electric potential at a very large distance from a point charge is usually chosen to be zero (as was done for the gravitational case of a point mass in Chapter 7). With this choice, the *electric potential* $V$ at a distance $r$ from a point charge is

$$V = \frac{kq}{r} \quad \text{(electric potential for a point charge)} \qquad (16.4)$$

Even though this expression is for the electric potential, $V$, keep in mind that only electric potential *differences* ($\Delta V$) are important, as Integrated Example 16.3 illustrates.

---

**INTEGRATED EXAMPLE 16.3: DESCRIBING THE HYDROGEN ATOM – POTENTIAL DIFFERENCES NEAR A PROTON**

According to the Bohr model of the hydrogen atom, the electron in orbit around the proton can exist only in certain sized circular orbits. The smallest orbit has a radius of 0.0529 nm, and the next largest has a radius of 0.212 nm. (a) How do the values of electric potential at each orbit compare: (1) the smaller orbit is at a higher potential, (2) the larger is at a higher potential, or (3) they have the same potential? Explain your reasoning. (b) Verify your answer to part (a) by calculating the values of the electric potential at the locations of the two orbits.

(a) **CONCEPTUAL REASONING.** The electron orbits in the field of a proton, whose charge is positive. Because electric potential increases with decreasing distance from a positive charge, the answer must be (1).

(b) **QUANTITATIVE REASONING AND SOLUTION.** The charge on the proton is known, so Equation 16.4 can be used to find the potential values. Listing the values,

***Given:***

$q_p = +1.60 \times 10^{-19}$ C
$r_1 = 0.0529 \text{ nm} = 5.29 \times 10^{-11} \text{ m}$
$r_2 = 0.212 \text{ nm} = 2.12 \times 10^{-10} \text{ m}$ (*Note:* $1 \text{ nm} = 10^{-9}$ m)

***Find:*** $V$ (the electric potential at each orbit)

Applying Equation 16.4 to the smaller orbit first:

$$V_1 = \frac{kq_p}{r_1} = \frac{(9.00 \times 10^9 \text{ N·m}^2/\text{C}^2)(+1.60 \times 10^{-19} \text{ C})}{5.29 \times 10^{-11} \text{ m}} = +27.2 \text{ V}$$

and then to the larger orbit:

$$V_2 = \frac{kq_p}{r_2} = \frac{(9.00 \times 10^9 \text{ N·m}^2/\text{C}^2)(+1.60 \times 10^{-19} \text{ C})}{2.12 \times 10^{-10} \text{ m}} = +6.79 \text{ V}$$

**FOLLOW-UP EXERCISE** In this Example, suppose the electron were moved from the smallest to the next orbit. (a) What would be the change in electrical potential? (b) What would be the change in electric potential energy?

---

## 16.1.4 Electric Potential Energy of Point Charge Configurations

In Chapter 7 the gravitational potential energy of systems of masses was explored. The expressions for electric force and gravitational force are mathematically similar, and so therefore are those for potential energy, except that charge takes the place of mass (however, remember that charge comes in two signs). In the case of two masses, the mutual gravitational potential energy is negative, because the force is always attractive. For electric potential energy, the result can be positive *or* negative, because the electric force can be repulsive or attractive.

For example, consider a positive point charge, $q_1$, fixed in space. Suppose a second positive charge $q_2$ is brought in from a very large distance (i.e., let its initial location $r \to \infty$) to a distance $r_{12}$ (▼ **Figure 16.7a**). In this case, the work to assemble them is positive. Therefore, this system (of two like charges) gains electric potential energy. The potential at a large distance ($V_\infty$) is, as is usual for point charges and masses, chosen as zero. (Recall the zero reference point is arbitrary.) Thus, from Equation 16.3, the change in potential energy is

$$\Delta U_e = q_2 \Delta V = q_2 (V_1 - V_\infty) = q_2 \left( \frac{kq_1}{r_{12}} - 0 \right) = \frac{kq_1 q_2}{r_{12}}$$

With $V_\infty$ (and therefore $U_\infty$) chosen as zero, then the change in electric potential energy is $\Delta U_e = U_{12} - U_\infty = U_{12}$. With this choice of zero reference, the conclusion is that the

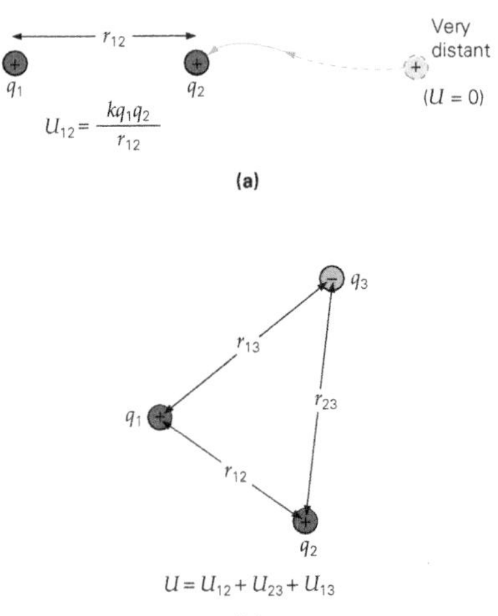


▲ **FIGURE 16.7 Mutual electric potential energy of point charges** **(a)** If a positive charge is moved from a large distance to a distance $r_{12}$ from another positive charge, there is an increase in potential energy because positive work must be done to bring the mutually repelling charges closer. **(b)** For more than two charges, the system's electric potential energy is the sum of the mutual potential energies of each pair.

electric potential energy associated with any two-charge system is

$$U_{12} = \frac{kq_1q_2}{r_{12}} \quad \text{(electric potential energy of two charges)} \quad (16.5)$$

Thus if the two same-sign charges are released, they will move apart, gaining kinetic energy as they lose potential energy. Conversely, it would take positive work to increase the separation of two opposite charges, such as the proton and the electron, much like stretching a spring. (See the Follow-Up Exercise in Integrated Example 16.3.)

Because energy is a scalar, for a configuration of any number of point charges (a multi-charge system), the total potential energy ($U$) is the sum of the mutual potential energies of all pairs of charges:

$$U = U_{12} + U_{23} + U_{13} + U_{14} \cdots \quad (16.6)$$

Only the first three terms of Equation 16.6 are needed for the configuration shown in Figure 16.7b. Note that the signs of the charges keep things straight mathematically, as the biomolecular situation in Example 16.4 shows.

---

### EXAMPLE 16.4: MOLECULE OF LIFE – THE ELECTRIC POTENTIAL ENERGY OF A WATER MOLECULE

The water molecule is the foundation of life as we know it. Many of its properties (such as the reason it is a liquid on the Earth's surface) are related to the fact that it is a permanent polar molecule (see Section 15.4 on electric dipoles). A simple picture of the water molecule, including the charges, is shown in ▶ **Figure 16.8.** The distance from each hydrogen atom to the oxygen atom is $9.60 \times 10^{-11}$ m, and the angle ($\theta$) between the two hydrogen-oxygen bond directions is 104°. What is the total electrostatic energy of the water molecule?

**THINKING IT THROUGH.** The model of this molecule involves trigonometry. The total electrostatic potential energy is the three charges. The charges are given, but the distance algebraic sum of the potential energies of the three pairs of between the hydrogen atoms must be calculated using charges (that is, Equation 16.6 will have three terms).

**SOLUTION**

The following data are taken from Figure 16.6.

***Given:***

$q_1 = q_2 = +5.20 \times 10^{-20}$ C

$q_3 = -10.4 \times 10^{-20}$ C

$r_{13} = r_{23} = 9.60 \times 10^{-11}$ m

$\theta = 104°$

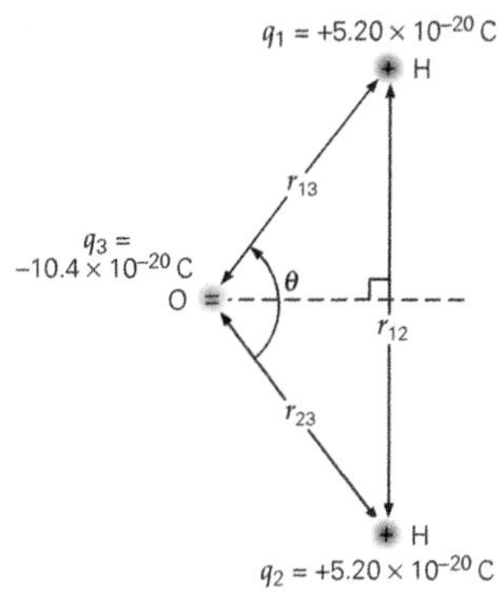


▲**FIGURE 16.8 Electrostatic potential energy of a water molecule** The charges shown on the water molecule are *average* charges because the atoms within the molecule share electrons. That is the reason why the charges on the water molecule ends can be smaller than the charge on the electron or proton.

***Find:*** $U$ (electrostatic potential energy of water molecule)

From the geometry: $\dfrac{(r_{12}/2)}{r_{13}} = \sin\left(\dfrac{\theta}{2}\right)$. Solving for $r_{12}$:

$$r_{12} = 2r_{13}\left(\sin\frac{\theta}{2}\right) = 2(9.60 \times 10^{-11}\,\text{m})(\sin 52°) = 1.51 \times 10^{-10}\,\text{m}$$

Now let's find the mutual electric potential energy each pair. Things are a bit easier, because $U_{13} = U_{23}$. (Can you explain why?) Applying Equation 16.5 pair by pair:

$$\begin{aligned} U_{12} &= \frac{kq_1q_2}{r_{12}} \\ &= \frac{(9.00 \times 10^9\,\text{N·m}^2/\text{C}^2)(+5.20 \times 10^{-20}\,\text{C})(+5.20 \times 10^{-20}\,\text{C})}{1.51 \times 10^{-10}\,\text{m}} \\ &= +1.61 \times 10^{-19}\,\text{J} \end{aligned}$$

and

$$\begin{aligned} U_{13} = U_{23} &= \frac{kq_2q_3}{r_{23}} \\ &= \frac{(9.00 \times 10^9\,\text{N·m}^2/\text{C}^2)(+5.20 \times 10^{-20}\,\text{C})(-10.4 \times 10^{-20}\,\text{C})}{9.60 \times 10^{-11}\,\text{m}} \\ &= -5.07 \times 10^{-19}\,\text{J} \end{aligned}$$

Thus the *total* electrostatic potential energy is

$$\begin{aligned} U &= U_{12} + U_{13} + U_{23} \\ &= (+1.61 \times 10^{-19}\,\text{J}) + (-5.07 \times 10^{-19}\,\text{J}) + (-5.07 \times 10^{-19}\,\text{J}) \\ &= -8.53 \times 10^{-19}\,\text{J} \end{aligned}$$

The negative result means that the molecule requires positive work to break it apart. (That is, it must be pulled apart.)

**FOLLOW-UP EXERCISE.** Another common neutral, but permanently polar, molecule is carbon monoxide (CO), a toxic gas commonly produced during incomplete hydrocarbon fuel combustion. The carbon atom is, on average, positively charged and the oxygen atom is, on average, negative. The distance between the carbon and oxygen atoms is 0.120 nm, and the total electrostatic potential energy of this molecule is $-3.27 \times 10^{-19}$ J. Determine the magnitude of the (average) charge on each end of the molecule.

## 16.2 Equipotential Surfaces and the Electric Field

### 16.2.1 Equipotential Surfaces

Suppose a positive charge is moved perpendicularly to an electric field (such as path I of ▼**Figure 16.9a**). As the charge moves from A to A′, *no work* is done on it, thus its potential energy does not change, or $\Delta U_{AA'} = 0$. Thus, points A and A′ – and indeed *all* points on path I – are at the same potential; that is,

$$\Delta V_{AA'} = V_{A'} - V_{A} = \frac{\Delta U_{AA'}}{q} = 0 \quad \text{or} \quad V_{A} = V_{A'}$$

This result actually holds for all points on the plane parallel to the plates and containing path I. A surface, like this plane, on which the potential is constant, is an **equipotential surface** (or

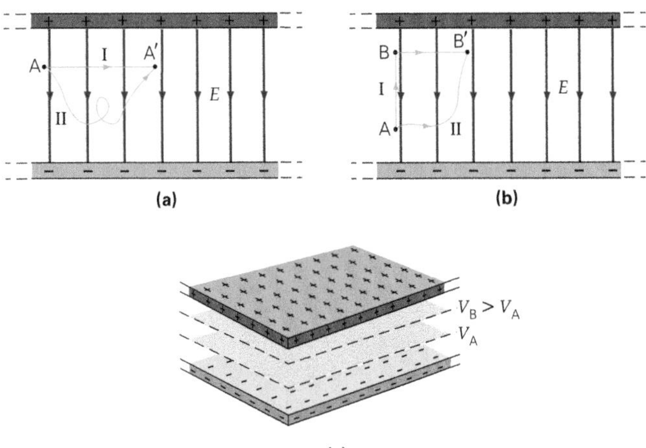


▲FIGURE 16.9 **Construction of equipotential surfaces between parallel plates (a)** The work done in moving a charge is zero as long as you start and stop on the same equipotential surface. (Compare paths I and II.) **(b)** Once the charge moves to a higher potential (for example, from point A to point B), it can stay on that new equipotential surface by moving perpendicularly to the electric field (B to B′). The change in potential is independent of the path, since the same change occurs whether path I or path II is used. (Why?) **(c)** The actual equipotential surfaces within the parallel plates are planes parallel to those plates. Two such plates are shown, with $V_B > V_A$.

simply an *equipotential*). The word equipotential means "same potential." Note that, unlike this special case, equipotential surfaces, are, in general, not planes.

Since no work is required to move a charge along an equipotential surface, it must be generally true that:

> Equipotential surfaces are always at right angles to the electric field.

Because the work done is independent of path, it is the same whether path I, path II, or any other path from $A$ to $A'$ is taken (Figure 16.9a). As long as the charge returns to the same equipotential surface from which it started, the work done on it is zero and the value of the electric potential is the same.

If the positive charge is moved opposite to the direction of $\vec{E}$ (e.g., path I in Figure 16.9b) – at right angles to the equipotential surfaces – the electric potential energy, and hence the electric potential, increases. When B is reached, the charge is on a different equipotential surface – one of a higher potential value than the surface of A. The work would be the same even if the charge was moved from A to B′; hence B and B′ are on the same equipotential surface. In summary, then, for parallel plates, the equipotentials are planes parallel to the plates (Figure 16.9c).

To help understand these equipotential surfaces, consider a gravitational analogy. If an object is lifted by an external force to a height $h = h_B - h_A$ (A to B in ▼**Figure 16.10**), the work done by that force ($mgh$) is positive. For horizontal movement, the potential energy does not change. Thus the plane at height $h_B$ is a gravitational equipotential surface – as is the plane at height $h_A$ – but at a lower value than the plane at $h_B$. Therefore, surfaces of constant gravitational potential energy are planes parallel to the Earth's surface. Topographic maps, which display land contours by plotting lines of constant elevation (relative to sea level), are also maps of constant gravitational potential (▶**Figure 16.11a** and **b**). Note how the equipotentials near a point charge (Figure 16.11c and d) are qualitatively similar to the gravitational contours due to a hill.

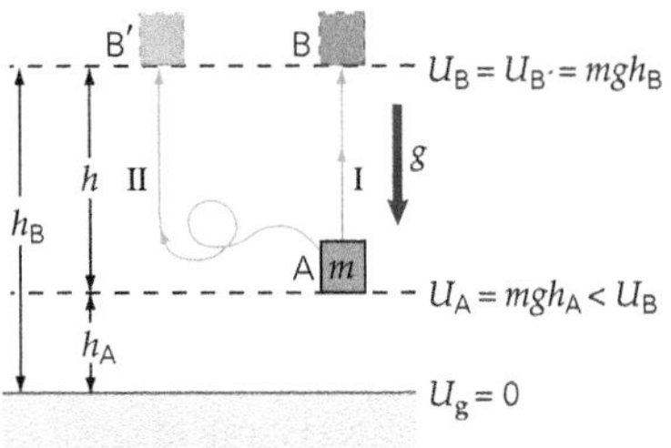


▲FIGURE 16.10 **Gravitational potential energy analogy** Raising an object in a uniform gravitational field results in an increase in gravitational potential energy; that is $U_B > U_A$. At a given height, the object's potential energy is constant as long as it remains on that (gravitational) equipotential surface.

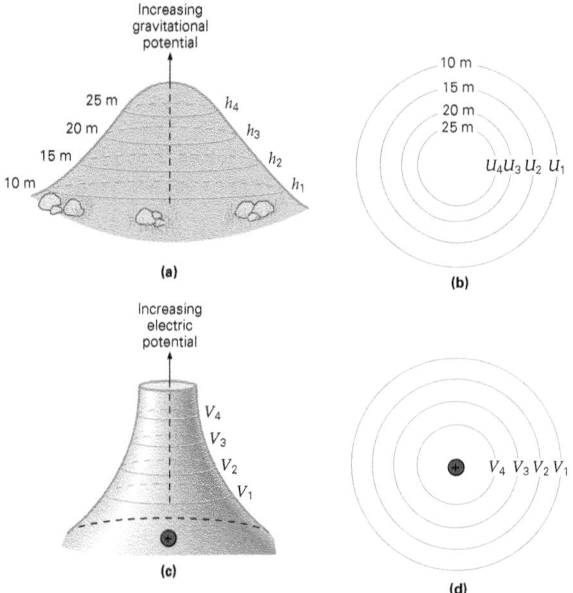


▲ **FIGURE 16.11 Topographic maps – a gravitational analogy to equipotential surfaces (a)** A symmetrical hill with slices at different elevations. Each slice is a plane of constant gravitational potential. **(b)** A topographic map of the slices in (a). The contours, where the planes intersect the surface, represent increasingly larger values of gravitational potential as one goes up the hill. **(c)** The electric potential $V$ near a point charge $q$ forms a similar symmetrical hill. $V$ is constant at fixed distances from $q$. **(d)** Electrical equipotentials around a point charge are spherical (in two dimensions they are circles) centered on the charge. The closer the equipotential to the positive charge, the larger its electric potential.

It is a useful skill to be able to visualize and sketch equipotential surfaces knowing the electric field pattern, and vice versa. ▼ **Figures 16.12** and **16.13** show examples of both these concepts. After studying these two figures, you should understand how these ideas were used to construct the equipotentials of an electric dipole in ▶ **Figure 16.14** knowing the electric field pattern from Chapter 15.

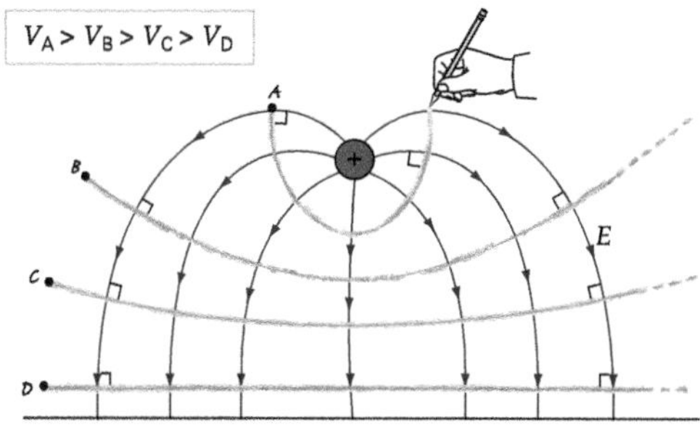


▲ **FIGURE 16.12 Sketching equipotentials from electric field lines** If the electric field line pattern is known, pick a point in the region of interest and move the pencil, being sure to plan ahead so as to cross the next field line at right angles.

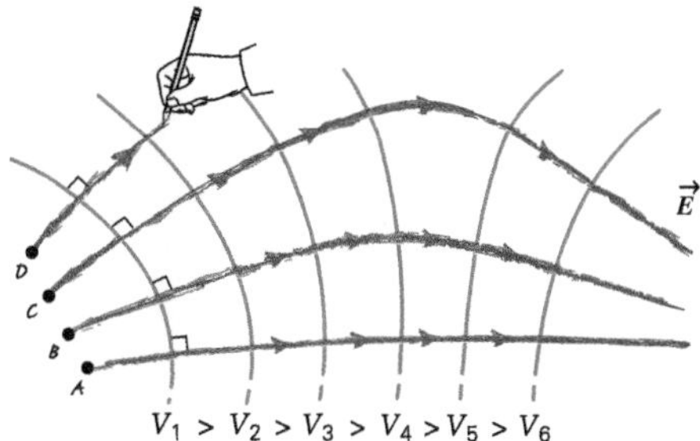


▲ **FIGURE 16.13 Mapping the electric field from equipotentials** If the equipotentials are known, then to determine the electric field line pattern, start at any convenient point, and, being careful to plan ahead, trace a line that crosses each equipotential at right angles.

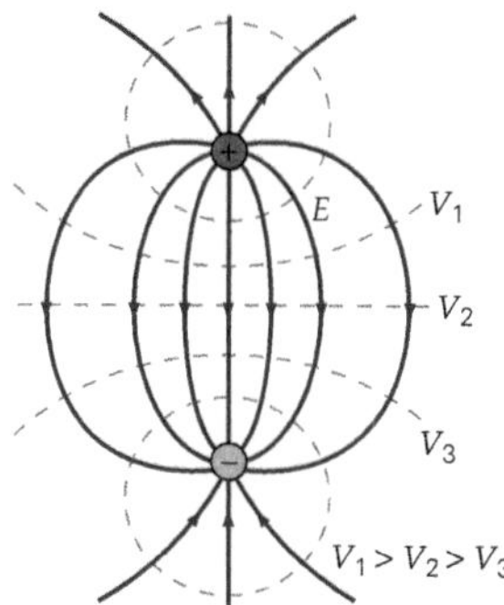


▲ **FIGURE 16.14 Equipotentials of an electric dipole** The field line pattern (solid) for an electric dipole is known (Chapter 15). Using the ideas in Figure 16.12, the equipotentials (dashed lines) can be determined.

To determine the mathematical relationship between electric field and electric potential, consider again the uniform electric field (▶ **Figure 16.15**). The potential difference ($\Delta V$) between any two equipotential planes ($V_1$ and $V_2$ in the figure) can be calculated with the same technique used to derive Equation 16.2. The result is

$$\Delta V = V_3 - V_1 = E \cdot \Delta x \tag{16.7}$$

Thus, if you start on equipotential surface 1 and move opposite to the electric field to surface 3, there is a potential increase ($\Delta V$) that depends on the electric field strength ($E$) and the distance ($\Delta x$). Furthermore, for a given distance $\Delta x$, moving perpendicular to the equipotential surfaces and opposite the electric field yields the maximum gain in potential. Think of moving a distance $\Delta x$ in any direction, starting from surface 1. To maximize the increase in $V$ would be to move to surface 3. Any step in a different direction (i.e., not perpendicular to surface 1 – for example, ending on surface 2) yields a smaller increase in potential.

This result can be summarized into a general principle:

The direction of $\vec{\mathbf{E}}$ is that in which the electric potential $V$ decreases most rapidly, or equivalently, opposite the direction in which $V$ increases most rapidly.

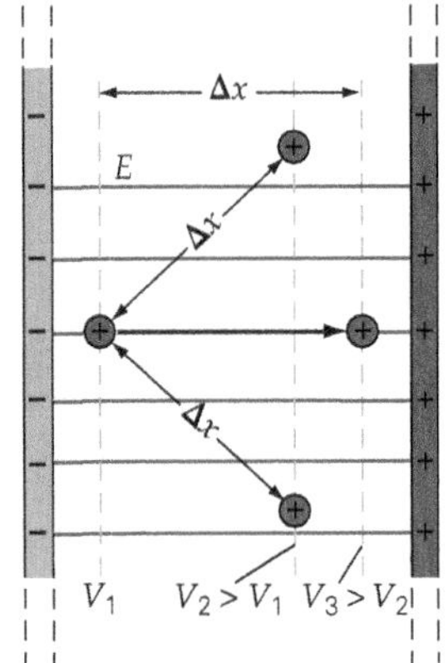


▲FIGURE 16.15 **Relationship between the electric potential change ($\Delta V$) and electric field ($\vec{E}$)** The electric field direction is that of maximum *decrease* in potential, or opposite the direction of maximum increase in potential. (Here, this maximum is in the direction of the black horizontal arrow, not the angled ones; why?) The electric field *magnitude* is given by the maximum rate at which the potential changes over distance.

Mathematically, then, at any location, the magnitude of $\vec{E}$ is the maximum rate of change of the potential with distance, or

$$E = \left|\frac{\Delta V}{\Delta x}\right|_{\max} \tag{16.8}$$

The unit of electric field is volts per meter (V/m). Previously, $E$ was expressed in newtons per coulomb (N/C; see Section 15.4). You should show, through dimensional analysis, that 1 V/m = 1 N/C.

A graphical depiction of the use of Equation 16.8 is shown in ▼**Figure 16.16**. Suppose an estimate of $\vec{E}$ at point $P$ in Figure 16.16 is desired. The potential values 1.0 cm on either side of $P$ must be known (shown). From these, it can be seen that the electric field points roughly from A to B, from high potential to low potential, and its approximate magnitude is the rate at which the potential changes with distance or

$$E = \left|\frac{\Delta V}{\Delta x}\right|_{\max} = \frac{(1000\,\text{V} - 950\,\text{V})}{2.0\times10^{-2}\,\text{m}}$$
$$= 2.5\times10^{3}\,\text{V/m}$$

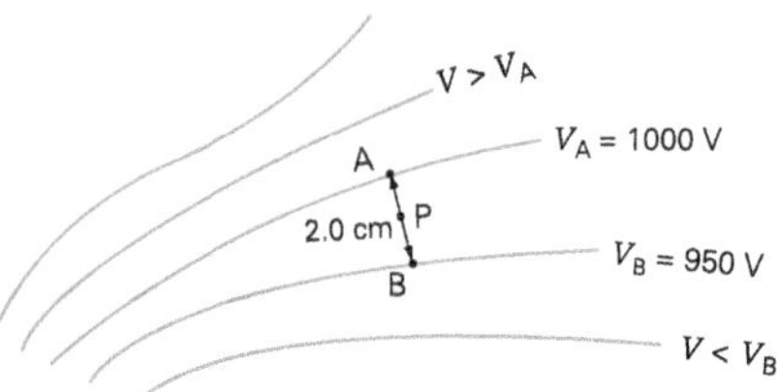


▲FIGURE 16.16 **Estimating the magnitude of the electric field** The magnitude of the potential change per meter at any point gives the strength of the electric field at that point. (See text for discussion and calculations.)

Equipotential surfaces can be useful for describing the field near a charged conductor, as Conceptual Example 16.5 shows.

---

### CONCEPTUAL EXAMPLE 16.5: THE EQUIPOTENTIAL SURFACES OUTSIDE A CHARGED CONDUCTOR

A solid conductor with an excess positive charge is shown in

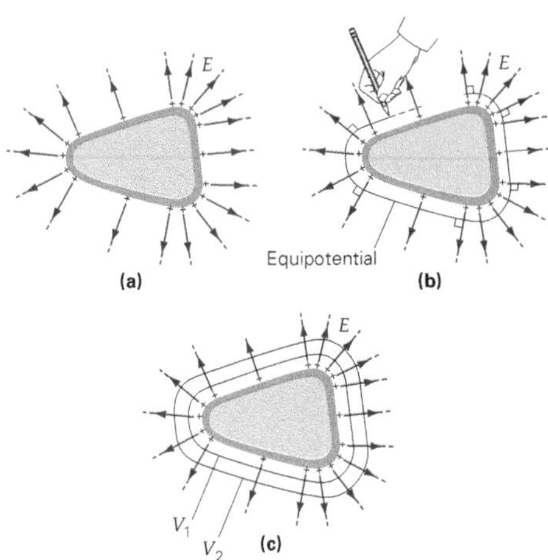


▲FIGURE 16.17 **Equipotential surfaces near a positively charged conductor.** See Conceptual Example 16.5.

▼**Figure 16.17a**. Which of the following best describes the shape of the equipotential surfaces just outside the conductor's surface: (a) flat planes, (b) spheres, or (c) approximately the shape of the conductor's surface? Explain your reasoning.

**REASONING AND ANSWER.** Choice (a) can be eliminated immediately, because flat (plane) equipotential surfaces are associated with flat plates. While it might be tempting to pick answer (b), a quick look at the electric field near the surface of a charged conductor (Section 15.5), shows that the correct answer must be (c). To verify that (c) is the correct answer, recall that near the surface the electric field is perpendicular to that surface. Since the equipotential surfaces are perpendicular to the field lines, they must follow the contour of the conductor's surface (Figure 16.17b).

**FOLLOW-UP EXERCISE.** In this Example, (a) which of the two equipotentials (1 or 2) shown in Figure 16.17c is at a higher potential? (b) What is the approximate shape of the equipotential surfaces very far from this conductor? Explain your reasoning. [*Hint*: What does the conductor "look like" from a large distance?]

---

In many practical situations, it is potential difference (commonly referred to as *voltage*), rather than electric field, that is useful. For example, a D-cell flashlight battery has a terminal voltage of 1.5 V, meaning that it can maintain a potential difference of 1.5 V between its terminals. Most automotive batteries have a terminal voltage of about 12 V. Some common potential differences, or voltages, are listed in ▶**Table 16.1**.

TABLE 16.1 Common Electric Potential Differences (Voltages)

| Source | Approximate Voltage ($\Delta V$) |
|---|---|
| Across nerve membranes | 100 mV |
| Small-appliance batteries | 1.5–9.0 V |
| Automotive batteries | 12 V |
| Household outlets (United States) | 110–120 V |
| Household outlets (Europe) | 220–240 V |
| Automotive ignitions (spark plug firing) | 10 000 V |
| Laboratory generators | 25 000 V |
| High-voltage electric power delivery lines | 300 kV or more |
| Cloud-to-Earth surface during thunderstorm | 100 MV or more |

### 16.2.2 The Electron-Volt Defined

The concept of electric potential provides a unit of energy, the **electron-volt (eV)** that is particularly useful in molecular, atomic, nuclear, and elementary particle physics. The electron-volt is defined as the kinetic energy acquired by an electron (or proton) accelerated through a potential difference, or voltage, of exactly 1 V. The gain in kinetic energy is equal but opposite to the loss in electric potential energy. Thus for an electron, the gain in kinetic energy in our SI unit, the joule, is:

$$\Delta K = -\Delta U_e = -(e\Delta V) = -(-1.60\times10^{-19}\ \text{C})(1.00\,\text{V})$$
$$= +1.60\times10^{-19}\ \text{J}$$

Thus the conversion factor between the electron volt and the joule (to three significant figures) is

$$1\ \text{eV} = 1.60\times10^{-19}\ \text{J}$$

The electron-volt is typical of energies on the atomic and nuclear scale. Hence at those distances, it is convenient to express energies in eV instead of joules. The energy of any charged particle accelerated through any potential difference is conveniently expressed in eV. For example, if a proton is accelerated through a potential difference of 1000 V, its gain in kinetic energy ($\Delta K$) is one thousand times that of a 1-eV proton, or $\Delta K = e\Delta V = (1\ \text{e})(1000\ \text{V}) = 1000\ \text{eV} = 1\ \text{keV}$. (Here keV stands for kiloelectron-volt.)

The electron-volt is defined in terms of particles with a charge of $\pm e$. However, the energy of a particle with any charge can be easily expressed in eV. For example, if a particle with a charge of $+2e$, such as an alpha particle, is accelerated through a potential difference of 1000 volts, it would gain a kinetic energy of $\Delta K = e\Delta V = (2e)(1000\ \text{V}) = 2000\ \text{eV} = 2\ \text{keV}$.

Occasionally, units larger than the eV are needed. For example, in nuclear and elementary particle physics, it is not uncommon to find particles with energies of *megaelectron-volts* (MeV) and *gigaelectron-volts* (GeV); 1 MeV = $10^6$ eV and 1 GeV = $10^9$ eV.

Care must be taken when using the eV since it is *not* an SI unit. For example, when using energies in calculations, the eV must first be converted to joules. Thus to determine the speed of an electron with a kinetic energy of 10.0 eV you must first convert) to joules:

$$K = (10.0\ \text{eV})(1.60\times10^{-19}\ \text{J/eV}) = 1.60\times10^{-18}\ \text{J}$$

In the SI system, mass is expressed in kilograms. Last, the speed is found:

$$v = \sqrt{2K/m} = \sqrt{2(1.60\times10^{-18}\ \text{J})/9.11\times10^{-31}\ \text{kg}} = 1.87\times10^6\ \text{m/s}$$

The electron-volt language will primarily be utilized in the later chapters of this book during our discussion of modern physics.

## 16.3 Capacitance

A pair of parallel plates, if charged, stores electrical energy (▼ **Figure 16.18**). Such an arrangement of conductors is called a **capacitor**. (Any pair of conductors qualifies – they need not be parallel plates.) The energy storage occurs because it takes work to transfer charge from one plate to the other. Imagine moving one electron between a pair of initially uncharged plates. Once that is done, transferring a *second* electron is more difficult, because it is repelled by the first electron on the negative plate, and attracted by the positive charge on the positive plate. Thus the separation of charges increasingly requires more work as more charge accumulates on the plates.

The work to charge parallel plates occurs quickly (usually in a few microseconds) by a battery. Although batteries won't be

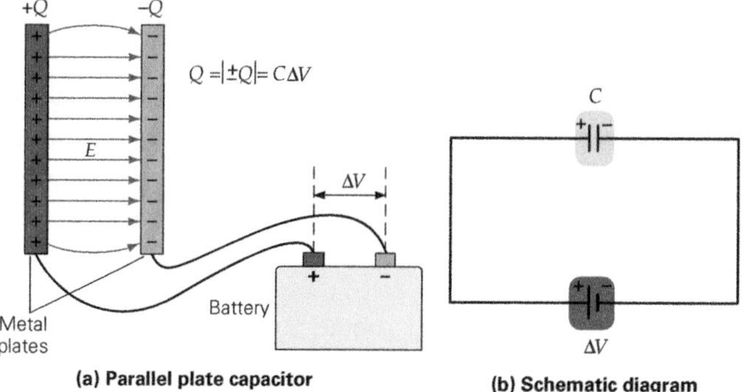


▲ **FIGURE 16.18 Capacitor and circuit diagram (a)** Two parallel plates are charged by a battery that moves electrons from the positive plate to the negative one through the wire. Work is done by the battery during the charging, and thus energy is stored in the capacitor. **(b)** This circuit diagram represents the charging situation in part (a). It also shows the symbols commonly used for a battery terminal voltage ($\Delta V$) and a capacitor (capacitance $C$). The longer line of the battery symbol is the positive terminal, and the shorter represents the negative terminal. Although the symbol for a capacitor is similar, its lines are of equal length.

discussed until Chapter 17, all you need to know now is that a battery removes electrons from the positive plate and transfers, or "pumps," them through a wire to the negative plate. (Take note of Figure 16.18b – our first electric circuit diagram – for future reference.) In the process of doing work, the battery loses some of its stored chemical energy. Of primary interest here is the result: a separation of charge and the creation of an electric field between the plates of the capacitor. The battery will continue to transfer charge until the potential difference between the plates is the same as the terminal voltage of the battery. When the capacitor is disconnected from the battery, the energy remains stored on it until the charge is removed.

Let us use $Q$ be the *magnitude* of the charge on either plate (of course the net charge on the capacitor is zero). For a capacitor, $Q$ is proportional to the voltage (electric potential difference) applied the plates, or $Q \propto \Delta V$. This proportionality can be made into an equation by using a constant, $C$, called **capacitance**:

$$Q = C \cdot \Delta V \quad \text{or} \quad C = \frac{Q}{\Delta V} \quad \text{(definition of capacitance)} \tag{16.9}$$

SI unit of capacitance is the coulomb per volt (C/V) or the farad (F)

The coulomb per volt is named the **farad** after Michael Faraday, a famous physicist of the nineteenth century, thus 1 C/V = 1 F. The farad is a large unit (see Example 16.6), so the *microfarad* ($1\ \mu\text{F} = 10^{-6}$ F), the *nanofarad* ($1\ \text{nF} = 10^{-9}$ F), and the *picofarad* ($1\ \text{pF} = 10^{-12}$ F) are commonly used.

Capacitance represents the charge $Q$ that can be stored *per volt*. If a capacitor has a large capacitance value, this means it is capable of holding a large charge per volt compared to one of smaller capacitance. Thus if you connect the same battery to two different capacitors, the one with the larger capacitance stores more charge and energy.

Capacitance depends *only* on the geometry (size, shape, and spacing) of the plates (and possibly any material between the plates – see Section 16.5) but specifically *not* the charge on the plates. To understand this, consider again a set of parallel plates which is now called a parallel plate capacitor. The electric field between the plates given by Equation 16.5:

$$E = \frac{4\pi kQ}{A}$$

The potential difference between the plates can be computed from Equation 16.2 as follows:

$$\Delta V = Ed = \frac{4\pi kQd}{A}$$

The capacitance of a parallel plate arrangement is then

$$C = \frac{Q}{\Delta V} = \left(\frac{1}{4\pi k}\right)\frac{A}{d} \quad \text{(parallel plates only)} \tag{16.10}$$

It is common to replace the constants in the parentheses in Equation 16.10 with a single constant called the **permittivity of free space ($\varepsilon_o$)**. Knowing $k$, the permittivity has a value of

$$\varepsilon_o = \frac{1}{4\pi k} = 8.85 \times 10^{-12} \frac{\text{C}^2}{\text{N}\cdot\text{m}^2} \quad \text{(permittivity of free space)} \tag{16.11}$$

$\varepsilon_o$ describes the electrical properties of free space (vacuum), but its value in air is only 0.05% larger. In our calculations, they will be taken to be the same.

It is common to rewrite Equation 16.10 in terms of $\varepsilon_o$:

$$C = \frac{\varepsilon_o A}{d} \quad \text{(parallel plates only)} \tag{16.12}$$

The next Example shows just how unrealistically large an air-filled capacitor with a capacitance of 1.0 F would be.

---

**EXAMPLE 16.6: PARALLEL PLATE CAPACITORS – HOW LARGE IS A FARAD?**

What must the plate area for an air-filled parallel plate capacitor be in order for it to have a capacitance of 1.0 F, if the plate separation is 1.0 mm? Would it be realistic to consider building such a capacitor?

**THINKING IT THROUGH.** The area can be calculated directly from Equation 16.12. Remember to keep all quantities in SI units, so that the answer will be in square meters. The vacuum value of $\varepsilon_o$ for air can be used without creating a significant error.

**SOLUTION**

Listing the data:

***Given:***

$C = 1.0$ F
$d = 1.0\ \text{mm} = 1.0 \times 10^{-3}$ m

***Find:*** $A$ (area of one of the plates)

Solving Equation 16.12 for area

$$A = \frac{Cd}{\varepsilon_o} = \frac{(1.0\ \text{F})(1.0 \times 10^{-3}\ \text{m})}{8.85 \times 10^{-12}\text{C}^2/(\text{N}\cdot\text{m}^2)} = 1.1 \times 10^8\ \text{m}^2$$

This would be a square more than $10^4$ m or 10 km or 6.2 mi on a side! It is clearly unrealistic to build a parallel-plate capacitor that big; 1.0 F is therefore a very large value of capacitance. There are ways, however, to make compact high-capacity capacitors, as will be seen in Section 16.4.

**FOLLOW-UP EXERCISE.** In this Example, what would the plate spacing have to be if you wanted the capacitor to have a plate area of 1 cm²? Compare your answer with a typical atomic diameter of $10^{-9}$–$10^{-10}$ m. Is it feasible to build this capacitor?

---

Deriving an expression for energy storage by a capacitor is based on the fact that the charge and the voltage across the plates are proportional since $\Delta V = (1/C)Q \propto Q$. Thus the work done by the battery $W_{batt}$ (and the energy stored in the capacitor $U_C$) can be found by considering the battery as transferring the total charge $Q$, across an average voltage $\overline{\Delta V}$. Because the voltage varies linearly with charge, assuming the capacitor initially uncharged, $\overline{\Delta V}$ is

$$\overline{\Delta V} = \frac{\Delta V_{final} + \Delta V_{initial}}{2} = \frac{\Delta V + 0}{2} = \frac{\Delta V}{2}$$

Since $U_C$ is equal to $W_{batt}$ and, by definition, $W_{batt} = Q\overline{\Delta V}$, an expression for $U_C$ is

$$U_C = W_{batt} = Q \cdot \overline{\Delta V} = \frac{1}{2} Q \cdot \Delta V$$

Since $Q = C \cdot \Delta V$, this can be rewritten in several equivalent forms:

$$U_C = \frac{1}{2} Q \cdot \Delta V = \frac{Q^2}{2C} = \frac{1}{2} C(\Delta V)^2 \quad \text{(energy storage in a capacitor)} \qquad (16.13)$$

Typically, the form $U_C = \frac{1}{2}C(\Delta V)^2$ is the most practical, since capacitance and voltage are usually the known quantities. A very important medical application of energy storage in a capacitor is the *cardiac defibrillator*, discussed in Example 16.7.

EXAMPLE 16.7: CAPACITORS TO THE RESCUE - ENERGY STORAGE IN A CARDIAC DEFIBRILLATOR

During a heart attack, the heart can beat in an erratic fashion, called *fibrillation*. One way to get it back to normal rhythm is to shock it with electrical energy supplied by a cardiac defibrillator (▼**Figure 16.19**). About 300 J of energy is required to produce the desired effect. Typically, a defibrillator stores this energy in a capacitor charged by a 5000-V power supply. (a) What capacitance is required? (b) What is the charge on the capacitor's plates?

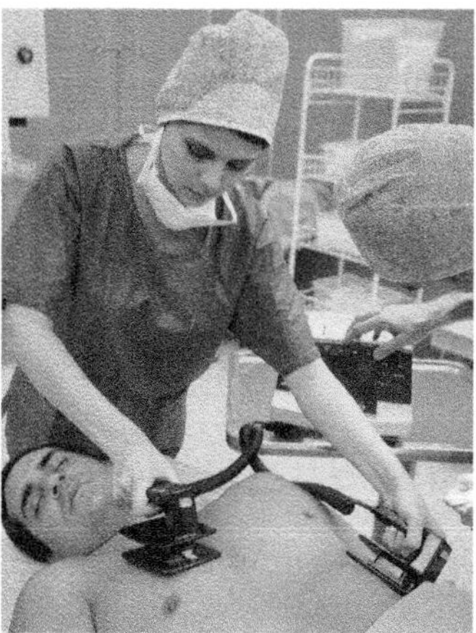

▲FIGURE 16.19 **Capacitors in action: The defibrillator** A flow of charge (and thus energy) from a discharging capacitor through the heart muscle may restore a normal heartbeat during a bout of cardiac fibrillation.

**THINKING IT THROUGH.** (a) To find the capacitance, solve for $C$ in Equation 16.13. (b) The charge then follows from the definition of capacitance Equation 16.9.

**SOLUTION**

Listing the data:

***Given:***

$U_C = 300\,\text{J}$

$\Delta V = 5000\,\text{V}$

***Find:***

**(a)** $C$ (the capacitance)

**(b)** $Q$ (charge on capacitor)

**(a)** The useful form of Equation 16.13 here is $U_C = \frac{1}{2}C(\Delta V)^2$ Solving for $C$,

$$C = \frac{2U_C}{(\Delta V)^2} = \frac{2(300\,\text{J})}{(5000\,\text{V})^2} = 2.40 \times 10^{-5}\,\text{F} = 24.0\,\mu\text{F}$$

**(b)** The charge (magnitude) on either plate is then

$$Q = C \cdot \Delta V = (2.40 \times 10^{-5}\,\text{F})(5000\,\text{V}) = 0.120\,\text{C}$$

**FOLLOW-UP EXERCISE.** For the capacitor in this Example, if the maximum allowable energy for any single defibrillation attempt is 750 J, what is the maximum voltage that should be used?

## 16.4 Dielectrics

In most capacitors, a sheet of insulating material, such as paper or plastic, lies between the plates. Such an insulating material, called a **dielectric**, serves several purposes. For one, it keeps the plates from coming into contact. Contact would allow the electrons to move back onto the positive plate, neutralizing the charge on the capacitor and thus the energy stored. A dielectric also allows flexible plates of metallic foil to be rolled into a cylinder, giving the capacitor a more compact (and therefore practical) size (▶**Figure 16.20**). Finally, a dielectric increases the charge storage capacity of the capacitor and therefore, under the right conditions, the energy stored in the capacitor. This capability depends on the type of material and is characterized by the **dielectric constant** ($\kappa$ - lower case Greek kappa). Values of $\kappa$ for some common materials are shown in ▶**Table 16.2**.

How a dielectric affects the electrical properties of a capacitor is illustrated in ▶**Figure 16.21**. Here the air-filled capacitor must be first fully charged (creating a field $\vec{E}_o$) and then disconnected from its battery, after which a dielectric is inserted (Figure 16.21a). In the dielectric material, the molecular dipoles are aligned with the field (Figure 16.21b). (This molecular polarization may be permanent or temporarily induced by the electric field. In either case, the effect is the same.)

Thus the dielectric creates a "reverse" electric field ($\vec{E}_d$ in Figure 16.21c) that partially cancels the original field. Hence the

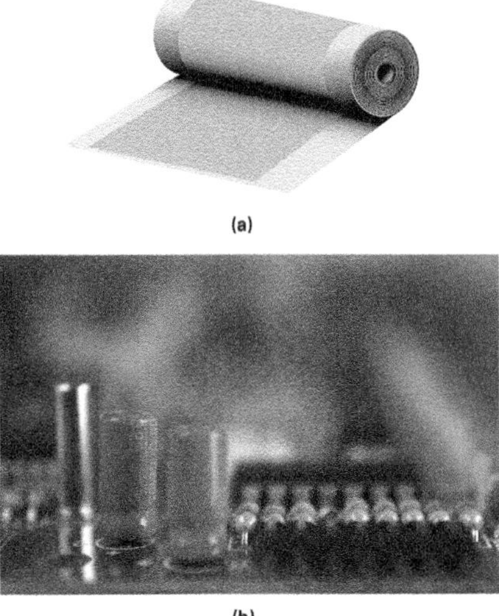

▲ **FIGURE 16.20** **Capacitor designs (a)** The dielectric material between capacitor plates enables the plates to be close together, thus increasing the capacitance. In addition, the plates can be rolled into a compact, more practical capacitor. **(b)** Cylindrical capacitors inserted into a printed circuit board, along with other circuit elements.

*net* field ($\vec{E}$) is reduced, and so is the voltage across the plates (since they are related by $\Delta V = Ed$). The dielectric constant $\kappa$ of the material is defined (with charge constant – i.e., battery disconnected) as the ratio of the vacuum voltage ($\Delta V_o$) to the voltage with the material in place ($\Delta V$). Because potential differences are proportional to the electric field, this is the same as the electric field ratio, or:

$$\kappa = \frac{\Delta V_o}{\Delta V} = \frac{E_o}{E} \quad \text{(at constant capacitor charge)} \qquad (16.14)$$

$\kappa$ is a dimensionless quantity and is greater than 1, since $\Delta V < \Delta V_o$. Because the battery was disconnected before insertion of the dielectric, and the capacitor remained isolated, the charge on the plates, $Q_o$, is unaffected. Since $\Delta V = \Delta V_o/\kappa$,

**TABLE 16.2** Dielectric Constants for Some Common Materials

| Material | Dielectric Constant ($\kappa$) |
|---|---|
| Vacuum | 1.0000 |
| Air | 1.00059 |
| Paper | 3.7 |
| Polyethylene | 2.3 |
| Polystyrene | 2.6 |
| Teflon | 2.1 |
| Glass (range) | 3–7 |
| Pyrex glass | 5.6 |
| Bakelite | 4.9 |
| Silicon oil | 2.6 |
| Water | 80 |
| Strontium titanate | 233 |

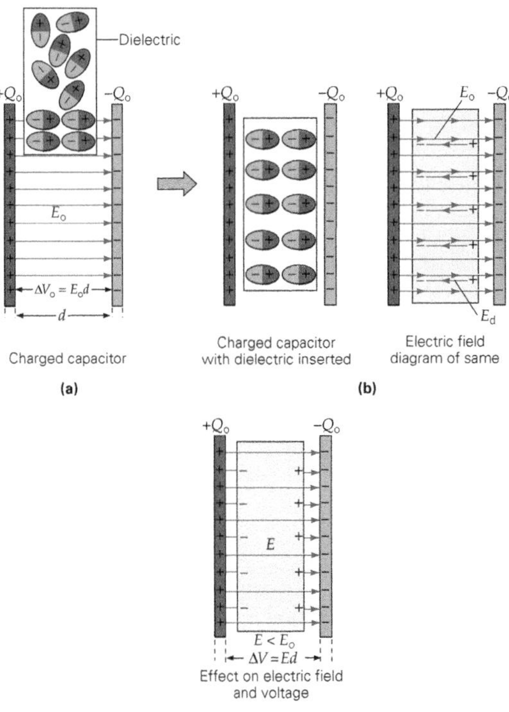


▲ **FIGURE 16.21** **The effects of a dielectric on an isolated capacitor (a)** A dielectric material with randomly oriented permanent molecular dipoles (or dipoles induced by the electric field) is inserted between the plates of an isolated charged capacitor. As the dielectric is inserted, the capacitor tends to pull it in, thus doing work on it. (Note the attractive forces between the plate charges and those induced on the dielectric surfaces.) **(b)** When the material is in the capacitor's electric field, the dipoles orient themselves with that field, thus creating its own, opposing, electric field $\vec{E}_d$. **(c)** The dipole field partially cancels the original field and the net effect is a decrease in both electric field and voltage (plate potential difference). Because the stored charge remains the same, but at a reduced voltage, the capacitance *value* increases.

the value of the capacitance with the dielectric is *larger* than the vacuum value by a factor of $\kappa$. In effect, the same charge is now stored at a lower voltage, and the result is an increase in capacitance. Mathematically this argument goes like this, applying the definition of capacitance:

$$C = \frac{Q}{\Delta V} = \frac{Q_o}{(\Delta V_o/\kappa)} = \kappa\left(\frac{Q_o}{\Delta V_o}\right) \quad \text{or} \quad C = \kappa C_o \qquad (16.15a)$$

So inserting a dielectric into an isolated capacitor results in a larger capacitance. But what about energy storage? Because there is no energy input (the battery is disconnected) and the capacitor does work by aligning (turning) the molecular dipoles, the stored energy actually *drops* by a factor of $\kappa$ (▶ **Figure 16.22a**), as the following shows:

$$U_C = \frac{Q^2}{2C} = \frac{Q_o^2}{2\kappa C_o} = \frac{(Q_o^2/2C_o)}{\kappa} = \frac{U_o}{\kappa} < U_o \quad \text{(battery disconnected)}$$

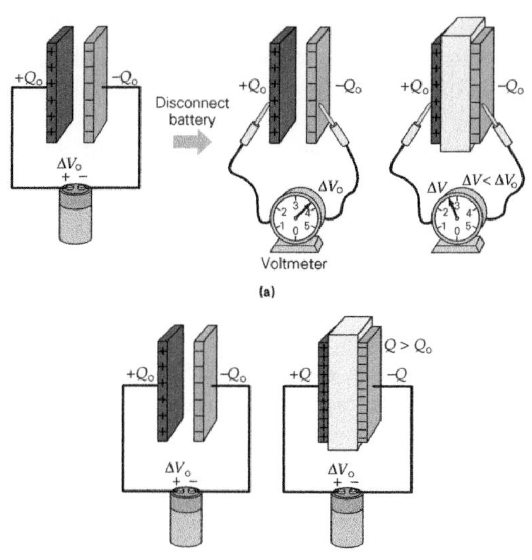


▲ FIGURE 16.22 **Dielectrics and capacitance (a)** A parallel plate capacitor in air is charged by a battery of voltage $\Delta V_o$ to a charge $Q_o$ (left). If the battery disconnected, the potential difference (voltage) remains at $\Delta V_o$ (center). If a dielectric is now inserted between the plates, the voltage drops (right) and the stored energy *decreases*. (Can you estimate the dielectric constant from the voltage readings?) **(b)** A capacitor is charged as in (a), but the battery is left connected. Thus with the dielectric inserted, the voltage is unchanged at $\Delta V_o$. However here the charge storage increases, and therefore the stored energy *increases*. Regardless, the capacitance increases by a factor of $\kappa$.

A different situation occurs, however, if the dielectric is inserted *and the battery remains connected*. In this case, the voltage stays constant and the battery supplies more charge to the capacitor (Figure 16.22b). Because the battery does work to provide more charge, the energy stored in the capacitor *increases*. With the battery remaining connected, the charge on the plates increases by a factor $\kappa$: $Q = \kappa Q_o$. Again the capacitance increases, but now it is because more charge is stored at the same voltage. Mathematically this argument goes like this, applying the definition of capacitance:

$$C = Q/\Delta V = (\kappa Q_o)/\Delta V_o = \kappa(Q_o/\Delta V_o) = \kappa C_o \quad \text{or} \quad C = \kappa C_o \qquad (16.15b)$$

In summary, then,

> The effect of a dielectric is to increase the capacitance by a factor of $\kappa$, regardless of the conditions under which the dielectric is inserted.

In the case of constant voltage, the energy storage increases at the expense of the battery since the battery does work to provide more charge on the plates. To see this, let's calculate the energy with the dielectric in place at constant voltage:

$$U_C = \frac{1}{2}C(\Delta V)^2 = \frac{1}{2}\kappa C_o(\Delta V_o)^2 = \kappa\left(\frac{1}{2}C_o(\Delta V_o)^2\right) = \kappa U_o > U_o \quad \text{(battery connected)}$$

For a parallel plate capacitor with a dielectric, the capacitance is given by Equation 16.12 increased by a factor of $\kappa$:

$$C = \kappa C_o = \frac{\kappa \varepsilon_o A}{d} \quad \text{(parallel plates)} \qquad (16.16)$$

This relationship is sometimes written as $C = (\varepsilon A)/d$, where $\varepsilon = \kappa\varepsilon_o$ is the **dielectric permittivity** of the material. Note that $\varepsilon > \varepsilon_o$.

Under constant voltage conditions, changes in capacitance can be used to indirectly monitor plate movement by monitoring the charge flowing onto or from it as Example 16.8 shows.

---

## EXAMPLE 16.8: A VARIABLE CAPACITOR – MOTION OF CHARGE AS A MOTION SENSOR?

Suppose a parallel a capacitor filled with a compressible dielectric (assume its dielectric constant does not change if it is compressed). The capacitor is permanently connected to a 12.0-V battery with an uncompressed plate separation of 3.00 mm and plate area of 0.750 cm$^2$. (a) What is the dielectric constant if its uncompressed capacitance is to be 1.10 pF? (b) How much charge is stored on the plates under uncompressed conditions? (c) If the capacitor is compressed to a plate separation of 2.00 mm, does charge flow onto or off of the plates? (d) Ammeters, as will be seen in Chapter 17, are instruments used to detect charge movement in wires. How much charge would be measured by such an instrument during this compression?

**THINKING IT THROUGH.** (a) The capacitance of air-filled plates can be found from Equation 16.12, and thus the dielectric constant can be determined from Equation 16.15. (b) The charge follows from Equation 16.9. (c) The compressed plate separation distance must be used to recompute the capacitance. It is expected that the capacitance will increase as the plates get closer, and at constant voltage, more charge should result, thus charge should flow onto the capacitor. (d) Then the new charge can be found as in (b) and the charge difference determined – it should be an increase if our reasoning in (c) is correct.

**SOLUTION**

The given data are as follows:

***Given:***

$\Delta V = 12.0\text{ V}$

$d = 3.00\text{ mm} = 3.00 \times 10^{-3}\text{ m}$

$A = 0.750\text{ cm}^2 = 7.50 \times 10^{-5}\text{ m}^2$

$C = 1.10\text{ pF} = 1.10 \times 10^{-12}\text{ F}$

$d' = 2.00\text{ mm} = 2.00 \times 10^{-3}\text{ m}$

***Find:***

**(a)** $\kappa$ (dielectric constant)
**(b)** $Q$ (initial capacitor charge)
**(c)** Explain if capacitor charge increases or decreases
**(d)** $\Delta Q$ (change in capacitor charge)

**(a)** From Equation 16.12, the capacitance in air is

$$C_o = \frac{\varepsilon_o A}{d} = \frac{(8.85\times10^{-12}\,\text{C}^2/\text{N}\cdot\text{m}^2)(7.50\times10^{-5}\,\text{m}^2)}{3.00\times10^{-3}\,\text{m}} = 2.21\times10^{-13}\,\text{F}$$

Because the dielectric increases the capacitance, the dielectric constant must be

$$\kappa = \frac{C}{C_o} = \frac{1.10\times10^{-12}\,\text{F}}{2.21\times10^{-13}\,\text{F}} = 4.98$$

**(b)** The initial charge is then

$$Q = C\cdot\Delta V = (1.10\times10^{-12}\,\text{F})(12.0\,\text{V}) = 1.32\times10^{-11}\,\text{C}$$

**(c)** Since the capacitance will increase as the plates get closer, and at constant voltage more charge will be stored, the change in charge stored is positive and the ammeter will detect a flow onto the capacitor.

**(d)** Under compressed conditions, the capacitance is

$$C' = \frac{\kappa\varepsilon_o A}{d'} = \frac{(4.98)(8.85\times10^{-12}\,\text{C}^2/\text{N}\cdot\text{m}^2)(7.50\times10^{-5}\,\text{m}^2)}{2.00\times10^{-3}\,\text{m}} = 1.65\times10^{-12}\,\text{F}$$

Since the voltage remains constant, $Q' = C'\cdot\Delta V = (1.65\times10^{-12}\,\text{F})\,(12.0\,\text{V}) = 1.98\times10^{-11}\,\text{C}$. Thus, the difference in charge is

$$\Delta Q = Q' - Q = (1.98\times10^{-11}\,\text{C}) - (1.32\times10^{-11}\,\text{C}) = +6.60\times10^{-12}\,\text{C}.$$

This is clearly an increase, as the + sign shows.

**FOLLOW-UP EXERCISE.** In this Example, suppose instead that the spacing between the plates were *increased* by 1.00 mm from the normal value of 3.00 mm. Would charge flow onto or away from the capacitor? How much charge would this entail?

## 16.5 Capacitors in Series and in Parallel

Capacitors can be connected in two basic ways: *series* or *parallel*. In series, the capacitors are connected head to tail (▼ **Figure 16.23a**). When connected in parallel, all the leads on one side of the capacitors have a common connection. (Think of all the "tails" connected together and all the "heads" connected together; Figure 16.23b.)

### 16.5.1 Capacitors in Series

When capacitors are in series, the charge $Q$ must be the same on all the plates. To see why this must be true, examine ▶ **Figure 16.24**. Note that only plates A and F are connected to

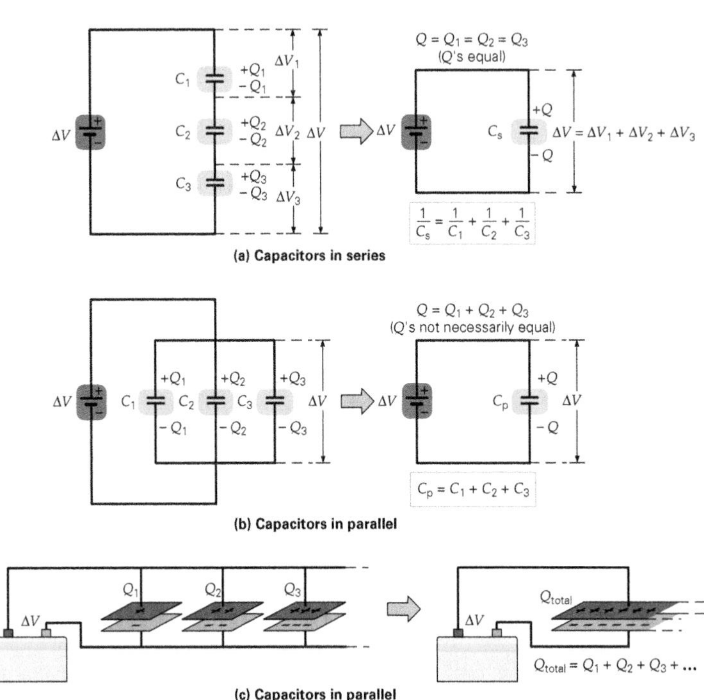


▲ **FIGURE 16.23** **Capacitors in series and in parallel** **(a)** All capacitors connected in series have the charge, and the sum of the voltage drops is equal to the voltage of the battery. The total series capacitance is equivalent to one capacitor of capacitance $C_s$. **(b)** When capacitors are in parallel, the voltage drops across the capacitors are the same as the battery voltage. The total charge is the sum of the charges on the individual capacitors. The total parallel capacitance is equivalent to one capacitor of capacitance $C_p$. **(c)** To visualize a parallel arrangement, think of the charge on each pair of plates.

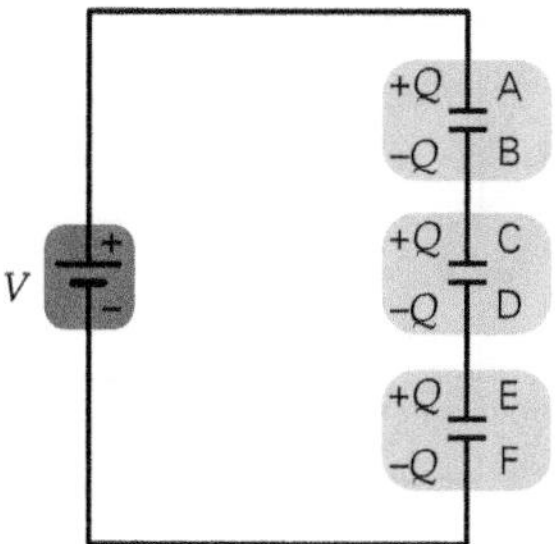


▲ FIGURE 16.24 **Charges on capacitors in series** Plates B and C together had zero net charge to start. When the battery placed $+Q$ on plate A, charge $-Q$ was induced on B; thus, C must have acquired $+Q$ for the BC combination to remain neutral. Continuing this way through the string, we see that all the charges must be the same in magnitude.

the battery. Because the plates labeled B and C are isolated, the total (net) charge on them must be zero. Thus if the battery puts a charge of $+Q$ on plate A, then $-Q$ is induced on B at the expense of plate C, which must acquire a charge of $+Q$. This charge in turn induces $-Q$ on D, and so on down the line.

When all the capacitor voltage drops are added (see Figure 16.24a), the total must be the same as the voltage of the battery. Thus, in series $\Delta V = \Delta V_1 + \Delta V_2 + \Delta V_3 + \cdots$. The **equivalent series capacitance ($C_s$)** is defined as the value of a single capacitor that could replace the series combination and store the same charge and energy at the same voltage. Because the combination of capacitors stores a charge of $Q$ at a voltage of $V$, it follows that $C_s = Q/\Delta V$, or $\Delta V = Q/C_s$. However, the individual voltages are related to the individual charges by $\Delta V_1 = Q/C_1$, $\Delta V_2 = Q/C_2$, $\Delta V_3 = Q/C_3$, etc. Substituting these expressions into the sum of the voltages,

$$\frac{Q}{C_s} = \frac{Q}{C_1} + \frac{Q}{C_2} + \frac{Q}{C_3} + \cdots$$

Canceling the common $Q$'s, the result is

$$\frac{1}{C_s} = \frac{1}{C_1} + \frac{1}{C_2} + \frac{1}{C_3} + \cdots \quad \text{(equivalent series capacitance)} \qquad (16.17)$$

This means that $C_s$ is *always smaller than the smallest capacitance* in series. For example, with $C_1 = 1.0\ \mu\text{F}$ and $C_2 = 2.0\ \mu\text{F}$, you should be able to show that $C_s = 0.67\ \mu\text{F}$, which is less than the smallest value, $1.0\ \mu\text{F}$.

The reasoning goes like this. In series, all capacitors have the same charge, so the charge stored by this arrangement is $Q = C_i \cdot \Delta V_i$ (where the subscript $i$ refers to any capacitor). Because $\Delta V_i < \Delta V$, a series arrangement must store less charge than any individual capacitor when connected by itself to the same battery. Note also that in series the smallest capacitance receives the largest voltage. This is because a small value of $C$ means less charge stored per volt. Since all the capacitors have the same charge, it must be that the smaller the capacitance, the larger the fraction of the total voltage the capacitor will have.

## 16.5.2 Capacitors in Parallel

With a parallel arrangement (Figure 16.23b), the voltage across each capacitor is the same as that of the battery, $\Delta V = \Delta V_1 = \Delta V_2 = \Delta V_3 = \cdots$, and the total charge stored is the sum of the charges on each capacitor (Figure 16.23c), $Q_{\text{total}} = Q_1 + Q_2 + Q_3 + \cdots$. The equivalent parallel capacitance should be larger than the largest capacitance in the string, because more charge per volt is stored than if any one capacitor were connected by itself to the battery. The individual charges are $Q_1 = C_1 \cdot \Delta V$, $Q_2 = C_2 \cdot \Delta V$, etc. One capacitor with the **equivalent parallel capacitance ($C_p$)** would hold this same total charge when connected to this battery, so $C_p = Q_{\text{total}}/\Delta V$ or $Q_{\text{total}} = C_p \cdot \Delta V$. Substituting these expressions into $Q_{\text{total}} = Q_1 + Q_2 + Q_3 + \cdots$ yields

$$C_p \cdot \Delta V = C_1 \cdot \Delta V + C_2 \cdot \Delta V + C_3 \cdot \Delta V + \cdots$$

and canceling the common $\Delta V$ gives

$$C_p = C_1 + C_2 + C_3 + \cdots \quad \text{(equivalent parallel capacitance)} \qquad (16.18)$$

In the parallel case, the equivalent capacitance $C_p$ is the sum of the individual capacitances and thus the equivalent capacitance is larger than the largest individual capacitance. Because capacitors in parallel have the same voltage, the largest capacitance will also store the most charge and energy). For a comparison of capacitors in series and in parallel, consider Example 16.9.

---

**EXAMPLE 16.9: CHARGING CAPACITORS IN SERIES AND IN PARALLEL**

Given two capacitors, one with a capacitance of $2.50\ \mu\text{F}$ and the other with that of $5.00\ \mu\text{F}$, what are the charges on each and the total charge stored if they are connected across a 12.0-V battery (a) in series and (b) in parallel?

**THINKING IT THROUGH.** (a) Capacitors in series have the same charge. From Equation 16.17 the equivalent capacitance can be found and then the charge on each capacitor can be determined. (b) Capacitors in parallel have the same voltage; from that, the charge on each can be easily determined because their individual capacitances are known.

**SOLUTION**

Listing the data:

***Given:***

$$C_1 = 2.50\ \mu\text{F} = 2.50 \times 10^{-6}\ \text{F}$$
$$C_2 = 5.00\ \mu\text{F} = 5.00 \times 10^{-6}\ \text{F}$$
$$\Delta V_{\text{batt}} = 12.0\ \text{V}$$

***Find:***

**(a)** $Q$ (on each capacitor in series) and $Q_{total}$ (total charge)
**(b)** $Q$ (on each capacitor in parallel) and $Q_{total}$ (total charge)

**(a)** In series, the equivalent capacitance is:

$$\frac{1}{C_s} = \frac{1}{2.50\times10^{-6}\,\text{F}} + \frac{1}{5.00\times10^{-6}\,\text{F}} = \frac{3}{5.00\times10^{-6}\,\text{F}}$$

Therefore

$$C_s = 1.67\times10^{-6}\,\text{F}$$

[*Note:* $C_s$ is less than the smallest capacitance, as expected.]
Because the charge on each capacitor is the same in series (and the same as the total), it follows

$$Q_{total} = Q_1 = Q_2 = C_s \cdot \Delta V_{batt} = (1.67\times10^{-6}\,\text{F})(12.0\,\text{V}) = 2.00\times10^{-5}\,\text{C}$$

**(b)** Here, the parallel equivalent capacitance relationship gives:

$$C_p = C_1 + C_2 = 2.50\times10^{-6}\,\text{F} + 5.00\times10^{-6}\,\text{F} = 7.50\times10^{-6}\,\text{F}$$

[*Note:* $C_p$ is larger than the largest capacitance, as expected.]
The total charge is

$$Q_{total} = C_p \cdot \Delta V_{batt} = (7.50\times10^{-6}\,\text{F})(12.0\,\text{V}) = 9.00\times10^{-5}\,\text{C}$$

In parallel, each capacitor has the full 12.0 V across it; hence,

$$Q_1 = C_1 \cdot \Delta V_{batt} = (2.50\times10^{-6}\,\text{F})(12.0\,\text{V}) = 3.00\times10^{-5}\,\text{C}$$
$$Q_2 = C_2 \cdot \Delta V_{batt} = (5.00\times10^{-6}\,\text{F})(12.0\,\text{V}) = 6.00\times10^{-5}\,\text{C}$$

As a final double check, notice that the total stored charge is the sum of the charges on the individual capacitors.

**FOLLOW-UP EXERCISE.** In this Example, determine which combination, series or parallel, stores the most energy by calculating how much is stored in each arrangement.

Capacitor arrangements generally involve *both* series and parallel connections, as shown in Integrated Example 16.10. In this situation, the circuit is simplified, using the equivalent parallel and series capacitance expressions, until it results in one single, overall equivalent capacitance. To find the results for each individual capacitor, the steps are undone until the original arrangement is reached.

### INTEGRATED EXAMPLE 16.10: FORWARD, THEN BACKWARD – A SERIES-PARALLEL COMBINATION OF CAPACITORS

Three capacitors are connected in a circuit as shown in ▶**Figure 16.25a**. (a) By looking at the capacitance values, how do the voltages across the various capacitors compare: (1) $\Delta V_3 > \Delta V_2 > \Delta V_1$, (2) $\Delta V_3 < \Delta V_2 = \Delta V_1$, (3) $\Delta V_3 = \Delta V_2 = \Delta V_1$, or (4) $\Delta V_3 = \Delta V_2 > \Delta V_1$? (b) How does the energy stored in $C_3$ compare to the total energy stored in $C_1$ plus $C_2$: (1) $U_3 > U_{1+2}$, (2) $U_3 = U_{1+2}$, or (3) $U_3 < U_{1+2}$? (c) Determine the voltages across each capacitor. (d) Determine the stored energy in each capacitor.

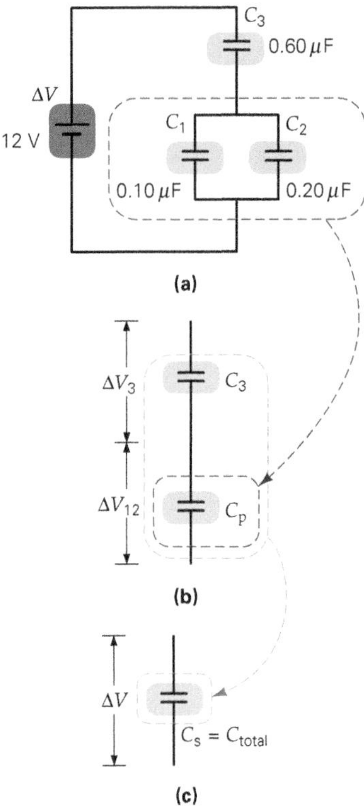


▲**FIGURE 16.25 Circuit reduction** When capacitances are combined, the combination of capacitors is reduced to a single equivalent capacitance.

**(a) CONCEPTUAL REASONING.** For part (a), since $C_1$ and $C_2$ are in parallel, they have the same voltage. Thus answers (1) and (4) cannot be correct. Their capacitances ($C_1$ plus $C_2$) add to 0.30 $\mu$F, which is less than $C_3$. Since the parallel combination ($C_1$ and $C_2$) is in series with $C_3$, most of the 12-V drop will occur across that pair. So the correct choice is (2) $\Delta V_3 < \Delta V_2 = \Delta V_1$. For (b), because $C_1$ and $C_2$ (as a parallel combination) are in series with $C_3$, the parallel combination has the same (total) charge, as $C_3$. In series, the lowest capacitance has the most energy storage, thus the correct choice is (3) $U_3 < U_{1+2}$.

**(b) QUANTITATIVE REASONING AND SOLUTION.** For (c), the voltage across each capacitor could be found from $\Delta V = Q/C$ if the charge on each capacitor was known. The total charge on the capacitors can be found by reducing the series-parallel combination down to a single equivalent capacitance. Two of the capacitors are in parallel. Their single equivalent capacitance ($C_p$) is itself in series with the last capacitor – a fact that enables the total capacitance to be found. Working backward will allow the voltage across each capacitor to be found. For (d), once the voltages are known, individual energy storages can be calculated using the most convenient for form this example, since capacitance and voltages are known: $U_C = \frac{1}{2}C(\Delta V)^2$ .

***Given:***

Values of capacitance and battery voltage given in Figure 16.25a

***Find:***

**(c)** $\Delta V_1$, $\Delta V_2$, and $\Delta V_3$ (voltages across each capacitor)
**(d)** $U_1$, $U_2$, and $U_3$ (energy stored in each capacitor)

**(c)** Starting with the parallel combination,

$$C_p = C_1 + C_2 = 0.10\,\mu F + 0.20\,\mu F = 0.30\,\mu F$$

Now the arrangement is partially reduced, as shown in Figure 16.25b. Next, considering $C_p$ in series with $C_3$, the total, or equivalent, capacitance of the arrangement is:

$$\frac{1}{C_s} = \frac{1}{C_3} + \frac{1}{C_p} = \frac{1}{0.60\,\mu F} + \frac{1}{0.30\,\mu F} = \frac{1}{0.60\,\mu F} + \frac{2}{0.60\,\mu F} = \frac{1}{0.20\,\mu F}$$

Therefore,

$$C_s = 0.20\,\mu F = 2.0 \times 10^{-7}\,F$$

This is the equivalent capacitance of the arrangement when the circuit is reduced to one equivalent capacitor in Figure 16.25c. Treating the problem as one single capacitor, the charge on that equivalent capacitor is:

$$Q = C_s \cdot \Delta V = (2.0 \times 10^{-7}\,F)(12\,V) = 2.4 \times 10^{-6}\,C$$

This is also the charge on $C_3$ and $C_p$ because they are in series. This fact can be used to calculate the voltage across $C_3$:

$$\Delta V_3 = \frac{Q}{C_3} = \frac{2.4 \times 10^{-6}\,C}{6.0 \times 10^{-7}\,F} = 4.0\,V$$

It is known that the sum of the voltages across the capacitors equals the voltage across the battery. The voltages across $C_1$ and $C_2$ are the same because they are in parallel. Because the voltage across $C_1$ (or $C_2$) plus the voltage across $C_3$ equals the battery voltage, $\Delta V = \Delta V_{12} + \Delta V_3 = 12$ V. (See Figure 16.22a.) Here, $V_{12}$ stands for the voltage across either $C_1$ or $C_2$. Thus

$$\Delta V_{12} = \Delta V - \Delta V_3 = 12\,V - 4.0\,V = 8.0\,V$$

**(d)** The individual energies are found as follows:

$$U_1 = \frac{1}{2} C_1 \cdot (\Delta V_1)^2 = \frac{1}{2}(0.10 \times 10^{-6}\,F)(8.0\,V)^2 = 3.2 \times 10^{-6}\,J = 3.2\,\mu J$$

$$U_2 = \frac{1}{2} C_2 \cdot (\Delta V_2)^2 = \frac{1}{2}(0.20 \times 10^{-6}\,F)(8.0\,V)^2 = 6.4 \times 10^{-6}\,J = 6.4\,\mu J$$

and

$$U_3 = \frac{1}{2} C_3 \cdot (\Delta V_3)^2 = \frac{1}{2}(0.60 \times 10^{-6}\,F)(4.0\,V)^2 = 4.8 \times 10^{-6}\,J = 4.8\,\mu J$$

The total energy stored in capacitors 1 and 2 is 9.6 $\mu$J, which is greater than that stored in capacitor 3.

**FOLLOW-UP EXERCISE.** In this Example, (a) what value of capacitance for capacitor 2 would make $\Delta V_3 = \Delta V_1$? (b) After the change is made in part (a), what is the ratio of energy stored in capacitor 3 to that stored in capacitor 1?

# Chapter 16 Review

- The **electric potential difference** (or *voltage*) between two points is the change in electric potential energy per unit positive charge moved between those points.

$$\Delta V = \frac{\Delta U_e}{q_+} \quad (16.1)$$

- The **electric potential difference between two parallel plates** is

$$\Delta V = Ed \quad \text{(parallel plates)} \quad (16.2)$$

- **Equipotential surfaces** (also *equipotentials*) are of constant electric potential. These surfaces are everywhere perpendicular to the electric field.
- The **electric potential difference** between two locations near a point charge is

$$\Delta V = \frac{kq}{r_B} - \frac{kq}{r_A} \quad (16.3)$$

- The **electric potential near a point charge** (choosing zero reference at infinite distance) is

$$V = \frac{kq}{r} \quad (16.4)$$

- The **electric potential energy of two point charges** (choosing $U = 0$ at $r = \infty$) is

$$U_{12} = \frac{kq_1q_2}{r_{12}} \quad (16.5)$$

- The **electric potential energy of more than two point charges** is the sum of point charge pair terms from Equation 16.5:

$$U = U_{12} + U_{23} + U_{13} + U_{14} \cdots \quad (16.6)$$

- The electric field $(\vec{E})$ is in the direction of the most rapid decrease in electric potential ($V$). The field magnitude ($E$) is the maximum rate of change of the potential with distance, or

$$E = \left|\frac{\Delta V}{\Delta x}\right|_{max} \quad (16.8)$$

- The **electron-volt (eV)** is the kinetic energy gained by an electron or a proton accelerated through a potential difference of 1 V. The conversion factor to SI units is $1\ eV = 1.60 \times 10^{-19}$ J.
- A **capacitor** is any arrangement of two metallic plates that can store charge and energy. Its **capacitance** is a measure of how effective it is in storing charge and is defined as the magnitude of the charge stored on either plate per volt:

$$C = \frac{Q}{\Delta V} \quad (16.9)$$

- The expression for the **capacitance of a parallel plate capacitor** (in air) is

$$C = \frac{\varepsilon_0 A}{d} \quad (16.12)$$

where $\varepsilon_0 = 8.85 \times 10^{-12}$ $C^2/(N{\cdot}m^2)$ is the **permittivity of free space**.
- The **energy stored in a capacitor** depends on its capacitance and the charge the capacitor stores (or, equivalently, the voltage across its plates). There are three equivalent expressions for this energy:

$$U_C = \frac{1}{2}Q{\cdot}\Delta V = \frac{Q^2}{2C} = \frac{1}{2}C(\Delta V)^2 \quad (16.13)$$

- A **dielectric** is a non-conducting material that increases capacitance. Its **dielectric constant** $\kappa$ describes its effect on capacitance.

$$C = \kappa C_o \quad (16.15a,b)$$

The **dielectric permittivity of material** ($\varepsilon$) is defined as $\varepsilon = \kappa\varepsilon_o$.
- Capacitors in series are equivalent to one capacitor, with a capacitance called the **equivalent series capacitance** $C_s$. The equivalent series capacitance is

$$\frac{1}{C_s} = \frac{1}{C_1} + \frac{1}{C_2} + \frac{1}{C_3} + \cdots \quad (16.17)$$

- Capacitors in parallel are equivalent to one capacitor, with a capacitance called the **equivalent parallel capacitance** $C_p$. The equivalent parallel capacitance is

$$C_p = C_1 + C_2 + C_3 + \cdots \quad (16.18)$$

# End of Chapter Questions and Exercises

## Multiple Choice Questions

**16.1 Electric Potential Energy and Electric Potential Difference**

1. The SI unit of electric potential difference is the (a) joule, (b) newton per coulomb, (c) newton-meter, (d) joule per coulomb.
2. How does the electrostatic potential energy of a system of two positive point charges change when the distance between them is tripled: (a) it is reduced to one-third its original value, (b) it is reduced to one-ninth its original value, (c) it is unchanged, or (d) it is increased to three times its original value?
3. An electron is moved from the positive plate to the negative plate of a charged parallel plate arrangement. How does the sign of the change in the system's electrostatic potential energy compare to the sign of the change in electrostatic potential the electron experiences: (a) both are positive, (b) the energy change is positive but the potential change is negative, (c) the energy change is negative but the potential change is positive, or (d) both are negative?
4. An isolated system consists of three point charges. Two are negative and one is positive. What can you say about the sign of this system's potential energy: (a) it is positive, (b) it is negative, (c) it is zero, or (d) you can't tell from the data given?
5. A positive point charge is fixed at the origin, and an electron is brought near to it from a large distance. If the magnitude of the work to move the electron is called $W$, then the change in the system's potential energy is (a) $+W$, (b) $-W$, (c) zero, (d) unrelated to $W$.
6. The spacing between two closely spaced oppositely charged parallel plates is decreased. What happens to the potential difference between them, assuming they are isolated: (a) it increases, (b) it decreases, (c) it stays the same, or (d) you can't tell from the data given?
7. The charge on each of two closely spaced oppositely charged parallel plates is increased equally. What happens to the potential difference between them: (a) it increases, (b) it decreases, (c) it stays the same, or (d) you can't tell from the information given?

**16.2 Equipotential Surfaces and the Electric Field**

8. On an equipotential surface (a) the electric potential is constant, (b) the electric field is zero, (c) the electric potential is zero, (d) there are equal negative and positive charges.
9. Equipotential surfaces (a) are parallel to the electric field, (b) are perpendicular to the electric field, (c) can be at any angle with respect to the electric field.
10. An electron is moved from an equipotential surface at +5.0 V to one at +10.0 V. It is moving generally in a direction (a) parallel to the electric field, (b) opposite

to the electric field, (c) you can't tell its direction relative to the electric field from the data given.

11. As an electron is moved perpendicularly away from a large uniformly charged plate, the system's electrostatic potential energy is observed to decrease. The charge on the plate must be (a) positive, (b) negative, (c) zero.
12. A proton with an initial kinetic energy of 9.50 eV is fired directly at another proton whose location is fixed. When the moving proton has reached its point of closest approach, by how much has the electric potential energy of this two-particle system changed: (a) +9.50 eV, (b) −9.50 eV, (c) zero, or (d) it depends on the distance of closest approach, so you can't tell from the data given?

**16.3 Capacitance**

13. A capacitor is first connected to a 6.0-V battery and then disconnected and connected to a 12.0-V battery. How does its capacitance change: (a) it increases, (b) it decreases, (c) it stays the same?
14. A capacitor is first connected to a 6.0-V battery and then disconnected and connected to a 12.0-V battery. How does the charge on either of its plates change: (a) it increases, (b) it decreases, (c) it stays the same?
15. A capacitor is first connected to a 6.0-V battery and then disconnected and connected to a 12.0-V battery. By how much does the electric field strength between its plates change: (a) it is two times greater, (b) it is four times greater, or (c) it stays the same?
16. The distance between the plates of a capacitor is cut in half. By what factor does its capacitance change: (a) it is cut in half, (b) it is reduced to one-fourth its original value, (c) it is doubled, or (d) it is quadrupled?
17. The area of the plates of a capacitor is reduced. How should the distance between those plates be adjusted to keep the capacitance constant: (a) increase it, (b) decrease it, or (c) changing the distance cannot make up for the plate area change?

**16.4 Dielectrics**

18. Putting a dielectric into a charged parallel plate capacitor that is not connected to a battery (a) decreases the capacitance, (b) decreases the voltage, (c) increases the charge, (d) causes the plates to discharge because the dielectric is a conductor.
19. A parallel plate capacitor is connected to a battery and remains so. If a dielectric is then inserted between the plates, (a) the capacitance decreases, (b) the voltage increases, (c) the voltage decreases, (d) the charge increases.
20. A parallel plate capacitor is connected to a battery for a while and then disconnected from it. If a dielectric is then inserted between the plates, what happens to the charge on its plates: (a) the charge decreases, (b) the charge increases, or (c) the charge stays the same?
21. A parallel plate capacitor is connected to a battery and remains so. If a dielectric is now inserted between the plates, what happens to the electric field there: (a) it decreases, (b) it increases, (c) it remains the same, (d) it may increase, decrease, or not change depending on the dielectric constant?
22. A parallel plate capacitor is first connected to a battery for a while and then disconnected from it. If a dielectric is now inserted between the plates, what happens to the electric field there: (a) it decreases, (b) it increases, (c) it remains the same, (d) the field may increase, decrease, or not change depending on the dielectric constant?

**16.5 Capacitors in Series and in Parallel**

23. Capacitors in series must have the same (a) voltage, (b) charge, (c) energy storage, (d) none of these.
24. Capacitors in parallel must have the same (a) voltage, (b) charge, (c) energy storage, (d) none of these.
25. Capacitors 1, 2, and 3 all have the same capacitance, $C$. 1 and 2 are in series and that combination is in parallel with 3. What is the total capacitance of this system: (a) $C$, (b) $1.5C$, (c) $3C$, or (d) $C/3$?

## Conceptual Questions

**16.1 Electric Potential Energy and Electric Potential Difference**

1. What is the difference between (a) electrostatic potential energy and electric potential and (b) electric potential difference and voltage?
2. When a proton approaches another fixed proton, what happens to (a) the kinetic energy of the approaching proton, (b) the electric potential energy of the system, and (c) the total energy of the system?
3. Using the language of electrical potential and energy (not forces), explain why positive charges speed up when they are released near negative charges.
4. An electron is released in a region where the electric potential decreases to the left. Which way will the electron begin to move? Explain.
5. An electron is released in a region where the electric potential is constant. Which way will the electron accelerate? Explain.
6. If two locations are at the same electrical potential, how much work does it take to move a charge from the first location to the second? Explain.

**16.2 Equipotential Surfaces and the Electric Field**

7. Sketch the topographic map as you walk away from the ocean up a gently sloping uniform beach. Label the gravitational equipotentials showing how gravitational potential energy changes. Show how to predict, using your map, which way a ball accelerates if it is initially rolled up the beach (away from the water).

8. Explain physically why two equipotential surfaces cannot intersect.
9. Suppose a charge starts at rest on an equipotential surface, is moved off that surface, and is eventually returned to the same surface at rest after a round trip. How much work did it take to do this? Explain.
10. What shape are the equipotential surfaces between two charged parallel plates?
11. What shape are the equipotential surfaces in the region between two concentric metallic spherical shells that have equal but opposite charge on them?
12. Near a fixed positive point charge, if you move from one equipotential surface to another with a smaller radius, (a) what happens to the value of the potential? (b) What was your general direction relative to the electric field?
13. If a proton is accelerated from rest by a potential difference of 1 million volts, (a) how much kinetic energy does it gain? (b) How would your answer to (a) change if the particle had twice the charge of the proton (same sign) and four times the mass?
14. (a) Can the electric field at a point be zero if there is a nonzero electric potential there? (b) Can the electric potential at a point be zero while there is also a nonzero electric field there? Explain. If your answer to either part is yes, give an example.

**16.3 Capacitance**

15. If the plates of an isolated parallel plate capacitor are pulled farther apart, does the energy storage of this capacitor increase, decrease, or remain the same? Explain.
16. If the potential difference across a capacitor is doubled, what happens to (a) the charge on the capacitor and (b) the energy stored in the capacitor?
17. A capacitor is connected to a 12-V battery. If the plate separation is tripled and the capacitor remains connected to the battery, (a) by what factor does the charge on the capacitor change? (b) By what factor does the energy stored in the capacitor change? (c) By what factor does the electric field between the plates of the capacitor change?

**16.4 Dielectrics**

18. Give several reasons why a conducting material is not be a good choice as a plate separator in a capacitor.
19. A parallel plate capacitor is connected to a battery and then disconnected. If a dielectric is inserted between the plates, what happens to (a) the capacitance, (b) the voltage across the capacitor's plates, and (c) the electric field between the plates?
20. Explain clearly why the electric field between two parallel plates of a capacitor decreases when a dielectric is inserted if the capacitor is not connected to a power supply but remains the same when it is connected to a power supply.

**16.5 Capacitors in Series and in Parallel**

21. Under what conditions would two capacitors in series have the same voltage across them? What if they were in parallel?
22. Under what conditions would two capacitors in parallel have the same charge on them? What if they were in series?
23. You are given two capacitors. How should you connect them to get (a) maximum equivalent capacitance and (b) minimum equivalent capacitance?
24. You have $N$ (an even number $\geq 2$) identical capacitors, each with a capacitance of $C$. In terms of $N$ and $C$, what is their total effective capacitance (a) if they are all in series? (b) If they are all in parallel? (c) If two halves ($N/2$ each) are connected in series and these two sets then connected in parallel?

## Exercises*

*Integrated Exercises* (**IEs**) *are two-part exercises. The first part typically requires a conceptual answer choice based on physical thinking and basic principles. The following part is quantitative calculations associated with the conceptual choice made in the first part of the exercise.*

**16.1 Electric Potential Energy and Electric Potential Difference**

1. • A pair of parallel plates is charged by a 12-V battery. If the electric field between the plates is 1200 N/C, how far apart are the plates?
2. • A pair of parallel plates is charged by a 12-V battery. How much work is required to move a particle with a charge of $-4.0\ \mu C$ from the positive to the negative plate?
3. • It takes $+1.6 \times 10^{-5}$ J to move a positively charged particle between two oppositely charged parallel plates. (a) What is the charge on the particle if the plates are connected to a 6.0-V battery? (b) Was it moved from the negative to the positive plate or from the positive to the negative plate? Explain.
4. • An electron is accelerated by a uniform electric field ($E = 1000$ V/m) vertically upward. Use Newton's laws to find the electron's velocity after it moves 0.10 cm from rest.
5. • (a) Repeat Exercise 4 using energy methods. Find the direction in which the electron is moving by considering electric potential energy changes. (b) Does the electron gain or lose electric potential energy?
6. **IE** • Consider two points at different distances from a positive point charge. (a) The point closer to the charge is at a (1) higher, (2) equal, (3) lower potential than the point farther away. (b) How much different is

* The bullets denote the degree of difficulty of the exercises: •, simple; ••, medium; and •••, more difficult.

the electric potential 20 cm from a charge of $+5.5\ \mu C$ compared to 40 cm from it?

7. IE •• (a) At one-third the original distance from a positive point charge, by what factor is the electric potential changed: (1) 1/3, (2) 3, (3) 1/9, or (4) 9? Why? (b) How far from a $+1.0$-$\mu C$ charge is a location that has an electric potential value of 10 kV? (c) How much of a change in potential would occur if the location was three times that distance?
8. IE •• According to the Bohr model of the hydrogen atom, the single electron can exist only in circular orbits of certain radii about a proton. (a) Will a larger orbit have (1) a higher, (2) an equal, or (3) a lower electric potential than a smaller orbit? Why? (b) Determine the potential difference between two orbits of radii 0.21 nm and 0.48 nm.
9. •• In Exercise 8, by how much does the potential *energy* of the atom change if the electron changes location (a) from the lower to the higher orbit, (b) from the higher to the lower orbit, and (c) from the larger orbit to a very large distance?
10. •• What is the least amount of work required to completely separate two charges (each $-1.4\ \mu C$) if they were initially 8.0 mm apart?
11. •• In Exercise 10, if the charges were, instead, released from rest at their initial separation, how much kinetic energy would each have when they are very distant from one another?
12. •• It takes $+6.0$ J of work to move two charges from a large distance apart to 1.0 cm from one another. If the charges have the same magnitude, (a) how large is each charge, and (b) what can you tell about their signs?
13. •• A $+2.0$-$\mu C$ charge is initially 0.20 m from a fixed $-5.0$-$\mu C$ charge and is then moved to 0.50 m from that fixed charge. (a) How much work is required to move the charge? (b) Does the work depend on the path through which the charge is moved?
14. •• An electron is moved from A to B and then to C along two legs of an equilateral triangle with sides of 0.25 m (▼ **Figure 16.26**). If the horizontal electric field is 15 V/m, (a) what is the magnitude of the work required? (b) What is the potential difference between A and C? (c) Which is at a higher potential?

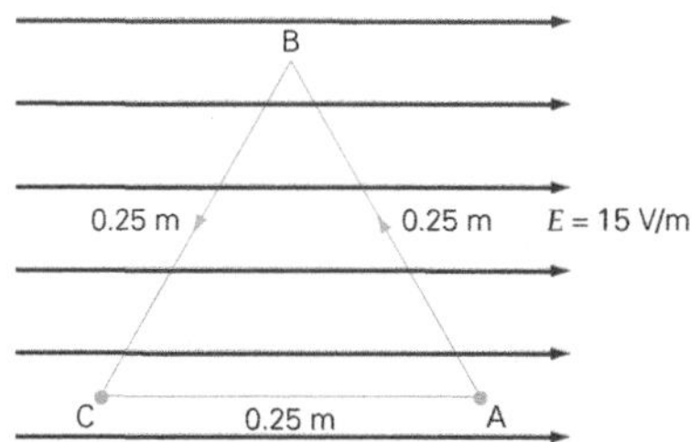


▲ FIGURE 16.26 **Work and energy** See Exercise 14.

15. •• Find the work necessary to bring together (from a large distance) the charges in ▼ **Figure 16.27.**

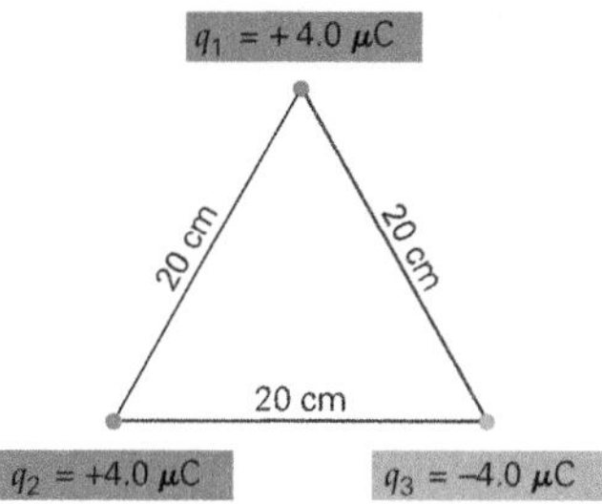


▲ FIGURE 16.27 **A charge triangle** See Exercises 15 and 17.

16. •• Find the work necessary to bring together (from a large distance) the charges in ▼ **Figure 16.28.**

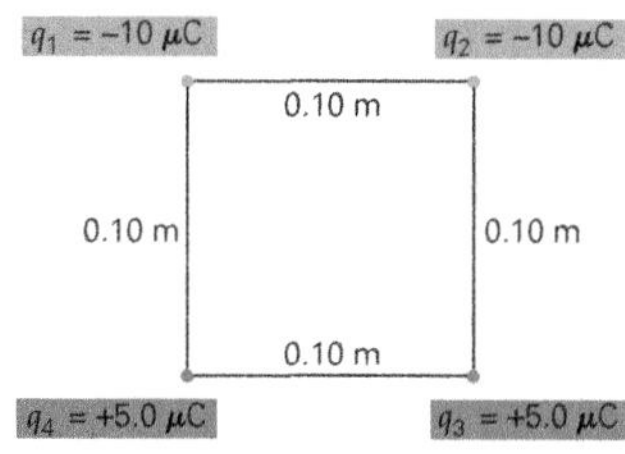


▲ FIGURE 16.28 **A charge rectangle** See Exercises 16 and 18.

17. ••• What is the electric potential at (a) the center of the triangle and (b) a point midway between $q_2$ and $q_3$ in Figure 16.27?
18. ••• What is the electric potential at (a) the center of the square and (b) a point midway between $q_1$ and $q_4$ in Figure 16.28?
19. IE ••• In older computer monitors, electrons were accelerated from rest through a potential difference in an "electron gun" arrangement (▼ **Figure 16.29**). (a) Should the left side of the gun be at (1) a higher, (2) an equal, or (3) a lower potential than the right side? Why? (b) If the potential difference across the gun is 5.0 kV, what is the "muzzle speed" of the electrons emerging from it? (c) If the gun is directed at a screen 25 cm away, how long do the electrons take to reach the screen?

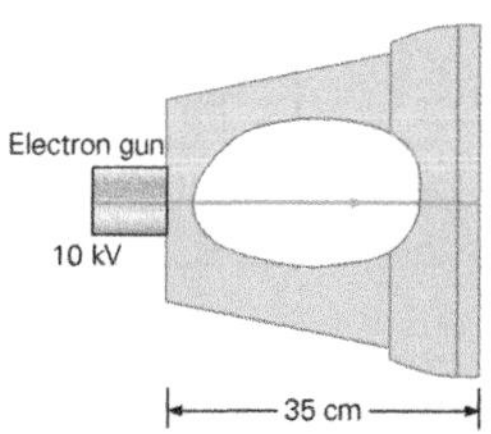


▲ FIGURE 16.29 **Electron speed** See Exercise 19.

## 16.2 Equipotential Surfaces and the Electric Field

20. • A uniform electric field of 10 kV/m points vertically upward. How far apart are the equipotential planes that differ by 100 V?
21. • In Exercise 20, if the ground is designated as zero potential, how far above the ground is the equipotential surface corresponding to 7.0 kV?
22. • Determine the potential 2.5 mm from the negative plate of a pair of parallel plates separated by 20.0 mm and connected to a 24-V battery. Assume the negative plate is designated as zero potential.
23. • Relative to the positive plate in Exercise 22, where is the point with a potential of 14 V?
24. • If the radius of the equipotential surface due to a point charge is 10.5 m and is at a potential of +2.20 kV (compared to zero at infinity), what are the magnitude and sign of the point charge?
25. **IE** • (a) The equipotential surfaces in the neighborhood of a positive point charge are spheres. Which sphere is associated with the higher electric potential: (1) the smaller one, (2) the larger one, or (3) they are associated with the same potential? (b) Calculate the amount of work (in electron-volts) it would take to move an electron from 12.6 m to 14.3 m away from a +3.50-$\mu$C point charge.
26. •• Calculate the voltage to accelerate a beam of protons initially at rest, and calculate their speed if they have kinetic energies of (a) 3.5 eV, (b) 4.1 keV, and (c) $8.0 \times 10^{-16}$ J.
27. •• Repeat the calculation in Exercise 26 for a beam of electrons.
28. ••• Two large parallel plates are separated by 3.0 cm and connected to a 12-V battery. Starting at the negative plate and moving 1.0 cm toward the positive plate at a 45° angle (▼**Figure 16.30**), (a) what value of potential would be reached, assuming the negative plate were defined as zero potential? (b) Repeat part (a) except move 1.0 cm directly toward the positive plate. Explain why your answers to (a) and (b) are different. (c) After the movement in (b), suppose you moved 0.50 cm parallel to the plane of the plates. What would be the electric potential value then?

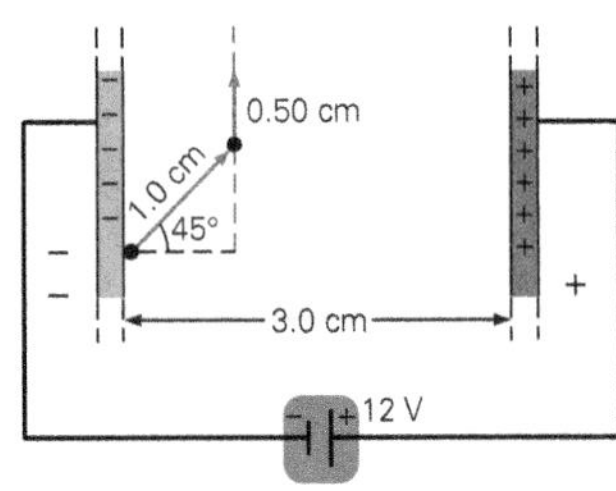


▲**FIGURE 16.30 Reaching our potential** See Exercises 28.

## 16.3 Capacitance

29. • How much charge flows from a 12-V battery when a 2.0-$\mu$F capacitor is connected across its terminals?
30. • A parallel plate capacitor has a plate area of 0.525 m$^2$ and a plate separation of 2.15 mm. What is its capacitance?
31. • What plate separation is required for a parallel plate capacitor to have a capacitance of 9.00 nF if the plate area is 0.425 m$^2$?
32. **IE** • (a) For a parallel plate capacitor with a fixed plate separation distance, a larger plate area results in (1) a larger capacitance value, (2) an unchanged capacitance value, (3) a smaller capacitance value. (b) A 2.50-nF parallel plate capacitor has a plate area of 0.514 m$^2$. If the plate area is doubled, what is the new capacitance value?
33. •• A 12.0-V battery remains connected to a parallel plate capacitor of plate area 0.224 m$^2$ and separation 5.24 mm. (a) What is the charge on the capacitor? (b) How much energy is stored in the capacitor? (c) What is the electric field between its plates?
34. •• If the plate separation of the capacitor in Exercise 33 changed to 10.48 mm after the capacitor is disconnected from the battery, what is the change in your answers?
35. •• Modern capacitors are now capable of storing many times the energy of previous ones. Such a capacitor, with a capacitance of 1.0 F, after being disconnected from its battery, is found to be able to light a small 0.50-W bulb at steady full power for 5.0 s before it quits. What is the terminal voltage of the battery?
36. ••• A modern large 1.50-F capacitor is connected to a 12.0-V battery for a long time, and then disconnected. The capacitor is able to briefly run a 1.00-W toy motor for 2.00 s. After this time, (a) by how much has the energy stored in the capacitor decreased? (b) What is the voltage across the plates? (c) How much charge is stored on the capacitor? (d) How much longer could the capacitor run the motor, assuming the motor ran at full power until the end?
37. ••• Two parallel plates have a capacitance of 0.17 $\mu$F when they are 1.5 mm apart. They are connected permanently to a 100-V power supply. If you pull the plates to a distance of 4.5 mm, (a) what is the electric field between them? (b) By how much has the capacitor's charge changed? (c) By how much has its energy storage changed? (d) Repeat these calculations if the power supply is disconnected before you pull the plates.

## 16.4 Dielectrics

38. • A capacitor has a capacitance of 50 pF, which increases to 150 pF when a dielectric material is between its plates. What is the dielectric constant of the material?

39. • A 50-pF capacitor is immersed in silicone oil, which has a dielectric constant of 2.6. When the capacitor is connected to a 24-V battery, (a) what will be the charge on the capacitor? (b) How much energy is stored in the capacitor?
40. •• The dielectric of a parallel plate capacitor is to be a slab of glass that completely fills the volume between the plates. The area of each plate is 0.50 m$^2$. (a) What thickness should the glass have if the capacitance is to be 0.10 $\mu$F? (b) What is the charge on the capacitor if it is connected to a 12-V battery? (c) How much more energy is stored in this capacitor compared to an identical one without the dielectric insert?
41. IE ••• A parallel plate capacitor has a capacitance of 1.5 $\mu$F with air between the plates. The capacitor is connected to a 12-V battery and charged. The battery is then removed. When a dielectric is placed between the plates, a potential difference of 5.0 V is measured across the plates. (a) What is the dielectric constant of the material? (b) What happened to the energy storage in the capacitor: (1) it increased, (2) it decreased, or (3) it stayed the same? (c) By how much did the energy storage of this capacitor change when the dielectric was inserted?
42. IE ••• An air-filled parallel plate capacitor has rectangular plates with dimensions of 6.0 cm × 8.0 cm. It is connected to a 12-V battery. While the battery remains connected, a sheet of 1.5-mm-thick Teflon (dielectric constant of 2.1) is inserted and completely fills the space between the plates. (a) While the dielectric was being inserted, (1) charge flowed onto the capacitor, (2) charge flowed off the capacitor, (3) no charge flowed. (b) Determine the change in the charge storage of this capacitor because of the dielectric insertion. (c) Determine the change in energy storage in this capacitor because of the dielectric insertion. (d) By how much was the battery's stored energy changed?

**16.5 Capacitors in Series and in Parallel**

43. • What is the equivalent capacitance of two capacitors with capacitances of 0.40 $\mu$F and 0.60 $\mu$F when connected (a) in series and (b) in parallel?
44. • Two identical capacitors are in series and their equivalent capacitance is 1.0 $\mu$F. What is each one's capacitance? Repeat the calculation if, instead, they were in parallel.
45. IE •• (a) Two capacitors can be connected to a battery in either a series or parallel combination. The parallel combination will require (1) more, (2) equal, (3) less energy from a battery than the series combination. Why? (b) Two uncharged capacitors, one with a capacitance of 0.75 $\mu$F and the other with that of 0.30 $\mu$F, are connected in series to a 12-V battery. Then the capacitors are disconnected, discharged, and reconnected to the same battery in parallel. Calculate the energy loss of the battery in both cases.
46. •• For the capacitors in ▼**Figure 16.31**, what value of $C_1$ would result in an equivalent capacitance of 1.7 $\mu$F?

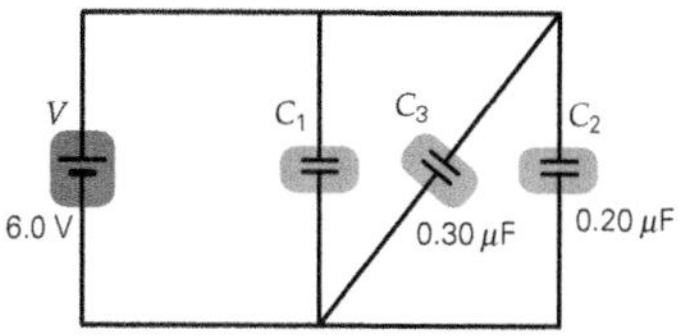


▲**FIGURE 16.31** **A capacitor triad** See Exercises 46 and 48.

47. IE •• (a) Three capacitors of equal capacitance are connected in parallel to a battery, and together they acquire a certain total charge $Q$ from that battery. Will the charge on each capacitor be (1) $Q$, (2) $3Q$, or (3) $Q/3$? (b) Three capacitors of 0.25 $\mu$F each are connected in parallel to a 12-V battery. What is the charge on each capacitor? (c) How much total charge was acquired from the battery?
48. ••• If the capacitance of $C_1$ is 0.10 $\mu$F, (a) what is the charge on each of the capacitors in the circuit in Figure 16.31? (b) How much energy is stored in each capacitor?
49. ••• Four capacitors are connected in a circuit, as illustrated in ▼**Figure 16.32**. Find the charge on, the voltage difference across, and the energy stored for each of the capacitors.

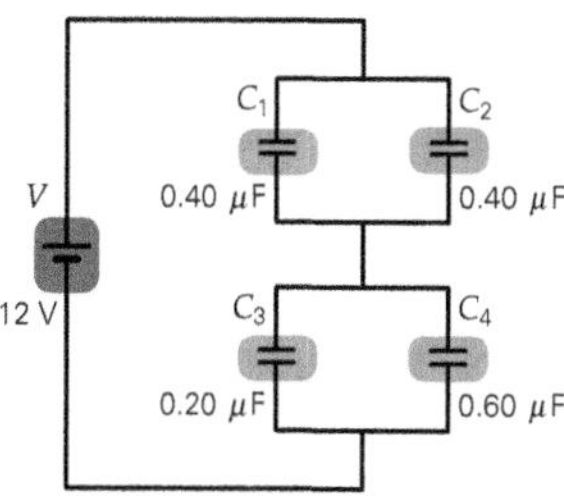


▲**FIGURE 16.32** **Double parallel in series** See Exercise 49.

# 17

# Electric Current and Resistance

**This artificial sparking is caused by an electric current in the air driven by an electric potential difference.**

If you were asked to think of electricity and its uses, many favorable images would probably come to mind, including such diverse applications as lamps, television sets, and computers. You might also think of unfavorable images, such as lightning, a shock, or sparks from an overloaded electric outlet.

Common to all of these images is the concept of electric energy. For electric appliances, electric current through wires supplies energy; for lightning or a spark, it is conducted through the air as in this chapter-opening photo. Here the light (energy) is emitted by air molecules. These molecules are excited by collisions with electrons moving through the air from one electrode to the other.

In this chapter, the fundamental principles governing electric circuits are our primary concern. These principles will enable us to answer questions such as: What is electric current and how does it travel? What causes an electric current to move through an appliance when it is switched on? Why does the electric current cause the filament in a bulb to glow brightly, but not affect the connecting wires in the same way? These basic electrical principles can be applied to enable an understanding of a wide range of phenomena, from the operation of household appliances to the workings of nature's spectacular fireworks – lightning.

## 17.1 Batteries and Direct Current

After studying electric force and energy, you can probably guess what is required to produce the flow of electric charge (commonly called *electric current*). To start, some analogies might help. Water naturally flows downhill, from higher to lower gravitational potential energy. It is this difference in gravitational potential energy that "drives" or causes the water current. Similarly, thermal energy (heat) flows naturally from hot to cold because of the temperature difference ($\Delta T$). Likewise, in electricity, a flow of electric charge is caused by an electric potential difference $\Delta V$ – commonly called "voltage."

In conductors, particularly metals, some of the outer electrons of atoms are relatively free to move. (In other materials such as liquid conductors and charged gases called *plasmas*, positive and negative ions as well as electrons may move.) Energy is required to move electric charge. This electric energy is typically generated through the conversion of other forms of energy, giving rise to a voltage. Any device that can produce and maintain such a potential difference is given the general designation as a *power supply.*

### 17.1.1 Battery Action

One common type of power supply is the **battery**. A battery converts stored chemical potential energy into electrical energy. The Italian scientist Alessandro Volta (1745–1827) constructed one of the first practical batteries in 1880. A simple battery consists of two unlike metal electrodes in an electrolyte, a solution that conducts electric charge. With the appropriate electrodes and electrolyte, a potential difference develops across the electrodes as a result of chemical action (▶ **Figure 17.1**).

The internal operation of the battery is explained in the caption to the figure. While the internal structure is interesting, for our purposes here a battery will simply be pictured as a "black box" that is able to maintain a constant potential difference (or voltage) across its terminals. Inserted into a circuit of wires and appliances, it is able to transfer energy to electrons in the wire (at the expense of its chemical energy), which in turn can deliver that energy to the circuit elements. There the energy is converted into other forms, such as mechanical motion (as in electric fans), heat (as in immersion heaters), and light (as in flashlights). Other sources of voltage (i.e., other types of power supplies), such as electric generators and photocells, will be considered later.

To help visualize the role of a battery, consider the gravitational analogy in ▶ **Figure 17.2**. A pump (analogous to the battery) lifts water, thus increasing the water's gravitational potential energy. The water then returns to the pump flowing down the trough (analogous to the wire) into the pond. On its way down, the water does work on the wheel resulting in rotational kinetic energy, analogous to the electrons transferring energy to an appliance such as a lightbulb or fan. Just as we can use a pump without knowing its inner workings, similarly we will not be concerned with the internals of a battery – only with its ability to cause charge movement.

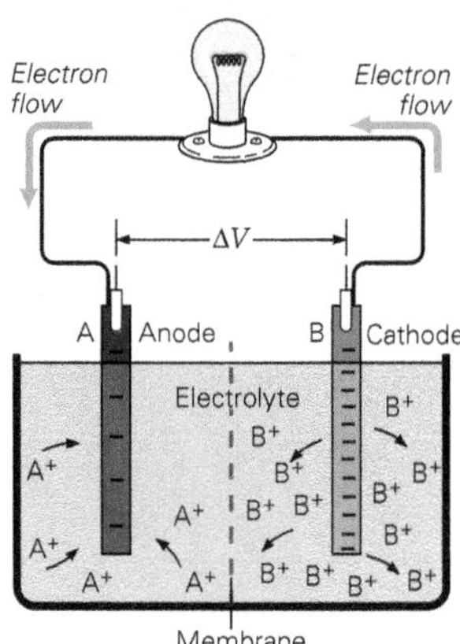


▲ **FIGURE 17.1 Battery action** Chemical processes involving an electrolyte and two unlike metal electrodes cause ions of both metals to dissolve into solution at different rates. Thus, one electrode (*cathode*) becomes more negatively charged than the other (*anode*) and therefore the anode is at a higher potential than the cathode. By convention, the anode is called the positive terminal and the cathode the negative one. This potential difference, or voltage, $\Delta V$, is continually maintained by this chemical action. This voltage can, in turn, produce a current, or a flow of charge (electrons here), in an attached wire or appliance. The positive ions in solution migrate as shown. A membrane – the dashed line – prevents mixing of the two types of ions.

▲ **FIGURE 17.2 Gravitational analogy to a battery and lightbulb** A pump lifts water from the pond, increasing the gravitational potential energy of the water. As the water flows downhill, it transfers energy to (or does work on) a waterwheel, causing the wheel to spin. This action is analogous to the delivery of energy to a lightbulb or fan.

### 17.1.2 Battery Emf and Terminal Voltage

The potential difference across the terminals of a battery when it is not connected to a circuit to a circuit is called the battery's **electromotive force (emf)**, symbolized by $\mathcal{E}$. The name is misleading, because emf is *not* a force, but a potential difference, or voltage.* To avoid confusion with force, the electromotive force is referred to as simply emf. Because it is a voltage and not a force, it represents the energy that could be transferred by the battery per coulomb of charge passing through it. Thus if a battery could do 12 joules of work on 1 coulomb of charge, its emf would be 12 joules per coulomb (12 J/C), or 12 volts (12 V).

* This voltage is sometimes called the battery's "open circuit" voltage.

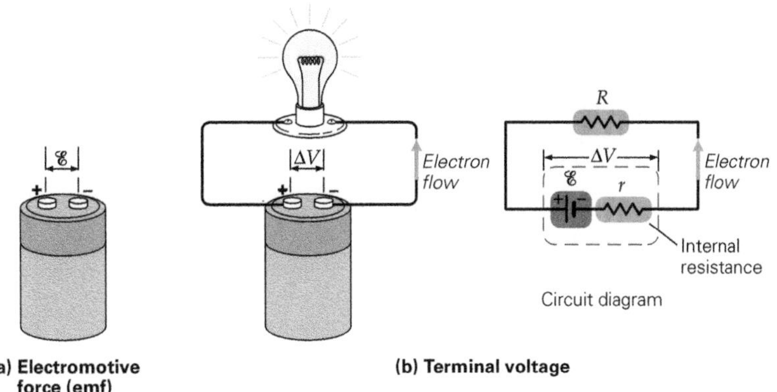


▲ **FIGURE 17.3 Electromotive force (emf) and terminal voltage (a)** The emf (ℰ) of a battery is the potential difference across its terminals when the battery is *not* connected to an external circuit. **(b)** Because of internal resistance ($r$), the terminal voltage ($\Delta V$) of the battery during operation is less than ℰ. (Here, $R$ represents the resistance of the lightbulb.) For our purposes, internal resistances will be assumed negligible, and thus the two voltages will be taken to be the same.

The emf actually represents the maximum potential difference across the terminals (▲ **Figure 17.3a**). Under practical circumstances, when a battery is in a circuit and charge flows, the voltage across the terminals is always slightly *less* than the emf. This "operating voltage" *($\Delta V$)* of a battery (the battery symbol is the pair of unequal-length parallel lines in Figure 17.3b) is called its **terminal voltage**. Because batteries in actual operation are of the most interest, it is the terminal voltage that is important.

Under many conditions, and for most practical purposes, the emf and terminal voltages of a battery (or any power supply) are the same and it will be assumed that this is the case unless stated otherwise. Any difference is due to the presence of the battery's *internal resistance* ($r$), shown explicitly in the circuit in Figure 17.3b. (Resistance, defined in Section 17.3, is a measure of the opposition to charge flow.) Since internal resistances are typically small, the terminal voltage is essentially the same as the emf.

There exists a wide variety of batteries. A common one is the 12-V automobile battery, consisting of six 2-V cells that are connected in *series*.* By series it is meant that the positive terminal of each cell (or battery) is connected to the negative terminal of the next (see the three batteries in ▶ **Figure 17.4a**). When batteries or cells are connected in this fashion, their voltages add to produce a total, larger voltage.

If, however, the batteries or cells are connected in *parallel*, the positive terminals are connected to each other, as are the negative ones (Figure 17.4b). When identical cells or batteries are connected this way, the potential difference or terminal voltage of the arrangement is equal to the voltage of any one.†

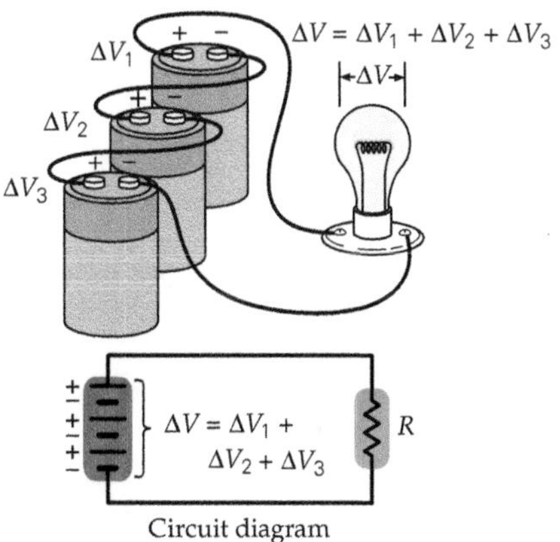


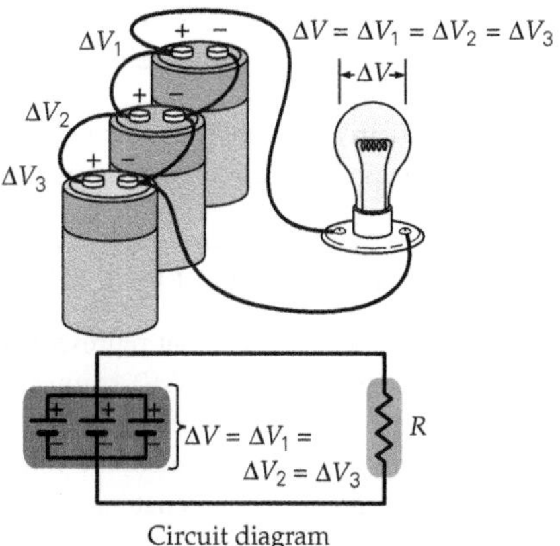


▲ **FIGURE 17.4 Batteries in series and in parallel (a)** When batteries or cells are in series, their voltages add, and the voltage across the lightbulb is the sum of the voltages. **(b)** When batteries *of the same voltage* are in parallel, the voltage across the bulb is that of one.

* Chemical energy is converted to electrical energy in a chemical *cell*. The term *battery* generally refers to a collection, or "battery," of cells.

† If the battery voltages differ then the higher voltage battery will supply, at its expense, energy to the lower voltage battery.

### 17.1.3 Circuit Diagrams and Symbols

To help analyze and visualize circuits, it is common to draw *circuit diagrams* that are schematic representations of the wires, batteries, appliances, and so on. In these diagrams, each element is represented by its own symbol. As seen previously (e.g., Figures 17.3b and 17.4) the battery symbol is two parallel lines, the longer one representing the positive terminal and the shorter one the negative terminal. Any element (such as a lightbulb or appliance) that opposes the flow of charge is represented by the symbol -\/\/\/-. (Electrical resistance is defined in Section 17.3; here the symbol is merely introduced.)

Connecting wires are unbroken lines and are assumed, unless stated otherwise, to have negligible resistance. Where lines cross, it is assumed that they do not contact one another, unless they have a dot at their intersection. Last, switches are shown as "drawbridges," capable of going up (to open the circuit and stop the current) and down (closed to complete the circuit and allow current). These symbols, along with that of the capacitor (Chapter 16), are summarized in ▼**Figure 17.5**. The use of these symbols and circuit diagrams to understand circuits conceptually is demonstrated in the next Example.

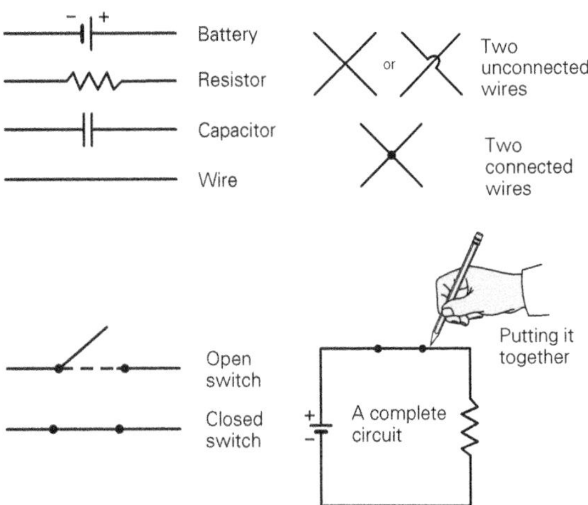


▲FIGURE 17.5 **Some circuit symbols** The symbols shows are the standard and common ones used in circuit diagraming.

---

CONCEPTUAL EXAMPLE 17.1: ASLEEP AT THE SWITCH?

▶**Figure 17.6** shows a circuit diagram that represents two identical batteries (each with a terminal voltage $\Delta V$) connected in parallel to a lightbulb (the resistor symbol). As always, assuming the wires offer no resistance, before switch $S_1$ is opened, the voltage across the lightbulb equals $\Delta V$, (i.e., $\Delta V_{AB} = \Delta V$). What happens to the voltage across the lightbulb when $S_1$ is opened: (a) it remains the same as before the switch was opened, (b) it drops to $\Delta V/2$, because only one battery is now connected to the bulb, or (c) the voltage drops to zero?

**REASONING AND ANSWER.** It might be tempting to choose answer (b), because there is now just one battery. But look again. The remaining battery is still connected to the lightbulb. This means that there must be some voltage across the lightbulb, so the answer certainly cannot be (c). But it also means that the answer cannot be (b), because the remaining battery itself will maintain a voltage of $\Delta V$ across the bulb. Hence, the answer is (a).

**FOLLOW-UP EXERCISE.** In this Example, what would the correct answer be if, in addition to $S_1$, switch $S_2$ were also opened? Explain your answer and reasoning.

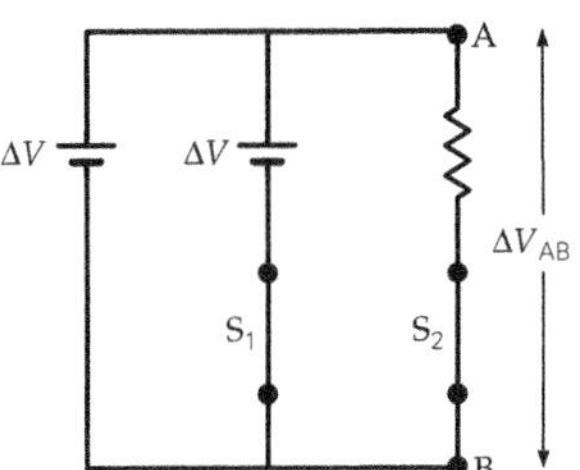


▲FIGURE 17.6 **What happens to the voltage?** See text for discussion.

---

## 17.2 Current and Drift Velocity

Maintaining an electric current requires a voltage source and a **complete circuit** – the name given to a continuous conducting path. Most practical circuits include a switch to "open" or "close" the circuit. An open switch eliminates the continuous part of the path, thereby stopping the flow of charge in the wires. (This is situation is called an *open circuit.*)

### 17.2.1 Electric Current

Because it is electrons that move in wires, the charge flow in a complete circuit leaves from the negative terminal of the battery. Historically, however, circuit analysis is done in terms of **conventional current**. The conventional current's direction is that in which positive charges flow or *appear* to flow. In wires, then, the conventional current direction is *opposite* the electron flow (▶**Figure 17.7**).*

It is said that a battery's job is to "deliver current" to a circuit, or to a component in that circuit. Alternatively, it is sometimes said that a circuit (or its components) "draws current" from the battery. In a complete circuit, the current returns to the battery. A battery, because its terminals do not change sign, is capable of producing current in only one direction. One-directional charge flow is called **direct current (dc)**. (If the current changes direction and/or magnitude, it is called *alternating current [ac]*, a type of current that will be studied in Chapter 21.)

---

* Conventional current is a historic remnant from Ben Franklin (1706–1790), in whose time there were no batteries and no knowledge of the existence of electrons. He advanced an "electric fluid" theory of electricity. When objects became electrostatically charged, he postulated that the fluid moved from a place of excess ("positive") to a place of deficit ("negative"). The idea or convention that electricity "flowed" from positive to negative was adopted and is applied even today.

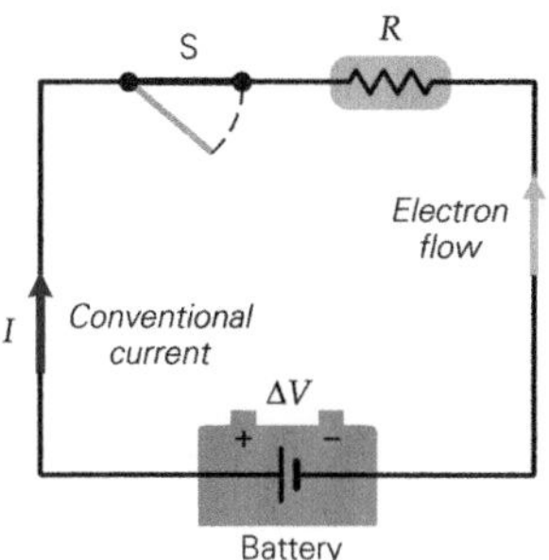


▲ FIGURE 17.7 **Conventional current** For historical reasons, circuit analysis is done with conventional current. Conventional current is in the direction in which positive charges flow, or appear to flow, which is the opposite to the electron flow in metallic wires.

Quantitatively, **electric current (*I*)** is defined as time rate of flow of net charge. Assuming dc conditions and thus a *steady* charge flow, if a net charge $\Delta q$ passes a point in a wire in a time interval $\Delta t$, then the electric current at that point is defined as

$$I = \frac{\Delta q}{\Delta t} \quad \text{(electric current)} \qquad (17.1)$$

SI unit of current: coulomb per second (C/s), or ampere (A)

The coulomb per second is named the **ampere (A)**, in honor of the French physicist André Ampère (1775–1836), an early investigator of electrical and magnetic phenomena. In everyday usage, the ampere is commonly shortened to *amp*. Thus, a current of 10 A is read as "ten amps." Smaller currents are expressed in *milliamperes* (mA, or $10^{-3}$ A), *microamperes* ($\mu$A, or $10^{-6}$ A), or *nanoamperes* (nA, or $10^{-9}$ A). These are also commonly shortened to *milliamps*, *microamps*, and *nanoamps*, respectively. In a typical household circuit, it is not unusual for wires to carry several amps of current. To understand the relationship between charge and current, consider Example 17.2.

---

EXAMPLE 17.2: COUNTING ELECTRONS - CURRENT AND CHARGE

Suppose a steady current of 0.50 A in a wire lasts for 2.0 min. How much charge moves past a point in this wire during this time? How many electrons does this represent?

**THINKING IT THROUGH.** The current and time are given; therefore, the definition of current (Equation 17.1) allows the calculation of the charge $\Delta q$. Each electron carries a charge of magnitude $1.6 \times 10^{-19}$ C, which can be used to find the electron number.

SOLUTION

Listing the data given and converting the time into seconds:

***Given:***

$I = 0.50\text{ A} = 0.50\text{ C/s}$
$\Delta t = 2.0\text{ min} = 1.2 \times 10^2\text{ s}$

***Find:***

$\Delta q$ (amount of charge)
$n$ (number of eletrons)

Solving Equation 17.1, the magnitude of the charge is

$$\Delta q = I\Delta t = (0.50\text{ C/s})(1.2 \times 10^2\text{ s}) = 60\text{ C}$$

Solving for the number of electrons ($n$) from $q = ne$,

$$n = \frac{q}{e} = \frac{60\text{ C}}{1.6 \times 10^{-19}\text{ C/electron}} = 3.8 \times 10^{20}\text{ electrons}$$

(That's a lot of electrons!)

**FOLLOW-UP EXERCISE.** Many sensitive laboratory instruments can measure currents in the nanoamp range or smaller. How long, in years, would it take for 1.0 C of charge to flow past a given point in a wire that carries a current of 1.0 nA?

---

### 17.2.2 Drift Velocity, Electron Flow, and Electric Energy Transmission

Although charge flow is sometimes described as analogous to water flow, electric charge traveling in a conductor does not move in the same way. In the absence of a potential difference across a wire, the free electrons move in random directions at high speeds, colliding many times per second with the metal atoms. As a result, there is no *net* charge flow, and thus no current.

However, when a potential difference (voltage) *is* applied across the wire (such as by a battery), a flow of electrons begins. This does not mean the electrons move directly from one end of the wire to the other. They still move randomly, but there is now a very small added component (toward the positive battery terminal) to their velocities. Thus their velocities are now directed, on average, toward the positive terminal (▼ **Figure 17.8**).

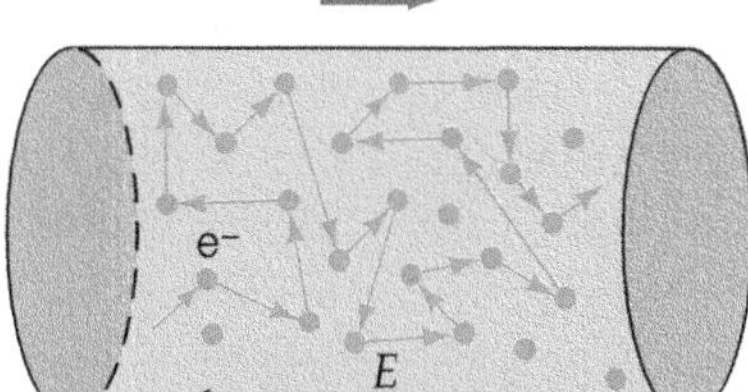


▲ FIGURE 17.8 **Drift velocity** In a metal (here a section of a wire) with no voltage across it, electron motion is random. However, when a battery's terminals are connected at each end of the wire to form a complete circuit, there is then a potential difference across the wire. This provides a tiny velocity component to each electron toward the high-potential (positive) terminal. The rate of this net motion is called the electron drift velocity.

This net electron flow is characterized by an average velocity called the **drift velocity**. The drift velocity is much smaller than the random electron velocities. Typically, the drift velocity is on the order of 1 mm/s. At that speed, it would take an electron about 17 min to travel 1 m along a wire. Yet a lamp comes on almost instantaneously when a switch is closed (completing the circuit), and electronic signals carrying telephone conversations travel almost instantaneously over miles of wire. How can that be?

When a potential difference is applied to a wire, for example, the electrons *throughout the conductor* are each affected almost instantaneously. This is because the voltage, and thus the force on each electron, travels at the speed of light in the wire. Thus, since this speed is so enormous, over everyday distances, the current everywhere in the wire seems to start instantaneously! This means that the current starts everywhere in the circuit essentially simultaneously. In other words, you don't have to wait for electrons to get there from the switch for example. Thus in the lightbulb example, the electrons that are already in its filament before the switch is thrown begin to move almost immediately, delivering energy and creating light with no noticeable delay.

## 17.3 Resistance and Ohm's Law

▲ FIGURE 17.9 **Resistors in use** A printed circuit board, typically used in many electrical devices, can include resistors of different values. The striped cylinders are resistors; their banded color code indicates their resistance value in ohms.

If a voltage (potential difference) is placed across any conducting material in a complete circuit, what factors determine the resulting current? As might be expected, usually the greater the voltage, the greater the current. However, another factor also influences current. Any object that offers significant resistance to electrical current is termed a *resistor* and is represented by the zigzag symbol (—\/\/\/—). This symbol represents all types of resistors, from the cylindrical color-coded ones on printed circuit boards (▼ **Figure 17.9**) to electrical devices and appliances such as hair dryers and lightbulbs.

But how is resistance quantified? For example, if a large voltage applied across an object produces only a small current, it is clear that the object has a high electrical resistance. Keeping this notion in mind, the electrical resistance of any object is defined as the ratio of the voltage applied across it to the resulting current in it, or

$$R = \frac{\Delta V}{I} \quad \text{(electrical resistance definition)} \tag{17.2}$$

SI unit of resistance: volt per ampere (V/A), or ohm (Ω)

Resistance units are volts per ampere (V/A), named the **ohm** (Ω) in honor of the German physicist Georg Ohm (1789–1854), who investigated the relationship between current and voltage. Large values of resistance are expressed as kilohms (kΩ) and even megohms (MΩ). A schematic circuit diagram showing how, in principle, resistance is determined is illustrated in ▼ **Figure 17.10**.

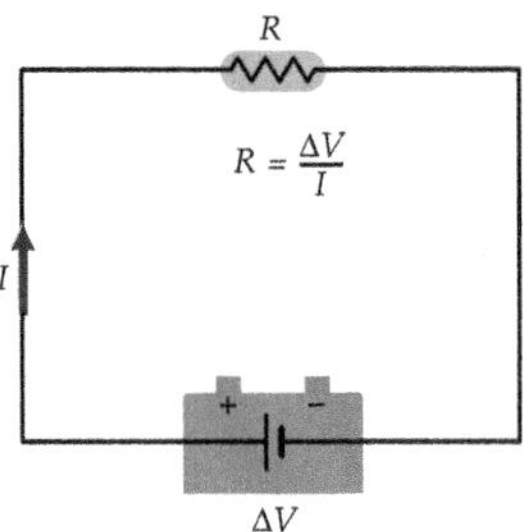


▲ FIGURE 17.10 **Measuring resistance** In principle, any object's electrical resistance can be determined by dividing the voltage across it by the resulting current through it.

For some materials, the resistance may be constant over a range of voltages. A resistor that exhibits constant resistance is said to obey **Ohm's law**, or to be *ohmic*. The law was named after Ohm, who found that many materials, particularly metals, possessed this property. A plot of voltage versus resulting current for an ohmic material is linear with a slope equal to its resistance *R* (▼ **Figure 17.11**). In most of our study, it will be assumed that the resistors are ohmic. Remember however, that current will be directly proportional to the applied voltage *only if R is constant.*

$$\Delta V = IR \quad \text{(Ohm's law: } R \text{ is constant)} \tag{17.3}$$

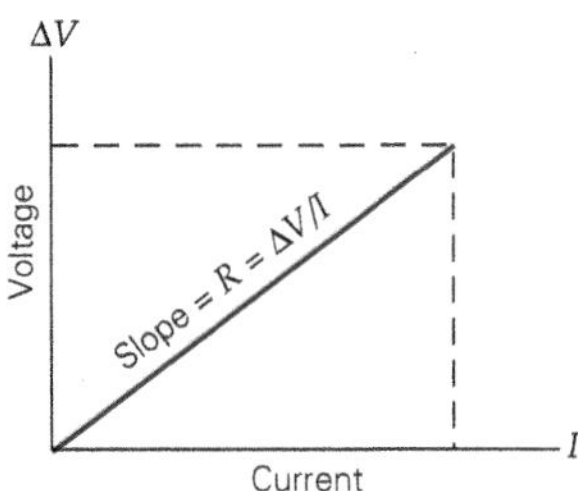


▲ FIGURE 17.11 **Ohm's law** If an object obeys Ohm's law (constant resistance), a plot of voltage versus current is linear with a slope equal to *R*, the object's resistance.

Keep in mind, then, that Ohm's "law" is not a fundamental law in the same sense as, for example, the law of conservation of energy. There is no "law" that states that materials must have constant resistance. Indeed, many advances in electronics are based on materials such as semiconductors, which have variable resistance or nonlinear (non-ohmic) voltage-current relationships.

Example 17.3 shows how the resistance of the human body can make the difference between life and death.

EXAMPLE 17.3: DANGER IN THE HOUSE – HUMAN RESISTANCE

Any room in a house that is exposed to water and electrical voltage can present hazards. (See the discussion of electrical safety in Section 18.5.) For example, suppose a person steps out of a shower and inadvertently touches an exposed 120-V wire (perhaps a frayed cord on a hair dryer) with a finger, thus creating a complete circuit though the body to the wet floor. The human body, when wet, can have an electrical resistance as low as 300 Ω. Using this value, estimate the current in that person's body.

**THINKING IT THROUGH.** The voltage across the body (from finger to feet) is 120 V. To determine the current, Equation 17.2, the definition of resistance can be used.

**SOLUTION**

Listing the data:

***Given:***

$\Delta V = 120\text{ V}$

$R = 300\,\Omega$

***Find:*** $I$ (current in the body)

From Equation 17.2,

$$I = \frac{\Delta V}{R} = \frac{120\text{ V}}{300\,\Omega} = 0.400\text{ A} = 400\text{ mA}$$

While this is a small current by everyday standards, it is a large current for the human body. A current over 10 mA can cause severe muscle contractions, and currents on the order of 100 mA can stop the heart. So this current is potentially deadly (see Section 18.5).

**FOLLOW-UP EXERCISE.** When the human body is dry, its resistance (over its length) can be as high as 100 kΩ. Under these conditions, what voltage would produce a current of 1.0 mA (the value that a person can barely feel)?

## 17.3.1 Factors That Influence Resistance

On the atomic level, resistance arises when electrons collide with the atoms of the material. Thus, resistance depends on the type of material involved. However, geometrical factors also influence resistance. To summarize, the electrical resistance of an object of uniform cross-section, such as a length of uniform wire, depends on four properties: (1) the type of material, (2) length, (3) cross-sectional area, and (4) temperature (▶ **Figure 17.12**).

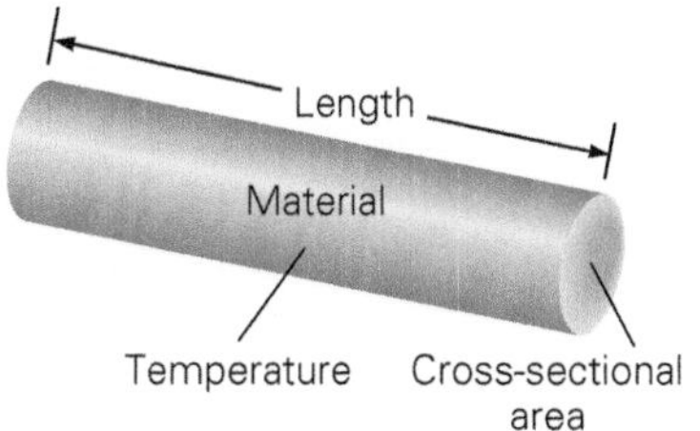


▲ FIGURE 17.12 **Resistance factors** The physical factors that directly affect the electrical resistance of a cylindrical conductor are: (a) type of material, (b) length ($L$), (c) cross-sectional area ($A$), and (d) temperature ($T$).

As might be expected, the resistance ($R$) of an object is inversely proportional to its cross-sectional area ($A$), and directly proportional to its length ($L$); that is, $R \propto L/A$. For example, a uniform metal wire 4.0 m long offers twice as much resistance as a similar wire 2.0 m long, but a wire with a cross-sectional area of 0.50 mm$^2$ has only half the resistance of one with an area of 0.25 mm$^2$. These geometrical resistance conditions are analogous to those for liquid flow in a pipe. The longer the pipe, the greater its resistance (drag), but the larger its cross-sectional area, the more liquid it can carry per second.

## 17.3.2 Resistivity

The resistance of an object is partly determined by its material's atomic properties, quantitatively described by the material's **resistivity ($\rho$)**. Instead of the proportionality relation ($R \propto L/A$), resistivity is the constant of proportionality, enabling the resistance to be written as an equation. Except for temperature dependence, which will be considered later, that expression is:

$$R = \rho\left(\frac{L}{A}\right) \quad \text{(uniform cross-section only)} \tag{17.4}$$

SI unit of resistivity: ohm-meter (Ω·m)

The unit of resistivity ($\rho$) is the ohm-meter (Ω·m). (You should show this.) Thus, from knowing its resistivity (i.e., material type) and using Equation 17.4, the resistance of any constant-area object can be calculated, as long as its length and cross-sectional area are known.

The values of the resistivities of some conductors, semiconductors, and insulators are given in ▶ **Table 17.1**. The values strictly apply only at 20 °C, because resistivity generally depends upon temperature. Most common wires are composed of copper or aluminum with cross-sectional areas on the order of $10^{-6}$ m$^2$ or 1 mm$^2$. You should be able to show that the resistance of a 1.5-m-long copper wire with this area is on the order of 0.025 Ω (25 mΩ). This explains why wire resistances are neglected in circuits – their values are much less than most household devices.

To get a feeling for the magnitudes of these quantities in living tissue, consider Example 17.4.

**TABLE 17.1** Resistivities (at 20 °C) and Temperature Coefficients of Resistivity for Various Materials[a]

| | $\rho$ (Ω·m) | $\alpha$ (1/°C) | | $\rho$ (Ω·m) | $\alpha$ (1/°C) |
|---|---|---|---|---|---|
| **Conductors** | | | **Semiconductors** | | |
| Aluminum | $2.82 \times 10^{-8}$ | $4.29 \times 10^{-3}$ | Carbon | $3.6 \times 10^{-5}$ | $-5.0 \times 10^{-4}$ |
| Copper | $1.70 \times 10^{-8}$ | $6.80 \times 10^{-3}$ | Germanium | $4.6 \times 10^{-1}$ | $-5.0 \times 10^{-2}$ |
| Iron | $10 \times 10^{-8}$ | $6.51 \times 10^{-3}$ | Silicon | $2.5 \times 10^{2}$ | $-7.0 \times 10^{-2}$ |
| Mercury | $98.4 \times 10^{-8}$ | $0.89 \times 10^{-3}$ | | | |
| Nichrome (alloy of nickel and chromium) | $100 \times 10^{-8}$ | $0.40 \times 10^{-3}$ | | | |
| | | | **Insulators** | | |
| Nickel | $7.8 \times 10^{-8}$ | $6.0 \times 10^{-3}$ | Glass | $10^{12}$ | |
| Platinum | $10 \times 10^{-8}$ | $3.93 \times 10^{-3}$ | Rubber | $10^{15}$ | |
| Silver | $1.59 \times 10^{-8}$ | $4.1 \times 10^{-3}$ | Wood | $10^{10}$ | |
| Tungsten | $5.6 \times 10^{-8}$ | $4.5 \times 10^{-3}$ | | | |

[a] Values for semiconductors are general ones, and resistivities for insulators are typical orders of magnitude.

### EXAMPLE 17.4: ELECTRIC EELS – COOKING WITH BIO-ELECTRICITY?

Suppose an electric eel touches the head and tail of a long approximately cylindrically shaped fish and applies a voltage of 600 V across it. If a current of 0.80 A results (likely killing the fish), estimate the average resistivity of the fish's flesh, assuming it is 20 cm long and 2.0 cm in diameter.

**THINKING IT THROUGH.** With cylindrical geometry, the fish's length and cross-sectional area can be found from its dimensions. From the voltage and current, its resistance can be determined. Last, its resistivity can be found from Equation 17.4.

**SOLUTION**

Listing the data:

***Given:***

$$L = 20\ \text{cm} = 0.20\ \text{m}$$
$$d = 2.0\ \text{cm} = 2.0 \times 10^{-2}\ \text{m}$$
$$\Delta V = 600\ \text{V}$$
$$I = 0.80\ \text{A}$$

***Find:*** $\rho$ (resistivity)

The cross-sectional area of the fish is

$$A = \pi r^2 = \pi\left(\frac{d}{2}\right)^2 = \frac{\pi\left(2.0\times10^{-2}\ \text{m}\right)^2}{4} = 3.1\times10^{-4}\ \text{m}^2$$

The fish's overall resistance is $R = \Delta V/I = 600\ \text{V}/0.80\ \text{A} = 7.5 \times 10^2\ \Omega$. Thus from Equation 17.4:

$$\rho = \frac{RA}{L} = \frac{(7.5\times10^2\ \Omega)(3.1\times10^{-4}\ \text{m}^2)}{0.20\ \text{m}} = 1.2\ \Omega\cdot\text{m} = 120\ \Omega\cdot\text{cm}$$

Comparing this to the values in Table 17.1, the fish's flesh is more resistive than metals (as expected). Its value is on the order of those of human tissues, such as cardiac muscle – about 175 Ω·cm.

**FOLLOW-UP EXERCISE**. Suppose for its next meal, the eel in this Example chooses a different species of fish. The next fish has twice the average resistivity, half the length, and half the diameter of the first fish. What current would be expected in this fish if the eel applied 400 V across its body?

For many materials, especially metals, the temperature dependence of resistivity is nearly linear if the temperature does not vary too far from room temperature. Under these conditions, the resistivity of a material at a temperature $T$, and thus a temperature difference $\Delta T = T - T_o$ from a reference temperature $T_o$ (usually 20 °C), is given by

$$\rho = \rho_o(1 + \alpha\Delta T) \quad \text{(temperature variation of resistivity)} \quad (17.5a)$$

where $\alpha$ is a constant called the **temperature coefficient of resistivity** of the material and $\rho_o$ is the material's reference resistivity at $T_o$.

Equation 17.5a can be rewritten as

$$\Delta\rho = \rho_o\alpha\Delta T \quad (17.5b)$$

where $\Delta\rho$ is the change in resistivity when the temperature changes by $\Delta T$. The ratio $\Delta\rho/\rho_o$ is dimensionless, hence $\alpha$ has units of inverse degrees Celsius, written as 1/°C or °C$^{-1}$. Physically, $\alpha$ represents the fractional change in resistivity ($\Delta\rho/\rho_o$) per degree Celsius. The temperature coefficients of resistivity for some materials are listed in Table 17.1. These coefficients will be assumed constant over normal temperature ranges. For semiconductors and insulators, the coefficients are given only as orders of magnitude and are usually not constant.

Since resistance is directly proportional to resistivity (Equation 17.4), an object's resistance has the same dependence on temperature as its resistivity. Thus the temperature dependence of resistance is given by:

$$R = R_o(1+\alpha\Delta T) \quad \text{or} \quad \Delta R = R_o\alpha\Delta T \quad \text{(temperature variation of resistance)} \tag{17.6}$$

Here, $\Delta R = R - R_o$, the change in resistance relative to its reference value $R_o$, again taken as 20 °C. The variation of resistance with temperature provides a means of measuring temperature in the form of an *electrical resistance thermometer*, as illustrated in Example 17.5.

---

**EXAMPLE 17.5: AN ELECTRICAL THERMOMETER – VARIATION OF RESISTANCE WITH TEMPERATURE**

A platinum wire has a resistance of 0.50 Ω at 0 °C. It is placed in a water bath, where, after coming to thermal equilibrium, its resistance comes to a value of 0.60 Ω. What is the temperature of the bath?

**THINKING IT THROUGH.** Using the temperature coefficient of resistivity for platinum from Table 17.1, $\Delta T$ can be found from Equation 17.6 and added to 0 °C, the initial temperature, to find the temperature of the bath.

**SOLUTION**

***Given:***

$T_1 = 0$ °C
$R_1 = 0.50\ \Omega$
$R_2 = 0.60\ \Omega$
$\alpha = 3.93 \times 10^{-3}$/°C (from Table 17.1)

***Find:*** $T_2$ (temperature of the bath)

Solving Equation 17.6 for $\Delta T$:

$$\Delta T = \frac{\Delta R}{\alpha R_o} = \frac{R - R_o}{\alpha R_o} = \frac{0.60\ \Omega - 0.50\ \Omega}{(3.93\times10^{-3}/°\text{C})(0.50\ \Omega)} = 51\ °\text{C}$$

Since the initial temperature was 0 °C, the bath must be at 51 °C.

**FOLLOW-UP EXERCISE.** In this Example, if the material had been copper with the same resistance at the ice point, rather than platinum, what would its resistance be at 51 °C? From this, you should be able to explain which material makes the more "sensitive" thermometer, one with a high temperature coefficient of resistivity or one with a low value.

---

### 17.3.3 Superconductivity

Many materials, including most metals, have positive temperature coefficients, which means that their resistance decreases as the temperature decreases. You might wonder just how far electrical resistance can be reduced by lowering temperature. It turns out that in certain cases, the resistance can reach zero – not just close to zero, but, as accurately as can be measured, *exactly* zero. This phenomenon is called **superconductivity** (discovered in 1911 by Heike Onnes, a Dutch physicist). Currently, the required temperatures are less than 100 K. Thus, at present, practical usage is mainly restricted to laboratory apparatus, medical instruments (such as MRI), as well as research and industrial equipment.

However, superconductivity does have the potential for important everyday applications, if materials can be found that are superconducting near room temperature. In the absence of resistance, high currents and very high magnetic fields are possible (Chapter 19). Used in motors or engines, superconducting electromagnets would be more efficient, providing more power for the same energy input. Superconductors might also be used as electrical transmission lines with no resistive losses. Some researchers envision superfast superconducting computer memories. The absence of electrical resistance opens almost endless possibilities. You are likely to hear much more about superconductor applications in the future as new materials are developed.

## 17.4 Electric Power

When a sustained current exists in a circuit, the electrons are given energy by the voltage source, such as a battery. As these charge carriers pass through circuit components, they collide with the atoms of the material, (i.e., they encounter resistance) and lose energy. The energy transferred in the collisions can result in an increase in temperature. In this way, electrical energy can be transformed, at least partially, into thermal energy.

However, electric energy can also be converted into other forms of energy such as light (as in lightbulbs) and mechanical motion (as in electric motors). According to conservation of energy, whatever forms the energy may take, the total energy delivered to the charge carriers by the battery must be completely transferred to the circuit elements (neglecting losses in the wires). That is, on return to the power supply or battery, the charge carriers must lose the energy gained from the source, and the process is then repeated.

The energy gained by an amount of charge $\Delta q$ from a voltage source (voltage $\Delta V$) is the same as the work done on it, or $W = \Delta q{\cdot}\Delta V$. Over a time interval of $\Delta t$, the rate at which energy is delivered (or work is done) may not be constant. Thus the average rate of energy delivery, or the **average electric power** is $\bar{P} = (W/\Delta t) = (\Delta q{\cdot}\Delta V)/\Delta t)$. In the special case when the current and voltage are constant (as with a battery), then the average power is constant. For steady (dc) currents, the current is

$I = \Delta q/\Delta t$ (Equation 17.1). Thus, under dc conditions the preceding equation can be rewritten as:

$$P = I \cdot \Delta V \quad \text{(dc electric power)} \tag{17.7a}$$

Recall that the SI unit of power is the watt (W). The ampere ($I$) times the volt ($\Delta V$) is the joule per second (J/s), or watt (W). (You should be able to show this.)

Because $R = \Delta V/I$, power can be written in three alternate (but equivalent) forms:

$$P = I \cdot \Delta V = \frac{(\Delta V)^2}{R} = I^2 R \quad \text{(dc electric power)} \tag{17.7b}$$

### 17.4.1 Joule Heat

The thermal power expended in a current-carrying resistor is referred to as **joule heat**, or **$I^2R$ losses** (pronounced "*I* squared *R*" losses). In many instances (such as in electrical transmission lines), joule heating is an undesirable side effect. However, in other situations, the conversion of electrical energy to thermal energy is the main purpose. Heating applications include the heating elements (burners) of electric stoves, hair dryers, immersion heaters, and toasters.

Electrical appliances typically display their electrical specifications (▼**Figure 17.13**). Power requirements of some household appliances are given in ▼**Table 17.2**. Even though most common appliances specify a nominal operating voltage (in the United States) of 120 V, household voltage can vary from 110 to 120 V and still be considered in "normal." Also, in many other areas of the world, the normal operating voltage is 220–240 V. Take a look at Example 17.6 for more insight on appliances and their operation.

▲**FIGURE 17.13 Appliance electrical specifications** Electrical specifications on a garbage disposal for a sink.

**TABLE 17.2** Typical Power Requirements for Household Appliances (at 120 V Except as Noted[a])

| Appliance | Power | Appliance | Power |
|---|---|---|---|
| Air conditioner, room | 1–3 kW | Heater, portable | 1–2 kW |
| Air conditioning, central | 4 kW[a] | Microwave oven | 1000 W |
| Blender | 500 W | Radio | 10–20 W |
| Clothes dryer | 3 kW[a] | Refrigerator | 500 W |
| Clothes washer | 800 W | Stove, top burners | 6 kW[a] |
| Coffeemaker | 1000 W | Stove, oven | 4.5 kW[a] |
| Dishwasher | 1200–1500 W | Television | 200 W |
| Electric blanket | 200 W | Toaster | 850 W |
| Hair dryer | 1500 W | Water heater | 4.5 kW[a] |

[a] A high-power appliance such as this is typically wired to a 240-V house supply to reduce the current. (See Section 18.5.)

**EXAMPLE 17.6: A POTENTIALLY DANGEROUS REPAIR – DON'T DO IT YOURSELF!**

A hair dryer has a power output of 1200 W when operating at 115 V. Its uniform wire filament breaks near one end, and the owner repairs it by removing the section near the break and simply reconnecting it. The resulting filament is 10.0% shorter than its original length. At this voltage, will be the heater's power output after this "repair?"

**THINKING IT THROUGH.** The wire always operates at 115 V. Thus shortening the wire, which decreases its resistance, will result in a larger current. With this increase in current, one would expect the power output to go up.

**SOLUTION**

Let's use a subscript 1 to indicate "before breakage" and a subscript 2 to mean "after the repair." Listing the given values:

***Given:***

$$P_1 = 1200 \text{ W}$$
$$\Delta V_1 = \Delta V_2 = 115 \text{V}$$
$$L_2 = 0.900 L_1$$

***Find:*** $P_2$ (power output after the repair)

After repair, the wire has 90.0% of its original resistance, because resistance is directly proportional to length (Equation 17.3). To show the reduction to 90% explicitly, let's express the resistance after repair ($R_2 = \rho L_2/A$) in terms of the original resistance ($R_1 = \rho L_1/A$):

$$R_2 = \rho \frac{L_2}{A} = \rho \frac{0.900 L_1}{A} = 0.900\left(\rho \frac{L_1}{A}\right) = 0.900\, R_1$$

The current will increase, because the voltage is the same, thus $\Delta V_2 = I_2 R_2 = \Delta V_1 = I_1 R_1$. Thus the new current can be expressed in terms of the original:

$$I_2 = \left(\frac{R_1}{R_2}\right) I_1 = \left(\frac{R_1}{0.900 R_1}\right) I_1 = (1.11) I_1$$

So the current after repair is about 11% larger than before.

The original power is $P_1 = I_1 \cdot \Delta V = 1200$ W and after the repair it is $P_2 = I_2 \cdot \Delta V$ (the voltages have no subscripts because they are the same). Forming a ratio of the powers gives

$$\frac{P_2}{P_1} = \frac{I_2 \cdot \Delta V}{I_1 \cdot \Delta V} = \frac{I_2}{I_1} = 1.11$$

Solving for $P_2$:

$$P_2 = 1.11 P_1 = 1.11(1200 \text{ W}) = 1330 \text{ W}$$

Thus the power output increased by 130 W. This is not the way a repair should be done, because the resulting power output may exceed the manufacturer's requirements and cause the dryer to melt or, worse, start a fire.

**FOLLOW-UP EXERCISE.** In this Example, determine the (a) initial and final resistances and (b) initial and final currents.

### 17.4.2 The Kilowatt-Hour: An Energy Unit

People often complain about their electric bills, but what are they actually paying for? What is being sold is electric energy, usually measured in units of the **kilowatt-hour (kWh)**. But what is a kWh? It is not power, because power is the *rate* at which work is done or energy delivered, so you don't pay for power – you pay for energy. Analogously, water bills are paid for water used, not the rate at which it is used.

Since electric power is defined by $P = \Delta E/\Delta t$, the energy delivered in a time $\Delta t$ is $\Delta E = P \cdot \Delta t$. Thus energy could be said to have units of watt-seconds (an alternate way of saying joule). But for billing purposes, since the energy delivered is a huge number expressed in joules, the choice of unit is the kWh. Thus the kWh is a larger (and thus more convenient for billing) energy unit. It is equivalent to 3.6 million joules, as can be seen below.

$$\begin{aligned}1\text{ kwh} &= (1\text{ kw})(1\text{ hr}) = (1000\text{ W})(3600\text{ s}) = (1000\text{ J/s})(3600\text{ s})\\ &= 3.6\times10^{6}\text{ J}\end{aligned}$$

Thus, consumers pay their "power" company for the electrical energy used to run their appliances.

The cost of electric energy varies with location and time. Currently in the United States, this cost ranges from a low of several cents per kilowatt-hour to several times that. Since the 1990s, electric energy rates have been deregulated. Coupled with increasing demand, this has given rise to higher rates in some areas of the country. Do you know the price of electricity in your locality? Check an electric bill to find out. Now let's look at a comparison of the electric energy costs to run some appliances in Integrated Example 17.7.

---

INTEGRATED EXAMPLE 17.7: A MODERN APPLIANCE DILEMMA – COMPUTING OR EATING

(a) Consider two appliances that operate at the same voltage. Appliance A has a higher power rating than B. How does the resistance of A compare with that of B: is it (1) larger, (2) smaller, or (3) the same? (b) Suppose your computer system has a power requirement of 200 W, whereas a countertop broiler/toaster oven is rated at 1500 W. What is the resistance of each if both are designed to run at 120 V? (c) If the computer system is on 15 h/day every day and the higher-powered broiler is only used for a half hour twice a week, estimate the electrical cost that each generates per month. Assume the electric energy rate is 15 cents per kWh and 30 days per month.

**(a) CONCEPTUAL REASONING.** Power depends on current and voltage. Because the two appliances operate at the same voltage, they must carry different currents. Therefore, answer (3) cannot be correct. Because both appliances operate at the same voltage, the one with higher power (A) must carry more current. For appliance A to carry more current at the same voltage as B, it must have less resistance than B. Therefore, the correct answer is (2); A has less resistance than B.

**(b,c) QUANTITATIVE REASONING AND SOLUTION.** For both appliances in (b) the definition of resistance ($R = \Delta V/I$) can be applied. However, this requires the current to be determined first, which can be done from the known power ratings (Equation 17.7). In (c), the energy used depends both on the rate (power) of usage and the time of usage. This will account for any differences in energies and resulting costs. Listing the data, using the subscript c for computer and b for broiler/toaster:

***Given:***

$P_c = 200\text{ W}$
$P_b = 1500\text{ W}$
$\Delta V = 120\text{ V}$

***Find:*** $R$ (resistance of each appliance)

**(b)** The computer current is

$$I_c = \frac{P_c}{\Delta V} = \frac{200\text{ W}}{120\text{ V}} = 1.67\text{ A}$$

and that in the broiler/toaster is

$$I_b = \frac{P_b}{\Delta V} = \frac{1500\text{ W}}{120\text{ V}} = 12.5\text{ A}$$

Thus the resistances are

$$R_c = \frac{\Delta V}{I_c} = \frac{120\text{ V}}{1.67\text{ A}} = 71.9\ \Omega$$

and

$$R_b = \frac{\Delta V}{I_b} = \frac{120\text{ V}}{12.5\text{ A}} = 9.60\ \Omega$$

Because all operate at the same voltage, resistance is inversely related to power requirement.

**(c)** The time of operation of each appliance (per month) is needed. This is determined as follows (employing the same subscripts as in part (b):

$$\Delta t_c = (15\text{ h/d})(30\text{ d/mo}) = 450\text{ h/mo}$$

and

$$\Delta t_b = (0.5\text{ h/d})(2\text{ d/wk})(4.3\text{ wk/mo}) = 4.3\text{ h/mo}$$

Thus the energy and its cost to run the computer for one month are

$$E_c = 450\text{ h}\times200\text{ W} = 90\times10^{3}\text{ Wh} = 90\text{ kWh}$$

for a monthly cost of

$$(90\text{ kWh})(\$0.15/\text{kWh}) = \$13.50$$

and for the broiler

$$E_b = 4.3\text{ h}\times1500\text{ W} = 6.45\times10^{3}\text{ Wh} = 6.45\text{ kWh}$$

for a monthly cost of

$$(6.45\text{ kWh})(\$0.15/\text{kWh}) = \$0.98$$

Thus the higher power (rate) associated with the broiler is more than offset by its much shorter time of usage, resulting in the computer system costing about 13–14 times more per month.

**FOLLOW-UP EXERCISE.** An immersion heater is a common appliance in most college dorms, useful for heating water for tea, coffee, or soup. Assuming that 100% of the heat produced goes into the water, (a) what must be the heater's resistance (operating at 120 V) to heat a cup of water (mass 250 g) from room temperature (20 °C) to boiling in 3.00 min? (b) How much will this usage increase your monthly electric bill if you prepare two cups of water per day this way?

### 17.4.3 Electrical Efficiency and Natural Resources

About 25% of the electric energy generated in the United States is used for lighting. Refrigerators use about 7% of the energy produced in the United States. This huge (and growing) demand for electric energy has prompted the establishment of efficiency standards for appliances such as refrigerators, freezers, air conditioners, and water heaters (▼**Figure 17.14**). Many households have converted from incandescent lightbulbs to LEDs (*L*ight *E*mitting *D*iodes) (▼**Figure 17.15**), which over time can easily recover their initial cost and more, while providing significant energy conservation.

The result of these and other measures has meant significant energy savings as the newer, more efficient appliances gradually replace the inefficient models. Energy saved translates directly into savings of fuels and natural resources, as well as a reduction in environmental hazards such as pollution and possibly climate change. To see the results that can be achieved by making a change to more efficient lighting, consider Example 17.8.

▲FIGURE 17.14 **Energy guide** These can make consumers aware of the efficiencies of appliances in terms of the average yearly costs in dollars and kWh.

▲FIGURE 17.15 **LED replacement floodlights** Can you tell from the label how many times more efficient this bulb is than an old-fashioned incandescent bulb with the same visible light output? See Example 17.8.

**EXAMPLE 17.8: ELECTRIC ENERGY EFFICIENCY – SAVING MONEY WHILE SAVING ENERGY**

Consider the LED floodlight in Figure 17.15. At a power input of 10 W, it claims, in the same time interval, to put out the same amount of light as an incandescent floodlight that needs a power input of 65 W. The manufacturer claims a lifetime of 25 000 hours. Assume the cost of electric energy is 15 cents/kWh and they operate for 250 days each year. (a) If they are both on for 2 hours/day (h/d), how much money will the LED save in a year? (b) How long will it take for the LED to "pay for itself" (the so-called *payback time*) assuming its purchase price was $9.00?

**THINKING IT THROUGH.** Once the yearly energy usages are determined, the monetary costs will follow. The yearly cost for the LED in comparison to the purchase price should tell how many years are needed to pay for the bulb.

**SOLUTION**

Energy bills are expressed in kWh, so the calculations should be done with this in mind. Listing the data and using the subscript L for LED and I for incandescent:

***Given:***

$$P_L = 10\text{ W}$$
$$P_I = 65\text{ W}$$
$$\text{Cost rate} = \$0.15/\text{kWh}$$
$$\Delta t = 2\text{ h/d}$$

***Find:***

**(a)** Money saved per year using the LED
**(b)** Payback time period

**(a)** For the incandescent bulb per year, the energy usage is:

$$E_I = P_I \cdot \Delta t = (65\text{ W})(2\text{ h/d})(200\text{ d/y}) = 2.60 \times 10^4\text{ Wh/y} = 26.0\text{ kWh/y}$$

at a cost of $(26.0\text{ kWh/y}) \times (\$0.15/\text{kWh}) = \$3.90/\text{y}$.
For the LED bulb per year, the energy usage is:

$$E_I = P_I \cdot \Delta t = (10\text{ W})(2\text{ h/d})(200\text{ d/y}) = 4.00 \times 10^3\text{ Wh/y} = 4.00\text{ kWh/y}$$

at a cost of $(4.00\text{ kWh/y}) \times (\$0.15/\text{kWh}) = \$0.60/\text{y}$.

Thus the savings is $3.30 per year for switching to the LED. While this may not sound like a lot, think of the savings over many years especially since the lifetime of a typical incandescent bulb is only 1500 hours. If, in any given house, many such bulbs were changed, the effect would be noticeable in your budget. Of course, if many buildings around the world made the switch, the savings in energy, and thus fuel, would be dramatic.

**(b)** Since the LED costs \$3.30 less to operate per year than the incandescent, it will take about 3 years to recoup the original LED cost (\$9.00).

**FOLLOW-UP EXERCISE.** A 50" plasma HDTV requires about 300 W to power it, while a similar-sized LCD/HDTV (*liquid crystal display/high-definition television*) requires only 150 W. Assuming they both are on for 5 hours per day, 250 days per year, how much less does it cost to run the LCD version for one year (using 15 cents/kWh for the rate cost)? If you did replace the plasma with the LCD version, how long would the payback period be, assuming the purchase price was \$900.

## Chapter 17 Review

- A **battery** produces an **electromotive force (emf)** ($\mathscr{E}$), or voltage difference, across its terminals. The high-voltage terminal is the **anode** (+), and the low-voltage one is the **cathode** (–). $\mathscr{E}$ is measured in volts and represents the number of joules of energy that a battery (or any power supply) gives to 1 coulomb of charge passing through it: 1 J/C = 1 V.
- **Electric current (*I*)** is the time rate of charge flow.

$$I = \frac{\Delta q}{\Delta t} \quad \text{(electric current)} \qquad (17.1)$$

  Its direction is that of the *conventional current*, that is the direction in which positive charge actually flows, or appears to flow. In metals, the charges are electrons, thus conventional current is opposite the direction of electron movement. The SI unit of current is the **ampere**: 1 A = 1 C/s.
- The electrical **resistance (*R*)** of an object is defined as the voltage difference across the object divided by the resulting current through it.

$$R = \frac{\Delta V}{I} \quad \text{(electrical resistance definition)} \qquad (17.2)$$

  The SI unit of resistance is the **ohm (Ω)**: 1 Ω = 1 V/A.
- A circuit element is said to obey **Ohm's law** if it has a constant electrical resistance. In this case, the current through it is proportional to the voltage applied to it.
- The resistance of an object depends on the **resistivity ($\rho$)** of its material (which depends on its atomic properties and possibly temperature), cross-sectional area *A*, and length *L*. For objects of uniform cross-section

$$R = \rho\left(\frac{L}{A}\right) \quad \text{(uniform cross-section only)} \qquad (17.4)$$

  The SI unit of resistivity is the ohm-meter: Ω·m.
- Resistivity for metals can vary with temperature according to

$$\rho = \rho_o(1 + \alpha\Delta T) \quad \text{(temperature variation of resistivity)} \qquad (17.5a)$$

  where $\alpha$ is the **temperature coefficient of resistivity** and $\rho_o$ is the material's reference resistivity at $T_o$ (usually 20 °C). The SI units of $\alpha$ are inverse degrees Celsius: 1/°C or °C$^{-1}$.
- **Electric power (*P*)** is the rate at which work is done by a battery (or any power supply), or the rate at which energy is transferred to the current passing through it.

$$P = I{\cdot}\Delta V = \frac{(\Delta V)^2}{R} = I^2R \quad \text{(dc electric power)} \qquad (17.7b)$$

# End of Chapter Questions and Exercises

## Multiple Choice Questions

**17.1 Batteries and Direct Current**

1. When four 1.5-V batteries are connected, the output voltage of the combination is measured as 1.5 V. These batteries therefore are connected (a) in series, (b) in parallel, (c) as a pair in series connected in parallel to the other pair in series, (d) you can't tell the connection from the data given.
2. When helping someone whose car has a "dead" battery, how should your car's battery be connected in relation to the dead battery: (a) in series, (b) in parallel, or (c) either in series or in parallel would work fine?
3. When several 1.5-V batteries are connected in series, the overall output voltage of the combination is measured to be 12 V. How many batteries are needed to achieve this voltage: (a) two, (b) ten, (c) eight, or (d) six?
4. To move 3.0 C of charge from one electrode to the other, a 12-V battery does much work: (a) 12 V, (b) 12 J, (c) 3.0 J, (d) 36 W, or (e) 36 J?
5. In a circuit diagram, a battery is represented by (a) two parallel equal-length lines, (b) a straight line in the direction of the wires, (c) two unequal-length lines, (d) a wiggly jagged symbol.
6. In the previous question, for which does the short line stand? (a) Anode, (b) a capacitor plate, (c) cathode, (d) the high voltage terminal of the battery.
7. In a circuit diagram if two wires cross it indicates (a) they are physically joined, making a junction, (b) they are not connected, they just pass over each other, or (c) either (a) or (b).

**17.2 Current and Drift Velocity**

8. In which of these current situations does more charge flow past a point in a wire? (a) 2.0 A for 1.0 min, (b) 4.0 A for 0.5 min, (c) 1.0 A for 2.0 min, or (d) all are the same.
9. Which of these situations represents the least current? (a) 1.5 C passing a point in 1.5 min, (b) 3.0 C passing a point in 1.0 min, or (c) 0.5 C passing a point in 0.10 min.
10. In a dental X-ray machine, the accelerated electrons move to east. The conventional current is in what direction: (a) east, (b) west, or (c) you can't tell from the data given?

11. In a current-carrying metal wire, the drift velocity of the electrons is on the order of (a) the speed of light, (b) the speed of sound, (c) millimeters per second.
12. When a light switch that controls a single lightbulb is thrown to the "on" position, the electric energy gets to the lightbulb at a speed on the order of (a) the speed of light, (b) the speed of sound, (c) a few millimeters per second, or (d) you can't tell from the data given since it depends on the power output of the bulb.

**17.3 Resistance and Ohm's Law**

13. The ohm is another name for the (a) volt/ampere, (b) ampere/volt, (c) watt, (d) volt.
14. Two ohmic resistors are connected across a 12-V battery, one at a time. The current in resistor A is twice that in B. What can you say about their resistances: (a) $R_A = 2R_B$, (b) $R_A = R_B$, (c) $R_A = R_B/2$, or (d) none of the preceding?
15. An ohmic resistor is placed across the terminals of two different batteries, one at a time. When the resistor is connected to battery A, the resulting current is three times the current when it is attached to B. What can you say about the battery voltages: (a) $\Delta V_A = 3 \cdot \Delta V_B$, (b) $\Delta V_A = \Delta V_B$, (c) $\Delta V_B = 3 \cdot \Delta V_A$, or (d) none of the preceding?
16. If you double the voltage across a resistor while cutting its resistance to one-third its original value, what happens to the current in the resistor: (a) it doubles, (b) it triples, (c) it increases by six times, or (d) you can't tell from the data given?
17. Both the length and diameter of cylindrical resistor are doubled. What happens to the resistance: (a) it doubles, (b) it is halved, (c) it is cut to one-fourth its original value, or (d) none of these?
18. Two wires are identical except that one is aluminum and the other copper. Which one's resistance will increase more rapidly as they are heated: (a) the aluminum wire, (b) the copper wire, (c) both would increase at the same rate, or (d) you can't tell?

**17.4 Electric Power**

19. The electric power unit, the watt, is equivalent to what combination of SI units: (a) $A^2 \cdot \Omega$, (b) J/s, (c) $V^2/\Omega$, or (d) all of the preceding?
20. If the voltage across an ohmic resistor is doubled, the power supplied to the resistor (a) doubles, (b) quadruples, (c) is cut by half, or (d) none of the preceding.
21. If the current through an ohmic resistor is halved, the power supplied the resistor (a) doubles, (b) quadruples, (c) is cut by half, or (d) decreases by a factor of 4.
22. A cylindrical resistor dissipates thermal energy at a rate $P$ when connected to a battery. It is disconnected and its length is cut in half. One of the halves is reconnected across the battery. The new power for the shortened resistor is (a) $P$, (b) $2P$, (c) $P/2$, (d) $P/4$.
23. Two wires are of the same length and thickness, but one is aluminum and the other copper. Both are connected, one at a time, to the terminals of the same battery. Which has a higher power output: (a) aluminum, (b) copper, (c) they have the same power output, or (d) you can't tell?

## Conceptual Questions

**17.1 Batteries and Direct Current**

1. Explain why electrode A (see battery in Figure 17.1) is labeled with a plus sign when it has an excess of electrons, which carry a negative charge.
2. Why does the battery design in Figure 17.1 require a chemical membrane?
3. The manufacturer's rating of a battery is 12 V. Does this mean that the battery will necessarily measure 12 V across its terminals when it is placed in a complete circuit? Explain.
4. Sketch the following *complete* circuits, using the symbols shown in Figure 17.5: (a) two ideal 6.0-V batteries in series wired to a capacitor followed by a resistor; (b) two ideal 12.0-V batteries in parallel, connected as a unit to two resistors in series with one another; (c) a nonideal battery (one with internal resistance) wired to two capacitors that are in parallel with each other, followed by two resistors in series with one another.

**17.2 Current and Drift Velocity**

5. In Figure 17.4a, what is the direction of (a) the electron flow in the resistor, (b) the conventional current in the resistor, and (c) the conventional current in the battery?
6. The drift speed of electrons in a complete circuit is a few millimeters per second. Yet a bulb 3.0 m from a switch will turn on instantaneously when the switch is flipped. Explain this apparent paradox.
7. In the battery design shown in Figure 17.1, how does the direction of current inside the battery compare to that in the wire connecting the electrodes?
8. To move charges through a wire, a voltage must be placed between the ends of the wire. If the left side of the wire is at a higher potential than the right, the direction of the conventional current in the wire is (a) left to right, (b) right to left, or (c) it could be either (a) or (b) depending on the size of the voltage applied.

**17.3 Resistance and Ohm's Law**

9. Voltage is plotted on the vertical axis of a graph and current is plotted horizontally. For two ohmic resistors with different resistances, how could you tell from the graph which one is less resistive?
10. Filaments in lightbulbs usually fail just after the bulbs are turned on rather than when they have already been on for a while. Can you explain why?

11. A wire is connected across a steady voltage source. (a) If that wire is replaced with one of the same material that is twice as long and has twice the cross-sectional area, how will the current in the wire be affected? (b) How will the current be affected if, instead, the new wire has the same length as the old one but half the diameter? In both cases, explain your reasoning.
12. A real battery always has some internal resistance $r$ that increases with the battery's age (▼**Figure 17.16**). Explain why, in a complete circuit connection, this results in a drop of the terminal voltage $\Delta V$ of the battery with time.

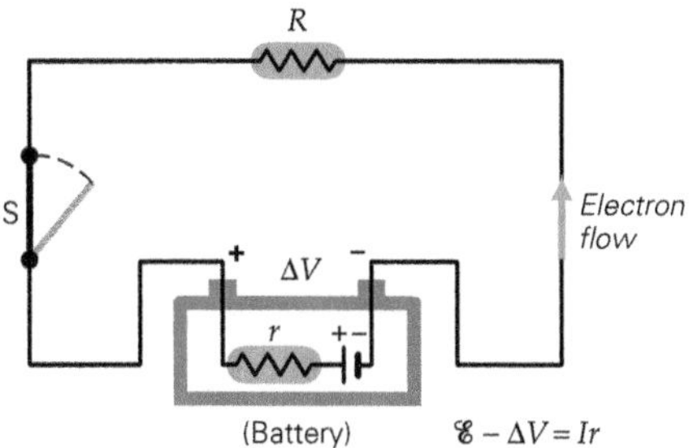


▲FIGURE 17.16 **Emf and terminal voltage** See Conceptual Question 12 and Exercise 15.

13. If you cut a long wire in half, what would you have to do to its diameter in order to keep its resistance at the same original value? Explain your reasoning.

**17.4 Electric Power**

14. Assuming the wire coils in your hair dryer obey Ohm's law, what would happen to its power output if you plugged it into a 240-V outlet in Europe? Remember, it is designed to be used with the 120-V outlets of the United States.
15. Most lightbulb filaments are made of tungsten and are about the same length. What then must be different about the tungsten filament in a 60-W bulb compared with the one in a 40-W bulb?
16. Which gives off more joule heat when connected across a 12-V battery: a 5.0-Ω resistor or a 10-Ω resistor? Explain.
17. From the electric power relationship $P = I^2R$, it would *appear* that increasing an appliance's resistance, would cause its power output to also increase. Yet a 60-W lightbulb has *less* resistance than its 40-W counterpart. Explain.
18. Explain clearly why, in terms of electric energy generation from nonrenewable fuels, an electric appliance can be, at most, only about 33% efficient when calculated in terms of the energy content of the original fuel. [*Hint*: See Chapter 12 on thermal cycle efficiencies and modern power plants. That is where most of our electric energy comes from.]

## Exercises*

*Integrated Exercises (IEs) are two-part exercises. The first part typically requires a conceptual answer choice based on physical thinking and basic principles. The following part is quantitative calculations associated with the conceptual choice made in the first part of the exercise.*

**17.1 Batteries and Direct Current**

*Assume all batteries are ideal unless told otherwise.*

1. • (a) Three 1.5-V batteries are connected in series. What is the total voltage of the combination? (b) What would be the total voltage if the cells were connected in parallel?
2. • What is the voltage across six 1.5-V batteries when they are connected (a) in series, (b) in parallel, (c) three in parallel with one another and this combination wired in series with the remaining three?
3. • Two 6.0-V batteries and one 12-V battery are connected in series. (a) What is the voltage across the whole arrangement? (b) What arrangement of these three batteries would give a total voltage of 12 V?
4. •• Given three batteries with voltages of 1.0, 3.0, and 12 V, what are the minimum and maximum voltages that could be achieved by connecting them in series?
5. IE •• You are given four AA batteries that are rated at 1.5 V each. The batteries are grouped in pairs. In arrangement A, the two batteries in each pair are in series, and then the pairs are connected in parallel. In arrangement B, the two batteries in each pair are in parallel, and then the pairs are connected in series. (a) Compared with arrangement B, will arrangement A have (1) a higher, (2) the same, or (3) a lower total voltage? (b) What are the total voltages for each arrangement?

**17.2 Current and Drift Velocity**

6. • How long does it take for a charge of 3.50 C to pass through the cross-sectional area of a wire carrying a current of 0.57 A?
7. • A net charge of 30 C passes through the cross-sectional area of a wire in 2.0 min. What is the current in the wire?
8. • (a) How long would it take for a net charge of 2.5 C to pass a location in a wire if it is to carry a steady current of 5.0 mA? (b) If the wire is actually connected directly to the two electrodes of a battery and the battery does 25 J of work on the charge during this time, what is the terminal voltage of the battery?

* The bullets denote the degree of difficulty of the exercises: •, simple; ••, medium; and •••, more difficult.

9. • A small toy car draws a 0.50-mA current from a 3.0-V NiCd (nickel-cadmium) battery. In 10 min of operation, (a) how much charge flows through the toy car, and (b) how much energy is lost by the battery?
10. • A car's starter motor draws 50 A from the car's battery during startup. If the startup time is 1.5 s, how many electrons pass a given location in the circuit during that time?
11. •• A net charge of 20 C passes a location in a wire in 1.25 min. How long does it take for a net 30-C charge to pass that location if the current in the wire is doubled?
12. ••• Car batteries are often rated in "ampere-hours" or A·h. (a) Show that the A·h has SI units of charge and specifically that 1 A·h = 3600 C. (b) A fully charged, heavy-duty battery is rated at 100 A·h and can deliver a current of 5.0 A steadily until depleted. What is the maximum time this battery can deliver that current, assuming it isn't being recharged? (c) How much charge will the battery deliver in this time?
13. IE ••• Imagine that some protons are moving to the left at the same time that some electrons are moving to the right past the same location. (a) Will the net current be (1) to the right, (2) to the left, (3) zero, or (4) none of the preceding? (b) In 4.5 s, 6.7 C of electrons flow to the right at the same time that 8.3 C of protons flow to the left. What are the direction and magnitude of the current due to the protons? (c) What are the direction and magnitude of the current due to the electrons? (d) What are the direction and magnitude of the total current?
14. ••• In a proton linear accelerator, a 9.5-mA proton current hits a target. (a) How many protons hit the target each second? (b) How much energy is delivered to the target each second if each proton has a kinetic energy of 20 MeV and loses all its energy in the target? (c) If the target is a 1.00-kg block of copper, at what rate will its temperature increase if it is not cooled?

### 17.3 Resistance and Ohm's Law

*Assume here that the temperature coefficients of resistivity in Table 17.1 apply over large temperature ranges.*

15. • A battery labeled 12.0 V supplies 1.90 A to a 6.00-Ω resistor (Figure 17.16). (a) What is the terminal voltage of the battery? (b) What is its internal resistance?
16. • How much current is drawn from an ideal 12-V battery when a 15-Ω resistor is connected across its terminals?
17. • What terminal voltage must an ideal battery (no significant internal resistance) have to produce a 0.50-A current through a 2.0-Ω resistor?
18. • What is the emf of a battery with a 0.15-Ω internal resistance if the battery delivers 1.5 A to an externally connected 5.0-Ω resistor?
19. IE • Some states allow the use of aluminum wire in houses in place of copper. (a) If you wanted the resistance of your aluminum wire to be the same as that of copper (assuming the same lengths), would the aluminum wire have to have (1) a greater diameter than, (2) a smaller diameter than, or (3) the same diameter as the copper wire? (b) Calculate the ratio of the thickness of aluminum to that of copper needed to make their resistances equal.
20. •• A material is formed into a long rod with a square cross-section 0.50 cm on each side. When a 100-V voltage is applied across a 20-m length of the rod, a 5.0-A current is carried. (a) What is the resistivity of the material? (b) Is the material a conductor, an insulator, or a semiconductor?
21. •• Two copper wires have equal cross-sectional areas and lengths of 2.0 and 0.50 m, respectively. (a) What is the ratio of the current in the shorter wire to that in the longer one if they are connected to the same power supply? (b) If you wanted the two wires to carry the same current, what would the ratio of their cross-sectional areas have to be? (Give your answer as a ratio of longer to shorter.)
22. IE •• Two copper wires have equal lengths, but the diameter of one is three times that of the other. (a) The resistance of the thinner wire is (1) 3, (2) 1/3, (3) 9, (4) 1/9 times that of the resistance of the thicker wire. (b) If the thicker wire has a resistance of 1.0 Ω, what is the resistance of the thinner wire?
23. •• The wire in a heating element of an electric stove burner has a 0.75-m length and a $2.0 \times 10^{-6}$ m$^2$ cross-sectional area. (a) If the wire is made of iron and operates at 380 °C, what is its operating resistance? (b) What is its resistance when the stove is "off?"
24. •• (a) What is the percentage variation of the resistivity of copper over the temperature range from 20 °C to 100 °C? (b) Assume a copper wire's resistance changes due to only resistivity changes over this temperature range. Further assume that it is connected to the same power supply. By what percentage would its current change? Would it be an increase or decrease?
25. •• A copper wire has a 25-mΩ resistance at 20 °C. When the wire is carrying a current, heat produced by the current causes the temperature of the wire to increase by 27 °C. (a) What is the change in the wire's resistance? (b) If its original current was 10.0 mA, what is its final current?
26. •• When a resistor is connected to a 12-V source, it draws a 185-mA current. The same resistor connected to a 90-V source draws a 1.25-A current. (a) Is the resistor ohmic? Justify your answer mathematically. (b) What is the rate of Joule heating in both cases?

27. •• A particular application requires a 20-m length of aluminum wire to have a 0.25-m Ω resistance at 20 °C. (a) What is the wire's diameter? (b) What would its resistance be if its length was halved and it was then placed in an ice water bath?
28. •• (a) If the resistance of the wire in Exercise 27 cannot vary by more than ±5.0% from its value at 20°C, to what operating temperature range should it be restricted? (b) What would be the operating range if the wire were, instead, made of copper?
29. IE •• As a wire is stretched out so that its length increases, its cross-sectional area decreases, while the total volume of the wire remains constant. (a) Will the resistance after the stretch be (1) greater than, (2) the same as, or (3) less than that before the stretch? (b) A 1.0-m length of copper wire with a 2.0-mm diameter is stretched out; its length increases by 25% while its cross-sectional area decreases but remains uniform. Compute the resistance ratio (final to initial).
30. •• ▼ **Figure 17.17** shows data on the dependence of the current through a resistor on the voltage across that resistor. (a) Is the resistor ohmic? Explain your reasoning. (b) What is the value of its resistance? (c) Use the data to predict what voltage would be needed to produce a 4.0-A current in the resistor.

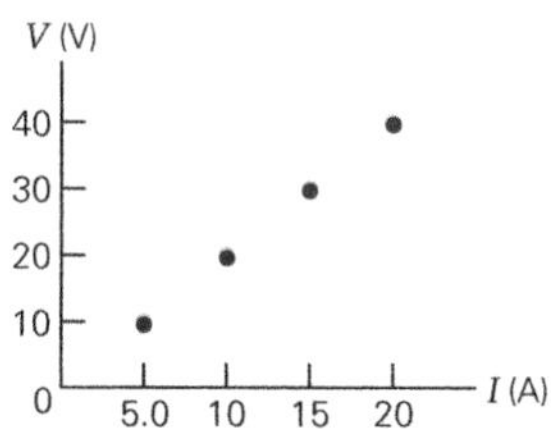


▲ **FIGURE 17.17 An ohmic resistor?** See Exercise 30.

## 17.4 Electric Power

31. • A DVD player is rated at 100 W at 120 V. What is its resistance?
32. • A freezer of resistance 10 Ω is connected to a 110-V source. What is the power delivered when this freezer is on?
33. • The current in a refrigerator with a resistance of 12 Ω is 13 A (when the refrigerator is on). What is the power delivered to the refrigerator?
34. • Show that the quantity volts squared per ohm ($V^2/\Omega$) has SI units of power.
35. • An electric water heater is designed to produce 50 kW of heat when it is connected to a 240-V source. What is its resistance?
36. •• If the heater in Exercise 35 is 90% efficient, how long would it take to heat 50 gal of water from 20 °C to 80 °C?
37. IE •• An ohmic resistor in a circuit is designed to operate at 120 V. (a) If you connect the resistor to a 60-V power source, will the resistor dissipate heat at (1) 2, (2) 4, (3) 1/2, or (4) 1/4 times the designed power? Why? (b) If the designed power is 90 W at 120 V, but the resistor is connected to a 30-V power supply, what is the power delivered to the resistor?
38. •• An electric toy with a resistance of 2.50 Ω is operated by a 3.00-V battery. (a) What current does the toy draw? (b) Assuming that the battery delivers a steady current for its lifetime of 4.00 h, how much charge passed through the toy in that time? (c) How much energy was delivered to the toy in that time?
39. •• A welding machine draws 18 A of current at 240 V. (a) What is its power rating? (b) What is its resistance? (c) When it is inadvertently connected to a 120 V outlet, the current in it is 10 A. Is the machine's resistance ohmic? Prove your answer.
40. •• On average, an electric water heater operates for 2.0 h each day. (a) If the cost of electricity is $0.15/kWh, what is the cost of operating the heater during a 30-day month? (b) What is the resistance of a typical water heater? [*Hint*: See Table 17.2.]
41. •• (a) What is the resistance of an immersion-type heating coil if it is to generate 15 kJ of heat per minute when it is connected to a 120-V source? (b) What would the coil's resistance have to be if instead 10 kJ of heat per minute was desired?
42. •• A 200-W computer power supply is on 10 h per day. (a) If the cost of electricity is $0.15/kWh, what is the energy cost if this computer is used like this every day for a year (365 days)? (b) If this power supply is replaced by a more efficient 100-W version and it costs $75, how long will it take the decreased operating cost to pay for the new supply?
43. •• A 120-V air conditioner unit draws 15 A of current. If it operates for 20 min, (a) how much energy in kilowatt-hours does it require in that time? (b) If the cost of electricity is $0.15/kWh, what is the cost of operating it for 20 min? (c) If the air conditioner initially cost $450 and it is operated, on average, 4 h per day, how long does it take before the operating costs equal the price?
44. •• Two resistors, 100 Ω and 25 kΩ, are rated for a maximum power output of 1.5 and 0.25 W, respectively. (a) What is the maximum voltage that can be safely applied to each resistor? (b) What is the maximum current that each resistor can have?
45. •• A wire 5.0 m long and 3.0 mm in diameter has a resistance of 100 Ω. A15-V potential difference is applied across the wire. Find (a) the current in the wire, (b) the resistivity of its material, and (c) the rate at which heat is being produced in the wire.

46. **IE** •• When connected to a voltage source, a coil of tungsten wire initially dissipates 500 W of power. In a short time, the temperature of the coil increases by 150 °C because of joule heating. (a) Will the dissipated power (1) increase, (2) remain the same, or (3) decrease? Why? (b) What is the corresponding change in the power? (c) What is the percentage change in its current?
47. •• A 20-Ω resistor is connected to four 1.5-V batteries. What is the joule heat loss per minute in the resistor if the batteries are connected (a) in series and (b) in parallel?
48. •• A 5.5-kW water heater operates at 240 V. (a) Should the heater circuit have a 20-A or a 30-A circuit breaker? (A circuit breaker is a safety device that "opens" the circuit – shutting down the current – at its rated current.) (b) Assuming 85% efficiency, how long will the heater take to heat the water in a 55-gal tank from 20 °C to 80 °C?
49. •• A student uses an immersion heater to heat 0.30 kg of water from 20 °C to 80 °C for tea. (a) If the heater is 75% efficient and takes 2.5 min to heat the water, what is its resistance? (b) How much current is in the heater? (Assume 120-V household voltage.)
50. •• An ohmic appliance is rated at 100 W when it is connected to a 120-V source. If the power company cuts the voltage by 5.0% to conserve energy, what is (a) the current in the appliance and (b) the power consumed by the appliance after the voltage drop?
51. •• A lightbulb's output is 60 W when it operates at 120 V. If the voltage is cut in half and the power dropped to 20 W during a brownout, what is the ratio of the bulb's resistance at full power to its resistance during the brownout?
52. •• To empty a flooded basement, a water pump must do work (lift the water) at a rate of 2.00 kW. If the pump is wired to a 240-V source and is 84% efficient, (a) how much current does it draw and (b) what is its resistance?
53. ••• (a) Find the individual monthly (30-day) electric energy costs (to the nearest dollar) for each of the following household appliances if the utility rate is $0.12/kWh: central air conditioning that runs 30% of the time; a blender that is used 0.50 h/month; a dishwasher that is used 8.0 h/month; a microwave oven that is used 15 min/day; the motor of a frost-free refrigerator that runs 15% of the time; a stove (burners plus oven) that is used a total of 10 h/month; and a television that is operated 120 h/month. (b) Determine the percentage that each appliance contributes to the total monthly electric cost. (c) What is the resistance of each appliance and current in each appliance when they are operating? (You may find useful information in Table 17.2.)
54. A computer hard drive that operates on 120 V is rated at 40 W when it is operating. (a) How much current does the drive draw? (b) What is the drive's resistance? (c) How much energy (in kWh) does this drive use per month assuming it operates 15 min per day? (d) Estimate the electric energy bill per month, assuming 15 cents per kWh.

# 18

# Basic Electric Circuits



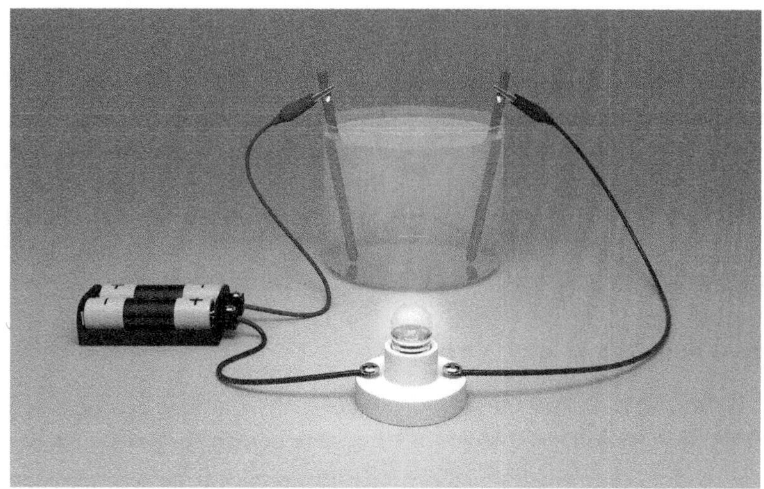

**This photo shows that material (here saltwater) containing charged particles can be used to carry current and thus complete a circuit.**

Metallic wires are usually thought of as the connectors between elements in a circuit. However, wires are not the only conductors of electricity, as the circuit in the chapter-opening photo shows. Notice that the bulb is lit. Hence the circuit, which consists of batteries, wires, electrodes, and a saltwater solution, must be complete. The conclusion is that the solution itself must be a conductor. Circuits containing liquid conductors such as these have practical applications in the laboratory and industry; for example, they can be used to synthesize or purify chemical substances.

Electric circuits are of many kinds and serve many specific purposes, from boiling water to providing light to restarting a heart. Building on the principles discussed previously, this chapter emphasizes the analysis and applications of electric circuits. Such analysis can be useful, for example, to check that a circuit will function properly as designed, or to be sure there would be no safety problems. As might be expected, our circuit analysis will rely heavily on circuit diagrams. (A few of these diagrams were employed in Chapter 17.) To begin, let's explore some of the ways that resistive elements, such as lightbulbs, can be connected.

## 18.1 Resistances in Series, Parallel, and Series–Parallel Combinations

In our discussions and diagrams, the resistance symbol —/\/\/\— can represent *any* type of circuit element, such as a toaster, lightbulb, or computer. For our purposes, all elements will be treated as if ohmic (constant resistance) and wire resistance will be neglected unless otherwise stated. There are two basic ways that circuit elements can be connected: in series and in parallel. (As shown for capacitors in Section 16.5.) Let's start with the series connection.

### 18.1.1 Resistors in Series

To begin, because voltage represents energy (per unit charge), to conserve energy the *sum of the voltages around a complete circuit loop must be zero.* Since the term voltage means "change in electrical potential," voltage gains and losses are represented by + and − signs, respectively.

Consider the circuit in ▼**Figure 18.1a**. These elements are connected in **series**, or head to tail. As a clockwise complete path is followed around circuit, summing the voltage gains (battery) and losses (resistors 1, 2, and 3) to zero gives: $+\Delta V_B - \Delta V_1 - \Delta V_2 - \Delta V_3 = 0$. Furthermore, in series, each element must have the same current because charge is conserved and cannot "pile up" or "leak out." Thus there is only one current ($I$), and each resistor's voltage is related to its resistance by $\Delta V_i = IR_i$ (the subscript $i$ refers to any of the three). The previous equation becomes:

$$\Delta V_B - \Sigma(\Delta V_i) = 0 \quad \text{or} \quad \Delta V_B = \Sigma(IR_i) \quad \text{(series)} \quad (18.1a)$$

► **Figure 18.2** shows a water current situation analogous to the circuit in Figure 18.1. In Figure 18.2, there is a flow of water ("current") over a smooth streambed ("wires") punctuated by a series of rapids ("resistance").

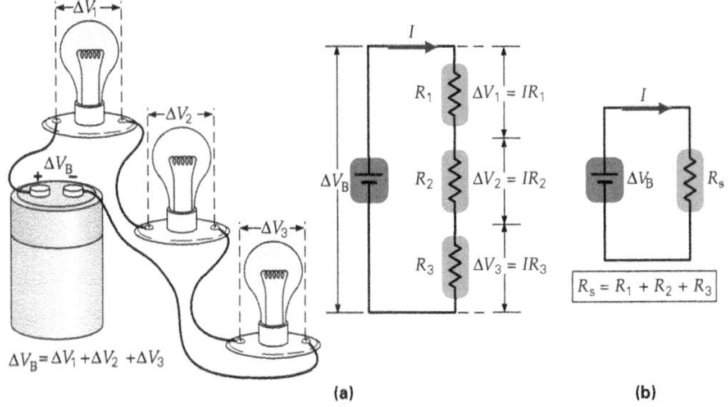


▲ **FIGURE 18.1 Resistors in series (a)** When resistors (representing the resistances of lightbulbs) are in series, the current in each is the same and $\Sigma\Delta V_i$, the sum of the voltage drops across the resistors, is equal to $\Delta V_B$, the battery voltage. **(b)** The equivalent resistance $R_s$ of the resistors in series is the sum of the resistances.

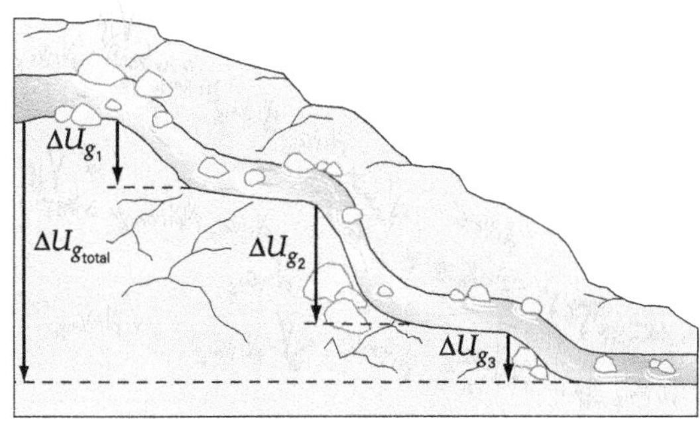


▲ **FIGURE 18.2 Water flow analogy to resistors in series** Even though, in general, a different amount of gravitational potential energy (per kilogram) is lost as the water flows down each set of rapids, the current of water is the same everywhere. The total loss of gravitational potential energy (per kilogram) is the sum of the losses. To make this a "complete" water circuit, an external power source, such as a pump, must do work on the water and take it to the top of the hill, producing a gain in its energy equal to the sum of the losses incurred on the way down.

For the special case of three series resistors, Equation 18.1a becomes:

$$\Delta V_B = IR_1 + IR_2 + IR_3 = I(R_1 + R_2 + R_3) \quad (18.1b)$$

The **equivalent series resistance ($R_s$)** is defined as the resistance value of a *single resistor* that could replace the actual resistors and yet maintain the same current. In Figure 18.1, this means that $\Delta V_B = IR_s$ or $R_s = \Delta V_B/I$. Hence, the equivalent resistor for three series resistors is

$$R_s = \Delta V_B/I = R_1 + R_2 + R_3$$

For example, if each resistor had a resistance of 10 Ω, then $R_s$ would be 30 Ω.

This result can be generalized to any number of **resistors in series**:

$$R_s = R_1 + R_2 + R_3 + \cdots = \Sigma R_i \quad \left(\begin{matrix}\text{equivalent series}\\ \text{resistance}\end{matrix}\right) \quad (18.2)$$

*Note that the equivalent series resistance is larger than the resistance of the largest resistor in the series.*

Series connections are not common in many circuits, such as in-house wiring, because there are two major disadvantages. The first is clear if you consider what happens if one of the bulbs in Figure 18.1a burns out (or is turned off). In this case, all the bulbs would go out, because the circuit would no longer be complete, or continuous. In this situation, the circuit is said to be *open*. Thus an *open circuit* has infinite equivalent resistance, because the current in it is zero, even though the battery voltage is not.

The second series disadvantage is that each resistor's operating power depends on the others. Consider, for example, what would happen in Figure 18.1a if a fourth bulb were added. The voltage across each of the original bulbs (as well as the current) would decrease, resulting in reduced light output from all bulbs. Thus none of them would glow at their rated output. Clearly, this type of connection (series) is not acceptable in a household setting. Compare these disadvantages to the advantages in the following parallel circuit discussion.

### 18.1.2 Resistors in Parallel

Resistors can also be connected in **parallel** (▼**Figure 18.3a**). In this case, all the resistors (here light bulbs) have common connections - that is, all the leads on one side of the resistors are attached together and then in turn to one terminal of the battery. The remaining leads are also attached together and then to the other terminal. *When resistors are connected in parallel to a source of emf, the voltage across each resistor is the same and equal to that of the source.* It may not surprise you that household circuits are wired in parallel. (See Section 18.5.) In parallel, each appliance operates at full voltage, and turning one appliance off or on does not affect the others.

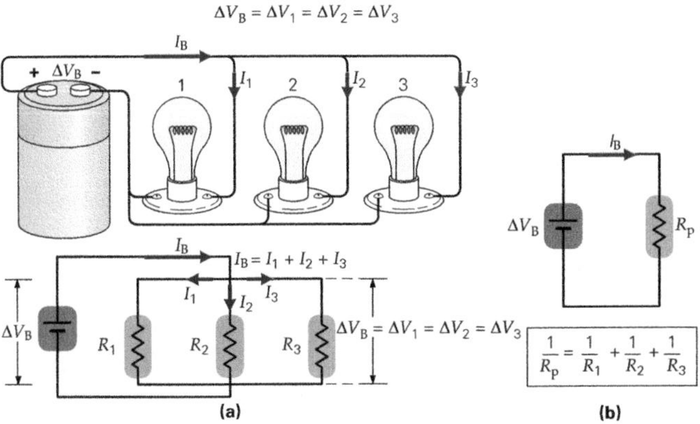


▲**FIGURE 18.3 Resistors in parallel (a)** When resistors (here bulbs) are in parallel, the voltage across each is the same. The current from the battery divides (generally unequally) among the resistors. **(b)** The expression for the parallel equivalent resistance, $R_p$, is of a reciprocal nature.

Unlike resistors in series, the current in a parallel circuit divides into the different paths (Figure 18.3a). This occurs at any *junction* (a location where several wires come together), much as traffic divides or merges together when it reaches a junction in the road (▶**Figure 18.4a**). Thus by charge conservation, the total current out of the battery ($I_B$) must be the sum of the separate currents. Specifically, for three resistors in parallel, this is written as $I_B = I_1 + I_2 + I_3$. In general, the current will divide among the resistors in inverse proportion to their resistances. Thus the largest current will take the path of least resistance, but no one resistor will carry the total.

(a)

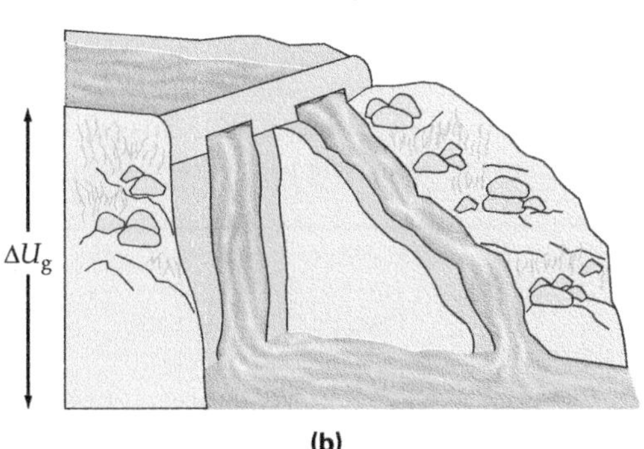

(b)

▲**FIGURE 18.4 Analogies for resistors in parallel (a)** When a road forks, the total number of cars entering the two branches each minute is equal to the number of cars arriving at the fork each minute. Movement of charge into and then out of a junction can be considered in the same way. **(b)** When water flows from a dam, the gravitational potential energy lost (per kilogram of water) in falling to the stream below is the same regardless of the path. This is analogous to voltages across parallel resistors.

The **equivalent parallel resistance ($R_p$)** is the value of a *single* resistor that could replace all the resistors and maintain the same total current. Thus, $R_p = \Delta V_B/I_B$ or $I_B = \Delta V_B/R_p$. As mentioned previously, the voltage must be the same across each resistor. To visualize this, imagine two separate water paths, each leading from the top of a dam to the bottom. The water loses the same amount of gravitational potential energy per gallon (analogous to $\Delta V_B$) regardless of the path (Figure 18.4b). For electricity, a given charge loses the same electrical potential energy, regardless of which parallel resistor it passes through.

Since all the voltages are the same and equal to the battery voltage, the current in each resistor is $I_i = \Delta V_B/R_i$ (the subscript $i$ represents *any* of the resistors). Substituting for each current into $I_B = I_1 + I_2 + I_3$:

$$I_B = I_1 + I_2 + I_3 = \frac{\Delta V_B}{R_1} + \frac{\Delta V_B}{R_2} + \frac{\Delta V_B}{R_3}$$

Since $I_B = \Delta V_B/R_p$,

$$\frac{\Delta V_B}{R_p} = \Delta V_B\left(\frac{1}{R_1} + \frac{1}{R_2} + \frac{1}{R_3}\right)$$

and thus expression for $R_p$ is

$$\frac{1}{R_p} = \frac{1}{R_1} + \frac{1}{R_2} + \frac{1}{R_3}$$

This result can be generalized to any number of **resistors in parallel***:

$$\frac{1}{R_p} = \frac{1}{R_1} + \frac{1}{R_2} + \frac{1}{R_3} + \cdots = \Sigma\left(\frac{1}{R_i}\right) \quad \left(\begin{array}{l}\text{equivalent parallel}\\ \text{resistance}\end{array}\right) \quad (18.3a)^*$$

A handy relationship for the special case of just two resistors in parallel is given by solving Equation 18.3. $R_p$. This can be handy for quick calculations, but you must be careful to use it only for two resistors.

$$R_p = \frac{R_1 R_2}{R_1 + R_2} \quad \text{(two parallel resistors)} \quad (18.3b)$$

The results show that $R_p$ *is always less than the smallest resistance* in the arrangement. For example, parallel resistors with values of 1.0, 6.0 and 12.0 Ω, are equivalent to a single resistance of 0.8 Ω (you should show this). But why does this occur?

To show that this makes sense, consider a 12-V battery (terminal voltage $\Delta V_B$) connected to a single 6.0-Ω resistor ($R_1$). The current in the circuit ($I_1$) will be 2.0 A after using $I_1 = \Delta V_B/R_1$. Now consider what happens when connecting a 12.0-Ω resistor ($R_2$) in parallel to the 6.0-Ω resistor. The current through the 6.0-Ω resistor will be unaffected and remain at 2.0 A. However, the added resistor will have a current of 1.0 A (using $I_2 = \Delta V_B/R_2$). Thus the total current in the circuit, before it splits, is 1.0 + 2.0 A = 3.0 A. Now consider the overall result of connecting the second resistor. When attached in parallel, the total current delivered by the battery increases. Since the voltage did not increase, the equivalent resistance of the circuit must have decreased (below its initial value of 6.0 Ω). In other words, every time an extra parallel path is added, the result is more total current at the same voltage. Thus when adding more resistors in parallel, the equivalent resistance always decreases. Conversely, removing parallel resistors always increases the equivalent parallel resistance.

This argument does not depend on the size of added resistance. All that matters is that another path to carry more current is added. (Try this using a 2.0-Ω or a 2.0-MΩ resistor in place of the 12-Ω resistor. A decrease in equivalent resistance, no matter how small, always happens.)

To see how these parallel and series connection calculations are typically employed, consider Example 18.1.

* Equation 18.3a gives $1/R_p$ *not* $R_p$. To solve for $R_p$, the reciprocal must be taken. Unit analysis shows that the units are not ohms until inverted. As usual, carrying units along with calculations makes errors less likely to occur.

### EXAMPLE 18.1: CONNECTIONS COUNT – RESISTORS IN SERIES AND IN PARALLEL

What is the equivalent resistance of three resistors (1.0, 2.0, and 3.0 Ω) when connected (a) in series (Figure 18.1a) and (b) in parallel (Figure 18.3a)? (c) How much total current will be delivered by a 12-V battery in each of these arrangements? (d) How much current will be in each resistor and what is the voltage across each resistor in each of these arrangements?

**THINKING IT THROUGH.** For (a) and (b), apply Equations 18.2 and 18.3, respectively. For the series arrangement in (c), the current is the total current and is determined by treating the battery as if connected to a single resistor – the series equivalent resistance. For the parallel arrangement, the total current is determined by doing the same with the parallel equivalent resistance. (d) In series, the total current is the same as the current in all the resistors. From the current and resistance values, voltages can be calculated. When in parallel, each resistor has 12 V across it; and the individual currents can be found by using the resistance values.

**SOLUTION**

Listing the data:

***Given:***

$$R_1 = 1.0\ \Omega$$
$$R_2 = 2.0\ \Omega$$
$$R_3 = 3.0\ \Omega$$
$$\Delta V_B = 12\ \text{V}$$

***Find:***

**(a)** $R_s$ (series resistance)
**(b)** $R_p$ (parallel resistance)
**(c)** $I_B$ (total current for each case)
**(d)** Current in and voltage across each resistor (for each case)

**(a)** The equivalent series resistance is

$$R_s = R_1 + R_2 + R_3 = 1.0\ \Omega + 2.0\ \Omega + 3.0\ \Omega = 6.0\ \Omega$$

(Larger than the largest resistance, as expected.)

**(b)** The equivalent parallel resistance is:

$$\frac{1}{R_p} = \frac{1}{R_1} + \frac{1}{R_2} + \frac{1}{R_3} = \frac{1}{1.0\ \Omega} + \frac{1}{2.0\ \Omega} + \frac{1}{3.0\ \Omega}$$
$$= \frac{6.0}{6.0\ \Omega} + \frac{3.0}{6.0\ \Omega} + \frac{2.0}{6.0\ \Omega} = \frac{11}{6.0\ \Omega}$$

thus, after inverting,

$$R_p = \frac{6.0\ \Omega}{11} = 0.55\ \Omega$$

(Less than the least resistance, as expected.)

**(c)** From the equivalent series resistance and the battery voltage the total current in series is:

$$I_B = \frac{\Delta V_B}{R_s} = \frac{12\text{ V}}{6.0\ \Omega} = 2.0\text{ A}$$

A similar calculation for the parallel arrangement, finds the total current:

$$I_B = \frac{\Delta V_B}{R_p} = \frac{12\text{ V}}{0.55\ \Omega} = 22\text{ A}$$

**(d)** The voltage across each resistor in series can be found, since the current in each is the same as the total (i.e., $I_B = I_1 = I_2 = I_3$):

$$\Delta V_1 = I_B R_1 = (2.0\text{ A})(1.0\ \Omega) = 2.0\text{ V}$$
$$\Delta V_2 = I_B R_2 = (2.0\text{ A})(2.0\ \Omega) = 4.0\text{ V}$$
$$\Delta V_3 = I_B R_3 = (2.0\text{ A})(3.0\ \Omega) = 6.0\text{ V}$$

Since current in each series resistor is the same, in series, the larger resistors require the larger percentage of the battery voltage. As a last check, the sum of the resistor voltages does equal the battery voltage.

The current in each parallel resistor can be determined, because each has 12 V across it (i.e., $\Delta V_B = \Delta V_1 = \Delta V_2 = \Delta V_3$):

$$I_1 = \frac{\Delta V_B}{R_1} = \frac{12\text{V}}{1.0\ \Omega} = 12\text{A}$$
$$I_2 = \frac{\Delta V_B}{R_2} = \frac{12\text{V}}{2.0\ \Omega} = 6.0\text{ A}$$
$$I_3 = \frac{\Delta V_B}{R_3} = \frac{12\text{ V}}{3.0\ \Omega} = 4.0\text{ A}$$

As a double check, the sum of the currents does equal the total current before it splits.

Since the voltage across each is the same, in parallel, the resistor with the smallest resistance gets the largest share (but never all) of the total current.

**FOLLOW-UP EXERCISE.** (a) Calculate the power delivered to each resistor for both arrangements in this Example. (b) What generalizations can you make? For instance, which resistor gets the most power in series? In parallel? (c) For each arrangement, verify the total power delivered to all the resistors is equal the power output of the battery.

### 18.1.3 Series–Parallel Resistor Combinations

Resistors may be connected in a circuit in a variety of series–parallel combinations. As shown in ▼ **Figure 18.5**, circuits with only one voltage source can sometimes be reduced to a single equivalent loop, containing just the voltage source and one equivalent resistance, by applying the series and parallel results.

A procedure for analyzing such combination circuits (i.e., for determining voltage, current, and power for each circuit element) is as follows:

1. Determine which groups of resistors are in series and which are in parallel, and reduce all groups to equivalent resistances, using Equations 18.2 and 18.3.
2. Reduce the circuit further by treating the separate equivalent resistances (from Step 1) as individual resistors. Proceed until you get to a single loop with one total (overall or equivalent) resistance value.
3. Find the total current delivered to the reduced circuit by the battery using $I_B = \Delta V_B / R_{total}$.
4. Expand the reduced circuit back to the actual circuit by reversing the reduction steps, one at a time. Use the current of the reduced circuit to find the currents and voltages for the resistors in each step.

To see this procedure in use, consider Example 18.2.

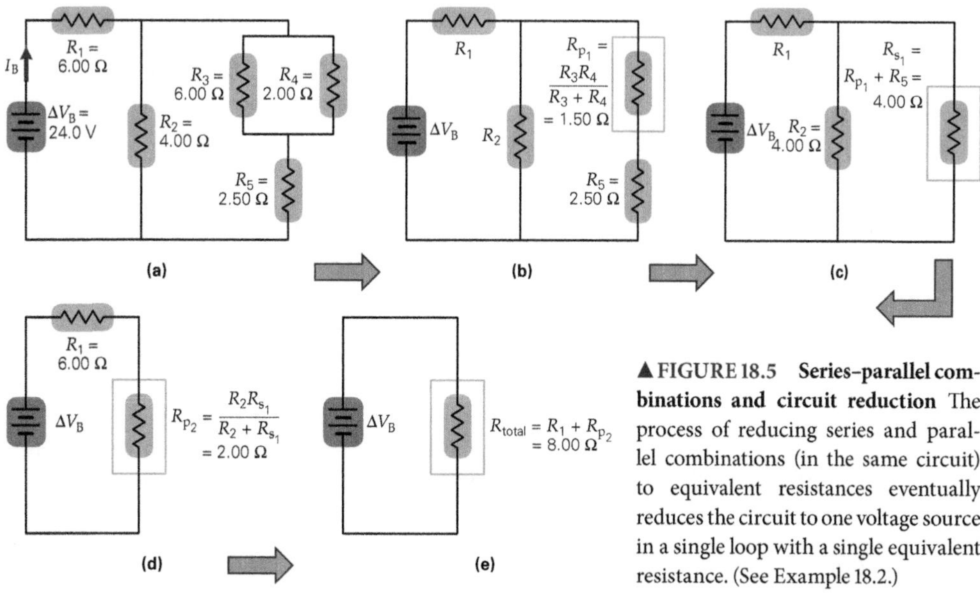


▲ **FIGURE 18.5 Series-parallel combinations and circuit reduction** The process of reducing series and parallel combinations (in the same circuit) to equivalent resistances eventually reduces the circuit to one voltage source in a single loop with a single equivalent resistance. (See Example 18.2.)

**EXAMPLE 18.2: SERIES-PARALLEL COMBINATION OF RESISTORS - SAME VOLTAGE OR SAME CURRENT?**

What are the voltages and the currents for resistors $R_1$ through $R_5$ in Figure 18.5a?

**THINKING IT THROUGH.** It is important to identify parallel and series combinations before starting. First, note that $R_3$ is in parallel with $R_4$ (written as $R_3||R_4$). This parallel *combination* is then in series with $R_5$. Furthermore, the $R_3||R_4 + R_5$ leg is itself, in parallel with $R_2$. Last, this parallel combination is in series with $R_1$. Combining the resistors step by step enables the determination of the total equivalent circuit resistance (Step 2). From that, the total current can be calculated. Working backward, the current in, and voltage of, each resistor can then be found.

**SOLUTION**
(To avoid rounding errors, results are carried to three significant figures.)

***Given:*** See values in Figure 18.5a.

***Find:*** Current and voltage for each resistor (Figure 18.5a)

The parallel combination at the right-hand side of the circuit diagram can be reduced to the equivalent resistance $R_{p1}$ (see Figure 18.5b), using Equation 18.3a as follows:

$$\frac{1}{R_{p1}}=\frac{1}{R_3}+\frac{1}{R_4}=\frac{1}{6.00\,\Omega}+\frac{1}{2.00\,\Omega}\quad \text{thus } R_{p1}=1.50\,\Omega$$

The series combination of $R_{p1}$ and $R_5$ can then be reduced to $R_{s1}$ using Equation 18.2 (Figure 18.5c):

$$R_{s1}=R_{p1}+R_5=1.5\,\Omega+2.50\,\Omega=4.00\,\Omega$$

Since $R_2$ and $R_{s1}$ are in parallel, they can be reduced (Equation 18.3a) to $R_{p2}$ (Figure 18.5d):

$$\frac{1}{R_{p2}}=\frac{1}{R_2}+\frac{1}{R_{s1}}=\frac{1}{4.00\,\Omega}+\frac{1}{4.00\,\Omega}\quad \text{thus } R_{p2}=2.00\,\Omega$$

This leaves $R_1$ and $R_{p2}$ in series. Thus the total equivalent resistance ($R_{\text{total}}$) of the circuit is (see Figure 18.5e):

$$R_{\text{total}}=R_1+R_{p2}=6.00\,\Omega+2.00\,\Omega=8.00\,\Omega$$

Therefore, the battery delivers a total current of

$$I_B=\frac{\Delta V_B}{R_{\text{total}}}=\frac{24.0\text{ V}}{8.00\,\Omega}=3.00\text{ A}$$

Now work backward and "rebuild" to the actual circuit. Note the battery (total) current is the same as the current through $R_1$ and $R_{p2}$. (In Figure 18.5d $I_B = I_1 = I_{p2} = 3.00$ A.) Therefore, the voltages across these resistors are

$$\Delta V_1=I_1R_1=(3.00\text{ A})(6.00\,\Omega)=18.0\text{ V}$$

and

$$\Delta V_{p2}=I_{p2}R_{p2}=(3.00\text{ A})(2.00\,\Omega)=6.00\text{ V}$$

Because $R_{p2}$ is composed of $R_2$ and $R_{s1}$ (Figures 18.5c and d), there must be a 6.00-V voltage across both of these. Thus the current in each is:

$$I_2=\frac{\Delta V_2}{R_2}=\frac{6.00\text{ V}}{4.00\,\Omega}=1.50\text{ A}$$

and

$$I_{s1}=\frac{\Delta V_{s1}}{R_{s1}}=\frac{6.00\text{ V}}{4.00\,\Omega}=1.50\text{ A}$$

Next, in Figure 18.5b notice that $I_{s1} = I_{p1} = I_5 = 1.50$ A because they are in series. The resistors' individual voltages are therefore

$$\Delta V_{p1}=I_{s1}R_{p1}=(1.50\text{ A})(1.50\,\Omega)=2.25\text{ V}$$

and

$$\Delta V_5=I_{s1}R_5=(1.50\text{ A})(2.50\,\Omega)=3.75\text{ V}$$

(As a check, note that the voltages do, in fact, add to 6.00 V.)

Finally, the voltages across $R_3$ and $R_4$ are the same as $\Delta V_{p1}$ (why?), and thus

$$\Delta V_{p1}=\Delta V_3=\Delta V_4=2.25\text{ V}$$

With these voltages and known resistances, the last two currents $I_3$ and $I_4$, are

$$I_3=\frac{\Delta V_3}{R_3}=\frac{2.25\text{ V}}{6.00\,\Omega}=0.375\text{ A}$$

and

$$I_4=\frac{\Delta V_4}{R_4}=\frac{2.25\text{ V}}{2.00\,\Omega}=1.13\text{ A}$$

Since the current $I_{s1}$ divides at the $R_3$–$R_4$ junction, a double-check is available: $I_3 + I_4$ does, in fact, equal $I_{s1}$, within rounding errors.

**FOLLOW-UP EXERCISE.** In this Example, verify that the total power delivered to all of the resistors is the same as the power output of the battery. Why must this be true?

## 18.2 Multiloop Circuits and Kirchhoff's Rules

Series–parallel circuits with a single voltage source can be reduced to a single loop, as seen in Example 18.2. However, circuits may contain several voltage sources as well as resistors that are neither in series nor parallel. As an example of this, a multiloop circuit which does not lend itself to the methods in Example 18.2 is in ▶ **Figure 18.6a**. Even though resistor groups can be replaced by their equivalent resistances (see Figure 18.6b), this circuit can be reduced only so far using parallel and series methods.

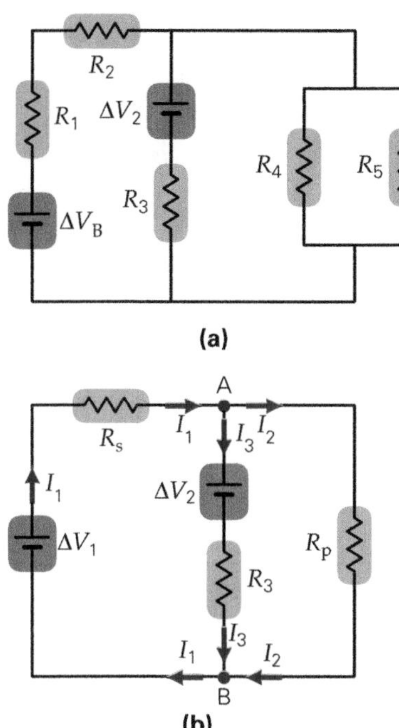


▲ FIGURE 18.6 **Multiloop circuit** In general, a circuit that contains voltage sources in more than one loop cannot be completely reduced by series and parallel methods alone. However, some reductions within each loop may be possible, such as from part **(a)** to part **(b)**. At a junction the current divides or comes together, as at junctions A and B in part (b), respectively. Any path between two junctions is called a *branch*. In part (b), there are three branches – that is, there are three different ways to get from junction A to junction B. (See Example 18.3.)

Analyzing these types of circuits requires a more general approach – that is, the application of **Kirchhoff's rules.*** These rules embody conservation of charge and energy. (Although not stated specifically, Kirchhoff's rules were applied to the parallel and series arrangements in Section 18.1.) First, it is useful to introduce some terminology:

- A point where three or more wires are joined is called a **junction** or **node** – for example, point A in Figure 18.6b.
- A path connecting two junctions is called a **branch**. A branch may contain one or more circuit elements, and there may be more than two branches between two junctions.

## 18.2.1 Kirchhoff's Junction Theorem

**Kirchhoff's first rule**, or **junction theorem**, states that the *algebraic sum* of the currents at any junction is zero:

$$\sum I_i = 0 \quad \text{(sum of currents at a junction)} \tag{18.4}$$

* Gustav Robert Kirchhoff (1824–1887) was a German scientist who made important contributions to electrical circuit theory and light spectroscopy.

Our sign convention is such that currents entering a junction are designated as positive and those leaving as negative. This rule is a really just a restatement of charge conservation – because of electrical repulsion, no charge can pile up at any junction. For the junction at point A in Figure 18.6b, for example, using the sign conventions, the algebraic sum of the currents is

$$I_1 - I_2 - I_3 = 0$$

or equivalently

$$I_1 = I_2 + I_3$$
$$\text{current in} = \text{current out}$$

(This rule was applied in analyzing parallel resistances in Section 18.1.)

Sometimes it may not be evident whether a current is directed into or out of a junction just from the circuit diagram. In this case, a direction (i.e., sign) is simply *assumed.* Then the currents are calculated, without worry about direction. If some of the assumed directions are opposite the actual resulting directions, then negative answers for these will result. In other words, the mathematical outcome will show that the directions of the actual currents are opposite to the directions initially assumed.

## 18.2.2 Kirchhoff's Loop Theorem

**Kirchhoff's second rule**, or **loop theorem**, states that the algebraic sum of the potential differences (voltages) across all of the elements of any closed loop is zero:

$$\sum \Delta V_i = 0 \quad \text{(sum of voltages around a closed loop)} \tag{18.5}$$

Thus the sum of the voltage gains (i.e., an increase in potential) equals the sum of the voltage drops (i.e., a decrease in potential) around a closed loop, which must be true if energy is conserved. (This rule was used in analyzing series resistances in Section 18.1.)

Notice that the direction chosen to traverse a circuit element will yield either a voltage rise or a voltage drop, depending on the direction. Thus, it is also important to establish a sign convention for voltages. The conventions used here are illustrated in ▶ **Figure 18.7.** The voltage across a battery is taken as positive (potential rise) if it is traversed from cathode (–) to anode (+) terminal (Figure 18.7a) and negative (potential drop) if it is traversed from anode to cathode. (Note that the direction of the current through the battery has nothing to do with these signs. The sign of the battery voltage depends only on the direction chosen to cross it.)

The voltage across a resistor is designated as negative (potential drop) if the resistor is traversed in the same direction as the assigned current, in essence going "downhill" potential-wise (Figure 18.7b). The voltage will be designated as positive (potential rise) if the resistor is traversed in the direction opposite the assigned current.

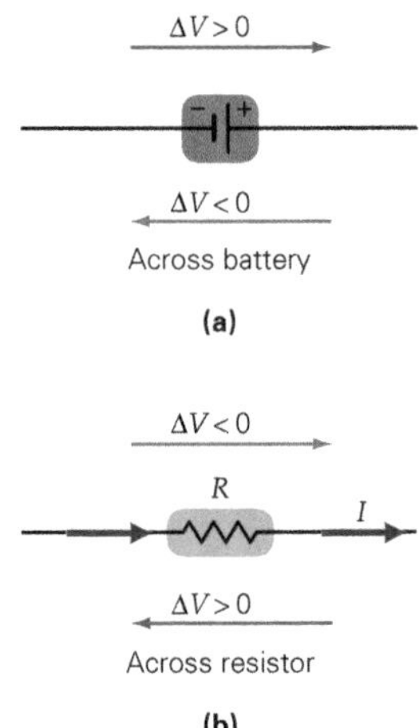


▲ **FIGURE 18.7 Sign convention for Kirchhoff's loop theorem** **(a)** The battery voltage is taken as positive (a gain) if it is traversed from the negative to the positive terminal. It is assigned a negative value if traversed from the positive to the negative (a drop). **(b)** The voltage across a resistor is taken as negative if the resistor is traversed in the direction of the assigned current ("downstream"). It is taken as positive if the resistance is traversed in the direction opposite that of the assigned current ("upstream").

Taken together, these sign conventions allow the summation of the voltages around a closed loop, regardless of the direction chosen to do that sum. It should be clear that Equation 18.5 is the same in either case. To see this, note that reversing the chosen loop direction simply amounts to multiplying Equation 18.5 by −1, which, of course, does not change the equation or the physics.

In applying the loop theorem, the sign of a voltage across a resistor is based on the direction of the current in that resistor. However, there can be situations in which the current direction is not obvious. How do you handle the voltage signs in such cases? The answer is simple – after assuming a direction for the current, follow the voltage sign convention based on this assumed direction. This guarantees that the signs of all the voltages are mathematically consistent. If it results that the actual direction is opposite your choice, the voltages will automatically reflect that.

The equation form of Kirchhoff's loop theorem has a geometrical visualization that may help develop better insight into its meaning. This graphical approach allows the visualization of how the potential changes in a circuit, either to anticipate the results of mathematical analysis or to qualitatively confirm the results. The idea is to make a three-dimensional plot based on the circuit diagram. The wires and elements of the circuit form the basis for the $x$–$y$ plane, or the diagram's "floor." Plotted perpendicularly to this plane, along the $z$-axis, is the electric potential with an appropriate choice for zero. Such a diagram is called a *Kirchhoff plot*, two examples of which are shown in ▶ **Figure 18.8** and elaborated upon in the caption.

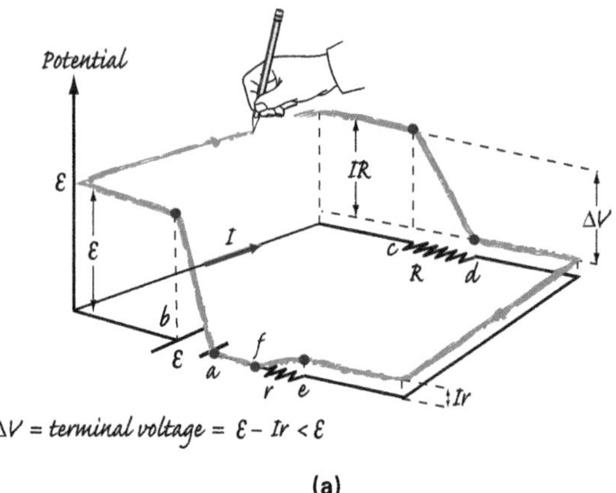


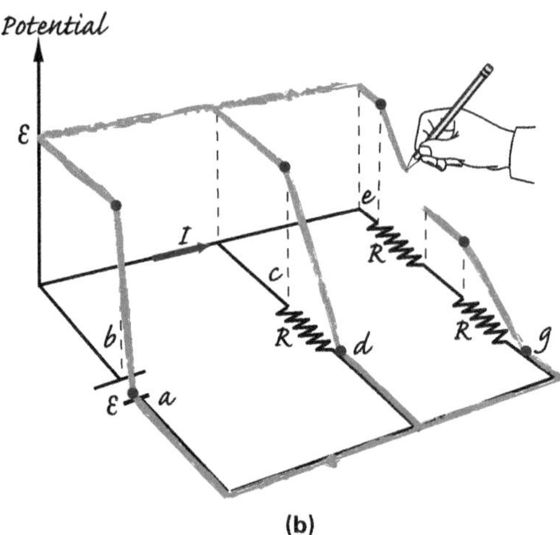


▲ **FIGURE 18.8 Kirchhoff plots: Two circuits, two examples** In preparing these plots, the circuit schematic is laid out in the $x$–$y$ plane, and electric potential is plotted along the $z$-axis. Usually, the zero potential is taken as cathode (−) of the battery. A direction for each current is assigned, and the potential is plotted around the circuit, following the rules for gains and losses. **(a)** This circuit's plot shows a rise in potential when the battery is traversed from cathode to anode (a–b), followed by a drop in potential across the external resistor $R$ (c–d), and a smaller drop in potential across the battery's internal resistance $r$ (e–f) back down to zero potential at the battery cathode. Thus the total voltage is shared between the two resistors. **(b)** This circuit's plot shows a rise in potential when the battery is traversed from cathode to anode (a–b), followed by two independent paths of resistors (c–d and e–g) over which the potential drops to zero upon return to the cathode. It can be seen why each parallel leg has the same voltage and how the resistors in one path are in series and share the total drop. In all of the diagrams there is no change in potential across the wires. Why?

### 18.2.3 Application of Kirchhoff's Rules

In this book, the following general steps will be used when applying these principles:

1. Assign a current and direction of current for each branch in the circuit. This assignment is done most conveniently at junctions.

2. Indicate the loops and the directions in which they are traversed (as in ▼**Figure 18.9**). Every branch *must* be in at least one loop.
3. Apply Kirchhoff's first rule (junction rule) at each junction. (This step gives a set of equations that includes *all* currents, but there may be redundant equations from two different junctions.)
4. Traverse the number of loops necessary to include all branches. In traversing a loop, apply Kirchhoff's loop theorem (using $\Delta V = IR$ to write the voltage for each resistor in terms of its current), using the proper sign conventions.

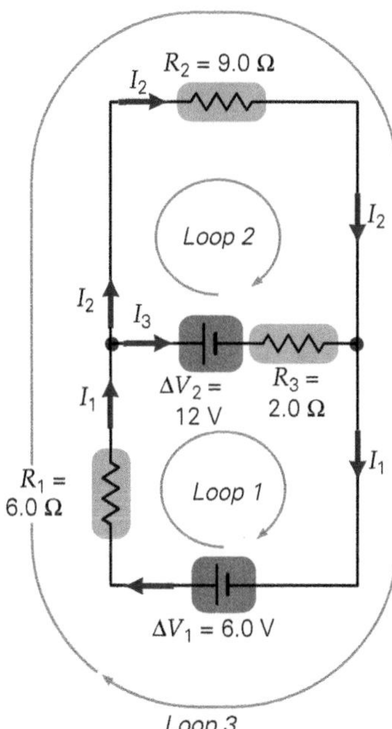


▲**FIGURE 18.9 Application of Kirchhoff's rules** (See Example 18.3.) To analyze this circuit, assume a current and its direction for each branch (most conveniently at junctions). Identify each loop and the direction of traversal. Then write current equations for each independent junction (using the junction theorem). Next write voltage equations (using the loop theorem) for as many loops as needed to include every branch. Always be careful to observe sign conventions.

If this procedure is applied properly, the result will be a set of equations in terms of the unknown currents. There will be enough equations to match the number of unknown currents – no more, no less. That is, the result is a set of $N$ equations if there are $N$ unknown currents. These equations can then be solved for the currents. The voltage drops and power for each element can then be found from these currents.

If more loops are traversed than necessary, one (or more) redundant loop equation(s) might appear. Only the number of loops that includes each branch *once* is needed. The actual algebra to solve all the equations may be daunting, but in principle, every circuit can be analyzed this way. This procedure is displayed in Integrated Example 18.3.

## INTEGRATED EXAMPLE 18.3: BRANCH CURRENTS – USING KIRCHHOFF'S RULES AND ENERGY CONSERVATION

Consider the circuit in Figure 18.9. (a) What can you say about the magnitude of the electric potential change between the two nodes if the path taken is through $R_2$ compared to the magnitude if instead the path is through $R_3$ and battery #2: (1) the change is greater through $R_2$, (2) the change is greater through $R_3$ and the battery, (3) the change is the same for both paths? Explain. (b) Find the current in each branch and determine the two changes in part (a). (c) Determine the power dissipated in each resistor and compare their total to the rate at which energy is lost (or gained) by the two batteries.

**(a) CONCEPTUAL REASONING.** The electric potential difference between the nodes must be the same. If not, the nodes would have different potential values, which cannot be the case. Hence the correct answer is (3), the change is the same for both paths.

**(b,c) QUANTITATIVE REASONING AND SOLUTION.** The solution begins by assigning current directions in each loop, and then applying the junction theorem and the loop theorem (twice – once for each inner loop) to generate three equations, because there are three currents. For the resistors in each loop, their voltages can be written in terms of currents and resistances. After solving the set of equations for all the currents, the power levels in each of the circuit element can be calculated.

***Given:*** Values in Figure 18.9.

***Find:***

**(b)** $I_1$, $I_2$, and $I_3$ (current in each branches) and $\Delta V_{\text{n-n}}$ (potential change between the two nodes via two paths)

**(c)** $P_1$, $P_2$, and $P_3$ (power dissipated in each resistor) and $P_{\text{B1}}$ and $P_{\text{B2}}$ (rate of energy lost or gained by each battery)

**(b)** The chosen current directions and loop traversal directions are shown in the figure. (Remember, these directions are not unique; choose them, and check the final current signs to see if they were correct.) There is a current in every branch, and every branch is in at least one loop. (Some branches are in more than one loop, which is acceptable.)

Applying Kirchhoff's first rule at the left-hand junction

$$I_1 - I_2 - I_3 = 0 \quad \text{or} \quad I_1 = I_2 + I_3 \tag{1}$$

Going around loop 1 and applying Kirchhoff's loop theorem with the sign conventions gives

$$\sum_{\text{loop 1}} \Delta V_i = +\Delta V_1 + (-I_1R_1) + (-\Delta V_2) + (-I_3R_3) = 0 \tag{2a}$$

Putting in the resistance values and rearranging yields

$$3I_1 + I_3 = -3 \tag{2b}$$

For convenience, units are omitted (they are all in amps and ohms, and are self-consistent).

For loop 2, applying loop theorem yields

$$\sum_{\text{loop 2}} \Delta V_i = +\Delta V_2 + (-I_2R_2) + (+I_3R_3) = 0 \quad (3a)$$

Putting in the resistance values and rearranging yields

$$9I_2 - 2I_3 = 12 \quad (3b)$$

Equations 1, 2b, and 3b are a set of three equations with three unknowns. The currents can be found in many ways. One way to begin is to substitute Equation 1 into Equation 2b and eliminate $I_1$:

$$3(I_2 + I_3) + I_3 = -3$$

Solving this for $I_2$ in terms of $I_3$

$$I_2 = -1 - \frac{4}{3}I_3 \quad (4)$$

Then, substituting Equation 4 into Equation 3b eliminates $I_2$:

$$9\left(-1 - \frac{4}{3}I_3\right) - 2I_3 = 12$$

Finishing the algebra and solving for $I_3$,

$$-14I_3 = 21 \quad \text{or} \quad I_3 = -1.5\,\text{A}$$

The minus sign means the wrong direction was assumed for $I_3$.

Putting the result for $I_3$ into Equation 4 gives $I_2$:

$$I_2 = -1 - \frac{4}{3}(-1.5\,\text{A}) = 1.0\,\text{A}$$

Then, from Equation 1,

$$I_1 = I_2 + I_3 = 1.0\,\text{A} - 1.5\,\text{A} = -0.5\,\text{A}$$

Again, the sign means the direction of $I_1$ was wrong as initially chosen.

The voltage, traversing from left to right node through battery #2, is determined by the sign conventions. Since the current through $R_3$ is actually to the left, its voltage is positive; however, the battery is traversed from anode to cathode, and its change is negative. Thus

$$\Delta V_{\text{n-n}} = -12\,\text{V} + |I_3|R_3 = -12\,\text{V} + |-1.5\,\text{A}|(2.0\,\Omega) = -9.0\,\text{V}$$

The voltage, going from the left node to the right node but through $R_2$, is negative, because the path is taken in the direction of the current. Its value is

$$\Delta V_{\text{n-n}} = -I_2R_2 = -(1.0\,\text{A})(9.0\,\Omega) = -9.0\,\text{V}$$

in agreement with the conceptual part (a).

**(c)** The resistor rates of energy dissipation (i.e., their power) are:

$$P_{R_1} = I_1^2R_1 = (0.50\,\text{A})^2(6.0\,\Omega) = 1.5\,\text{W}$$
$$P_{R_2} = I_2^2R_2 = (1.0\,\text{A})^2(9.0\,\Omega) = 9.0\,\text{W}$$
$$P_{R_3} = I_3^2R_3 = (1.50\,\text{A})^2(2.0\,\Omega) = 4.5\,\text{W}$$

for a total dissipation rate of 15 W.

Batteries can gain or lose energy, depending upon the direction of their current. Here #2 is losing energy because the current leaves its anode. Its rate of energy loss (power) is

$$P_2 = I_3 \cdot \Delta V_2 = (1.5\,\text{A})(12\,\text{V}) = 18\,\text{W} \quad \text{(a loss)}$$

#1 is having its stored energy increased (it is being "recharged") because current enters its anode.

$$P_1 = I_1 \cdot \Delta V_1 = (-0.5\,\text{A})(6.0\,\text{V}) = -3.0\,\text{W} \quad \text{(a gain)}$$

The results are consistent with conservation of energy, since the net battery rate (a loss rate of 18–3 W = 15 W) equals the total resistor loss rate of 15 W.

**FOLLOW-UP EXERCISE.** Find the currents in this Example by using the junction theorem and loops 3 and 1 instead of loops 1 and 2.

## 18.3 RC Circuits

Until now, only circuits that have constant currents have been considered. In some direct-current (dc) circuits, the current can vary with time while maintaining a constant direction. Such is the case in **RC circuits**, which consist of resistors, capacitors, and power supplies in various combinations.

### 18.3.1 Charging a Capacitor through a Resistor

The charging of an uncharged capacitor by a battery is shown in ▶ **Figure 18.10**. Here the maximum capacitor charge is $Q_o$ and the battery voltage is $\Delta V_o$. The time sequence of the charging process is shown in Figures 18.10a through c. At $t = 0$, there is no charge on the capacitor and no voltage across it. By the loop theorem, this means that the full battery voltage initially appears across the resistor, resulting in an initial (and maximum) current of $I_o = \Delta V_o/R$. As the capacitor charges, the voltage across its plates increases, thus reducing the resistor's share of the total voltage and therefore the current. Eventually, when the capacitor is charged to its maximum, the current drops to zero. At this time, the resistor's voltage is now zero and the capacitor's voltage is at a maximum value of $\Delta V_o$. Thus (see Chapter 16) the maximum capacitor charge is $Q_o = C \cdot \Delta V_o$.

The resistance value is one factor that determines how quickly a capacitor charges, because the larger its value, the lower the current and the longer it will take. The capacitance is the other factor that influences the charging speed – it takes longer to charge a larger capacitor (more capacity or capacitance). Analysis of this type of circuit requires mathematics beyond the scope of this

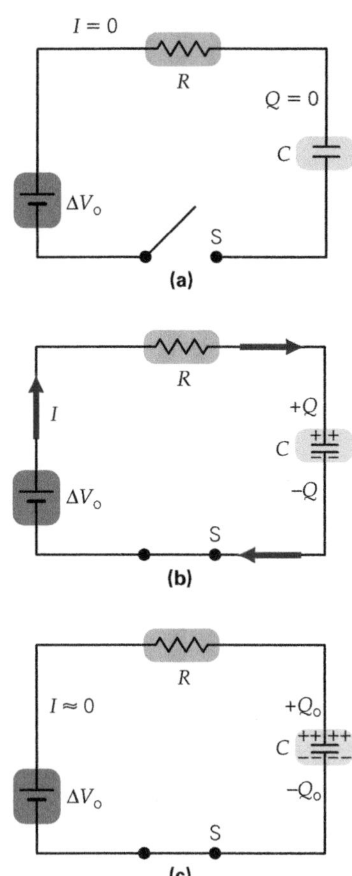


▲ **FIGURE 18.10 Charging a capacitor in a series RC circuit (a)** Initially there is no current and no charge on the capacitor. **(b)** When the switch is closed, there is a current in the circuit until the capacitor is charged to its maximum value. The rate of charging depends on the circuit's time constant, $\tau (= RC)$. **(c)** For times much longer than $\tau$, the current becomes very close to zero, and the capacitor is essentially fully charged.

book. However, it can be shown that as a capacitor is charged, the voltage across it increases exponentially with time according to

$$\Delta V_C = \Delta V_o[1 - e^{-t/(RC)}] \quad \text{(charging capacitor voltage in an RC circuit)} \tag{18.6}$$

where $e$ has an approximate value of 2.718. (Recall that the irrational number $e$ is the base of the system of natural logarithms.*) A graph of $\Delta V_C$ versus $t$ is shown in ▶ **Figure 18.11a**. As expected, $\Delta V_C$ starts at zero and approaches $\Delta V_o$, the capacitor's maximum voltage, after a "long" time.

Figure 18.11b shows the current decreasing with time and detailed analysis gives the following equation to describe the current delay

$$I = I_o e^{-t/(RC)} \quad \text{(current in a charging RC circuit)} \tag{18.7}$$

Note the current decreases exponentially with time and has its largest value initially, as expected.

* For a review of exponential functions, see Appendix I.

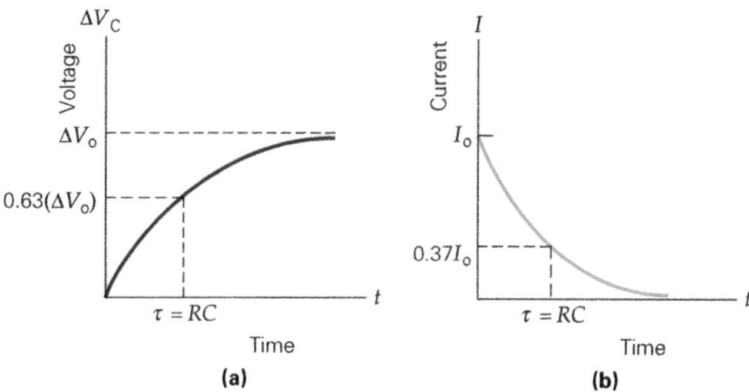


▲ **FIGURE 18.11 Capacitor charging in a series RC circuit (a)** In a series RC circuit, as the capacitor charges, the voltage across it increases nonlinearly, reaching 63% of its maximum voltage ($\Delta V_o$) in one time constant. **(b)** The current in this circuit is initially a maximum ($I_o = \Delta V_o/R$) and decays exponentially, falling to 37% of its initial value in one time constant.

According to Equation 18.6, it would theoretically take an infinite time for a capacitor to become fully charged. However, in practice, most capacitors become "close" to completely charged in relatively short times. It is customary to use a special quantity to express "charging time." This quantity is named the **time constant ($\tau$)** and is defined as follows.

$$\tau = RC \quad \text{(time constant for RC circuits)} \tag{18.8}$$

(You should be able to show that $RC$ has units of seconds.) After an elapsed time of one time constant, that is, $t = \tau = RC$, the voltage across a charging capacitor has risen to 63% of the maximum. This can be seen by evaluating $\Delta V_C$ (Equation 18.6), after replacing $t$ with $RC$:

$$\Delta V_C = \Delta V_o(1 - e^{-\tau/\tau}) = \Delta V_o(1 - e^{-1}) \approx \Delta V_o\left(1 - \frac{1}{2.718}\right)$$
$$= 0.63(\Delta V_o)$$

Because $Q \propto \Delta V_C$, the capacitor also has 63% of its maximum charge after one time constant. You should show that after one time constant, the current has dropped to 37% of its initial (maximum) value, $I_o$.

After an elapsed time of two time constants (i.e., $t = 2\tau = 2RC$), the capacitor is charged to more than 86% of its maximum value; after $t = 3\tau = 3RC$, the capacitor is charged to 95% of its maximum value, and so on. (You should check these results.) As a general rule, a capacitor is said to be "fully charged" after "several time constants" have elapsed.

### 18.3.2 Discharge of a Capacitor through a Resistor

▶ **Figure 18.12a** shows a capacitor being discharged through a resistor. In this case, the voltage across the capacitor, starting from an initial value of $\Delta V_o$, decreases exponentially with time, as does the current. The expression for the decay of the capacitor's voltage is

$$\Delta V_C = \Delta V_o e^{-t/(RC)} \quad \text{(discharging capacitor voltage – RC circuit)} \tag{18.9}$$

After one time constant, the capacitor voltage is at 37% of its original value (Figure 18.12b). The current in the circuit decays exponentially as in Equation 18.7. In an important medical use, the capacitor in a heart defibrillator will discharge its stored energy (as current) to the heart (of resistance $R$) in a discharge time of several time constants) or about 0.1 s. RC circuits are also an integral part of cardiac pacemakers, which charge a capacitor, transfer the energy to the heart, and repeat this at a rate determined by the time constant. Another practical application related to cameras is considered in Example 18.4.

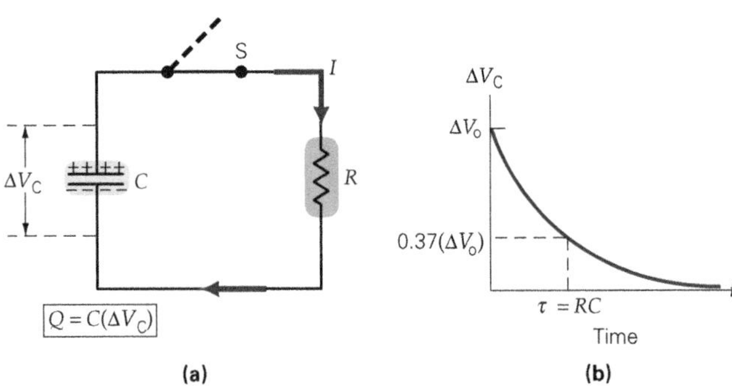


▲FIGURE 18.12 **Capacitor discharging in a series RC circuit (a)** The capacitor is initially fully charged. When the switch is closed, current appears in the circuit as the capacitor begins to discharge. **(b)** In this case, the voltage across the capacitor (and the current in the circuit) decays exponentially with time, falling to 37% of its initial value in one time constant, $\tau$.

---

**EXAMPLE 18.4: RC CIRCUITS IN CAMERAS – FLASH PHOTOGRAPHY IS AS EASY AS FALLING OFF A LOG(ARITHM)**

In many cameras, a built-in flash gets its energy from that stored in a capacitor. The capacitor is charged using long-life batteries with voltages of typically 9.00 V. Once the bulb is fired, the capacitor must recharge quickly through an internal RC circuit. If the capacitor has a value of 0.100 F, what must the resistance be so that the capacitor is charged to 80% of its maximum charge (the minimum charge to fire the bulb again) in 5.00 s?

**THINKING IT THROUGH.** After one time constant, the capacitor is charged to 63% of its maximum charge. Because it needs 80%, the time constant must be *less* than 5.00 s. Equation 18.6 can be used to determine the time constant. From that, the required resistance can be found.

**SOLUTION**

The data given include the final voltage across the capacitor, $V_C$, which is 80% of the battery's voltage, which means that $Q$ is 80% of the maximum charge.

***Given:***

$$C = 0.100\ \text{F}$$
$$\Delta V_B = \Delta V_o = 9.00\ \text{V}$$
$$\Delta V_C = 0.80(\Delta V_o) = 7.20\ \text{V}$$
$$t = 5.00\ \text{s}$$

***Find:*** $R$ (required resistance for 80% charge)

For a charging capacitor, $\Delta V_C = \Delta V_o(1 - e^{-t/\tau})$, so putting in the data yields

$$7.20 = 9.00(1 - e^{-5.00/\tau})$$

Rearranging this yields $e^{-500/\tau} = 0.20$ and taking the reciprocal gives a positive exponent:

$$e^{5.00/\tau} = 5.00$$

To solve for the time constant, recall that if $e^a = b$, then $a$ is the *natural logarithm* (ln) of $b$. So for us this means that $5.00/\tau$ is the natural logarithm of 5.00. A calculator shows that ln 5.00 = 1.61. Therefore

$$\frac{5.00}{\tau} = \ln 5.00 = 1.61 \quad \text{thus } \tau = RC = \frac{5.00}{1.61} = 3.11\ \text{s}$$

Solving for $R$

$$R = \frac{\tau}{C} = \frac{3.11\ \text{s}}{0.100\ \text{F}} = 31.1\ \Omega$$

As expected, the time constant is less than 5.00 s, because achieving 80% of the maximum voltage requires a time interval longer than one time constant.

**FOLLOW-UP EXERCISE.** (a) In this Example, how does the energy stored in the capacitor (after 5.00 s) compare with the maximum energy storage? Explain why it isn't 80%. (b) If you waited 10.00 s to charge the capacitor, what would its voltage be? Why isn't it twice the voltage that exists across the capacitor after 5.00 s?

---

## 18.4 Ammeters and Voltmeters

In previous circuit analyses the voltages and currents were given – assumed to have been measured. But how is this done? What instruments are used and how are they connected in the circuit? In this section the discussion will focus on the two primary instruments: ammeters and voltmeters.

Until recently, these instruments were based on sensitive mechanical (analog) movements of a galvanometer that uses the magnetic force to measure current (▶**Figure 18.13**). Today electronic digital *multimeters* (see ▶**Figure 18.14**) are commonplace. They are typically capable of measuring voltage, current, and often resistance. In place of mechanical galvanometers, these use electronic circuits to analyze digital signals and calculate voltages, currents, and resistances, which are then displayed on a small screen.

Since our focus is on what they measure and how to place them in a circuit to make such measurements, the internal workings and details will not concern us.

▲ FIGURE 18.13 **The galvanometer – the historical basis for ammeters/voltmeters** A galvanometer is a current-sensitive device whose needle deflection (analog) is proportional to the current in it. Enclosed in a protective case, the ammeter/voltmeter of the recent past looked like this.

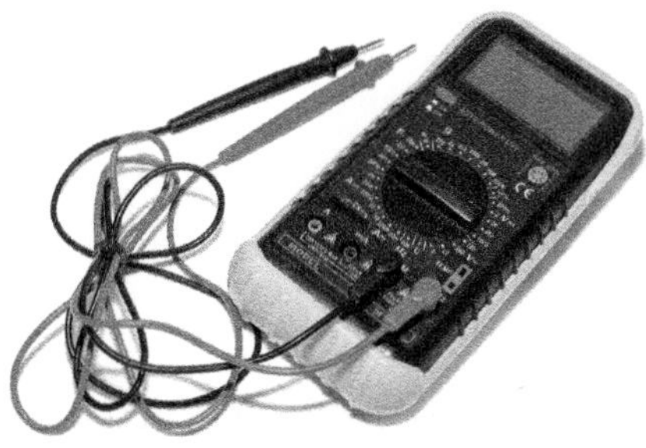

▲ FIGURE 18.14 **A modern digital multi-meter** This instrument is capable of measuring voltages, currents and resistances over a wide range of values by virtue of its internal digital circuitry.

### 18.4.1 The Ammeter

An ammeter is a device that measures the current in a circuit element. Since current flows *through* circuit elements, it must be placed in series with the element of interest, as shown in ▼ **Figure 18.15**. (The symbol for an ammeter is an A inside a circle.) The wire directly before or directly after the element of interest can be thought of as being cut and the ammeter inserted to perform this measurement. It should be clear that to not affect the current appreciably, the ammeter should have no significant resistance. The *ideal ammeter*, in fact, has zero resistance. Unless stated otherwise, this will be our assumption. From a practical viewpoint, it is important to note that if an ammeter is connected in parallel across a circuit element, since it has very little resistance, it will draw a huge current through it and likely be destroyed or damaged (unless protected by a fuse – see household safety Section 18.5).

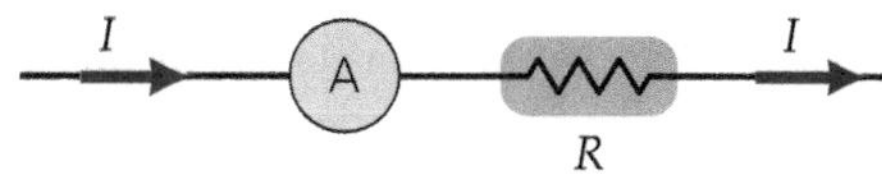


▲ FIGURE 18.15 **A dc ammeter** $R$ is the resistance of a resistor whose current is being measured. (The symbol for an ammeter is a circle with an A inside.) Since it must be placed in series with the resistor, the ammeter's resistance must be very low so as not to change the current it is measuring.

### 18.4.2 The Voltmeter

A voltmeter is a device that measures the potential difference (voltage) across circuit element. Thus it must be wired in *parallel* with the element of interest, as shown in ▼ **Figure 18.16**. (The symbol for a voltmeter is a V inside a circle.) In practice, to measure a difference in potential, the voltmeter is connected just before and just after the element of interest. It should be clear that to make an accurate measurement, the voltmeter should not draw any current away from the element – something that would change the voltage being measured. Hence it should have a very large resistance. The *ideal voltmeter*, in fact, has infinite resistance. Unless stated otherwise, this will be our assumption. From a practical viewpoint, if a voltmeter is incorrectly wired in series with a circuit element, since its resistance is very large, the current in the circuit will essentially drop to zero, making the voltage measurement useless, but unlike the ammeter, the voltmeter will not be in any immediate damage.

For some applications and details on the correct usage of voltmeters and ammeters, consider the next two examples.

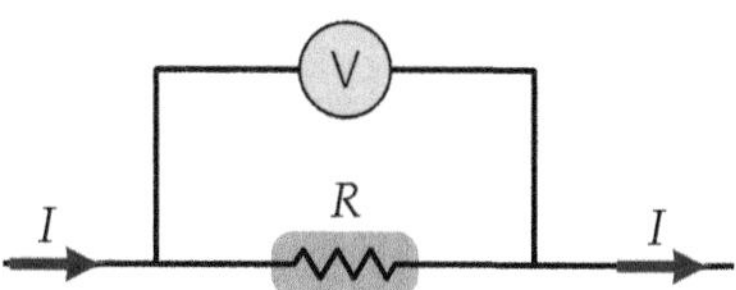


▲ FIGURE 18.16 **A dc voltmeter** $R$ is the resistance of a resistor whose voltage (potential difference across) is being measured. (The symbol for a voltmeter is a circle with an V inside.) Since it must be placed in parallel with the resistor, the voltmeter's resistance must be very high. This ensures that it will not draw current away from the resistor – something that would change its voltage.

---

**INTEGRATED EXAMPLE 18.5: AMMETERS – REAL AND IDEAL**

An (ideal) 12.0-V battery has a series arrangement of two resistors connected across its terminals. Starting at the battery anode, the first one ($R_1$) has a resistance of 2.00 Ω and the last one ($R_2$) has resistance of 4.00 Ω (a) To determine the current in $R_1$, what is the correct way to wire the ammeter? (1) In series but on the high potential side of $R_1$, (2) in series but in between the resistors, (3) in series but after $R_2$, (4) anywhere in the circuit, or (5) in parallel across either one of them. (b) Determine the current measured by an ideal ammeter in this circuit. (c) If the ammeter was real and had an internal resistance of 0.50 Ω, would the current be less or more than the ideal measurement? By what percentage would they differ? (*Note:* Practical ammeter resistances are actually very much smaller than the one in this example. The relatively large resistance value here is used so that the effects are obvious.)

**(a) CONCEPTUAL REASONING.** Due to its very low resistance, placing an ammeter in parallel as suggested by choice (5) suggested will lead a huge current through it and likely damage and so cannot be correct. Since the circuit is one complete loop with no junctions, the current is the same everywhere.

Hence choice (4) is the most correct - the ammeter can be placed anywhere. Thus it simultaneously measures the current passing through the battery as well as both resistors.

**(b) QUANTITATIVE REASONING AND SOLUTION.**

**SOLUTION**

Listing the data:

***Given:***

$$R_1 = 2.00\ \Omega$$
$$R_2 = 4.00\ \Omega$$
$$R_A = 0.50\ \Omega$$
$$\Delta V_B = 12.0\ \text{V}$$

***Find:***

**(b)** Ideal circuit current ($I_o$)
**(c)** Circuit current with "real" ammeter ($I$) and % difference

**(b)** For the ideal ammeter, the series resistance of the circuit is just the sum of the two resistors, or 6.00 Ω. Treating the battery as if it were connected to this one resistor, the ideal current is

$$I_o = \frac{\Delta V_B}{R_s} = \frac{12.0\ \text{V}}{6.00\ \Omega} = 2.00\ \text{A}$$

**(c)** If a real ammeter is used, the equivalent resistance across the battery would be increased because the three elements are in series. Thus the current in the circuit would be reduced. To find it, consider the three resistances (ammeter and two resistors) in series to find the equivalent resistance of 6.50 Ω. In this case, the measured current is

$$I = \frac{\Delta V_B}{R_s} = \frac{12.0\ \text{V}}{6.50\ \Omega} = 1.85\ \text{A}$$

This represents a reduction of 11% as shown below:

$$\%\ \text{difference} = \frac{(I_o - I)}{I_o} \times 100 = \frac{(2.00\ \text{A} - 1.85\ \text{A})}{2.00\ \text{A}} \times 100$$
$$= 7.50\%$$

This is a significant error and it should be clear that the ammeter resistance must be much less than the resistances in the circuit.

**FOLLOW-UP EXERCISE.** In this Example, (a) describe how you would connect a voltmeter to measure the voltage across $R_2$. (b) What voltage would an ideal voltmeter measure? (c) If the voltmeter actually had a resistance of 50.0 Ω, would you expect it to measure a larger or smaller voltage than the ideal case? Explain. (d) To back up your answer to (c), find the voltage that would be measured by the nonideal voltmeter. (*Note:* Voltmeter resistances actually are in the MΩ range to ensure they are larger than any normally encountered resistance. The low value here is used so that the effect on the currents and voltages is obvious.)

**EXAMPLE 18.6: USING AN AMMETER AND VOLTMETER TO DETERMINE RESISTANCE**

You have two resistors: $R_1$ has an unknown resistance, and $R_2$ has resistance of 6.00 Ω They are connected in series across a battery. (a) Describe how you would place an ideal voltmeter to measure both the voltage across the battery and $R_2$. (b) Suppose the battery measures 12.0 V and the voltage across $R_2$ measures at 3.00 V. What is the unknown resistance?

**THINKING IT THROUGH.** (a) Voltmeters are connected "across" elements, so for the battery the voltmeter should be connected from anode and cathode. For the resistor, the voltmeter is connected to both ends in parallel.

**SOLUTION**

Listing the data:

***Given:***

$$R_2 = 6.00\ \Omega$$
$$\Delta V_B = 12.0\ \text{V}$$
$$\Delta V_2 = 3.00\ \text{V}$$

***Find:*** $R_1$ (unknown resistance)

(b) The current in this single loop arrangement can be found because all the elements carry the same current. The current through $R_2$ (and thus through $R_1$ and the battery) is

$$I_2 = \frac{\Delta V_2}{R_2} = \frac{3.00\ \text{V}}{6.00\ \Omega} = 0.50\ \text{A}$$

To determine $R_1$ requires both its voltage and current. The voltage is found by applying the loop theorem, summing the voltages around the loop to zero: $\sum_i \Delta V_i = +\Delta V_B - \Delta V_1 - \Delta V_2 = 0$, and putting in the numbers gives

$$+12.0\ \text{V} - \Delta V_1 - 3.00\ \text{V} = 0 \quad \text{thus} \quad \Delta V_1 = 9.00\ \text{V}$$

The unknown resistance is then

$$R_1 = \frac{\Delta V_1}{I_2} = \frac{9.00\ \text{V}}{0.50\ \text{A}} = 18.0\ \Omega$$

**FOLLOW-UP EXERCISE.** You have two resistors: $R_1$ has an unknown resistance and $R_2$ has resistance of 6.00 Ω. They are connected in *parallel* across a battery. (a) Describe how you would make one measurement with an ideal voltmeter to measure all of the voltages at one time. (b) Suppose the voltmeter, when properly placed in (a), measures 12.0 V. After removing the voltmeter, show how you would place an ideal ammeter to measure the total current in the circuit. (c) If the current in (b) is 4.00 A, use your results to determine the unknown resistance.

## 18.5 Household Circuits and Electrical Safety

Although household circuits use alternating current, their operation (and many applications) can be understood using the circuit principles just studied. For example, wouldn't you expect the elements in a household circuit (lamps, appliances, and so on) to be in parallel? When a bulb in a lamp in your kitchen burns out, other appliances on that circuit, such as the coffee maker, would continue to work – which would not happen if the wiring were series. Moreover, appliances are generally designed to operate at approximately 120 V. If appliances were in series, then each would have only a fraction of the full 120 V.

Electrical power is supplied to a house by a three-wire system (▼**Figure 18.17**). There is an average difference in potential of 240 V between the two "hot," or high-potential, wires. Each of these "hot" wires has an average 120-V difference in potential with respect to the ground. The two "hot" wires are at opposite polarities. The third wire (the *ground*) is located at the point where the wires enter the house, usually by a metal rod driven into the ground. This wire is defined to be at zero of potential.

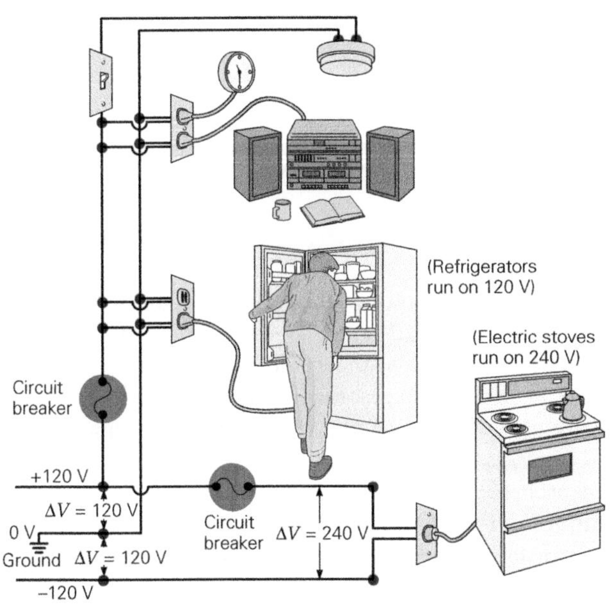


▲FIGURE 18.17 **Household wiring schematic** A120-V circuit is obtained by connecting either of the "hot" lines to the ground line. A voltage of 240 V (for appliances that require a lot of power such as electric stoves) can be obtained by connecting the two "hot" lines of opposite polarity.

The 120 V needed by most appliances is obtained by connecting them between ground and either high-potential wire. The result is the same in either case, $\Delta V = 120$ V (see Figure 18.17). Note that even though the ground wire is at zero potential, it *is* a current-carrying wire, because it is part of the complete circuit. High-power appliances such as ovens and water heaters typically operate at 240 V to reduce current and thus heating losses. The 240 V is obtained by connecting such appliances between the two hot wires: $\Delta V = 120\text{ V} - (-120\text{ V}) = 240$ V. The current needed for the operation of an appliance may specified on its label. If not, it can be determined from the power rating. For example, a stereo rated at 180 W would require a current of 1.50 A, since $I = P/\Delta V = 180\text{ W}/120\text{ V} = 1.50$ A.

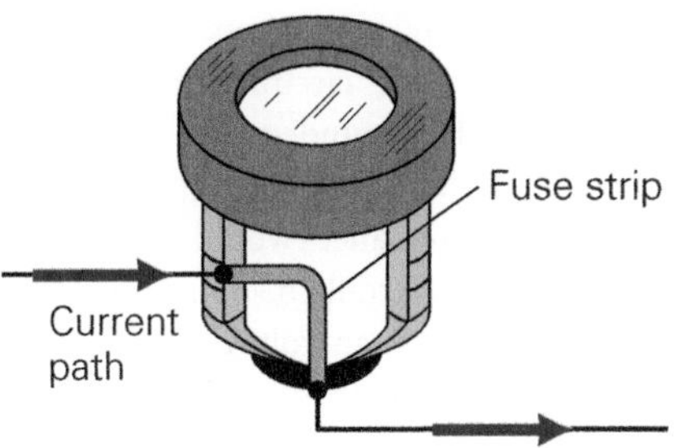


▲FIGURE 18.18 **Fuses** A fuse contains a metallic strip that melts when the current exceeds a rated value. This action opens the circuit and prevents overheating.

There are limitations on the *total* current in the wires of any circuit. Specifically, joule heating of the wires must be considered. Since appliances are connected in parallel, adding appliances (by turning them on or plugging them in) increases the total current in their circuit. Since real wires have some resistance, they can be subject to significant joule heating if the current is large enough. By adding too many appliances, it is possible to overload a circuit and thus produce too much heat in the wires. This could melt insulation and perhaps start a fire.

This potential overloading is prevented by limiting the current using one of two common devices: fuses and circuit breakers. **Fuses** can be seen mostly in older homes (▲**Figure 18.18**). Inside a fuse is a metal strip designed to melt when the current is larger than a rated value (typically 15 A for a 120-V circuit). The melting of the strip opens the circuit, and the current drops safely to zero.

**Circuit breakers** are used exclusively in newer homes. One type (▼**Figure 18.19**) uses a bimetallic strip (see Chapter 10). As the current in the strip increases, the strip warms and bends. At the rated value of current, the strip will bend sufficiently to open the circuit. The strip then cools, so the breaker can be reset. It should always be remembered that a blown fuse or a tripped circuit breaker indicates that the circuit is attempting to carry too much current. The problem should be found and repaired before replacing a fuse or resetting a circuit breaker.

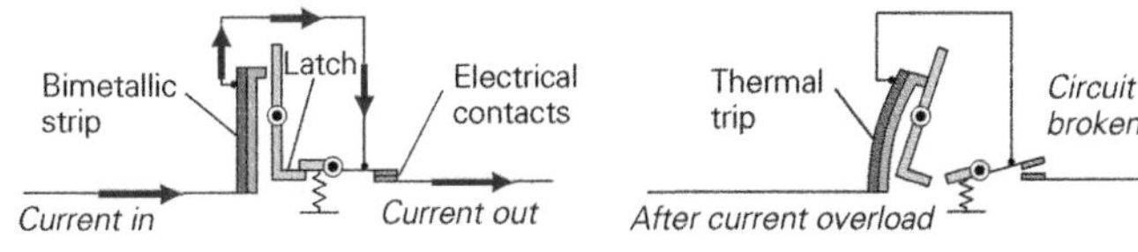


▲FIGURE 18.19 **Circuit breakers** A diagram of a thermal trip element. With increased current and joule heating, the element bends until it opens the circuit at some preset current value.

Switches, fuses, and circuit breakers are placed in the "hot" (high-potential) side of the circuit. They would, of course, also work if placed in the grounded side. However if they were installed there, even if the switch were open, the fuse blown, or the breaker tripped, the appliances would all remain connected to a high voltage wire – which could be dangerous if a person made electrical contact with the high voltage; for example, if a worn wire comes in contact with a metallic case (▼ **Figure 18.20a**).

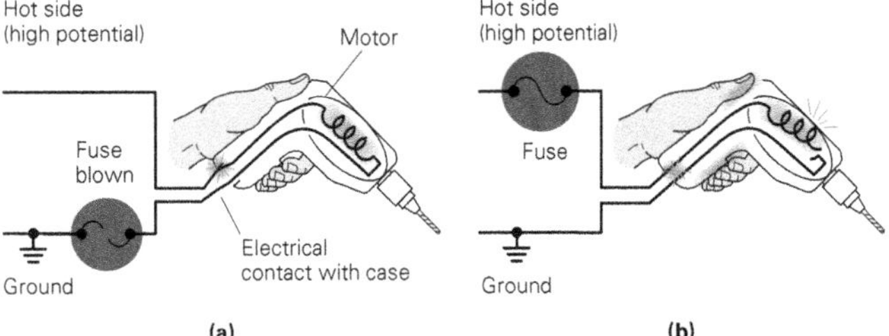


▲ FIGURE 18.20 **Electrical safety (a)** Switches and fuses or circuit breakers are always wired in the hot side of the line, *not* in the grounded side as shown. If these elements are wired in the grounded side, the line (and potentially the metallic case of the appliance) could remain at a high voltage even when the fuse is blown or a switch is open. **(b)** Even if the fuse or circuit breaker is wired correctly to the hot side, a potentially dangerous situation exists. If an internal wire comes in contact with the metal casing, a person touching the casing, which is at high voltage, can get a shock. To prevent this possibility, a third dedicated ground line runs from the case to ground (see Figure. 18.21).

Even with fuses or circuit breakers wired correctly into the "hot" side of the circuit, there is still a possibility of receiving an electrical shock from a defective appliance that has a metal casing, such as a hand drill. For example, if a wire comes loose inside, it could make contact with the casing, which would then be at a high voltage (Figure 18.20b). If the case were touched, the person's body would carry current to ground – a potential electrical shock.

To prevent such a shock, a third, dedicated grounding wire is usually provided with the appliance power cord (▶ **Figure 18.21**). If a hot wire does come in contact with the casing, the circuit is completed through this grounding wire, not you, thus avoiding any harm to you. This type of appliance cord ends in the familiar **three-prong grounded plugs**, where the large, round prong connects with the grounding wire (▶ **Figure 18.22**).

In the unlikely event that an electrical shock *does* result from exposure to high voltage, the extent of injury depends on the current in the body and on its pathway through the body. If the body as a whole is subjected to a voltage (say shoulder to foot), the current in the body ($I = \Delta V/R_{\text{body}}$) clearly depends on the body's resistance. If the skin is dry, total body resistance can be as high as 0.50 MΩ ($0.50 \times 10^6\ \Omega$) or more. For a voltage of 120 V with that resistance, the current would be about one-quarter of a milliamp:

$$I = \frac{\Delta V}{R_{\text{body}}} = \frac{120\ \text{V}}{0.50 \times 10^6\ \Omega} = 0.24 \times 10^{-3}\ \text{A} = 0.24\ \text{mA}.$$

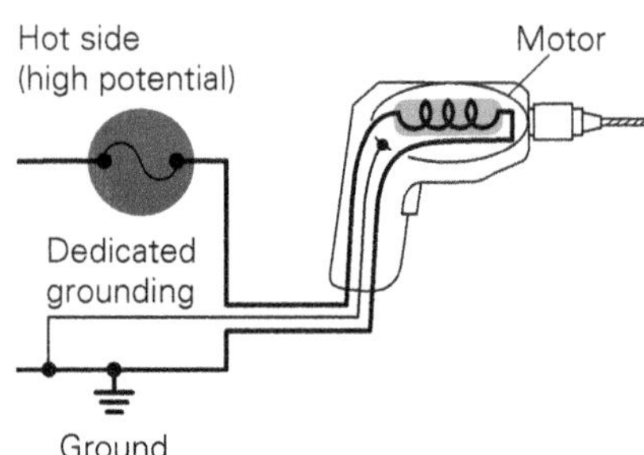


▲ FIGURE 18.21 **Dedicated grounding** For safety, a third wire is connected from the appliance to ground. This dedicated grounding wire normally carries no current. If the hot wire should somehow come in contact with the metal case, the current would follow the ground wire (the path of least resistance) rather than go through the body of the operator holding the case. The plug used for this is shown in Figure 18.22.

▲ FIGURE 18.22 **Plugging into ground** To accommodate the dedicated ground wire (Figure 18.21), the familiar three-prong plug is used.

This current is almost too weak to be felt (▼ **Table 18.1**). However, if the skin is wet, $R_{\text{body}}$ can be as low as 5.0 kΩ and the current would be 24 mA, a value that could be dangerous.

The physical damage resulting from high-voltage contact clearly also depends on the current path through the body. If the path is from finger to thumb on one hand, a large current probably would result in only a burn. However, if the path is from hand to hand through the chest (and therefore likely through the heart), the effect could be much worse (see Table 18.1 again).

**TABLE 18.1** Effects of Electric Current on the Human Body[a]

| Current (Approximate) | Effect |
|---|---|
| 2.0 mA (0.002 A) | Mild shock or heating |
| 10 mA (0.01 A) | Paralysis of motor muscles |
| 20 mA (0.02 A) | Paralysis of chest muscles, causing respiratory arrest; fatal in a few minutes |
| 100 mA (0.1 A) | Ventricular fibrillation, preventing coordination of the heart's beating; fatal in a few seconds |
| 1000 mA (1 A) | Serious burns; fatal almost instantly |

[a] The effect of a given current depends on a variety of conditions. This table gives only general and relative descriptions and assumes a circuit path that includes the upper chest.

Injury results because the current interferes with muscle function and/or causes burns. Muscle function is regulated by electrical nerve impulses, which can be influenced by external currents. Muscle reaction and pain can occur from a current of just a few milliamperes. At about 10 mA, muscle paralysis can prevent a person from releasing the conductor. At about 20 mA, contraction of the chest muscles occurs, which can cause impairment or stoppage of breathing. Death can occur in a few minutes. At 100 mA, rapid uncoordinated movements of the heart muscles (*ventricular fibrillation*) prevent proper heart pumping action and can be fatal in seconds. Working safely with electricity requires a knowledge of electrical principles as well as common sense. Electricity must be treated with respect.

## Chapter 18 Review

- Resistors in series all carry the same current. The **equivalent series resistance** of resistors in series is

$$R_s = R_1 + R_2 + R_3 + \cdots = \sum R_i \quad \text{(equivalent series resistance)} \tag{18.2}$$

- Resistors in parallel all experience the same voltage. The **equivalent parallel resistance** of resistors in parallel is

$$\frac{1}{R_p} = \frac{1}{R_1} + \frac{1}{R_2} + \frac{1}{R_3} + \cdots = \sum\left(\frac{1}{R_i}\right) \quad \text{(equivalent parallel resistance)} \tag{18.3a}$$

- **Kirchhoff's junction theorem** states that the total current into any **junction** equals the total current out of that junction and is written as:

$$\sum I_i = 0 \quad \text{(sum of currents at a junction)} \tag{18.4}$$

- **Kirchhoff's loop theorem** states that the algebraic sum of the voltage gains and losses around a complete circuit loop is zero; that is, the sum of the voltage gains equals the sum of the voltage losses, and is written as

$$\sum \Delta V_i = 0 \quad \text{(sum of voltages around a closed loop)} \tag{18.5}$$

- The voltage across a **charging capacitor** in an **RC circuit** is

$$\Delta V_C = \Delta V_o[1 - e^{-t/(RC)}] \quad \text{(charging capacitor – RC circuit)} \tag{18.6}$$

- The voltage across a **discharging capacitor** in an **RC circuit** is

$$\Delta V_C = \Delta V_o e^{-t/(R/C)} \quad \text{(discharging capacitor – RC circuit)} \tag{18.9}$$

- The **time constant** ($\tau$) for an RC circuit is a characteristic time by which the capacitor's charging and discharging rate is measured.

$$\tau = RC \tag{18.8}$$

- An **ammeter** is a device for measuring the current in a circuit element. Ammeters are wired in series with that element. Ideally, they have zero resistance.
- A **voltmeter** is a device for measuring the voltage across a circuit element. Voltmeters are wired in parallel across that element. The ideal voltmeter has an infinite resistance.

# End of Chapter Questions and Exercises

## Multiple Choice Questions

### 18.1 Resistances in Series, Parallel, and Series–Parallel Combinations

1. Which of the following must be the same for resistors in series: (a) voltage, (b) current, (c) power, or (d) energy?
2. Which of the following must be the same for resistors in parallel: (a) voltage, (b) current, (c) power, or (d) energy?
3. Resistors A and B are connected in series to a 12-V battery. A has 9 V across it. Which has the least resistance: (a) A, (b) B, (c) both have the same, or (d) can't tell from the data?
4. Resistors A and B are connected in parallel to a 12-V battery. A has 2.0 A of current and the total current in the battery is 3.0 A. Which has the most resistance: (a) A, (b) B, (c) both have the same, or (d) can't tell from the data given?
5. Two resistors (one of resistance 2.0 Ω and the other of 6.0 Ω) are connected in parallel to a battery. Which one produces the most joule heating: (a) the 2.0-Ω resistor, (b) the 6.0-Ω resistor; (c) both produce the same, or (d) you can't tell from the data given?
6. Two resistors (one of resistance 2.0 Ω and the other of 6.0 Ω) are connected in series to a battery. Which one produces the most joule heating: (a) the 2.0-Ω resistor, (b) the 6.0-Ω resistor; (c) both produce the same, or (d) you can't tell from the data given?
7. Resistors A and B are wired in parallel, and that combination is in turn connected in series to resistor C. The whole network is then connected across the terminals of a battery. Which of the following statements is true: (a) the current in C must be less than the current in either A or B; (b) the current in C must be more than the current in either A or B; (c) the current in C must equal the sum of the currents in A and B; or (d) the current in C must exceed the sum of the currents in A and B?
8. For the circuit in Question 7, which of the following statements is true: (a) the voltage across C must be less than that across either A or B; (b) the voltage across

C must be more than the voltage across either A or B; (c) the voltage across C must equal the sum of the voltages across A and B; or (d) the voltages across A and B must be equal?

**18.2 Multiloop Circuits and Kirchhoff's Rules**

9. A multiloop circuit has just one battery. After leaving the battery, the current encounters a junction into two wires. One wire carries 1.5 A and the other 1.0 A. What is the current in the battery: (a) 2.5 A, (b) 1.5 A, (c) 1.0 A, (d) 5.0 A, or (e) it can't be determined from the given data?
10. By our sign conventions, if a resistor is traversed in the direction opposite of the current in it, what can you say about the sign of the change in electric potential (the voltage) (a) it is negative, (b) it is positive, (c) it is zero, or (d) you can't tell from the data given?
11. By our sign conventions, if a battery is traversed in the actual direction of the current in it, what can you say about the sign of the change in electric potential (the battery's terminal voltage) (a) it is negative, (b) it is positive, (c) it is zero, or (d) you can't tell from the data given?
12. A multiloop circuit has just one battery with a terminal voltage of 12 V. After leaving the positive terminal of the battery, a wire takes you to a junction where the current (and circuit) splits into three wires, each containing two resistors. Later on, the three arms of the circuit rejoin and then are connected to the negative terminal of the battery. As you traverse each wire/resistor arm separately leading to the final junction, what can you say about the sum of the voltages across the two resistors in each wire: (a) they all total +12 V, (b) they all total −12 V, (c) they all total less than 12 V in magnitude.

**18.3 RC Circuits**

13. A fully charged capacitor stores 2.5 mJ of electric energy. Then it has a resistor connected across its oppositely charged plates. What can you say about the total heat generated in the resistor (ignore wire resistance): (a) it is greater than 2.5 mJ, (b) it is equal to 2.5 mJ, (c) it is less than 2.5 mJ.
14. As a capacitor discharges through a resistor, the voltage across the resistor is a maximum (a) at the beginning of the process, (b) near the middle of the process, (c) at the end of the process, (d) after one time constant.
15. When a capacitor discharges through a resistor, the current in the circuit is a minimum (a) at the beginning of the process, (b) near the middle of the process, (c) at the end of the process, (d) after one time constant.
16. A charged capacitor discharges through a resistor (situation #1). The resistance is doubled and the identically charged capacitor allowed to discharge again (situation #2). How do the time constants compare? (a) $\tau_1 = 2\tau_2$, (b) $\tau_1 = \tau_2$, (c) $\tau_1 = (1/2)\tau_2$.
17. An uncharged capacitor is charged by a battery through a resistor (situation #1). The capacitor is then recharged (using a different battery but the same resistor) to twice the final charge in situation #1 (charging situation #2). How do the time constants compare? (a) $\tau_1 = 2\tau_2$, (b) $\tau_1 = \tau_2$, or (c) $\tau_1 = 1/2\tau_2$?

**18.4 Ammeters and Voltmeters**

18. To accurately measure the voltage across a 1-kΩ resistor, a voltmeter should have a resistance that is (a) much larger than 1 kΩ, (b) much smaller than 1 kΩ, (c) about the same as 1 kΩ, (d) as close to zero as possible.
19. To accurately measure the current in a 1.0-kΩ resistor, an ammeter should have a resistance that is (a) much larger than 1.0 kΩ, (b) much smaller than 1.0 kΩ, (c) about the same as 1.0 kΩ, (d) as large as possible.
20. To correctly measure the voltage across a circuit element, a voltmeter should be connected (a) in series with it, (b) in parallel with it, (c) between the high potential side of the element and ground, (d) in none of the preceding.
21. Which of the following enables an ammeter to make a correct measurement of current in a circuit element (there may be more than one correct answer): (a) connect the ammeter in series with the element and just before it; (b) connect the ammeter in series with the element and just after it; (c) connect the ammeter in parallel with the element; (d) connect the ammeter between the high potential side of the element and ground; and/or (e) none of these is the correct way to use an ammeter.

**18.5 Household Circuits and Electrical Safety**

22. The circuit breaker in a normal 120-V household circuit limits the total current in its circuit to (a) about 15–20 A, (b) about 1 A, (c) about 120 V, (d) about 100 A.
23. Two identical appliances are on the same household circuit and the circuit breaker trips when someone tries to turn on both at the same time. Which of the following is true (there may be more than one correct answer): (a) if the breaker is reset and only one appliance is on, the breaker *will not* trip; (b) if the breaker is reset, having only one appliance on *might not* trip the breaker again; (c) each appliance had too low a resistance to work with a 120-V circuit breaker; and/or (d) the appliances together have an equivalent resistance too low to work with a 120-V circuit breaker.

## Conceptual Questions

**18.1 Resistances in Series, Parallel, and Series–Parallel Combinations**

1. Are the voltage drops across resistors in series generally the same? If not, under what circumstance(s) could they be the same?
2. Are the joule heating rates for resistors in series generally the same? If not, under what circumstance(s) could they be the same?

3. Are the currents in resistors in parallel generally the same? If not, under what circumstance(s) could they be the same?
4. Are the joule heating rates in resistors in parallel generally the same? If not, under what circumstance(s) could they be the same?
5. If a large resistor and a small resistor are connected in series, will the value of the effective resistance be closer to that of the large resistance or that of the small one? What if they are connected in parallel?
6. Lightbulbs are labeled with their power output. For example, when a lightbulb is labeled 60 W, it is assumed that the bulb is connected to a 120-V source. Suppose you have two bulbs. A 60-W bulb is followed by a 40-W bulb in series to a 120-V source. Which one glows brighter? Why? What happens if you switch the order of the bulbs? Are either of them at full power rating? [*Hint*: Consider their relative resistance values.]
7. Three identical resistors are connected to a battery. Two are wired in parallel, and that combination is followed in series by the third resistor. Which resistor has (a) the largest current, (b) the largest voltage, and (c) the largest power output?
8. Three resistors have values of 5.0, 2.0, and 1.0 Ω. The first one is followed in series by the last two wired in parallel. When this arrangement is connected to a battery, which resistor has (a) the largest current, (b) the largest voltage, and (c) the largest power output?

**18.2 Multiloop Circuits and Kirchhoff's Rules**

9. Must currents always leave from the positive terminal of a battery that is in a complete circuit? Explain. If not, give an example in which the current can enter at the positive terminal.
10. Use Kirchhoff's junction theorem to explain why the total equivalent resistance of a circuit is reduced, not increased, by connecting a second resistor in parallel to another resistor.
11. Use Kirchhoff's loop theorem to explain why a 60-W lightbulb produces *more* light than one rated at 100 W when they are connected in series to a 120-V source. [*Hint*: Recall that the power ratings are meaningful only at 120 V.]
12. Use both of Kirchhoff's theorems to explain why a 60-W lightbulb produces less light than one rated at 100 W when they are connected in parallel to a 120-V source. [*Hint*: Recall that the power ratings are meaningful only at 120 V.]
13. Use Kirchhoff's loop theorem to explain why, in a series connection, the largest resistance has the greatest voltage drop across it.

**18.3 RC Circuits**

14. An alternative way to describe the discharge/charge time of an RC circuit is to use a time interval called the *half-life*, which is defined as the time for the capacitor to lose *half* its initial charge. Based on this definition, is the time constant longer or shorter than the half-life? Explain your reasoning.
15. Is the time it takes to charge a capacitor in an RC circuit to 25% of its maximum value longer or shorter than one time constant? Is the time it takes to discharge a capacitor to 25% of its initial charge longer or shorter than one time constant? Explain your answers.
16. Use Kirchhoff's loop theorem to explain why the current in an RC circuit that is discharging a capacitor decreases as time goes on. Use the loop theorem to explain why the current in a charging RC circuit also decreases with time. [*Hint*: The loop theorem will tell you about the voltage across the resistor, which is directly related to the current in the circuit.]

**18.4 Ammeters and Voltmeters**

17. (a) What would happen if an ammeter were connected in parallel with a current-carrying circuit element? (b) What would happen if a voltmeter were connected in series with a current-carrying circuit element?
18. Explain clearly, using Kirchhoff's laws, why the resistance of an ideal voltmeter is infinite.
19. If designed properly, a good ammeter should have a very small resistance. Why? Explain clearly, using Kirchhoff's laws.
20. Draw the circuit diagrams indicating the correct placement for the ammeter in the following situations. (Use a circle with an "A" in it to represent the ammeter.) (a) Three resistors are wired in parallel and you want to measure the total current through all of them with just one measurement. (b) Three resistors are wired in parallel and you want to measure the current of just one of them. (c) Three resistors are wired in series and you want to measure the total current through all of them. (d) Three resistors are wired in series and you want to measure the current through just one of them.
21. Draw the circuit diagrams indicating the correct placement for the voltmeter in the following situations. (Use a circle with a "V" in it to represent the voltmeter.) (a) Three resistors are wired in parallel and you want to measure the voltage across all of them with just one measurement. (b) Three resistors are wired in parallel and you want to measure the voltage across just one of them. (c) Three resistors are wired in series and you want to measure the total voltage across them. (d) Three resistors are wired in series and you want to measure the voltage across just one of them.

**18.5 Household Circuits and Electrical Safety**

22. In terms of electrical safety, explain clearly what is wrong with the circuit in ▶ **Figure 18.23**, and why.
23. The severity of injury from electrocution depends on the magnitude of the current and its path, yet we commonly see signs that warn "Danger: High Voltage." Shouldn't such signs be changed to refer to high current? Explain.

▲**FIGURE 18.23 A safety problem?** (The dark red circular element represents a fuse or circuit breaker.) See Conceptual Question 22.

24. Explain why it is safe for birds to perch with both feet on the same high-voltage wire, even if the insulation is worn through.
25. After a collision with a power pole, you are trapped in your car, with a high-voltage line (with frayed insulation) in contact with the hood of the car. If you must get out before help arrives, is it safer to step out of the car one foot at a time or to jump with both feet leaving the car at the same time? Explain your reasoning.

## Exercises*

*Integrated Exercises (IEs) are two-part exercises. The first part typically requires a conceptual answer choice based on physical thinking and basic principles. The following part is quantitative calculations associated with the conceptual choice made in the first part of the exercise.*

### 18.1 Resistances in Series, Parallel, and Series–Parallel Combinations

1. • Three resistors that have values of 10, 20, and 30 Ω are to be connected. (a) How should you connect them to get the maximum equivalent resistance, and what is this maximum value? (b) How should you connect them to get the minimum equivalent resistance, and what is this minimum value?
2. • Two identical resistors (each with resistance $R$) are connected together in series and then this combination is wired in parallel to a 20-Ω resistor. If the total equivalent resistance is 10 Ω, what is the value of $R$?
3. • Two identical resistors ($R$) are connected in parallel and then wired in series to a 40-Ω resistor. If the total equivalent resistance is 55 Ω, what is the value of $R$?
4. IE • (a) In how many different ways can three 4.0-Ω resistors, be wired: (1) three, (2) five, or (3) seven? (b) Sketch the different ways you found in part (a) and determine the equivalent resistance for each.

* The bullets denote the degree of difficulty of the exercises: •, simple; ••, medium; and •••, more difficult.

5. • Three resistors with values of 5.0, 10, and 15 Ω are connected in series in a circuit with a 9.0-V battery. (a) What is the total equivalent resistance? (b) What is the current in each resistor? (c) At what rate is energy delivered to the 15-Ω resistor?
6. • Three resistors with values 1.0 Ω, 2.0 Ω, and 4.0 Ω are connected in parallel in a circuit with a 6.0-V battery. What are (a) the total equivalent resistance, (b) the voltage across each resistor, and (c) the power delivered to the 4.0-Ω resistor?
7. IE •• A length of wire with a resistance $R$ is cut into two equal-length segments. These segments are then twisted together (i.e., in parallel) to form a conductor half as long as the original wire. (a) The resistance of the shortened conductor is (1) $R/4$, (2) $R/2$, (3) $R$. Explain your reasoning. (b) If the resistance of the original wire is 27 $\mu\Omega$ and the wire is, instead, cut into *three* equal segments and then twisted together (i.e., in parallel), what is the resistance of the shortened conductor?
8. •• You are given four 5.00-Ω resistors. (a) Show how to connect all the resistors so as to produce an effective total resistance of 3.75 Ω. (b) If this network were then connected to a 12-V battery, determine the current in and voltage across each resistor.
9. •• Two 8.0-Ω resistors are connected in parallel, as are two 4.0-Ω resistors. These two combinations are then connected in series in a circuit with a 12-V battery. What is the current in each resistor and the voltage across each resistor?
10. •• What is the equivalent resistance of the resistors in ▼**Figure 18.24**?

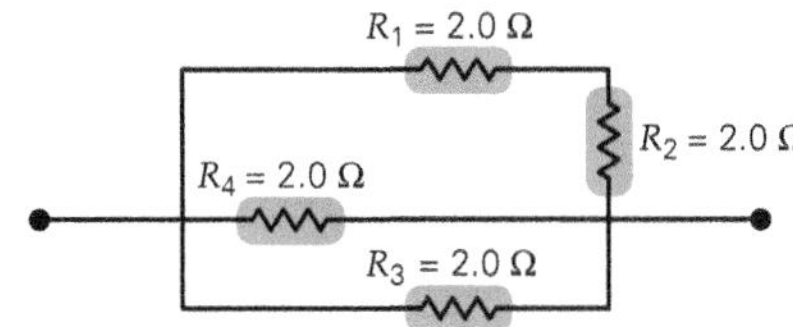


▲**FIGURE 18.24 Series-parallel combination** See Exercises 10 and 14.

11. •• What is the equivalent resistance between points A and B in ▼**Figure 18.25**?

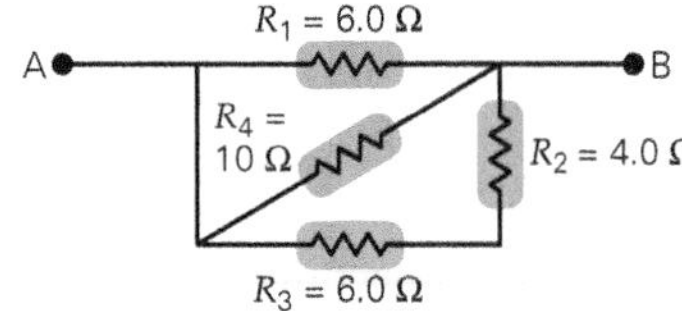


▲**FIGURE 18.25 Series-parallel combination** See Exercises 11 and 16.

12. •• Find the current in and voltage across the 10-Ω resistor shown in ▶**Figure 18.26**.

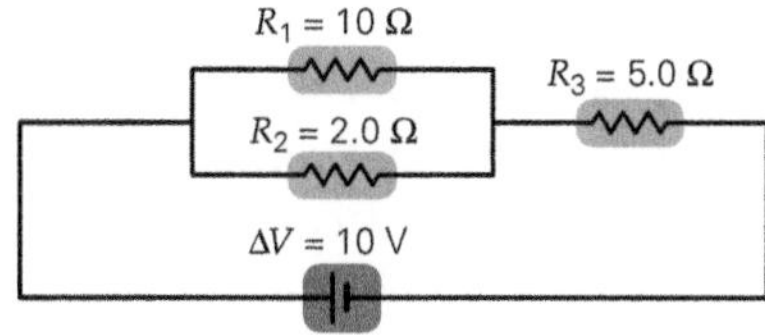


▲ **FIGURE 18.26 Current and voltage drop of a resistor** See Exercise 12.

13. •• For the circuit in ▼ **Figure 18.27**, find (a) the current in each resistor, (b) the voltage across each resistor, and (c) the total power delivered.

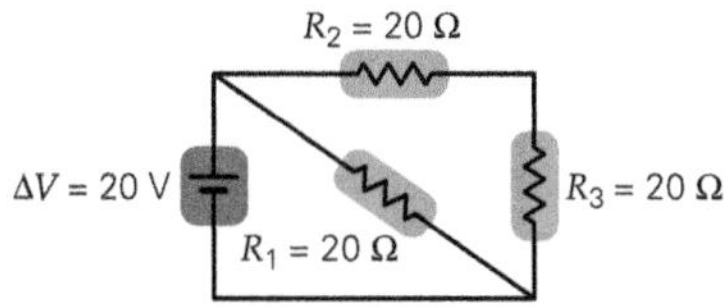


▲ **FIGURE 18.27 Circuit reduction** See Exercises 13 and 21.

14. •• Suppose that the resistor arrangement in Figure 18.24 is connected to a 12-V battery. What will be (a) the current in each resistor, (b) the voltage drop across each resistor, and (c) the total power delivered?
15. **IE**•• Suppose, in Exercise 14, that another 2.0-Ω resistor is in series with one in the lower branch. (a) Redraw the circuit and predict how the currents in the three different arms will change (increase, decrease, or stay the same) compared to those in the original circuit. (b) Calculate the new currents in each arm of the new circuit.
16. ••• The terminals of a 6.0-V battery are connected to points A and B in Figure 18.25. (a) How much current is in each resistor? (b) How much power is delivered to each? (c) Compare the sum of the individual powers with the power delivered to the equivalent resistance for the circuit.
17. ••• Lightbulbs with the power ratings (in watts) given in ▼ **Figure 18.28** are connected in a circuit as shown. (a) What current does the voltage source deliver to the circuit? (b) Find the power delivered to each bulb. (Take the bulbs' resistances to be the same as at their normal operating voltage.)

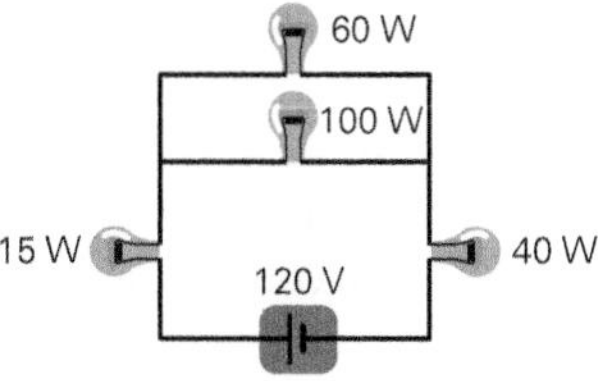


▲ **FIGURE 18.28 Watt's up?** See Exercise 17.

18. ••• Two resistors $R_1$ and $R_2$ are in series with a 7.0-V battery. If $R_1$ has a resistance of 2.0 Ω and $R_2$ receives energy at the rate of 6.0 W, what is (are) the value(s) for the circuit's current(s)? (There may be more than one answer.)
19. ••• For the circuit in ▼ **Figure 18.29**, find (a) the current in each resistor, (b) the voltage across each resistor, (c) the power delivered to each resistor, and (d) the total power delivered by the battery.

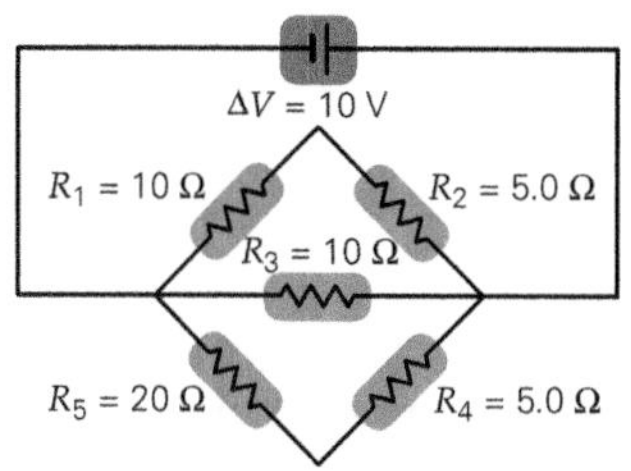


▲ **FIGURE 18.29 Resistors and currents** See Exercise 19.

20. ••• (a) Determine the equivalent resistance of the circuit in ▼ **Figure 18.30**. Find (b) the current in each resistor, (c) the voltage across each resistor, and (d) the total power delivered to the circuit.

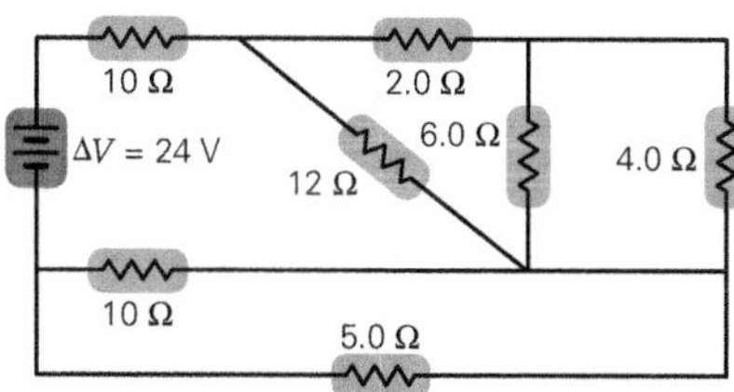


▲ **FIGURE 18.30 Power dissipation** See Exercise 20.

## 18.2 Multiloop Circuits and Kirchhoff's Rules

21. • Apply Kirchhoff's rules to the circuit in Figure 18.27 to find the current in each resistor.
22. **IE** •• Two batteries with terminal voltages of 10 and 4 V are connected with their positive terminals together. A 12-Ω resistor is wired between their negative terminals. (a) The current in the resistor is (1) 0 A, (2) between 0 and 1.0 A, (3) greater than 1.0 A. Explain your choice. (b) Use Kirchhoff's loop theorem to find the current in the circuit and the power delivered to the resistor. (c) Compare this result with the power output of each battery. Do both batteries lose stored energy? Explain.
23. •• Using Kirchhoff's rules, find the current in each resistor in ▼ **Figure 18.31**.

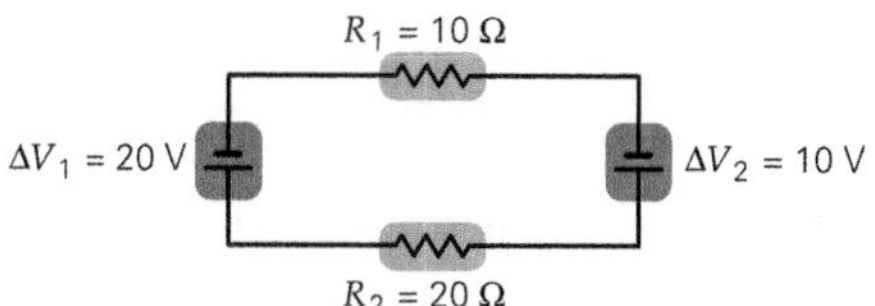


▲ **FIGURE 18.31 Single-loop circuit** See Exercise 23.

24. •• Apply Kirchhoff's rules to the circuit in ▼ **Figure 18.32**, and find (a) the current in each resistor and (b) the rate at which energy is being delivered to the 8.0-Ω resistor.

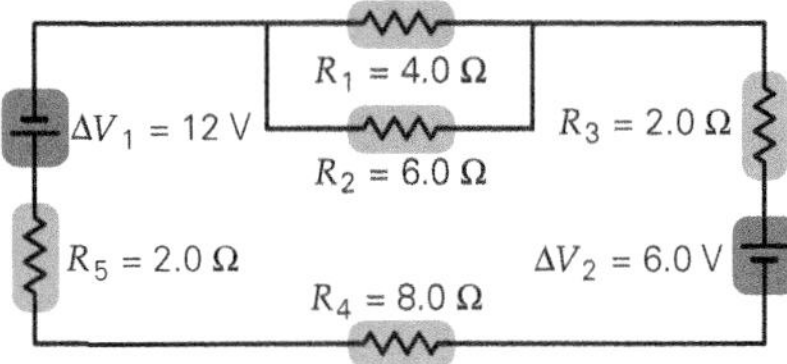


▲ **FIGURE 18.32 A loop in a loop** See Exercise 24.

25. ••• Find the current in each resistor in the circuit shown in ▼ **Figure 18.33**.

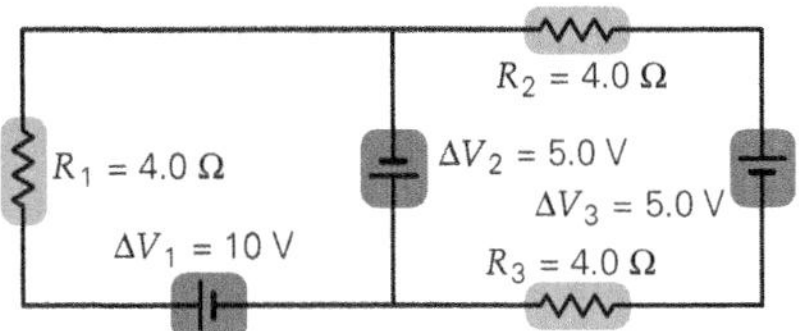


▲ **FIGURE 18.33 Double-loop circuit** See Exercise 25.

26. ••• Find the currents in the circuit branches in ▼ **Figure 18.34**.

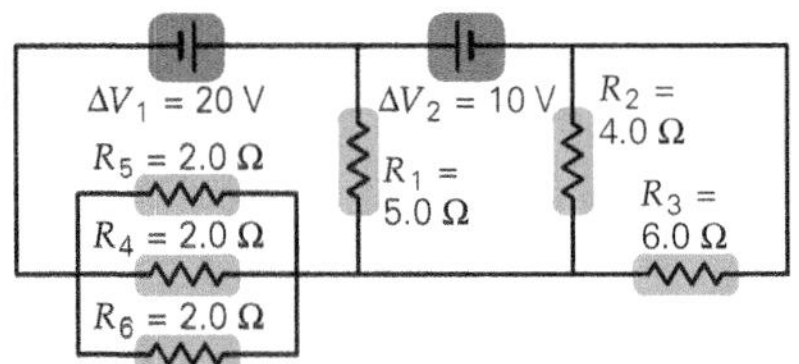


▲ **FIGURE 18.34 How many loops?** See Exercise 26.

27. ••• For the multiloop circuit shown in ▼ **Figure 18.35**, what is the current in each branch?

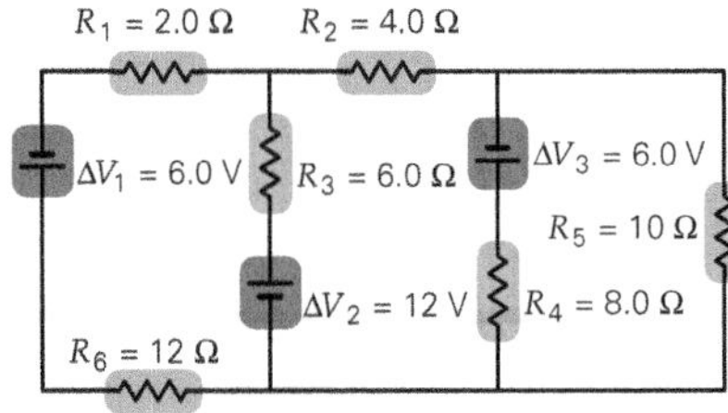


▲ **FIGURE 18.35 Triple-loop circuit** See Exercise 27.

## 18.3 RC Circuits

28. • A capacitor in a single-loop RC circuit is charged to 63% of its final voltage in 1.5 s. Find (a) the time constant for the circuit and (b) the percentage of the circuit's final voltage after 3.5 s.
29. •• How many time constants will it take for a charged capacitor to be discharged to one-fourth of its initial stored energy?
30. •• A 1.00-$\mu$F capacitor, initially charged to 12 V, discharges when connected in series with a resistor. (a) What resistance is necessary to cause the capacitor to have only 37% of its initial charge 1.50 s after starting? (b) What is the voltage across the capacitor at a time $t = 3\tau$ if the capacitor is instead charged by the same battery through the same resistor?
31. •• A 3.00-$\mu$F capacitor, initially charged to 24 V, discharges when connected in series with a resistor. (a) How much energy does this capacitor store when fully charged? (b) What is the capacitor's voltage when it has only half of its maximum energy? Is it 12 V? Why or why not? (c) What resistance is necessary to cause the capacitor to have only 50% of its energy left after 0.50 s of discharge? (d) What is the current in the resistor at this time?
32. •• A series RC circuit with $C = 40\ \mu$F and $R = 6.0\ \Omega$ has a 24-V source in it. With the capacitor initially uncharged, an open switch in the circuit is closed. (a) What is the voltage across the resistor immediately afterward? (b) What is the voltage across the capacitor at that time? (c) What is the current in the resistor at that time?
33. •• (a) For the circuit in Exercise 32, after the switch has been closed for a time $t = 4\tau$, what is the charge on the capacitor? (b) After a very long time, what are the voltages across the capacitor and the resistor?
34. ••• A series RC circuit consisting of a 5.0-MΩ resistor and a 0.40-$\mu$F capacitor is connected to a 12-V battery. If the capacitor is initially uncharged, (a) what is the change in voltage across it between $t = 2\tau$ and $t = 4\tau$? (b) By how much does the capacitor's stored energy change in the same time interval?
35. ••• A 3.0-MΩ resistor is connected in series with an initially uncharged 0.28-$\mu$F capacitor. This arrangement is then connected across four 1.5-V batteries (also in series). (a) What is the maximum current in the circuit and when does it occur? (b) What percentage of the maximum current is in the circuit after 4.0 s? (c) What is the maximum charge on the capacitor and when does it occur? (d) What percentage of the maximum charge is on the capacitor after 4.0 s? (e) How much energy is stored in the capacitor after one time constant has elapsed?

**18.4 Ammeters and Voltmeters**

36. •• An ammeter has a resistance of 1.0 mΩ. Sketch the circuit diagram and find (a) the current in the ammeter and (b) the voltage drop across a 10-Ω resistor that is in series with a 6.0-V ideal battery when the ammeter is properly connected to that 10-Ω resistor. (Express your answer to five significant figures to show how the current differs from 0.60 A and the voltage differs from 6.0 V, which are the expected values when no ammeter is in place.)

37. •• A voltmeter has a resistance of 30 kΩ. (a) Sketch the circuit diagram and find the current in a 10-Ω resistor that is in series with a 6.0 V ideal battery when the voltmeter is properly connected across that 10-Ω resistor. (b) Find the voltage across the 10-Ω resistor under the same conditions. (Express your answer to five significant figures to show how the current differs from 0.60 A and the voltage differs from 6.0 V, which are the expected values when no voltmeter is in place.)

**18.5 Household Circuits and Electrical Safety**

38. • Suppose you are using a drill that is incorrectly wired as in Figure 18.20a, and you make electrical contact with an ungrounded metal case. (a) Explain why this is a dangerous situation for you. (b) Estimate the current in you, assuming an overall body resistance of 300 Ω between your hand and feet. Would this be dangerous?

39. • In Exercise 38, suppose instead the case had been properly wired and grounded as shown in Figure 18.21. (a) If the grounding wire had a resistance of 0.10 Ω, what is the ratio of the current in you to the current in the ground wire, assuming that the fuse/circuit breaker does not "trip?" (b) Show that there is enough current to "trip" it, thus reducing the current to zero (and probably saving you from injury).

40. •• One day your electric stove does not turn on. You decide to check the 240-V outlet to see if it is the problem. You use two voltmeter probes inserted into the outlet slots, but because of cramped conditions, you accidentally touch the metal part of both probes, one with each hand. (a) How much current is in you during the time you are touching the probes, assuming that the outlet was actually operating properly and there is a resistance of 100 Ω between your hands? (b) Is this enough current to be dangerous to you? (c) Is there enough current to "trip" the circuit breaker and save the day? Comment on the relative sizes of your answers to parts (b) and (c).

# 19

# Magnetism

**Magnetic forces are responsible for suspending this high-speed train thus eliminating friction with rails.**

When thinking about magnetism, most people tend to envision an attractive force, because it is well known that certain materials can be picked up with a magnet. For example, paper clips are attracted to a magnet, and you have probably seen souvenir magnets that can stick to a refrigerator. It is less likely, however, that a repulsive magnetic force would come to mind. But there are repulsive magnetic forces, and they can be as useful as attractive ones. In this regard, the chapter-opening photo shows an interesting example. At first glance, the vehicle looks like an ordinary train. But where are its wheels? In fact, it isn't a conventional train, but a high-speed, magnetically levitated one. It doesn't physically touch the rails; rather, it "floats" above them, supported by repulsive forces produced by powerful magnets. The advantages are obvious: with no wheels, there is no rolling friction and few moving parts.

But where do these magnetic forces come from? For centuries, the properties of magnets were attributed to the supernatural. The original "natural" magnets, called *lodestones*, were found in the ancient Greek province of Magnesia. Magnetism was later discovered to be associated with electricity in an area of study that is now called *electromagnetism*. The forces described by electromagnetism are routinely used in such applications as motors and generators.

It is now known that electricity and magnetism are actually different manifestations of the same force. However, initially it is instructive to consider them individually.

## 19.1 Permanent Magnets, Magnetic Poles, and Magnetic Field

One of the features of a common bar magnet is that it has two "centers of force," called *poles*, near each end (▼**Figure 19.1**). To avoid confusion with the ± designation used for electric charge, magnetic poles are instead labeled north (N) and south (S). This terminology stems from the early use of the magnetic compass to determine direction. The north pole of a compass needle was historically defined as the north-seeking pole – that is, the end that points north on the Earth. The other end is south-seeking – that is, a south pole. A confusing result of this definition can result because the north pole of a compass needle is attracted to the Earth's north polar region (that is, *geographic* north), which means that the geographic north area is actually a *south magnetic* pole. (See Section 19.8 for more details about the Earth's magnetic field.)

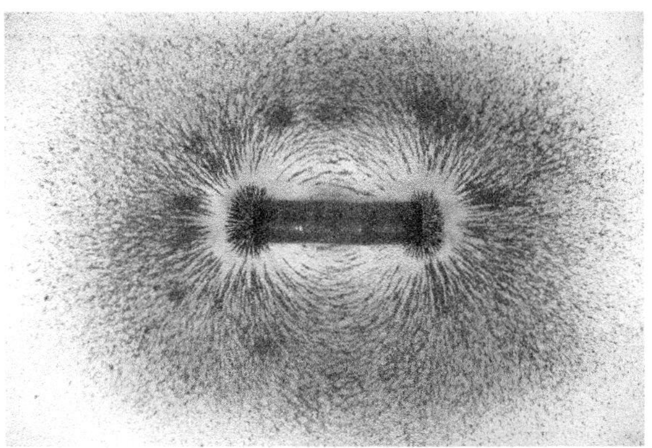

▲ FIGURE 19.1 **Bar magnet** The iron filings (acting as little compass needles) indicate the poles, or centers of force, of a common bar magnet. Determination of the type of pole at each end is found using a compass (see Figure 19.3).

By using two bar magnets, the forces between magnetic poles can be studied. The experimental findings are that each pole is attracted to the opposite pole of the other magnet and repelled by the same pole of the other magnet. This result is called the **pole-force law**, or **law of poles** (▼**Figure 19.2**):

> Like magnetic poles repel each other, and unlike magnetic poles attract each other.

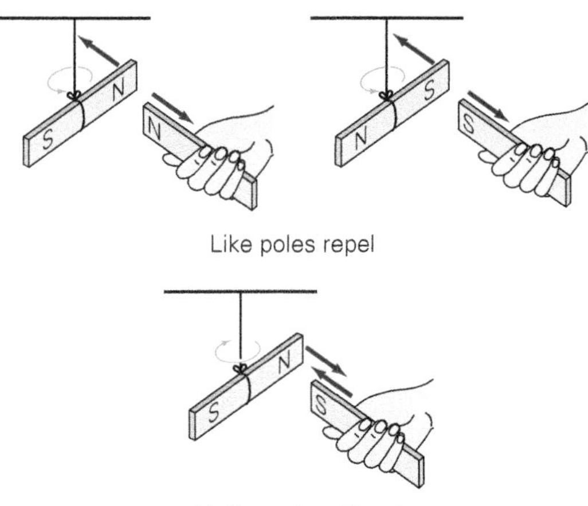


▲ FIGURE 19.2 **The pole-force law, or law of poles** Like poles (N-N or S-S) repel, and unlike poles (N-S) attract.

Two opposite magnetic poles, such as those of a bar magnet, form a *magnetic dipole*. At first glance, a bar magnet's field might appear to be the magnetic analog of the electric dipole. There are, however, fundamental differences between the two. For example, permanent magnets always have two poles occurring together, never one by itself. You might think that breaking a bar magnet in half would yield two isolated poles. However, the resulting pieces of the magnet always turn out to be two shorter magnets, each with its own set of north and south poles. While a single isolated magnetic pole (a *magnetic monopole*) could exist in theory, it has yet to be found experimentally.

Thus there is no magnetic analog to electric charge. This fact provides a strong hint about the differences between electric and magnetic fields. For example, while magnetic fields are produced by electric charges, this only occurs when they are *in motion*, such as electric currents and orbiting (or spinning) atomic electrons. As we will see, the latter is actually the source of the bar magnet's field.

### 19.1.1 Magnetic Field Direction

The historical approach to analyzing a bar magnet's field was to try to express the magnetic force between poles in a mathematical form similar to Coulomb's law for the electric force (Section 15.3). In fact, Coulomb developed such a law, using magnetic pole strengths in place of electric charge. However, this approach is rarely used, because it does not fit our modern understanding – since single magnetic poles, unlike single electric charges, do not exist. Instead, the modern description uses the concept of the magnetic field.

Recall that electric charges produce electric fields, which can be represented by electric field lines. The electric field (vector) is defined as the force per unit charge at a given location ($\vec{\mathbf{E}} = \vec{\mathbf{F}}_e/q_o$).

Similarly, magnetic interactions can be described in terms of the **magnetic field**, a vector quantity represented by the symbol $\vec{\mathbf{B}}$. Just as electric fields exist near electric charges, magnetic fields occur near permanent magnets. The magnetic field pattern surrounding a magnet can be made visible by sprinkling iron filings near it as in Figure 19.1.

Because the magnetic field is a vector field, both magnitude (or "strength") and direction must be specified. The direction $\vec{\mathbf{B}}$ at a given location is defined using a compass that has been calibrated (for direction) using the Earth's magnetic field as follows:

> The direction of $\vec{\mathbf{B}}$ at any location is the direction that the north pole of a compass would point if placed there.

This provides a convenient method for mapping a magnetic field by moving a small compass to various locations in the field. At any location, the compass needle will point in the direction of the "B-field" there. If the compass is then moved in the direction in which it points, the path it traces out is a *magnetic field line*, as illustrated in ▼**Figure 19.3a**. Note that the north end of a compass will point away from the north pole of a bar magnet, thus the field lines point away from the north pole and toward the south pole.

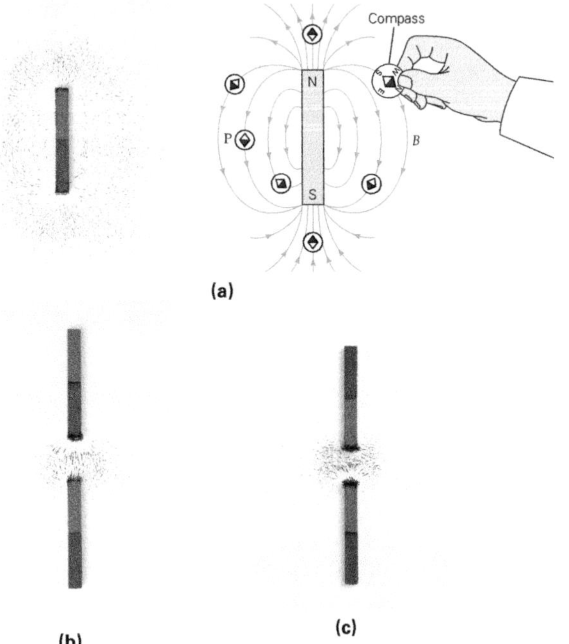


▲FIGURE 19.3 **Magnetic fields (a)** The magnetic field lines can be traced and the overall field pattern determined by using iron filings or a compass, as shown. The filings behave as tiny compasses and line up with the field. The closer together the field lines, the stronger the magnetic field. **(b)** Iron filing pattern for the magnetic field between unlike poles; the field lines converge. **(c)** Iron filing pattern for the magnetic field between like poles; the field lines diverge.

In summary, the rules that govern the interpretation of the magnetic field lines are very similar to those that apply to electric field lines:

> The closer together the $B$ field lines, the stronger the field. The direction $\vec{\mathbf{B}}$ is tangent to the field lines, or, equivalently, in the direction that the north end of a compass points.

Notice the dense concentration of iron filings in the pole regions of the bar magnets in Figure 19.3. This indicates closely spaced field lines and therefore a relatively strong magnetic field compared with other locations. As an example of field direction, observe that just outside the middle of the magnet, the field is downward, tangent to the field line at point P in the sketch in Figure 19.3a.

Two nearby bar magnets produce a net magnetic field that is the vector sum of their individual fields. For example, in Figure 19.3b the area between the nearby opposite poles shows a strong magnetic field since the two fields are in the same direction. Figure 19.3c shows a weak resultant field when two nearby north poles produce opposing fields that tend to cancel.

The magnitude of $\vec{\mathbf{B}}$ is defined in terms of the magnetic force exerted on a moving electric charge, as will be discussed in the following section.

## 19.2 Magnetic Field Strength and Magnetic Force

Experiments indicate that one determining factor of the magnetic force on a particle is its *electric* charge. That is, there is a connection between electrical properties of objects and how they respond to magnetic fields. As stated before, the study of these interactions is called **electromagnetism**. Consider the following electromagnetic interaction. Suppose a positively charged particle is moving at a constant velocity as it enters a uniform magnetic field. For simplicity, it is assumed that the particle's velocity is perpendicular to the magnetic field. (A fairly uniform field exists between the poles of a "horseshoe" magnet, as shown in ▶**Figure 19.4a**.) When the charged particle enters the field, it experiences a magnetic force and is deflected into an upward curved path, which is actually an arc segment of a circular path (assuming the field is uniform), as shown in Figure 19.4b.

From the study of circular motion (Section 7.3), for a particle to move in a circular arc, a centripetal force must act on it. Recall that this centripetal ("center-seeking") force is perpendicular to the particle's velocity. But what provides this force here? No electric field is present. The gravitational force would deflect the particle into a downward parabolic arc, not an upward circular one. Evidently, the force is a magnetic one, due to the interaction of the moving charge and the magnetic field. *Thus experiments indicate that a magnetic field can exert a force on a moving charged particle.*

From detailed measurements, the magnitude of this magnetic force $F_{\mathrm{m}}$ is proportional to the particle's charge and speed. When the particle's velocity ($\vec{\mathbf{v}}$) is perpendicular to the magnetic

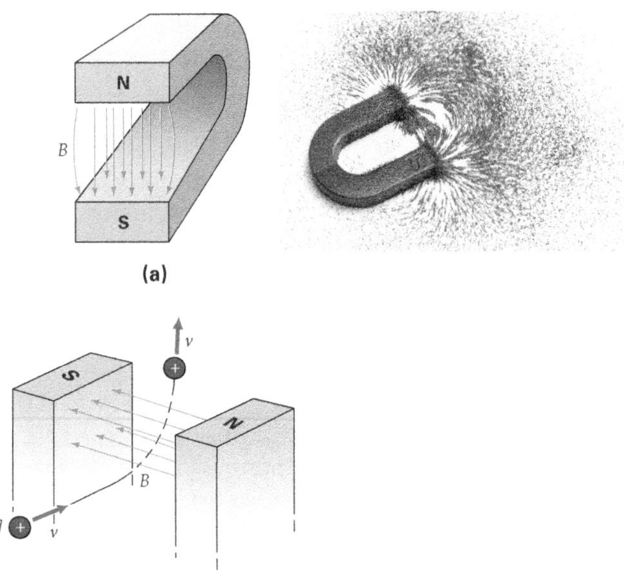


▲ FIGURE 19.4 **Force on a moving charged particle (a)** A horseshoe magnet, created by bending a permanent bar magnet, produces a fairly uniform field between its poles as shown in the sketch and also indicated by the iron filings in the photograph. **(b)** When a charged particle enters a magnetic field, the particle is acted on by a magnetic force whose direction is determined by observing the deflection of the particle.

field ($\vec{\mathbf{B}}$), the magnitude of the field, or the field strength, $B$ is defined as

$$B = \frac{F_{\mathrm{m}}}{qv} \quad \text{(valid only if } \vec{\mathbf{v}} \text{ is perpendicular to } \vec{\mathbf{B}}) \qquad (19.1)$$

SI unit of magnetic field: newtons per ampere-meter [N/(A·m), or tesla (T)].

Thus the magnitude of $\vec{\mathbf{B}}$ represents the magnetic force exerted on a charged particle per unit charge (coulomb) and per unit speed (m/s). From this definition, the units of $B$ are N/(C·m/s) or N/(A·m) since 1 A = 1 C/s. This combination of units is named the **tesla** (**T**) after Nikola Tesla (1856–1943), an early researcher in magnetism, and 1 T = 1 N/(A·m).

Most everyday magnetic field strengths, such as those from permanent magnets, are much smaller than 1 T. In such situations, it is common to express magnetic field strengths in milliteslas (1 mT $= 10^{-3}$ T) or microteslas (1 $\mu$T $= 10^{-6}$ T). A non-SI unit commonly used by geologists and geophysicists, the *gauss* (G), is defined as one ten-thousandth of a tesla, thus 1 G $= 10^{-4}$ T = 0.1 mT. The Earth's field is on the order of tenths of a G, or hundredths of an mT. Many laboratory and medical applications, such as MRI machines, utilize fields on the order of 3 T, while superconducting magnets can create fields of 25 T or higher.

If the field strength is known, the magnitude of the magnetic force on *any* charged particle moving at *any* speed can be found by rearranging Equation 19.1:

$$F_{\mathrm{m}} = qvB \quad \text{(valid only if } \vec{\mathbf{v}} \text{ is perpendicular to } \vec{\mathbf{B}}) \qquad (19.2)$$

More generally, a particle's velocity may not be perpendicular to the field. In that case the magnitude of the force depends the (smallest) angle ($\theta$) between the velocity vector and the magnetic field vector as follows:

$$F_{\mathrm{m}} = qvB\sin\theta \quad \text{(magnetic force on a charged particle)} \qquad (19.3)$$

Thus the magnetic force is zero when $\vec{\mathbf{v}}$ and $\vec{\mathbf{B}}$ are parallel ($\theta = 0°$) or oppositely directed ($\theta = 180°$) since $\sin 0° = \sin 180° = 0$. The force has its maximum value when the vectors are perpendicular. With $\theta = 90°$ ($\sin 90° = 1$), the maximum value is $qvB$.

### 19.2.1 The Right-Hand Force Rule

The direction of the magnetic force on a moving charged particle is determined by the orientation of its velocity relative to the magnetic field. Experiment shows that the magnetic force direction is given by the **right-hand force rule** as stated below and shown in ▶ **Figure 19.5a**.

> When the fingers of the right hand are pointed in the direction of a charged particle's velocity $\vec{\mathbf{v}}$ and then curled (through the smallest angle) toward the field $\vec{\mathbf{B}}$, the extended thumb points in the direction of the magnetic force $\vec{\mathbf{F}}_{\mathrm{m}}$ that would act on a *positive* charge. If the particle is *negatively* charged, the magnetic force is in the direction opposite to that of the thumb.

It is sometimes convenient to imagine the fingers of the right hand as physically turning or rotating the vectors $\vec{\mathbf{v}}$ into $\vec{\mathbf{B}}$ until they are aligned, much like rotating a right-hand screw thread. Several common physically equivalent alternatives are shown in Figure 19.5c and d. It does not matter which one you use, just choose the one that is the most comfortable for you and remember that if the particle is negatively charged, you must reverse the result.

Notice that the magnetic force is always *perpendicular to the plane formed by* $\vec{\mathbf{v}}$ *and* $\vec{\mathbf{B}}$ (see Figure 19.5b). Thus the magnetic force cannot do work on the particle. (This follows from the definition of work in Section 5.1, with a right angle between the direction of the force and displacement, then $W = Fd$ (cos) 90° = 0.) Therefore, a magnetic field does not change the speed (that is, kinetic energy) of a particle – only its direction.

To see how the right-hand force rule is applied to both charge types, consider the following Conceptual Example.

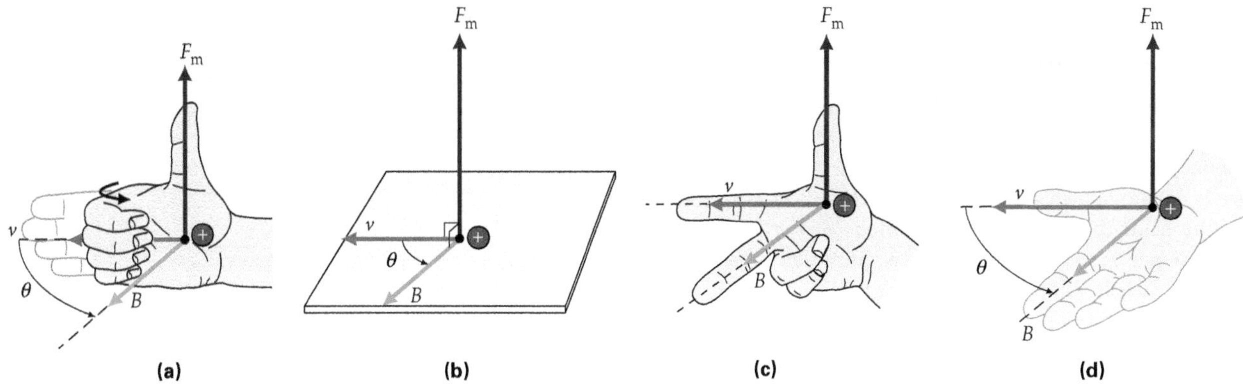


▲ FIGURE 19.5 **Equivalent right-hand force rules for determining magnetic force direction** **(a)** If the fingers of the right hand are pointed in the direction of $\vec{\mathbf{v}}$ and then curled in the direction of $\vec{\mathbf{B}}$, the extended thumb points in the direction of the magnetic force $\vec{\mathbf{F}}_m$ on a positive charge. **(b)** The magnetic force is always perpendicular to the plane formed by $\vec{\mathbf{B}}$ and $\vec{\mathbf{v}}$ and thus is always perpendicular to the particle's velocity. **(c)** When the extended forefinger of the right hand points in the direction of the particle's velocity and the middle finger points in the direction of the field, the extended right thumb points in the direction of the magnetic force on a positive charge. **(d)** When the fingers of the right hand are in the direction of the field and the thumb in the velocity direction, the palm points in the direction of the magnetic force on a positive charge. (Regardless of the rule you employ, remember to use your right hand and, at the end, reverse the direction for a negative charge.)

## CONCEPTUAL EXAMPLE 19.1: EVEN "LEFTIES" USE THE RIGHT-HAND RULE

Suppose in a linear particle accelerator, protons are traveling horizontally northward. To deflect them eastward using a uniform magnetic field, which direction should the magnetic field point: (a) vertically downward, (b) west, (c) vertically upward, or (d) south?

**REASONING AND ANSWER.** Because the force is perpendicular to the plane formed by $\vec{\mathbf{v}}$ and $\vec{\mathbf{B}}$, the field *cannot* be horizontal. If it were, it would deflect the protons down or up. Thus, (b) and (d) are eliminated. Now let's use the right-hand force rule with positive charges to see if $\vec{\mathbf{B}}$ could be downward (answer a). You should verify (make a sketch) that for a downward magnetic field, the force would be westward. Hence, the answer is (c). The magnetic field must point upward to deflect the protons east. You should verify this answer with a sketch.

**FOLLOW-UP EXERCISE.** What direction would the particles in this Example deflect if they were electrons moving horizontally southward in the same upward magnetic field?

Because the right-hand force rule involves thinking in three dimensions, it is common to designate vectors that point into the plane of the page by a letter ×. Vectors that point out of the plane are indicated by a dot •. These symbols are chosen so as to graphically mimic the feathered end of an arrow pointing away as ×, and the tip of that arrow pointing toward as •. This notation is shown in ▶ **Figure 19.6** and used in Example 19.2.

According to the previous discussion, charged particles entering uniform magnetic fields at right angles follow circular arcs (that is, uniform circular motion). To see this more explicitly, consider Example 19.2 carefully.

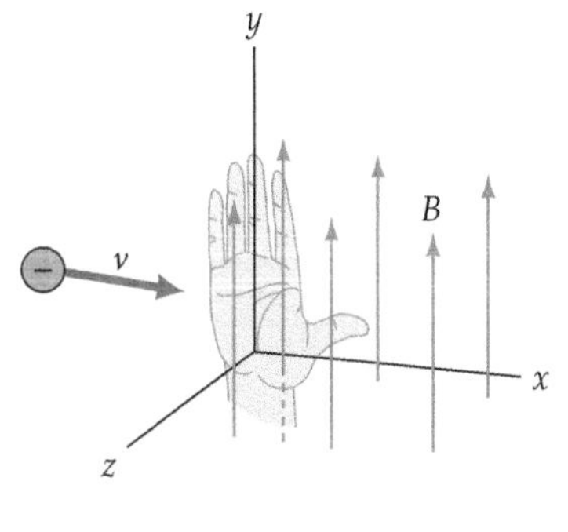


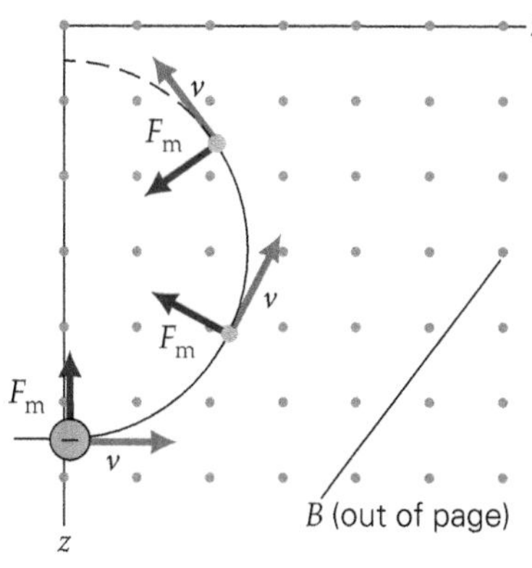


▲ FIGURE 19.6 **Path of a charged particle in a magnetic field** **(a)** A charged particle entering a uniform magnetic field will be deflected in the *z* direction by the right-hand rule, because the charge is negative. **(b)** In the field, the force is always perpendicular to the particle's velocity. The particle moves in a circular path if the field is constant and the particle enters the field perpendicularly to its direction. (See Example 19.2.)

**EXAMPLE 19.2: GOING AROUND IN CIRCLES – FORCE ON A MOVING CHARGE**

A particle with a charge of $-5.0 \times 10^{-4}$ C and a mass of $2.0 \times 10^{-9}$ kg moves at $1.0 \times 10^{3}$ m/s in the $+x$-direction. It enters a uniform magnetic field of 0.20 T that points in the $+y$-direction (see Figure 19.6a). (a) Which way will the particle deflect as it initially enters the field? (b) What is the magnitude of the force on the particle when it is in the field? (c) What is the radius of the circular arc that the particle will travel while in the field?

**THINKING IT THROUGH.** The initial deflection direction is in the direction of the initial magnetic force. In Figure 19.6a, the right-hand force rule in Figure 19.5d is being employed – fingers in the field direction and thumb in the velocity direction. Since the particle is negative, care must be taken in applying this force rule. A circular arc is expected, as the magnetic force is always perpendicular to the particle's velocity. The magnitude of the magnetic force on a charged particle is given by Equation 19.3. This is the only significant force on the electron; therefore, it is also the net force. From this, Newton's second law applied to centripetal forces should enable the determination of the circular orbit radius.

**SOLUTION**

Listing the given data:

***Given:***

$$q = -5.0 \times 10^{-4}\text{ C}$$
$$v = 1.0 \times 10^{3}\text{ m/s } (+x \text{ direction})$$
$$m = 2.0 \times 10^{-9}\text{ kg}$$
$$B = 0.20\text{ T } (+y \text{ direction})$$

***Find:***

(a) Initial deflection direction
(b) $F_m$ (magnitude of the magnetic force)
(c) $r$ (radius of the orbit)

(a) By the right-hand force rule, the force on a positive charge would be in the $+z$-direction (see direction of palm, Figure 19.6a). Because the charge is negative, the force is actually opposite this; thus, the particle will begin to deflect in the $-z$-direction.
(b) The force's magnitude can be determined from Equation 19.3. Because only the magnitude is of interest, the sign of $q$ can be dropped.

$$\begin{aligned} F_m &= qvB\sin\theta \\ &= (5.0\times10^{-4}\text{ C})(1.0\times10^{3}\text{ m/s})(0.20\text{ T})(\sin 90°) \\ &= 0.10\text{ N} \end{aligned}$$

(c) The magnetic force is the only force on the particle, thus it is the net force (Figure 19.6b). This net force points toward the circle's center and is the centripetal force acting on the particle. Recall that Newton's second law as applied to circular motion reads: $\vec{F}_c = m\vec{a}_c$. Here the magnetic force is $qvB$ from Equation 19.2 because $\theta = 90°$. Substituting this for the magnitude of the net force and using the expression for centripetal acceleration magnitude ($a_c = v^2/r$ from Section 7.3):

$$qvB = \frac{mv^2}{r} \quad \text{or} \quad r = \frac{mv}{qB}$$

Finally, inserting the numerical values,

$$\begin{aligned} r = \frac{mv}{qB} &= \frac{(2.0\times10^{-9}\text{ kg})(1.0\times10^{3}\text{ m/s})}{(5.0\times10^{-4}\text{ C})(0.20\text{ T})} \\ &= 2.0\times10^{-2}\text{ m} = 2.0\text{ cm} \end{aligned}$$

(You should check that the SI unit for the combination of $mv/qB$ is the meter.)

**FOLLOW-UP EXERCISE.** In this Example, if the particle had been a proton traveling initially in the $+z$-direction, (a) in what direction would it be initially deflected? (b) If the radius of its circular path is 5.0 cm and its speed is $1.0 \times 10^{5}$ m/s, what is the magnetic field strength?

## 19.3 Applications: Charged Particles in Magnetic Fields

The previous discussion showed that charged particles moving in magnetic fields can experience a magnetic force. Let's take a look at the role this magnetic force plays in some interesting applications.

### 19.3.1 The Velocity Selector and the Mass Spectrometer

Have you ever thought about how the mass of an atom or molecule is measured? Electric and magnetic fields provide a way in an instrument known as a **mass spectrometer** ("mass spec" for short). Mass spectrometers perform many functions in modern laboratories. For example, they can be used to analyze the composition of complex mixtures, such as a sample of smog-laden air. Forensic scientists use the mass spectrometer to identify trace materials, such as a streak of paint in a car accident. In archaeology, they are used to separate atoms to establish ages of ancient rocks and human artifacts. In hospitals, mass spectrometers are essential for maintaining the proper balance of anesthetic gases administered during an operation.

In actuality, it is the masses of *ions*, or charged molecules, that are measured in a mass spectrometer.* As shown in the schematic sketch in ▶ **Figure 19.7**, a beam of ions of charge $+q$ are produced by removing electrons from atoms and molecules.

* Removing or adding electrons from or to an atom or molecule produces an *ion*. However, an ion's mass is negligibly different from that of its neutral atom, because the electron's mass is very small.

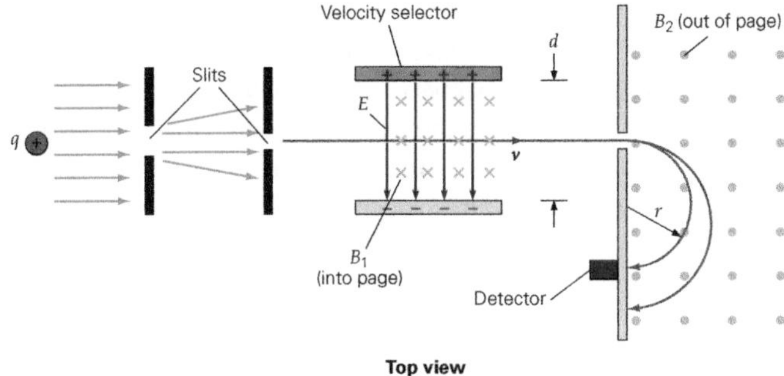


▲FIGURE 19.7 **Principle of the mass spectrometer** Ions pass through the velocity selector; only those with a particular velocity ($v = E/B_1$) then enter a magnetic field ($B_2$). These ions are deflected, with the radius of the circular path depending on the mass and charge of the ion. Paths of two different radii indicate that the beam contains ions of two different masses (assuming that they have the same charge).

The beam is then narrowed by a series of slits. At this point, the ions in the beam have a distribution of speeds. Thus, if they all entered the spectrometer, then ions with different speeds would take different paths. So, before the ions enter the spectrometer, a specific ion speed must be selected. This can be accomplished by using a *velocity selector*. This instrument consists of an electric field and a magnetic field crossed at right angles.

A velocity selector allows only particles traveling at a unique speed to pass through undeflected. To see how this works, consider a positive ion approaching the crossed-field arrangement at right angles to both fields as in Figure 19.7. Here the electric field produces a downward force ($F_e = qE$), and the magnetic field produces an upward force ($F_m = qvB_1$). (You should verify each force's direction is correct.) If the ions are not to be deflected, these two forces must cancel. Therefore, they must be equal in magnitude and oppositely directed. Equating their magnitudes gives $qE = qvB_1$, which can be solved for the "selected" speed:

$$v = \frac{E}{B_1} \quad \text{(velocity selector result)} \tag{19.4a}$$

If the plates are parallel, the electric field between them is $E = \Delta V/d$, where $\Delta V$ is the voltage across the plates and $d$ is their spacing. Thus a more practical version Equation 19.41 is

$$v = \frac{\Delta V}{B_1 d} \quad \text{(alternative velocity selector result)} \tag{19.4b}$$

The desired speed is usually selected by varying the plate voltage $\Delta V$.

Once through the velocity selector, the beam passes through a slit into a second magnetic field ($\vec{\mathbf{B}}_2$) that is perpendicular to the velocity. At this point, the ions begin to travel in a circular arc. The analysis is identical to that in Example 19.2:

$$F_c = ma_c \quad \text{or} \quad qvB_2 = m\frac{v^2}{r}$$

Using Equation 19.4, the mass of the particle is given by

$$m = \left(\frac{qdB_1B_2}{\Delta V}\right)r \quad \text{(mass determined with a mass spectrometer)} \tag{19.5}$$

The quantity in the parentheses is a constant (assuming all ions have the same charge). Hence, the larger the ion mass, the larger the radius of its circular path. Two paths of different radii are shown in Figure 19.7. This figure shows the beam actually contains ions of two different masses. Once the radius is measured by recording the position where the ions hit the detector after a semicircular trajectory, ion mass can be determined from Equation 19.5.

In a mass spectrometer of slightly different design, the detector is kept at a fixed position. This design employs a time-varying magnetic field ($B_2$), and a computer records and stores the detector reading as a function of time. In this design, $m$ is proportional to $B_2$. To see this, rewrite Equation 19.5 as $m = (qdB_1r/\Delta V)B_2$. Since the quantity in the parentheses is a constant, then $m \propto B_2$. Thus as $B_2$ is varied, the detector data recorded by a high-speed computer enables determination of both ion masses and relative number (that is, percentage) of each type of ion.

Regardless of design, the result - called a *mass spectrum* (the number of ions plotted against their mass) - is typically displayed on a computer screen and digitized for storage and analysis (▼**Figure 19.8**). Consider the following Example of a mass spectrometer arrangement.

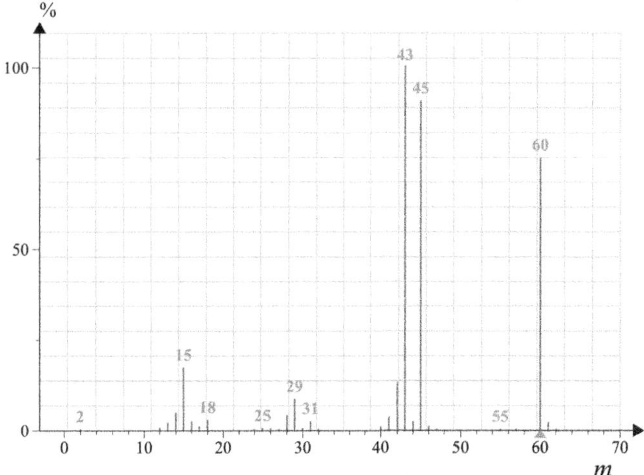

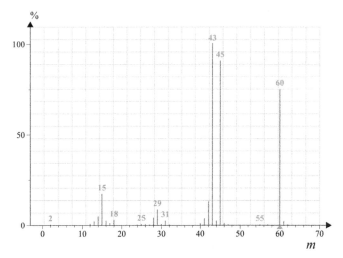

▲FIGURE 19.8 **Mass spectrometer** Display of a mass spectrometer, with the relative percentage of molecules plotted vertically and the molecular mass horizontally. Such patterns, *mass spectra*, help determine the composition and structure of molecules. The mass spectrometer can also be used to identify tiny amounts of a molecule in a complex mixture.

**EXAMPLE 19.3: THE MASS OF A MOLECULE – A MASS SPECTROMETER**

One electron is removed from a methane molecule before it enters the mass spectrometer in Figure 19.7. After passing through the velocity selector, the ion has a speed of $1.0 \times 10^3$ m/s. It then enters the main magnetic field region of field strength $6.70 \times 10^{-3}$ T. From there it follows a circular path and lands 5.00 cm from the field entrance. Determine the mass of this molecule. (Neglect the mass of the electron that is removed.)

**THINKING IT THROUGH.** The centripetal force is provided by the magnetic force on the ion. Since one electron has been removed, the ion's charge is $+e$. Because the velocity and magnetic field are at right angles, the magnetic force is given by Equation 19.2. By applying Newton's second law to circular motion, the molecule's mass can be determined.

**SOLUTION**

First, list the given data:

***Given:***

$$q = +e = 1.60 \times 10^{-19}\,\text{C}$$
$$r = (5.00\,\text{cm})/2 = 0.0250\,\text{m}$$
$$B_2 = 6.70 \times 10^{-3}\,\text{T}$$
$$v = 1.00 \times 10^3\,\text{m/s}$$

***Find:*** $m$ (mass of a methane molecule)

The centripetal force on the ion ($F_c = mv^2/r$) is provided by the magnetic force ($F_m = qvB_2$):

$$\frac{mv^2}{r} = qvB_2$$

Solving this equation for $m$ and putting in the numerical values:

$$m = \frac{qB_2r}{v} = \frac{(1.60 \times 10^{-19}\,\text{C})(6.70 \times 10^{-3}\,\text{T})(0.0250\,\text{m})}{1.00 \times 10^3\,\text{m/s}}$$
$$= 2.68 \times 10^{-26}\,\text{kg}$$

**FOLLOW-UP EXERCISE.** In this Example, if the magnetic field between the velocity selector's parallel plates (10 mm apart) is $5.00 \times 10^{-2}$ T, what is the voltage across the plates?

### 19.3.2 Silent Propulsion: Magnetohydrodynamics

In a search for quiet and efficient methods of propulsion at sea, engineers have invented a system based on *magnetohydrodynamics* – the study of the interactions of moving fluids and magnetic fields. This method of propulsion relies on the magnetic force and does not require moving parts so as to avoid detection – a feature that is of importance in modern submarine design.

Basically, seawater enters the front of the unit and is accelerated and expelled at high speeds out the rear (▼**Figure 19.9**). A superconducting electromagnet (see Section 19.7) is used to produce a large magnetic field. At the same time, an electric generator produces a large dc voltage, moving charged ions through the seawater. Seawater has a large concentration of sodium ($Na^+$) and chlorine ($Cl^-$) ions. Figure 19.9 shows what happens to the $Na^+$ ions, and you should show that the $Cl^-$ ions are also pushed rearward. The magnetic force on these ions pushes the water backward. By Newton's third law, a reaction force acts in the forward direction on the submarine, enabling it to move silently.

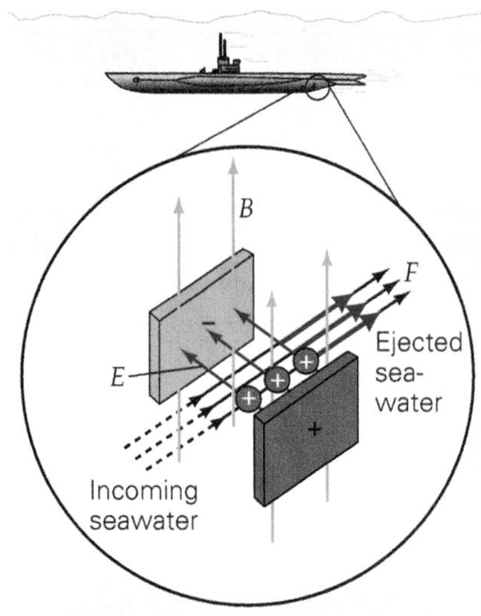


▲**FIGURE 19.9 Magnetohyrodynamic propulsion** In magnetohydrodynamic propulsion, ions (here only the $Na^+$ ions are shown, but the $Cl^-$ ions behave the same way) are accelerated by an electric field created by a dc voltage. A magnetic field exerts a force on both types of ions, pushing the water out of the submarine. The reaction force pushes the vessel in the opposite direction.

## 19.4 Magnetic Forces on Current-Carrying Wires

A charged particle moving in a magnetic field will generally experience a magnetic force. Because an electric current is composed of moving charges, it should be expected that a current-carrying wire, when placed in a magnetic field, would also experience such a force. The sum of the magnetic forces on each of the charges that make up the current should give the total magnetic force on the wire.

Recall that the direction of the "conventional current" assumes that electric current in a wire is due to the motion of positive charges, as depicted in ▶**Figure 19.10**. You should use the right-hand force rule to convince yourself that electrons that actually travel to the left would experience a magnetic force in the same direction as the conventional current carriers.

In the orientation depicted, the magnetic force is a maximum, because $\theta = 90°$. In a time $t$, an average charge will move a length

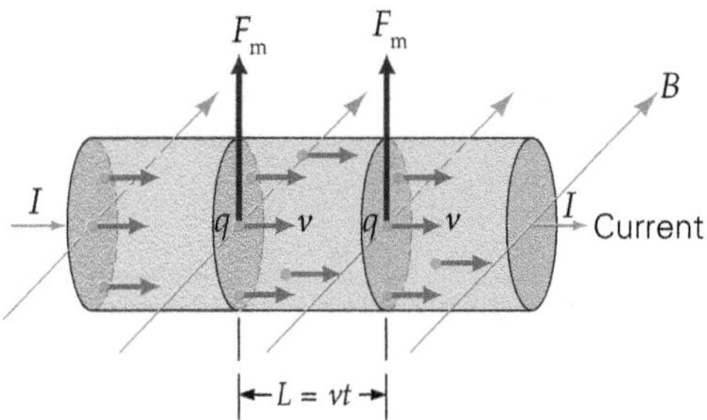


▲ FIGURE 19.10 **Force on a wire segment** Magnetic fields exert forces on current-carrying wires because electric current is composed of moving charged particles. The maximum magnetic force on such a wire is shown, when the angle between the charge velocity and the field is 90°.

$L = vt$, where $v$ is the average drift speed of the charge carriers. All the moving charges (total charge $= \sum_i q_i$) in this length of wire are acted on by a magnetic force in the same direction. Thus the total force on this length of wire can be determined by summing Equation 19.2 over all the charges. Assuming the field to be uniform, both the field ($B$) and the drift velocity ($v$) can be factored out of the sum, yielding

$$F_m = \Sigma (q_i vB) = vB(\Sigma q_i)$$

or, after replacing $v$ by $L/t$:

$$F_m = (\Sigma q_i)\left(\frac{L}{t}\right)B = \left(\frac{\Sigma q_i}{t}\right)LB$$

But $(\Sigma q_i)/t$ is, by definition, the current in the wire ($I$). Therefore, this result can be expressed in terms of the current as

$$F_m = ILB \quad \text{(valid only if current and magnetic field are perpendicular)} \tag{19.6}$$

This result is the *maximum* force on the wire. If the current direction makes an angle $\theta$ with respect to the field direction, then the force on the wire will be less. In general, the force on a length of current-carrying wire in a uniform magnetic field is

$$F_m = ILB \sin\theta \quad \text{(magnetic force on a current-carrying wire)} \tag{19.7}$$

Note that if the conventional current is in the direction of, or directly opposite to, the field, then the magnetic force on the wire is zero.

The direction of the magnetic force on a current-carrying wire is also given by a right-hand rule. As was the case for individual charged particles, there are several equivalent versions of the **right-hand force rule for a current-carrying wire**, one of the most common being:

> When the fingers of the right hand are pointed in the direction of the (conventional) current $I$ and imagine to be curled toward the magnetic field $\vec{\mathbf{B}}$, the extended thumb gives the direction of the magnetic force on the wire.

This version is illustrated in ► **Figure 19.11a** and **b**.

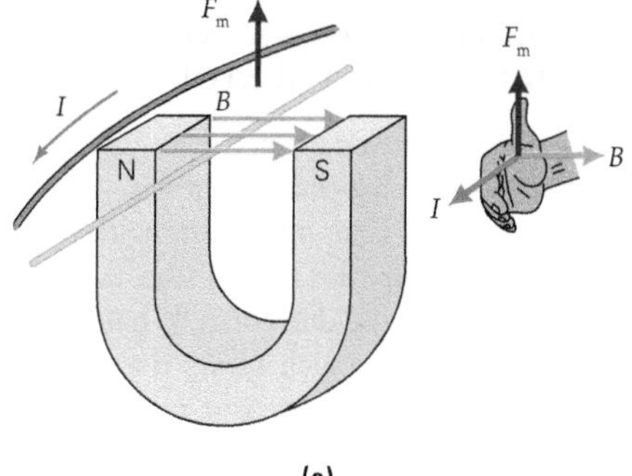


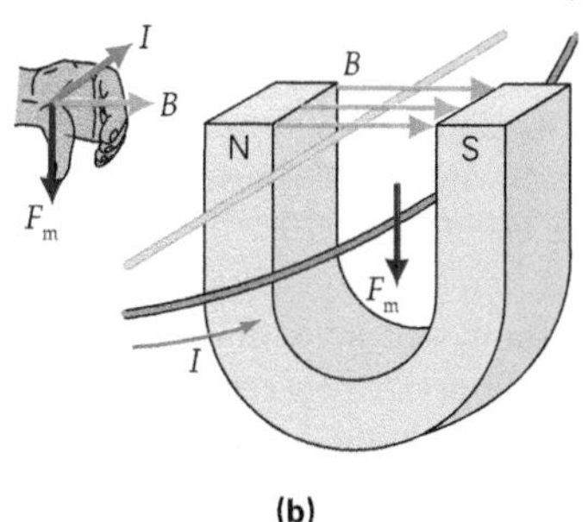


▲ FIGURE 19.11 **A right-hand force rule for current-carrying wires** The direction of the force is given by pointing the fingers of the right hand in the direction of the conventional current $I$ and then curling them toward $\vec{\mathbf{B}}$. The extended thumb points in the direction of the magnetic force on the wire. In **(a)** the force is upward, and then **(b)** downward when the direction of the current is reversed.

An equivalent alternative technique is shown in ▼ **Figure 19.12**. Here, when the fingers of the right hand are extended in the direction of the magnetic field and the thumb is in the direction of the (conventional) current in the wire, the palm of the right hand points in the direction of the magnetic force on the wire. Both techniques, of course, yield the same direction, because they are extensions of the right-hand force rules for individual charges. To see how current-carrying wires can interact magnetically with the Earth's field, consider Integrated Example 19.4.

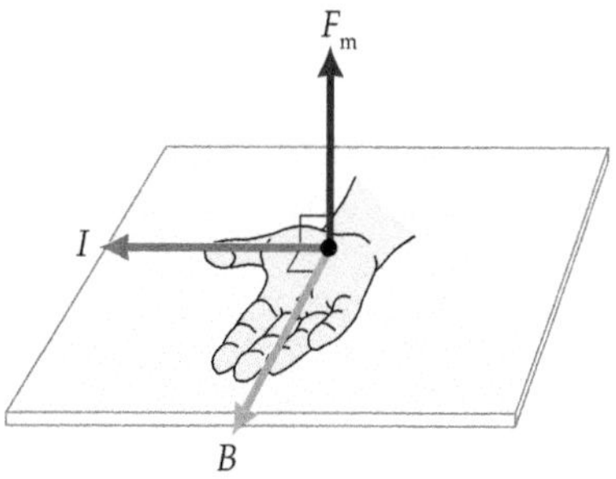


▲ FIGURE 19.12 **An alternative right-hand force rule for current-carrying wires** When the fingers of the right hand are extended in the direction of the magnetic field and the thumb is pointed in the direction of the conventional current, the palm points in the direction of the magnetic force on the wire. You should check that this gives the same direction as the technique in Figure 19.11.

## INTEGRATED EXAMPLE 19.4: MAGNETIC FORCES ON WIRES AT THE EQUATOR

Because a current-carrying wire is acted on by a magnetic force, it would seem possible to suspend such a wire above the ground using the Earth's magnetic field. (a) Assuming this could be done, consider long, straight wire located at the equator where the magnetic field is horizontal and due north. What would the current direction have to be to perform such a feat: (1) up, (2) down, (3) east, or (4) west? Draw a sketch to show your reasoning. (b) Calculate the current required to suspend the wire, assuming that the Earth's magnetic field is 0.40 G (gauss) at the equator and the wire is 1.0 m long with a mass of 30 g.

(a) **CONCEPTUAL REASONING.** The magnetic force direction must be upward (that is, away from the Earth's center), because gravity acts downward (▼**Figure 19.13**). The Earth's magnetic field at the equator is parallel to the ground and north. Because the magnetic force is perpendicular to both current and field, the current cannot be up or down, which eliminates the first two choices. To decide between east and west, simply choose one and see if it works (or doesn't work). Suppose the current is to the west. The right-hand force rule shows that the magnetic force acts downward (toward the Earth's center). This is incorrect; thus the correct answer is (3), east. You should verify that this is correct by using the force right-hand rule.

(b) **QUANTITATIVE REASONING AND SOLUTION.** The mass of the wire is known, and therefore its weight can be found. To balance the wire, this must be equal and opposite to the magnetic force. The current and the field are at right angles to each other; hence, the magnetic force is given by Equation 19.6, From that the current can be determined.

Listing the data and converting to SI units:

***Given:***

$m = 30\text{ g} = 3.0 \times 10^{-2}\text{ kg}$
$B = (0.40\text{ G})(10^{-4}\text{ T/G})$
$= 4.0 \times 10^{-5}\text{ T}$
$L = 1.0\text{ m}$

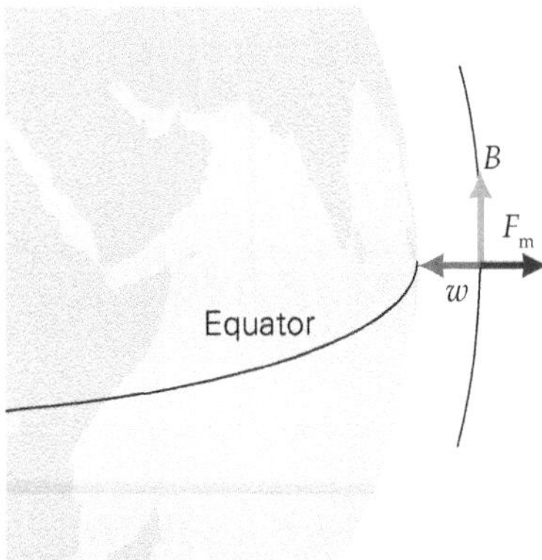


▲ FIGURE 19.13 **Defying gravity by using a magnetic field?** Near the Earth's equator it is theoretically possible to oppose the pull of gravity with an upward magnetic force on a wire. See Integrated Example 19.4 for details.

***Find:*** $I$ (current required to suspend the wire)

The wire's weight is $w = mg = (3.0 \times 10^{-2}\text{ kg})(9.8\text{ m/s}^2) = 0.29\text{ N}$. To suspend the wire, this must equal the magnitude of the magnetic force, or $w = ILB$. Solving for the current:

$$I = \frac{w}{LB} = \frac{0.29\text{ N}}{(1.0\text{ m})(4.0 \times 10^{-5}\text{ T})} = 7.4 \times 10^{3}\text{ A}$$

(As always, you should verify that the units are consistent.) This is a huge current, so suspending the wire in this manner is probably not a practical idea.

**FOLLOW-UP EXERCISE.** (a) Using the right-hand force rule, show that the idea of suspending a wire, as in this Example, could not work at either the south or north magnetic poles of the Earth. (b) In this Example, what would the wire's mass have to be for it to be suspended when carrying a more reasonable current of 10 A? Does this seem like a reasonable mass for a 1-m length of wire?

### 19.4.1 Torque on a Current-Carrying Loop

Besides exerting forces on wires, the magnetic field can exert torques on current-carrying loops, such as the rectangular one in a uniform field shown in ▼**Figure 19.14a**. (The voltage source that provides the current is not shown.) Suppose the loop is free to rotate about an axis passing through opposite sides, as

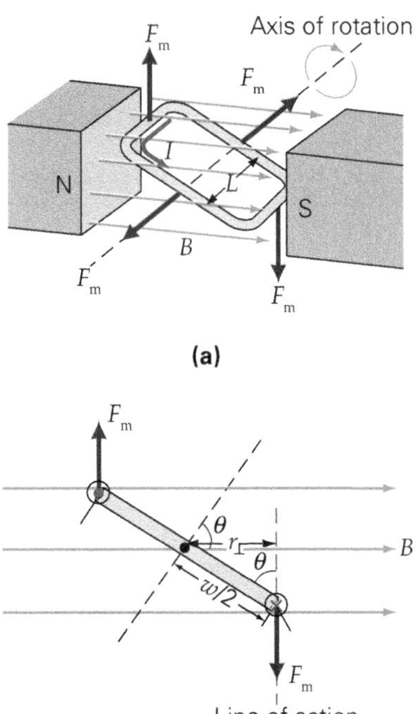


▲ FIGURE 19.14 **Force and torque on a current-carrying pivoted loop (a)** A current-carrying rectangular loop oriented in a magnetic field as shown is acted on by a magnetic force on each side. However only the forces on the sides parallel to the rotation axis produce a torque on the loop. **(b)** A side view shows the geometry for determining the torque. (See text for details.)

shown. There is no net force or torque due to the forces acting on its pivoted sides (the sides through which the rotation axis passes). The forces on these sides are equal and opposite and in the plane of the loop. Therefore, they produce no net torque or force. However, the equal and opposite forces on the two sides of the loop *parallel* to the axis of rotation, while not exerting a net force, do create a net torque (see Section 8.2).

To see how this works, consider Figure 19.14b, which is a side view of Figure 19.14a. The magnitude of the magnetic force on one of the non-pivoted sides (length $L$) is $F_m = ILB$. The torque produced by this force (Section 8.2) is $\tau = r_\perp F_m$, where $r_\perp$ is the perpendicular distance (or lever arm) from the axis of rotation to the line of action of the force. From Figure 19.16b, $r_\perp = \frac{1}{2} w \sin\theta$, where $w$ is the width of the loop and $\theta$ is the angle between the normal to the loop's plane and the direction of the field. The *net* torque exerted on the loop is due to the torques from both forces and thus is twice this result, or

$$\tau = 2r_\perp F_m = 2\left(\tfrac{1}{2} w \sin\theta\right) F_m = wF_m \sin\theta$$
$$= w(ILB)\sin\theta$$

But $wL$ is the area ($A$) of the loop. Thus the torque on one pivoted, current-carrying loop is

$$\tau = IAB\sin\theta \quad \text{(torque on one current-carrying loop)} \qquad (19.8)$$

(Although derived for a rectangular loop, Equation 19.8 is actually valid for a flat loop of *any* shape and area.)

A *coil* is composed of $N$ individual loops wired in series (where $N = 2, 3, \ldots$). Thus, for a coil, the torque is $N$ times that on one loop (because each loop carries the same current). Therefore the torque on a *coil* is

$$\tau = NIAB\sin\theta \quad \text{(torque on current-carrying coil of } N \text{ loops)} \qquad (19.9)$$

It is convenient to define the magnitude of a coil's **magnetic moment** vector, $\vec{\mathbf{m}}$, as

$$m = NIA \quad \text{(magnitude of a coil's magnetic moment)} \qquad (19.10)$$

SI units of magnetic moment: ampere·meter$^2$ or A·m$^2$

The *direction* of $\vec{\mathbf{m}}$ is determined by circling the fingers of the right hand in the direction of the (conventional) current in the coil. The thumb then points in the direction of $\vec{\mathbf{m}}$. Thus $\vec{\mathbf{m}}$ is always perpendicular to the plane of the coil (▶ **Figure 19.15a**). Equation 19.10 can be rewritten in terms of the magnetic moment:

$$\tau = mB\sin\theta \quad \text{(magnetic torque on a coil)} \qquad (19.11)$$

The magnetic torque tends to align the magnetic moment vector ($\vec{\mathbf{m}}$) with the magnetic field direction. To see this, notice that a loop or coil is subject to a torque until $\sin\theta = 0$ (that is, $\theta = 0°$), at which point the forces producing the torque are parallel to the plane of the loop (see Figure 19.15b). This situation exists when the plane of the loop is perpendicular to the field. Thus if the loop is started from rest with its magnetic moment at some angle with the field, it will experience an angular acceleration that will tend to rotate it toward the zero angle position (that is, when the magnetic moment is in the direction of the field).

Rotational inertia will then carry it through this equilibrium position (zero angle, Figure 19.15c) and on to the other side. On that side, the torque will then slow the loop, eventually stopping it, and then act to reaccelerate it back toward equilibrium. In other words, the torque on the loop is *restoring* and tends to cause the magnetic moment to oscillate about the field direction, much like a compass needle as it settles down (due to friction on its pivot) to eventually point north.

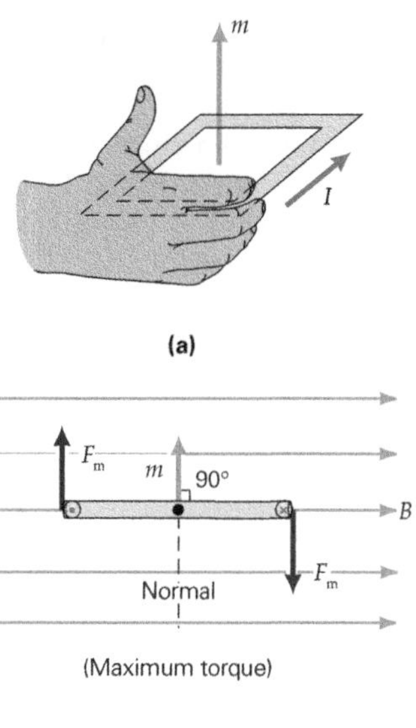


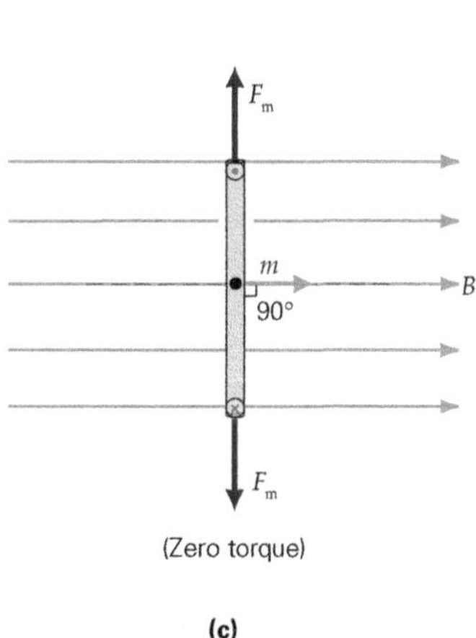


▲ FIGURE 19.15 **Magnetic moment of a current-carrying loop** (a) A right-hand rule determines the direction of a loop's magnetic moment vector $\vec{\mathbf{m}}$. The fingers wrap around the loop in the direction of the current, and the thumb gives the direction of $\vec{\mathbf{m}}$. **(b)** Condition of maximum torque. **(c)** Condition of zero torque. Thus if the loop is free to rotate, its magnetic moment vector will tend to align with the direction of the external magnetic field.

### EXAMPLE 19.5: MAGNETIC TORQUE – DOING THE TWIST?

A laboratory technician makes a circular coil out of 100 loops of thin copper wire with a resistance of 0.50 Ω. The coil diameter is 10 cm and the coil is connected to a 6.0-V battery. (a) Determine the magnetic moment (magnitude) of the coil. (b) Determine the maximum torque (magnitude) on the coil if it were placed in a uniform magnetic field of 0.40 T.

**THINKING IT THROUGH.** The magnetic moment includes the number of loops, the coil area, and the current in the wires. Ohm's law can be used to find that current. The maximum torque occurs when the angle between the magnetic moment vector and the magnetic field is 90°, as given by Equation 19.11.

**SOLUTION**

Listing the given data with the radius of the circle expressed in SI units:

***Given:***

$N = 100$ loops
$r = d/2 = 5.0\ \text{cm} = 5.0 \times 10^{-2}\ \text{m}$
$R = 0.50\ \Omega$
$\Delta V = 6.0\ \text{V}$

***Find:***

**(a)** $m$ (coil magnetic moment)
**(b)** $\tau$ (maximum torque on the coil)

**(a)** The magnetic moment is given by Equation 19.10, so the area and current are needed:

$$A = \pi r^2 = (3.14)(5.0 \times 10^{-2}\ \text{m})^2 = 7.9 \times 10^{-3}\ \text{m}^2$$

and

$$I = \frac{\Delta V}{R} = \frac{6.0\ \text{V}}{0.50\ \Omega} = 12\ \text{A}$$

Therefore the magnetic moment magnitude is

$$m = NIA = (100)(12\ \text{A})(7.9 \times 10^{-3}\ \text{m}^2) = 9.5\ \text{A}\cdot\text{m}^2$$

**(b)** The magnitude of the maximum torque Equation 19.11 is:

$$\tau = mB \sin\theta = (9.5\ \text{A}\cdot\text{m}^2)(0.40\ \text{T})(\sin 90^\circ) = 3.8\ \text{m}\cdot\text{N}$$

**FOLLOW-UP EXERCISE.** In this Example, (a) show that if the coil were rotated so that its magnetic moment vector was at 45° with respect to the field direction, the torque would *not* be half the maximum torque. (b) At what angle would the torque be half the maximum torque?

## 19.5 Applications: Current-Carrying Wires in Magnetic Fields

With the principles of electromagnetic interactions learned so far, the operation of the motor and digital balance can be understood.

### 19.5.1 The DC Motor

An *electric motor* is a device that converts electrical energy into mechanical energy. To understand the basics of a dc motor, consider a current-carrying coil in a uniform magnetic field. Because of the restoring nature of the magnetic torque, it will not be able to make a complete revolution (see Figure 19.16b for an example). To provide continuous rotation of such a coil, its design must be altered. Somehow the current must be reversed every half-turn so that the torque-producing forces are reversed, thus producing continual acceleration in the same direction. For example, this can be done by using a *split-ring commutator*, which is an arrangement of two metal half-rings insulated from each other (▶ **Figure 19.16a**).

Here the ends of the coil's wire are fixed to the half-rings so they rotate together. The current is supplied to the coil through the commutator by means of *contact brushes*. Then, with one half-ring electrically positive and the other negative, the coil and ring will start to rotate. When they have gone through half a rotation, the half-rings come in contact with the opposite brushes. Because their polarity is now reversed, the current in the coil also reverses. In turn, this action reverses the directions of the magnetic forces, keeping the torque in the same direction (clockwise in Figure 19.16b). Even though the torque is zero at the equilibrium position, the coil is in unstable equilibrium and has enough rotational momentum to continue through the equilibrium point, whereupon the torque takes over and angularly accelerates the coil through the next half-cycle. The process repeats in continuous operation providing the desired smooth rotation of the motor shaft.

### 19.5.2 The Electronic Balance

Traditional laboratory balances measure masses by balancing the weight of an unknown mass against a known one. Digital electronic balances (▶ **Figure 19.17a**) work on a different principle. In one type of design, there is still a suspended beam with a pan on one end that holds the object to be weighed, but no known mass is needed. Instead, the balancing downward force is supplied by current-carrying coils of wire in the field of a permanent magnet (Figure 19.17b). The coils move up and down in the cylindrical gap of the magnet, and the downward force is proportional to the current in the coils. The weight of the object in the pan is determined from the coil current, which produces a force just sufficient to balance the beam. From the weight, the balance's software determines the object's mass from $m = w/g$.

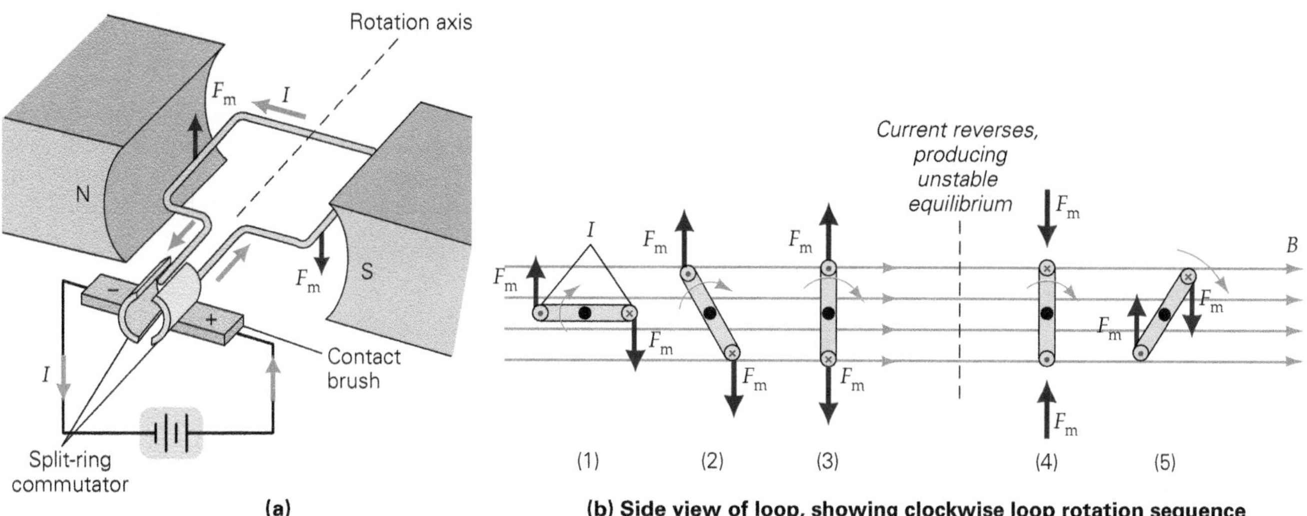


▲ **FIGURE 19.16** **A dc motor (a)** A split-ring commutator reverses the polarity and current each half-cycle, so the coil rotates continuously. **(b)** An end view shows the forces on the coil and its orientation during a half-cycle. For simplicity, only a single loop is depicted. Note the current reversal (shown by dot and cross notation) between situations (3) and (4).

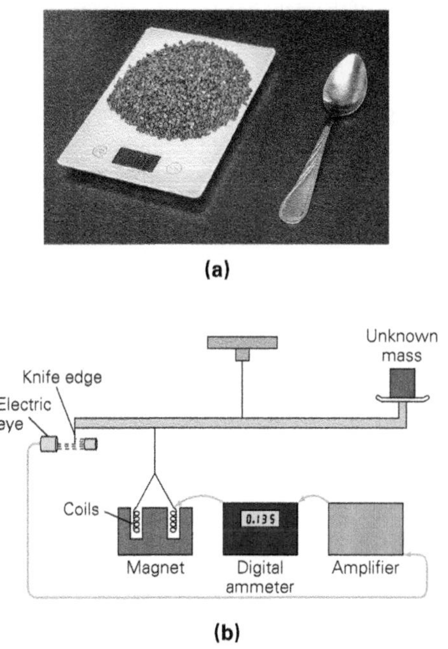


▲ **FIGURE 19.17** **Electronic balance (a)** A digital electronic balance. **(b)** Diagram of the principle of an electronic balance. The balance force is supplied by electromagnetism. See text for a detailed description.

Photo sensors and an electronic feedback loop typically control the current required for balance. When the beam is balanced and horizontal, a knife-edge obstruction cuts off part of the light from a source that falls on a photosensitive "electric eye," the resistance of which depends on the amount of light falling on it. This resistance controls the current in the coil. Thus if the beam tilts so the knife edge rises and more light strikes the sensor, the current in the coil is increased to counterbalance the tilting. In this manner, the beam is electronically maintained in nearly horizontal equilibrium.

## 19.6 Electromagnetism: Currents as a Magnetic Field Source

Electric and magnetic phenomena, although apparently quite different, are actually closely and fundamentally related. As has been seen, the magnetic force on a particle depends on its electric charge. But what is the source of the magnetic field? Danish physicist Hans Christian Oersted discovered the answer in 1820, when he found that *electric currents produce magnetic fields*. His studies marked the beginnings of the discipline called **electromagnetism** – the study of the relationship between electricity and magnetism.

In particular, Oersted first noted that an electric current could produce a deflection of a compass needle. This property can be demonstrated with an arrangement such as that shown in ▶ **Figure 19.18**. When the circuit is open and there is no current, the compass needle points, as usual, in the northerly direction. However, when the switch is closed and there is current in the circuit, the compass needle points in a different direction, indicating that an additional magnetic field (due to the current) is affecting the needle.

Developing expressions for the magnetic field created by various configurations of current-carrying wires requires mathematics beyond the scope of this book. Here we present the results for the magnetic fields for several common current configurations.

### 19.6.1 Magnetic Field Near a Long, Straight, Current-Carrying Wire

At a perpendicular distance $d$ from a long, straight wire carrying a current $I$ (▶ **Figure 19.19**), the magnitude of its magnetic field is

$$B = \frac{\mu_o I}{2\pi d} \quad \text{(magnetic field due to a long, straight wire)} \tag{19.12}$$

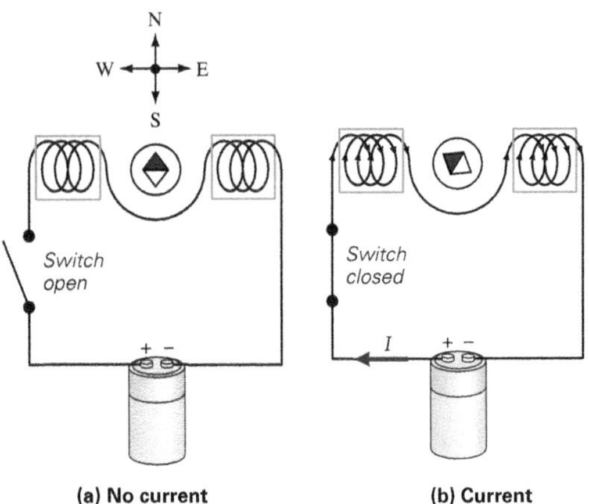


▲ **FIGURE 19.18 Electric current and magnetic field** (**a**) With no current in the wire, the compass needle points north. (**b**) With a current in the wire, the compass needle is deflected, indicating the presence of an additional magnetic field added (vectorially) to the Earth's. In this sketch, the strength of the additional field is roughly equal in magnitude to that of the Earth. How can you tell?

where $\mu_o = 4\pi \times 10^{-7}\ \mathrm{T \cdot m/A}$ is a constant – **the magnetic permeability of free space.**

For long, straight wires, the magnetic field lines are closed circles centered on the wire (Figure 19.19a). As seen in **Figure 19.19b**, the direction of $\vec{\mathbf{B}}$ due to current in a long, straight wire is given by a **right-hand source rule**:

> If a long, straight, current-carrying wire is grasped with the right hand and the extended thumb points in the direction of the current ($I$), then the curled fingers indicate the circular sense of the magnetic field lines.

### EXAMPLE 19.6: COMMON FIELDS – MAGNETIC FIELD FROM A CURRENT-CARRYING WIRE

The maximum household current in a wire is about 15 A. Assume that this current exists in a long, straight, horizontal wire in a west-to-east direction (▼**Figure 19.20**). What are the magnitude and direction of the magnetic field 1.0 cm directly below the wire?

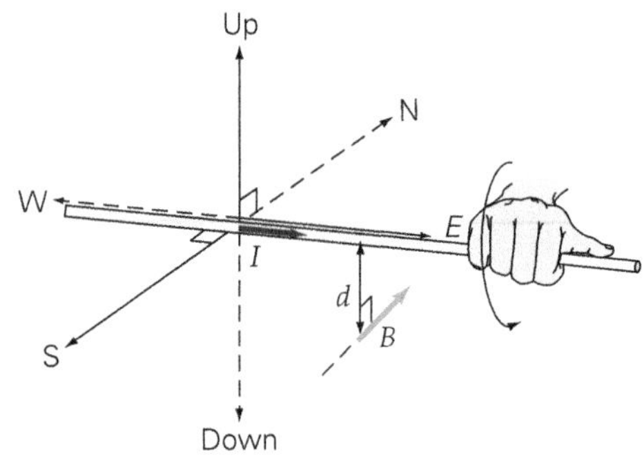


▲ **FIGURE 19.20 Magnetic field** Finding the magnitude and direction of the magnetic field produced by a straight current-carrying wire using one version of the right-hand source rule (see Figure 19.21a). See Example 19.6 for details.

**THINKING IT THROUGH.** Equation 19.12 allows for the determination of the magnetic field magnitude. The direction of the field is given by the right-hand source rule.

**SOLUTION**

***Given:***

$I = 15\ \mathrm{A}$
$d = 1.0\ \mathrm{cm} = 0.010\ \mathrm{m}$

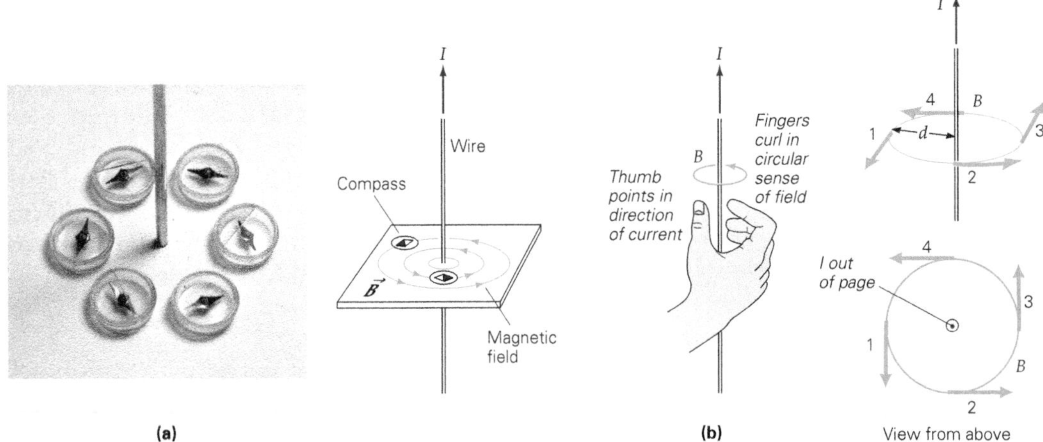


▲ **FIGURE 19.19 Magnetic field near a long, straight, current-carrying wire** (**a**) The field lines form concentric circles around the wire, as revealed by the pattern of small compasses. (**b**) The circular sense of the field lines is given by the right-hand source rule, and, as usual, the magnetic field vector (direction) is tangent to a circular field line at any point.

***Find:*** $\vec{\mathbf{B}}$ (magnitude and direction)

From Equation 19.12, the magnitude of the field 1.0 cm below the wire is

$$B = \frac{\mu_o I}{2\pi d} = \frac{(4\pi \times 10^{-7}\ [\text{T}\cdot\text{m}]/\text{A})(15\ \text{A})}{2\pi(0.010\ \text{m})} = 3.0 \times 10^{-4}\ \text{T}$$

By the source rule (see Figure 19.19), the field direction directly below the wire is north. [Note: The fingers of the right hand, below the wire, point north.]

**FOLLOW-UP EXERCISE.** (a) In this Example, what is the field direction 5.0 cm above the wire? (b) What current is needed to produce a magnetic field at this location with one-half the field strength in the Example?

### 19.6.2 Magnetic Field at the Center of a Circular Coil

At the center of a circular coil of wire of $N$ loops, each of radius $r$ and each carrying the same current $I$ (▼ **Figure 19.21a** shows one such loop), the magnitude of the magnetic field is

$$B = \frac{\mu_o N I}{2r} \quad \text{(magnetic field at center of circular coil of } N \text{ loops)} \qquad (19.13\text{a})$$

In this case (and for all *circular* current arrangements, such as solenoids, discussed next), it is convenient to determine the magnetic field direction using the right-hand source rule that is slightly different from (but equivalent to) the one for straight wires (see Figure 19.21a):

> If a current-carrying coil/loop is grasped with the right hand so the fingers are curled in the direction of the current, the magnetic field direction *inside* the area of the coil/loop is the direction in which the extended thumb points.

In all cases, the magnetic field lines are always closed curves, the direction of which is determined by the right-hand source rule.

As seen before, the direction of $\vec{\mathbf{B}}$ is tangent to the field line and will depend upon location. Notice that the overall field pattern of the loop is geometrically similar to that of a bar magnet (shown overlaid and shadowy in Figure 19.21b). This is not a coincidence and will be discussed later in the chapter.

More generally, on the central axis of a circular coil of wire consisting of $N$ loops (Figure 19.21c shows this location $P$ for one loop), each of radius $r$ and each carrying a current $I$, the magnitude of $\vec{\mathbf{B}}$ varies with the distance $x$ from the loops' center according to

$$B = \frac{\mu_o N I r^2}{2(r^2 + x^2)^{3/2}} \quad \text{(magnetic field on central axis of circular coil of } N \text{ loops)} \qquad (19.13\text{b})$$

### 19.6.3 Magnetic Field Inside a Current-Carrying Solenoid

A *solenoid* is constructed by wrapping a long wire to form a coil, or *helix*, with many circular loops, as in ▼ **Figure 19.22**. If the solenoid's radius is small compared to its length ($L$), the interior magnetic field is parallel to the solenoid's longitudinal axis and constant in magnitude. Notice how the solenoid's external field resembles that of a permanent bar magnet.

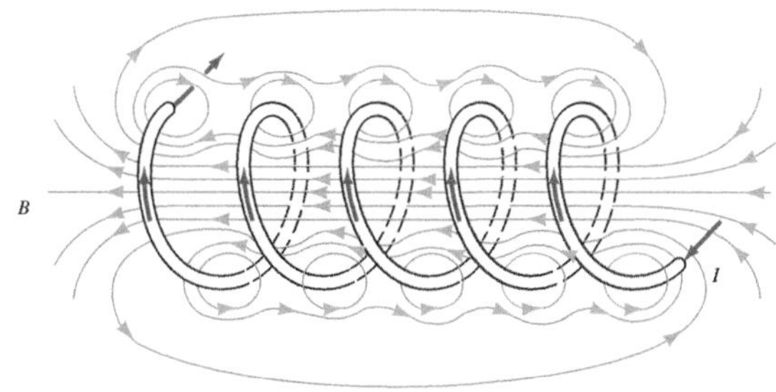


▲ FIGURE 19.22 **Magnetic field of a solenoid** The magnetic field of a solenoid is fairly uniform near the central axis of the solenoid. The direction of the field in the interior can be determined by applying the right-hand source rule to any of the loops.

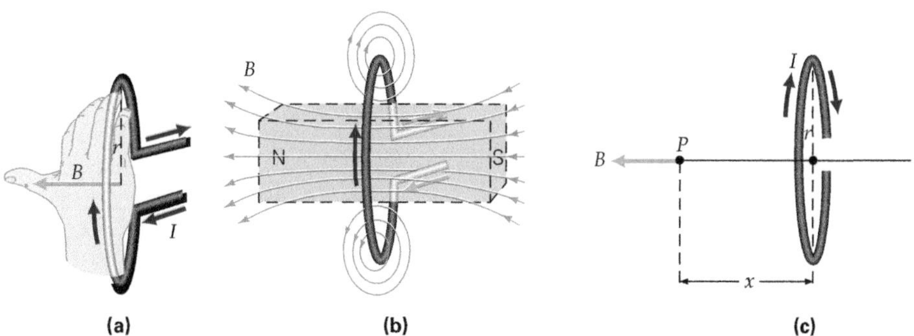


▲ FIGURE 19.21 **Magnetic field due to a circular current-carrying loop (a)** The direction of the field in the interior of the loop is given by a right-hand source rule. With the fingers wrapped around the loop in the direction of the conventional current, the thumb indicates the direction of $\vec{\mathbf{B}}$ inside the loop. Note that the magnetic field at the loop's center is perpendicular to the loop's plane. **(b)** The overall magnetic field of a current-carrying circular loop is similar to that of a bar magnet. **(c)** The magnetic field on the central axis of a current-carrying loop. See text discussion for more details.

As usual, the direction of the interior field is given by the right-hand source rule for circular geometry. If the solenoid has $N$ turns and each carries a current $I$, the magnitude of the magnetic field near its center is given by

$$B = \frac{\mu_o NI}{L} \quad \text{(magnetic field near the center of a solenoid)} \tag{19.14}$$

Note that the solenoid's interior field depends on how close the coils are together, or the coil density, $N/L$. To quantify this, the *linear turn density* $n$ is defined as $n = N/L$. The units of $n$ are turns per meter (in SI terms, this is $m^{-1}$). Equation 19.14 is sometimes rewritten in terms of the turn density as $B = \mu_o nI$.

To see how the solenoid might be well suited for magnetic applications requiring a large and fairly uniform magnetic field, consider Example 19.7.

### EXAMPLE 19.7: WIRE VERSUS SOLENOID – CONCENTRATING THE MAGNETIC FIELD

A solenoid is 0.30 m long with 300 turns and carries a current of 15.0 A. (a) What is the magnitude of the magnetic field near its center? (b) Compare your result with the field near the single wire in Example 19.6, which carries the same current, and comment.

**THINKING IT THROUGH.** The field strength depends on the number of turns ($N$), the solenoid length ($L$), and the current ($I$). This is a direct application of Equation 19.14.

#### SOLUTION

***Given:***

$I = 15.0$ A
$N = 300$ turns
$L = 0.30$ m

***Find:***

**(a)** $B$ (magnitude of magnetic field near the solenoid center)
**(b)** Compare the answer in part (a) with that from a long, straight wire in Example 19.6

**(a)** From Equation 19.14,

$$B = \frac{\mu_o NI}{L} = \frac{(4\pi \times 10^{-7}\ \text{T}\cdot\text{m/A})(300)(15.0\ \text{A})}{0.30\,\text{m}}$$
$$= 6\pi \times 10^{-3}\ \text{T} \approx 18.8\ \text{mT}$$

**(b)** This field is more than sixty times larger than the field produced by the wire in Example 19.6. Winding many loops close together in a helix fashion increases the field while employing the same current. This is because the solenoid's field is the vector sum of the fields from 300 loops, and since individual loop field directions are all approximately the same, this results in a larger net field for the same current.

**FOLLOW-UP EXERCISE.** In this Example, if the current was reduced to 1.0 A and the solenoid shortened to 10 cm, how many turns would be needed to create the same magnetic field?

Integrated Example 19.8 uses most of the aspects of electromagnetism discussed so far: the production of magnetic fields by electric currents and the resulting forces those fields can exert on other electric currents. Study the example carefully, especially the use of the appropriate right-hand rules (source and force).

### INTEGRATED EXAMPLE 19.8: ATTRACTION OR REPULSION? MAGNETIC FIELDS OF, AND FORCES ON, PARALLEL WIRES

Two long, parallel wires carry currents in the same direction, as illustrated in ▼**Figure 19.23a**. (a) Is the magnetic force between these wires (1) attractive or (2) repulsive? Make a sketch to show how you obtained your result. (b) Wire 1 has a current of 5.0 A and the current in wire 2 is 10 A. Both have a length of 50 cm, and they are separated by 3.0 mm. Determine the magnitude of the magnetic field created by each wire at the other wire's location. (c) Determine the magnitude of the magnetic force that each wire exerts on the other.

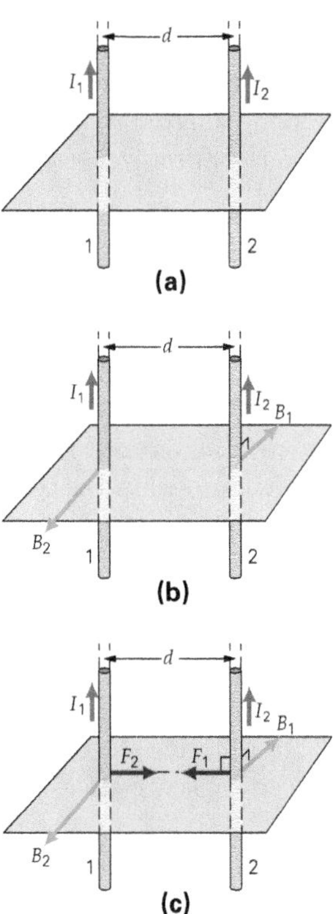


▲FIGURE 19.23 **Mutual interaction of parallel current-carrying wires (a)** Two parallel wires carry current in the same direction. **(b)** Wire 1 creates a magnetic field at the site of wire 2, and vice versa. **(c)** The wires exert equal but opposite (attractive) forces on each other. (See Integrated Example 19.8 for details.)

(a) **CONCEPTUAL REASONING.** Choose one wire at a time and determine the direction of the magnetic field it produces at the location of the other wire using the source right-hand rule. In Figure 19.23b, the fields from both wires ($\vec{\mathbf{B}}_1$ and $\vec{\mathbf{B}}_2$), *at the location of the other wire,* are shown. [Note that $\vec{\mathbf{B}}_2$ is represented by a longer arrow. Why?]

Next, use the right-hand *force* rule on each wire to determine the direction of the force on each. The result, in Figure 19.23c, is an attractive force on each wire. So (1) is the correct answer. [In keeping with Newton's third law, these forces are represented by arrows of the same length and oppositely directed. The magnitudes should be the same and are found in (c).]

(b, c) **QUANTITATIVE REASONING AND SOLUTION.** To find the magnitude of the field produced by each wire, Equation 19.12 is used. Because both fields are at right angles to the currents in the wires, the magnitude of the force on each wire is a maximum and is given by *ILB*. Be careful to use the appropriate field and current, that is, watch the subscripts carefully! The symbols are in Figure 19.23. Listing the data and converting to SI units:

***Given:***

$I_1 = 5.0$ A
$I_2 = 10$ A
$d = 3.0$ mm $= 3.0 \times 10^{-3}$ m
$L = 50$ cm $= 0.50$ m

***Find:***

(b) $B_1$ and $B_2$ (the magnitude of the fields due to each wire)
(c) $F_1$ and $F_2$ (the magnitude of the magnetic forces on each wire)

(b) The magnitude of the magnetic field due to wire 1 at the location of wire 2 is

$$B_1 = \frac{\mu_o I_1}{2\pi d} = \frac{(4\pi \times 10^{-7}\ \text{T·m/A})(5.0\ \text{A})}{2\pi(3.0 \times 10^{-3}\ \text{m})} = 3.3 \times 10^{-4}\ \text{T}$$

The magnitude of the magnetic field due to wire 2 at the location of wire 1 is

$$B_2 = \frac{\mu_o I_2}{2\pi d} = \frac{(4\pi \times 10^{-7}\ \text{T·m/A})(10\ \text{A})}{2\pi(3.0 \times 10^{-3}\ \text{m})} = 6.6 \times 10^{-4}\ \text{T}$$

(c) The magnitude of the magnetic force on wire 1 due to the field created by wire 2 is

$$F_1 = I_1 L B_2 = (5.0\ \text{A})(0.50\ \text{m})(6.6 \times 10^{-4}\ \text{T}) = 1.7 \times 10^{-3}\ \text{N}$$

The magnitude of the magnetic force on wire 2 due to the field created by wire 1 is

$$F_2 = I_2 L B_1 = (10\ \text{A})(0.50\ \text{m})(3.3 \times 10^{-4}\ \text{T}) = 1.7 \times 10^{-3}\ \text{N}$$

As expected, the forces are the same magnitude, in keeping with Newton's third law.

(A question about these results: wire 2 is subject to a weaker field than wire 1. How can it then experience the same force as wire 1?)

**FOLLOW-UP EXERCISE.** (a) In this Example, determine the force direction if the current in just one of the wires is reversed. (b) If the magnitude of the force on the wires is to be kept the same as in the Example, but the current in each tripled, how far apart must the wires be?

The magnetic force between parallel wires such as in Integrated Example 19.8 provides the modern basis for the definition of the ampere. The National Institute of Standards and Technology (NIST) defines the ampere* as follows.

One ampere is a current that, if maintained in each of two long parallel wires separated by a distance of exactly 1 m in free space, produces a magnetic force between the wires of exactly $2 \times 10^{-7}$ N per meter of wire.

## 19.7 Magnetic Materials

Why is it that some materials are magnetic or easily magnetized and others are not? How can a bar magnet create a magnetic field when it carries no obvious current? To answer these questions, let's start with some basics. It is known that an electric current produces a magnetic field. If the magnetic fields of a bar magnet and a long solenoid are compared (see Figures 19.1 and 19.23), it seems logical that the magnetic field of the bar magnet might be due to *internal* currents. Perhaps these "invisible" currents are due to electrons orbiting the atomic nucleus or electron spin. Detailed analysis of atomic structure shows that the net magnetic field produced by *orbital motion* is usually zero or very small.

What then *is* the source of the magnetism produced by magnetic materials? Modern atomic quantum theory tells us that the permanent type of magnetism, like that exhibited by an iron bar magnet, is produced by *electron spin.* Classical physics likens a spinning electron to the Earth rotating on its axis. However, this mechanical analog is *not* actually the case, since electron spin is a quantum mechanical concept with no direct classical analog. Nonetheless, the picture of spinning electrons creating magnetic fields is useful for qualitative thinking and reasoning. In effect, each "spinning" electron produces a field similar to a current loop (Figure 19.21b). This pattern, resembling that produced by a small bar magnet, enables us to treat electrons, magnetically speaking, as tiny bar magnets or compass needles.

In multi-electron atoms, the electrons usually are arranged in pairs with their spins oppositely aligned. In this case, their fields will effectively cancel, and the material will not be magnetic. Aluminum is such a material.

However, in certain materials, known as **ferromagnetic materials**, the fields due to electron spins in individual atoms do not cancel. Thus each atom possesses a magnetic moment. There is a strong interaction between these neighboring moments that leads to the formation of regions called **magnetic domains.**

* While in the final production stages of this book (2019), the International Committee for Weights and Measures has adopted a new definition of the ampere based on a fundamental constants, such as *e* the elementary charge unit.

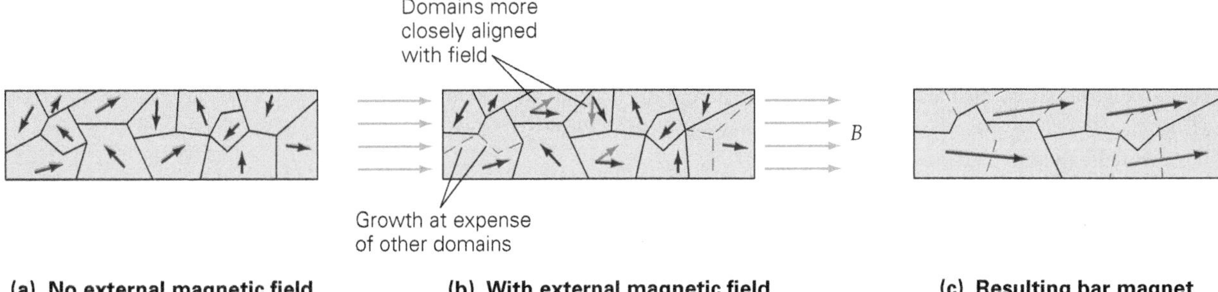


▲ FIGURE 19.24 **Magnetic domains (a)** With no external magnetic field, the magnetic domains of a ferromagnetic material are randomly oriented and the material is unmagnetized. **(b)** In an external magnetic field, domains with orientations parallel to the field may grow at the expense of other domains, and the orientations of some domains may become more aligned with the field. **(c)** As a result, the material becomes magnetized.

In a given domain, the electron spin moments are aligned in approximately the same direction, producing a relatively strong (net) magnetic field within that domain. Not many ferromagnetic materials occur naturally. The most common are iron, nickel, and cobalt. Gadolinium and certain manufactured alloys, such as neodymium and other rare earth alloys, are also ferromagnetic.

In an unmagnetized ferromagnetic material, the domains are randomly oriented and there is no net magnetization (▲ **Figure 19.24a**). But when a ferromagnetic material (such as an iron bar) is placed in an external magnetic field, the domains change their orientation and size (Figure 19.24b). Remembering the picture of the electron spin acting as a small compass, they begin to "line up" in the external field. As the external field and the iron bar interact, the iron exhibits the following two effects:

1. Domain boundaries change as some of the domains with magnetic orientations in the direction of the external field grow at the expense of the others.
2. The magnetic orientation of some domains may change slightly so as to be more aligned with the field.

Upon removal of the external field, the iron domains will remain more or less aligned in the original external field direction, thus creating an overall "permanent" magnetic field of their own.

Now you can also understand why an unmagnetized piece of iron is attracted to a magnet and why iron filings line up with a magnetic field. Essentially, the pieces of iron become induced magnets (Figure 19.24c).

## 19.7.1 Electromagnets and Magnetic Permeability

Ferromagnetic materials are used to make electromagnets, usually by wrapping a wire around an iron core (▶ **Figure 19.25a**). When the current is on, it induces magnetism in ferromagnetic materials (for example, the iron sliver in Figure 19.25b) and, if the forces are large enough, it can be used to pick up large amounts of scrap iron (Figure 19.25c). When an electromagnet is on (lower drawing in Figure. 19.25a), the iron core is magnetized and adds to the field of the solenoid. The total field can be written as

$$B = \frac{\mu NI}{L} \quad \text{(field at center of the iron core solenoid)} \qquad (19.15)$$

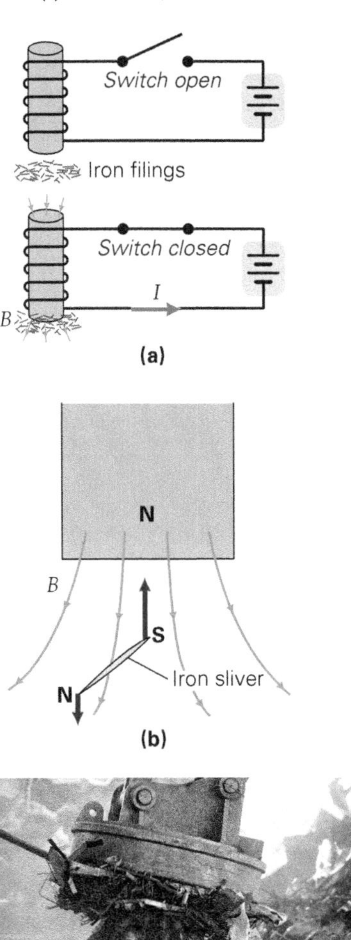


▲ FIGURE 19.25 **Electromagnet (a)** (top) With no current in the circuit, there is no magnetic force. (bottom) However, with a current in the coil, there is a magnetic field and the iron core becomes magnetized. **(b)** Detail of the lower end of the electromagnet in part (a). The sliver of iron is attracted to the end of the electromagnet. **(c)** An electromagnet picking up scrap metal.

This expression is nearly identical to that for the field of an air core solenoid Equation 19.14, *except* that it contains $\mu$ instead of $\mu_o$, the permeability of free space. Here, $\mu$ represents the **magnetic permeability** of the core material, not free space. The role permeability plays in magnetism is similar to that of electric permittivity for electric fields (Chapter 16). Magnetic permeability is defined in comparison to the free space value as:

$$\mu = \kappa_m \mu_o \tag{19.16}$$

where $\kappa_m$ is the *relative* permeability (dimensionless) in analogy to the dielectric constant $\kappa$.

For ferromagnetic materials, the total magnetic field far exceeds that from the wire wrapping alone in free space. Thus for ferromagnetic materials, $\mu \gg \mu_o$ and $\kappa_m \gg 1$. Typically, a ferromagnetic core in an electromagnet can enhance its field thousands of times compared with an air core alone. In other words, ferromagnetic materials have values of $\kappa_m$ on the order of thousands. To see the effect of ferromagnetic materials, refer to Example 19.9.

EXAMPLE 19.9: MAGNETIC ADVANTAGE – USING FERROMAGNETIC MATERIALS

A laboratory solenoid with 200 turns in a length of 30 cm is limited to carrying a maximum current of 2.0 A. The scientists in the lab need an interior magnetic field strength of at least 2.0 T and are debating whether they need to employ a ferromagnetic core. (a) Is their field possible if no material fills its core? (b) If not, determine the minimum magnetic permeability of the ferromagnetic material that should comprise the core in order to meet their needs.

**THINKING IT THROUGH.** The field strength depends upon the number of turns ($N$), solenoid length ($L$), current ($I$), and the permeability of the core material ($\mu$). This is a direct application of Equations 19.14 and 19.15.

SOLUTION

***Given:***

$I = 2.0$ A
$N = 200$ turns
$L = 0.30$ m

***Find:***

**(a)** $B$ (is 2.0 $T$ possible with no core material?)
**(b)** $\mu$ (magnetic permeability required to attain $B = 2.0$ $T$)

**(a)** From Equation 19.14, with no core, the interior field clearly would not be large enough, since:

$$B = \frac{\mu_o NI}{L} = \frac{(4\pi \times 10^{-7}\ \text{T·m/A})(200)(2.0\ \text{A})}{0.30\ \text{m}}$$
$$= 1.7 \times 10^{-3}\ \text{T} = 1.7\ \text{mT}$$

**(b)** The required field is 2.0 T/($1.7 \times 10^{-3}$ T) or about 1200 times larger than the answer to (a). Thus, since $B \propto \mu$ (if everything else is kept constant), attaining a field of at least 2.0 T requires a permeability of $\mu \geq 1200\mu_o$ or $\mu \geq 1.5 \times 10^{-3}$ T·m/A.

**FOLLOW-UP EXERCISE.** In this Example, if the scientists found a way for the solenoid to handle up to 5.0 A, what would be the minimum relative permeability?

The type of iron that retains some of its magnetism after being in an external magnetic field is called *hard iron* and is used to make so-called permanent magnets. You may have noticed that a paper clip or a screwdriver blade becomes slightly magnetized after being near a magnet. Permanent magnets are produced by heating pieces of ferromagnetic material in an oven and then allowing them to cool while in a strong external magnetic field. In permanent magnets, the domains remain largely aligned when the external field is removed.

However, a *permanent* magnet is not truly permanent, because its magnetism can be diminished or destroyed. Hitting such a magnet or dropping it on the floor can cause a loss of some domain alignment, reducing the magnet's field. Another mechanism for loss of magnetism is to increase the thermal motions of atoms and thus their spin directions. This tends to disrupt domain alignment. One of the worst things you can do to a credit card's magnetic strip is to leave the card in the direct sun or a hot environment. The increased thermal motion of the electron spins can partially destroy the magnetic pattern on the strip. Above a certain critical temperature, called the **Curie temperature** (or Curie point), domain coupling is destroyed by these increased thermal oscillations, and a ferromagnetic material loses its "permanent" magnetism. This effect was first discovered by French physicist Pierre Curie (1859–1906 and husband to Madame Curie of nuclear radioactivity fame). For example, the Curie temperature for iron is 770 °C.

Ferromagnetic domain alignment plays an important role in geology and geophysics. For example, it is well known that when cooled, lava flows that initially contained iron above its Curie temperature can retain some residual magnetism due to being in the Earth's field as it existed when the lava cooled below the Curie temperature. Measuring the strength and orientation of older lava flows at various locations has enabled geophysicists to map the changes in the Earth's magnetic field and polarity over time.

This technique provided geophysicists with some of the first evidence to support the theory of plate tectonic motion. They discovered that the seafloor near the mid-Atlantic ridge, for example, is composed of a series of lava flows from underwater volcanoes. These solidified flows were found to exhibit permanent magnetism, but with polarity that varied with time (older samples are farther out from the ridge) as the Earth's magnetic polarity changed (reversed).

## 19.8 Geomagnetism

The magnetic field of the Earth was used for centuries before people had any clues about its origin. It is known that in ancient times, navigators used lodestones or magnetized needles to locate north. An early study of magnetism was first carried out by the English scientist Sir William Gilbert in the 1600s. He theorized that a large body of permanently magnetized material within the Earth might produce its field.

In fact, the Earth's external magnetic field, known as the *geomagnetic field*, does have a configuration similar to that which would be produced by a large interior bar magnet with the south pole of the magnet near the north geographic pole (▼**Figure 19.26**). The magnitude of the Earth's (horizontal) magnetic field at the equator is on the order of $10^{-5}$ T (about 0.4 G), and the (vertical) field at the poles is about $10^{-4}$ T (roughly, 1 G). The idea of a solid permanent ferromagnet of within the Earth may not seem unreasonable at first, but this cannot be correct, because it is known that the temperatures deep inside the Earth are well above the Curie temperatures of iron and nickel, which are the ferromagnetic materials believed to be the most abundant in the Earth's interior. So the existence of a permanent solid internal Earth magnet is not possible.

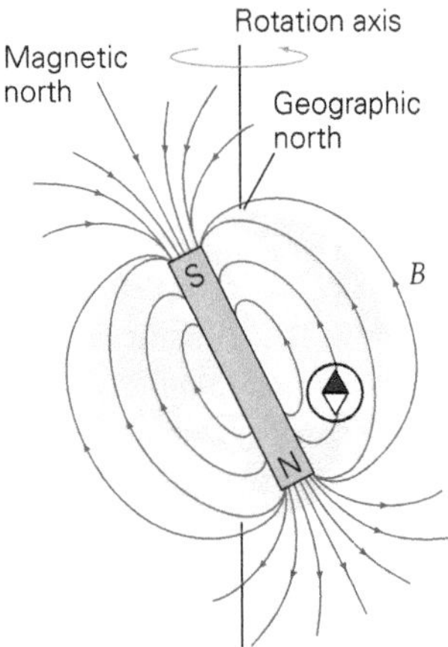


▲**FIGURE 19.26 Geomagnetic field** The Earth's magnetic field has a similar structure to that of a bar magnet. However, a permanent solid magnet could not exist within the Earth because of the high temperatures there. The Earth's magnetic field is believed to be associated with motions in the liquid outer core deep within the planet.

Knowing that electric currents produce magnetic fields has led scientists to believe that the Earth's field is associated with motions in the liquid outer core, which, in turn, may be connected in some way with the Earth's rotation. It is known, for example, that Jupiter, a planet that rotates rapidly, has a magnetic field much larger than Earth's. On the other hand, Mercury and Venus have relatively weak fields; these planets are more like Earth and rotate relatively slowly. Several models have been proposed to explain the Earth's magnetic field, but the details of its mechanism are still not clear.

It is important to note that the axis of the Earth's magnetic field does *not* coincide with the its rotational axis, which defines the geographic poles. Hence, the Earth's (south) magnetic pole and the geographic North Pole do not coincide (see Figure 19.26). Currently the south magnetic pole is about a thousand kilometers south of the geographic North Pole (true north). The Earth's north magnetic pole is displaced even more from its south geographic pole, meaning that the magnetic axis does not even pass through the center of the Earth.

A compass indicates the direction of magnetic north, not "true," or geographic, north. The angular difference in these two directions is called the magnetic declination (▼**Figure 19.27**). The magnetic declination varies with location, thus knowing these variations has historically been particularly important for accurate navigation. Most recently, with the advent of accurate GPSs (global positioning systems), high-tech travelers no longer have to depend on compasses.

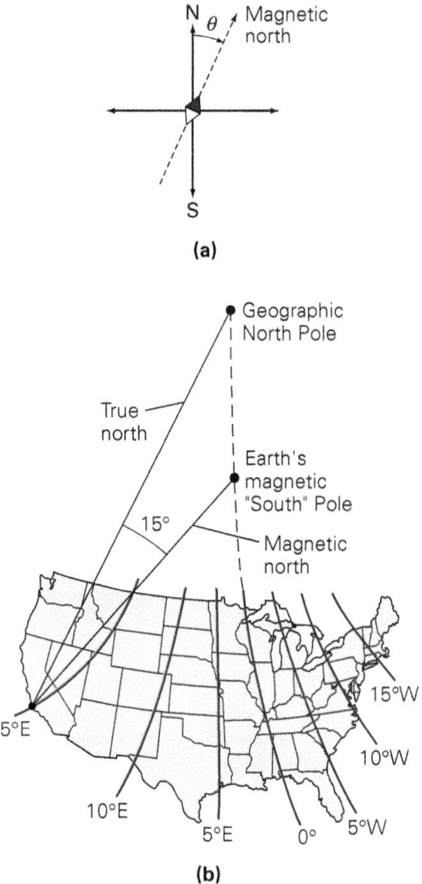


▲**FIGURE 19.27 Magnetic declination (a)** The angular difference between magnetic north and "true," or geographic, north is called the magnetic declination. **(b)** The magnetic declination varies with location and time. The map shows *isogonic* lines (lines with the same magnetic declination) for the continental United States. For locations on the 0° line, magnetic north is in the same direction as true (geographic) north. On either side of this line, a compass has an easterly or westerly variation. For example, on a 15°E line, a compass has an easterly declination of 15°. (Magnetic north is 15° east of true north.)

The Earth's magnetic field exhibits a variety of fluctuations with time. For example, it is known that Earth's magnetic poles have switched polarity at various times in the past, most recently about 700,000 years ago. The mechanism for periodic magnetic polarity reversal is not clearly understood, and scientists are still investigating it.

On a much shorter time scale, the magnetic poles tend to "wander," or change location. For example, the Earth's south magnetic pole (the north polar region) has recently been moving about 1° in latitude per decade. This long-term drift means that the magnetic declination map (Figure 19.27b) must be updated periodically.

## Chapter 19 Review

- The **pole-force law**, or **law of poles**: opposite magnetic poles attract and like poles repel.
- **The magnetic field** ($\vec{B}$) has SI units of the **tesla** (T), where $1T = 1N/(A{\cdot}m)$ and can exert forces on moving charged particles and electric currents.
- **The magnitude of the magnetic force** ($F_m$) on a charged **particle** where $\theta$ is the smallest angle between its velocity and the field direction is

$$F_m = qvB\sin\theta \tag{19.3}$$

- **The magnitude** of the magnetic force on a current-carrying wire where $\theta$ is the smallest angle between the conventional current direction and the field direction is

$$F_m = ILB\sin\theta \tag{19.7}$$

- **Right-hand force rules** determine the direction of the magnetic force on moving charged particles and current-carrying wires.

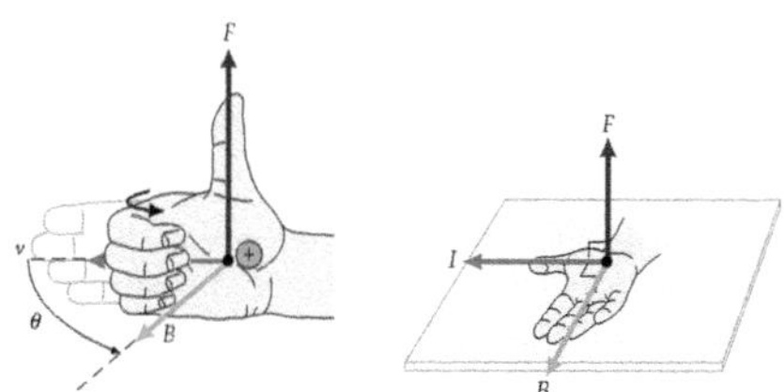


- A series connection of $N$ current-carrying circular loops, each with plane area $A$ and carrying a current $I$ can experience a **magnetic torque** when placed in a magnetic field. If $\theta$ is the smallest angle between the normal to the loop area and field, then the magnitude of such a torque is

$$\tau = NIAB\sin\theta \tag{19.9}$$

- The magnitude of the **magnetic field produced by a long, straight wire** is

$$B = \frac{\mu_o I}{2\pi d} \tag{19.12}$$

where $\mu_o = 4\pi \times 10^{-7}\,(\text{T·m})/\text{A}$ is the **magnetic permeability of free space.**
- The magnitude of the **magnetic field at the center of *N* circular current-carrying loops** of radius $r$ is

$$B = \frac{\mu_o NI}{2r} \tag{19.13a}$$

- The magnitude of the **magnetic field** on the central axis of ***N* circular current-carrying loops** of radius $r$ is

$$B = \frac{\mu_o NIr^2}{2(r^2 + x^2)^{3/2}} \tag{19.13b}$$

- The magnitude of the **magnetic field** produced near the center of the interior of a **solenoid** with $N$ windings and a length $L$ is

$$B = \frac{\mu_o NI}{L} \tag{19.14}$$

- **Right-hand source rules** determine the direction of the magnetic field due to various current configurations.

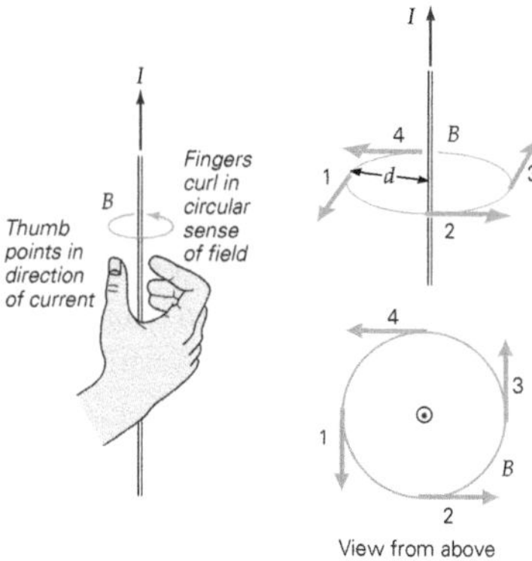


- In **ferromagnetic materials**, the electron spins align, creating **domains**. When an external field is applied, the effect is to increase the size of those domains that already point in the direction of the field at the expense of the others. When the external magnetic field is removed, a **permanent magnet** remains.

## End of Chapter Questions and Exercises

### Multiple Choice Questions

**19.1 Permanent Magnets, Magnetic Poles, and Magnetic Field Direction**

1. When the ends of two bar magnets are close, they attract. These ends must be (a) one north, one south, (b) both north, (c) both south, (d) either (b) or (c).
2. A compass is just above the end of a permanent bar magnet, and points away from that end. It can be concluded that this end acts as (a) a north magnetic pole, (b) a south magnetic pole, (c) you can't conclude anything about the magnetic properties of that end.
3. If you look directly at the south pole of a bar magnet, its magnetic field points (a) to your right, (b) to your left, (c) away from you, (d) toward you.
4. Which way would a compass point if placed midway between the ends of the two bar magnets shown in Figure 19.3b: (a) up, (b) down, (c) left, or (d) right?

5. Which way would a compass point if placed just to the right of the midway point between the ends of the two magnets in Figure 19.3c: (a) up, (b) down, (c) left, or (d) right?

### 19.2 Magnetic Field Strength and Magnetic Force

6. A proton moves vertically upward and perpendicular to a uniform magnetic field. It deflects to the left as you watch it. What is the magnetic field direction: (a) directly away from you, (b) directly toward you, (c) to the right, or (d) to the left?
7. An electron moves horizontally east in a vertical uniform magnetic field. It is found to deflect north. What direction is the field: (a) up, (b) down, or (c) the direction can't be determined from the given data?
8. If a negatively charged particle were moving upward along the right edge of this page, which way should a magnetic field (perpendicular to the plane of the paper) be oriented so that the particle would initially be deflected left: (a) out of the page, (b) in the plane of the page, or (c) into the page?
9. An electron passes through a magnetic field without being deflected. What can you conclude about the magnetic field direction relative to the electron's velocity, assuming that no other forces act: (a) it could be in the same direction as the velocity, (b) it could be perpendicular to the velocity, (c) it could be directly opposite the velocity, or (d) both (a) and (c) are possible?

### 19.3 Applications: Charged Particles in Magnetic Fields

10. In a mass spectrometer two ions with identical charge and speed form two different semicircular arcs. Ion A's arc has a radius of 25.0 cm and ion B's arc has a radius of 50.0 cm. What can you say about their relative masses: (a) $m_A = m_B$, (b) $m_A = 2m_B$, (c) $m_A = \frac{1}{2}m_B$, or (d) you can't say anything about their relative masses from this data?
11. In a mass spectrometer two ions with identical mass and speed are accelerated form two different semicircular arcs. Ion A's arc has a radius of 25.0 cm and ion B's arc has a radius of 50.0 cm. What can you say about their net charges: (a) $q_A = q_B$, (b) $q_A = 2q_B$, (c) $q_A = \frac{1}{2}q_B$, or (d) you can't say anything about their relative charges from this data?
12. In the velocity selector shown in Figure 19.9, which way will an ion be deflected if its velocity is less than $E/B_1$: (a) up, (b) down, or (c) there will be no deflection?

### 19.4 Magnetic Forces on Current-Carrying Wires and

### 19.5 Applications: Current-Carrying Wires in Magnetic Fields

13. A long, straight, horizontal wire located on the equator carries a current directed west. What is the direction of the force on the wire due to the Earth's magnetic field: (a) east, (b) west, (c) south, or (d) downward?
14. A long, straight, horizontal wire located on the equator carries a current. In what direction should the current be if the purpose is to balance the wire's weight with the magnetic force on it (a) east, (b) west, (c) south, or (d) upward?
15. You are looking horizontally due west directly and perpendicularly at the circular plane of a current-carrying coil. The coil is in a uniform vertically upward magnetic field. When released, the top of the coil starts to rotate away from you as the bottom rotates toward you. Which direction is the current in the coil: (a) clockwise, (b) counterclockwise, or (c) you can't tell from the data given?
16. For the current-carrying loop shown in Figure 19.16, the maximum torque occurs for what value of the angle $\theta$: (a) 0°, (b) 90°, (c) 180°, or (d) the torque is the same for all angles?

### 19.6 Electromagnetism: Currents as a Magnetic Field Source

17. A long, straight wire is parallel to the ground and carries a steady current to the west. At a point directly below the wire, what is the direction of the magnetic field the wire produces: (a) north, (b) east, (c) south, or (d) west?
18. You are looking directly into one end of a long solenoid. The magnetic field at its center points directly away from you. What is the direction of the current in the solenoid, as viewed by you: (a) clockwise, (b) counterclockwise, (c) directly toward you, or (d) directly away from you?
19. A current-carrying loop of wire is in the plane of this paper. Outside the loop, its magnetic field points into the paper. What is the direction of the current in the loop: (a) clockwise, (b) counterclockwise, or (c) you can't tell from the data given?
20. Consider a current-carrying circular loop of wire. On its central axis (that is, the line perpendicular to the area of the loop and passing through its center), which location has the least magnetic field strength: (a) the center of the loop, (b) 10 cm above the center of the loop, or (c) 20 cm above the center of the loop?

### 19.7 Magnetic Materials

21. The main source of magnetism in magnetic materials is (a) electron orbits, (b) electron spin, (c) magnetic poles, (d) nuclear properties.
22. When a ferromagnetic material is placed in an external magnetic field, (a) the domain orientation may change, (b) the domain boundaries may change, (c) new domains are created, (d) both (a) and (b).
23. Heating a permanent magnet can significantly reduce the magnetic field it produces. The reduction in field strength by this method is mainly due to (a) domain orientation change, (b) domain boundary change, (c) increase thermal motion of electron spin directions, (d) all of these.

**19.8 Geomagnetism: The Earth's Magnetic Field**

24. The Earth's magnetic field (a) has poles that coincide with the geographic poles, (b) only exists at the poles, (c) reverses polarity every few hundred years, (d) none of these.
25. If a proton orbits just above the Earth's equator, which way must it be moving for the magnetic force to help the force of gravity: (a) west, (b) east, or (c) either direction?
26. If the direction of your compass pointed straight up, where would you be: (a) near the Earth's north geographic pole, (b) near the equator, or (c) near the Earth's south geographic pole?

## Conceptual Questions

**19.1 Permanent Magnets, Magnetic Poles, and Magnetic Field Direction**

1. Given two identical iron bars, one of which is a permanent magnet and the other unmagnetized, how could you tell which is which by using only the two bars?
2. The direction of any magnetic field is taken to be in the direction that a compass points. Explain why this means that magnetic field lines must leave from the north pole of a permanent bar magnet and enter its south pole.
3. (a) As you start in the very middle of Figure 19.3b and move horizontally to the right, what happens to the magnetic field line spacing as indicated by the iron filing pattern? What does this imply for the field strength? (b) What is the direction of the field in this region of the figure? Can you tell the direction from the iron filing pattern alone? Explain.

**19.2 Magnetic Field Strength and Magnetic Force**

4. A proton and electron are moving at the same velocity perpendicularly to a constant magnetic field. (a) How do the magnitudes and directions of the magnetic forces on them compare? (b) What about the magnitudes of their accelerations?
5. If a charged particle moves in a straight line and there are no other forces on it except possibly from a magnetic field, can you say with certainty that no magnetic field is present? Explain.
6. Three particles with the same velocity enter the same uniform magnetic field as shown in ▶ **Figure 19.28a**. What can you say about (a) the charges of the particles and (b) their masses?
7. You want to deflect a positively charged particle in an S-shaped path, as shown in Figure 19.28b, using only magnetic fields. (a) Explain how this could be done by using magnetic fields perpendicular to the plane of the page. (b) How would the emerging particle's kinetic energy compare with its initial kinetic energy?
8. A magnetic field can be used to determine the *sign* of charge carriers in a current-carrying wire. Consider a wide conducting strip in a magnetic field oriented as shown in ▼ **Figure 19.29**. The charge carriers are deflected by the magnetic force and accumulate on one side of the strip, giving rise to a measurable voltage across it. (This is known as the *Hall effect* and the voltage is the *Hall voltage*.) If the sign of the charge carriers is unknown (they are either positive charges moving as indicated by the arrows in the figure or negative charges moving in the opposite direction), discuss how the polarity of the Hall voltage allows the sign of the charge to be determined. Assume that only one type of charge carrier is responsible for the current and consider each case.

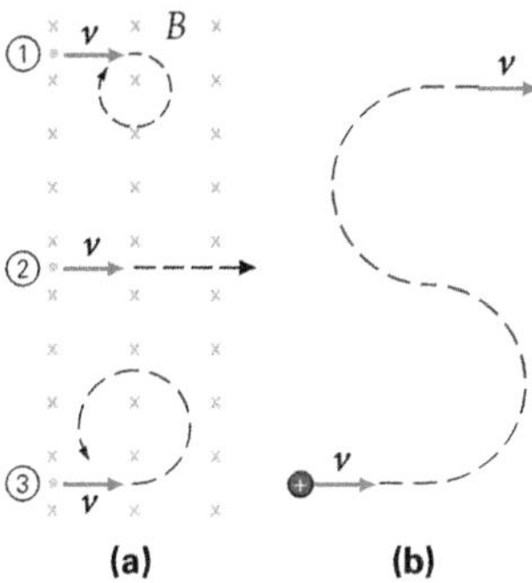


▲ FIGURE 19.28 **Charges in motion** See Conceptual Questions 6 and 7.

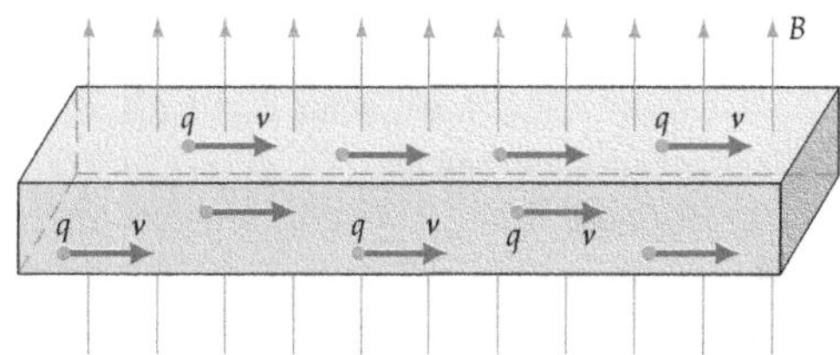


▲ FIGURE 19.29 **The Hall effect** See Conceptual Question 8.

**19.3 Applications: Charged Particles in Magnetic Fields**

9. Explain how particle accelerator operators utilize magnetic fields to "steer" various charged particles in the correction to a specific target for a particular experimental setup.
10. The circular inset in Figure 19.11 shows how positive sodium ($Na^+$) ions in seawater are accelerated out the rear of the submarine to provide a propulsive force. But what about the negative chlorine ($Cl^-$) ions? Because they have the opposite charge, aren't they accelerated *forward*, resulting in a net force of zero on the submarine? Explain.
11. Explain clearly why the speed selected in a velocity selector setup such as in Figure 19.9 does *not* depend on the charges of the ions passing through.
12. Redraw the charged particle path in the apparatus diagrammed in Figure 19.9 if the ions were negatively charged instead of positively charged.

13. (a) Redraw the charged particle path in the apparatus diagrammed in Figure 19.9 if the electric field of the velocity selector were reduced in magnitude. (b) Redraw the charged particle path in the apparatus diagrammed in Figure 19.9 if the magnetic field of the velocity selector were reduced in magnitude. In both cases, explain your reasoning.

**19.4 Magnetic Forces on Current-Carrying Wires and**
**19.5 Applications: Current-Carrying Wires in Magnetic Fields**

14. Two parallel straight wires carry different currents in the same direction. Do they attract or repel each other? How do the magnitudes of these forces on each wire compare?
15. Predict what should happen to the length of a metal spring when a large current passes through it. [*Hint:* Consider the direction of the current in the neighboring spring coils.]
16. (a) How would you orient a square current loop in a uniform magnetic field so that there is no torque on the loop? (b) How would the orientation change to maximize the torque on the loop? (c) In each case, is there a net magnetic force on this loop? Explain.
17. (a) Show that the SI unit for magnetic moment multiplied by the SI unit for magnetic field yields the SI unit for torque. (b) If you are looking down onto the area of a current-carrying loop of wire and the current is counterclockwise, what is the direction of the loop's magnetic moment?
18. Suppose a long, straight, current-carrying wire had a current from west to east. If it were immersed in a vertically upward magnetic field that was stronger on its west side than on its east side, what would be the initial motion of the wire if released from rest? Explain.

**19.6 Electromagnetism: Currents as a Magnetic Field Source**

19. A circular current-carrying loop is lying flat on a table. A calibrated compass, when placed at the center of the loop, points toward the ground. If you look straight down on the loop, what is the direction of the current? Explain your reasoning.
20. If you doubled your distance from a long current-carrying wire, what changes would need to be made to the current to keep the magnetic field strength the same as at the nearer position but reversed direction?
21. Given two solenoids, one with 100 turns and the other with 200 turns, if both carry the same current, will the one with more turns necessarily produce a stronger magnetic field at its center? Explain.
22. Wires carrying current to and from an appliance are typically wrapped close together, insulated from each other, and formed into the power cord. Explain how this works to reduce the magnetic field created external to the cord.
23. Two circular wire loops are coplanar (their areas are in the same plane) and have a common center. The outer one carries 10 A in the clockwise direction. To create a zero magnetic field at their common center, what should be the direction of the current in the inner loop? Should its current be 10 A, larger than 10 A, or smaller than 10 A? Explain.

**19.7 Magnetic Materials**

24. If you are looking down on the orbital plane of the electron in a hydrogen atom and the electron orbits counterclockwise, what is the direction of the magnetic field the electron produces at the proton?
25. What is the purpose of the iron core often used at the center of a solenoid?
26. Discuss several ways that the magnetic field of a permanent magnet can be destroyed or reversed.
27. Suppose the lava flow from the Kilauea volcano on the "Big Island" of Hawaii cools below its Curie temperature as it moves, and finally solidifies. What is the direction of the *remnant* magnetism in the resulting lava rocks? Explain.

**19.8 Geomagnetism: The Earth's Magnetic Field**

28. Determine the direction of the force due to the Earth's magnetic field on an electron near the equator when the electron's velocity is (a) due south, (b) northwest, and (c) upward.
29. In a relatively short time, geologically speaking, data indicate that the Earth's magnetic field direction will reverse. After that, what would be the polarity of the magnetic pole near the Earth's geographic North Pole?
30. Based on Figure 19.27, approximately how far off (in angle and direction) would your compass direction be from geographic north if you were located in (a) Phoenix, (b) Chicago, and (c) New Orleans?

## Exercises*

*Integrated Exercises (IEs) are two-part exercises. The first part typically requires a conceptual answer choice based on physical thinking and basic principles. The following part is quantitative calculations associated with the conceptual choice made in the first part of the exercise.*

**19.1 Permanent Magnets, Magnetic Poles, and Magnetic Field Direction**

1. • A permanent bar magnet (#1) is vertically oriented so that its north end is below its south end. The north end of this magnet feels an upward magnetic force of 1.5 m N from an identical vertically oriented magnet (#2) that has one end located 2.5 cm directly below the north end of #1. (a) Make a sketch showing the

* The bullets denote the degree of difficulty of the exercises: •, simple; ••, medium; and •••, more difficult.

orientation of the second magnet's poles. (b) A third identical magnet (#3) is brought in and oriented horizontally so that its north end is closest to #1 and 2.5 cm directly to the right of its north pole. What is the magnetic force on the north end of magnet #1 now?

2. • Two identical bar magnets of negligible width are located in the $x$-$y$ plane. Magnet #1 lies on the $x$-axis and its north end is at $x = +1.0$ cm, while its south end is at $x = +5.0$ cm. Magnet #2 lies on the $y$-axis and its north end is at $y = +1.0$ cm, while its south end is at $y = +5.0$ cm. (a) In what direction would a compass point if it were located at the origin? (b) Repeat part (a) for the situation where magnet #1 is reversed in polarity. [*Hint*: Make a sketch of the two magnets and their individual fields at the origin.]

3. •• Two bar very narrow magnets are located in the $x$-$y$ plane. Magnet #1 lies on the $x$-axis and its north end is at $x = +1.0$ cm, while its south end is at $x = +5.0$ cm. Magnet #2 lies on the $y$-axis and its north end is at $y = +1.0$ cm, while its south end is at $y = +5.0$ cm. Magnet #2 produces a magnetic field that is only one-half the magnitude of magnet #1. (a) In what direction would a compass point if it were located at the origin? (b) Repeat part (a) for the situation where magnet #1 is reversed in polarity.

### 19.2 Magnetic Field Strength and Magnetic Force

4. • A positive charge moves horizontally to the right across this page and enters a magnetic field directed vertically downward in the plane of the page. (a) What is the direction of the magnetic force on the charge: (1) into the page, (2) out of the page, (3) downward in the plane of the page, or (4) upward in the plane of the page? Explain. (b) If the charge is 0.25 C, its speed is $2.0 \times 10^2$ m/s, and it is acted on by a force of 20 N, what is the magnetic field strength?

5. • A charge of 0.050 C moves vertically in a field of 0.080 T that is oriented 45° from the vertical. What speed must the charge have such that the force acting on it is 10 N?

6. • A charge of 0.250 C moves vertically in a field of 0.500 T that is oriented some angle from the vertical. If the charge's speed is $2.0 \times 10^2$ m/s, what field angle(s) will ensure that the force acting on the charge is 5.0 N?

7. •• A beam of protons is accelerated to a speed of $5.0 \times 10^6$ m/s in a particle accelerator and emerges horizontally into a uniform magnetic field. What magnetic field (direction and magnitude) oriented perpendicularly to the velocity of the protons would cancel the force of gravity and keep the beam moving horizontally?

8. • An electron travels in the $+x$-direction in a magnetic field and is acted on by a magnetic force in the $-y$-direction. (a) In which of the following directions could the magnetic field be oriented: (1) $-x$, (2) $+y$, (3) $+z$, or (4) $-z$? Explain. (b) If the electron speed is $3.0 \times 10^6$ m/s and the magnitude of the force is $5.0 \times 10^{-19}$ N, what is the field strength?

9. • An electron travels at a speed of $2.0 \times 10^4$ m/s through a uniform magnetic field of magnitude is $1.2 \times 10^{-3}$ T. What is the magnitude of the magnetic force on the electron if its velocity and the field (a) are perpendicular, (b) at angle of 45°, (c) are parallel, and (d) are exactly opposite?

10. •• (a) What angle(s) does a particle's velocity have to make with the magnetic field direction for it to be subjected to half the maximum possible magnetic force, $F_{max}$? (b) Express the magnetic force on a charged particle in terms of $F_{max}$ if the angle between its velocity and the field direction is (i) 10°, (ii) 80°, and (iii) 100°. (c) If the particle's velocity makes an angle of 50° with respect to the field direction, at what other angle(s) would the magnetic force be the same? Would the direction be the same? Explain.

11. ••• A beam of protons exits a particle accelerator due east at a speed of $3.0 \times 10^5$ m/s. They then enter a uniform magnetic field of magnitude 0.50 T that is oriented at 37° above the horizontal relative to the beam direction. (a) What is the initial acceleration of a proton as it enters the field? (b) What if the magnetic field were angled at 37° below the horizontal instead? (c) If the beam were instead electrons at the same speed and the field were angled upward at 37°, would there be any difference in the force on the electrons compared to protons? Explain. (d) In part (c), what would be the ratio of the acceleration of an electron to that of a proton?

### 19.3 Applications: Charged Particles in Magnetic Fields

12. • An ionized deuteron (a bound proton-neutron system with a net $+e$ charge) passes through a velocity selector whose perpendicular magnetic and electric fields have magnitudes of 40 mT and 8.0 kV/m, respectively. Find the speed of the ion.

13. • In a velocity selector, the uniform magnetic field of 1.5 T is produced by a large magnet. Two parallel plates with a separation of 1.5 cm produce the perpendicular electric field. What voltage should be applied across the plates so that (a) a singly charged ion traveling at $8.0 \times 10^4$ m/s will pass through undeflected and (b) a doubly charged ion traveling at the same speed will pass through undeflected?

14. • In a velocity selector, a charged particle travels undeflected through perpendicular electric and magnetic fields whose magnitudes are 3000 N/C and 30 mT, respectively. Find the speed of the particle if it is (a) a proton and (b) an alpha particle. (An alpha particle is a helium nucleus – a positive ion with a double positive charge of $+2e$.)

15. •• In an experimental technique for treating deep tumors, unstable positively charged pions ($\pi+$, elementary particles with a mass of $2.25 \times 10^{-25}$ kg) penetrate the flesh and disintegrate at the tumor site, releasing energy to kill cancer cells. If pions with a kinetic energy of 10 keV are required and if a velocity selector with an electric field strength of $2.0 \times 10^3$ V/m is used, what must be the magnetic field strength?
16. •• In a mass spectrometer, a singly charged ion having a particular velocity is selected by using a magnetic field of 0.10 T perpendicular to an electric field of $1.0 \times 10^3$ V/m. A magnetic field of this same magnitude is then used to deflect the ion, which moves in a circular path with a radius of 1.2 cm. What is the mass of the ion?
17. •• In a mass spectrometer, a doubly charged ion having a particular velocity is selected by using a magnetic field of 100 mT perpendicular to an electric field of 1.0 kV/m. This same magnetic field is then used to deflect the ion in a circular path with a radius of 15 mm. Find (a) the mass of the ion and (b) the kinetic energy of the ion. (c) Does the kinetic energy of the ion increase in the circular path? Explain.
18. ••• In a mass spectrometer, a beam of protons enters a magnetic field. Some protons make exactly a one-quarter circular arc of radius 0.50 m. If the field is always perpendicular to the proton's velocity, (a) what is the field's magnitude if exiting protons have a kinetic energy of 10 keV? (b) How long does it take the proton to complete the quarter circle? (c) Find the net force (magnitude) on a proton while it is in the field.

**19.4 Magnetic Forces on Current-Carrying Wires and**

**19.5 Applications: Current-Carrying Wires in Magnetic Fields**

19. • (a) Find the direction of the current in the wires in ▼**Figure 19.30**. In each case, the magnetic force direction is shown. (b) If in each case the wire is a straight segment 15 cm long carrying a current of 5.5 A, and is in a field whose strength is 1.0 mT, determine the magnitude of the magnetic force.

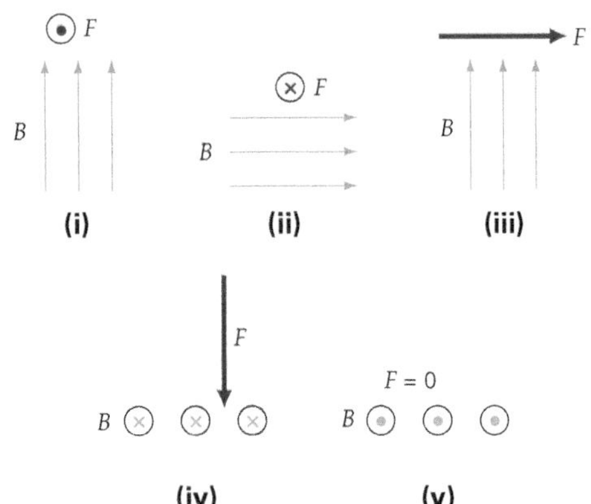


▲**FIGURE 19.30 The right-hand force rule** See Exercise 19.

20. • A straight, horizontal segment of wire carries a current in the $+x$-direction in a magnetic field that is directed in the $-z$-direction. (a) Is the magnetic force on the wire directed in the (1) $-x-$, (2) $+z-$, (3) $+y-$, or (4) $-y$-direction? Explain. (b) If the wire is 1.0 m long and carries a current of 5.0 A and the magnetic field is 0.30 T, what is the magnitude of the force on the wire?
21. • A 2.0-m length of straight wire carries a current of 20 A in a uniform magnetic field of 50 mT at an angle of 37° from the direction of the current. Find the force on the wire.
22. •• A horizontal magnetic field of $1.0 \times 10^{-4}$ T is at 30° to the current in a straight, horizontal wire 75 cm long. If the wire carries a current of 15 A, (a) what is the magnitude of the force on the wire? (b) What angle(s) would be required for the force to be half the value found in (a), assuming nothing else changed?
23. •• A wire carries a current of 10 A in the $+x$-direction in a uniform magnetic field of 0.40 T. Find the magnitude of the force per unit length and the direction of the force on the wire if the magnetic field is (a) in the $+x$-direction, (b) in the $+y$-direction, (c) in the $+z$-direction, (d) in the $-y$-direction, (e) in the $-z$-direction, and (f) at an angle of 45° above the $+x$-axis and in the $x$-$y$ plane.
24. •• A straight current-carrying wire 25 cm long is at right angles to a uniform horizontal magnetic field of 0.30 T in the $-x$-direction. (a) Along which of the $x$-$y$-$z$ axes would the current direction have to be to cause the wire to be subject to a force of (a) 0.050 N in the $+y$-direction, (b) 0.025 N in the $+z$-direction, and (c) 0.020 N in the $+x$-direction?
25. •• A wire carries a current of 10 A in the $+x$-direction. (a) Find the force per unit length on the wire if it is in a magnetic field that has components of $B_x = 0.020$ T, $B_Y = 0.040$ T, and $B_z = 0$. (b) Find the force per unit length on the wire if only the field's $x$-component is changed to 0.050 T. (c) Find the force per unit length on the wire if only the field's $y$-component is changed to $-0.050$ T.
26. •• A horizontal dc power line carries a current of 1000 A east. If the Earth's field at that location is northward with a magnitude of $5.0 \times 10^{-5}$ T at 45° below the horizontal, what are the magnitude and direction of the magnetic force on a 15-m section of this line?
27. •• A wire is bent as shown in ▶**Figure 19.31** and placed in a magnetic field with a magnitude of 1.0 T in the indicated direction. Find the net force on the whole wire if $x = 50$ cm and it carries a current of 5.0 A in the direction shown.

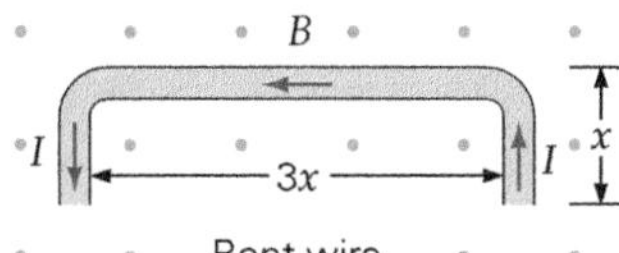


▲ **FIGURE 19.31 Current-carrying wire in a magnetic field** See Exercise 27.

28. IE ••• A loop of current-carrying wire is in a 1.6-T magnetic field. (a) For the magnetic torque on the loop to be at a maximum, should the plane of the coil be (1) parallel, (2) perpendicular, or (3) at a 45° angle to the magnetic field? Explain. (b) If the loop is rectangular with dimensions 20 cm by 30 cm and carries a current of 1.5 A, what is the magnitude of the magnetic moment of the loop, and what is the maximum torque? (c) What would be the angle(s) between the magnetic moment vector and the magnetic field direction if the loop felt only 20% of its maximum torque?
29. ••• A rectangular wire loop with a cross-sectional area of 0.20 $m^2$ carries a current of 0.25 A. The loop is free to rotate about an axis that is perpendicular to a uniform magnetic field of 0.30 T. The plane of the loop is at 30° to the direction of the field. (a) What is the magnitude of the torque on the loop? (b) How would you change the field to double the torque in (a)? (c) Instead, how would you change only the current to double the torque in (a)? (d) Instead, how would you double the torque by changing only the loop area, and what would the new area be? (e) Could you double the torque in (a) by changing only the angle? Explain.

## 19.6 Electromagnetism: Currents as a Magnetic Field Source

30. • The magnetic field at the center of a 50-turn coil of radius 15 cm is 0.80 mT. Find the current in the coil.
31. • In Exercise 30, if you wanted to double the field strength while keeping the current and turn count the same, what would the coil area have to be?
32. • (a) Show that the right-hand side of Equation 19.13b gives the correct SI units for magnetic field. (b) Show that Equation 19.13b reduces to Equation 19.13a when the central axis location is at the center of the loop.
33. • The magnetic field 7.5 cm directly above the center of a 25-turn coil of radius 15 cm is 0.80 mT. Find the current in the coil.
34. • A long, straight wire carries a current of 2.5 A. Find the magnitude of the magnetic field 25 cm from the wire.
35. • In a physics lab, a student discovers that the magnitude of the magnetic field at a certain distance from a long wire is 4.0 $\mu$T. If the wire carries a current of 5.0 A, what is the distance from the wire?
36. • A solenoid is 0.20 m long and consists of 100 turns of wire. At its center, the solenoid produces a magnetic field of 1.5 mT. Find the current in the coil.
37. •• Two long, parallel wires carry currents of 8.0 and 2.0 A (▼ **Figure 19.32**). (a) What is the magnitude of the magnetic field midway between the wires? (b) Where on a line perpendicular to and joining the wires is the magnetic field zero?

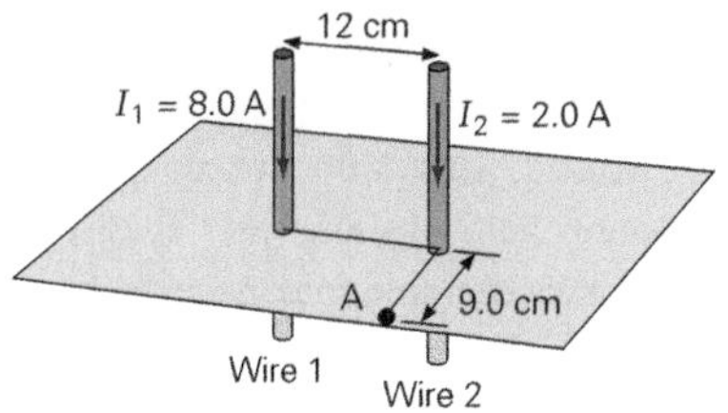


▲ **FIGURE 19.32 Parallel current-carrying wires** See Exercises 37 and 40.

38. •• Two long, parallel wires separated by 50 cm each carry currents of 4.0 A in a horizontal direction. Find the magnetic field midway between the wires if the currents are (a) in the same direction and (b) in opposite directions.
39. •• Two long, parallel wires separated by 0.20 m carry equal currents of 1.5 A in the same direction. Find the magnitude and direction of the magnetic field 0.15 m away from each wire on the side opposite the other wire (▼ **Figure 19.33**).

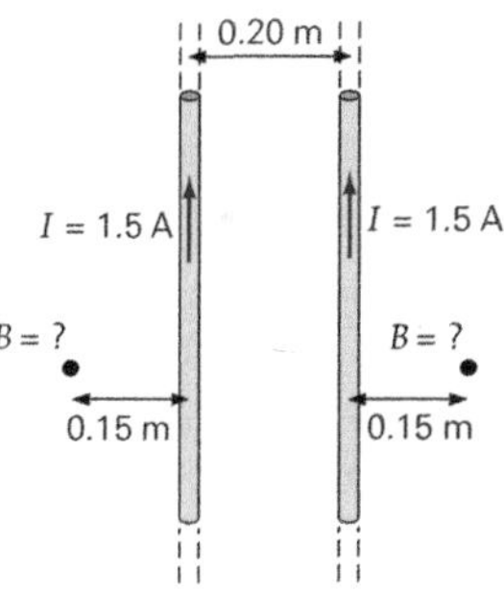


▲ **FIGURE 19.33 Magnetic field summation** See Exercise 39.

40. •• In Figure 19.32, find the magnetic field (magnitude and direction) at point A, located 9.0 cm away from wire 2 on a line perpendicular to the line joining the wires.
41. •• A coil of four circular loops of radius 5.0 cm carries a current of 2.0 A clockwise, as viewed from above the coil's plane. What is the magnetic field at the center of the coil?
42. IE •• A circular loop of wire in the horizontal plane carries a counterclockwise current, as viewed from above. (a) Use the right-hand source rule to determine whether the direction of the field at the center of the loop is (1) toward or (2) away from the observer. (b) If the diameter of the loop is 12 cm and the current is 1.8 A, what is the magnitude of the magnetic field at the center of the loop?

43. •• A circular loop of wire of radius 5.0 cm carries a current of 1.0 A. Another circular loop of wire is concentric with (that is, has a common center with) the first and has a radius of 10 cm. The field at their common center is double what it would be from the first alone, but oppositely directed. What is the current in the second loop?

44. •• A current-carrying solenoid is 10 cm long and is wound with 1000 turns of wire. It produces a magnetic field of $4.0 \times 10^{-4}$ T at its center. (a) How long would the solenoid have to be to produce a field of $6.0 \times 10^{-4}$ T at its center without changing the number of turns? (b) Adjusting only the windings and not the length, what number would be needed to produce a field of $8.0 \times 10^{-4}$ T at the center? (c) What current in the original solenoid is needed to produce a field of $9.0 \times 10^{-4}$ T in the opposite direction?

45. •• A solenoid is wound with 200 turns/cm. An outer layer of insulated wire with 180 turns/cm is wound over the first layer of wire. When the solenoid is operating, the inner coil carries a current of 10 A and the outer carries 15 A in the direction opposite to that in the inner coil (▼ **Figure 19.34**). (a) What is the direction of the field at the center for this configuration? (b) What is the magnitude of the field at the center of this arrangement?

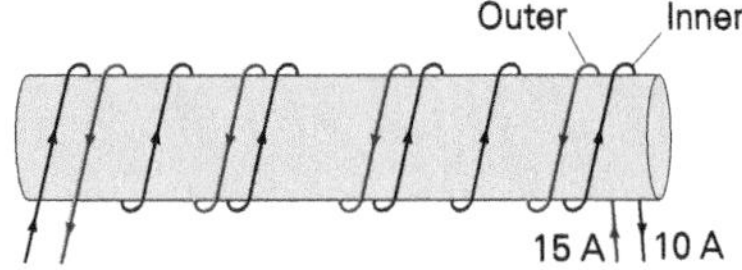


▲ FIGURE 19.34 **Double it up?** See Exercise 45.

46. •• A set of jumper cables used to start a car from another car's battery is connected to the terminals of both batteries. If 15 A of current exists in the cables at some instant during the starting procedure and the cables are parallel and 15 cm apart, (a) do the wires repel or attract? Explain. (b) What is the magnetic field strength that each wire produces at the location of the other? (c) What is the force per unit length on the cables?

47. •• Two long, straight, parallel wires carry the same current in the same direction. (a) Use both the right-hand source and force rules to determine whether the forces between the wires are (1) attractive or (2) repulsive. (b) If the wires are 24 cm apart and experience a force per unit length of 24 mN/m, find the current in each. (c) What is the field strength midway between the two?

48. ••• Four wires running through the corners of a square with sides of length a, as shown in ▼ **Figure 19.35**, carry equal currents $I$. Calculate the magnetic field at the center of the square in terms of $I$ and $a$. [*Hint:* Take careful note of the direction of the currents – some are into the paper and some are out of the paper.]

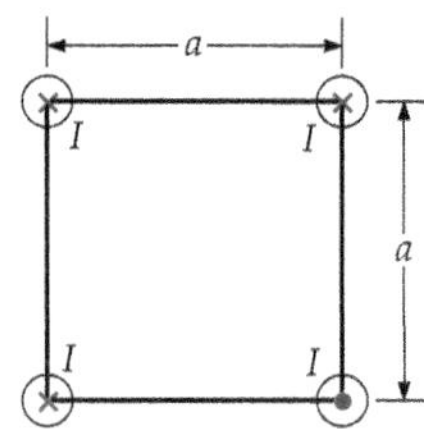


▲ FIGURE 19.35 **Current-carrying wires in a square array** See Exercise 48.

## 19.7 Magnetic Materials

49. •• A 50-cm-long solenoid has 100 turns of wire and carries a current of 0.95 A. It has a ferromagnetic core completely filling its interior where the field is 0.71 T. Determine the (a) magnetic permeability and (b) relative magnetic permeability of the material.

50. ••• A circular wire (planar) coil consists of 100 turns and is wound around a very long iron cylinder of radius 2.5 cm and a relative permeability 2200. The loop carries a current of 7.5 A. Find the field strength (a) at the center of the circular coil and (b) on the central axis of the iron cylinder 5.0 cm above the center of the coil.

## 19.8 Geomagnetism: The Earth's Magnetic Field

51. •• (a) Exiting from a small linear accelerator near Washington, DC, a proton is moving parallel to the ground. What should the direction(s) of its velocity be in order to maximize the upward component of the Earth's magnetic force on it? (b) Assuming a value of 0.05 mT the horizontal component of the Earth's field, what is the upward acceleration of the proton if it exits at 2000 m/s?

52. ••• A crude classical (and wrong!) model for the production of the Earth's magnetic field consists of a circular loop of current with a radius of 500 km with its center coincident with the Earth's. Assuming the plane of the loop to be in the equatorial plane, and using a value of 0.10 mT for the Earth's field at the poles, (a) estimate the current in this theoretical loop. (b) What would be the magnetic moment (magnitude and direction) of this supposed loop?

# 20

# Electromagnetic Induction and Waves

**These windmills provide renewable electric energy from wind using the principles of electromagnetic induction.**

As was seen in Chapter 19, an electric current produces a magnetic field. But the relationship between electricity and magnetism does not stop there. In this chapter, it will be shown that under the right conditions, a magnetic field can produce an electric field and electric current. How is this done? Chapter 19 considered only constant magnetic fields. No current is produced in a loop of wire that is stationary in a constant magnetic field. However, if the magnetic field changes with time, or if the wire loop moves into or out of, or is rotated in, the field, a current *is* produced in the wire.

One of the most useful applications of this interrelationship provides alternative and renewable sources of electric energy. At wind farms such as that in the chapter-opening photo, one of the oldest and simplest energy sources on Earth - wind - is used to generate "clean" electric energy. The windmill generators convert some of the air's kinetic energy into electric energy.

But how does this last step take place? Regardless of the source of the energy - the burning of fossil fuels; heat from a nuclear reactor, wind, waves, or falling water - the actual conversion to electric energy is accomplished by using magnetic fields and the concept of electromagnetic induction. This chapter not only examines the underlying principles of induction, but also discusses several practical applications. Last, it will be shown that the creation and propagation of electromagnetic radiation are intimately related to electromagnetic induction.

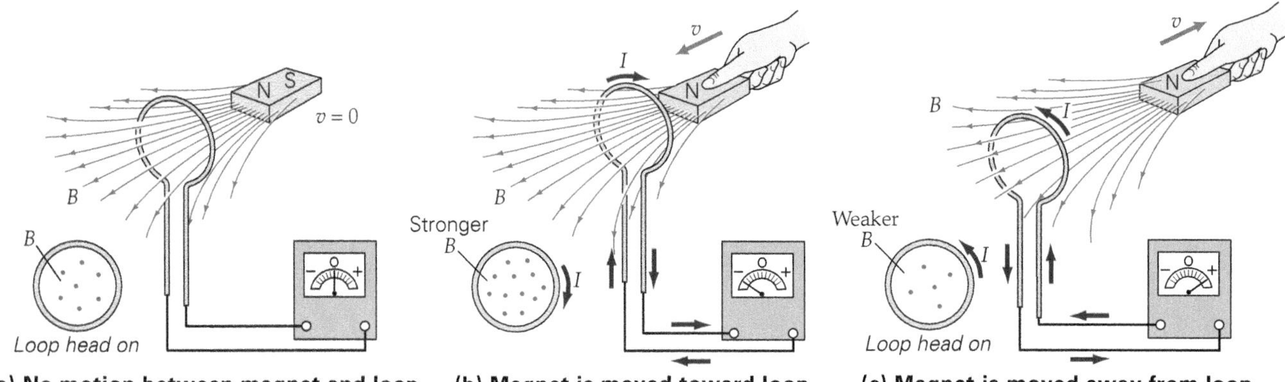


▲ FIGURE 20.1 **Electromagnetic induction (a)** When there is no relative motion between the magnet and the wire loop, the number of field lines through the loop (in this case, seven) is constant, and the galvanometer (see text footnote) shows no deflection, indicating no induced current. **(b)** However, moving the magnet toward the loop increases the number of field lines passing through the loop (now twelve), and an induced current is detected. **(c)** Moving the magnet away from the loop decreases the number of field lines passing through the loop (to five). The result is that the induced current is now in the opposite direction. (Note the needle deflection.)

## 20.1 Induced emf: Faraday's Law and Lenz's Law

Recall from Section 17.1 that the term *emf* stands for electromotive force, which is an electric potential difference capable of causing an electric current. It is observed experimentally that a magnet, when held stationary near a conducting wire loop, does *not* induce an emf (and therefore induces no current) in that loop (▲ **Figure 20.1a**). If the magnet is *moved* toward the loop, however, as in Figure 20.1b, the deflection of the galvanometer* needle indicates that current is created in the loop, but only during the motion of the magnet. Moreover, if the magnet is then moved away from the loop, as in Figure 20.1c, the galvanometer needle deflects in the opposite direction, which indicates a reversal of the current direction but, again, this only occurs during the motion of the magnet.

Further experiments indicate that deflection of the galvanometer needle, indicating the presence of *induced currents*, also occurs if the loop is moved toward or away from a stationary magnet. Thus the inducing effect depends on the relative motion of loop and magnet. It has also been determined that the magnitude of the induced current depends on the speed of that motion. Experimentally, there is a noteworthy exception. If a loop is moved (but not rotated) in a uniform magnetic field, as shown in ▶ **Figure 20.2**, no current is induced. This situation will be discussed later in this section.

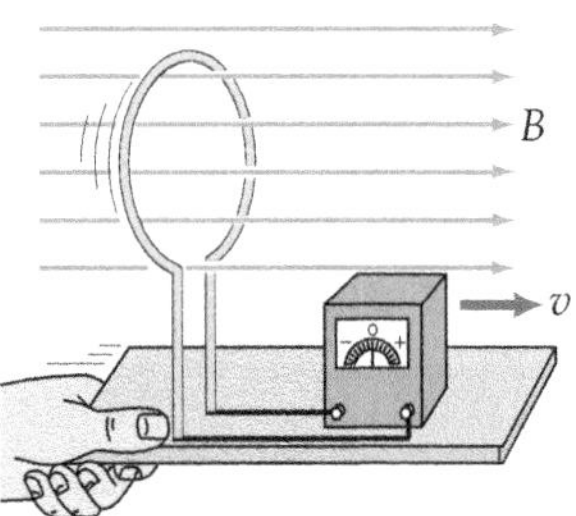


▲ FIGURE 20.2 **Relative motion and no induction** When a loop is moved parallel to a uniform magnetic field, there is no change in the number of field lines passing through the loop, and there is no induced current.

Yet another method to induce a current in a stationary wire loop is to vary the current in another, nearby loop. When the switch in the battery-powered circuit in ▶ **Figure 20.3a** is closed, the current in the loop on the right goes from zero to some constant value in a short time. However only during the buildup time does the magnetic field caused by the current in this loop increase in the region of the loop on the left. During the buildup, the galvanometer needle deflects, indicating an induced current is present in the left loop. When the current in the right loop attains a steady value, the field it produces becomes constant, and the current in the left loop drops to zero. Similarly, when the switch in the right loop is opened (Figure 20.3b), its current and field decrease to zero, and the galvanometer deflects in the opposite direction, indicating a reversal in direction of the current induced in the left loop.

There is a convenient way of summarizing the process of **electromagnetic induction** that is happening in both Figures 20.1 and 20.3:

> To induce emfs and currents in a loop or complete circuit, all that matters is whether the magnetic field through the loop or circuit is changing.

Detailed experiments by Michael Faraday in the mid-1800s investigating electromagnetic induction found that an important factor in electromagnetic induction was the

* A galvanometer is an electromechanical meter that can detect electric current and its direction by noting which way its needle moves. In Figure 20.1 the galvanometer is wired in series with the loop of interest.

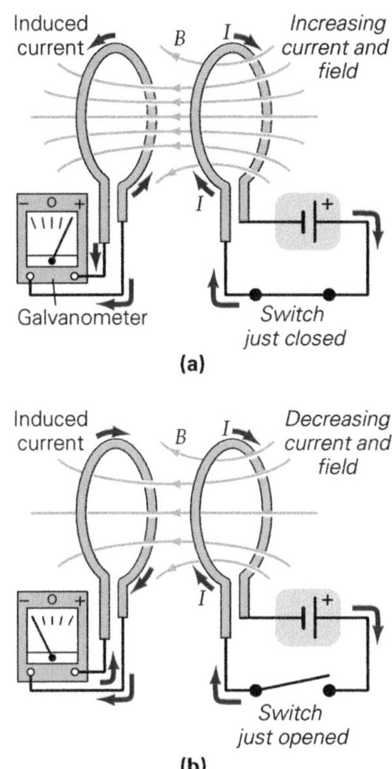


▲FIGURE 20.3 **Mutual induction** **(a)** When the switch is closing in the right-loop circuit, the buildup of current produces a changing magnetic field in the other loop, inducing a current in it. **(b)** When the switch is opened, the magnetic field collapses, and the magnetic field in the left loop decreases. The induced current in this loop is then in the opposite direction. The induced currents occur only when the magnetic field passing through a loop changes and vanish when the field reaches a constant value.

*time rate of change* of the number of magnetic field lines passing through the loop or circuit area. That is, an induced emf (and resulting induced current) is produced in a loop or complete circuit whenever the number of magnetic field lines passing through the plane of the loop or circuit changes with time, and its magnitude is related to how rapidly that change occurs.

## 20.1.1 Magnetic Flux

Because of Faraday's discovery, determining induced emf requires that the number of field lines through the loop be quantified. Consider a loop of wire in a uniform magnetic field (▶ **Figure 20.4a**). The number of field lines through the loop depends on the loop's area, its orientation relative to the field, and the strength of that field. To describe the loop's orientation, the concept of an *area vector* ($\vec{\mathbf{A}}$) is employed. Its direction is normal to the loop's plane, and its magnitude is equal to the loop area. The angle between the field ($\vec{\mathbf{B}}$) and the area vector ($\vec{\mathbf{A}}$), $\theta$, is a measure of their relative orientation. For example, in Figure 20.4a, $\theta = 0°$ meaning that the vectors

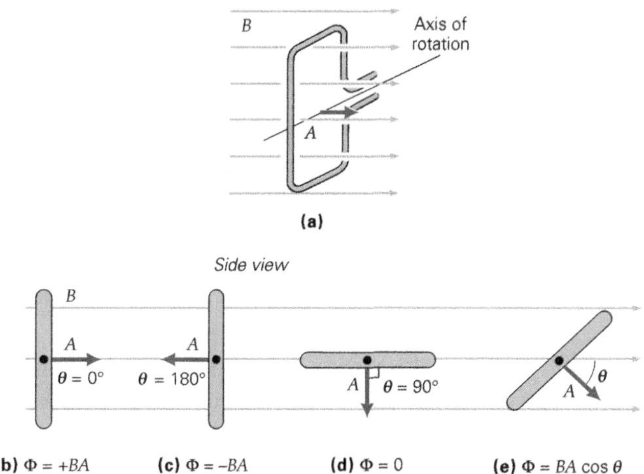


▲FIGURE 20.4 **Magnetic flux** **(a)** Magnetic flux ($\Phi$) is a measure of the number of field lines passing through an area ($A$). The area can be represented by a vector $\vec{\mathbf{A}}$ perpendicular to the plane of the area. **(b)** When the plane of a loop is perpendicular to the field and $\theta = 0°$, then $\Phi = \Phi_{max} = +BA$. **(c)** When $\theta = 180°$, the magnetic flux has the same magnitude, but is opposite in direction: $\Phi = -\Phi_{max} = -BA$. **(d)** When $\theta = 90°$, then $\Phi = 0$. **(e)** As the loop's plane is changed from being perpendicular to the field to being more parallel to the field, less area is available to the field lines, and therefore the flux decreases. In general, $\Phi = BA\cos\theta$.

are in the same direction, or, alternatively, that the area plane is perpendicular to the field.

In a magnetic field that does *not* vary over the area, the number of magnetic field lines passing through a particular area (the loop area here) is proportional to the **magnetic flux ($\Phi$)**, which is defined as

$$\Phi = BA\cos\theta \quad \text{magnetic flux (constant magnetic field)} \qquad (20.1)$$

SI unit of magnetic flux: tesla-meter squared (T·m²), or weber (Wb)

The SI unit of magnetic flux is T·m², and this combination was named the weber (in honor of Wilhelm Eduard Weber*) with 1 Wb = 1 T·m².

Note that the *orientation* of the loop with respect to the field affects the number of field lines passing through it, and this factor is accounted for by the cosine term in Equation 20.1. Let us consider several possible orientations:

- If $\vec{\mathbf{B}}$ and $\vec{\mathbf{A}}$ are parallel ($\theta = 0°$), then the magnetic flux is positive and has a maximum value of $\Phi_{max} = \Phi_{0°} = BA\cos 0° = +BA$ indicating that the maximum number of field lines pass through the loop in this orientation (Figure 20.4b).

* Wilhelm Eduard Weber (1804–1891), a German physicist, was noted for his work in magnetism and electricity, particularly terrestrial magnetism.

- If $\vec{\mathbf{B}}$ and $\vec{\mathbf{A}}$ are oppositely directed ($\theta = 180°$), then the *magnitude* of the flux is a maximum again, but of opposite sign since $\Phi_{180°} = BA \cos 180° = -BA = -\Phi_{max}$ (Figure 20.4c).
- If $\vec{\mathbf{B}}$ and $\vec{\mathbf{A}}$ are perpendicular, there are no field lines passing through the plane of the loop, and the flux is zero: $\Phi_{90°} = BA \cos 90° = 0$ (Figure 20.4d).
- For situations at intermediate angles, the flux is less than the maximum value, but nonzero (Figure 20.4e).

$A \cos \theta$ can be interpreted as the *effective area* of the loop perpendicular to the field lines (▼ **Figure 20.5a**). Alternatively, $B \cos \theta$ can be viewed as the perpendicular component of the field through the full area of the loop, $A$, as in Figure 20.5b. In other words, Equation 20.1 can be viewed mathematically grouped as $\Phi = (B \cos \theta)A$ or $\Phi = B(A \cos \theta)$. Of course, either way, the answer is the same.

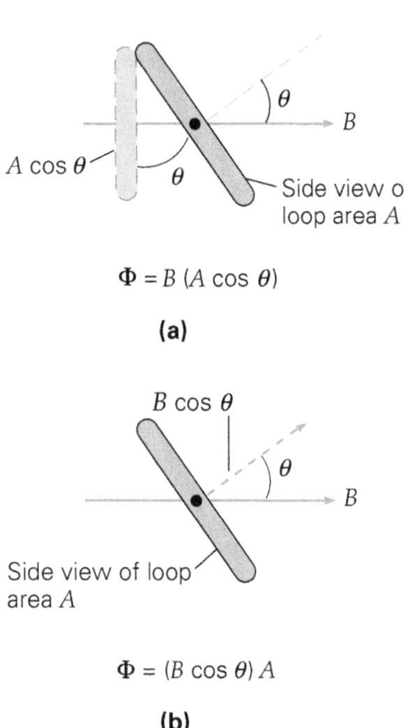


▲ FIGURE 20.5 **Magnetic flux through a loop: Two alternative interpretations** Instead of defining the flux ($\Phi$) **(a)** in terms of the magnetic field magnitude ($B$) passing through a reduced area ($A \cos \theta$), we can define it **(b)** in terms of the perpendicular component of the magnetic field ($B \cos \theta$) passing through A. Either way, $\Phi$ is a measure of the number of field lines passing through $A$ and is given by $\Phi = BA \cos \theta$.

## 20.1.2 Faraday's Law of Induction and Lenz's Law

From quantitative experiments, Faraday determined that the emf ($\mathscr{E}$) induced in a coil (a coil, by definition, consists of a series connection of $N$ individual loops) depends on the time rate of change of the magnetic field lines through all the loops, or the time rate of change of the magnetic flux through all the loops (total flux). This dependence, known as Faraday's law of induction, is expressed mathematically as

$$\mathscr{E} = -\frac{\Delta(N\Phi)}{\Delta t} = -N\frac{\Delta\Phi}{\Delta t} \quad \text{(Faraday's law for induced emf)} \qquad (20.2)$$

where $\Delta\Phi$ is the change in flux through one loop. Thus in a coil consisting of $N$ loops, the total change in flux is $N\Delta\Phi$.

The minus sign is included in Equation 20.2 to give an indication of the *direction* of the induced emf, which has not been mentioned until now. The Russian physicist Heinrich Lenz (1804–1865) discovered the law that governs the direction of the induced emf. **Lenz's law** is stated as follows:

> An induced emf in a wire loop or coil has a direction such that the current it creates produces its own magnetic field that opposes the *change* in magnetic flux through that loop or coil.

This means that the magnetic field *due to the induced current* is in such a direction to attempt to keep the flux through the loop from changing. For example, if the flux increases in the $+x$-direction, the magnetic field due to the induced current will be in the $-x$-direction (▶ **Figure 20.6a**). This effect tends to cancel the increase in the flux, or *oppose the change.*

Essentially, the magnetic field due to the induced current tries to maintain the existing magnetic flux. This effect is sometimes called "electromagnetic inertia," by analogy to the tendency of objects to resist changes in their velocity. In the long run, the induced current cannot prevent the magnetic flux from changing. However, during the time that the flux is changing, the induced magnetic field will oppose that change.

The direction of the induced current is given by the **induced-current right-hand rule**: With the thumb of the right hand pointing in the direction of the induced field, the fingers curl in the direction of the induced current. (See Figure 20.6b and Integrated Example 20.1.) This rule is a version of the right-hand rule used to find the direction of a magnetic field produced by a current (Chapter 19). Here it is used in reverse. Typically, the induced field direction is known (e.g., $-x$ in Figure 20.6b) and the direction of the current that produces it is to be determined. An application of Lenz's law is illustrated in Integrated Example 20.1.

### INTEGRATED EXAMPLE 20.1: LENZ'S LAW AND INDUCED CURRENTS

(a) The south end of a bar magnet is pulled far away from a small wire coil. (See ▶ **Figure 20.7a.**) Looking from behind the coil toward the south end of the magnet (Figure 20.7b), what is the direction of the induced current? (1) Counterclockwise, (2) clockwise, or (3) there is

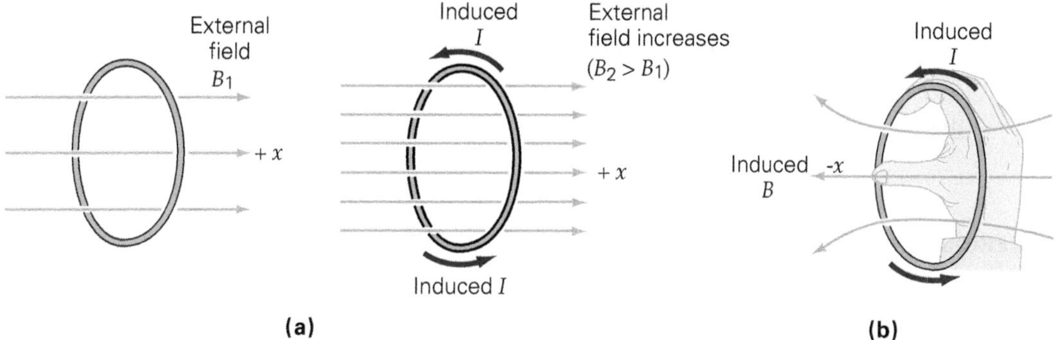


▲ **FIGURE 20.6 Finding the direction of the induced current (a)** Here an external magnetic field is increasing to the right through a wire loop. The induced current in that loop creates its own magnetic field to try to counteract the flux change that is occurring in it. **(b)** The (induced) current right-hand (source) rule determines the direction of the induced current. Here the direction of the induced field must be to the left. With the thumb of the right hand pointing left, the fingers give the induced current direction.

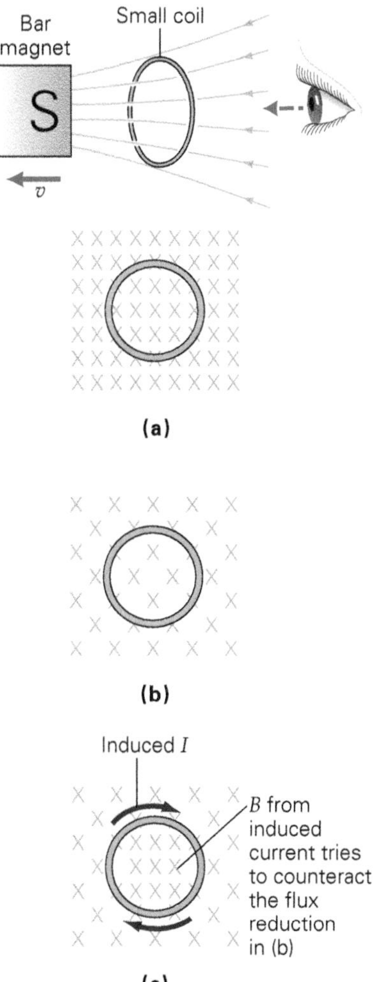


▲ **FIGURE 20.7 Using a bar magnet to induce currents (a)** The south end of a bar magnet is pulled away from a wire loop. **(b)** The view from the right of the loop shows the magnetic field pointing away from the observer, or into the page, and decreasing. **(c)** To counteract this loss of flux into the page, current is induced in the clockwise direction, so as to provide its own field into the page. See Integrated Example 20.1.

no induced current. (b) Suppose that the magnetic field over the area of the coil is initially constant at 40.0 mT, the coil's radius is 2.0 mm, and there are 100 loops in the coil. Determine the magnitude of the average induced emf in the coil if the bar magnet is removed in 0.750 s.

(a) **CONCEPTUAL REASONING.** There is initially magnetic flux into the plane of the coil (Figure 20.7b), and later, when the magnet is far away from the coil, there is no flux; and thus the flux has changed. Therefore, there must be an induced emf, so answer (3) cannot be correct. As the bar magnet is pulled away, the field weakens but maintains the same direction. The induced emf will produce an (induced) current that, in turn, will produce a magnetic field into the page so as to try to prevent this decrease in flux. Therefore, the induced emf and current are in the clockwise direction, as found using the induced current right-hand rule (Figure 20.7c), and the correct answer is (2), clockwise.

(b) **QUANTITATIVE REASONING AND SOLUTION.** This Example is a straightforward application of Equation 20.2. The initial flux is the maximum possible and the final flux is zero. The data are listed and converted to SI units:

***Given:***

$B_i = 40.0\text{ mT} = 0.0400\text{ T}$

$B_f = 0$

$r = 2.00\text{ mm} = 2.00 \times 10^{-3}\text{ m}$

$N = 100\text{ loops}$

$\Delta t = 0.750\text{ s}$

***Find:*** $\mathscr{E}$ (magnitude of average induced emf)

The initial magnetic flux through the coil is found by using Equation 20.1 with $\theta = 0°$. (Why?) The area is $A = \pi r^2 = \pi(2.00 \times 10^{-3}\text{ m})^2 = 1.26 \times 10^{-5}\text{ m}^{-2}$. Therefore, the initial flux, $\Phi_i$, through one loop is positive (why?) and given by

$$\Phi_i = B_i A \cos\theta = (0.0400\text{ T})(1.26 \times 10^{-5}\text{ m}^2)\cos 0°$$
$$= 5.03 \times 10^{-7}\text{ T}\cdot\text{m}^2$$

Since the final flux is zero, $\Delta\Phi = \Phi_f - \Phi_i = 0 - \Phi_i = -\Phi_i$. Therefore, from Equation 20.2 the magnitude of the average induced emf is

$$|\mathscr{E}| = N\frac{|\Delta\Phi|}{\Delta t} = (100\text{ loops})\frac{[5.03\times10^{-7}(\text{T}\cdot\text{m}^2)/\text{loop}]}{(0.750\text{ s})}$$
$$= 6.70\times10^{-5}\text{ V}$$

**FOLLOW-UP EXERCISE.** In this Example, (a) in which direction is the induced current if instead a north magnetic pole approaches the coil quickly? Explain. (b) In this Example, what would be the average induced current if the coil had a total resistance of 0.200 Ω?

Lenz's law incorporates the principle of energy conservation. Consider a situation in which a wire loop experiences an increasing magnetic flux through its area. Contrary to Lenz's law, suppose instead that the magnetic field from the induced current *added* to the flux instead of keeping it at its original value. This increased flux would then lead to an even greater induced current. In turn, this greater induced current would produce a still greater magnetic flux, which in turn would give a greater induced current, and so on. Such a something-for-nothing energy situation would violate conservation of energy and thus cannot happen.

To understand the *direction* of the induced emf in terms of forces, consider the case of the moving magnet (e.g., Figure 20.1b). The induced current sets in that loop creates a magnetic field similar in character to that of a bar magnet with a polarity that will oppose the motion of the real bar magnet (▼ **Figure 20.8**). You should be able to show that if the bar magnet is pulled away from the loop, the loop then will exert a magnetic *attraction* to try to keep the magnet from leaving – electromagnetic inertia in action.

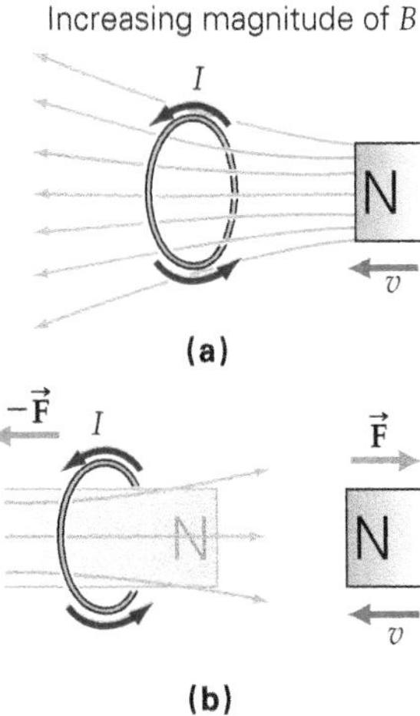


▲ FIGURE 20.8 **Lenz's law in terms of forces (a)** If the north end of a bar magnet is moved rapidly toward a wire loop, current is induced in the direction shown. **(b)** While the induced current exists, the loop acts like a bar magnet with its "north end" close to the north end of the real magnet. Thus there is a magnetic repulsion. This is an alternative way of viewing Lenz's law: Induce a current so as to try to keep the flux from changing – in this case, to try to keep the bar magnet away from the loop and maintain its initial value of flux, which is zero.

When the expression for the magnetic flux Φ (Equation 20.1) is substituted into Equation 20.2, the result is

$$\mathscr{E} = -N\frac{\Delta\Phi}{\Delta t} = -N\frac{\Delta(BA\cos\theta)}{\Delta t} \tag{20.3}$$

From this expression, it can be concluded that an induced emf in a loop/coil will result if:

1. the magnitude of the magnetic field through the loop/coil changes, and/or
2. the loop/coil area changes, and/or
3. the orientation between the loop area and the field direction changes.

In (1), a flux change is created by a time-varying field, such as that from a time-varying current in a nearby circuit or that created by moving a magnet near or away from a loop/coil. In (2), a flux change results because of a varying loop/coil area. Finally, in (3), a change in flux can result from a change in orientation of the loop/coil.

Situation (3) can occur, for example, if a coil is rotated in a magnetic field. The change in the number of field lines through a single loop is evident in the sequential views in Figure 20.4. Rotating a coil in a field is a common way of inducing an emf and will be considered on its own in Section 20.2. The emfs that result from changing the field strength and loop area are analyzed in the next two Examples.

### CONCEPTUAL EXAMPLE 20.2: FIELDS IN THE FIELDS – ELECTROMAGNETIC INDUCTION

In rural areas where electric power lines carry electricity to big cities, it is possible to generate small electric currents by means of induction in a conducting loop. The overhead power lines carry alternating currents that reverse direction sixty times per second. How would you orient the plane of the loop to maximize the induced current if the power lines run north to south? (a) Parallel to the Earth's surface, (b) perpendicular to the Earth's surface in the north-south direction, or (c) perpendicular to the Earth's surface in the east-west direction. (See ▼ **Figure 20.9a.**)

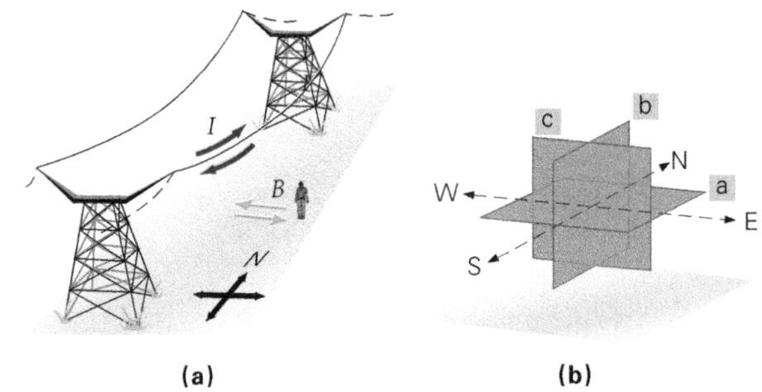


▲ FIGURE 20.9 **Induced emfs below power lines (a)** If current-carrying wires run in the north-south direction, then directly below the alternating current produces a magnetic field that oscillates between pointing east and west. **(b)** These are the three choices for loop orientation in Conceptual Example 20.2.

**REASONING AND ANSWER.** Magnetic field lines from long wires are circular. (See Figure 19.23.) By the source right-hand rule, the magnetic field direction at ground level is parallel to the Earth's surface and alternates in direction. The orientation choices are shown in Figure 20.9b. Neither answer (a) nor (c) can be correct, because in these orientations there would *never* be any magnetic flux passing through the loop. In the situation in this Example, the flux would be constant and there would be no induced emf. Hence, the answer is (b). If the loop is oriented perpendicular to the Earth's surface with its plane in the north-south direction, the flux through it would vary from zero to its maximum value and back sixty times per second, thus maximizing the induced emf and resulting current in the loop.

**FOLLOW-UP EXERCISE.** Suggest possible ways of increasing the induced current in this Example by changing only properties of the loop and not those of the overhead wires.

---

### EXAMPLE 20.3: INDUCED CURRENTS – A POTENTIAL HAZARD TO EQUIPMENT?

Electrical instruments can be damaged or destroyed if they are in a rapidly changing magnetic field; for example, near an electromagnet operating under ac conditions. If the induced currents are large enough, they could damage the instrument. Consider an audio speaker's coil that is near such an electromagnet (▼ **Figure 20.10**). Suppose an electromagnet exposes this coil to a maximum magnetic field of 1.00 mT that reverses direction every 1/120 s.

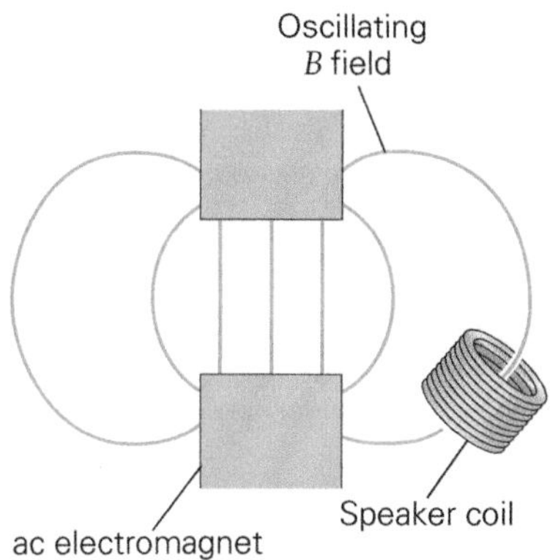


▲ FIGURE 20.10 **Instrument hazard?** The coil of a speaker is close to an alternating current electromagnet. The changing flux in the coil produces an induced emf and thus an induced current that depends on the resistance of the coil.

Assume that the speaker's coil consists of 100 circular loops (each with a radius of 3.00 cm) and has a total resistance of 1.00 Ω. According to the manufacturer of the speaker, the average current in the coil should not exceed 25.0 mA. (a) Calculate the magnitude of the average induced emf in the coil during the 1/120-s interval. (b) Is the induced current likely to damage the speaker coil?

**THINKING IT THROUGH.** (a) The flux goes from a (maximum) positive to a (maximum) negative value in 1/120 s. The magnetic flux change can be determined from Equation 20.1 with $\theta = 0°$ and $\theta = 180°$. The average induced emf can then be calculated from Equation 20.2. (b) Once the emf is determined, the average induced current can be calculated from $I = \mathscr{E}/R$.

**SOLUTION**

Listing the data and converting to SI units:

***Given:***

$$B_i = +1.00\,\text{mT} = +1.00\times10^{-3}\,\text{T (pointing one way)}$$
$$B_f = -1.00\,\text{mT} = -1.00\times10^{-3}\,\text{T (pointing the opposite way)}$$
$$\Delta t = 1/120\ \text{s} = 8.33\times10^{-3}\,\text{s}$$
$$N = 100\ \text{loops}$$
$$R = 1.00\,\Omega$$
$$r = 3.00\ \text{cm} = 3.00\times10^{-2}\,\text{m}$$

***Find:***

**(a)** $\mathscr{E}$ (magnitude of average induced emf)
**(b)** $I$ (magnitude of average induced current)

**(a)** The loop area is $A = \pi r^2 = \pi(3.00\times10^{-2}\,\text{m})^2 = 2.83\times10^{-3}\,\text{m}^2$. Thus the initial flux through one loop is (Equation 20.1):

$$\begin{aligned}\Phi_i &= B_i\,A\cos\theta\\ &= (1.00\times10^{-3}\,\text{T})(2.83\times10^{-3}\,\text{m}^2/\text{loop})(\cos 0°)\\ &= 2.83\times10^{-6}\,\text{T·m}^{-2}/\text{loop}\end{aligned}$$

Because the final flux is the negative of this, the change in flux through one loop is

$$\begin{aligned}\Delta\Phi &= \Phi_f - \Phi_i = -\Phi_i - \Phi_i = -2\Phi_i\\ &= -5.66\times10^{-6}\,\text{T·m}^2/\text{loop}\end{aligned}$$

Therefore, for all 100 loops that make up the coil, the average induced emf is (Equation 20.2)

$$\begin{aligned}\mathscr{E} &= N\frac{|\Delta\Phi|}{\Delta t} = (100\ \text{loops})\left[\frac{5.66\times10^{-6}(\text{T·m}^2/\text{loop})}{8.33\times10^{-3}\,\text{s}}\right]\\ &= 6.79\times10^{-2}\,\text{V}\end{aligned}$$

**(b)** This voltage is small by everyday standards, but keep in mind that the speaker coil's resistance is small. The average induced current in the coil is:

$$I = \frac{\mathscr{E}}{R} = \frac{6.79\times10^{-2}\,\text{V}}{1.00\ \Omega} = 6.79\times10^{-3}\,\text{A} = 67.9\ \text{mA}$$

This exceeds the allowed average speaker current and therefore the speaker's coil could possibly be damaged.

**FOLLOW-UP EXERCISE.** In this Example, if the speaker's coil were moved farther from the electromagnet, it could reach a point where the induced current would be below the "dangerous" level of 25.0 mA. Determine the maximum field strength at this point.

---

Emfs and currents can also be induced in conductors as they are moved through a magnetic field. In this situation, the induced emf is called a *motional emf.* To see how this works, consider the situation in ▼**Figure 20.11a**. As the bar moves upward, the circuit area increases by $\Delta A = L\Delta x$ (Figure 20.11a) At constant speed, the distance traveled by the bar in a time $\Delta t$ is $\Delta x = v\Delta t$, hence $\Delta A = Lv\Delta t$. The angle between the magnetic field and the normal to the area ($\theta$) is always 0°. However, the area is changing, so the flux varies. Since $\Phi = BA \cos 0° = BA$, it follows that $\Delta\Phi = B\Delta A$, or $\Delta\Phi = BLv\Delta t$. Therefore, from Faraday's law, the magnitude of this "motional" (induced) emf, $\mathscr{E}$, is $|\mathscr{E}| = |\Delta\Phi/\Delta t| = BLv\Delta t/t = BLv$. This is the fundamental principle behind *electric energy generation*: movement of a conductor in a magnetic field results in the conversion of the work done on it into electrical energy. To see some of the details, consider the following Integrated Example.

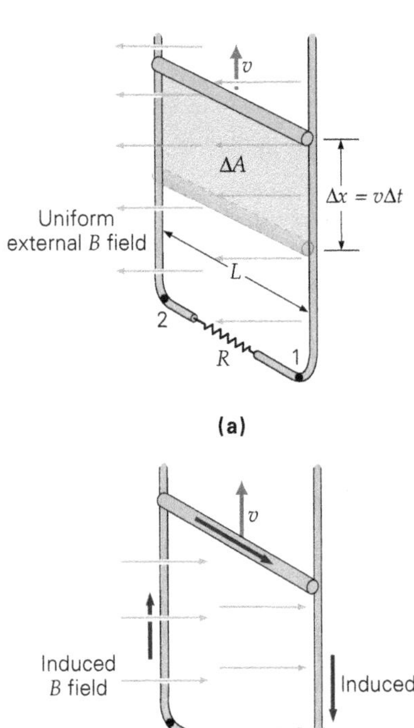


▲**FIGURE 20.11 Motional emf (a)** As the metal rod is pulled on the metal frame, the area of the rectangular loop varies with time. A current is induced in the loop as a result of the changing flux. **(b)** To counteract the increase of flux to the left, an induced current creates its own magnetic field to the right. See Integrated Example 20.4.

### INTEGRATED EXAMPLE 20.4: THE ESSENCE OF ELECTRIC ENERGY GENERATION - MECHANICAL WORK INTO ELECTRICAL CURRENT

Consider the situation in Figure 20.11a but with the bar moving *down* instead. An external force does work as the bar moves, and this work is converted to electrical energy. Because the "circuit" (wires, resistor, and bar) is in a magnetic field, the flux through it changes with time, inducing a current. (a) What is the direction of the induced current in the resistor: (1) from 1 to 2 or (2) from 2 to 1? (b) If the bar is 20 cm long and is moving at a steady speed of 10 cm/s, what is the induced current if the resistor has a value of 5.0 Ω and the circuit is in a uniform magnetic field of 0.25 T?

**(a) CONCEPTUAL REASONING.** In Figure 20.11a, the magnetic flux points left and is decreasing. According to Lenz's law, the field due to the induced current must be to the left to make up for the reduced flux. Using the induced-current right-hand rule, the direction of the induced current must be from 1 to 2 (Figure 20.11b), and the correct answer is (2).

**(b) QUANTITATIVE REASONING AND SOLUTION.** The flux change here is due to an area change. The expression for motional emfs has been derived in the preceding text. Once this emf has been found, the induced current can be found from Ohm's law.

Listing the data and converting to SI units:

***Given:***

$B = 0.25$ T
$L = 20$ cm $= 0.20$ m
$v = 10$ cm/s $= 0.10$ m/s
$R = 5.0\ \Omega$

***Find:*** $I$ (induced current in the resistor)

The induced emf is $|\mathscr{E}| = BLv = (0.25\text{ T})(0.20\text{ m})(0.10\text{ m/s}) = 5.0\times10^{-3}$ V. Hence the induced current is

$$I = \frac{\mathscr{E}}{R} = \frac{5.0\times10^{-3}\text{ V}}{5.0\ \Omega} = 1.0\times10^{-3}\text{ A}$$

Clearly this arrangement isn't a practical way to generate large amounts of electrical energy. Here the power dissipated in the resistor is only $5.0\times10^{-6}$ W. (You should verify this.)

**FOLLOW-UP EXERCISE.** In this Example, if the field were three times larger and the bar's width changed to 45 cm, what bar speed would be needed to induce a current of 0.10 A?

Electromagnetic induction is used in our daily lives, in many cases without our realizing it. For example, in the field of air travel safety, induction is used to prevent dangerous metallic objects (such as knives and guns) from being carried onto airplanes. As a passenger walks through the arch of an airport metal detector, a series of large "spiked" currents is periodically delivered to a coil (solenoid) in one of the nonmagnetic sides of the archway. In the most common system, called PI (*pulsed induction*), these current spikes occur hundreds of times per second. As the current rises and falls, a changing magnetic field is created in the region of the passenger. If the passenger has a metal object on their person, a current will be induced in that object, which in turn produces its own (induced) magnetic field that can be sensed as a "magnetic echo." Sophisticated electronics measure the emf induced by the "echo" field and can be set to trigger a warning for further inspection of that passenger.

## 20.2 Electric Generators and Back emf

One way to induce an emf in a loop is through a change in the loop's orientation in its magnetic field (Figure 20.4). This is the operational principle behind electric generators.

### 20.2.1 Electric Generators

An *electric generator* is a device that converts mechanical energy into electrical energy. Basically, the function of a generator is the reverse of that of a motor.

Recall that a battery supplies direct current (dc). That is, the voltage polarity (and therefore current direction) do not change. However, most generators produce *alternating current* (ac), named because the polarity of the voltage (and therefore current direction) change periodically. The electric energy used in homes and industry is delivered in the form of alternating voltage and current. (See Chapter 21 for analysis of ac circuits and Chapter 18 for household wiring diagrams.)

An **ac generator** is sometimes called an *alternator*, particularly in automobiles. The elements of a simple ac generator are shown in ▼ **Figure 20.12**. A wire loop called an *armature* is mechanically rotated in a magnetic field by some external force, such as water flow or steam hitting turbine blades. The rotation of the blades in turn causes a rotation of the loop. This results in a change in the loop's magnetic flux and therefore an induced emf in the loop. The ends of the loop are connected to an external circuit by means of slip rings and brushes. The result is an alternating current in that circuit. Note that in practice, generators have many loops, or windings, on their armatures.

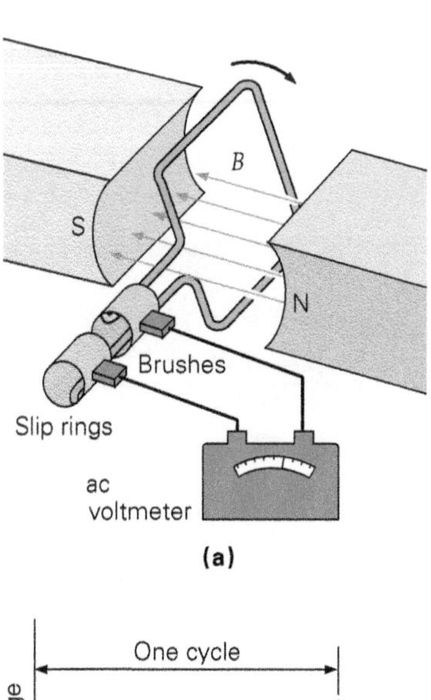


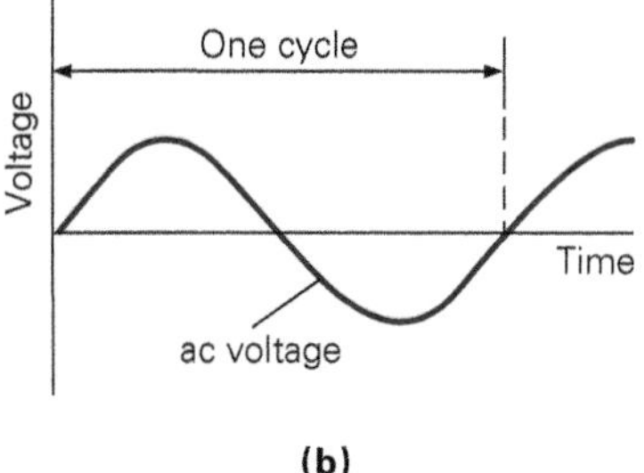


▲ FIGURE 20.12 **A simple ac generator** (**a**) The rotation of a wire loop in a magnetic field produces (**b**) a voltage output whose polarity reverses with each half-cycle. This alternating voltage is picked up by a brush/slip ring arrangement as shown.

When the loop is rotated at a constant angular speed ($\omega$), the angle ($\theta$) between the magnetic field vector and the area vector of the loop changes with time, since $\Delta\theta = \omega\Delta t$. As a result, the magnetic flux through the loop changes with time, resulting in an induced emf. From Equation 20.1, the flux (one loop and assuming $\theta = 0$ at $t = 0$) varies with time as $\Phi = BA\cos\theta = BA\cos\omega t$. Using this result, the time-varying induced emf can be computed from Faraday's law (assuming $N$ loops)

$$\mathscr{E} = -N\frac{\Delta\Phi}{\Delta t} = -NBA\left(\frac{\Delta(\cos\omega t)}{\Delta t}\right)$$

$B$ and $A$ have been removed from the time rate of change, because they are constant. By using methods beyond the scope of this book, it can be shown that the this can be rewritten as $\mathscr{E} = (NBA\omega)\sin\omega t$. The product of terms in the parentheses, $NBA\omega$, is the magnitude of the maximum emf, which occurs whenever $\sin\omega t = \pm 1$. If $NBA\omega$ is labeled $\mathscr{E}_{max}$, the maximum value of the emf, then the previous equation can be rewritten more compactly as

$$\mathscr{E} = (NBA\omega)\sin\omega t = \mathscr{E}_{max}\sin\omega t \quad \text{(ac generator output)} \quad (20.4a)$$

As can be seen in Equation 20.4, the sine function varies between $\pm 1$, and thus the polarity of the emf changes with time (▼ **Figure 20.13**). Note that the emf has its maximum value when $\theta = 90°$ or $\theta = 270°$. That is, at the instants when the plane of the loop is parallel to the field, and the magnetic flux is zero, the emf will be at its largest (magnitude). The rate of change in flux is greatest at these angles, because although the flux itself is momentarily zero, it is changing rapidly due to a *sign* change. Near the angles that produce the flux's largest value ($\theta = 0°$ and $\theta = 180°$), the flux is approximately constant and thus the induced emf is zero at those angles.

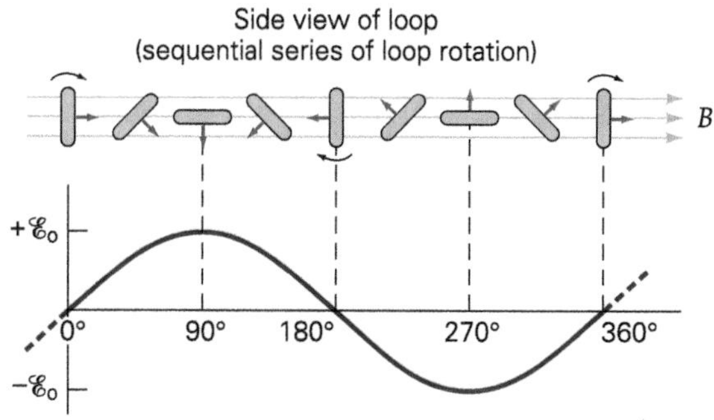


▲ FIGURE 20.13 **An ac generator output** A graph of the sinusoidal output of a generator, with a side view of the corresponding loop orientations during a cycle, showing the flux variation with time. Note that the emf is a maximum when the flux changes most rapidly, as it passes through zero and changes in sign.

In a circuit connected to such a generator, the induced current will also change direction periodically. In everyday applications, it is common to refer to the frequency ($f$) of the armature [in hertz (Hz)], rather than its angular frequency ($\omega$). Since they are related by $\omega = 2\pi f$, Equation 20.4a can be rewritten as

$$\mathscr{E} = (NBA\omega)\sin(2\pi ft) = \mathscr{E}_{max}\sin(2\pi ft) \quad \text{(ac generator output)} \tag{20.4b}$$

The ac frequency in the United States and most of the western hemisphere is 60 Hz. A frequency of 50 Hz is common in Europe and other areas of the world. It is important to keep in mind that Equations 20.4 and 20.5 give the *instantaneous* value of the emf. Thus the generator's output voltage varies from $+\mathscr{E}_{max}$ to $-\mathscr{E}_{max}$ over half of a rotational period. However, for most practical ac circuits, the time-averaged values of ac voltage and current are more important. This concept will be developed in detail in Chapter 21. To see how various factors influence the generator's output, examine Example 20.5 closely.

### EXAMPLE 20.5: AN AC GENERATOR – RENEWABLE ELECTRIC ENERGY

A farmer decides to use a waterfall to create a small hydroelectric power plant for his farm. He builds a coil consisting of 1500 circular loops of wire with a radius of 20 cm, which rotates on the generator's armature at 60 Hz in a magnetic field. To generate a time-averaged voltage of 120 V, he will need to create a *maximum* emf of 170 V (this concept will be discussed in more detail in Chapter 21). What is the magnitude of the generator's magnetic field necessary for this to happen?

**THINKING IT THROUGH.** The magnetic field can be determined from the expression for $\mathscr{E}_o$.

**SOLUTION**

***Given:***

$\mathscr{E}_{max} = 170\text{ V}$
$N = 1500\text{ loops}$
$r = 20\text{ cm} = 0.20\text{ m}$
$f = 60\text{ Hz}$

***Find:*** $B$ (magnitude of the generators magnetic field)

The generator's maximum emf is $\mathscr{E}_{max} = NBA\omega$. Because $\omega = 2\pi f$ and, for a circle, $A = \pi r^2$, this can be rewritten as

$$\mathscr{E}_{max} = NB(\pi r^2)(2\pi f) = 2\pi^2 NBr^2 f$$

Solving for $B$

$$B = \frac{\mathscr{E}_{max}}{2\pi^2 Nr^2 f} = \frac{170\text{ V}}{2\pi^2(1500)(0.20\text{ m})^2(60\text{ Hz})} = 2.4\times10^{-3}\text{ T}$$

**FOLLOW-UP EXERCISE.** In this Example, suppose the farmer lived in France where they use a time-averaged emf of 240 V and a maximum emf of 340 V. If he does this by changing only the size of the coil, what would the new radius have to be?

Clearly, electromagnetic induction plays a crucial role in our lives by supplying our electric energy. In most power plants the armature is actually stationary, and magnets revolve about it. The revolving magnetic field produces a time-varying flux through the coils of the armature and thus an ac output. A turbine supplies the mechanical energy required to spin the magnets. The turbines are typically powered by steam generated from the heat of combustion of fossil fuels or from nuclear fission.

Recently, as supplies of fossil fuels dwindle and concerns about their carbon emissions related to climate change arise, there has been increased interest in electric generation using renewable energy sources. For example, *hydroelectricity* generates electric energy using falling water to rotate the turbine, as shown in ▼**Figure 20.14a**. In Figure 20.14b a "wind farm" in a mountain pass near Palm Springs, California generates as much electric energy as a medium-sized power plant. Intense winds driven by desert heating and the narrowing of the pass rotate the blades that turn the turbines to generate electric energy.

(a)

(b)

▲FIGURE 20.14 **Electrical energy generation (a)** Gravitational potential energy of water is converted into electric energy. **(b)** Wind kinetic energy is converted into electric energy by turning the blades on these windmills in a "wind farm" near the San Gorgonio Pass in California.

Keep in mind, however, that in terms of basic physics, the only difference between these types of electric energy generation is the source of the energy that turns the turbines. In all cases, some form of energy is converted into electric energy using Faraday's law of induction.

### 20.2.2 Back emf

Although their main job is to convert electric energy into mechanical energy, motors also generate (induced) emfs at the same time. Like a generator, a motor has a rotating armature in a magnetic field. For motors, this induced emf (labeled as $\mathscr{E}_b$) is called a **back emf** (or **counter emf**), because its direction is opposite that of the line voltage that runs the motor and tends to reduce the that voltage and thus the current in the armature coils.

If $\Delta V$ is the line voltage, then the net voltage driving the motor is less than $\Delta V$ since the line voltage and the back emf ($\mathscr{E}_b$) are of opposite polarity. Thus the net voltage driving the motor is $\Delta V_{net} = \Delta V - \mathscr{E}_b$. If the motor's armature's resistance is $R$, the current in the motor while in operation is $I = \Delta V_{net}/R = (\Delta V - \mathscr{E}_b)/R$. Solving for the back emf yields

$$\mathscr{E}_b = \Delta V - IR \quad \text{(back emf of a motor)} \tag{20.5}$$

The back emf of a motor depends on the rotational speed of the armature and increases from zero to some maximum value as the armature goes from rest to normal operating speed. On startup, the back emf is zero (why?). Therefore the starting current is a maximum (Equation 20.5 with $\mathscr{E}_b = 0$). Ordinarily, a motor is trying to turn something, such as a drill bit; that is, it has a mechanical load. Without a load, the armature speed will increase until the back emf almost equals the line voltage. The result is a small current in the coils, just enough to overcome friction and joule heat losses. Under normal load conditions, the back emf is less than the line voltage. The larger the load, the slower the motor rotates and the smaller the back emf. If a motor is overloaded and turns very slowly, the back emf may be reduced so much that the current becomes very large (since $\Delta V_{net}$ increases as $\mathscr{E}_b$ decreases) and may burn out the coils. Thus back emf plays a vital role in the regulation of a motor's operation by limiting the current in it.

Schematically, a back emf in a dc motor circuit can be represented as an "induced battery" with polarity opposite that of the driving voltage (▶ **Figure 20.15**). To see how the back emf affects the current in a motor, consider the next Example.

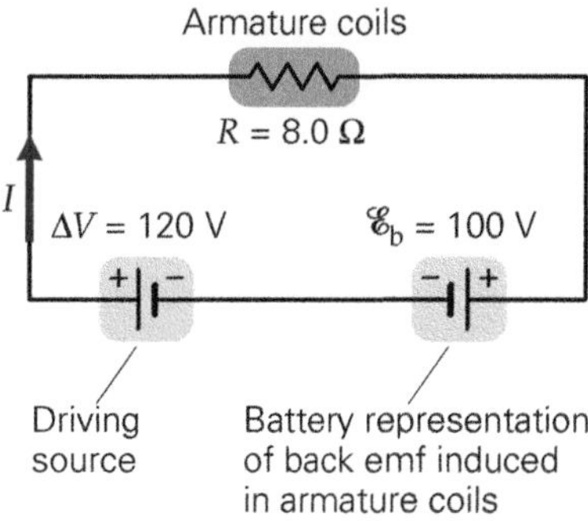


▲ FIGURE 20.15 **Back emf** The back emf in the armature of a dc motor can be represented as a battery with a polarity opposite that of the driving voltage. See Example 20.6.

---

**EXAMPLE 20.6: GETTING UP TO SPEED – BACK EMF IN A DC MOTOR**

A dc motor has windings that have a resistance of 8.00 Ω and operates at a line voltage of 120 V. With a normal load, there is a back emf of 100 V when the motor reaches full speed. (See Figure 20.15.) Determine (a) the starting current drawn by the motor and (b) the armature current at operating speed under a normal load.

**THINKING IT THROUGH.** (a) The only difference between startup and full speed is that there is no back emf at startup. The net voltage and resistance determine the current, so Equation 20.5 can be applied. (b) At operating speed, the back emf increases and is opposite in polarity to the line voltage. Equation 20.6 can again be used to determine the current.

**SOLUTION**

Listing the data as usual:

***Given:***

$$R = 8.00\ \Omega$$
$$\Delta V = 120\ \text{V}$$
$$\mathscr{E}_b = 100\ \text{V}$$

***Find:***

**(a)** $I_s$ (starting current)
**(b)** $I$ (operating current)

**(a)** From Equation 20.5, the current in the windings is

$$I_s = \frac{\Delta V}{R} = \frac{120\ \text{V}}{8.00\ \Omega} = 15.0\ \text{A}$$

**(b)** When the motor is at full speed, the back emf is 100 V; thus, the current is less.

$$I = \frac{\Delta V - \mathscr{E}_b}{R} = \frac{120\ \text{V} - 100\ \text{V}}{8.00\ \Omega} = 2.50\ \text{A}$$

With little or no back emf, the *starting* current is relatively large. Thus when a big motor, such as that of a central air conditioning unit, starts up, the lights in a building might momentarily dim. In some designs, resistors are temporarily connected in series with a motor's coil to protect the windings from burning out as a result of large starting currents.

**FOLLOW-UP EXERCISE.** In this Example, (a) how much energy is required to bring the motor to operating speed if it takes 10 s and the back emf averages 50 V during that time? (b) Compare this amount with the amount of energy required to keep the motor running for 10 s once it reaches operating conditions.

---

## 20.3 Transformers and Power Transmission

Electric energy is typically distributed long distances by power lines. It is desirable to minimize $I^2R$ losses (joule heating) that can occur in these lines. Because the resistance of a line is fixed, reducing $I^2R$ losses means reducing current. However, the power output of a generator is determined by its output current and voltage ($P = IV$), and for a fixed voltage, such as 120 V, a reduction in current would mean reduced power output. It might appear that there is no way to reduce the current while maintaining the power level. Fortunately, electromagnetic induction enables the reduction of transmission losses by increasing voltage while simultaneously reducing current such that the delivered power is essentially unchanged. This is done using devices called **transformers**.

A simple transformer consists of two coils of insulated wire wound on the same iron core (▼ **Figure 20.16a**). When ac voltage is applied to the input coil, or *primary coil*, the alternating current produces an alternating magnetic flux concentrated in the iron core, without any significant leakage of flux outside the core. Under these conditions, the same changing flux also passes through the output coil, or *secondary coil*, inducing an alternating voltage and current in it. (Note that it is common in transformer design to refer to emfs as "voltages," as was done in Chapter 18. Thus for this discussion the symbol $\Delta V$ will be used in place of $\mathscr{E}$.)

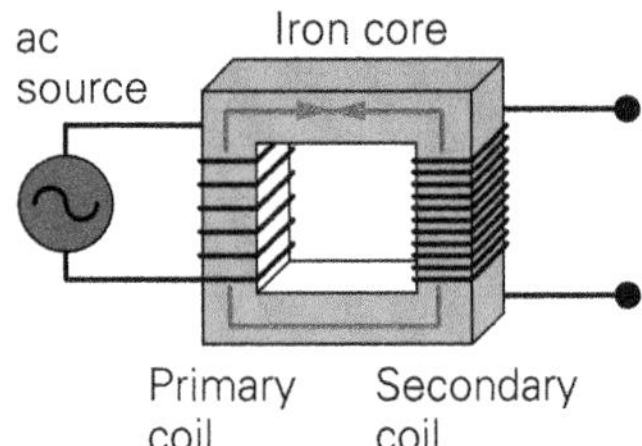


**(a) Step-up transformer: high-voltage (low-current) output**

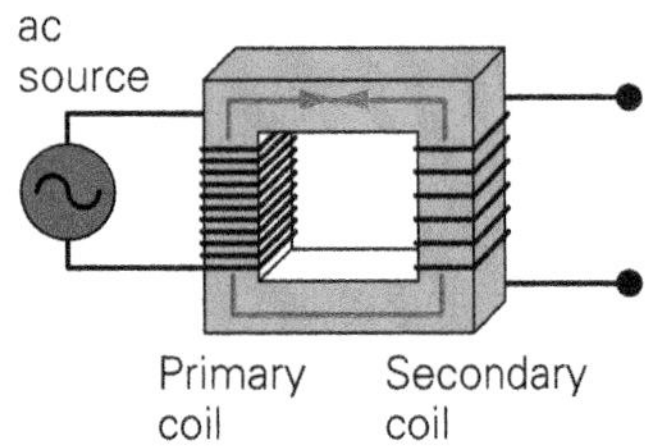


**(b) Step-down transformer: low-voltage (high-current) output**

▲ FIGURE 20.16 **Transformers (a)** A step-up transformer has more turns in the secondary coil than in the primary coil. **(b)** A step-down transformer has more turns in the primary coil than in the secondary coil.

The ratio of the induced voltage in the secondary coil to the voltage in the primary depends on the ratio of the number of turns in the coils. By Faraday's law, the induced voltage in the secondary (subscript "s") is $\Delta V_s = -N_s(\Delta\Phi/\Delta t)$ where $N_s$ is the number of turns in the secondary coil. The changing flux in the primary coil (subscript "p") produces a back emf of $\Delta V_p = -N_p(\Delta\Phi/\Delta t)$ where $N_p$ is the number of turns in the primary coil.

If the resistance of the primary coil is neglected, then the back emf is equal in magnitude to the external voltage applied to the primary (why?). Forming a ratio of output voltage (secondary) to input voltage (primary) yields

$$\frac{\Delta V_s}{\Delta V_p} = \frac{-N_s(\Delta\Phi/\Delta t)}{-N_p(\Delta\Phi/\Delta t)}$$

which can finally be rewritten as

$$\frac{\Delta V_s}{\Delta V_p} = \frac{N_s}{N_p} \quad \text{(voltage ratio in a transformer)} \qquad (20.6)$$

If the transformer is ideal, that is 100% efficient (no energy losses), then the power input is equal to the power output: $P_p = I_p\Delta V_p = P_s = I_s\Delta V_s$. Although some energy is always lost to joule heat, this is a good assumption, since a well-designed transformer has efficiency greater than 95%. (The details of transformer energy losses will be discussed shortly.) Assuming the ideal case, from Equation 20.6, the transformer currents and voltages are related to the turn ratio by

$$\frac{I_p}{I_s} = \frac{\Delta V_s}{\Delta V_p} = \frac{N_s}{N_p} \quad \text{(current ratio in a transformer)} \quad (20.7)$$

Thus in practice the transformer action in terms of voltage and current output can be summarized as

$$\Delta V_s = \left(\frac{N_s}{N_p}\right)\Delta V_p \quad \text{(output transformer voltage)} \quad (20.8a)$$

and

$$I_s = \left(\frac{N_p}{N_s}\right)I_p \quad \text{(output transformer current)} \qquad (20.8b)$$

If the secondary coil has more windings than the primary ($N_s/N_p > 1$) as in Figure 20.16a, the voltage is "stepped up," because that is $\Delta V_s > \Delta V_p$. In this case it is called a *step-up transformer*. Notice that while the voltage is increased, to keep the power constant there is *reduced* current in the secondary. If, however, the secondary coil has fewer turns than the primary, the transformer is called a *step-down transformer* (Figure 20.16b), because the voltage is "stepped down," and the current, therefore, increased. Depending on the design details, a step-up transformer may be used as a step-down transformer by reversing output and input connections.

### INTEGRATED EXAMPLE 20.7: TRANSFORMER ORIENTATION – STEP-UP OR STEP-DOWN CONFIGURATION?

An ideal 600-W transformer has 50 turns on its primary coil and 100 turns on its secondary coil. (a) Is this transformer acting as (1) a step-up or (2) a step-down transformer? (b) If the primary coil is connected to a 120-V source, what are its output voltage and current?

**(a) CONCEPTUAL REASONING.** The terms step-up and step-down refer to what happens to voltage, not current. Because the voltage is proportional to the number of turns, in this case the secondary voltage is greater than the primary voltage. Thus the correct answer is (1), a step-up transformer.

**(b) QUANTITATIVE REASONING AND SOLUTION.** The output voltage can be determined from Equation 20.8a once the turn ratio is established. From the power, the current can be determined.

***Given:***

$$N_p = 50$$
$$N_s = 100$$
$$\Delta V_p = 120 \text{ V}$$

***Find:*** $\Delta V_s$ and $I_s$ (secondary voltage and current)

The secondary voltage can be found using Equation 20.8a with a turn ratio of 2, because $N_s = 2N_p$

$$\Delta V_s = \left(\frac{N_s}{N_p}\right)\Delta V_p = (2)(120 \text{ V}) = 240 \text{ V}$$

Since the transformer is ideal, the input power equals the output power. On the primary side, the input power is $P_p = I_p\Delta V_p = 600$ W, so the input current is

$$I_p = \frac{600 \text{ W}}{\Delta V_p} = \frac{600 \text{ W}}{120 \text{ V}} = 5.00 \text{ A}$$

Because the voltage is stepped up by a factor of two, the output current must thus be stepped down by a factor of two. From Equation 20.8b, or

$$I_s = \left(\frac{N_p}{N_s}\right)I_p = \left(\frac{1}{2}\right)(5.00 \text{ A}) = 2.50 \text{ A}$$

**FOLLOW-UP EXERCISE.** (a) When a European visitor (the average ac voltages are 240 V in Europe) visits the United States, what type of transformer should be used to enable her hair dryer to work properly? Explain. (b) For a 1500-W hair dryer (assumed ohmic), what would be the transformer's input current in the United States, assuming it to be ideal?

Many factors combine to determine how close a real transformer comes to being "ideal." First, there is flux leakage, since not all of the flux makes it to the secondary coil. Second, the ac current in the primary means there is a changing magnetic flux in those coils. In turn, this gives rise to an induced emf in the primary. This effect is called *self-induction*. This self-induced emf opposes, and thus limits, the primary current in a similar way to the back emf in a motor. A third thing that prevents transformers from being ideal is joule heating in its wires. Last, consider the effect of induction on the core material itself. To increase magnetic flux, the core is made of a highly permeable material (such as iron), but such materials are also good conductors. The changing magnetic flux in the core induces emfs there, which in turn create *eddy (or "swirling") currents* in the core material. These eddy currents cause energy loss by heating the core ($I^2R$ losses again). To reduce the loss of energy due to eddy currents, transformer cores are made of thin sheets of material (usually iron) laminated with a non-conducting glue between them. The insulating layers between the sheets break up the eddy currents, thus greatly reducing energy loss.

## 20.3.1 Power Transmission and Transformers

For power transmission over long distances, transformers provide a way to increase the voltage while reducing the current output of the generating power plant, thus cutting down the joule heating losses in the transmission wires. The voltage output of the generating plant is first stepped up, reducing the current, and in that form the energy is transmitted over long distances to an area substation near consumers. At that point, the voltage is stepped down, increasing the current. There are further step-downs at distributing substations and utility poles before the electricity is supplied to homes and businesses at the normal voltage and current. The following Example illustrates the benefits of being able to step up the voltage (while stepping down the current) for electrical power transmission.

### EXAMPLE 20.8: CUTTING YOUR LOSSES – POWER TRANSMISSION AT HIGH VOLTAGE

A small hydroelectric power plant produces energy in the form of electric current at 10 A and a voltage of 440 V. The voltage is stepped up to 4400 V (by an ideal transformer) for transmission over 40 km of power line, which has a total resistance of 20 Ω. (a) What percentage of the produced power would have been lost in transmission if the voltage had not been stepped up? (b) What percentage of the produced power is actually lost when the voltage is stepped up?

**THINKING IT THROUGH.** (a) The power output can be computed from $P = I\Delta V$ and compared with the power lost in the wire, $P = I^2R$. (b) Equations 20.8a and 20.8b enable the determination of the stepped-up voltage and stepped-down currents, respectively. Then the calculation is repeated, and the results are compared with those of part (a).

**SOLUTION**

***Given:***

$$I_p = 10 \text{ A}$$
$$\Delta V_p = 440 \text{ V}$$
$$\Delta V_s = 4400 \text{ V}$$
$$R = 20\ \Omega$$

***Find:***

**(a)** Percentage power loss without voltage step-up
**(b)** Percentage power loss with voltage step-up

**(a)** The power output by the generator is

$$P_{out} = I_p \Delta V_p = (10 \text{ A})(440 \text{ V}) = 4400 \text{ W}$$

The power loss in the wire if transmitting a current of 10 A is

$$P_{loss} = I^2 R = (10 \text{ A})^2 (20\ \Omega) = 2000 \text{ W}$$

Thus, the percentage of power lost to joule heat in the wires is

$$\% \text{ power loss} = \frac{P_{loss}}{P_{out}} \times 100\% = \frac{2000 \text{ W}}{4400 \text{ W}} \times 100\% = 45\%$$

**(b)** When the voltage is stepped up to 4400 V, the output current is reduced by a factor of 10 from its value in part (a). Thus the secondary current is

$$I_s = \left(\frac{\Delta V_p}{\Delta V_s}\right) I_p = \left(\frac{440 \text{ V}}{4400 \text{ V}}\right)(10 \text{ A}) = 1.0 \text{ A}$$

The power (loss) is thus reduced by a factor of 100, since it varies as the square of the current:

$$P_{loss} = I^2 R = (1.0 \text{ A})^2 (20\ \Omega) = 20 \text{ W}$$

Thus the percentage of power lost is also reduced by a factor of 100 for a considerable saving:

$$\% \text{ power loss} = \frac{P_{loss}}{P_{out}} \times 100\% = \frac{20 \text{ W}}{4400 \text{ W}} \times 100\% = 0.45\%$$

**FOLLOW-UP EXERCISE.** Some heavy-duty electrical appliances, such as water pumps, can be wired to operate at either 240 V or 120 V. Their power rating is the same regardless of the voltage at which they run. (a) Explain the efficiency advantage of operating such appliances at the higher voltage. (b) For a 1.00-hp pump (746 W), estimate the ratio of the power lost in the wires at 240 V to the power lost at 120 V (assuming the pump and connecting wire resistances are ohmic and do not change).

## 20.4 Electromagnetic Waves

Electromagnetic waves (or *electromagnetic radiation*) were considered as a means of heat transfer in Section 11.4. The production and characteristics of electromagnetic radiation can now be understood, because these waves are composed of electric and magnetic fields.

Scottish physicist James Clerk Maxwell (1831–1879) is credited with first bringing together, or *unifying*, electric and magnetic phenomena. Using mathematics beyond the scope of this book, he took the equations that governed both electric and magnetic fields and predicted the existence of electromagnetic waves. In fact, he went further and calculated their speed in a vacuum, and his prediction agreed with the experiment. Because of these contributions, the set of equations is collectively known as *Maxwell's equations*, although they were, for the most part, developed by others (e.g., Faraday's law of induction).

Essentially, the electric and magnetic fields are really just two parts of a single *electromagnetic field*. The two apparently separate fields are symmetrically related in the sense that either one can create the other under the proper conditions. This symmetry was evident to Maxwell during his analysis of the equations (not shown). A qualitative summary of the results is sufficient:

> A time-varying magnetic field produces a time-varying electric field.
>
> A time-varying electric field produces a time-varying magnetic field.

The first statement restates, in field language, our observations in Section 20.1: A changing magnetic flux gives rise to an induced emf. The second statement (which will not be studied in detail) is crucial to the self-propagating characteristic of electromagnetic waves. Together, these two phenomena enable these waves to travel through a vacuum, whereas all other waves, such as water waves, require a supporting medium.

According to Maxwell's theory, *accelerating* electric charges, such as an oscillating electron, produce electromagnetic waves. The electron in question could, for example, be one of the many electrons in the metal antenna of a radio transmitter, driven by an electrical (voltage) oscillator at a frequency of $10^6$ Hz (1 MHz). As each electron oscillates, it continually accelerates and decelerates and thus radiates an electromagnetic wave (▶ **Figure 20.17a**). The oscillations of these electrons produce time-varying electric and magnetic fields in the vicinity of the antenna. The electric field, shown in dark red in Figure 20.17a, is in the plane of the page and continually changes direction, as does the magnetic field (shown in light red and pointing into and out of the page).

Both the electric and the magnetic fields carry energy and propagate outward at the speed of light, symbolized by the letter *c*. To three significant figures, the speed of light in a vacuum is $3.00 \times 10^8$ m/s. At large distances from the source, these waves become planar. (Shown in Figure 20.17b at an instant in time.)

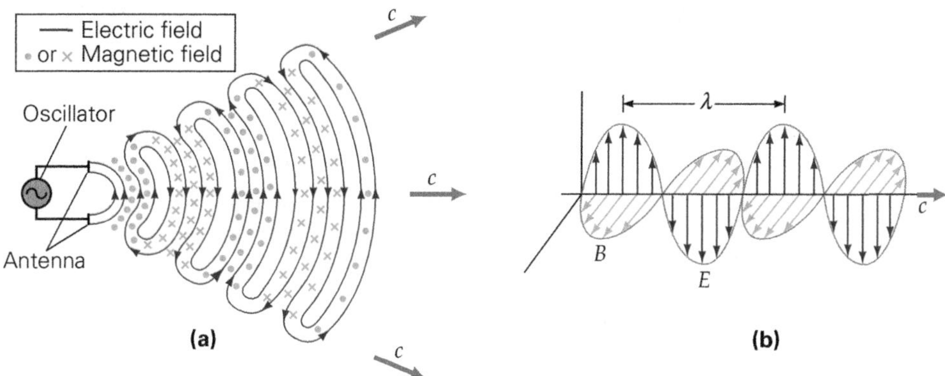


▲ FIGURE 20.17 **Source of electromagnetic waves** Electromagnetic waves are produced, fundamentally, by accelerating electric charges. **(a)** Here charges (electrons) in a metal antenna are driven by an oscillating voltage source. As the antenna polarity and current direction periodically change, alternating electric and magnetic fields propagate outward. The electric and magnetic fields are perpendicular to the direction of wave propagation. Thus, electromagnetic waves are transverse waves. **(b)** At large distances from the source, the initially curved wave fronts become planar.

Here, the electric field ($\vec{\mathbf{E}}$) is perpendicular to the magnetic field ($\vec{\mathbf{B}}$) as each varies sinusoidally with time. Both field vectors are perpendicular to the direction of wave propagation, thus electromagnetic waves are *transverse*. According to Maxwell's theory, as one field changes, it creates the other. This process, repeated again and again, gives rise to the traveling electromagnetic wave that we call light and explains why no medium is needed to support this wave. An important result of Maxwell's analysis is:

> In a vacuum, all electromagnetic waves, regardless of frequency or wavelength, travel at the same speed or $c$, or approximately $3.00 \times 10^8$ m/s.

This extremely high value speed of light (nearly 186 000 mi/s) means that for everyday distances, unlike sound, for example, the time delay due to the transmission of light can be neglected. However, for interplanetary trips, this delay can be a problem. Consider the following Example.

---

**EXAMPLE 20.9: LONG-DISTANCE GUIDANCE – THE SPEED OF ELECTROMAGNETIC WAVES IN A VACUUM**

Probes have successfully landed on Mars for many decades now, receiving control signals from Earth as well as sending data back. How much longer does it take for a signal to reach us when Mars is farthest from Earth than when it was closest? The average distances of Mars and the Earth from the Sun are 229 million km (call it $d_M$) and 150 million km ($d_E$), respectively. Assume both planets are in circular orbits and take these average distances for their orbital radii to estimate your answer.

**THINKING IT THROUGH.** This situation calls for a time–distance calculation. The planets are farthest apart when they are on opposite sides of the Sun and separated by a distance of $d_M + d_E$. (This arrangement requires signals to be sent through the Sun, which is not possible. However, it does serve to determine an upper limit on time.) The planets are closest when they are aligned on the same side of the Sun. In this case, their separation distance is at a minimum value of $d_M - d_E$. (Draw a diagram to help visualize this.) Because the speed of electromagnetic waves in a vacuum is known, the travel times can be found from $\Delta t = d/c$, and their difference is what we are asked for.

**SOLUTION**

Listing the data and converting the distances to meters:

***Given:***

$$d_M = 229 \times 10^6 \text{ km} = 2.29 \times 10^{11} \text{ m}$$
$$d_E = 150 \times 10^6 \text{ km} = 1.50 \times 10^{11} \text{ m}$$

***Find:*** $\Delta t$ (difference in time for light to travel the longest and shortest distances)

The longest travel time $\Delta t_L$ is

$$\Delta t_L = \frac{d_M + d_E}{c} = \frac{3.79 \times 10^{11} \text{ m}}{3.00 \times 10^8 \text{ m/s}} = 1.26 \times 10^3 \text{ s (or 21.1 min)}$$

The shortest travel time $t_s$ is

$$\Delta t_S = \frac{d_M - d_E}{c} = \frac{7.90 \times 10^{10} \text{ m}}{3.00 \times 10^8 \text{ m/s}} = 2.63 \times 10^2 \text{ s (or 4.39 min)}$$

The difference in the times is

$$\Delta T = \Delta t_L - \Delta t_S = 1.00 \times 10^3 \text{ s (or 16.7 min)}$$

**FOLLOW-UP EXERCISE.** Assume that a Martian exploratory rover is heading for a collision with a rock 2.0 m ahead of it. When it is at that distance, the vehicle sends a picture of the rock to controllers on Earth. If Mars is at the closest point to the Earth, what is the maximum speed that the vehicle could have and still avoid a collision? Assume that the video signal from the vehicle reaches Earth and that the signal for it to stop is sent back with no time delay.

---

### 20.4.1 Types of Electromagnetic Waves

Electromagnetic waves are classified by the range of frequencies (or wavelengths) they encompass. Recall from Chapter 13 that wave frequency and wavelength are inversely related. For electromagnetic waves this relationship is $\lambda = c/f$, where the speed of light, $c$, has been substituted for a general wave speed, $v$. Hence the higher the frequency, the shorter the wavelength, and vice versa. The electromagnetic spectrum is continuous, so the limits of the various types of radiation are approximate (▼**Figure 20.18**). ▶**Table 20.1** lists these ranges for the various types of electromagnetic waves. Note that in everyday language, the word "light" is taken to mean "visible electromagnetic waves" but in physics, the word "light" is taken to mean *any* electromagnetic wave type, since they have the same fundamental structure, differing only in frequency and wavelength.

#### 20.4.1.1 Power Waves

Electromagnetic waves with a frequency of 60 Hz result from alternating currents in electric power lines. These power waves have a wavelength of about 5000 km. Waves of such low frequency are of little practical use. They may occasionally produce a so-called 60-Hz hum on your speakers or introduce, via induction, unwanted electrical noise in delicate instruments. More serious concerns have been expressed about the possible effects of these waves on health. Some early research tended to suggest that very low-frequency fields may have potentially harmful biological effects on cells and tissues. However, recent surveys indicate that this is not the case.

#### 20.4.1.2 Radio and TV Waves

Radio and TV waves are generally in the frequency range from 500 kHz to about 1000 MHz. The AM (*amplitude-modulated*) band runs from 530 to 1710 kHz (1.71 MHz). Higher frequencies, up to 54 MHz, are used for "short-wave" bands. TV bands range from 54 MHz to 890 MHz. The FM (*frequency-modulated*) radio band runs from 88 to 108 MHz. Cellular phones use radio waves to transmit voice communication in the ultrahigh-frequency (UHF) band.

#### 20.4.1.3 Microwaves

Microwaves, with frequencies in the gigahertz (GHz) range, are produced by special vacuum tubes (called *klystrons* and *magnetrons*). Microwaves are commonly used in communications, ovens, and radar applications. In addition to roles in navigation and guidance, radar provides the basis for the speed guns used to time such things as baseball pitches and motorists using the Doppler effect (Section 14.5).

#### 20.4.1.4 Infrared Radiation

The infrared (IR) region of the electromagnetic spectrum lies adjacent to the low-frequency, or long-wavelength, end of the visible spectrum. A warm body, for example, emits IR radiation, which depends on that body's temperature. (See Chapter 27.) An object at or near room temperature emits radiation in the far IR region. ("Far" means relative to the visible region.) IR radiation is sometimes called "heat radiation." As such, IR lamps are widely used in therapeutic applications, such as easing pain in strained muscles, as well as to keep food warm.

#### 20.4.1.5 Visible Light

The region of visible light occupies only a small portion of the electromagnetic spectrum and covers a frequency range from about $4 \times 10^{14}$ Hz to about $7 \times 10^{14}$ Hz. In terms of wavelengths, the range is from about 700 to 400 nm (see inset in Figure 20.18). [1 nanometer (nm) = $10^{-9}$ m.] Only the radiation in this region activates the receptors on the retina of human eyes. Visible light emitted or reflected from objects provides us with visual information about our world. Visible light and optics will be discussed in Chapters 22 through 25. It is interesting to note that not all animals are sensitive to the same range of

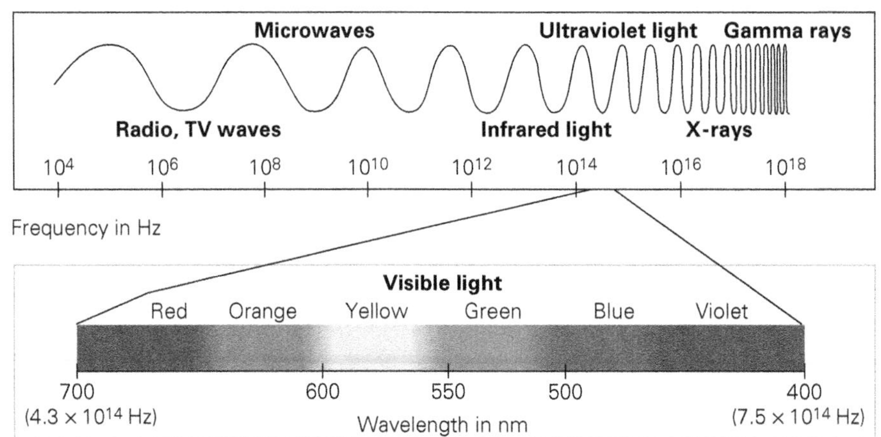


▲FIGURE 20.18 **The electromagnetic spectrum** The spectrum of frequencies or wavelengths is divided into various regions, or ranges. The visible light region is a very small part of the total spectrum. For visible light, wavelengths are usually expressed in nanometers (1 nm = $10^{-9}$ m). (The relative sizes of the wavelengths at the top of the figure are not to scale.)

**TABLE 20.1** Classification of Electromagnetic Waves

| Type of Wave | Approximate Frequency Range (Hz) | Approximate Wavelength Range (m) | Some Typical Sources |
|---|---|---|---|
| Power waves | 60 | $5.0 \times 10^6$ | Electric currents |
| Radio waves - AM | $0.53 \times 10^6$–$1.7 \times 10^6$ | 570–186 | Electric circuits/antennae |
| Radio waves - FM | $88 \times 10^6$–$108 \times 10^6$ | 3.4–2.8 | Electric circuits/antennae |
| TV | $54 \times 10^6$–$890 \times 10^6$ | 5.6–0.34 | Electric circuits/antennae |
| Microwaves | $10^9$–$10^{11}$ | $10^{-1}$–$10^{-3}$ | Special vacuum tubes |
| Infrared radiation | $10^{11}$–$10^{14}$ | $10^{-3}$–$10^{-7}$ | Warm and hot bodies, stars |
| Visible light | $4.0 \times 10^{14}$–$7.0 \times 10^{14}$ | $10^{-7}$ | The Sun and other stars, lamps |
| Ultraviolet radiation | $10^{14}$–$10^{17}$ | $10^{-7}$–$10^{-10}$ | Very hot bodies, stars, special lamps |
| X-rays | $10^{17}$–$10^{19}$ | $10^{-10}$–$10^{-12}$ | High-speed electron collisions, atomic processes |
| Gamma rays | Above $10^{19}$ | Below $10^{-12}$ | Nuclear reactions, nuclear decay processes |

wavelengths. For example, snakes can visually detect infrared radiation, and the visible range of many insects extends well into the ultraviolet range.

### 20.4.1.6 Ultraviolet Radiation

The Sun's spectrum has a component of ultraviolet (UV) light, whose frequency range lies beyond the violet end of the visible region. UV is also produced artificially by special lamps and very hot objects. In addition to tanning the skin, UV radiation can cause sunburn and/or skin cancer if exposure is too high. Upon its arrival at Earth, most of the Sun's UV light is absorbed in the ozone ($O_3$) layer of the atmosphere, at an altitude of about 30–50 km. Because this layer plays a protective role, there is concern about its depletion due to chlorofluorocarbon gases (such as Freon, once commonly used in refrigerators) that drift upward and react with the ozone.

### 20.4.1.7 X-Rays

Beyond the UV region of the electromagnetic spectrum is the important X-ray region. X-rays primarily bring to mind medical applications. They were discovered accidentally in 1895 by the German physicist Wilhelm Roentgen (1845–1923) when he noted the glow of a piece of fluorescent paper, evidently caused by some mysterious radiation coming from a cathode ray tube. Because of their apparent mystery, these rays were called *X-radiation*, or *X-rays* for short.

The basic elements of an X-ray tube are shown in ▶ **Figure 20.19**. An accelerating voltage, typically several thousand volts, is applied across the electrodes in a sealed, evacuated tube. Electrons emitted from the heated negative electrode (cathode) are accelerated toward the positive electrode (anode). When striking the anode they decelerate, and some of their kinetic energy is converted to electromagnetic energy in the form of X-rays.

As will be discussed in Chapter 27, the energy carried by electromagnetic radiation depends on its frequency. High-frequency X-rays have very high energies and can cause cancer, skin burns, and other harmful effects. However, at low intensities, X-rays can be used with relative safety to view the internal structure of the human body and other opaque objects.

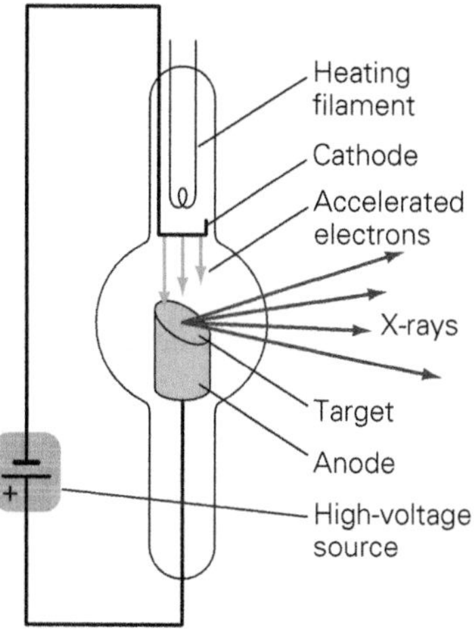


▲ **FIGURE 20.19 The X-ray tube** Electrons accelerated through a large voltage strike a target electrode. There they slow down and interact with the electrons of the target material. Energy is emitted in the form of X-rays during this "braking" (deceleration) process.

### 20.4.1.8 Gamma Rays

The electromagnetic waves of the uppermost frequency range of the electromagnetic spectrum are *gamma rays* ($\gamma$-rays). This high-frequency radiation is produced in nuclear reactions, particle accelerators, and certain types of nuclear decay (radioactivity). Gamma rays will be discussed in more detail in Chapter 29.

## Chapter 20 Review

- Magnetic flux is a measure of the number of magnetic field lines that pass through an area. For a single wire loop of area $A$ it is defined as

$$\Phi = BA\cos\theta \quad \text{(magnetic flux)}$$

where $B$ is the magnetic field strength (assumed constant) and $\theta$ is the angle between the direction of the field and the normal to the loop's plane.

- **Faraday's law of induction** relates the induced emf $\mathscr{E}$ in a loop (or coil composed of $N$ loops in series) to the time rate of change of the magnetic flux through that loop (or coil).

$$\mathscr{E} = -\frac{\Delta(N\Phi)}{\Delta t} = -N\frac{\Delta\Phi}{\Delta t} \quad (20.2)$$

  where $\Delta\Phi$ is the change in flux through one loop and there are $N$ total loops.
- **Lenz's law** states that when a change in magnetic flux induces an emf in a coil, loop, or circuit, the resulting, or induced, current direction is such as to create a magnetic field to oppose the change in flux.
- An **ac generator** converts mechanical energy into electrical energy. The generator's emf as a function of time is

$$\mathscr{E} = (NBA\omega)\sin(2\pi ft) = \mathscr{E}_{\text{max}}\sin(2\pi ft) \quad \text{(ac generator output)} \quad (20.4b)$$

  where $\mathscr{E}_{\text{max}}$ is the maximum emf and the generator's frequency is $f$.
- A **transformer** is a device that changes the voltage supplied to it by means of induction. The voltage at the input, or primary (p), side of the transformer is changed into the output, or secondary (s), voltage. The current and voltage relationships for a transformer are

$$\Delta V_s = \left(\frac{N_s}{N_p}\right)\Delta V_p \quad (20.8a)$$

$$I_s = \left(\frac{N_p}{N_s}\right)I_p \quad (20.8b)$$

- An **electromagnetic wave** (light) consists of time-varying electric and magnetic fields that propagate at a speed of light $c = (3.00 \times 10^8\ \text{m/s})$ in a vacuum. The various types of electromagnetic radiation (such as UV, radio waves, and visible light) have the same electromagnetic structure and basis but differ in frequency and wavelength.

# End of Chapter Questions and Exercises

## Multiple Choice Questions

### 20.1 Induced emf: Faraday's Law and Lenz's Law

1. Which of the following is an SI unit of magnetic flux (there may be more than one correct answer): (a) Wb, (b) T·m$^2$, (c) (T·m)/A, or (d) T?
2. The magnetic flux through a loop can change due to a change in which of the following (there may be more than one correct answer): (a) the area of the coil, (b) the strength of the magnetic field, (c) the orientation of the loop with respect to a fixed field direction, or (d) the direction of the field relative to a fixed loop?
3. For a current to be induced in a wire loop, (a) there must be a large magnetic flux through the loop, (b) the loop's plane must be parallel to the magnetic field, (c) the loop's plane must be perpendicular to the magnetic field, (d) the magnetic flux through the loop must vary with time.
4. Identical single loops A and B are oriented so they initially have the maximum amount of flux in a magnetic field. Loop A is then quickly rotated so that its normal is perpendicular to the magnetic field, and in the same time, B is rotated so its normal makes an angle of 45° with the field. How do their induced emfs compare: (a) they are the same; (b) A's is larger than B's; (c) B's is larger than A's; or (d) you can't tell the relative emf magnitudes from the data given?
5. Identical single loops A and B are oriented so they have the maximum amount of flux when placed in a magnetic field. Both loops maintain their orientation relative to the field, but in the same amount of time A is moved to a region of stronger field, while B is moved to a region of weaker field. How do their induced emfs compare: (a) they are the same; (b) A's is larger than B's; (c) B's is larger than A's; or (d) you can't tell the relative emf magnitudes from the data given?
6. A bar magnet is thrust toward the center of a circular metallic loop. The magnet approaches with its length perpendicular to the coil's plane. As the bar recedes from your view and approaches the coil, a clockwise current is induced in the loop. What polarity is that end of the bar magnet nearest the coil: (a) north, (b) south, (c) you can't tell from the data given?
7. The north end of a bar magnet is quickly pulled away from the center of a circular metallic loop. The magnet's length is always perpendicular to the coil's plane. As the south end of the magnet approaches you, what would be the induced current direction in the coil: (a) clockwise, (b) counterclockwise, (c) the induced current would be zero, or (d) you can't tell from the data given?

### 20.2 Electric Generators and Back emf

8. Increasing only the coil area in an ac generator would result in (a) an increase in the frequency of rotation, (b) a decrease in the maximum induced emf, (c) an increase in the maximum induced emf, (d) no change in the generator output.
9. In an ac generator, the maximum emf output occurs when the magnetic flux through the coil is (a) zero, (b) maximum, (c) not changing, (d) it does not depend on the flux in any way.

10. In an ac generator, the maximum emf output occurs when the magnetic flux through the coil is (a) changing most rapidly, (b) not changing, (c) at its maximum value, (d) it does not depend on the flux in any way.
11. The back emf of an electric motor depends on which of the following (there may be more than one correct answer): (a) the line voltage, (b) the current in the motor, (c) the armature's rotational speed, or (d) none of the preceding?

**20.3 Transformers and Power Transmission**

12. A transformer at the local substation of the delivery system just before your house has (a) more windings in the primary coil, (b) more windings in the secondary coil, (c) the same number of windings in the primary and secondary coils.
13. The output power delivered by a realistic step-down transformer is (a) greater than the input power, (b) less than the input power, (c) the same as the input power.
14. A transformer located just outside a power plant before the energy is delivered over the wires would have (a) more windings in the primary coil than in the secondary, (b) more windings in the secondary coil than in the primary, (c) the same number of windings in the primary and secondary coils.
15. A transformer located just outside a power plant before the energy is delivered over the wires would have (a) more current in the primary coil than in the secondary, (b) more current in the secondary coil than in the primary, (c) the same current in the primary and secondary coils.

**20.4 Electromagnetic Waves**

16. Relative to the blue end of the visible spectrum, the yellow and green regions have light waves of (a) higher frequencies, (b) longer wavelengths, (c) shorter wavelengths, (d) both (a) and (c).
17. Which of these electromagnetic waves has the lowest frequency: (a) UV, (b) IR, (c) X-ray, or (d) microwave?
18. Which of these electromagnetic waves travels slowest in a vacuum: (a) green light, (b) infrared light, (c) gamma rays, (d) radio waves, or (e) they all have the same speed?
19. If the frequency of an orange source of light was halved, what kind of light would it then put out: (a) red, (b) blue, (c) violet, (d) UV, (e) X-ray, or (f) IR?

## Conceptual Questions

**20.1 Induced emf: Faraday's Law and Lenz's Law**

1. A bar magnet is dropped through a coil of wire as shown in ▶ **Figure 20.20**. (a) Describe what is observed on the galvanometer by sketching a graph of induced emf versus t. (b) Does the magnet fall freely? Explain.

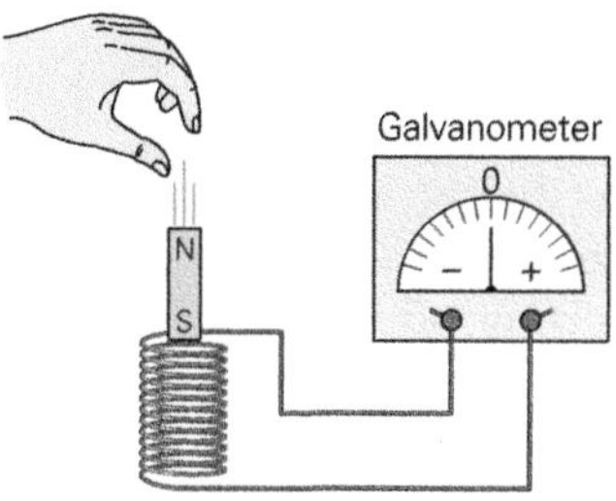


▲FIGURE 20.20 **A time-varying magnetic field** What will the galvanometer measure? See Conceptual Question 1.

2. In Figure 20.1b, what would be the direction of the induced current in the loop if the south pole of the magnet were approaching instead of the north pole?
3. In Figure 20.7a, how could you move the coil so as to prevent any current from being induced in it? Explain.
4. Two identical strong magnets are dropped simultaneously by two students into two vertical tubes of the same dimensions (▼**Figure 20.21**). One tube is made of copper, and the other is made of plastic. From which tube will the magnet emerge first? Why?

▲FIGURE 20.21 **Free fall?** See Conceptual Question 4.

5. By using the appropriate SI units, show that the units on both sides of Equation 20.2 (Faraday's law) are the same.
6. If a fixed-area metal loop is kept entirely within a uniform magnetic field and a current is induced in it, what must its motion be? Explain.
7. A circular metal loop is kept entirely within a magnetic field but moved to a region of higher field strength while *not* rotated. What could you do to its diameter to prevent an induced current in it? Explain.
8. (a) In Figure 20.10, what would happen to the direction of the induced current if the metal rod were moving downward instead? (b) What would happen to the direction of the induced magnetic field? Is the induced magnetic field necessarily opposite to the external field direction? [*Hint:* Compare your answer to part (b) with the direction in the figure.]

**20.2 Electric Generators and Back emf**

9. (a) Explain why the maximum emf produced by an ac generator occurs when the flux through its armature coil is zero. (b) Explain why the emf produced by an ac generator is zero when the flux through its armature coil is at its maximum.
10. A student has a bright idea for a generator that apparently generates electric energy without a corresponding loss of energy somewhere else. His suggested arrangement is shown in ▼**Figure 20.22.** The magnet is initially pulled down and released. When the magnet is attached to a spring, the student thinks that there should be a continuous electrical output as the magnet oscillates. What is wrong with this idea? [*Hint:* Check the total energy stored in the spring – magnet system and the forces on the magnet as it oscillates.]

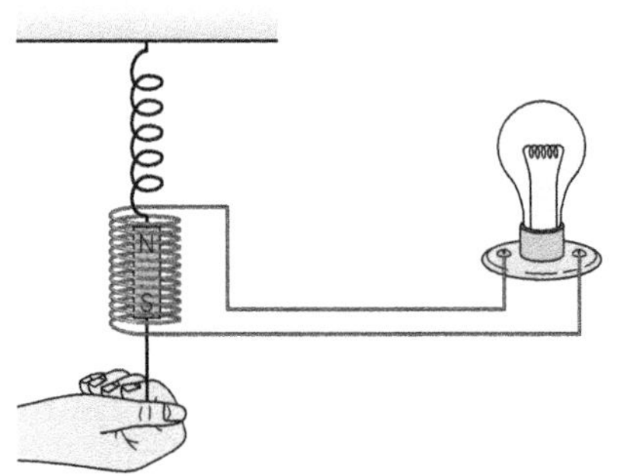


▲FIGURE 20.22 **Inventive genius?** See Conceptual Question 10.

11. In a dc motor, if the armature is jammed or turns very slowly under a heavy load, the coils in the motor may burn out. Explain why this can happen.
12. If you wanted to make a more compact ac generator (operating at the same frequency) by reducing the area of the coils, how would you compensate by changing its other physical characteristics in order to maintain the same emf output? Explain how each characteristic would change (larger, smaller, etc.) and why the change compensates.

**20.3 Transformers and Power Transmission**

13. Explain why large-scale electric energy delivery systems operate at such high voltages when such voltages can be dangerous.
14. For an emergency project in his automotive workshop, a mechanic needs a step-down transformer, but has quick access to only step-up transformers in the shop. Show how he might be able to use a step-up transformer as a step-down one.
15. A metallic plate is in the plane of this paper. A uniform magnetic field is perpendicular to the plane of the paper and creates a magnetic flux in the plate. Give the direction of the eddy currents in the plate (clockwise, counterclockwise, or zero) for each of the following cases. Assume you are looking down onto the paper's plane. (a) The field points away from you and is decreasing. (b) The field points toward you and is increasing. (c) The field points toward you and is decreasing. (d) The field points away from you and does not change with time.

**20.4 Electromagnetic Waves**

16. (a) An antenna has been connected to a car battery. Under these conditions, will the antenna emit electromagnetic radiation? Why or why not? Explain. (b) Repeat part (a) for the time when the battery is disconnected and the current in the antenna drops to zero.
17. On a cloudy summer day, you work outside and feel cool, yet that evening you find that you are sunburned. Explain how this is possible.
18. Electromagnetic energy can exert a force on surfaces that it strikes (this phenomenon is called *radiation pressure*). (a) Will this force be greater on a shiny or a dark surface? (b) For the same surface, would you expect the force on it to be greater using a bright source rather than a fainter source of the same color? Explain. (c) For a given surface, would you expect the force on it to vary with distance from a light source? For all parts, clearly explain your reasoning.
19. (a) When police radar waves bounce off an oncoming car and are received by the transmitting radar "gun," they have a different frequency than the emitted waves. Explain. (b) Is the frequency in part (a) higher or lower than the original frequency? What about wavelength? What about wave speed? [*Hint:* Remember the Doppler effect.]

## Exercises*

*Integrated Exercises* (**IEs**) *are two-part exercises. The first part typically requires a conceptual answer choice based on physical thinking and basic principles. The following part is quantitative calculations associated with the conceptual choice made in the first part of the exercise.*

**20.1 Induced emf: Faraday's Law and Lenz's Law**

1. • What is the diameter of a circular wire loop if a magnetic field of 0.15 T oriented perpendicular to its area produces a magnetic flux of $1.2\times10^{-2}\,T{\cdot}m^2$?
2. • A circular loop with an area of 0.015 $m^2$ is in a uniform magnetic field of 0.30 T. What is the flux through the loop's plane if it is (a) parallel to the field, (b) at an angle of 37° to the field, and (c) perpendicular to the field?

* The bullets denote the degree of difficulty of the exercises: •, simple; ••, medium; and •••, more difficult.

3. • A circular loop (radius of 20 cm) is in a uniform magnetic field of 0.15 T. What angle(s) between the normal to the plane of the loop and the field would result in a flux with a magnitude of $1.4 \times 10^{-2}$ T·m²?
4. • The plane of a conductive loop with an area of 0.020 m² is perpendicular to a uniform magnetic field of 0.30 T. If the field drops to zero in 0.0045 s, what is the magnitude of the average emf induced in the loop?
5. • An ideal solenoid with a current of 1.5 A has a radius of 3.0 cm and a turn density of 250 turns/m. (a) What is the magnetic flux (due to its own field) through only one of its loops at its center? (b) What current is required to double the flux in part (a)?
6. •• A uniform magnetic field is at right angles to the plane of a wire loop. If the field decreases by 0.20 T in $1.0 \times 10^{-3}$ s and the magnitude of the average emf induced in the loop is 80 V, (a) what is the area of the loop? (b) What would be the value of the average induced emf if the field change was the same but took twice as long to decrease? (c) What would be the value of the average induced emf if the field decrease was twice as much and also took twice as long to change?
7. •• (a) A square loop of wire with sides of length 40 cm is in a uniform magnetic field perpendicular to its area. If the field's strength is initially 100 mT and it decays to zero in 0.010 s, what is the magnitude of the average emf induced in the loop? (b) What would be the average emf if the sides of the loop were only 20 cm?
8. •• The magnetic flux through one loop of wire is reduced from 0.35 Wb to 0.15 Wb in 0.20 s. The average induced current in the coil is 10 A. (a) Can you determine the area of the loop from the data given? Explain. (b) Find the resistance of the wire.
9. •• When the magnetic flux through a single loop of wire increases by 30 T · m², an average current of 40 A is induced in the wire. Assuming that the wire has a resistance of 2.5 Ω, (a) over what period of time did the flux increase? (b) If the current had been only 20 A, how long would the flux increase have taken?
10. •• In 0.20 s, a coil of wire with 50 loops experiences an average induced emf of 9.0 V due to a changing magnetic field perpendicular to the plane of the coil. The radius of the coil is 10 cm, and the initial strength of the magnetic field is 1.5 T. Assuming that the strength of the field decreased with time, (a) what is the final strength of the field? (b) If the field strength had, instead, increased, what would its final value have been? (c) Explain a method whereby you could, in principle, tell whether the field was increasing or decreasing in magnitude.
11. IE •• A boy is traveling due north at a constant speed while carrying a metal rod. The rod's length is oriented in the east-west direction and is parallel to the ground. (a) There will be no induced emf when the rod is (1) at the equator, (2) near the Earth's magnetic poles, (3) somewhere between the equator and the poles. Why? (b) Assume that the Earth's magnetic field is $1.0 \times 10^{-4}$ T near the North Pole and $1.0 \times 10^{-5}$ T near the equator. If the boy runs with a speed of 1.3 m/s northward near each location, and the rod is 1.5 m long, calculate the induced emf in the rod in each location.
12. •• A metal airplane with a wingspan of 30 m flies horizontally along a north-south route in the northern hemisphere at a constant speed of 320 km/h in a region where the vertical component of the Earth's magnetic field is $5.0 \times 10^{-5}$ T. (a) What is the magnitude of the induced emf between the tips of its wings? (b) If the easternmost wing tip is negatively charged, is the plane flying due north or due south? Explain.
13. •• Suppose that the metal rod in Figure 20.11 is 20 cm long and is moving at a speed of 10 m/s in a magnetic field of 0.30 T and that the metal frame is covered with an insulating material. Find (a) the magnitude of the induced emf across the rod and (b) the current in the rod. (c) Repeat these calculations if the wire were not covered and the total resistance of the circuit (rod plus frame) were 0.15 Ω.
14. ••• The flux through a loop of wire changes uniformly from +40 Wb to −20 Wb in 1.5 ms. (a) What is the significance of the negative number attached to the final flux value? (b) What is the average induced emf in the loop? (c) If you wanted to double the average induced emf by changing only the time, what would the new time interval be? (d) If you wanted to double the average induced emf by changing only the final flux value, what would it be?
15. ••• A fixed coil of wire with 10 turns and an area of 0.055 m² is placed in a perpendicular magnetic field. This field oscillates in direction and magnitude at a frequency of 10 Hz and has a maximum value of 0.12 T. (a) What is the average emf induced in the coil during the time it takes for the field to go from its maximum value in one direction to its maximum value in the other direction? (b) Repeat part (a) for a time interval of one complete cycle. (c) At what time(s) during a complete magnetic field cycle would you expect the induced emf to have its maximum magnitude? What about its minimum value? Explain both answers.

### 20.2 Electric Generators and Back emf

16. • A hospital emergency room ac generator operates at a frequency of 60 Hz. If the output voltage is at a maximum value (magnitude) at $t = 0$, when is it next (a) a maximum (magnitude), (b) zero, and (c) at its initial value (direction and magnitude)?
17. • A student makes a simple ac generator by using a single square wire loop 10 cm on a side. The loop is then rotated at a frequency of 60 Hz in a magnetic field of 0.015 T. (a) What is the maximum emf output? (b) If she wanted to make the maximum emf output ten times larger by adding loops, how many should she use in total?

18. •• A simple ac generator consists of a coil with 10 turns (each turn has an area of 50 $cm^2$). The coil rotates in a uniform magnetic field of 350 mT with a frequency of 60 Hz. (a) Write an expression in the form of Equation 20.5 for the generator's emf variation with time. (b) Compute the maximum emf.
19. •• An ac generator operates at a rotational frequency of 60 Hz and produces a maximum emf of 100 V. Assume that its output at $t = 0$ is zero. What is the instantaneous emf (a) at $t = 1/240$ s? (b) at $t = 1/120$ s? (c) Find an expression for the instantaneous emf at any time $t$. (d) How much time elapses between successive 0-volt outputs? (e) What maximum emf would this generator produce if it were operated instead at 120 Hz?
20. •• The armature of a simple ac generator has 20 circular loops of wire, each with a radius of 10 cm. It is rotated with a frequency of 60 Hz in a uniform magnetic field of 800 mT. (a) What is the maximum emf induced in the loops? (b) How often is this value attained? (c) If the time period in part (b) were cut in half, what would be the new maximum emf value?
21. •• The armature of an ac generator has 100 turns. Each turn is a rectangular loop measuring 8.0 cm by 12 cm. The generator has a sinusoidal voltage output with an amplitude of 24 V. (a) If the magnetic field of the generator is 250 mT, with what frequency does the armature turn? (b) If the magnetic field was doubled and the frequency cut in half, what would be the amplitude of the output?
22. **IE** •• (a) To increase the output of an ac generator, a student has the choice of doubling *either* the generator's magnetic field *or* its frequency. To maximize the increase in emf output, (1) he should double the magnetic field, (2) he should double the frequency, (3) it does not matter which one he doubles. Explain. (b) Two students display their ac generators at a science fair. The generator made by student A has a loop area of 100 $cm^2$ rotating in a magnetic field of 20 mT at 60 Hz. The one made by student B has a loop area of 75 $cm^2$ rotating in a magnetic field of 200 mT at 120 Hz. Which one generates the largest maximum emf? Justify your answer mathematically.
23. **IE** •• A motor has a resistance of 2.50 Ω and is connected to a 110-V line. (a) Is the operating current of the motor (1) higher than 44 A, (2) 44 A, or (3) lower than 44 A? Why? (b) If the back emf of the motor at operating speed is 100 V, what is its operating current?
24. ••• The starter motor in an automobile has a resistance of 0.40 Ω in its armature windings. The motor operates on 12 V and has a back emf of 10 V when running at normal operating speed. How much current does the motor draw (a) when running at its operating speed, (b) when running at half its final rotational speed, and (c) when starting up?
25. ••• A 240-V dc motor has an armature whose resistance is 1.50 Ω. When running at its operating speed, it draws a current of 16.0 A. (a) What is the back emf of the motor when it is operating normally? (b) What is the starting current? (Assume that there is no additional resistance in the circuit.) (c) What series resistance would be required to limit the starting current to 25 A?

**20.3 Transformers and Power Transmission**

26. **IE** • The secondary coil of an ideal transformer has 450 turns, and the primary coil has 75 turns. (a) Is this transformer a (1) step-up or (2) step-down transformer? Explain your choice. (b) What is the ratio of the current in the primary coil to the current in the secondary coil? (c) What is the ratio of the voltage across the primary coil to the voltage in the secondary coil?
27. • An ideal transformer steps 8.0 V up to 2000 V, and the 4000-turn secondary coil carries 2.0 A. (a) Find the number of turns in the primary coil. (b) Find the current in the primary coil.
28. • The primary coil of an ideal transformer has 720 turns, and the secondary coil has 180 turns. If the primary coil carries 15 A at a voltage of 120 V, what are (a) the voltage and (b) the output current of the secondary coil?
29. •• The transformer in the power supply for a computer's external hard drive changes a 120-V input voltage (from a regular house line) to the 5.0-V output voltage that is required to power the drive. (a) Find the ratio of the number of turns in the primary coil to the number of turns in the secondary coil. (b) If the drive is rated at 10 W when running and the transformer is ideal, what is the current in the primary and secondary when the drive is in operation?
30. •• The primary coil of an ideal transformer is connected to a 120-V source and draws 1.0 A. The secondary coil has 800 turns and supplies an output current of 10 A to run an electrical device. (a) What is the voltage across the secondary coil? (b) How many turns are in the primary coil? (c) If the maximum power allowed by the device (before it is destroyed) is 240 W, what is the maximum input current to this transformer?
31. • An ideal transformer has 840 turns in its primary coil and 120 turns in its secondary coil. If the primary coil draws 2.50 A at 110 V, what are (a) the current and (b) the output voltage of the secondary coil?
32. •• The efficiency $e$ of a transformer is defined as the ratio of the power output to the power input, or $e = P_s/P_p$. (a) Show that for an ideal transformer, this expression gives an efficiency of 100% ($e = 1.00$). (b) Suppose a step-up transformer increased the line voltage from 120 to 240 V, while at the same time the output current was reduced to 5.0 A from 12 A. What is the transformer's efficiency? Is it ideal?

33. IE •• The specifications of a transformer used with a small appliance read as follows: input, 120 V, 6.0 W; output, 9.0 V, 300 mA. (a) Is this transformer (1) an ideal or (2) a non-ideal transformer? Why? (b) What is its efficiency? (See Exercise 32.)
34. •• An ac generator supplies 20 A at 440 V to a 10 000-V power line through a step-up transformer that has 150 turns in its primary coil. (a) If the transformer is 95% efficient (see Exercise 32), how many turns are in the secondary coil? (b) What is the current in the power line?
35. •• The electricity supplied in Exercise 34 is transmitted over a line 80.0 km long with a resistance of 0.80 Ω/km. (a) How many kilowatt-hours are saved in 5.00 h by stepping up the voltage? (b) At $0.15/kWh, how much of a savings (to the nearest $10) is this to all the consumers the line supplies in a 30-day month, assuming that the energy is supplied continuously?
36. •• A small plant produces electric energy and, through a transformer, sends it out over the transmission lines at 50 A and 20 kV. The line reaches a small town over 25-km-long transmission lines whose resistance is 1.2 Ω/km. (a) What is the power loss in the lines if the energy is transmitted at 20 kV? (b) What should be the output voltage of the transformer to decrease the power loss by a factor of 15? Assume the transformer is ideal. (c) What would be the current in the lines in part (b)?
37. •• Electrical power from a generator is transmitted through a power line 175 km long with a resistance of 1.2 Ω/km. The generator's output is 50 A at its operating voltage of 440 V. This output is increased by a single step-up for transmission at 44 kV. (a) How much power is lost as joule heat during the transmission? (b) What must be the turn ratio of a transformer at the delivery point in order to provide an output voltage of 220 V? (Neglect the voltage drop in the line.)

## 20.4 Electromagnetic Waves

38. • Find the frequencies of electromagnetic waves with wavelengths of (a) 3.0 cm, (b) 650 nm, and (c) 1.2 fm. (d) Classify the type of light in each case.
39. • In a small town there are only two AM radio stations, one at 920 kHz and one at 1280 kHz. What are the wavelengths of the radio waves transmitted by each station?
40. • A meteorologist for a TV station is using radar to determine the distance to a cloud. He notes that a time of 0.24 ms elapses between the sending and the return of a radar pulse. How far away is the cloud?
41. • How long does a laser beam take to travel from the Earth to a reflector on the Moon and back? Take the distance from the Earth to the Moon to be $2.4 \times 10^5$ mi. (This experiment was done when the *Apollo* flights of the early 1970s left laser reflectors on the lunar surface.)
42. •• Orange light has a wavelength of 600 nm, and green light has a wavelength of 510 nm. (a) What is the difference in frequency between these two types of light? (b) If you doubled the wavelength of both, what type of light would they become?
43. •• A certain type of radio antenna is called a *quarter-wavelength antenna*, because its length is equal to one-quarter of the wavelength to be received. If you were going to make such antennae for the AM and FM radio bands by using the middle frequencies of each band, what lengths of wire would you use?
44. IE ••• Microwave ovens can have cold spots and hot spots due to standing electromagnetic waves, analogous to standing wave nodes and antinodes in strings (▼ **Figure 20.23** shows such spots in the lasagna cheese topping – some melted, some not). (a) The longer the distance is between cold spots, (1) the higher the frequency of the waves, (2) the lower the frequency of the waves, (3) the frequency of the waves is independent of this distance. Why? (b) In your microwave the cold spots (nodes) occur approximately every 5.0 cm, but your neighbor's microwave produces them at every 6.0 cm. Which microwave operates at a higher frequency and by how much?

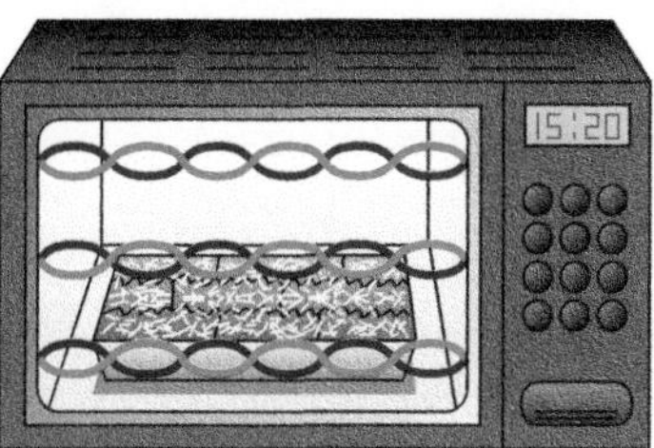


▲ FIGURE 20.23 **Cold spots?** See Exercise 44.

# 21

# AC Circuits

**Almost every instrument in this photo depends on an alternating voltage source to operate properly.**

Direct current (dc) circuits have many uses, but the airport control tower in the chapter-opening photo operates many devices that use alternating current (ac). The electric power delivered to our homes and offices is also ac, and most everyday devices and appliances require alternating current.

There are several reasons for our reliance on alternating current. For one thing, almost all electric energy generators produce electric energy using electromagnetic induction, and thus produce ac outputs (Chapter 20). Furthermore, electrical energy produced in ac fashion can be transmitted economically over long distances through the use of transformers. But perhaps the most important reason is that ac currents produce electromagnetic effects that can be exploited in a variety of devices. For example, when a radio is tuned to a station, it takes advantage of a special *resonance* property of ac circuits.

To determine currents in dc circuits, resistance values are of main concern (Chapter 18). There is, of course, resistance present in ac circuits as well, but additional factors can affect the flow of charge. For instance, a capacitor in a dc circuit is equivalent to an infinite resistance (an open circuit). However, in an ac circuit the alternating voltage continually charges and discharges a capacitor. Under such conditions, current can exist in a circuit even if it contains a capacitor. Moreover, wrapped coils of wire can oppose an ac current through the principle of electromagnetic induction (Lenz's law; Section 20.1).

In this chapter, the principles of ac circuits will be studied. More generalized forms of Ohm's law and expressions for power, applicable to ac circuits, will be developed. Finally, the phenomenon and uses of circuit resonance is explored.

## 21.1 Resistance in an AC Circuit

An ac circuit contains an ac voltage source (such as a generator or a household outlet) and one or more elements. An ac circuit with a single resistive element is shown in ▼**Figure 21.1**. If the source's voltage (electric potential difference) varies sinusoidally at a maximum voltage of $\Delta V_m$, for example as from a generator (Section 20.2), then the voltage across the resistor $\Delta V_R$ will also vary sinusoidally as

$$\Delta V_R = \Delta V_m \sin(\omega t) = \Delta V_m \sin(2\pi ft) \tag{21.1}$$

where $\omega$ is the angular frequency of the voltage (in rad/s) and is related to its frequency $f$ (in Hz) by $\omega = 2\pi f$. Thus the voltage across the resistor oscillates between $\pm\Delta V_m$, where $\Delta V_m$ is the **peak** (or *maximum*) **voltage**, and represents the amplitude of the voltage.

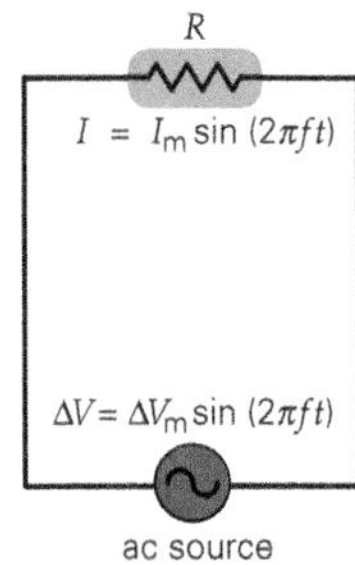


▲**FIGURE 21.1 A purely resistive circuit** The ac source supplies a sinusoidal voltage across a single resistor. The voltage across, and current in, the resistor vary sinusoidally at the frequency of the source's voltage.

### 21.1.1 AC Current and Power for Resistors

Under ac conditions, the current in the resistor oscillates both in direction and magnitude. From Ohm's law, the ac current in the resistor, as a function of time, is

$$I = \frac{\Delta V_R}{R} = \left[\frac{\Delta V_m}{R}\right]\sin(2\pi ft)$$

Because $\Delta V_m$ represents the peak voltage across the resistor, the expression in the parentheses represents the maximum current $I_m$ in the resistor. Thus this expression can be rewritten as

$$I = I_m \sin(2\pi ft) \tag{21.2}$$

where $I_m = \Delta V_m/R$ is the peak (*maximum*) **current**.

►**Figure 21.2** shows both current and voltage as functions of time for a resistor. Note that they are in step, or in phase. That is, both reach their zero, minimum, and maximum values at the same time. The current oscillates and takes on both positive and negative values, indicating its directional changes during each cycle. Because the current spends equal time in both directions, *the average current is zero*. Mathematically, this is because the time-averaged value of the sine function over one or more *complete* (360°) cycles is zero. Using overbars to denote a time-averaged value, then $\overline{\sin\theta} = \overline{\sin 2\pi ft} = 0$. Similarly, $\overline{\cos\theta} = 0$.

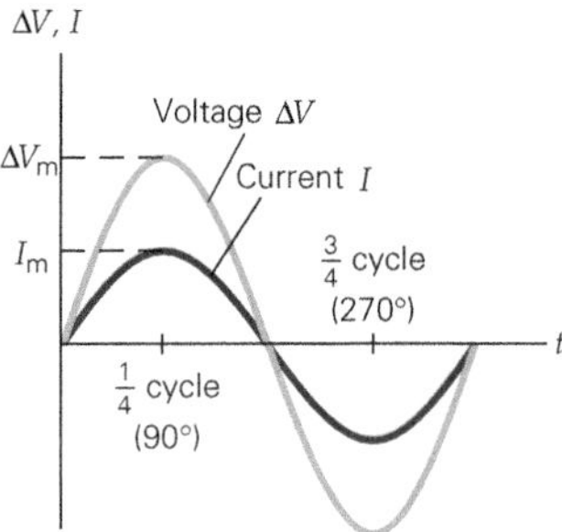


▲**FIGURE 21.2 Voltage and current in phase** In a purely resistive ac circuit, the voltage and current are in step, or in phase.

Even though the average current is zero, this does not mean that there is no joule heating ($I^2R$ losses). This is because the dissipation of electrical energy in a resistor does not depend on the current's direction. The instantaneous power as a function of time is obtained from the instantaneous current (Equation 21.2). Thus,

$$P = I^2R = \left(I_m^2R\right)\sin^2(2\pi ft) \tag{21.3}$$

Even though the current changes sign, the square of the current is always positive. Hence the time-averaged value of $I^2R$ is *not* zero. The time-averaged value of $I^2$ is $\overline{I^2} = I_m^2\overline{\sin^2(2\pi ft)}$. Using the trigonometric identity $\sin^2\theta = (1/2)(1 - \cos[2\theta])$ it follows that $\overline{\sin^2\theta} = (1/2)\left(1 - \overline{\cos[2\theta]}\right)$. However, $\overline{\cos[2\theta]} = 0$ (just as $\overline{\cos\theta} = 0$). Therefore $\overline{\sin^2\theta} = (1/2)$. Thus $\overline{I^2} = I_m^2\overline{\sin^2(2\pi ft)} = (1/2)I_m^2$. The average power is therefore

$$\bar{P} = \overline{I^2}R = \frac{1}{2}I_m^2R \tag{21.4}$$

Under ac conditions, it is customary to work with average power and a special kind of "average" current defined as follows:

$$I_{rms} = \sqrt{\overline{I^2}} = \sqrt{\frac{1}{2}I_m^2} = \frac{I_m}{\sqrt{2}} \approx 0.707I_m \tag{21.5}$$

$I_{rms}$ is called the **rms current, or effective current**. (Here, rms stands for *root-mean-square*, indicating the square *root* of the *mean* value of the *square* of the current.) The rms current represents the value of a steady (dc) current that would produce the same power as its ac current counterpart, hence the name effective current. When squared, Equation 21.5 yields $I_m^2 = 2I_{rms}^2$. Then the average power (Equation 21.4) can be rewritten as

$$\bar{P} = \frac{1}{2}I_m^2R = I_{rms}^2R \quad \text{(time-averaged power for a resistor)} \tag{21.6}$$

Thus the average power (joule heating) expended in a resistor is just the time-varying (oscillating) power averaged over time, which is half the maximum power as illustrated in ►**Figure 21.3**.

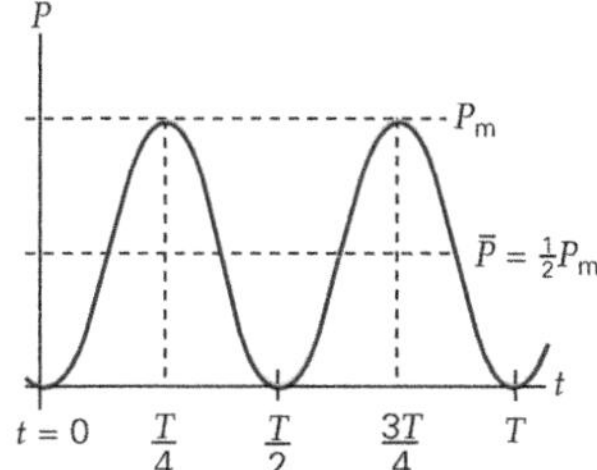


▲ **FIGURE 21.3 Power variation with time in a resistor** Although both current and voltage oscillate in direction (sign), their product (power) is always a positive oscillating quantity. The average power is one-half the peak power, $P_m$.

### 21.1.2 AC Voltage

The peak values of voltage and current for a resistor are related by $\Delta V_m = I_m R$. Using a development similar to that for rms current, the **rms voltage**, or **effective voltage**, is

$$\Delta V_{rms} = \frac{\Delta V_m}{\sqrt{2}} \approx 0.707 V_m \quad (21.7)$$

For resistors under ac conditions, then, dc-like relationships can be used – as long as it is kept in mind that the quantities represent rms values. Thus for ac situations involving only a resistor, the relationship between rms values of current and voltage is

$$\Delta V_{rms} = I_{rms} R \quad \text{(rms voltage across resistor)} \quad (21.8)$$

Combining Equations 21.8 and 21.6 results in three expressions for ac power in a resistor:

$$\bar{P} = I_{rms}^2 R = I_{rms}\Delta V_{rms} = \frac{(\Delta V_{rms})^2}{R} \quad \text{(ac power for a resistor)} \quad (21.9)$$

It is customary to measure and specify rms values when dealing with ac quantities. For example, a household voltage of 120 V is really the rms value of the voltage. Household voltage actually has a peak value of 170 V since $(\Delta V_m) = \sqrt{2}\Delta V_{rms} \approx 1.414(120\text{ V}) = 170\text{ V}$. See Examples 21.1 and 21.2 for usage of these concepts.

---

#### EXAMPLE 21.1: A BRIGHT LIGHTBULB – RMS AND PEAK VALUES

A lamp with an old-fashion incandescent 60-W bulb is plugged into a 120-V outlet. (a) What are the rms and peak currents in the lamp? (b) What is the resistance of the bulb under these conditions?

**THINKING IT THROUGH.** (a) Because the average power and rms voltage are known, the rms current can be found from one of the equivalent forms in Equation 21.9. From the rms current, Equation 21.5 can be used to calculate the peak current. (b) The resistance is found from Equation 21.8.

**SOLUTION**

The average power and the rms voltage of the source are given.

***Given:***

$$\bar{P} = 60\text{ W}$$
$$\Delta V_{rms} = 120\text{ V}$$

***Find:***

**(a)** $I_{rms}$ and $I_m$ (rms and peak currents)
**(b)** $R$ (bulb resistance)

**(a)** The rms current is

$$I_{rms} = \frac{\bar{P}}{\Delta V_{rms}} = \frac{60\text{ W}}{120\text{ V}} = 0.50\text{ A}$$

and the peak current is determined by rearranging Equation 21.5:

$$I_m = \sqrt{2} I_{rms} = (1.41)(0.50\text{ A}) = 0.71\text{ A}$$

**(b)** The resistance of the bulb is

$$R = \frac{\Delta V_{rms}}{I_{rms}} = \frac{120\text{ V}}{0.50\text{ A}} = 240\ \Omega$$

**FOLLOW-UP EXERCISE.** What would be the (a) rms current and (b) peak current in a 60-W lightbulb in Great Britain, where the house rms voltage is 240 V at 50 Hz? (c) How would the resistance of a 60-W bulb in Great Britain compare with one designed for operation at 120 V? Why are the two resistances different?

---

---

#### CONCEPTUAL EXAMPLE 21.2: ACROSS THE POND – BRITISH VERSUS AMERICAN ELECTRICAL SYSTEMS

In many countries, the line voltage is 240 V. If a British tourist visiting the United States plugged in a hair dryer from home (where the voltage is 240 V), you would expect it (a) not to operate, (b) to operate normally, (c) to operate poorly, or (d) to burn out.

**REASONING AND ANSWER.** British appliances operate at 240 V. At a decreased voltage of 120 V there would be decreased current and reduced joule heating. If the resistance of the appliance were constant, then at half the voltage, there would be only one-fourth the power output since both current and voltage are halved ($P = I\Delta V$). Thus the heating element of the dryer might get warm, but it would not work as expected, so the answer is (c).

When traveling to a foreign country, most people do not make this mistake, because the shape of plugs and sockets varies from country to country. If you are traveling with appliances from home, a conversion kit can be useful (▶ **Figure 21.4**). This typically contains a selection of plugs for adapting to foreign sockets, as well as a voltage converter. The converter is a device that converts 240 V to 120 V for U.S. travelers and vice versa for tourists visiting the United States. (There are appliances available that can be switched between

▲**FIGURE 21.4** **Converter and adapters** In countries that have 240-V line voltages, U.S. tourists need converters that convert to 120 V to operate normal U.S. appliances properly. Note also the different types of plugs for different countries.

120 V and 240 V, negating the need for voltage converters. However, plug socket adaptors may still be needed.)

**FOLLOW-UP EXERCISE.** What happens if an American tourist inadvertently plugs a 120-V appliance into a British 240-V outlet without a converter? Explain.

## 21.2 Capacitive Reactance

Discussions in Chapter 16 considered situations (such as RC circuits) in which a capacitor is connected to a dc voltage source. In these situations, current exists only for the short time required to charge or discharge the capacitor. As charge accumulates on the capacitor's plates, the voltage across them increases, opposing the external voltage and reducing the current. When the capacitor is fully charged, the current drops to zero.

Things are different when a capacitor is driven by an ac voltage source (▶**Figure 21.5a**). Under these conditions, the capacitor limits the current, but doesn't completely prevent the flow of charge. This is because the capacitor is alternately charged and discharged as the current and voltage reverse each half-cycle. Plots of ac current and voltage versus time for a circuit with just a capacitor are shown in Figure 21.5b. Let's look at how the conditions of the capacitor change with time (▶**Figure 21.6**).

- In Figure 21.6a, $t = 0$ is arbitrarily chosen as the time of maximum voltage (thus $\Delta V = \Delta V_m$). At the start, the capacitor is fully charged ($Q_m = C{\cdot}\Delta V_m$) with polarity shown. Because the plates are full, there is no current.
- As the voltage decreases ($0 < \Delta V < \Delta V_m$), the capacitor begins to discharge, giving rise to a counterclockwise current (labeled negative; compare Figures 21.5b through 21.6b).
- The current reaches its maximum when the voltage drops to zero and the plates are completely discharged (Figure 21.6c). This occurs one-quarter of the way through the cycle (at $t = T/4$).

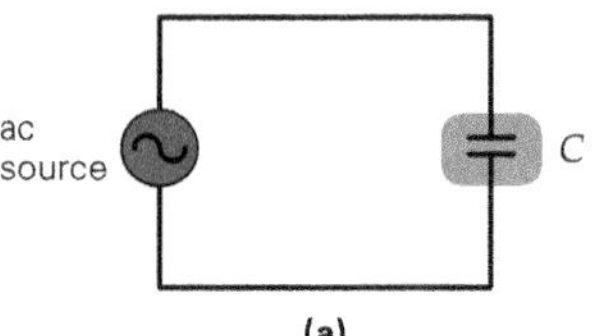


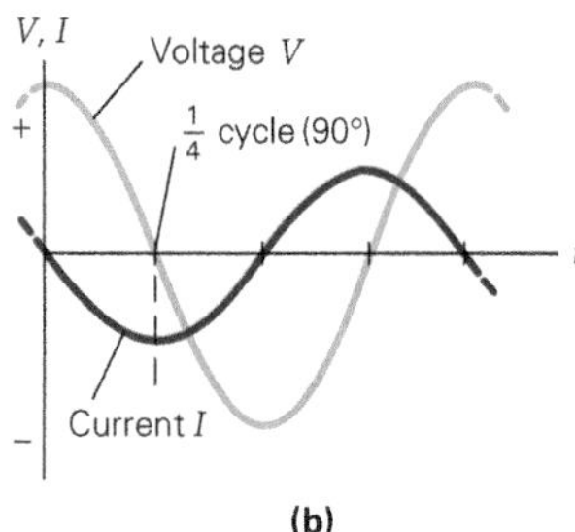


▲**FIGURE 21.5** **A purely capacitive circuit (a)** An ac circuit with only capacitance. **(b)** In this case, the current leads the voltage by 90°, or one quarter-cycle. The details of a half-cycle are shown in Figure 21.6.

- The ac voltage source now reverses polarity and starts to increase in magnitude. The capacitor begins to charge, this time with the opposite polarity (Figure 21.6d). With the plates uncharged, there is no opposition to the current, so the current is at its maximum. However, as the plates accumulate charge, the capacitor begins to inhibit the current, and thus it decreases.
- Halfway through the cycle (at $t = T/2$), the capacitor is fully charged, but is opposite in polarity to its starting condition (Figure 21.6e). The current is zero and the voltage is at its maximum, but opposite its initial polarity.

During the next half-cycle (not shown), the process is reversed, and the circuit returns to its initial condition. Note that the current and voltage are *not* in step (i.e., not in phase). The current reaches its maximum a quarter cycle ahead of the voltage. The phase relationship between the current and the voltage for a capacitor is commonly stated this way:

*In a purely capacitive ac circuit, the current leads the voltage by 90°, or one-quarter of a cycle.*

Thus in an ac situation, a capacitor provides some opposition to the charging process, but it is not totally limiting as it would be under dc conditions when it behaves as an open circuit. The quantitative measure of this "capacitive opposition" to current is called **capacitive reactance ($X_C$)**. Under ac conditions, the capacitive reactance of a capacitor is

$$X_C = \frac{1}{\omega C} = \frac{1}{2\pi f C} \quad \text{(capacitive reactance)} \qquad (21.10)$$

SI unit of capacitive reactance:
ohm (Ω), or seconds per farad (s/F)

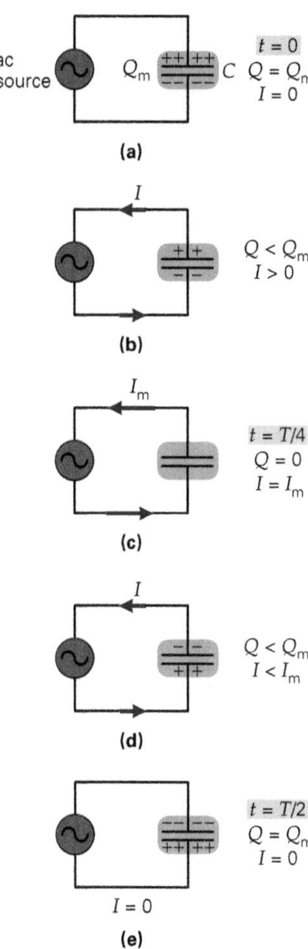


▲**FIGURE 21.6 A capacitor under ac conditions** This sequence (abcde) shows the voltage, charge, and current in a circuit containing only a capacitor and an ac voltage source. All five circuit diagrams taken together represent physically what is plotted in the first half of the cycle (from $t = 0$ to $t = T/2$) in the graph shown in Figure 21.5b.

where, as usual, $\omega = 2\pi f$, $C$ is the capacitance (in farads), and $f$ is the frequency (in Hz). Like resistance, reactance is measured in ohms (Ω). (Using unit analysis, you should show that the ohm is equivalent to seconds per farad.)

Equation 21.10 shows that the reactance is inversely proportional to both the capacitance ($C$) and the voltage frequency ($f$). Both of these dependencies can be understood as follows.

To understand the dependence on $C$, recall that capacitance means "charge stored per volt" ($C = Q/\Delta V$). Therefore, for a given source frequency and voltage, the greater the capacitance, the more charge the capacitor can accommodate. This requires a larger charge flow rate, or current. Thus a larger capacitance offers less opposition to charge flow (i.e., a reduced capacitive reactance) at a given frequency.

To understand the frequency dependence, consider the fact that the greater the frequency of the voltage, the shorter the time available for charging each cycle. A shorter charging time means less charge is able to accumulate on the plates and thus there is less opposition to current. So increasing the frequency results in a decrease in capacitive reactance.

It is always good to check a general relationship to see whether it gives a result known to be true in a special case. As a special case for a capacitor, note that as $f$ approaches zero (i.e., non-oscillating dc conditions), the capacitive reactance becomes infinite. This is as expected because under dc conditions, there is no current.

Capacitive reactance is related to the voltage across the capacitor and current by a relationship similar in form to that for resistances:

$$\Delta V_{\text{rms}} = I_{\text{rms}} X_{\text{C}} \quad \text{(voltage across a capacitor)} \qquad (21.11)$$

For a look at capacitive reactance, consider Example 21.3.

---

**EXAMPLE 21.3: CAPACITIVE REACTANCE AND CAPACITOR CURRENT UNDER AC CONDITIONS**

A 15.0-μF capacitor is connected to a 120-V, 60-Hz source. What are (a) the capacitive reactance and (b) the current (rms and peak) in the circuit?

**THINKING IT THROUGH.** (a) The capacitive reactance can be found via Equation 21.10. (b) The rms current can then be determined from the reactance and rms voltage using Equation 21.11. Finally, Equation 21.5 will enable the peak current to be found.

**SOLUTION**

Assuming 60 Hz is exact, answers will be shown to three significant figures.

***Given:***

$$C = 15.0\ \mu\text{F} = 15.0 \times 10^{-6}\ \text{F}$$
$$\Delta V_{\text{rms}} = 120\ \text{V}$$
$$f = 60\ \text{Hz}$$

***Find:***

**(a)** $X_{\text{C}}$ (capacitive reactance)
**(b)** $I_{\text{m}}$ (peak current), $I_{\text{rms}}$ (rms current)

**(a)** The capacitive reactance is

$$X_{\text{C}} = \frac{1}{2\pi f C} = \frac{1}{2\pi(60\ \text{Hz})(15.0 \times 10^{-6}\ \text{F})} = 177\ \Omega$$

**(b)** Then, the rms current is

$$I_{\text{rms}} = \frac{\Delta V_{\text{rms}}}{X_{\text{C}}} = \frac{120\ \text{V}}{177\ \Omega} = 0.678\ \text{A}$$

and therefore, the peak current is

$$I_{\text{m}} = \sqrt{2} I_{\text{rms}} = \sqrt{2}(0.678\ \text{A}) = 0.959\ \text{A}$$

Thus the current oscillates at 60 cycles per second with a maximum value of 0.959 A.

**FOLLOW-UP EXERCISE.** In this Example, (a) what is the peak voltage and (b) what frequency would give the same current if the capacitance were cut in half?

## 21.3 Inductive Reactance

Inductance is a measure of the opposition a circuit element presents to a time-varying current by virtue of Lenz's law. In principle, all circuit elements (even resistors) have some inductance. However, a coil of wire with negligible resistance has, in effect, only inductance. When placed in a circuit with a time-varying current, such a coil, called an **inductor**, exhibits a reverse voltage, or back emf, in opposition to the changing current. The changing current through the coil produces a changing magnetic field and flux. The back emf is the induced emf in opposition to this changing flux. Because the back emf is induced in the inductor as a result of its own changing magnetic field, this phenomenon is called *self-induction.*

The self-induced emf for a coil consisting of $N$ loops is given by Faraday's law (Equation 20.2): $\mathscr{E} = -N(\Delta\Phi/\Delta t)$. But time rate of change of the flux through the coil, $N(\Delta\Phi/\Delta t)$, is proportional to the time rate of change of the current in the coil, $\Delta I/\Delta t$. This is because the current produces the magnetic field responsible for the changing flux. Thus the back emf is proportional to, and oppositely directed to, the rate of current change. This relationship is expressed using a proportionality constant, $L$:

$$\mathscr{E} = -L\frac{\Delta I}{\Delta t} \tag{21.12}$$

where $L$ is the **inductance** of the coil (more properly, its self-inductance). You should be able to show, using unit analysis, that the units of inductance are volt-second per ampere (V·s/A). This combination is called a *henry* (**H**, 1 H = 1 V·s/A), in honor of Joseph Henry (1797–1878), an American physicist and early investigator of electromagnetic induction. Smaller units, such as the millihenry (mH), are commonly used (1 mH = $10^{-3}$ H).

The opposition presented to current by an inductor under ac conditions depends on its inductance value and the voltage frequency. This is expressed quantitatively by its **inductive reactance ($X_L$)**, which is

$$X_L = \omega L = 2\pi fL \quad \text{(inductive reactance)} \tag{21.13}$$

SI unit of inductive reactance:
ohm (Ω), or henrys per second (H/s)

where $f$ is the frequency of the driving ac voltage source, as usual $\omega = 2\pi f$, and $L$ is the inductance. Like capacitive reactance, inductive reactance is measured in ohms (Ω), which is equivalent to henrys per second.

Note that the inductive reactance is directly proportional to both the coil inductance ($L$) and the voltage frequency ($f$). Inductance is a property of the coil that depends on the number of turns, the coil's diameter and length, and the material in the coil (if any). The frequency of the voltage plays a role because the more rapidly the current in the coil changes, the greater the rate of change of its magnetic flux. This implies a larger self-induced (back) emf to oppose the changes in current.

In terms of $X_L$, the voltage across an inductor is related to the current and its inductive reactance by the following:

$$\Delta V_{rms} = I_{rms}X_L \quad \text{(voltage across an inductor)} \tag{21.14}$$

A circuit including only an ac voltage source connected to one inductor (with the inductor symbol shown) and the time-variation graphs of voltage across the inductor and current in the circuit are both shown in ▼ **Figure 21.7.**

When an inductor is connected to an ac voltage source, maximum voltage corresponds to zero current. When the

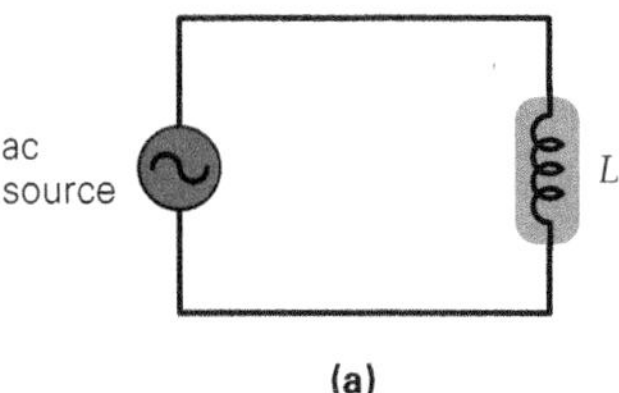


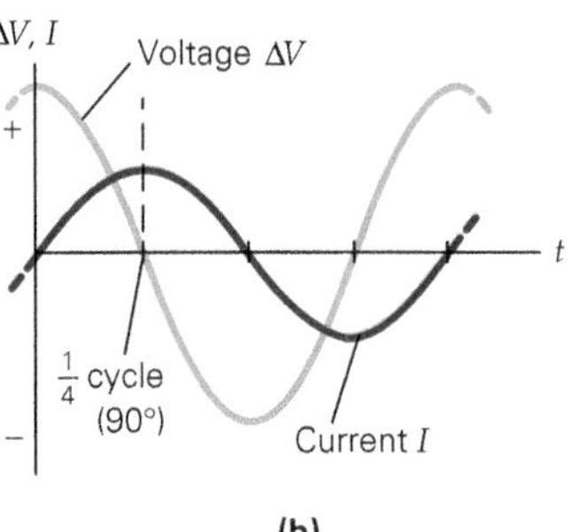


▲ **FIGURE 21.7 A purely inductive circuit (a)** An ac circuit with only inductance. **(b)** The voltage leads the current by 90°, or one quarter-cycle.

voltage drops to zero, the current is maximum. This happens because as the voltage changes polarity (causing the magnetic flux through the inductor to drop to zero), the inductor acts to prevent the change in accordance with Lenz's law, so the induced emf creates a current. In an inductor, the current *lags* one quarter-cycle behind the voltage, a relationship expressed as follows:

*In a purely inductive ac circuit, the voltage leads the current by 90°, or a one-quarter cycle.*

Because the phase relationships between current and voltage for purely inductive and purely capacitive circuits are opposite, there is a phrase that may help you remember the difference:

*EL̲I* the *IC̲E* man.

Here *E* represents voltage (for *emf*) and *I* represents current. The three letters *ELI* then indicate that for inductance (*L*), voltage leads current – reading the acronym from left to right. Similarly, *ICE* means that for capacitance (*C*), current leads voltage.

---

EXAMPLE 21.4: CURRENT OPPOSITION WITHOUT RESISTANCE – INDUCTIVE REACTANCE

A 125-mH inductor is connected to a 120-V, 60-Hz source. What are (a) the inductive reactance and (b) the rms current in the circuit?

**THINKING IT THROUGH.** Because the inductance and frequency are known, the inductive reactance can be calculated from Equation 21.13 and the current from Equation 21.14.

SOLUTION

Listing the given data:

***Given:***

$L = 125\ \text{mH} = 0.125\ \text{H}$

$\Delta V_{\text{rms}} = 120\ \text{V}$

$f = 60\ \text{Hz}$

***Find:***

**(a)** $X_L$ (inductive reactance)

**(b)** $I_{\text{rms}}$

**(a)** The inductive reactance is

$$X_L = 2\pi fL = 2\pi(60\,\text{Hz})(0.125\,\text{H}) = 47.1\,\Omega$$

**(b)** The rms current is then

$$I_{\text{rms}} = \frac{\Delta V_{\text{rms}}}{X_L} = \frac{120\ \text{V}}{47.1\,\Omega} = 2.55\ \text{A}$$

**FOLLOW-UP EXERCISE.** In this Example, (a) what is the peak current? (b) What voltage frequency would yield the same current if the inductance were reduced to one-third the value in this Example?

---

## 21.4 Impedance: RLC Circuits

In the previous sections, purely capacitive or purely inductive circuits were considered separately and without resistance present. However, in practical situations, it is impossible to have purely reactive circuits because there is always some resistance – at a minimum, from the connecting wires. Thus resistances, capacitive reactances, and inductive reactances combine to impede the current in ac circuits. The following circuits illustrate how reactances and resistances are handled when in combination.

### 21.4.1 Series RC Circuit

Suppose an ac circuit consists of an ac voltage source, a resistor, and a capacitor connected in series (▼ **Figure 21.8a**). The phase relationship between the current and the voltage is different for each circuit element. As a result, a special graphical method is needed to find the overall impedance to the current in the circuit. This method uses a visual aid called a *phase diagram.*

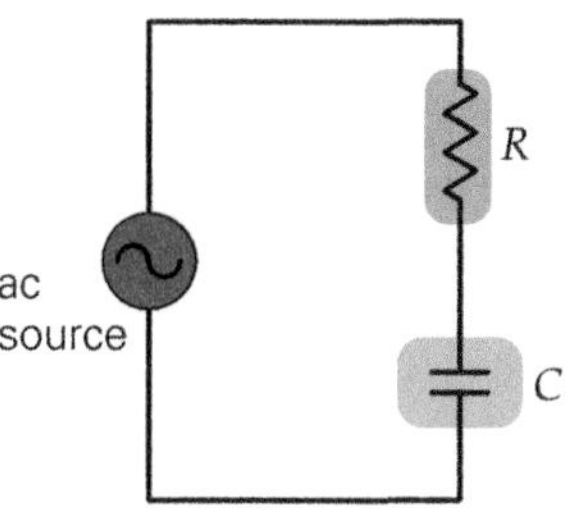


**(a) RC circuit diagram**

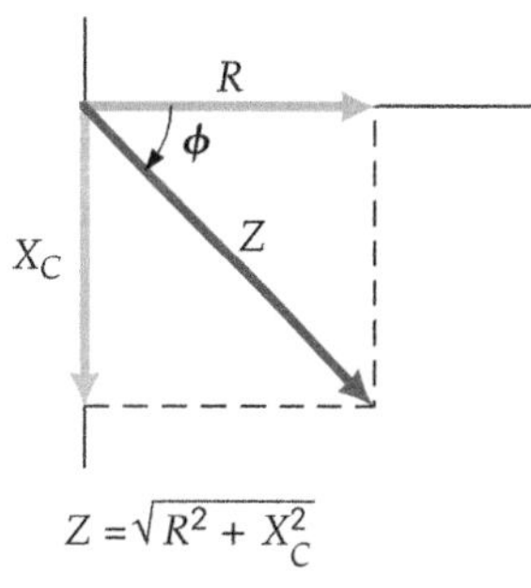


**(b) Phase diagram**

▲FIGURE 21.8 **A series RC circuit (a)** in a series RC circuit, **(b)** the impedance *Z* is the phasor (vector-like) sum (using Pythagorean theorem) of the resistance *R* and the capacitive reactance $X_C$. The phase angle $\phi$ is negative when the voltage lags the current as is true for RC circuits. See text for full description.

In the phase diagram for an RC circuit, such as in Figure 21.8b, the resistance and reactance in the circuit are endowed with vector-like properties and their magnitudes are represented by arrows called *phasors.* On a set of *x*–*y* coordinate axes, resistance is plotted on the positive *x*-axis (i.e., at 0°), because the voltage-current phase difference for a resistor is zero. The capacitive reactance is plotted along the negative *y*-axis, to reflect a voltage-current phase difference of −90° for a capacitor.

The *phasor sum* is the effective, or net, impedance to the current, and is called the circuit's **impedance (*Z*)**. Here the two phasors are combined as vectors to yield the overall impedance.

The angle $\phi$ is called the **phase angle** and is a measure of the phase difference between the voltage and current. For this RC circuit it is negative (below the $+x$-axis) because the voltage leads the current. Applying the Pythagorean theorem to find $Z$ in Figure 21.8b yields

$$Z = \sqrt{R^2 + X_C^2} \quad \text{(series RC circuit impedance)} \quad (21.15)$$

The SI unit of impedance is the ohm.

Ohm's law, as applied to ac circuits, uses impedance $Z$ in place of resistance $R$. It is a generalization to $\Delta V = IR$ which is used for circuits containing only resistances, and is written as

$$\Delta V_{\text{rms}} = I_{\text{rms}} Z \quad \text{(Ohm's law for ac circuits)} \quad (21.16)$$

To illustrate how phasors are used to analyze RC circuits, consider Example 21.5. Take particular note of part (b), in which there is an apparent violation of Kirchhoff's loop theorem – explained by phase differences between the voltages across the circuit elements.

---

### EXAMPLE 21.5: RC IMPEDANCE AND KIRCHHOFF'S LOOP THEOREM

A series RC circuit has a resistance of 100 Ω and a capacitance of 15.0 μF. (a) What is the (rms) current in the circuit when it is driven by a 120-V, 60-Hz source? (b) Compute the (rms) voltage across each circuit element and the two elements combined. Compare it with that of the voltage source. Is Kirchhoff's loop theorem satisfied? Explain your reasoning.

**THINKING IT THROUGH.** (a) Note that the voltage and capacitor values are the same as in Example 21.3 but a resistor has been added in series. Then, using phasors, the capacitive reactance and the resistance can be combined to determine the overall impedance (Equation 21.16). From Equation 21.17, the impedance and voltage can be used to find the current. (b) Because the current is the same everywhere at any given time in a series circuit, the result of part (a) can be used to calculate the voltages. The rms voltage across both elements together is found by recalling that the individual voltages are out of phase by 90°. What this means physically is that they do not reach their peak values at the same time, but rather are one-fourth of a period apart. Thus, the voltages cannot simply be added as plain numbers would.

#### SOLUTION

***Given:***

$$R = 100\,\Omega$$
$$C = 15.0\,\mu\text{F} = 15.0 \times 10^{-6}\,\text{F}$$
$$\Delta V_{\text{rms}} = 120\,\text{V}$$
$$f = 60\,\text{Hz}$$

***Find:***

**(a)** $I$ (rms current)
**(b)** $\Delta V_C$ (rms voltage across capacitor)
$\Delta V_R$ (rms voltage across resistor)
$\Delta V_{(R+C)}$ (combined rms voltage)

**(a)** In Example 21.3, the reactance for the capacitor at this frequency was $X_C = 177\,\Omega$. Now Equation 21.15 can be used to find the circuit impedance:

$$Z = \sqrt{R^2 + X_C^2} = \sqrt{(100\,\Omega)^2 + (177\,\Omega)^2} = 203\,\Omega$$

Using $\Delta V_{\text{rms}} = I_{\text{rms}} Z$, the rms current is

$$I_{\text{rms}} = \frac{\Delta V_{\text{rms}}}{Z} = \frac{120\,\text{V}}{203\,\Omega} = 0.591\,\text{A}$$

**(b)** Using Equation 21.17 first for the rms voltage across the resistor alone ($Z = R$),

$$\Delta V_R = I_{\text{rms}} R = (0.591\,\text{A})(100\,\Omega) = 59.1\,\text{V}$$

and then for the capacitor alone ($Z = X_C$),

$$\Delta V_C = I_{\text{rms}} X_C = (0.591\,\text{A})(177\,\Omega) = 105\,\text{V}$$

The algebraic sum of these two rms voltages is 164 V, which is not the same as the rms value of the voltage source (120 V). However, this does not mean that Kirchhoff's loop theorem has been violated. In fact, the source voltage does equal the combined voltages across the capacitor and resistor if you account for phase differences. The combined voltage must be calculated properly to take into account the 90° phase difference between the two voltages. Using the Pythagorean theorem to get the total voltage

$$\Delta V_{(R+C)} = \sqrt{(\Delta V_R)^2 + (\Delta V_C)^2} = \sqrt{(59.1\,\text{V})^2 + (105\,\text{V})^2} = 120\,\text{V}$$

Thus when the individual voltages are combined properly (taking into account that the voltages do not peak at the same time), Kirchhoff's laws are still valid. But care must be taken to add the voltages in a "vector-like" way because they are generally out of phase.

**FOLLOW-UP EXERCISE.** (a) How would the result in part (a) of this Example change if the circuit were driven by a voltage source with the same rms voltage, but oscillating at 120 Hz? (b) Is the resistor or the capacitor responsible for the change?

---

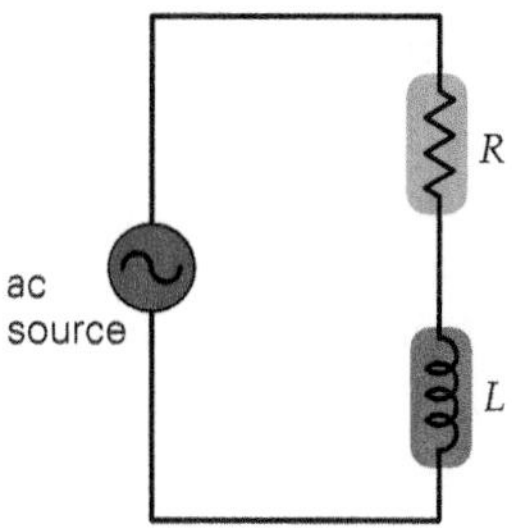


(a) RL circuit diagram

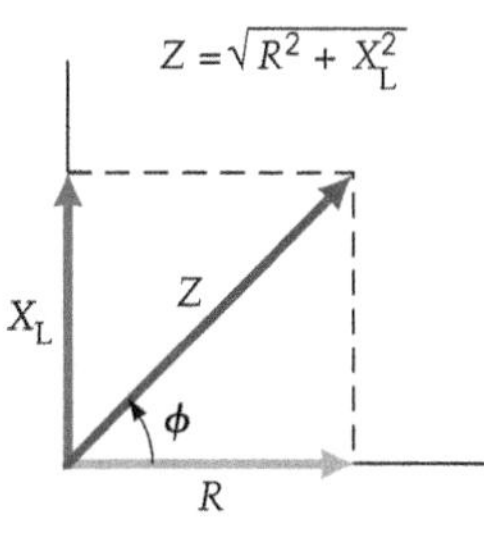


(b) Phase diagram

▲FIGURE 21.9 **A series RL circuit (a)** In a series RL circuit, **(b)** the impedance $Z$ is the phasor sum of the resistance $R$ and the inductive reactance $X_L$.

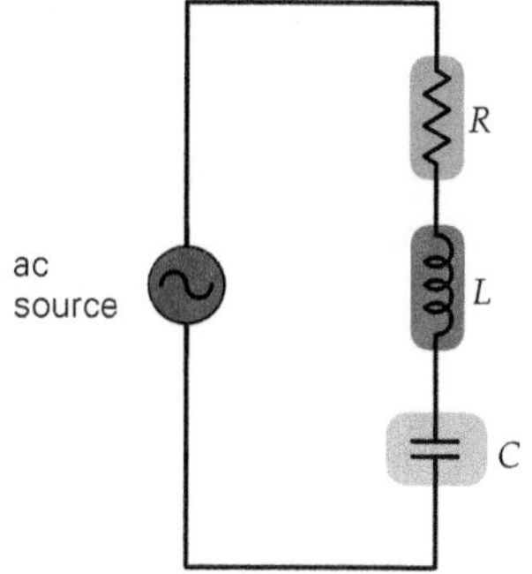


(a) RLC circuit diagram

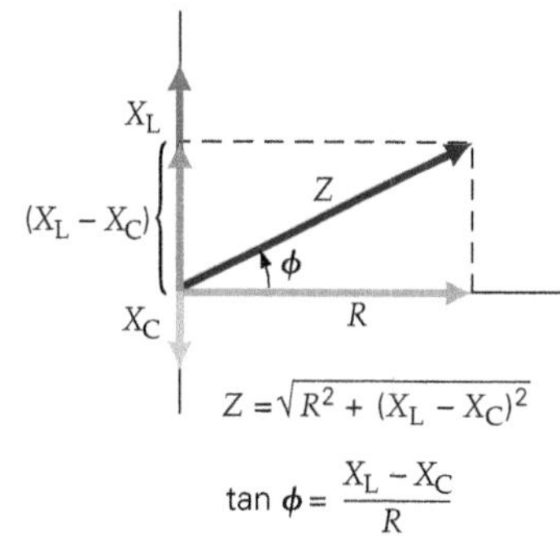


(b) Phase diagram

▲FIGURE 21.10 **A series RLC circuit (a)** In a series RLC circuit, **(b)** the impedance $Z$ is the phasor sum of the resistance $R$ and the total (or net) reactance $(X_L - X_C)$. Note that this phasor diagram is drawn for the case of $X_L > X_C$. Thus, $\phi$ is positive here (i.e., above the $+x$-axis).

## 21.4.2 Series RL Circuit

The analysis of a series RL circuit (▶**Figure 21.9**) is similar to that of a series RC circuit. However, the inductive reactance is now plotted on the positive $y$-axis in the phase diagram to reflect a phase difference of $+90°$ with respect to the resistance. Remember that a positive phase angle $\phi$ means that the voltage leads the current, as is true for an inductor. Thus the impedance in an RL series circuit is given by

$$Z = \sqrt{R^2 + X_L^2} \quad \text{(series RL circuit impendance)} \qquad (21.17)$$

## 21.4.3 Series RLC Circuit

In general, an ac circuit may contain all three circuit elements – a resistor, an inductor, and a capacitor – as shown in the series in ▶**Figure 21.10**. Again, phasor addition methods must be used to determine the overall circuit impedance. Combining the vertical components (i.e., inductive and capacitive reactances) gives the *total reactance*, $X_L - X_C$. Subtraction is used because the phase difference between $X_L$ and $X_C$ is 180°. The overall circuit impedance is the phasor sum of the resistance and the total reactance. Employing the Pythagorean theorem on the phasor diagram gives the expression for the RLC series circuit impedance

$$Z = \sqrt{R^2 + (X_L - X_C)^2} \quad \text{(series RLC impendance)} \qquad (21.18)$$

The phase angle $\phi$ between the source voltage and the current in the circuit is the angle between the overall impedance phasor ($Z$), and the $+x$-axis is shown in Figure 21.10b.

$$\tan\phi = \frac{X_L - X_C}{R} \quad \text{(phase angle in series RLC circuit)} \qquad (21.19)$$

Notice that if $X_L$ is greater than $X_C$ (as in Figure 21.10b), the phase angle is positive, and the circuit is said to be *inductive*, because the non-resistive part of the impedance (i.e., the reactance) is dominated by the inductor. If $X_C$ is greater than $X_L$, the phase angle is negative, and the circuit is said to be *capacitive*, because capacitive reactance dominates over inductive reactance. Consider Example 21.6 that shows phasor analysis applied to a series RLC circuit.

---

**EXAMPLE 21.6: ALL TOGETHER NOW – IMPEDANCE IN AN RLC CIRCUIT**

A series RLC circuit contains an element with resistance of 25.0 Ω, another with capacitance of 50.0 μF, and last one with an inductance of 0.300 H. If the circuit is driven by a 120-V, 60-Hz source, what are (a) the total impedance of the circuit, (b) the rms current in the circuit, and (c) the phase angle between the current and the voltage?

**THINKING IT THROUGH.** (a) To calculate the overall impedance from Equation 21.19, the individual reactances must first be determined. (b) The current is computed from the generalization of Ohm's law, $\Delta V_{rms} = I_{rms}Z$ (Equation 21.16). (c) The phase angle can be found from Equation 21.19.

**SOLUTION**

All the necessary data are given:

***Given:***

$$R = 25.0\,\Omega$$
$$C = 50.0\,\mu\text{F} = 5.00\times10^{-5}\text{ F}$$
$$L = 0.300\text{ H}$$
$$\Delta V_{rms} = 120\text{ V}$$
$$f = 60\text{ Hz}$$

***Find:***

(a) $Z$ (overall circuit impedance)
(b) $I_{rms}$
(c) $\phi$ (phase angle)

(a) The individual reactances are found as follows:

$$X_C = \frac{1}{2\pi fC} = \frac{1}{2\pi(60\text{ Hz})(5.00\times10^{-5}\text{ F})} = 53.1\,\Omega$$

$$X_L = 2\pi fL = 2\pi(60\text{ Hz})(0.300\text{ H}) = 113\,\Omega$$

$$Z = \sqrt{R^2 + (X_L - X_C)^2} = \sqrt{(25.0\,\Omega)^2 + (113\,\Omega - 53.1\,\Omega)^2} = 64.9\,\Omega$$

(b) $I_{rms} = \dfrac{\Delta V_{rms}}{Z} = 1.85\text{ A}$

(c) Solving for the phase angle gives

$$\phi = \tan^{-1}\left(\frac{X_L - X_C}{R}\right) = \tan^{-1}\left(\frac{113\,\Omega - 53.1\,\Omega}{25.0\,\Omega}\right) = +67.3^\circ$$

A positive phase angle was to be expected because the inductive reactance is greater than the capacitive reactance [see (a)] and this circuit is *inductive* in nature.

**FOLLOW-UP EXERCISE.** (a) Consider the RLC circuit in this Example, but with the driving frequency doubled. Reasoning conceptually, should the phase angle $\Phi$ be greater or less than the $+67.3°$ after the increase? (b) Compute the new phase angle to show that your reasoning is correct.

By now, you should appreciate the usefulness of phasor diagrams in determining impedances, voltages, and currents in ac circuits. However, you might still be wondering about the use and meaning of $\phi$. To illustrate its importance, let's examine the power loss in an RLC circuit. Note carefully how this analysis depends on the use of phasor diagrams.

### 21.4.4 Power Factor for a Series RLC Circuit

In considering power in an RLC circuit, a crucial fact to remember is that any power loss (joule heating) can take place only in the resistor. *There are no power losses associated with capacitors and inductors.* This is because, ideally, neither has resistance, and neither would exhibit joule heating. Thus capacitors and inductors simply store energy and give it back, without loss.

The average (rms) power dissipated by a resistor is $P_{rms} = I_{rms}^2 R$. This rms power can also be expressed in terms of the rms current and voltage, but the appropriate voltage is that across the resistor ($\Delta V_R$), because it is the only dissipative element. Thus the average power dissipated in a series RLC circuit is equal to that dissipated by the resistor: $\bar{P} = P_R = I_{rms}\Delta V_R$.

The voltage across the resistor can be found by creating a *voltage triangle* that corresponds to the phasor triangle (▼**Figure 21.11**). The rms voltages across the individual components in an RLC circuit are: $\Delta V_R = I_{rms}R$, $\Delta V_L = I_{rms}X_L$, and $\Delta V_C = I_{rms}X_C$. Combining the last two voltages yields $(\Delta V_L - \Delta V_C) = I_{rms}(X_L - X_C)$. Thus if each leg of the (impedance) phasor triangle (Figure 21.11a) is multiplied by a circuit's rms current, an equivalent voltage phasor triangle results (Figure 21.11b). As can be seen, the resistor voltage is

$$\Delta V_R = \Delta V_{rms}\cos\phi \tag{21.20}$$

The term $\cos\phi$ is called the **power factor**. Again from Figure 21.11, it can be seen that

$$\cos\phi = \frac{R}{Z} \quad \text{(series RLC power factor)} \tag{21.21}$$

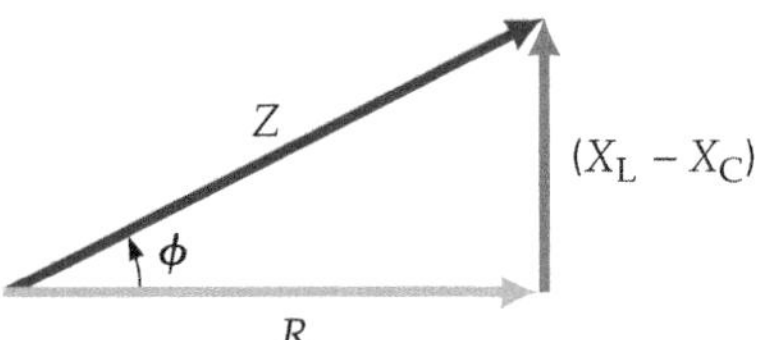


**(a) Phasor triangle**

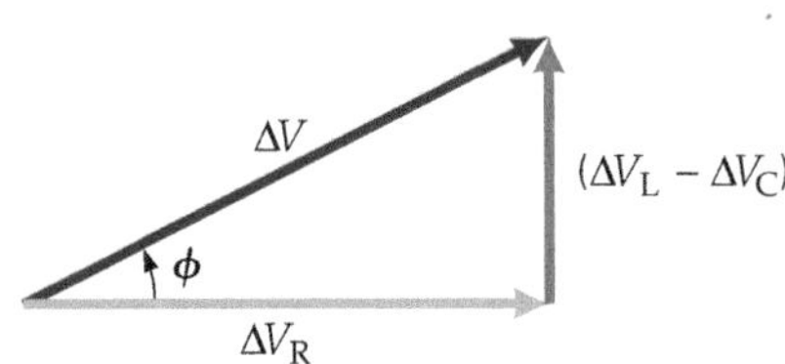


**(b) Equivalent voltage triangle**

▲FIGURE 21.11 **Phasor and voltage triangles** In series, the rms current is the same in each element, thus each element voltage is proportional to its impedance (reactance or resistance). Therefore the impedance phasor triangle **(a)** can be converted into an equivalent voltage phasor triangle **(b)**. Both diagrams are for the case of $X_L > X_C$.

The circuit's average power can be rewritten in terms of the power factor

$$\bar{P} = I_{rms} \cdot \Delta V_{rms} \cdot \cos\phi \quad \text{(series RLC average power)} \qquad (21.22)$$

Because power is dissipated only in the resistance $\left(\bar{P} = I_{rms}^2 R\right)$, Equation 21.22 can be rewritten alternatively as

$$\bar{P} = I_{rms}^2 \, Z \cos\phi \quad \text{(series RLC average power)} \qquad (21.23)$$

Recall that $\cos\phi$ has a maximum of $+1$ ($\phi = 0°$) and a minimum of zero ($\phi = \pm 90°$). When $\phi = 0°$, the circuit is *completely resistive* and there is maximum power dissipation (as if the circuit contained only a resistor – although it may actually include inductors and capacitors). The power factor decreases as the phase angle increases in either direction [since $\cos(-\phi) = \cos\phi$] – in other words, as the circuit becomes inductive or capacitive. If the angle is either $\pm 90°$, the circuit must contain no resistance – only capacitors and/or inductors. Thus for these extreme angles, there is no (joule heating) power dissipated in the circuit.

In practice, because there is always some resistance, a circuit can never be completely inductive or capacitive. It is possible, however, for an RLC circuit to appear to be completely resistive even if it contains a capacitor and an inductor, as will be seen in Section 21.5. For now, let's look at our previous RLC example with an emphasis on power.

---

EXAMPLE 21.7: POWER FACTOR REVISITED

What is the average power dissipated in the circuit described in Example 21.6?

**THINKING IT THROUGH.** The power factor can be found since the resistance ($R$) and impedance ($Z$) are known. Once this factor is known, the average power can be calculated.

**SOLUTION**

***Given:***

See Example 21.6.

***Find:*** $\bar{P}$ (average power)

In Example 21.6, the circuit had an impedance of $Z = 64.9\ \Omega$, and a resistance of $R = 25.0\ \Omega$.

Therefore, its power factor is

$$\cos\phi = \frac{R}{Z} = \frac{25.0\ \Omega}{64.9\ \Omega} = 0.385$$

Using the other data from Example 21.6 and Equation 21.22 gives

$$\bar{P} = I_{rms} \cdot \Delta V_{rms} \cdot \cos\phi = (1.85\,\text{A})(120\,\text{V})(0.385) = 85.5\,\text{W}$$

This is less than the power that would be dissipated without a capacitor and an inductor. (Can you determine what that value would be?)

**FOLLOW-UP EXERCISE.** If the frequency were doubled and the capacitor removed from this Example, what would be the rms power?

---

## 21.5 RLC Circuit Resonance

From the previous discussion, it can be seen that when the power factor ($\cos\phi$) of an RLC series circuit is equal to unity, maximum power is transferred to the circuit. In this situation, the current in the circuit must be at a maximum because the impedance is at its minimum. This occurs at a unique frequency when the inductive and capacitive reactances *effectively cancel* – that is, they are equal in magnitude and 180° out of phase.

The key to finding this unique frequency is to realize that because inductive and capacitive reactances are both frequency dependent, so is the overall impedance. From the expression for the RLC series impedance, $Z = \sqrt{R^2 + (X_L - X_C)^2}$, it can be seen that the impedance is a minimum at a frequency labeled $f_o$, found by setting $X_L = X_C$. From the expressions for the reactances this means $2\pi f_o L = 1/(2\pi f_o C)$, which can be solved for $f_o$:

$$f_o = \frac{1}{2\pi\sqrt{LC}} \quad \text{(series RLC resonance frequency)} \qquad (21.24)$$

This frequency satisfies the condition of minimum impedance – and therefore maximizes the current and power dissipation in the circuit, a situation analogous to pumping a swing at just the right frequency to create maximum amplitude. $f_o$ is called the circuit's **resonance frequency**. A plot of capacitive and inductive reactances versus frequency is shown in ▶ **Figure 21.12a**. The graphs for $X_C$ and $X_L$ as a function of source frequency intersect at $f_o$, that unique frequency when their values become equal.

The physical explanation of series RLC circuit resonance is worth exploring. Recall that capacitor and inductor voltages are *always* 180° out of phase, or of opposite polarity. In other words, they tend to cancel out but usually don't do so completely. If they only partially cancel, then the voltage across the resistor is less than that of the source voltage. This is because under these conditions, there remains a net voltage across the combination of capacitor and inductor, thus leaving less than the maximum for the resistor (Kirchhoff's loop theorem). In this case, the power dissipated in the resistor is less than its maximum value – the value it attains at resonance.

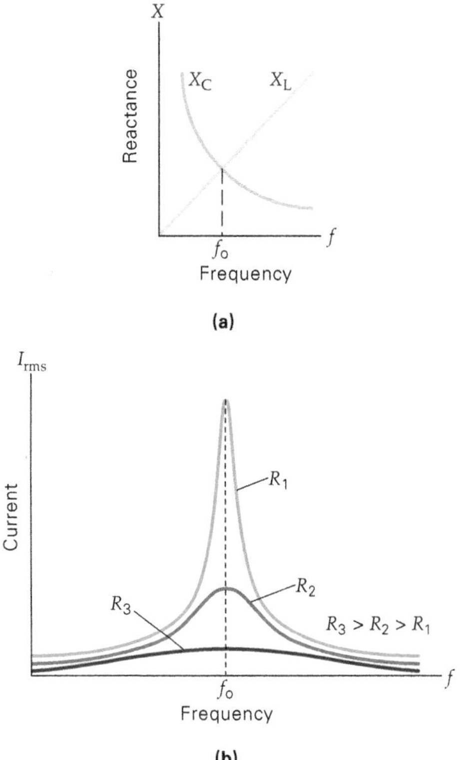


▲FIGURE 21.12 **Resonance frequency for a series RLC circuit** **(a)** At the resonance frequency ($f_o$), the capacitive and inductive reactances are equal. A graph of both reactances versus frequency illustrates that this is where the plots of $X_C$ and $X_L$ intersect. **(b)** On a graph of rms circuit current versus frequency, the current is a maximum at $f_o$. The curve becomes sharper and narrower as the resistance in the circuit decreases.

If the circuit is driven at its resonance frequency, however, then the capacitive and inductive voltages do cancel, and the full source voltage does appear across the resistor. In this special case, the power factor becomes 1, and the resistor dissipates the maximum possible power. Note that details of this power transfer depend on the value of the resistor. A graph of rms current versus driving frequency is shown in Figure 21.12b for several different values of resistance. As expected, the maximum current occurs at a driving frequency equal to the resonance (natural) frequency of the circuit, $f_o$, however the curve becomes sharper (narrower in frequency and larger in amplitude) as the resistance in the circuit decreases.

## 21.5.1 Resonance Applications

Resonant circuits have a variety of applications. One common application is as a **radio wave receiver**. Each broadcast station has an assigned frequency at which its waves are transmitted. When these are received at the antenna, their oscillating electric and magnetic fields set the electrons in the antenna into oscillatory motion. In other words, they produce an ac current in the receiver circuit, just as a normal ac voltage source would do.

In a given geographic area, usually several different radio signals reach an antenna at once, but a good receiver circuit selectively picks up only the signal with a frequency at or near its resonance frequency. Most radios allow you to alter this resonance frequency so as to "tune in" to different stations. In the early days of radio, variable air capacitors were used to change the capacitive reactance for this purpose (▼**Figure 21.13**). More compact variable capacitors in smaller radios and electronic devices with a polymer dielectric between thin plates were later used. The dielectric sheets increased the capacitance, allowing manufacturers to create much smaller capacitors. Today most modern electronic devices that require frequency tuning employ solid-state devices instead of variable capacitors.

▲FIGURE 21.13 **Variable air capacitor** Rotating the movable plates between the fixed plates changes the overlap area and thus the capacitance. Such capacitors were common in the tuning circuits of older radios.

On the other hand, if you wish to *emit* high-frequency electromagnetic waves, such as those in radio communications and television, current must oscillate at high frequencies in the emitting antenna. This can be accomplished by connecting the antenna to an RLC circuit. In this situation, the circuit is called a **broadcast circuit**. In such circuits, the current oscillates at a frequency determined by their inductive and capacitive elements. When the resistance in an RLC circuit is small, the circuit is essentially an LC circuit and the current oscillates at the circuit's "natural" frequency given by Equation 21.24. The oscillating electrons in the wire then emit electromagnetic radiation at that frequency, which is adjustable by altering the capacitive and/or inductive reactances. For a practical example of what happens when a radio changes reception frequencies, consider the next example.

### INTEGRATED EXAMPLE 21.8: AM VERSUS FM – RESONANCE IN RADIO RECEPTION

(a) Suppose you were listening to news on an AM station broadcasting at a frequency of 920 kHz and switched to a music station on the FM band at 99.7 MHz (the word *band* refers to a specific frequency range). When you switched, you effectively changed the capacitance of the receiving circuit, assuming constant inductance. Did the capacitance (1) increase or (2) decrease when you made this change? (b) By what factor did the capacitance of the receiving circuit in the radio change, assuming constant inductance?

(a) **CONCEPTUAL REASONING.** Because FM stations broadcast at significantly higher frequencies than AM stations, the resonance frequency of the receiver must be increased to receive signals on the FM band. An increase of the resonance frequency requires reducing the capacitance since the inductance is fixed. Thus the correct answer is (2).

(b) **QUANTITATIVE REASONING AND SOLUTION.** The resonance frequency (Equation 21.24) depends on the inductance and the capacitance. Because the question asks for a "factor," it is asking for a ratio of the new capacitance to the original. The frequencies must be in the same units. So let's convert MHz into kHz and use unprimed quantities to refer to AM and primed quantities to mean FM.

***Given:***

$$f_o = 920\text{ kHz}$$
$$f_o' = 99.7\text{ MHz}$$
$$= 99.7\times10^3\text{ kHz}$$

***Find:*** $C'/C$ (ratio of FM capacitance to AM capacitance)

From Equation 21.24, the two resonant frequencies can be written as

$$f_o = \frac{1}{2\pi\sqrt{LC}} \quad\text{and}\quad f_o' = \frac{1}{2\pi\sqrt{LC'}}$$

Solving for the frequency ratio yields

$$\frac{f_o}{f_o'} = \frac{2\pi\sqrt{LC'}}{2\pi\sqrt{LC}} = \sqrt{\frac{C'}{C}}$$

The capacitance ratio is found by squaring and substituting the given values.

$$\frac{C'}{C} = \left(\frac{f_o}{f_o'}\right)^2 = \left(\frac{920\text{kHz}}{99.7\times10^3\text{kHz}}\right)^2 = 8.51\times10^{-5}$$

Thus, $C' = 8.51\times10^{-5}\,C$ and the capacitance was decreased by a factor of almost one ten-thousandth ($8.51\times10^{-5} \approx 10^{-4}$).

**FOLLOW-UP EXERCISE.** (a) Based on the resonance curves shown in Figure 21.12b, explain how it is possible to pick up two radio stations *simultaneously* on your radio. (Stations are sometimes granted licenses for broadcasting at closely spaced frequencies under the assumption that they both won't be received by the same radio. However, under certain atmospheric conditions, this may not be true.) (b) In part (b) of this Example, if you next increased the capacitance by a factor of two (starting with the news at 920 kHz) to listen to a hockey game, to what new frequency on the AM band would you now be tuned?

# Chapter 21 Review

- An **ac voltage** is described by

$$\Delta V_R = \Delta V_m \sin(\omega t) = \Delta V_m \sin(2\pi ft) \tag{21.1}$$

- For a sinusoidally varying **ac current**, the peak (maximum) current $I_m$ and the rms current $I_{rms}$ are related by

$$I_{rms} = \frac{I_m}{\sqrt{2}} \approx 0.707 I_m \tag{21.5}$$

- For an **ac voltage**, the peak (maximum) voltage $\Delta V_m$ is related to its rms (root-mean-square) voltage $\Delta V_{rms}$ by

$$\Delta V_{rms} = \frac{\Delta V_m}{\sqrt{2}} \approx 0.707 V_m \tag{21.7}$$

- The current in a resistor is in phase with the voltage across it. For a capacitor, the current is 90° (one-quarter of a cycle) ahead of the voltage. For an inductor, the current lags behind the voltage by 90°.
- In ac circuits, joule heating is due entirely to the resistive elements, and the time-averaged **power dissipation** is

$$\bar{P} = I_{rms}^2 R = I_{rms}\Delta V_{rms} = \frac{(\Delta V_{rms})^2}{R} \tag{21.9}$$

- In ac circuits, capacitors and inductors can both allow and oppose current but without joule heating (dissipative losses). The opposition is characterized by two frequency-dependent quantities: **capacitive reactance** ($X_C$) and **inductive reactance** ($X_L$). The capacitive reactance is given by

$$X_C = \frac{1}{\omega C} = \frac{1}{2\pi fC} \tag{21.10}$$

and the expression for inductive reactance is

$$X_L = \omega L = 2\pi fL \tag{21.13}$$

- Ohm's law, as applied to each type of circuit element, is a generalization from dc circuits. The relationship between rms current and the rms voltage for a resistor is

$$\Delta V_{rms} = I_{rms} R \tag{21.8}$$

The relationship between rms current and the rms voltage for a capacitor is

$$\Delta V_{rms} = I_{rms} X_C \tag{21.11}$$

The relationship between rms current and the rms voltage for an inductor is

$$\Delta V_{rms} = I_{rms} X_L \tag{21.14}$$

- **Phasors** are vector-like quantities that allow resistances and reactances to be represented and combined graphically.
- **Impedance** ($Z$) is the total, or effective, opposition to current that takes into account both resistances and reactances. Impedance is related to current and circuit voltage by a generalization of Ohm's law:

$$\Delta V_{\text{rms}} = I_{\text{rms}} Z \quad (21.16)$$

- The **impedance for a series RLC circuit** is

$$Z = \sqrt{R^2 + (X_L - X_C)^2} \quad (21.18)$$

- The **phase angle** ($\phi$) between the rms voltage and the rms current in a series RLC circuit is given by

$$\tan\phi = \frac{X_L - X_C}{R} \quad (21.19)$$

- The **power factor** ($\cos\phi$) for a series RLC circuit is a measure of how close to the maximum power dissipation the circuit is. The power factor is

$$\cos\phi = \frac{R}{Z} \quad (21.21)$$

- The **average power dissipated** in a series RLC circuit (joule heating in the resistor) is

$$\bar{P} = I_{\text{rms}} \cdot \Delta V_{\text{rms}} \cos\phi \quad (21.22)$$

or

$$\bar{P} = I_{\text{rms}}^2 Z \cos\phi \quad (21.23)$$

- The **resonance frequency** ($f_o$) of an RLC circuit is the frequency at which the circuit dissipates maximum power and is given by

$$f_o = \frac{1}{2\pi\sqrt{LC}} \quad (21.24)$$

# End of Chapter Questions and Exercises

## Multiple Choice Questions

**21.1 Resistance in an AC Circuit**

1. Which of the following is true for a sinusoidally varying ac voltage: (a) $\Delta V_m = \Delta V_{\text{rms}}$, (b) $\Delta V_m \approx (0.707)\Delta V_{\text{rms}}$, (c) $\Delta V_{\text{rms}} \approx (0.707)\Delta V_m$, or (d) $\Delta V_{\text{rms}} \approx (0.5)\Delta V_{\text{rms}}$?
2. During one ac voltage cycle (in the United States) how long does the direction of the current stay constant in a resistor: (a) 1/60 s, (b) 1/120 s, or (c) 1/30 s?
3. Over seven complete ac voltage cycles (in the United States) what is the average voltage: (a) 0 V, (b) 60 V, (c) 120 V, or (d) 170 V?
4. During five complete ac voltage cycles (in the United States) how many times does the power dissipation in a resistor reach its maximum: (a) once, (b) five times, (c) ten times, or (d) twice?
5. In ac operation in the United States, how much time elapses between successive maximum power values in a resistor: (a) 1/60 s, (b) 1/120 s, or (c) 1/30 s?

**21.2 Capacitive Reactance**
**and**
**21.3 Inductive Reactance**

6. In a purely capacitive ac circuit, (a) the current and voltage are in phase, (b) the current leads the voltage, (c) the current lags behind the voltage, or (d) none of the preceding.
7. A single capacitor is connected to an ac voltage source. When the voltage across the capacitor is at a maximum, then the charge on it is (a) zero, (b) at a maximum, (c) neither of the preceding, but somewhere in between.
8. A single capacitor is connected to an ac voltage source. When the current in the circuit is at a maximum, then the energy stored in the capacitor is (a) zero, (b) at a maximum, (c) neither of the preceding, but somewhere in between.
9. A single inductor is connected to an ac voltage source. When the voltage across the inductor is at a maximum, the current in it is not changing. Is this statement (a) true, (b) false, or (c) cannot be determined from the given information?
10. A single inductor is connected to an ac voltage source. When the current in the inductor is at a maximum, the voltage across the inductor is not changing. Is this statement (a) true, (b) false, or (c) cannot be determined from the given information?

**21.4 Impedance: RLC Circuits**
**and**
**21.5 Circuit Resonance**

11. The impedance of an RLC circuit depends on (a) frequency, (b) inductance, (c) capacitance, (d) all of the preceding.
12. If the capacitance of a series RLC circuit is decreased, (a) the capacitive reactance increases, (b) the inductive reactance increases, (c) the current remains constant, (d) the power factor remains constant.
13. When a series RLC circuit is driven at its resonance frequency, (a) energy is dissipated only by the resistive element, (b) the power factor has a value of one, (c) there is maximum power delivered to the circuit, (d) all of the preceding.
14. When a series RLC circuit is not driven at its resonance frequency, energy may be dissipated as joule heat in either the capacitor or the inductor. Is this statement (a) true, (b) false, or (c) cannot be determined from the given information?

## Conceptual Questions

**21.1 Resistance in an AC circuit**

1. The average current in a resistor in an ac circuit is zero. Explain why the average power delivered to a resistor isn't zero.
2. The voltage and current associated with a resistor in an ac circuit are *in phase*. What does that mean physically?
3. A 60-W lightbulb designed to work at 240 V in England is instead connected to a 120-V source. Discuss the changes in the bulb's rms current and power when it is at 120 V compared with 240 V. Assume the bulb is ohmic.
4. If the ac voltage and current for a particular circuit element are given by $\Delta V = (120\text{ V})\sin(120\pi t)$ and $I = (30\text{ A})\sin(120\pi t)$, respectively, could the circuit element be a resistor? What is the frequency of the source?
5. A 25-Ω resistor is wired directly across a 120-V ac source. What happens to the time-average power, rms voltage, and rms current when the resistor's value changes to 50 Ω? Do their values increase, decrease, or remain the same? Explain.
6. A 25-Ω resistor is wired directly across a 120-V ac source. What happens to the time-average power, rms voltage, and rms current when the ac source is changed to 240 V? Do their values increase, decrease, or remain the same? Explain.

**21.2 Capacitive Reactive Reactance and**
**21.3 Inductive Reactance**

7. Explain why, under very low-frequency ac conditions, a capacitor acts almost as an open circuit while an inductor acts almost as a short circuit.
8. Can an inductor oppose dc current? What about a capacitor? Explain each and why they are different.
9. The current in an ac circuit containing only an ac voltage source and a 10-$\mu$F capacitor is described by $I = (2.0\text{ A})\ \sin(120\pi t + \pi/2)$. Explain why the instantaneous voltage across it at $t = 0$ *is* zero whereas the current at that time is *not*.
10. An inductor is connected by itself to a 60-Hz ac voltage source. To ensure it has the same inductive reactance when it is connected to a 240-Hz source, would you decrease or increase its inductance value? By what factor should it change? Explain your reasoning.
11. A capacitor is connected by itself to a 60-Hz ac voltage source. To ensure the circuit has the same capacitive reactance when the capacitor is replaced with one of twice the capacitance, would you decrease or increase the source frequency? By what factor should it change? Explain your reasoning.

**21.4 Impedance: RLC Circuits and**
**21.5 Circuit Resonance**

12. An RLC circuit consists of a 25-Ω resistor, a 1.00-$\mu$F capacitor, and a 250-mH inductor. If it is driven by an ac voltage source whose frequency is 60 Hz, (a) what is the impedance of an RLC circuit at resonance? (b) How does its impedance compare to that in part (a) if the source frequency is doubled: is it more, less, or the same value? (c) How does its impedance compare to that in part (a) if the source frequency is halved: is it more, less, or the same value? Explain your reasoning for all parts.
13. For each of the following cases for an RLC circuit, does the resonance frequency increase, decrease, or stay the same? If it changes, tell by what factor. (a) Only the capacitance is changed – it is quadrupled. (b) Only the inductance is changed – it is increased by nine times. (c) Only the resistance is changed – it is increased by three times. (d) Both the inductance and capacitance are doubled; nothing else changes.
14. An RLC circuit is driven by an ac voltage source and is at resonance. Then the driving frequency changes and the power output drops. Can you tell if the frequency increased or decreased? Explain.
15. If a driven RLC circuit has an inductive reactance of 250 Ω and a capacitive reactance of 150 Ω, is the driving frequency exactly at, above, or below the circuit's resonant frequency? Explain your reasoning.

## Exercises*

*Integrated Exercises (IEs) are two-part exercises. The first part typically requires a conceptual answer choice based on physical thinking and basic principles. The following part is quantitative calculations associated with the conceptual choice made in the first part of the exercise.*

**21.1 Resistance in an AC Circuit**

1. • What are the peak and rms voltages of a 120-V ac line and a 240-V ac line?
2. • An ac circuit has an rms current of 5.0 A. What is the peak current? What is the average current?
3. • How much ac rms current must be in a 10-Ω resistor to produce an average power of 15 W?
4. • An ac circuit contains a resistor with a resistance of 5.0 Ω. The resistor has an rms current of 0.75 A. (a) Find its rms voltage and peak voltage. (b) Find the average power delivered to the resistor.
5. • A hair dryer is rated at 1200 W when plugged into a 120-V outlet. Find (a) its rms current, (b) its peak current, and (c) its resistance.
6. IE •• The voltage across a resistor varies as $\Delta V = (170\text{ V})\sin(100\pi t)$. (a) Is the current in the resistor (1) in phase with the voltage, (2) ahead of the voltage by 90°, or (3) lagging behind the voltage

* The bullets denote the degree of difficulty of the exercises: •, simple; ••, medium; and •••, more difficult.

by 90°? (b) Write the expression for the current in the resistor as a function of time and determine the voltage frequency.

7. •• An ac voltage is applied to a 25.0-Ω resistor so that it dissipates 500 W of power. Find the resistor's (a) rms and peak currents and (b) rms and peak voltages.
8. **IE** •• An ac voltage source has a peak voltage of 85 V and a frequency of 60 Hz. The voltage at $t = 0$ is zero. (a) If a student measures the voltage at $t = 1/240$ s, how many possible results are there: (1) one, (2) two, or (3) three? Why? (b) Determine all possible voltages the student might measure.
9. •• An ac voltage source has an rms voltage of 120 V. Its voltage goes from zero to its maximum positive value in 4.20 ms. Write an expression for the voltage as a function of time.
10. •• What are the resistance, peak current, and power level of a computer monitor that draws an rms current of 0.833 A when connected to a 120-V outlet?
11. •• Find the rms and peak currents in a 40-W, 120-V lightbulb. What is its resistance?
12. •• A 50-kW electric heater is designed to run using a 240-V ac source. Find its (a) peak current and (b) peak voltage. (c) How much energy will you be billed for in a 30-day month if it operates 2.0 h per day?
13. •• The current in a resistor is given by $I = (8.0\text{ A})\sin(40\pi t)$ when an ac voltage given by $\Delta V = (60\text{ V})\sin(40\pi t)$ is applied to it. (a) What is the resistance value? (b) What are the frequency and period of the voltage source? (c) What is the average power delivered to the resistor?
14. •• The current and voltage outputs of an operating ac power supply have peak values of 2.5 A and 16 V, respectively. (a) What is the average power output of the generator? (b) What is the effective resistance of the circuit it is in?
15. ••• The current in a resistor is given by $I = (12.0\text{ A})\sin(380t)$. (a) What is the frequency of the current? (b) What is the rms current? (c) How much average power is delivered to the resistor? (d) Write an equation for the voltage across the resistor as a function of time. (e) Write an equation for the power delivered to the resistor as a function of time. (f) Show that the rms power obtained in part (e) is the same as your answer to part (c).

**21.2 Capacitive Reactance**
**and**
**21.3 Inductive Reactance**

16. • At what frequency does a 25-μF capacitor have a reactance of 25 Ω?
17. • A single 2.0-μF capacitor is connected across the terminals of a 60-Hz voltage source, and a current of 2.0 ma results. What is the capacitive reactance of the capacitor?
18. • What capacitance value would create a reactance of 100 Ω in a 60-Hz ac circuit?
19. • How much current is in a circuit containing only a 50-μF capacitor connected to an ac generator with an output of 120 V and 60 Hz?
20. •• A single 50-mH inductor is connected to an ac voltage source at 120 V and 60 Hz. (a) What is its inductive reactance? (b) What is the current in this circuit? (c) What is the phase angle between the current and applied voltage? (Assume negligible resistance.)
21. •• A variable capacitor in a circuit with a 120-V, 60-Hz source initially has a capacitance of 0.25 μF. The capacitance is then increased to 0.40 μF. (a) What is the percentage change in the capacitive reactance? (b) What is the percentage change in the current in the circuit?
22. •• (a) An inductor has a reactance of 90 Ω in a 60-Hz ac circuit. What is its inductance? (b) What frequency would be required to double its reactance?
23. •• (a) Find the frequency at which a 250-mH inductor has a reactance of 400 Ω. (b) At what frequency would a 0.40 μF capacitor have the same reactance?
24. **IE** •• A capacitor is connected to a variable-frequency ac voltage source. (a) If the frequency increases by a factor of 3, the capacitive reactance will be (1) 3, (2) 1/3, (3) 9, (4) 1/9 times the original reactance. Explain. (b) If the capacitive reactance of a capacitor at 120 Hz is 100 Ω, what is its reactance if the frequency is changed to 60 Hz?
25. •• A 150-mH inductor is in a circuit with a 60-Hz voltage source, resulting in a current of 1.6 A. (a) What is the rms voltage of the source? (b) What is the phase angle between the current and the source voltage?
26. •• (a) What inductance has the same reactance in a 120-V, 60-Hz circuit as a capacitance of 10 μF? (b) What would be the ratio of inductive reactance to capacitive reactance if the frequency were changed to 120 Hz?
27. •• A circuit with a single capacitor is connected to a 120-V, 60-Hz source. (a) What is its capacitance if there is a current of 0.20 A in the circuit? (b) What would be the current if the source frequency were halved?
28. **IE** •• An inductor is connected to a variable-frequency ac voltage source. (a) If the frequency decreases by a factor of 2, the rms current will be (1) 2, (2) 1/2, (3) 4, (4) 1/4 times the original rms current. Explain. (b) If the rms current in an inductor at 40 Hz is 9.0 A, what is its rms current if the frequency is changed to 120 Hz?

**21.4 Impedance: RLC Circuits**
**and**
**21.5 Circuit Resonance**

29. • A coil in a 60-Hz circuit has a resistance of 100 Ω and an inductance of 0.45 H. Calculate (a) the coil's reactance and (b) the circuit's impedance.
30. •• A series RC circuit has a resistance of 200 Ω and a capacitance of 25 μF and is driven by a 120-V, 60-Hz source. (a) Find the capacitive reactance and impedance of the circuit. (b) What is the current is in the circuit?

31. •• A series RL circuit has a resistance of 100 Ω and an inductance of 100 mH. It is driven by a 120-V, 60-Hz source. (a) Find the inductive reactance and the impedance of the circuit. (b) What is the current is in the circuit?
32. •• A series RC circuit has a resistance of 250 Ω and a capacitance of 6.0 $\mu$F. If the circuit is driven by a 60-Hz source, find (a) the capacitive reactance and (b) the impedance of the circuit.
33. IE •• A series RC circuit has a resistance of 100 Ω and a capacitive reactance of 50 Ω. (a) Is the phase angle (1) positive, (2) zero, or (3) negative? Explain. (b) What is the phase angle of this circuit?
34. •• A series RLC circuit contains a resistance of 25 Ω, an inductance of 0.30 H, and a capacitance of 8.0 $\mu$F. (a) At what frequency should the circuit be driven for the maximum power to be transferred from the driving source? (b) What is the impedance at that frequency?
35. IE •• In a series RLC circuit, $R = X_C = X_L = 40\ \Omega$ at a certain driving frequency. (a) This circuit is (1) inductive, (2) capacitive, (3) in resonance. Explain your reasoning. (b) If the driving frequency is doubled, what will be the impedance of the circuit?
36. IE •• (a) A series RLC circuit is in resonance. Which one of the following can you change without upsetting the resonance: (1) resistance, (2) capacitance, (3) inductance, or (4) frequency? Explain. (b) A resistor, an inductor, and a capacitor have resistance/reactance values of 500 Ω, 500 mH, and 3.5 $\mu$F, respectively. They are connected in series to a power supply of 240 V with a frequency of 60 Hz. What values of resistance and inductance would be required for this circuit to be in resonance (without changing the capacitor)?
37. •• (a) How much power is dissipated in the circuit described in Exercise 36b using the initial values of resistance, inductance, and capacitance? (b) How much power is dissipated in the same circuit at resonance?
38. •• (a) What is the resonance frequency of an RLC circuit with a resistance of 100 Ω, an inductance of 100 mH, and a capacitance of 5.00 $\mu$F? (b) What is the resonance frequency if all the values in part (a) are doubled?
39. •• A tuning circuit in a radio receiver has a fixed inductance of 0.50 mH and a variable capacitor. (a) If the circuit is tuned to a radio station broadcasting at 980 kHz, what is the capacitance value? (b) What value of capacitance is required to tune into a station broadcasting at 1280 kHz?
40. IE •• A coil with a resistance of 30 Ω and an inductance of 0.15 H is connected to a 120-V, 60-Hz source. (a) Is the phase angle of this circuit (1) positive, (2) zero, or (3) negative? Explain. (b) What is the phase angle of the circuit? (c) How much rms current is in the circuit? (d) What is the average power delivered to the circuit?
41. •• A small welding machine uses a voltage source of 120 V at 60 Hz. When the source is operating, it requires 1200 W of power, and the power factor is 0.75. (a) What is the machine's circuit impedance? (b) Find the rms current in the machine while operating.
42. •• A series circuit is connected to a 220-V, 60-Hz power supply. The circuit has the following components: a 10-Ω resistor, a coil with an inductive reactance of 120 Ω, and a capacitor with a reactance of 120 Ω. Compute the rms voltage across (a) the resistor, (b) the inductor, and (c) the capacitor.
43. •• A series RLC circuit has a resistance of 25 Ω, a capacitance of 0.80 $\mu$F, and an inductance of 250 mH. The circuit is connected to a variable-frequency source with a fixed rms voltage output of 12 V. If the source frequency is set at the circuit's resonance frequency, what is the rms voltage across each of the circuit elements?
44. •• (a) In Exercises 42 and 43, determine the numerical (scalar) sum of the rms voltages across the three circuit elements and explain why it is larger than the source voltage. (b) Determine the sum of these voltages using the proper phasor techniques and show that your result is equal to the source voltage.
45. IE ••• (a) If the circuit in ▶ **Figure 21.14** is in resonance, the impedance of the circuit is (1) greater than 25 Ω, (2) equal to 25 Ω, (3) less than 25 Ω. Explain. (b) If the driving frequency is 60 Hz, what is the circuit's impedance?

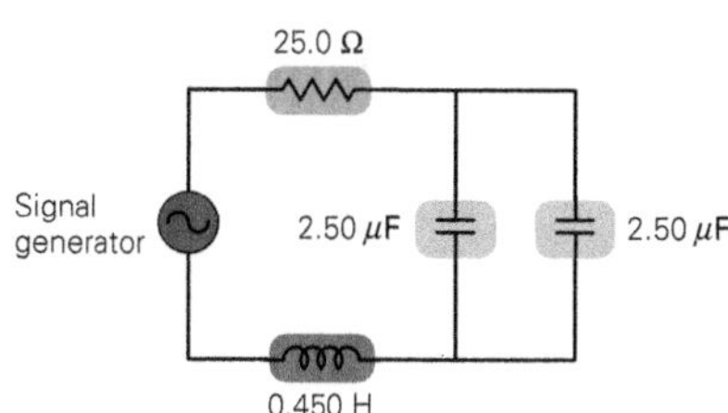


▲ **FIGURE 21.14** **Tune to resonance** See Exercise 45.

46. ••• A series RLC circuit with a resistance of 400 Ω has capacitive and inductive reactances of 300 and 500 Ω, respectively. (a) What is the power factor of the circuit? (b) If the circuit operates at 60 Hz, what additional capacitance should be connected to the original capacitance to give a power factor of 1.00? How should the capacitors be connected, in series or in parallel?
47. ••• A series RLC circuit has components with resistive/reactance values of: $R = 50\ \Omega$, $L = 0.15$ H, and $C = 20\ \mu$F. The circuit is driven by a 120-V, 60-Hz source. (a) What is the current in the circuit, expressed as a percentage of the maximum possible current? (b) What is the power delivered to the circuit, expressed as a percentage of the power delivered when the circuit is in resonance?

# 22

# Reflection and Refraction of Light

**The mirrorlike water surface forms a perfect reflection image of a gliding bald eagle with its talons out.**

We live in a visual world, surrounded by eye-catching images such as that reflective image of the bald eagle with its talons out gliding over the mirrorlike water surface of a lake, shown in the chapter-opening photograph. How these images are formed is something taken largely for granted – until something is seen that can't be easily explained. Optics is the study of light and vision. Human vision requires *visible light* of wavelengths from 400 to 700 nm (see Figure 20.18). Optical phenomena, such as reflection and refraction, are shared by all electromagnetic waves.

In this chapter, the basic optical phenomena of reflection, refraction, total internal reflection, and dispersion will be investigated. The principles that govern reflection explain the behavior of mirrors, while those that govern refraction explain the properties of lenses. With the aid of these and other optical principles, we can understand many optical phenomena experienced every day – why a glass prism spreads light into a spectrum of colors, what causes mirages, how rainbows are formed, and why the legs of a person standing in a swimming pool seem to shorten. Some less familiar but increasingly useful subjects, including the fascinating field of fiber optics, will also be explored.

A simple geometrical approach involving straight lines and angles can be used to investigate many aspects of the properties of light, especially how light propagates. For these purposes, we need not be concerned with the physical (wave) nature of electromagnetic waves described in Chapter 20. The principles of geometrical optics will be introduced here and applied in greater detail in the study of mirrors and lenses in Chapter 23.

## 22.1 Wave Fronts and Rays

Waves, electromagnetic or otherwise, are conveniently described in terms of wave fronts. A **wave front** is the line or surface defined by adjacent portions of a wave that are in phase (Section 13.3.1). If an arc is drawn along one of the crests of a circular water wave moving out from a point source, all particles on the arc will be in phase (▼ **Figure 22.1a**). An arc along a wave trough would work equally well. For a three-dimensional spherical wave, such as a sound or light wave emitted from a point source, the wave front is a spherical surface rather than a circle.

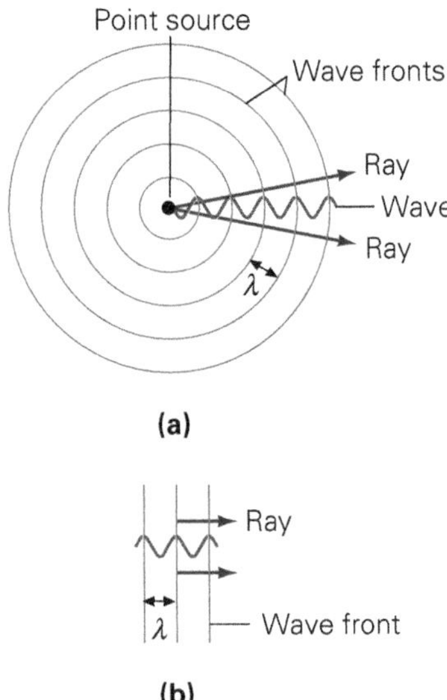


▲ FIGURE 22.1 **Wave fronts and rays** A wave front is defined by adjacent points on a wave that are in phase, such as those along wave crests or troughs. A line perpendicular to a wave front in the direction of the wave's propagation is called a ray. **(a)** Near a point source, the wave fronts are circular in two dimensions and spherical in three dimensions. **(b)** Very far from a point source, the wave fronts are approximately linear or planar and the rays nearly parallel.

Very far from the source, the curvature of a short segment of a circular or spherical wave front is extremely small. Such a segment may be approximated as a linear wave front (in two dimensions) or a **plane wave front** (in three dimensions), just as we take the surface of the Earth to be locally flat (Figure 22.1b). A plane wave front can also be produced directly by a large, luminous flat surface. In a uniform medium, wave fronts propagate outward from the source at a speed characteristic of the medium. This was seen for sound waves in Section 14.2, and the same occurs for light, although at a much faster speed. The speed of light is greatest in a vacuum: $c = 3.00 \times 10^8$ m/s. The speed of light in air, for all practical purposes, is the same as that in vacuum.

The geometrical description of a wave in terms of wave fronts tends to neglect the fact that the wave is actually oscillating, like those studied in Chapter 13. This simplification is carried a step further with the concept of a ray. As illustrated in Figure 22.1, a line drawn perpendicular to a series of wave fronts and pointing in the direction of propagation is called a **ray**. Note that a ray points in the direction of the energy flow of a wave. A plane wave is assumed to travel in a straight line in a medium in the direction of its rays, perpendicular to its plane wave fronts. A beam of light can be represented by a group of rays or simply as a single ray (▼ **Figure 22.2**). The representation of light as rays is adequate and convenient for describing many optical phenomena.

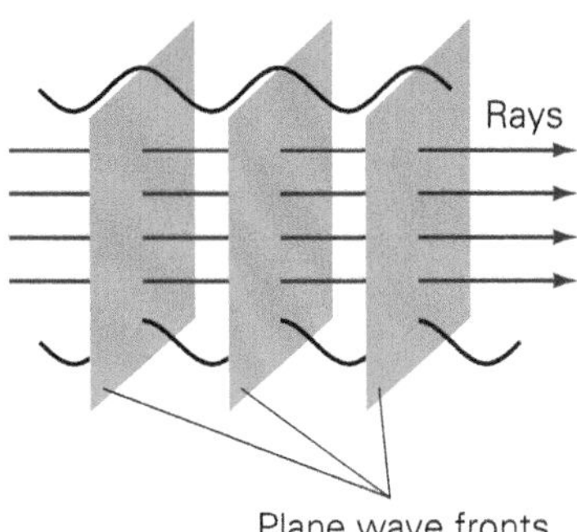


▲ FIGURE 22.2 **Light rays** A plane wave travels in a direction perpendicular to its wave fronts. A beam of light can be represented by a group of parallel rays (or by a single ray).

How do we see objects around us? They are seen because rays from the objects, or rays that appear to come from the objects, enter our eyes (▼ **Figure 22.3**) and form images of the objects on the retina (Chapter 23). The rays could be coming directly from the objects, as in the case of light sources, or could be reflected or refracted by the objects or other optical systems. Our eyes and brain working together, however, cannot tell whether the rays actually come from the objects or only *appear* to come from the objects. This is one way magicians can fool our eyes with seemingly impossible illusions.

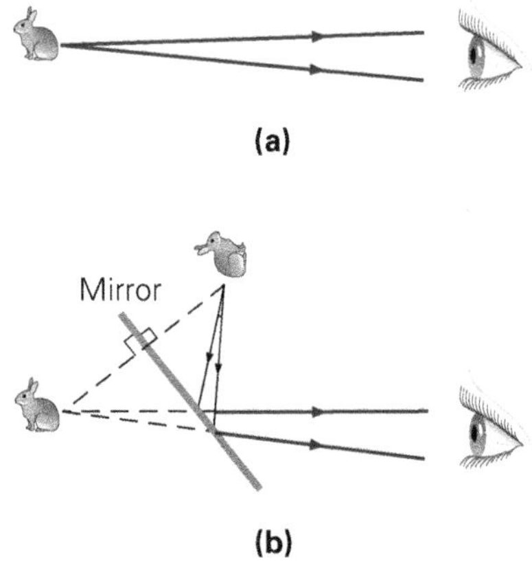


▲ FIGURE 22.3 **How things are seen** We see things because **(a)** rays from the objects or **(b)** rays appearing to come from the objects enter our eyes.

The use of the geometrical representations of wave fronts and rays to explain phenomena such as the reflection and refraction of light is called **geometrical optics**. However, certain other phenomena, such as the interference of light, cannot be treated in this manner and must be explained in terms of actual wave characteristics. These phenomena will be considered in Chapter 24.

## 22.2 Reflection

The reflection of light is an optical phenomenon of enormous importance: If light were not reflected by objects around us to our eyes, we wouldn't see the objects at all. **Reflection** involves the absorption and re-emission of light by means of complex electromagnetic oscillations in the atoms of the reflecting medium. However, the phenomenon is easily described by using rays.

A light ray incident on a surface is described by an **angle of incidence ($\theta_i$)**. This angle is measured relative to a *normal* – a line perpendicular to the reflecting surface (▼**Figure 22.4**). Similarly, the reflected ray is described by an **angle of reflection ($\theta_r$)**, also measured from the normal. The relationship between these angles is given by the **law of reflection**: *the angle of incidence is equal to the angle of reflection*, or

$$\theta_i = \theta_r \quad \text{(law of reflection)} \tag{22.1}$$

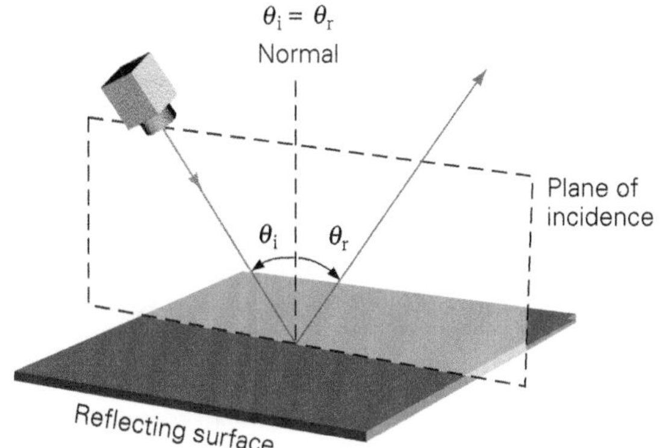


▲**FIGURE 22.4 The law of reflection** According to the law of reflection, the angle of incidence ($\theta_i$) is equal to the angle of reflection ($\theta_r$). Note that the angles are measured relative to a normal (a line perpendicular to the reflecting surface). The normal and the incident and reflected rays always lie in the same plane.

Two other attributes of reflection are that the incident ray, the reflected ray, and the normal all lie in the same plane, which is sometimes called the **plane of incidence**, and that the incident ray and the reflected ray are on opposite sides of the normal. When the reflecting surface is smooth and flat, the reflected rays from parallel incident rays are also parallel (▶**Figure 22.5a**). This type of reflection is called **specular**, or **regular, reflection**. The reflection from a smooth water surface is an example of specular (regular) reflection (Figure 22.5b). If the reflecting surface is rough, however, the reflected rays are not parallel, because of the irregular nature of the surface (▶**Figure 22.6**). This type of reflection is termed **diffuse**, or **irregular, reflection**. The reflection of light from this page you are reading is an example of diffuse reflection because the paper is microscopically rough. If the paper were perfectly smooth, there would have been so much glare (images of various things) that you would not have been able to read the words on the page.

Note in Figures 22.5a and 22.6 that the law of reflection still applies locally to both specular and diffuse reflection. However,

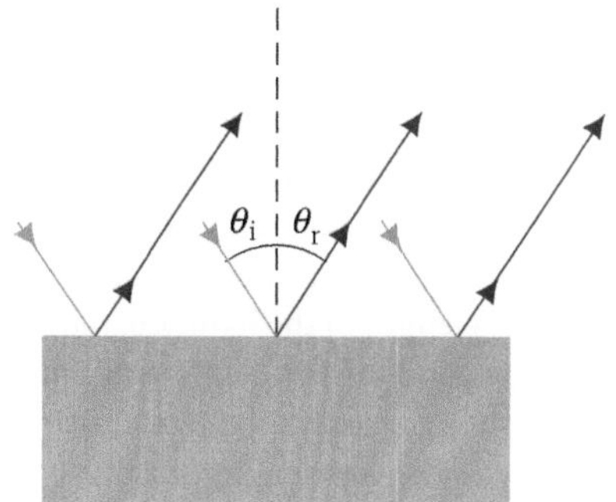


**(a) Specular (regular) reflection (diagram)**

**(b) Specular (regular) reflection (photo)**

▲**FIGURE 22.5 Specular (regular) reflection (a)** When a light beam is reflected from a smooth surface and the reflected rays are parallel (in the plane of incidence), the reflection is said to be specular or regular. **(b)** Specular (regular) reflection from a smooth water surface produces an almost perfect mirror image of a mountaineer looking to the Taranaki Volcano in the North Island of New Zealand, with the reflection of the mountain and the climber in a lake.

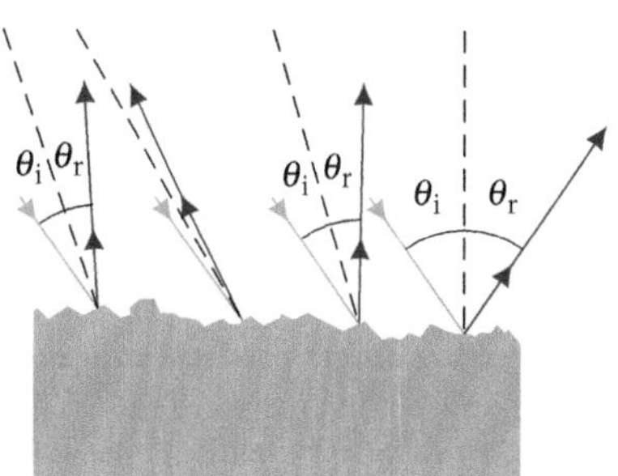


▲**FIGURE 22.6 Diffuse (irregular) reflection** Reflected rays from a relatively rough surface, such as this page, are not parallel; the reflection is said to be diffuse or irregular. (Note that the law of reflection still applies locally to each individual ray.)

the type of reflection involved determines whether we see images from a reflecting surface. In specular reflection, the reflected, parallel rays produce an image when they are viewed by an optical system such as an eye or a camera. Diffuse reflection does not produce an image, because the light is reflected in various directions.

In general, if the dimensions of the surface irregularities are greater than the wavelength of the light, the reflection is diffuse. Therefore, to make a good mirror, glass (with a metal coating) or metal must be polished at least until the surface irregularities are about the same size as the wavelength of light.

You probably have had the (bad) experience of driving on a dark and rainy night when you can hardly see the road ahead. Often nothing can be seen except the reflective glare of various things such as oncoming car headlights or building lights. What causes these conditions? When the road surface is dry, the reflection of light off the road is diffuse (irregular) because the surface is rough. However, when the road surface is wet, water fills the crevices, turning the road into a relatively smooth reflecting surface (▼**Figure 22.7a**) and the normally diffuse reflection is replaced by specular reflection. Reflected images of lighted buildings and road lights form, blurring the view of the surface, and the specular reflection of oncoming cars' headlights may make it difficult for you to see the road (Figure 22.7b).

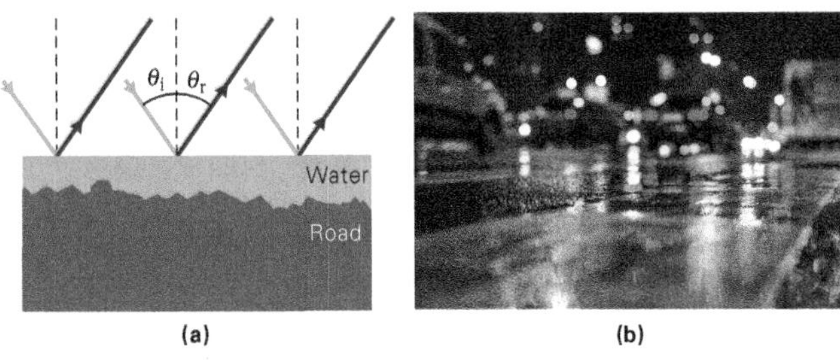


▲FIGURE 22.7 **Diffuse to specular (a)** The diffuse reflection from a dry road is turned into specular reflection by water on the road's surface. **(b)** Instead of seeing the road, a driver sees the reflected images of lights, buildings, and so on.

### EXAMPLE 22.1: TRACING THE REFLECTED RAY

Two mirrors, $M_1$ and $M_2$, are perpendicular to each other, with a light ray incident on one of the mirrors as shown in ▼**Figure 22.8**. (a) Sketch a diagram to trace the path of the reflected light ray. (b) Find the direction of the ray after it is reflected by $M_2$.

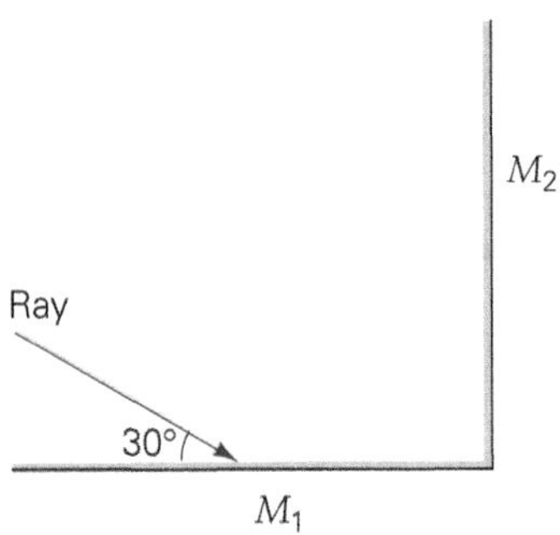


▲FIGURE 22.8 **Trace the ray** See Example 22.1.

**THINKING IT THROUGH.** The law of reflection can be used to determine the direction of the ray after it leaves the first and then the second mirror.

**SOLUTION**

***Given:***

$\theta = 30°$ (angle relative to $M_1$)

***Find:***

**(a)** Sketch a diagram tracing the light ray
**(b)** $\theta_{r_2}$ (angle of reflection from $M_2$)

Using the law of reflection to trace the ray for both mirrors in four steps (▼**Figure 22.9**):

**(a)** 1. Since the incident and reflected rays are measured from the normal (a line perpendicular to the reflecting surface), we draw the normal to mirror $M_1$ at the point where the incident ray hits $M_1$. From geometry, it can be seen that the angle of incidence on $M_1$ is $\theta_{i_1} = 60°$.

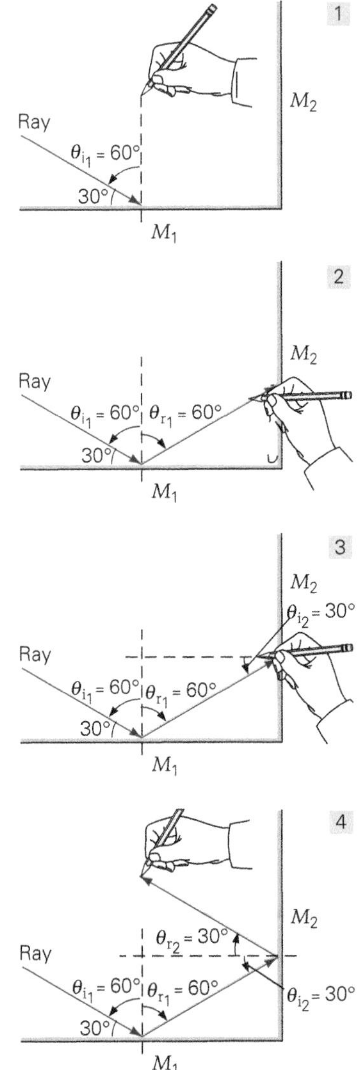


▲FIGURE 22.9 **Tracing the reflected rays** See Example 22.1.

2. According to the law of reflection, the angle of reflection from $M_1$ is also $\theta_{r_1} = 60°$. Next, draw this reflected ray with an angle of reflection of 60°, and extend it until it hits $M_2$.
3. Draw another normal to $M_2$ at the point where the ray hits $M_2$. Also from geometry (focus on the triangle in the diagram), the angle of incidence on $M_2$ is $\theta_{i_2} = 30°$. (Why?)

**(b)** The angle of reflection off $M_2$ is $\theta_{r_2} = \theta_{i_2} = 30°$ (step 4). This is the final direction of the ray reflected after both mirrors.

What if the directions of the rays are reversed? In other words, if a ray is first incident on $M_2$, in the direction opposite that of the one drawn in step 4, will all the rays reverse their directions? Draw another diagram to prove that this is indeed the case. Light rays are reversible.

**FOLLOW-UP EXERCISE.** When following an eighteen-wheel semitrailer, there may be a sign on the back stating, "If you can't see my mirror, I can't see you." What does this mean? (*Answers to all Follow-Up Exercises are given in Appendix V at the back of the book.*)

## 22.3 Refraction

**Refraction** refers to the change in direction of a wave at a boundary where the wave passes from one medium into another. In general, when a wave is incident on a boundary between media, some of the wave's energy is reflected and some is transmitted. For example, when light traveling in air is incident on a transparent material such as glass, it is partially reflected and partially transmitted (▼**Figure 22.10**). But the direction of the transmitted light is different from the direction of the incident beam, so the light is said to have been refracted; in other words, it has changed direction.

This change in direction is caused by the fact that light travels with different speeds in different media. Intuitively, you might expect the passage of light to take longer through a medium with more atoms per volume, and the speed of light is, in fact, generally less in denser media. For example, the speed of light in water is about 75% of that in air or a vacuum.

▲FIGURE 22.10 **Reflection and refraction** A beam of light is incident on a rectangular glass from the left. Part of the beam is reflected, and part is refracted at the air–glass surface. The refracted beam is again partially reflected and partially refracted at the bottom glass–air surface (right).

The change in the direction of wave propagation is described by the **angle of refraction**. In ▼**Figure 22.11**, refraction of light at a medium boundary, $\theta_1$ is the angle of incidence and $\theta_2$ is the angle of refraction. The symbols $\theta_1$ and $\theta_2$ are used for the angles of incidence and refraction so as to avoid confusion with $\theta_i$ and $\theta_r$, the angles of incidence and reflection. Willebrord Snell (1580–1626), a Dutch physicist, discovered a relationship between the angles ($\theta$) and the speeds ($v$) of light in two media:

$$\frac{\sin\theta_1}{\sin\theta_2} = \frac{v_1}{v_2} \quad \text{(Snell's law)} \tag{22.2}$$

This expression is known as **Snell's law** or the **law of refraction**. Note that $\theta_1$ and $\theta_2$ are always taken with respect to the normal.

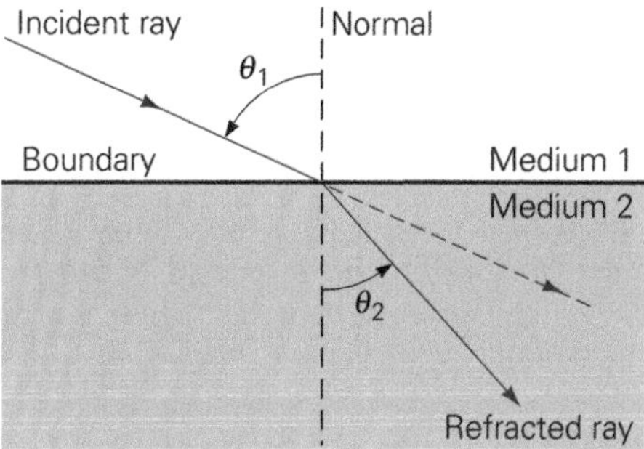


▲FIGURE 22.11 **Refraction** Light changes direction on entering a different medium. The angle of refraction $\theta_2$, describing the direction of the refracted ray, is different from the angle of incidence $\theta_1$. (Both angles are measured from the normal.)

Thus, light is refracted when passing from one medium into another because the speed of light is different in the two media. The speed of light is greatest in a vacuum, and it is therefore convenient to compare the speed of light in other media with this constant value ($c$). This is done by defining a ratio called the **index of refraction (*n*)**:

$$n = \frac{c}{v} \quad \left(\frac{\text{speed of light in a vacuum}}{\text{speed of light in a medium}}\right) \quad \text{(index of refraction)} \tag{22.3}$$

As a ratio of speeds, the index of refraction is a unitless quantity. The indices of refraction of several substances are given in ▶**Table 22.1**. Note that these values are for a specific wavelength of light. The wavelength is specified because $v$, and consequently $n$, are slightly different for different wavelengths within a particular medium. (This is the cause of dispersion, to be discussed in Section 22.5.) The values of $n$ given in the table will be used in Examples and Exercises in this chapter for all wavelengths of light in the visible region, unless otherwise noted. Observe that $n$

TABLE 22.1 Indices of Refraction (at $\lambda = 590$ nm)[a]

| Substance | $n$ |
|---|---|
| Air | 1.000 29 |
| Water | 1.33 |
| Ice | 1.31 |
| Ethyl alcohol | 1.36 |
| Fused quartz | 1.46 |
| Human eye | 1.336 – 1.406 |
| Polystyrene | 1.49 |
| Oil (typical value) | 1.50 |
| Glass (by type)[b] | 1.45 – 1.70 |
| Crown | 1.52 |
| Flint | 1.66 |
| Zircon | 1.92 |
| Diamond | 2.42 |

[a] One nanometer (nm) is $10^{-9}$ m.
[b] Crown glass is a soda-lime silicate glass; flint glass is a lead-alkali silicate glass. Flint glass is more dispersive than crown glass (Section 22.5).

is always greater than 1, because the speed of light in a vacuum is always greater than the speed of light in any material ($c > v$).

The frequency ($f$) of light does not change when the light enters another medium, but the wavelength of light in a material ($\lambda_m$) differs from the wavelength of that light in a vacuum ($\lambda$), as can be easily shown:

$$n = \frac{c}{v} = \frac{\lambda f}{\lambda_m f}$$

or

$$n = \frac{\lambda}{\lambda_m} \tag{22.4}$$

The wavelength of light in the medium is then $\lambda_m = \lambda/n$. Since $n > 1$, it follows that $\lambda_m < \lambda$. Thus, the wavelength is the longest in a vacuum.

### EXAMPLE 22.2: THE SPEED OF LIGHT IN WATER - INDEX OF REFRACTION

Light from a He–Ne laser with a wavelength of 632.8 nm travels from air into water. What are the speed and wavelength of the laser light in water?

**THINKING IT THROUGH.** If the index of refraction ($n$) of a medium is known, the speed and wavelength of light in the medium can be obtained from Equations 22.3 and 22.4.

**SOLUTION**

***Given:***

$n = 1.33$ (Table 22.1)
$\lambda = 632.8$ nm
$c = 3.00 \times 10^8$ m/s (speed of light in air)

***Find:*** $v$ and $\lambda_m$ (speed and wavelength in water)

Since $n = c/v$,

$$v = \frac{c}{n} = \frac{3.00 \times 10^8 \text{ m/s}}{1.33} = 2.26 \times 10^8 \text{ m/s}$$

Note that $1/n = v/c = 1/1.33 = 0.75$; therefore, $v$ is 75% of the speed of light in a vacuum. Also, $n = \lambda/\lambda_m$, so

$$\lambda_m = \frac{\lambda}{n} = \frac{362.8 \text{ nm}}{1.33} = 476 \text{ nm}$$

**FOLLOW-UP EXERCISE.** The speed of light of wavelength 500 nm (in air) in a particular liquid is $2.40 \times 10^8$ m/s. What are the index of refraction of the liquid and the wavelength of light in the liquid?

For practical purposes, the index of refraction is measured in air rather than in a vacuum, since the speed of light in air is very close to $c$, and

$$n_{air} = \frac{c}{v_{air}} \approx \frac{c}{c} = 1$$

(From Table 22.1, $n_{air} = 1.00$ 29, but for calculations $n_{air} = 1$ may be used.)

A more practical form of the law of refraction can be rewritten as

$$\frac{\sin\theta_1}{\sin\theta_2} = \frac{v_1}{v_2} = \frac{c/n_1}{c/n_2} = \frac{n_2}{n_1}$$

or

$$n_1 \sin\theta_1 = n_2 \sin\theta_2 \quad \text{(law of refraction)} \tag{22.5}$$

where $n_1$ and $n_2$ are the indices of refraction for the first and second media, respectively.

Note that Equation 22.5 can be used to measure the index of refraction. If the first medium is air, then $n_1 \approx 1$ and $n_2 \approx \sin\theta_1/\sin\theta_2$. Thus, only the angles of incidence and refraction need to be measured to determine the index of refraction of a material experimentally. On the other hand, if the index of refraction of a material is known, then the law of refraction can be used to find the angle of refraction for any angle of incidence.

Note also that the sine of the refraction angle is inversely proportional to the index of refraction: $\sin\theta_2 = n_1 \sin\theta_1/n_2$. Hence, for a given angle of incidence, the greater the index of refraction of the second medium, the smaller the $\sin\theta_2$ and the smaller the angle of refraction, $\theta_2$.

More generally, the following relationships hold:

- If the second medium has a higher index of refraction than the first medium ($n_2 > n_1$), the ray is refracted toward the normal ($\theta_2 > \theta_1$), as illustrated in ▶ **Figure 22.12a**.

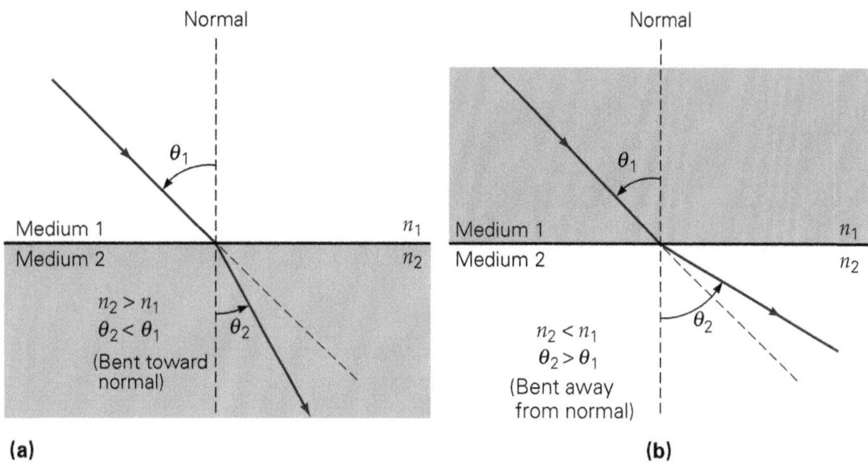


▲FIGURE 22.12 **Index of refraction and ray deviation (a)** When the second medium has a higher index of refraction than the first ($n_2 > n_1$), the ray is refracted toward the normal, as in the case of light entering water from air. **(b)** When the second medium has a lower index of refraction than the first ($n_2 < n_1$), the ray is refracted away from the normal. [This is the case if the ray in part (a) is traced in reverse, going from medium 2 to medium 1.]

- If the second medium has a lower index of refraction than the first medium ($n_2 < n_1$), the ray is refracted away from the normal ($\theta_2 < \theta_1$), as illustrated in Figure 22.12b.

---

## INTEGRATED EXAMPLE 22.3: ANGLE OF REFRACTION - THE LAW OF REFRACTION

Light in water is incident on a piece of crown glass at an angle of 37° (relative to the normal). (a) Will the transmitted ray be (1) bent toward the normal, (2) bent away from the normal, or (3) not bent at all? Use a diagram to illustrate. (b) What is the angle of refraction?

**(a) CONCEPTUAL REASONING.** The indices of refraction of water and crown glass can be found from Table 22.1. According to the law of refraction (Equation 22.5), $n_1 \sin\theta_1 = n_2 \sin\theta_2$, (1) is the correct answer. Since ($n_2 > n_1$), the angle of refraction must be smaller than the angle of incidence ($\theta_2 < \theta_1$).

Because both $\theta_1$ and $\theta_2$ are measured from the normal, the refracted ray will bend toward the normal. The ray diagram in this case is identical to Figure 22.12a.

**(b) QUANTITATIVE REASONING AND SOLUTION.** Again, the law of refraction (Equation 22.5) is most practical in this case. (Why?) Listing the given quantities,

***Given:***

$\theta_1 = 37°$
$n_1 = 1.33$ (water)
$n_2 = 1.52$ (crown glass*)

***Find:***

**(b)** $\theta_2$ (angle of refraction)

The angle of refraction is found by using Equation 22.5,

$$\sin\theta_2 = \frac{n_1 \sin\theta_1}{n_2} = \frac{(1.33)\sin 37°}{1.52} = 0.53$$

and

$$\theta_2 = \sin^{-1}(0.53) = 32°$$

**FOLLOW-UP EXERCISE.** It is found experimentally that a beam of light entering a liquid from air at an angle of incidence of 37° exhibits an angle of refraction of 29° in the liquid. What is the speed of light in the liquid?

---

## EXAMPLE 22.4: A GLASS TABLETOP - MORE ABOUT REFRACTION

A beam of light traveling in air strikes the glass top of a coffee table at an angle of incidence of 45° (▶**Figure 22.13**). The glass has an index of refraction of 1.5. (a) What is the angle of refraction for the light transmitted into the glass? (b) Prove that the beam emerging from the other side of the glass is parallel to the incident beam - that is, $\theta_4 = \theta_1$. (c) If the glass is 2.0 cm thick, what is the lateral displacement between the ray entering and the ray emerging from the glass (the perpendicular distance between the two rays - $d$ in the figure)?

**THINKING IT THROUGH.** Since two refractions are involved in this Example, the law of refraction is used in parts (a) and (b), and then some geometry and trigonometry in part (c).

**SOLUTION**

Listing the data:

***Given:***

$\theta_1 = 45°$
$n_1 = 1.0$ (air)
$n_2 = 1.5$
$y = 2.0$ cm

---

* Glass made without lead or iron. Formerly used in windows, it is now used as optical glass of low refractive index.

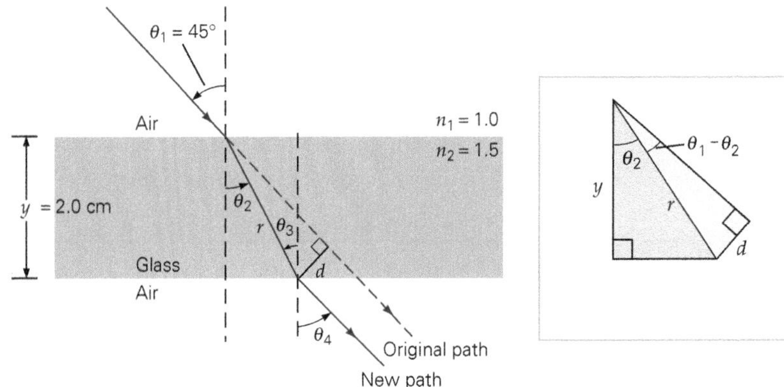


▲ **FIGURE 22.13** **Two refractions** In the glass, the refracted ray is displaced laterally (sideways) a distance $d$ from the incident ray, and the emergent ray is parallel to the original ray. See Example 22.4.

***Find:***

**(a)** $\theta_2$ (angle of refraction)
**(b)** Show that $\theta_4 = \theta_1$
**(c)** $d$ (lateral displacement)

**(a)** Using the law of refraction, Equation 22.5, with $n_1 = 1.0$ for air gives

$$\sin\theta_2 = \frac{n_1 \sin\theta_1}{n_2} = \frac{(1.0)\sin 45°}{1.5} = \frac{0.707}{1.5} = 0.47$$

Thus,

$$\theta_2 = \sin^{-1}(0.47) = 28°$$

Note that the beam is refracted toward the normal.

**(b)** If $\theta_1 = \theta_4$, then the emergent ray is parallel to the incident ray. Applying the law of refraction to the ray at both surfaces,

$$n_1 \sin\theta_1 = n_2 \sin\theta_2$$

and $(n_3 = n_2;\ n_4 = n_1)$

$$n_2 \sin\theta_3 = n_1 \sin\theta_4$$

From the figure, $\theta_2 = \theta_3$. Therefore,

$$n_1 \sin\theta_1 = n_1 \sin\theta_4$$

or

$$\theta_1 = \theta_4$$

Thus, the emergent beam is parallel to the incident beam but displaced laterally or perpendicularly to the incident direction at a distance $d$.

**(c)** It can be seen from the inset in Figure 22.13 that, to find $d$, we need to first find $r$ from the known information in the pink right triangle. Then,

$$\frac{y}{r} = \cos\theta_2 \quad \text{or} \quad r = \frac{y}{\cos\theta_2}$$

In the light-shade triangle, $d = r\sin(\theta_1 - \theta_2)$. Substituting $r$ from the previous step yields

$$d = \frac{y\sin(\theta_1 - \theta_2)}{\cos\theta_2} = \frac{(2.0\text{ cm})\sin(45° - 28°)}{\cos 28°} = 0.66\text{ cm}$$

**FOLLOW-UP EXERCISE.** If the glass in this Example had $n = 1.6$, would the lateral displacement be the same, larger, or smaller? Explain your answer conceptually, and then calculate the actual value to verify your reasoning.

---

### EXAMPLE 22.5: THE HUMAN EYE – REFRACTION AND WAVELENGTH

A simplified representation of the crystalline lens in a human eye shows it to have a cortex (an outer layer) of $n_{cortex} = 1.386$ and a nucleus (core) of $n_{nucleus} = 1.406$. (See Figure 25.1b.) If a beam of monochromatic (single-frequency or -wavelength) light of wavelength 590 nm is directed from air through the front of the eye and into the crystalline lens, qualitatively compare and list the frequency, speed, and wavelength of light in air, in the cortex, and in the nucleus. First do the comparison without numbers, and then calculate the actual values to verify your reasoning.

**REASONING AND ANSWER.** First, the relative magnitudes of the indices of refraction are needed, where $n_{air} < n_{cortex} < n_{nucleus}$. As learned earlier in this section, the frequency ($f$) of light is the same in all three media: air, the cortex, and the nucleus. Thus, the frequency can be calculated by using the speed and the wavelength of light in any of these materials, but it is easiest in air. (Why?) From the wave relationship (Equation 13.22),

$$f = f_{air} = f_{cortex} = f_{nucleus} = \frac{c}{\lambda} = \frac{3.00 \times 10^8\text{ m/s}}{590 \times 10^{-9}\text{ m}} = 5.08 \times 10^{14}\text{ Hz}$$

The speed of light in a medium depends on its index of refraction, since $v = c/n$. The smaller the index of refraction, the higher the speed. Therefore, the speed of light is the highest in air ($n = 1.00$) and lowest in the nucleus ($n = 1.406$).

The speed of light in the cortex is

$$v_{\text{cortex}} = \frac{c}{n_{\text{cortex}}} = \frac{3.00 \times 10^8 \text{ m/s}}{1.386} = 2.16 \times 10^8 \text{ m/s}$$

and the speed of light in the nucleus is

$$v_{\text{nucleus}} = \frac{3.00 \times 10^8 \text{ m/s}}{1.406} = 2.13 \times 10^8 \text{ m/s}$$

We also know that the wavelength of light in a medium depends on the index of refraction of the medium ($\lambda_m = \lambda/n$). The smaller the index of refraction, the longer the wavelength. Therefore, the wavelength of light is the longest in air ($n = 1.00$ and $\lambda = 590$ nm) and the shortest in the nucleus ($n = 1.406$).

The wavelength in the cortex can be calculated from Equation 22.4:

$$\lambda_{\text{cortex}} = \frac{\lambda}{n_{\text{cortex}}} = \frac{590 \text{ nm}}{1.386} = 426 \text{ nm}$$

and the wavelength in the nucleus is

$$\lambda_{\text{nucleus}} = \frac{590 \text{ nm}}{1.406} = 420 \text{ nm}$$

Finally, a table is constructed to compare the index of refraction, frequency, speed, and wavelength of light in the three media:

| | $n$ | $f$ (Hz) | $v$ (m/s) | $\lambda$ (nm) |
|---|---|---|---|---|
| Air | 1.00 | $5.08 \times 10^{14}$ | $3.00 \times 10^8$ | 590 |
| Cortex | 1.386 | $5.08 \times 10^{14}$ | $2.16 \times 10^8$ | 426 |
| Nucleus | 1.406 | $5.08 \times 10^{14}$ | $2.13 \times 10^8$ | 420 |

**FOLLOW-UP EXERCISE.** A light source of a single frequency is submerged in water in a fish tank. The beam travels in the water, through the glass pane at the side of the tank, and into the air. In general, what happens to (a) the frequency and (b) the wavelength of the light when it emerges into the outside air?

Refraction is common in everyday life and can be used to explain many phenomena.

**Mirage:** A common example of this phenomenon sometimes occurs on a highway on a hot summer day. The refraction of light is caused by layers of air that have different temperatures (the layer closer to the road has a higher temperature, lower density, and therefore lower index of refraction). This variation in indices of refraction gives rise to the observed "wet" spot and an inverted image of an object such as a car (▼ **Figure 22.14a**). The term *mirage* generally brings to mind a thirsty person in the desert "seeing" a pool of water that really isn't there. This optical illusion plays tricks on the mind, with the image usually seen as a pool of water and our eye's past experience unconsciously leading us to conclude that there is water on the road.

As shown in Figure 22.14b, there are two ways for light to get to our eyes from the car. First, the horizontal rays come directly from the car to our eyes, so we see the car above the ground. Also, the rays from the car that travel toward the road surface will be gradually refracted by the layered air. (See the inset in the figure.) After hitting the road, these rays will be refracted again and travel toward our eyes.

Cooler air has a higher density and thus a higher index of refraction. A ray traveling toward the road surface will be gradually refracted with an increasing angle of refraction until it hits the surface. It will then be refracted with a decreasing angle of refraction, going toward our eyes. As a consequence, we also see an inverted image of the car appearing below the road surface. In other words, the surface of the road acts almost as a mirror. The "pool of water" is actually skylight being refracted – an image of the sky. This layering of air of different temperatures, creating different indices of refraction, causes us to "see" the rising hot air as a result of continually changing refraction.

The opposite of this is the mirage at sea (the looming effect). At sea, the air above is warmer than that below. This causes the light to be refracted opposite to what is in Figure 22.14b, causing objects to be seen in the air above the water.

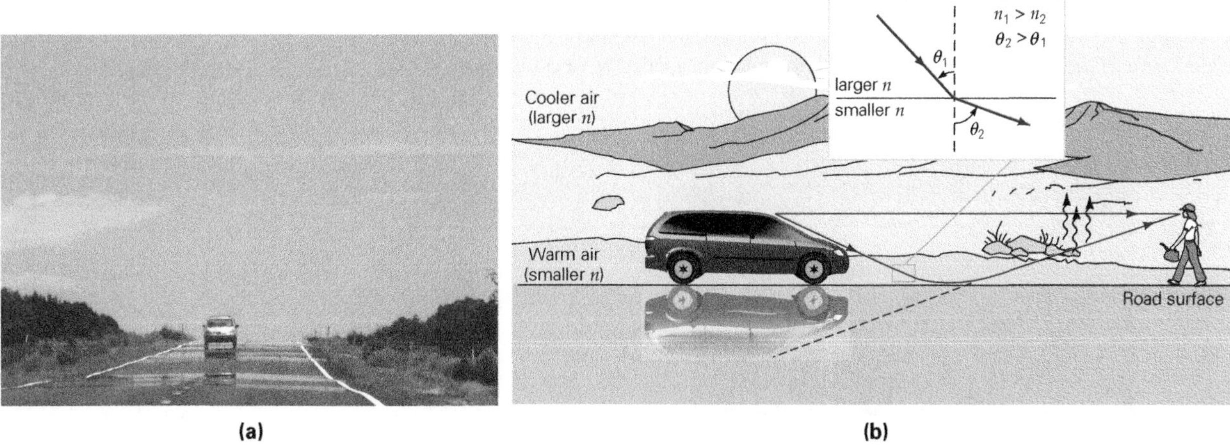


▲ FIGURE 22.14 **Refraction in action (a)** An inverted car on a "wet" road – a mirage. **(b)** The mirage is formed when light from the object is refracted by layers of air at different temperatures near the surface of the road.

**Not where it should be:** You may have experienced a refractive effect while trying to reach for something underwater, such as a fish (▼**Figure 22.15a**). We are used to light traveling in straight lines from objects to our eyes, but the light reaching our eyes from a submerged object has had a directional change at the water–air interface. (Note in the figure that the ray is refracted away from the normal.) As a result, the object appears closer to the surface than it actually is, therefore we tend to miss the object when reaching for it. For the same reason, a chopstick in a cup appears bent (Figure 22.15b), a coin in a glass of water appears closer than it really is (Figure 22.15c), and the legs of a person standing in water seem shorter than their actual length. The relationship between the true depth and the apparent depth can be calculated. (See Exercise 25.)

**Atmospheric effects:** The Sun on the horizon sometimes appears flattened, with its horizontal dimension greater than its vertical dimension (▼**Figure 22.16a**). This effect is the result of temperature and density variations in the denser air along the horizon. These variations occur predominantly vertically, so light from the top and bottom portions of the Sun are refracted differently as the two sets of beams pass through different atmospheric densities with different indices of refraction.

Atmospheric refraction lengthens the day, so to speak, by allowing us to see the Sun (or the Moon, for that matter) just before it actually rises above the horizon and just after it actually sets below the horizon (lengthening the day by as much as 20 min on both ends). The denser air near the Earth refracts the light over the horizon toward us (Figure 22.16b).

The twinkling of stars is due to atmospheric turbulence, which distorts the light from the stars. The turbulence refracts light in random directions and causes the stars to appear to "twinkle." Stars on the horizon appear to twinkle more than stars directly overhead because the light from those stars has to pass through more of the Earth's atmosphere.

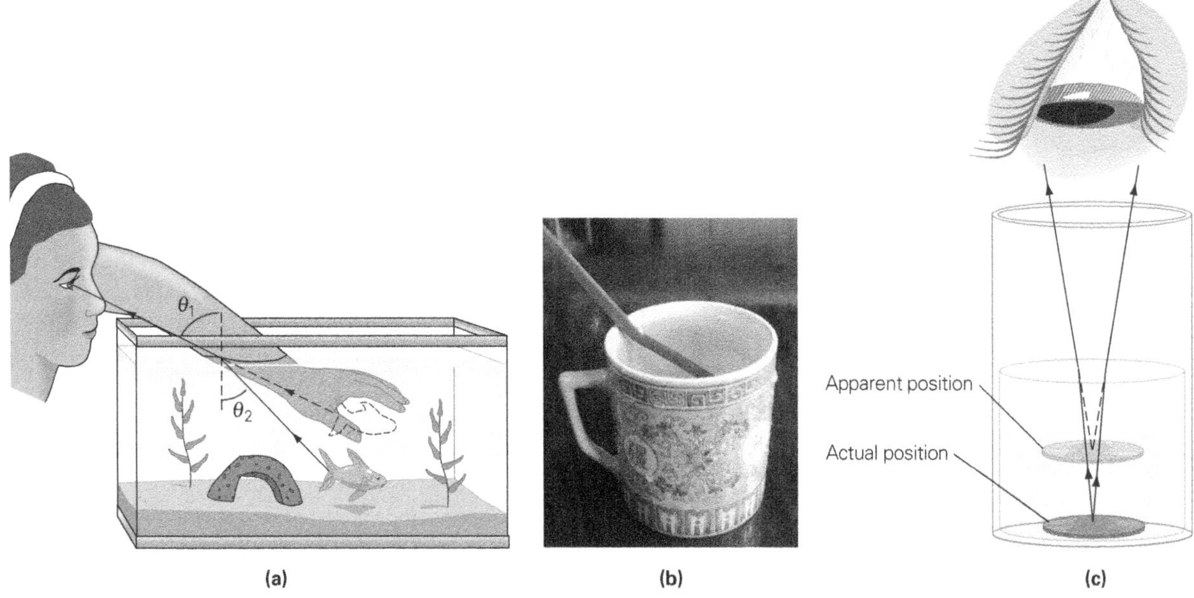


▲ FIGURE 22.15 **Refractive effects (a)** The light is refracted, and because we tend to think of light as traveling in straight lines, the fish is below where we think it is. **(b)** The chopstick appears bent at the air–water boundary. If the cup is transparent, we see a different refraction. (See Figure 22.25, Conceptual Question 7.) **(c)** Because of refraction, the coin appears to be closer than it actually is.

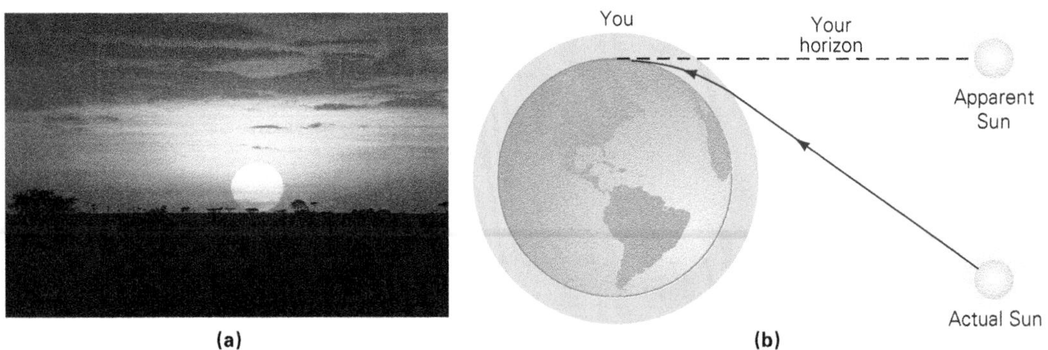


▲ FIGURE 22.16 **Atmospheric effects (a)** The Sun on the horizon commonly appears flattened as a result of atmospheric refraction. **(b)** Before rising and after setting, the Sun can be seen briefly also because of atmospheric refraction. (Exaggerated for illustration.)

## 22.4 Total Internal Reflection and Fiber Optics

An interesting phenomenon occurs when light travels from a medium with a higher index of refraction into a medium with a lower one, such as when light goes *from* water *into* air. As you know, in such a case a ray will be refracted away from the normal. (The angle of refraction is larger than the angle of incidence.) Furthermore, the law of refraction states that the greater the angle of incidence, the greater the angle of refraction. That is, as the angle of incidence increases, the refracted ray diverges farther from the normal.

However, there is a limit. For a certain angle of incidence called the **critical angle ($\theta_c$)**, the angle of refraction is 90° and the refracted ray is directed along the boundary between the media. But what happens if the angle of incidence is even larger? If the angle of incidence is greater than the critical angle ($\theta_1 > \theta_c$), the light isn't refracted at all, but is internally reflected (▼**Figure 22.17**). This condition is called **total internal reflection.** The reflection process is about 100% efficient. Because of total internal reflection, glass prisms can be used as mirrors (▼**Figure 22.18**). In summary, where $n_1 > n_2$, reflection and refraction occur at all angles for $\theta_1 \leq \theta_c$, but the refracted or transmitted ray disappears for $\theta_1 > \theta_c$.

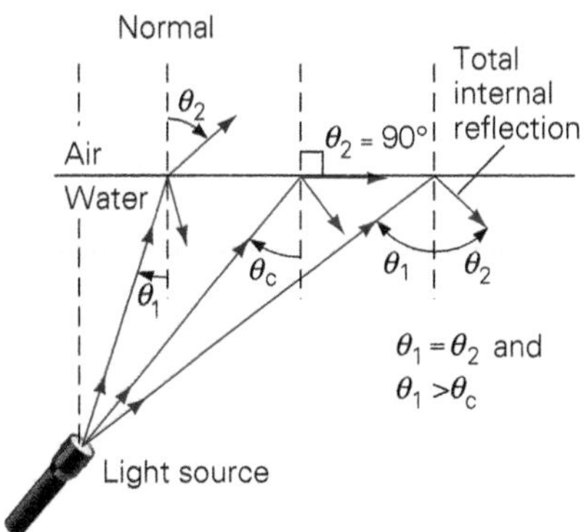


▲**FIGURE 22.17 Internal reflection** When light enters a medium with a lower index of refraction, it is refracted away from the normal. At a critical angle ($\theta_c$), the light is refracted along the interface (common boundary) of the media. At an angle greater than the critical angle ($\theta_1 > \theta_c$), there is total internal reflection.

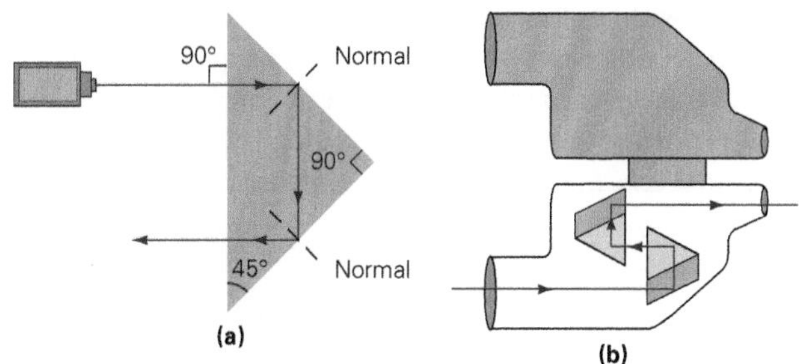


▲**FIGURE 22.18 Internal reflection in a prism (a)** Because the critical angle of glass is less than 45°, prisms with 45° and 90° angles can be used to reflect light through. **(b)** Internal reflection of light by prisms in binoculars makes this instrument much shorter than a telescope because the rays are "folded" by the prisms.

An expression for the critical angle can be obtained from the law of refraction. If $\theta_1 = \theta_c$ in the medium with a higher index of refraction, $\theta_2 = 90°$ and it follows that

$$n_1 \sin\theta_1 = n_2 \sin\theta_2 \quad \text{or} \quad n_1 \sin\theta_c = n_2 \sin 90°$$

Since $\sin 90° = 1$,

$$\sin\theta_c = \frac{n_2}{n_1} \quad \text{(critical angle, where } n_1 > n_2\text{)} \tag{22.6}$$

If the second medium is air, $n_2 \approx 1$, then the critical angle at the boundary from a medium into air is given by $\sin\theta_c = 1/n$, where $n$ is the index of refraction of the first medium.

---

**EXAMPLE 22.6: A VIEW FROM THE POOL – CRITICAL ANGLE**

(a) What is the critical angle for light traveling in water and incident on a water–air boundary? (b) If a diver submerged in a pool looked up at the surface of the water at an angle of $\theta < \theta_c$, what would she see?

**THINKING IT THROUGH.** (a) The critical angle is given by Equation 22.6. (b) As shown in Figure 22.17, $\theta_c$ forms a cone of vision for viewing from below the water.

**SOLUTION**

***Given:***

$n_1 = 1.33$ (water, from Table 22.1)
$n_2 \approx 1$ (air)

***Find:***

**(a)** $\theta_c$ (critical angle)
**(b)** View for $\theta_1 < \theta_c$

**(a)** The critical angle is

$$\theta_c = \sin^{-1}\left(\frac{n_2}{n_1}\right) = \sin^{-1}\left(\frac{1}{1.33}\right) = 48.8°$$

**(b)** Using Figure 22.17, trace the rays in reverse for light coming from all angles outside the pool. Light coming from the above-water 180° panorama could be viewed only in a cone with a half-angle of 48.8°. As a result, objects above the surface would also appear distorted. An underwater panoramic view is seen in ▶**Figure 22.19**. Now can you explain why wading birds like herons usually keep their bodies low while trying to catch a fish?

**FOLLOW-UP EXERCISE.** What would the diver see when looking up at the water surface at an angle of $\theta_1 > \theta_c$?

---

Internal reflections enhance the brilliance of cut diamonds. (Brilliance is a measure of the amount of light returning back to the viewer. Brilliance is reduced if light leaks out the back of a diamond – that is, if the reflection is not total.) The critical angle for a diamond–air surface is

▲ FIGURE 22.19 **Panoramic and distorted** An underwater view of the surface of a swimming pool in Hawaii.

$$\theta_c = \sin^{-1}\left(\frac{1}{n}\right) = \sin^{-1}\left(\frac{1}{2.42}\right) = 24.4°$$

A so-called brilliant-cut diamond has many facets, or faces (fifty-eight in all – thirty-three on the upper face and twenty-five on the lower). Light from above hitting the lower facets at angles greater than the critical angle is internally reflected in the diamond. The light then emerges from the upper facets, giving rise to the diamond's brilliance (▼ **Figure 22.20**).

(a)

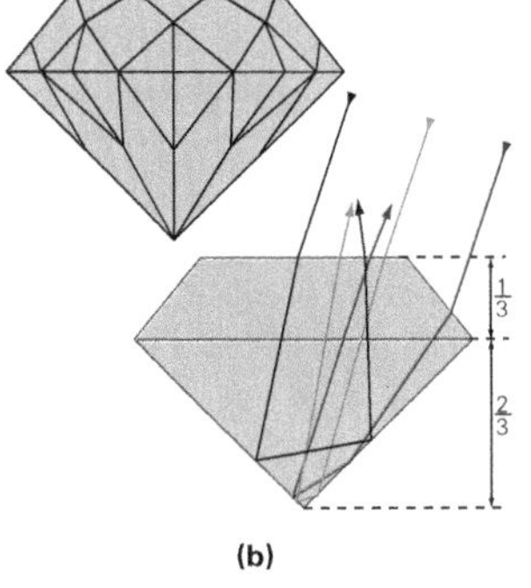


(b)

▲ FIGURE 22.20 **Diamond brilliance (a)** Internal reflection gives rise to a diamond's brilliance. **(b)** The "cut," or the depth proportions, of the facets is critical. If a stone is too shallow or too deep, light will be lost (refracted out) through the lower facets.

## 22.4.1 Fiber Optics

When a fountain is illuminated from below, the light is transmitted along the curved streams of water. This phenomenon was first demonstrated in 1870 by the Irish scientist John Tyndall (1820–1893), who showed that light was "conducted" along the curved path of a stream of water flowing from a hole in the side of a container. The phenomenon is observed because light undergoes total internal reflection along the stream.

Total internal reflection forms the basis of **fiber optics**, a fascinating technology centered on the use of transparent fibers to transmit light. Multiple total internal reflections make it possible to "pipe" light along a transparent rod (as in streams of water), even if the rod is curved (▶ **Figure 22.21**). Note from the figure that the smaller the diameter of the light pipe, the more total internal reflections it has. A small fiber can produce as many as several hundred total internal reflections per centimeter.

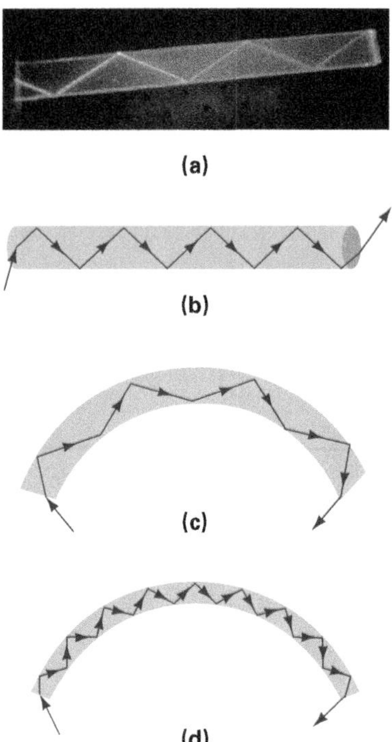


▲ FIGURE 22.21 **Light pipes (a)** Total internal reflection in an optical fiber. **(b)** When light is incident on the end of a cylindrical form of transparent material such that the internal angle of incidence is greater than the critical angle of the material, the light undergoes total internal reflection down the length of the light pipe. **(c)** Light is also transmitted along curved light pipes by total internal reflection. **(d)** As the diameter of the rod or fiber becomes smaller, the number of reflections per unit length increases.

Total internal reflection is an exceptionally efficient process. Compared with copper electrical wires, optical fibers can be used to transmit light over very long distances with much less signal loss. These losses are due primarily to impurities in the fiber, which scatter the light. Transparent materials have different degrees of transmission. Fibers are made of special plastics and glasses for maximum transmission efficiency. The greatest efficiency is achieved with infrared radiation, because there is less scattering, as will be learned in Section 24.5.

The greater efficiency of multiple total internal reflections compared with multiple mirror reflections can be illustrated by a good reflecting plane mirror, which has a typical reflectivity of about 95%. After each reflection, the beam intensity is 95% of that of the incident beam from the preceding reflection ($I_1 = 0.95I_o$, $I_2 = 0.95I_1 = 0.95^2 I_o, \ldots$). Therefore, the intensity $I$ of the reflected

$$I = 0.95^n I_o$$

where $I_o$ is the initial intensity of the beam before the first reflection. Thus, after fourteen reflections,

$$I = 0.95^{14} I_o = 0.49 I_o$$

In other words, after fourteen reflections, the intensity is reduced to less than half (49%). For 100 reflections, $I = 0.006I_o$ and the intensity is only 0.6% of the initial intensity! When you compare

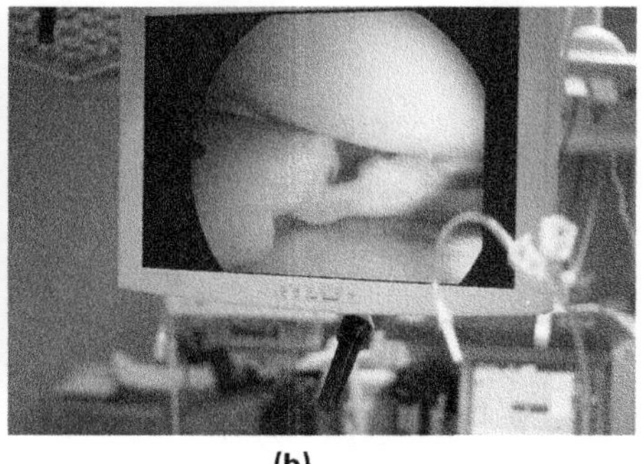

(a) (b)

▲ FIGURE 22.22 **Fiber-optic bundle (a)** Hundreds or even thousands of extremely thin fibers are grouped together to make an optical fiber. **(b)** A fiber-optic arthroscope used to perform keyhole microscopic surgery. The screen shows an arthroscopic view of an internal body part.

this to an intensity of about 75% of the initial intensity in optical fibers over a kilometer in length with *thousands* of reflections, you can see the advantage of total internal reflection.

Fibers whose diameters are about 10 $\mu$m ($10^{-5}$ m) are grouped together in flexible bundles that are 4–10 mm in diameter, depending on the application (▲ **Figure 22.22a**). A fiber bundle with a cross-sectional area of 1 $cm^2$ can contain as many as 50 000 individual fibers. (A coating on each fiber is needed to keep the fibers from touching each other.)

There are many important and interesting applications of fiber optics, including communications, computer networking, and medical instruments (Figure 22.22b). Light signals, converted from electrical signals, are transmitted through optical telephone lines and computer networks. At the other end, they are converted back to electrical signals. Optical fibers have lower energy losses than electric current–carrying wires, particularly at higher frequencies, and can carry far more data. Also, optical fibers are lighter than metal wires, have greater flexibility, and are not affected by electromagnetic disturbances (electric and magnetic fields), because they are made of materials that are electrical insulators.

## 22.5 Dispersion

Light of a single frequency, and consequently a single wavelength, is called monochromatic light (from the Greek *mono*, meaning "one," and *chroma*, meaning "color"). Visible light that contains all the component frequencies, or colors, at about the same intensities (such as sunlight) is called white light. When a beam of white light passes through a glass prism, as shown in ▼ **Figure 22.23a**, it is spread out, or dispersed, into a spectrum of colors. This phenomenon led Newton to believe that sunlight is a mixture of colors. When the beam enters the prism, the component colors, corresponding to different wavelengths of light, are refracted at slightly different angles, so they spread out into a spectrum.

The emergence of a spectrum indicates that the index of refraction of glass is slightly different for different wavelengths, which is true for many transparent media (Figure 22.23b). The reason has to do with the fact that in a dispersive medium the speed of light is slightly different for different wavelengths. Since the index of refraction of a medium is a function of the speed of light in that medium, the index of refraction is

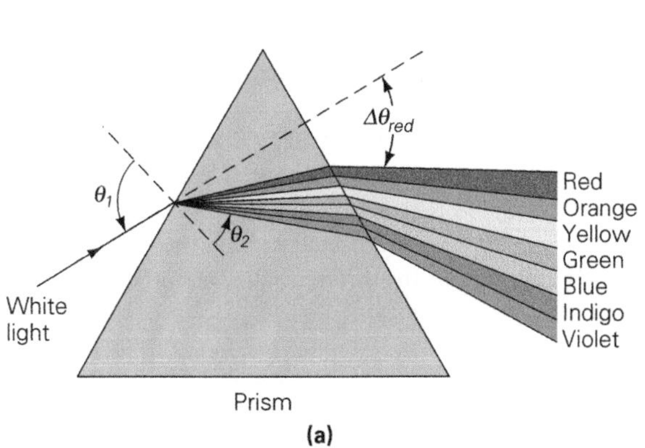


(a)

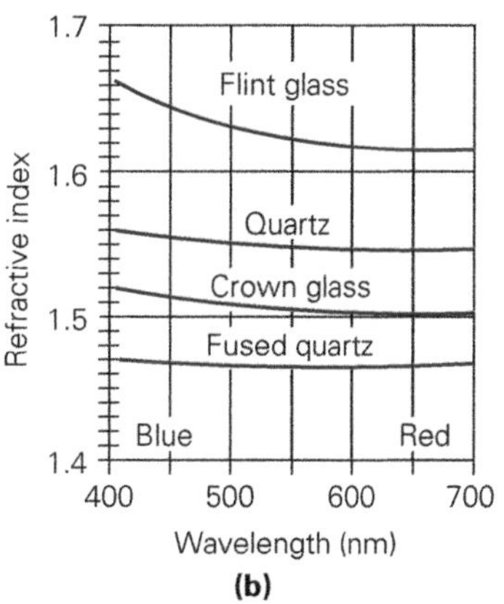


(b)

▲ FIGURE 22.23 **Dispersion (a)** White light is dispersed into a spectrum of colors by glass prisms. In a dispersive medium, the index of refraction varies slightly with wavelength. Red light, longest in wavelength, has the smallest index of refraction and is refracted least. The angle between the incident beam and an emergent ray is the angle of deviation for that ray. (The angles are exaggerated for clarity.) **(b)** Variation in the index of refraction with wavelength for some common transparent media.

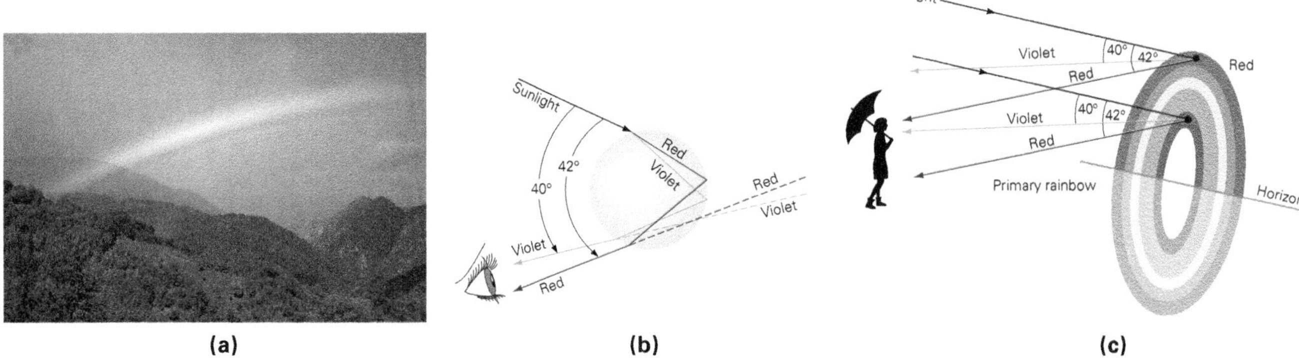


▲ **FIGURE 22.24** **Rainbow (a)** Rainbows are created by the refractions, dispersion, and internal reflection of sunlight within water droplets. The colors of the primary rainbow run vertically from red (top) to violet (bottom). **(b)** Light of different colors emerges from the droplet in different directions. **(c)** An observer sees red light at the top of the rainbow and violet at the bottom.

different for different wavelengths. It follows from the law of refraction that light of different wavelengths will be refracted at different angles of refraction even if the angle of incidence is the same.

The preceding discussion can be summarized by saying that in a transparent material with different indices of refraction for different wavelengths of light, refraction causes a separation of light according to wavelength, and the material is said to be dispersive and exhibit **dispersion**. Dispersion varies with different media (Figure 22.23b). Also, because the differences in the indices of refraction for different wavelengths are small, a representative value at some specified wavelength can be used for general purposes. (See Table 22.1.)

A good example of a dispersive material is diamond, which is about five times as dispersive as glass. In addition to revealing the brilliance resulting from internal reflections off many facets, a cut diamond shows a display of colors, or "fire," resulting from the dispersion of the refracted light.

Dispersion is a cause of chromatic aberration in lenses, which is described more fully in Section 23.5.2. Optical systems in cameras often consist of several lenses to minimize this problem.

Another dramatic example of dispersion is the production of the beautiful array of colors in a rainbow (▲ **Figure 22.24a**). A rainbow is formed by (two) refractions, dispersion, and reflection of light within water droplets. The light that forms the rainbow is first refracted and dispersed in each water droplet, then reflected once at the back surface of each droplet. Finally, it is refracted and dispersed again upon exiting each droplet, resulting in the light being spread out in different directions into a spectrum of colors (Figure 22.24b). However, because of the conditions for refraction and reflection in water, the angles between incoming and outgoing rays for violet to red light lie within a narrow range of 40°–42°.

Red appears on the top of the rainbow because light of shorter wavelengths from those water droplets will pass over our eyes (Figure 23.34c). Similarly, violet is at the bottom of the rainbow because light of longer wavelengths passes under our eyes.

The main, or primary, rainbow is sometimes accompanied by a fainter and higher secondary rainbow. The secondary rainbow is caused by two reflections within the water.

### INTEGRATED EXAMPLE 22.7: REFRACTION AND DISPERSION

The index of refraction of a particular transparent material is 1.4503 for the red end of the ($\lambda_r = 640$ nm) visible spectrum and 1.4698 for the blue end ($\lambda_b = 434$ nm) White light is incident on a prism of this material, as in Figure 22.23a, at an angle of incidence $\theta_1$ of 45.00°. (a) Inside the prism, the angle of refraction of the red light is (1) larger than, (2) smaller than, or (3) the same as the angle of refraction of the blue light? Explain. (b) What is the angular separation of the visible spectrum inside the prism?

**(a)** **CONCEPTUAL REASONING.** This example illustrates the essence of the cause of dispersion, that is, different colors have different indices of refraction in a material. Since red light has a smaller index of refraction than blue light, the angle of refraction of red light is larger than that of blue light for the same angle of incidence. Sometimes it is also said that red light is "refracted less" than blue light because the larger angle of refraction of red light means that it is closer to the direction of the original incident ray. So the answer is (1).

**(b)** **QUANTITATIVE REASONING AND SOLUTION.** The law of refraction is used to compute the angle of refraction for the red and blue ends of the visible spectrum. The angular separation of the visible spectrum inside the prism is the difference between the angles of refraction of the two colors.

***Given:***

(red) $n_r = 1.4503$ for $\lambda_r = 640$ nm
(blue) $n_b = 1.4698$ for $\lambda_b = 434$ nm
$\theta_1 = 45.00°$

***Find:*** $\Delta\theta_2$ (angular separation)

Using Equation 22.5 with $n_1 = 1.00$ (air),

$$\sin\theta_{2_r} = \frac{\sin\theta_1}{n_{2_r}} = \frac{\sin 45.00°}{1.4503} = 0.48756 \quad \text{and} \quad \theta_{2_r} = 29.180°$$

Similarly,

$$\sin\theta_{2_b} = \frac{\sin\theta_1}{n_{2_b}} = \frac{\sin 45.00}{1.4698} = 0.48109 \quad \text{and} \quad \theta_{2_b} = 28.757°$$

The angle of refraction of the red light is indeed larger than that of the blue light, as discussed in (a). The angular separation is

$$\Delta\theta_2 = \theta_{2_r} - \theta_{2_b} = 29.180° - 28.757° = 0.423°$$

This is not much of a deviation, but as the light travels to the other side of the prism, it is refracted and dispersed again by the second boundary. Thus the colors spread out even farther. When the light emerges from the prism, the dispersion becomes evident (Figure 22.23a).

**FOLLOW-UP EXERCISE.** In the prism in this Example, if the green light exhibits an angular separation of 0.156° from the red light, what is the index of refraction for green light in the material? Will the green light refract more or less than the red light? Explain.

## Chapter 22 Review

- **Law of reflection**: The angle of incidence equals the angle of reflection (as measured from the normal to the reflecting surface):

$$\theta_i = \theta_r \tag{22.1}$$

- The **index of refraction (*n*)** of any medium is the ratio of the speed of light in a vacuum to its speed in that medium:

$$n = \frac{c}{v} = \frac{\lambda}{\lambda_m} \tag{22.3, 22.4}$$

- The refraction of light as it enters a medium from another is given by **Snell's law** or **the law of refraction**:

$$\frac{\sin\theta_1}{\sin\theta_2} = \frac{v_1}{v_2} \tag{22.2}$$

$$n_1 \sin\theta_1 = n_2 \sin\theta_2 \tag{22.5}$$

If the second medium has a higher index of refraction ($n_2 > n_1$), the ray is refracted toward the normal; if the second medium has a lower index of refraction ($n_2 < n_1$), the ray is refracted away from the normal.

- **Total internal reflection** occurs if the second medium has a lower index of refraction than the first and the angle of incidence exceeds the critical angle, which is given by

$$\sin\theta_c = \frac{n_2}{n_1} \quad (n_1 > n_2) \tag{22.6}$$

- **Dispersion** of light occurs in a medium because different wavelengths have slightly different speeds and hence different indices of refraction in the medium. This results in slightly different refraction angles for different wavelengths.

## End of Chapter Questions and Exercises

### Multiple Choice Questions

**22.1 Wave Fronts and Rays**
**and**
**22.2 Reflection**

1. A ray is (a) perpendicular to the direction of energy flow, (b) parallel to the direction of energy flow, (c) always parallel to other rays, (d) parallel to a series of wave fronts.
2. The angle of reflection is the angle between (a) the reflected ray and the reflecting surface, (b) the incident ray and the normal to the surface, (c) the reflected ray and the incident ray, (d) the reflected ray and the normal to the reflecting surface.
3. For both specular (regular) and diffuse (irregular) reflections, (a) the angle of incidence equals the angle of reflection, (b) the incident and reflected rays are on opposite sides of the normal, (c) the incident ray, the reflected ray, and the local normal lie in the same plane, (d) all of the preceding.

**22.3 Refraction**
**and**
**22.4 Total Internal Reflection and Fiber Optics**

4. Refraction is caused by the fact that (a) different media have different speeds of light, (b) a given medium has different speeds of light for different wavelengths, (c) the angle of incidence is greater than the critical angle, (d) a given medium has different indices of refraction for different angles of incidence.
5. Light refracted at the boundary of two different media (a) is bent toward the normal when $n_1 > n_2$, (b) is bent away from the normal when $n_1 > n_2$, (c) is bent away from the normal when $n_1 < n_2$, (d) has the same angle of refraction as the angle of incidence.
6. The index of refraction (a) is always greater than or equal to 1, (b) is inversely proportional to the speed of light in a medium, (c) is inversely proportional to the wavelength of light in the medium, (d) all of the preceding.

7. Which of the following must be satisfied for total internal reflection to occur: (a) $n_1 > n_2$, (b) $n_2 > n_1$, (c) $\theta_1 > \theta_c$, or (d) $\theta_c < \theta_1$?

**22.5 Dispersion**

8. Dispersion can occur only if the light is (a) monochromatic, (b) polychromatic, (c) white light, (d) both b and c.
9. Dispersion can occur only during (a) specular reflection, (b) diffuse reflection, (c) refraction, (d) total internal reflection.
10. Dispersion is caused by (a) the difference in the speed of light in different media, (b) the difference in the speed of light for different wavelengths of light in a given medium, (c) the difference in the angle of incidence for different wavelengths of light in a given medium, (d) the difference in the indices of refraction of light in different media.

## Conceptual Questions

**22.1 Wave Fronts and Rays and 22.2 Reflection**

1. Under what circumstances is the angle of reflection not equal to the angle of incidence?
2. The book you are reading is not itself a light source, so it must be reflecting light from other sources. What type of reflection is this?
3. After rain, what kind of reflection can take place off of the road surface?
4. It is sometimes difficult to read magazines using high-gloss paper due to glare. Why?

**22.3 Refraction and 22.4 Total Internal Reflection and Fiber Optics**

5. Two hunters, one with a bow and arrow and the other with a laser gun, see a fish under water. They both aim directly for the location where they see the fish. Does the arrow or the laser beam have a better chance of hitting the fish? Explain.
6. As light travels from one medium to another, does its wavelength change? Its frequency? Its speed?
7. Explain why the pencil in ▶**Figure 22.25** appears almost severed. Also, compare this figure with Figure 22.15b and explain the difference.
8. If total internal reflection is taking place between two transparent materials, what can you conclude about the indices of refraction of the two materials? How about the angle of incidence?
9. When light enters medium 2 from medium 1, the angle of incidence is always greater than the angle of refraction. Could total internal reflection take place from medium 1 to medium 2? Explain.

▲FIGURE 22.25 **Refraction effect** See Conceptual Question 7.

**22.5 Dispersion**

10. Both refraction and dispersion are caused by the difference in the speed of light. What is the difference in the physical cause (reason)?
11. Why is dispersion more prominent when using a triangular prism rather than a square block?
12. If white light is incident on a square block of dispersive material, will there be a spectrum? How about the angles of the colors when they exit the block?
13. A glass prism disperses white light into a spectrum. Can a second glass prism be used to recombine the spectral components? Explain.
14. A light beam consisting of two colors, A and B, is sent through a prism. Color A is refracted more than color B. Which color has a longer wavelength? Explain.
15. If glass is dispersive, why don't we normally see a spectrum of colors when sunlight passes through a glass window? Explain.

## Exercises*†

*Integrated Exercises* (IEs) *are two-part exercises. The first part typically requires a conceptual answer choice based on physical thinking and basic principles. The following part is quantitative calculations associated with the conceptual choice made in the first part of the exercise.*

**22.1 Wave Fronts and Rays and 22.2 Reflection**

1. • The angle of incidence of a light ray on a mirrored surface is 30°. What is the angle between the incident and reflected rays?
2. • A beam of light is incident on a plane mirror at an angle of 35° relative to the normal. What is the angle between the reflected ray and the surface of the mirror?

* The bullets denote the degree of difficulty of the exercises: •, simple; ••, medium; and •••, more difficult.

† Assume all angles to be exact.

3. IE • A beam of light is incident on a plane mirror at an angle $\alpha$ relative to the surface of the mirror. (a) Will the angle between the reflected ray and the normal be (1) $\alpha$, (2) $90° - \alpha$, or (3) $2\alpha$? (b) If $\alpha = 33°$, what is the angle between the reflected ray and the normal?
4. IE •• Two upright plane mirrors touch along one edge, where their planes make an angle of $\alpha$. A beam of light is directed onto one of the mirrors at an angle of incidence of $\beta$ ($\beta < \alpha$) and is reflected onto the other mirror. (a) Will the angle of reflection of the beam from the second mirror be (1) $\alpha$, (2) $\beta$, (3) $\alpha + \beta$, or (4) $\alpha - \beta$? (b) If $\alpha = 60°$ and $\beta = 40°$, what will be the angle of reflection of the beam from the second mirror?
5. IE •• Two identical plane mirrors of width $w$ are placed a distance $d$ apart with their mirrored surfaces parallel and facing each other. (a) A beam of light is incident at one end of one mirror so that the light just strikes the far end of the other mirror after reflection. Will the angle of incidence be (1) $\sin^{-1}(w/d)$, (2) $\cos^{-1}(w/d)$, or (3) $\tan^{-1}(w/d)$? (b) If $d = 50$ cm and $w = 25$ cm, what is the angle of incidence?
6. •• Two people stand 5.0 m apart and 3.0 m away from a large plane mirror in a dark room. At what angle of incidence should one of them shine a flashlight on the mirror so that the reflected beam directly strikes the other person?
7. •• A beam of light is incident on a plane mirror at an angle of incidence of 35°. If the mirror rotates through a small angle of $\theta$, through what angle will the reflected ray rotate?
8. ••• Two plane mirrors, $M_1$ and $M_2$, are placed together, as illustrated in ▼**Figure 22.26**. (a) If the angle $\alpha$ between the mirrors is 70° and the angle of incidence, $\theta_{i_1}$, of a light ray incident on $M_1$ is 35°, what is the angle of reflection, $\theta_{i_2}$, from $M_2$? (b) If $\alpha = 115°$ and $\theta_{i_1} = 60°$, what is $\theta_{i_2}$?

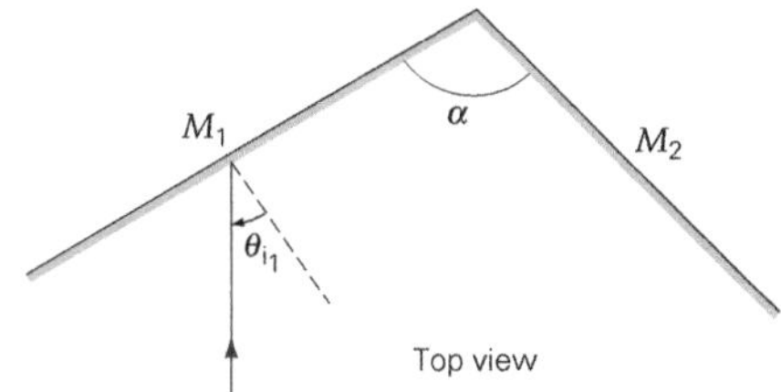


▲**FIGURE 22.26** **Plane mirrors together** See Exercises 8 and 9.

9. ••• For the plane mirrors in Figure 22.26, what angles $\alpha$ and $\theta_{i_1}$ would allow a ray to be reflected back in the direction from which it came (parallel to the incident ray)?

### 22.3 Refraction and 22.4 Total Internal Reflection and Fiber Optics

10. • The index of refraction in a certain precious transparent stone is 2.42. What is the speed of light in that stone?
11. • The speed of light in the core of the crystalline lens in a human eye is $2.13 \times 10^8$ m/s. What is the index of refraction of the core?
12. IE • The indices of refraction for zircon and fused quartz can be found in Table 22.1. (a) The speed of light in fused quartz is (1) greater than, (2) less than, (3) the same as the speed of light in zircon. Explain. (b) Compute the ratio of the speed of light in fused quartz to that in zircon.
13. • A beam of light traveling in air is incident on a transparent plastic material at an angle of incidence of 50°. The angle of refraction is 35°. What is the index of refraction of the plastic?
14. IE • A beam of light enters water from air. (a) Will the angle of refraction be (1) greater than, (2) equal to, or (3) less than the angle of incidence? Explain. (b) If the angle of incidence is 50°, find the angle of refraction.
15. IE • Light passes from a crown glass container into water. (a) Will the angle of refraction be (1) greater than, (2) equal to, or (3) less than the angle of incidence? Explain. (b) If the angle of refraction is 20°, what is the angle of incidence?
16. • The critical angle for a certain type of glass in air is 40.2°. What is the index of refraction of the glass?
17. IE • (a) For total internal reflection to occur, should the light be directed from (1) air to water or (2) water to air? Explain. (b) What is the critical angle of water in air?
18. •• What is the critical angle of a diamond in water?
19. •• A beam of light in air is incident on the surface of a slab of fused quartz. Part of the beam is transmitted into the quartz at an angle of refraction of 30° relative to a normal to the surface, and part is reflected. What is the angle of reflection?
20. •• A beam of light is incident from air onto a flat piece of polystyrene at an angle of 50° relative to a normal to the surface. What angle does the refracted ray make with the plane of the surface?
21. •• The laser used in cornea surgery to treat corneal disease is the *excimer laser*, which emits ultraviolet light at a wavelength of 193 nm in air. The index of refraction of the cornea is 1.376. What are the wavelength and frequency of the light in the cornea?
22. IE •• Light passes from material A, which has an index of refraction of $\frac{4}{3}$, into material B, which has an index of refraction of $\frac{5}{4}$. (a) The speed of light in material A is

(1) greater than, (2) the same as, (3) less than the speed of light in material B. Explain. (b) Find the ratio of the speed of light in material A to the speed of light in material B.

23. **IE** •• In Exercise 22, (a) the wavelength of light in material A is (1) greater than, (2) the same as, (3) less than the wavelength of light in material B. Explain. (b) What is the ratio of the light's wavelength in material A to that in material B?
24. •• The critical angle between two materials is 41°. If the angle of incidence is 35°, what is the angle of refraction? (Consider that light can travel to the interface from either material.)
25. •• (a) An object immersed in water appears closer to the surface than it actually is. What is the cause of this illusion? (b) Using ▼**Figure 22.27**, show that the apparent depth for small angles of refraction is $d' = d/n$, where $n$ is the index of refraction of the water. [*Hint*: Recall that for small angles, $\tan\theta \approx \sin\theta$.]

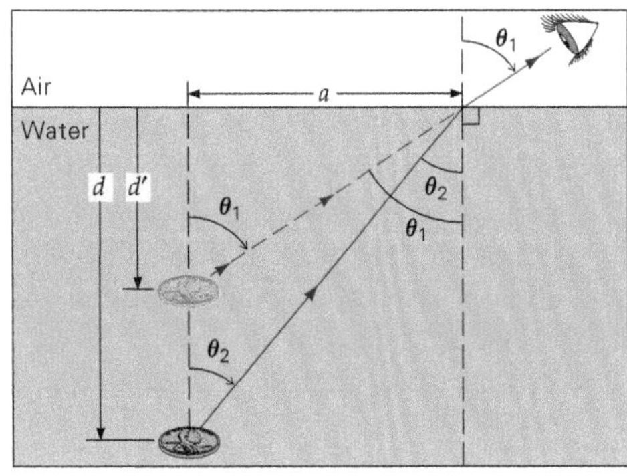


▲FIGURE 22.27 **Apparent depth?** See Exercise 25. (For small angles only; angles enlarged for clarity.)

26. •• A person looks over the edge of a pool and sees a bottle cap on the bottom directly below, where the depth is 3.2 m. How far below the water surface does the bottle cap appear to be? (See Exercise 25b.)
27. •• What percentage of the actual depth is the apparent depth of an object submerged in oil if the observer is looking almost straight downward? (See Exercise 25b.)
28. •• A light ray in air is incident on a glass plate 10.0 cm thick at an angle of incidence of 40°. The glass has an index of refraction of 1.65. The emerging ray on the other side of the plate is parallel to the incident ray but is laterally displaced. What is the perpendicular distance between the original direction of the ray and the direction of the emerging ray? [*Hint*: See Example 22.4.]
29. **IE** •• To a submerged diver looking upward through the water, the altitude of the Sun (the angle between the Sun and the horizon) appears to be 45°. (a) The actual altitude of the Sun is (1) greater than, (2) the same as, (3) less than 45°. Explain. (b) What is the Sun's actual altitude?
30. •• At what angle to the surface must a diver submerged in a lake look toward the surface to see the setting Sun just along the horizon?
31. •• A submerged diver shines a light toward the surface of a body of water at angles of incidence of 40° and 50°. Can a person on the shore see a beam of light emerging from the surface in either case? Justify your answer mathematically.
32. •• Light in air is incident on a transparent material. It is found that the angle of reflection is twice the angle of refraction. What is the *range* of the index of refraction of the material?
33. **IE** •• A beam of light is to undergo total internal reflection through a 45°–90°–45° prism (▼**Figure 22.28**). (a) Will this arrangement depend on (1) the index of refraction of the prism, (2) the index of refraction of the surrounding medium, or (3) the indices of refraction of both? Explain. (b) Calculate the minimum index of refraction of the prism if the surrounding medium is air. Repeat if it is water.

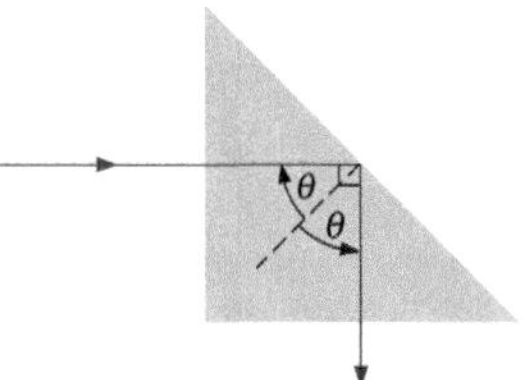


▲FIGURE 22.28 **Total internal reflection in a prism** See Exercises 33 and 34.

34. •• A prism (Figure 22.28) is made of a material with an index of refraction of 1.85. Can the prism be used to deflect a beam of light by (a) in air? (b) What about in water?
35. •• A coin lies on the bottom of a pool under 1.5 m of water and 0.90 m from the side wall (▼**Figure 22.29**). If a light beam is incident on the water surface at the wall, at what angle relative to the wall must the beam be directed so that it will illuminate the coin?

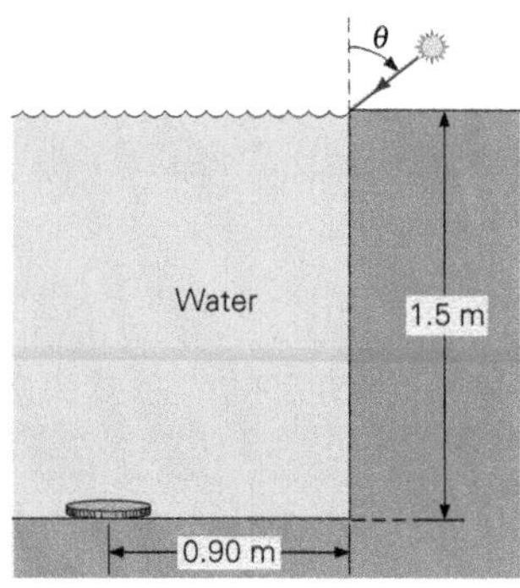


▲FIGURE 22.29 **Find the coin** See Exercise 35. (Not drawn to scale.)

36. •• A flint glass plate 3.5 cm thick is placed over a newspaper. How far beneath the top surface of the plate would the print appear to be if you were looking almost vertically downward through the plate? (See Exercise 25b.)
37. •• Yellow-green light of wavelength 550 nm in air is incident on the surface of a flat piece of crown glass at an angle of 40°. (a) What is the angle of refraction of the light? (b) What is the speed of the light in the glass? (c) What is the wavelength of the light in the glass?
38. IE •• A light beam traveling upward in a plastic material with an index of refraction of 1.60 is incident on an upper horizontal air interface. (a) At certain angles of incidence, the light is not transmitted into air. The cause of this is (1) reflection, (2) refraction, (3) total internal reflection. Explain. (b) If the angle of incidence is 45°, is some of the beam transmitted into air? (c) Suppose the upper surface of the plastic material is covered with a layer of liquid with an index of refraction of 1.20. What happens in this case?

**22.5 Dispersion**

39. • White light is incident from air onto a transparent material at an angle of incidence of 40°. The angles of refraction for the red and blue colors are 28.15° and 27.95°, respectively. What are the indices of refraction for the two colors?
40. IE •• The index of refraction of crown glass is 1.515 for red light and 1.523 for blue light. (a) If light of both colors is incident on crown glass from air, the blue color will be refracted (1) more, (2) less, or (3) the same amount as the red color. Explain. (b) Find the angle separating rays of the two colors in a piece of crown glass if their angle of incidence is 37°.
41. •• A beam of light with red and blue components of wavelengths 670 and 425 nm, respectively, strikes a slab of fused quartz at an incident angle of 30°. On refraction, the different components are separated by an angle of 0.001 31 rad. If the index of refraction of the red light is 1.4925, what is the index of refraction of the blue light?
42. •• In Exercise 40, if the angle of incidence is 41.15°, which color(s) of light will be refracted out into the air?

# 23

# Mirrors and Lenses



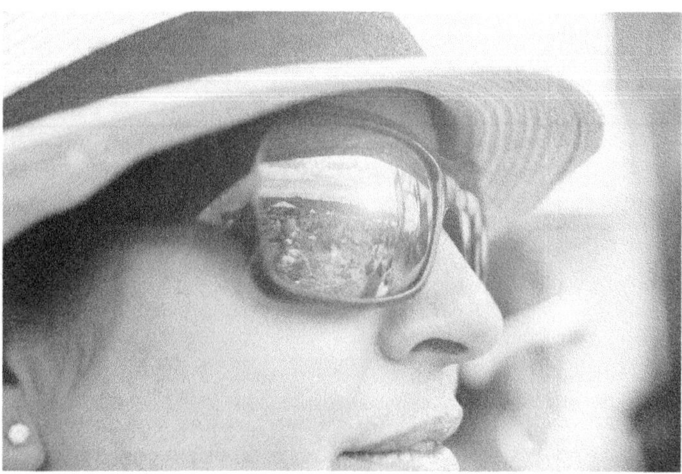

**Upright and reduced images are formed by the convex surface of the sunglasses.**

What would life be like if there were no mirrors in bathrooms or cars, and if eyeglasses did not exist? Imagine a world without optical images of any kind – no photographs, no movies, no TV. Think about how little we would know about the universe if there were no telescopes to observe distant planets and stars – or how little we would know about biology and medicine if there were no microscopes to see bacteria and cells. It is often forgotten how dependent we are on mirrors and lenses.

The first mirror was probably the reflecting surface of a pool of water. Later, people discovered that polished metals and glass also have reflective properties. They must also have noticed that when they looked at things through glass, the objects looked different than when viewed directly, depending on the shape of the glass. In some cases, the objects appeared to be reduced, as is the case in the interesting chapter-opening photograph. It shows the reflected images of people by the surface of sunglasses. In time, people learned to shape glass into lenses, paving the way for the eventual development of the many optical devices we now take for granted.

The optical properties of mirrors and lenses are based on the principles of reflection and refraction of light, as introduced in Chapter 22. In this chapter, the principles of mirrors and lenses will be discussed. Among other things, you'll discover why the image in the photo is upright and reduced, whereas your image in an ordinary flat mirror is the same size as you – but the image doesn't seem to comb your hair with the same hand you use.

## 23.1 Plane Mirrors

Mirrors are smooth reflecting surfaces, usually made of polished metal or glass that has been coated with some metallic substance. As you know, even an uncoated piece of glass, such as a window pane, can act as a mirror. However, when one side of a piece of glass is coated with a compound of tin, aluminum, or silver, the reflectivity is increased, as light is not transmitted through the coating. A mirror may be front coated or back coated, but most mirrors are back coated.

When you look directly into a mirror, you see the images of yourself and objects around you (apparently on the other side of the surface of the mirror). The geometry of a mirror's surface affects the size, orientation, and type of image. In general, an *image* is the visual counterpart of an object, produced by reflection (mirrors) or refraction (lenses).

A mirror with a flat surface is called a **plane mirror**. How images are formed by a plane mirror is illustrated by the ray diagram in ▼**Figure 23.1**. An image appears to be behind or "inside" the mirror. This is because when the mirror reflects a ray of light from the object to the eye (Figure 23.1a), the ray appears to originate from behind the mirror. Reflected rays from the top and bottom of an object are shown in Figure 23.1b. In actuality, light rays coming from all points on the side of the object facing the mirror are reflected, and an image of the complete object is observed.

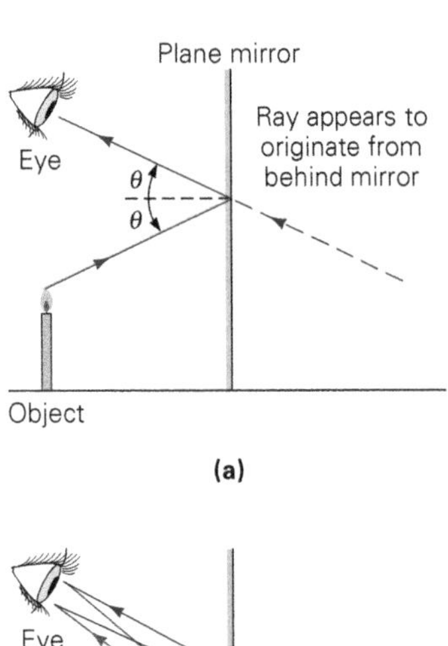


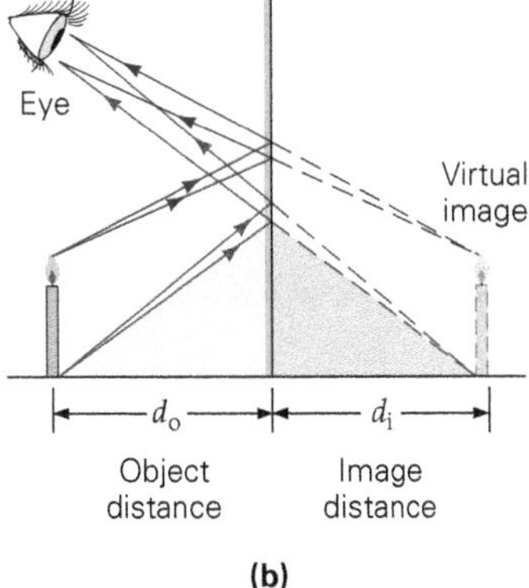


▲FIGURE 23.1 **Image formed by a plane mirror** (a) A ray from a point on the object is reflected in the mirror according to the law of reflection. (b) Rays from various points on the object produce an image. Because the two shaded triangles are identical, the image distance $d_i$ (the distance of the image from the mirror) is equal to the object distance $d_o$. That is, the image appears to be the same distance behind the mirror as the object is in front of the mirror. The rays appear to diverge from the image position. In this case, the image is said to be virtual.

The image formed in this way *appears* to be behind the mirror. Such an image is called a **virtual image**. Light rays appear to diverge from virtual images, but do not actually do so. No light rays actually come from or pass through the image; therefore virtual images cannot be formed on screens. However, spherical mirrors (discussed in Section 23.2) can project images in front of the mirror where light actually converges at and then diverges from the image. This type of image is called a **real image**. An example of a real image is the image produced on a screen by an overhead projector in a classroom.

Notice in Figure 23.1b the distances of the object and image from the mirror. Quite logically, the distance of an object from a mirror is called the *object distance* ($d_o$), and the distance its image *appears* to be behind the mirror is called the *image distance* ($d_i$). By geometry of identical triangles and the law of reflection, $\theta_i = \theta_r$ (Section 22.2), it can be shown that $d_o = |d_i|$, which means that *the image formed by a plane mirror appears to be at a distance behind the mirror that is equal to the distance between the object and the front of the mirror.* (See Exercise 9.) The absolute value sign on $d_i$ will be discussed in detail later, it is because of a sign convention.

We are interested in various characteristics of images. Two of these features are the height and orientation of an image compared with those of its object. Both are expressed in terms of the **lateral magnification factor** (***M***), which is defined as a ratio of heights of the image ($h_i$) and object ($h_o$):

$$M = \frac{\text{image height}}{\text{object height}} = \frac{h_i}{h_o} \quad \text{(magnification factor)} \tag{23.1}$$

A lighted candle used as an object allows us to address an important image characteristic: orientation. That is, whether the image is upright or inverted with respect to the orientation of the object. (In sketching ray diagrams, an arrow is a convenient object for this purpose.) For a plane mirror, the image is always upright. This means that the image is oriented in the same direction as the object. We say that $h_i$ and $h_o$ have *the same sign*, so $M$ is positive. Note that $M$ is a dimensionless quantity, as it is a ratio of heights.

In ▶**Figure 23.2**, you should also be able to see that the image and object have the same sizes (heights). Therefore, $M = (h_i/h_o) = +1$ for a plane mirror, the image is upright ($M$ is positive), and there is no magnification. That is, the object and the image in a plane mirror are the same size. In general, the sign of $M$ tells us the orientation of the image relative to the object, and the value of $M$ gives the magnification, for all optic elements.

With other types of mirrors, such as spherical mirrors (which will be considered shortly), it is possible to have real and inverted images where $M$ is negative and magnified, or reduced images where absolute value of $M$ is either greater than or less than 1.

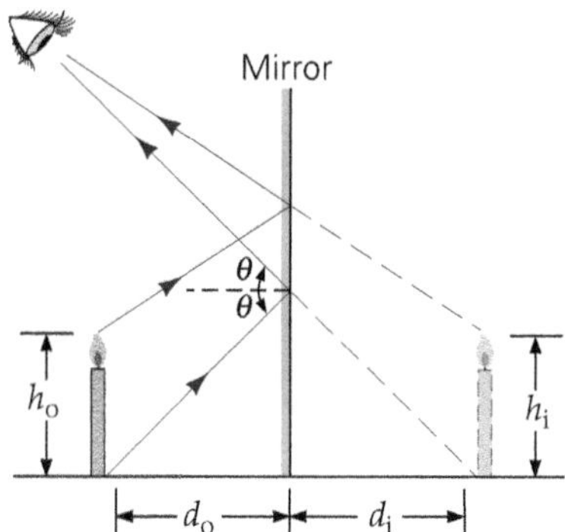


▲ **FIGURE 23.2** **Magnification** The lateral, or height magnification, factor is given by $M = h_i/h_o$. For a plane mirror, $M = +1$, which means that $h_i = h_o$; that is, the image is the same height as the object, and the image is upright.

Another characteristic of reflected images of plane mirrors is the so-called *right–left reversal*. When you look at yourself in a mirror and raise your right hand, it appears that your image raises its left hand. However, this right–left reversal is actually caused by the front–back reversal. For example, if your front faces south, then your back "faces" north. Your image, on the other hand, has its front to the north and back to the south – a front–back reversal. You can demonstrate this reversal by asking one of your friends to stand facing you (without a mirror). If your friend raises his right hand, you can see that that hand is actually on your left side.

The main characteristics of an image formed by a plane mirror are summarized in ▼**Table 23.1**.

TABLE 23.1 Characteristics of Images Formed by Plane Mirrors

| | |
|---|---|
| $d_o = \lvert d_i \rvert$ | The object distance is equal to the image distance; that is, the image appears to be as far behind the mirror as the object is in front. |
| $M = +1$ | The image is virtual, upright, and unmagnified. |

## EXAMPLE 23.1: ALL OF ME – MINIMUM MIRROR LENGTH

What is the minimum vertical length of a plane mirror needed for a woman to be able to see a complete (head-to-toe) image of herself (▶ **Figure 23.3**)?

**THINKING IT THROUGH.** Applying the law of reflection, it can be seen in the figure that two triangles are formed by the rays needed for the image to be complete. These triangles relate the woman's height to the minimum mirror length.

### SOLUTION

To determine this length, consider the situation shown in Figure 23.3. With a mirror of minimum length, a ray from the top of the woman's head would be reflected at the top of the mirror, and a ray from her feet would be reflected at the bottom of the mirror, to her eyes. The length $L$ of the mirror is then the distance between the dashed horizontal lines perpendicular to the mirror at its top and bottom.

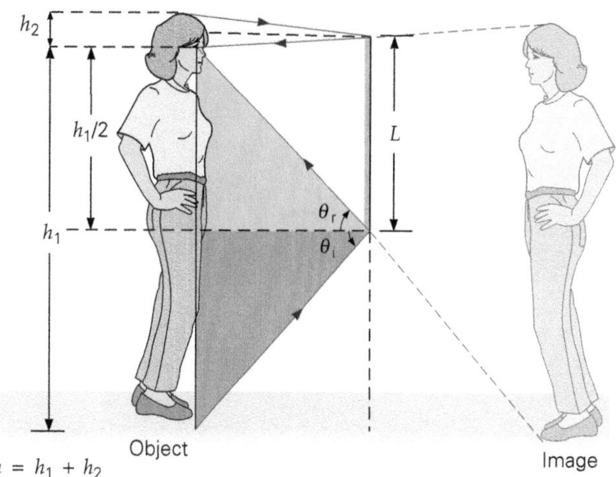


▲ **FIGURE 23.3** **Seeing it all** The minimum height, or vertical length, of a plane mirror needed for a person to see his or her complete (head-to-toe) image turns out to be half the person's height. (See Example 23.1 for description.)

However, these lines are also the normals for the ray reflections. By the law of reflection, they bisect the angles between incident and reflected rays; that is, $\theta_i = \theta_r$. Then, because their respective triangles on each side of the dashed normal are identical, the length of the mirror from its bottom to a point even with the woman's eyes is $h_1/2$ where $h_1$ is the woman's height from her feet to her eyes. Similarly, the small upper length of the mirror is $h_2/2$ (the vertical distance between the woman's eyes and the top of mirror). Then,

$$L = \frac{h_1}{2} + \frac{h_2}{2} = \frac{h_1 + h_2}{2} = \frac{h}{2}$$

where $h$ is the woman's total height.

Hence, for the woman to see her complete image (head-to-toe) in a plane mirror, the minimum height, or vertical length, of the mirror must be half her height.

You can do a simple experiment to prove this conclusion. Get some newspaper and tape, and find a full-length mirror. Gradually cover parts of the mirror with the newspaper until you can't see your complete image. You will find you need a mirror length that is only half your height to see a complete image.

**FOLLOW-UP EXERCISE.** What effect does a woman's distance from the mirror have on the minimum mirror length required to produce her complete image? *(Answers to all Follow-Up Exercises are given in Appendix V at the back of the book.)*

Mirrors are the primary equipment for magicians to "fool" the audience and make things "disappear." One of the most famous such illusions was the "Vanishing Elephant" by Harry Houdini performed on the stage of the Hippodrome Theater in New York City in 1918 (▶ **Figure 23.4a**). When the time for the elephant's disappearance came, two large plane mirrors at right angles to each other were slid quickly into place (Figure 23.4b). A strobe light was used to conceal the brief motion of the mirrors. When properly aligned, the mirrors

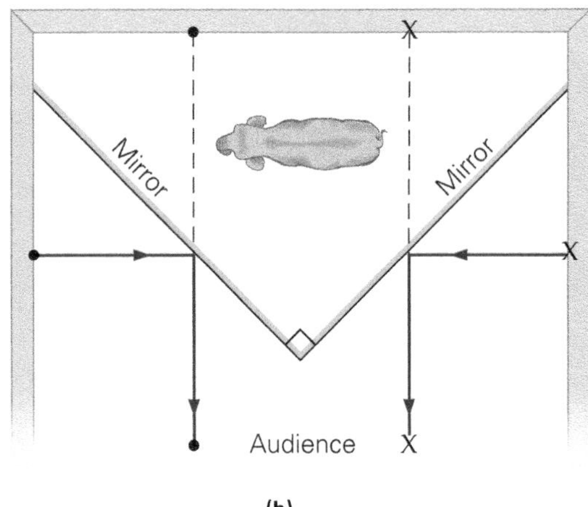


(a) (b)

▲ FIGURE 23.4 **It's all done with mirrors (a)** The elephant vanished from view when Houdini fired a pistol. **(b)** Two large mirrors at right angles to each other were used to conceal the elephant.

reflected light (dark red lines) from the side walls of the stage to form virtual images that matched the pattern of the stage backdrop. Thus the audience apparently saw the stage with no elephant visible. Unseen by the audience, the elephant was quickly led off stage.

## 23.2 Spherical Mirrors

As the name implies, a **spherical mirror** is a reflecting surface with spherical geometry. ▼ **Figure 23.5** shows that if a portion of a sphere of radius $R$ is sliced off along a plane, the severed section has the shape of a spherical mirror. Either the inside or the outside of such a section can be the reflecting surface. For reflections on the inside surface, the section behaves as a **concave mirror**. (Think of looking into a cave in order to help yourself remember that a concave mirror has a recessed surface.) For reflections from the outside surface, the section behaves as a **convex mirror**.

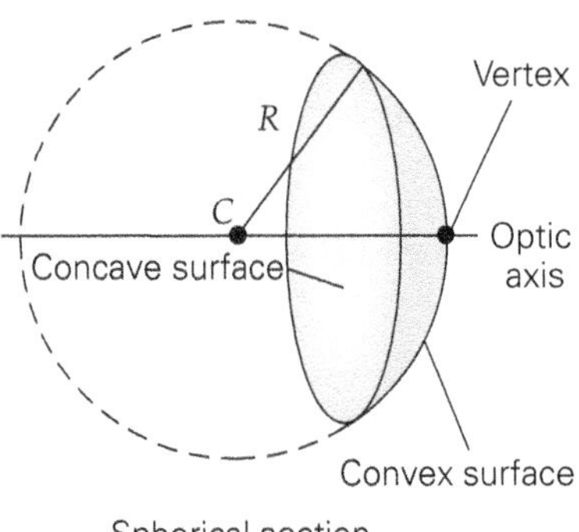


▲ FIGURE 23.5 **Spherical mirrors** A spherical mirror is a section of a sphere. Either the outside (convex) surface or the inside (concave) surface of the spherical section may be the reflecting surface.

The radial line through the center of the spherical mirror that intersects the surface of the mirror at the vertex of the spherical section is called the *optic axis* (Figure 23.5). The point on the optic axis that corresponds to the center of the sphere of which the mirror forms a section is called the **center of curvature (*C*)**. The distance between the vertex and the center of curvature is equal to the radius of the sphere and is called the **radius of curvature (*R*)**.

When rays parallel and close to the optic axis are incident on a concave mirror, the reflected rays intersect, or converge, at a common point called the **focal point (*F*)**. As a result, a concave mirror acts as a **converging mirror** (▼ **Figure 23.6a**). Note that the law of reflection, $\theta_i = \theta_r$ is satisfied for each ray.

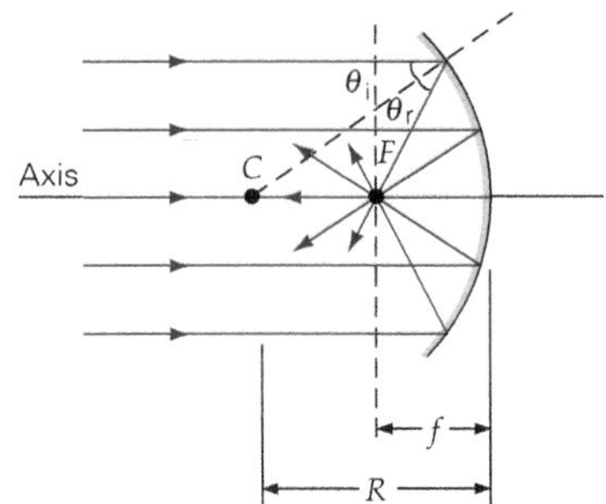


(a) Concave, or converging, mirror

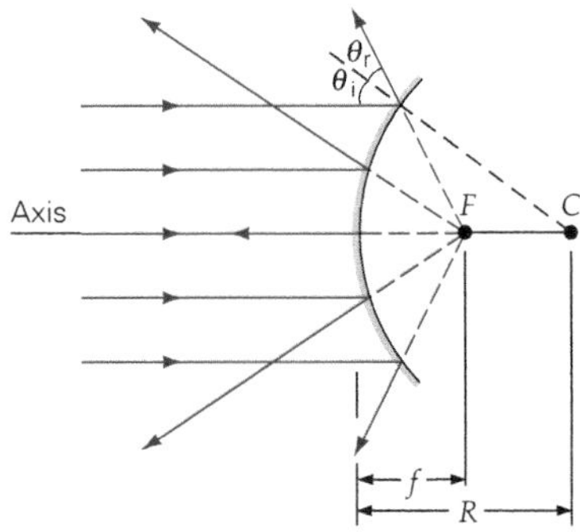


(b) Convex, or diverging, mirror

▲ FIGURE 23.6 **Focal point (a)** Rays parallel and close to the optic axis of a concave spherical mirror converge at the focal point $F$. **(b)** Rays parallel and close to the optic axis of a convex spherical mirror are reflected along paths as though they diverge from a focal point behind the mirror. Note that the law of reflection, $\theta_i = \theta_r$, is always satisfied for each ray.

Similarly, rays parallel and close to the optic axis of a convex mirror diverge on reflection, as though the reflected rays came from a focal point behind the mirror's surface (Figure 23.6b). Thus, a convex mirror acts as a **diverging mirror** (▼ **Figure 23.7**). When you see diverging rays, your brain assumes that there is an image from which the rays appear to diverge, even though no such image is actually there.

▲ FIGURE 23.7 **Diverging mirror** Note by reverse-ray tracing in Figure 23.6b that a diverging (convex) spherical mirror gives an expanded, although distorted, field of view, as can be seen from this convex mirror showing a dock scene.

The distance from the vertex to the focal point ($F$) is called the **focal length** ($f$). (See Figure 23.6.) The focal length can be shown to be half the radius of curvature:

$$f = \frac{R}{2} \quad \text{(focal length of spherical mirror)} \qquad (23.2)$$

The preceding result is valid only when the rays are close to the optic axis – that is, for small-angle approximation. Rays far away from the optic axis will focus at different focal points, resulting in some image distortion. In optics, this type of distortion is an example of a *spherical aberration*. Some telescope mirrors are parabolic in shape, rather than spherical, in which case *all* rays parallel to the optic axis are focused at the same focal point, thus eliminating spherical aberration.

## 23.2.1 Ray Diagrams

The characteristics of images formed by spherical mirrors can be determined from geometrical optics (Chapter 22). The method involves drawing rays emanating from one or more points on an object. The law of reflection $\theta_i = \theta_r$ applies, and three key rays are defined with respect to the mirror's geometry as follows (see ▶ **Figure 23.8** and also Example 23.2).

① A **parallel ray** is a ray that is incident along a path parallel to the optic axis and is reflected through (or appears to go through) the focal point $F$ (as do all rays near and parallel to the axis).

**Converging (concave) mirror**

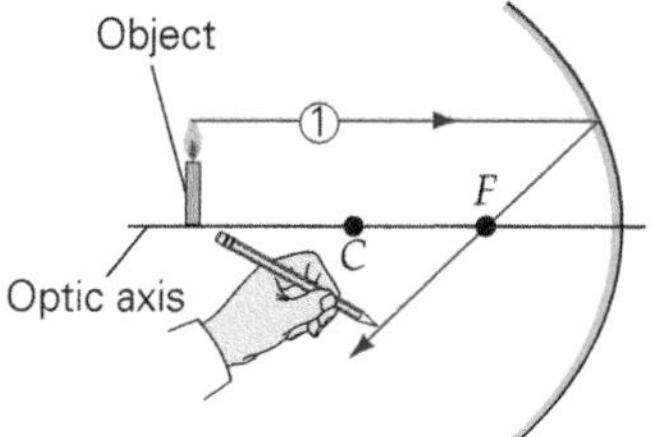


1 **Parallel ray**

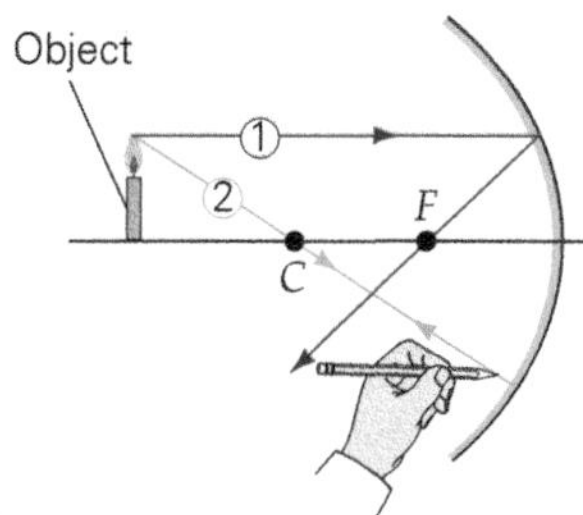


2 **Chief (radial) ray**

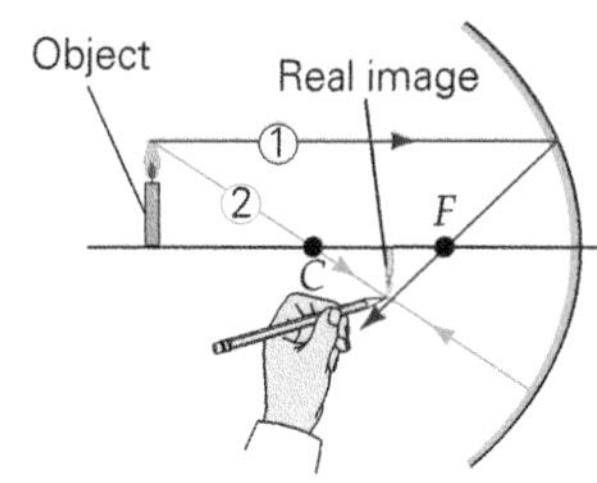


3 **Locating image**

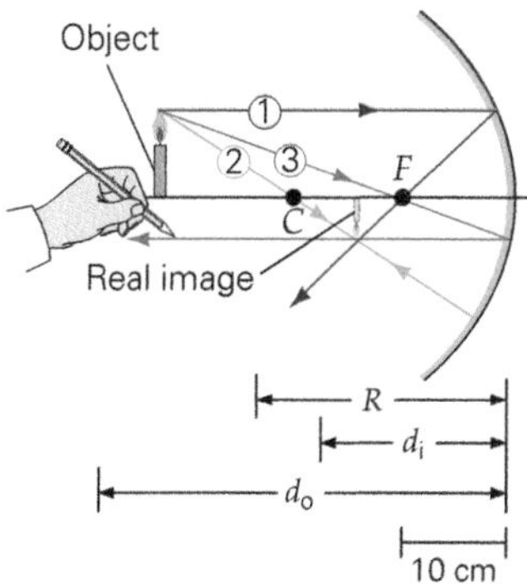


4 **Can also use focal ray to confirm image**

▲ FIGURE 23.8 **A mirror ray diagram** Rays can be used to determine image characteristics. See Example 23.2 for a description.

② A **chief ray**, or **radial ray**, is a ray that is incident through the center of curvature ($C$) of the spherical mirror. Since the chief ray is incident normal to the mirror's surface, this ray is reflected back along its incident path, through point $C$.

③ A **focal ray** is a ray that passes through (or appears to go through) the focal point $F$ and is reflected parallel to the optic axis. (It is a reversed parallel ray, so to speak.)

Using any two of these three rays, the image can be located to find the image distance and determine its type (real or virtual), orientation (upright or inverted), and size (magnified or reduced). It is customary to use the tip of an asymmetrical object (e.g., the head of an arrow or the flame of a candle) as the origin point of the rays. The corresponding point of the image is at the point of intersection of the rays. This makes it easy to see whether the image is upright or inverted.

Keep in mind, however, that properly traced rays from any point on the object can be used to find the image. Every point on a visible object acts as an emitter of light. For example, for a candle, the flame emits its own light, and many other points on the candle surface reflect light.

### EXAMPLE 23.2: A MIRROR RAY DIAGRAM

An object is placed 39 cm in front of a converging spherical mirror of radius 24 cm. (a) Use a ray diagram to locate the image formed by this mirror. (b) Discuss the characteristics of the image.

**THINKING IT THROUGH.** A ray diagram, drawn accurately, can by itself provide "quantitative" information about image location and image characteristics that might otherwise be determined mathematically.

**SOLUTION**

***Given:***

$R = 24$ cm
$d_o = 39$ cm

***Find:***

**(a)** Image location
**(b)** Image characteristics

**(a)** Since a ray diagram (drawing) is to be used to locate the image, a scale for the drawing is needed. If a scale of 1 cm (on the drawing) represents 10 cm, the object would be drawn 3.9 cm in front of the mirror.

First draw the optic axis (Figure 23.8, A mirror ray diagram), the mirror, the object (a lighted candle), and the center of curvature ($C$). From Equation 23.2, the focal length $f = 24\text{ cm}/2 = 12$ cm, so the focal point ($F$) is halfway from the vertex to the center of curvature.

To locate the image, follow steps 1–4 in Figure 23.8.

1. The first ray drawn is the parallel ray (① in the drawing). From the tip of the flame, draw a ray parallel to the optic axis. After reflecting, this ray goes through the focal point, $F$.
2. Then draw the chief ray (② in the drawing). From the tip of the flame, draw a ray going through the center of curvature, $C$. This ray will be reflected back along the original direction. (Why?)
3. Locate the image. It can be seen that these two rays, after reflection, actually intersect. The point of intersection is the tip of the *image* of the candle. From this point, draw the image by extending the tip of the flame to the optic axis. The image distance $d_i = 17$ cm as measured from the diagram.
4. Only two rays are needed to locate the image. However, the third ray can be drawn as a double check. In this case, the focal ray (③ in the drawing) should go through the same image point at which the other two rays intersect (if drawn carefully). The focal ray from the tip of the flame going through the focal point, $F$, after reflection, will travel out parallel to the optic axis.

**(b)** From the ray diagram drawn in part (a), it can clearly be seen that the image is real (because the reflected rays intersect). As a result, the real image could be seen on a screen (e.g., a piece of paper) that is positioned at the image point. The image is also inverted (the image of the candle points downward) and is reduced; that is, it is smaller in size than the object.

**FOLLOW-UP EXERCISE.** In this example, what would the characteristics of the image be if the object were 15 cm in front of the mirror? Locate the image and discuss its characteristics.

An example of a ray diagram using the same three rays for a convex (diverging) mirror will be shown in Integrated Example 23.4.

A converging mirror does not always form a real image. For a converging spherical mirror, the characteristics of the image change with the object distance. Dramatic changes take place at two points: $C$ (the center of curvature) and $F$ (the focal point). These points divide the optic axis into three regions: $d_o > R$, $R > d_o > f$, and $d_o < f$, where $R = 2f$ (▶ **Figure 23.9a**).

Let's start with an object in the region farthest from the mirror ($d_o > R$) and move toward the mirror:

- The case of $d_o > R$ was shown in Example 23.2.
- When $d_o = R = 2f$, the image is real, inverted, and the same size as the object.
- When $R > d_o > f$, a real, inverted, and magnified image is formed (Figure 23.9b). The image is magnified when the object is inside the center of curvature, $C$.
- When $d_o = f$, the object is at the focal point (Figure 23.9c). The reflected rays are parallel, and the image is said to be "formed at infinity." The focal point $F$ is a special "crossover" point between real and virtual images.
- When $d_o < f$, the object is inside the focal point (between the focal point and the mirror's surface). A virtual, upright, and magnified image is formed (Figure 23.9d).

When $d_o > f$, (Figure 23.9a and b) the image is real; when $d_o < f$ (Figure 23.9d) the image is virtual. For $d_o = f$, the image is said to be formed at infinity (Figure 23.9c). When an object is at "infinity" – when it is so far away that rays emanating from it and falling on the mirror are essentially parallel – its image is formed at the focal plane. This fact provides an easy method for determining the focal length of a converging mirror.

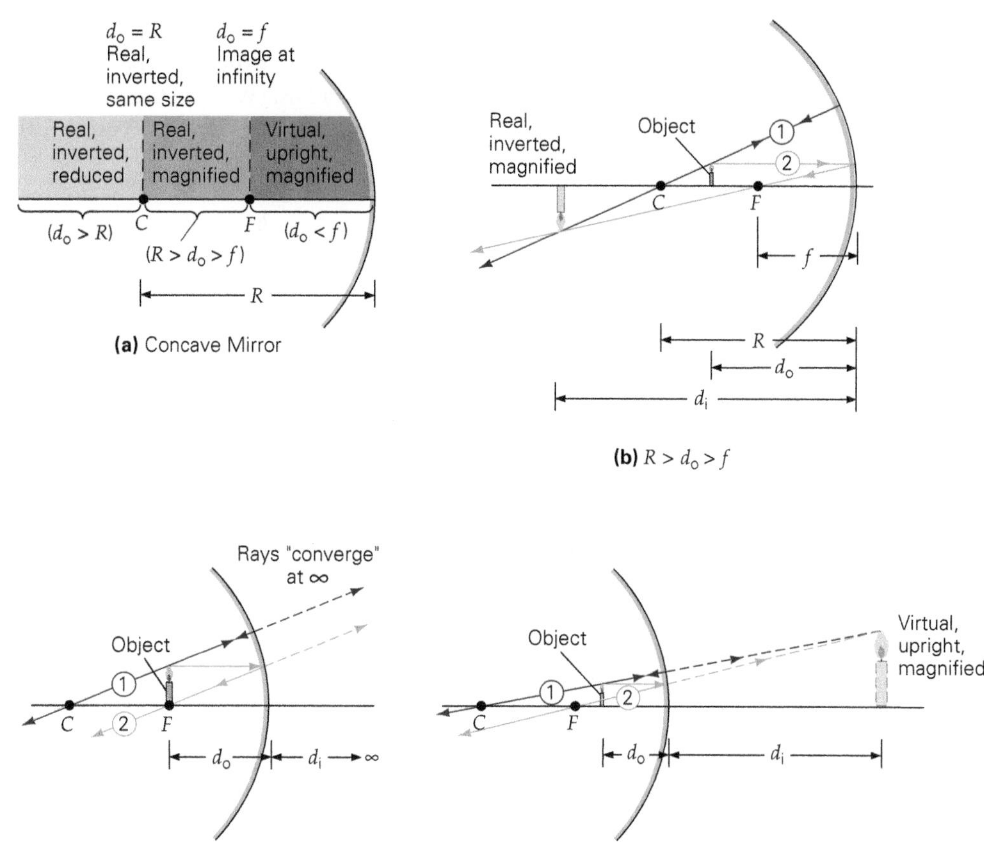


▲ **FIGURE 23.9 Converging mirrors (a)** For a converging (concave) mirror, the object is located within one of three regions defined by the center of curvature ($C$) and the focal point ($F$), or at one of these two points. For $d_o > R$, the image is real, inverted, and reduced, as shown by the ray diagrams in Example 23.2. **(b)** For $R > d_o > f$, the image will also be real and inverted but magnified. **(c)** For an object at the focal point $F$, or $d_o = f$, the image is said to be formed at infinity. **(d)** For $d_o < f$, the image will be virtual, upright, and magnified.

As we have seen, the position, orientation, and size of the image can be approximately determined graphically from ray diagrams drawn to scale. However, these characteristics can be determined more accurately by analytical methods. It can be shown by means of geometry that the object distance ($d_o$), the image distance ($d_i$), and the focal length ($f$) are related through the **spherical mirror equation**:

$$\frac{1}{d_o} + \frac{1}{d_i} = \frac{1}{f} = \frac{2}{R} \quad \text{(spherical mirror equation)} \tag{23.3}$$

Note that this equation can be written in terms of either the radius of curvature, $R$, or the focal distance, $f$, since by Equation 23.2, $f = R/2$. Both $R$ and $f$ can be either positive or negative, as will be discussed shortly.

If $d_i$ is the quantity to be found for a spherical mirror, it may be convenient to use an alternative form of the spherical mirror equation:

$$d_i = \frac{d_o f}{d_o - f} \tag{23.3a}$$

The signs of the various quantities are very important in the application of Equation 23.3. The sign conventions summarized in ▶ **Table 23.2** will be used. For example, for a real object, a positive $d_i$ indicates a real image and a negative $d_i$ corresponds to a virtual image.

The **lateral magnification factor ($M$)**, also called the *magnification factor*, or simply *magnification*, defined in Equation 23.1 can also be found analytically for a spherical mirror. Again, by using geometry, it can be expressed in terms of the image and object distances:

$$M = \frac{h_i}{h_o} = -\frac{d_i}{d_o} \quad \text{(magnification factor)} \tag{23.4}$$

(The derivations of Equations 23.3 and 23.4 follow as optional content.)

A positive value for $M$ indicates an upright image, whereas a negative $M$ implies an inverted image. Also, if $|M| > 1$, the image is magnified; if $|M| < 1$, the image is reduced; if $|M| = 1$, the image is the same size as the object.

**TABLE 23.2** Sign Conventions for Spherical Mirrors

| | |
|---|---|
| *Focal length (f)* | |
| Converging (concave) mirror | $f$ (or $R$) is positive |
| Diverging (convex) mirror | $f$ (or $R$) is negative |
| *Object distance ($d_o$)* | |
| Object is in front of the mirror (real object) | $d_o$ is positive |
| Object is behind the mirror (virtual object)[a] | $d_o$ is negative |
| *Image distance ($d_i$) and image type* | |
| Image is formed in front of the mirror (real image) | $d_i$ is positive |
| Image is formed behind the mirror (virtual image) | $d_i$ is negative |
| *Image orientation (M)* | |
| Image is upright with respect to the object | $M$ is positive |
| Image is inverted with respect to the object | $M$ is negative |

[a] In a combination of two (or more) mirrors, the image formed by the first mirror is the object of the second mirror (and so on). If this image–object falls behind the second mirror, it is referred to as a *virtual* object, and the object distance is taken to be negative. This concept is more important for lens combinations, as we will see in Section 23.3, and is mentioned here only for completeness.

**Derivation of the Spherical Mirror Equation (Optional)** You might wonder from where Equations 23.3 and 23.4 originate. The spherical mirror equation can be derived with the aid of a little geometry. Consider the ray diagram in ▼ **Figure 23.10**. The object and image distances ($d_o$ and $d_i$) and the heights of the object and image ($h_o$ and $h_i$) are shown. Note that these lengths make up the bases and heights of triangles formed by the ray reflected at the vertex ($V$). These triangles ($O'VO$ and $I'VI$) are similar, since, by the law of reflection, their angles at $V$ are equal. Hence,

$$\frac{h_i}{h_o} = -\frac{d_i}{d_o} \tag{1}$$

This equation is Equation 23.4, from the definition of Equation 23.1. The negative sign inserted here signifies the fact that the image is inverted, so $h_i$ is negative.

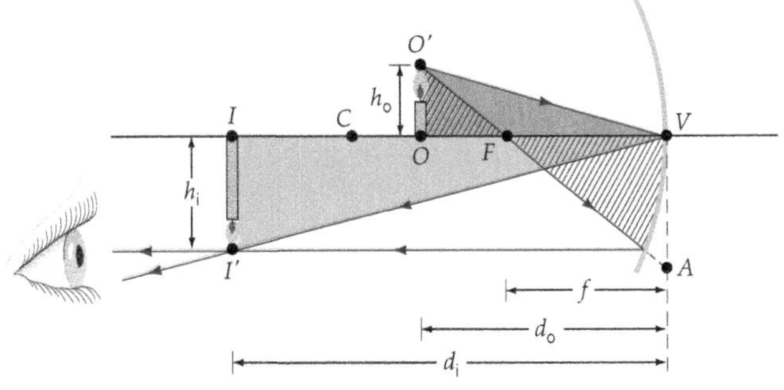


▲ **FIGURE 23.10** **Spherical mirror equation** The rays provide the geometry, through similar triangles, for the derivation of the spherical mirror and lateral magnification equations.

The (focal) ray through $F$ also forms similar triangles, $O'FO$ and $AVF$ in the approximation that the mirror is small compared with its radius. (Why are the triangles similar?) The bases of these triangles are $VF = f$ and $OF = d_o - f$. Then, if $VA$ is taken to be $h_i$,

$$\frac{h_i}{h_o} = -\frac{VF}{OF} = -\frac{f}{d_o - f} \tag{2}$$

Again, the negative sign inserted here signifies the fact that the image is inverted, so $h_i$ is negative.

Equating Equations 1 and 2,

$$\frac{d_i}{d_o} = \frac{f}{d_o - f} \tag{3}$$

Algebraic manipulation yields

$$\frac{1}{d_o} + \frac{1}{d_i} = \frac{1}{f}$$

which is the spherical mirror equation (Equation 23.3).

Example 23.3 and Integrated Example 23.4 show how these equations and sign conventions are used for spherical mirrors. In general, this approach usually involves finding the image of an object; you will be asked where the image is formed ($d_i$) and what the image characteristics are ($M$). These characteristics tell whether the image is real or virtual, upright or inverted, and larger or smaller than the object (i.e., magnified or reduced).

### EXAMPLE 23.3: WHAT KIND OF IMAGE? CHARACTERISTICS OF IMAGES FROM A CONVERGING MIRROR

A converging mirror has a radius of curvature of 30 cm. If an object is placed (a) 45 cm, (b) 20 cm, and (c) 10 cm from the mirror, where is the image formed, and what are its characteristics? (Specify whether each image is real or virtual, upright or inverted, and magnified or reduced.)

**THINKING IT THROUGH.** Here the radius $R$ is given so the focal length $f$ is $R/2$. Also given are three different object distances, which can be used in Equations 23.3 and 23.4 to determine the image location and characteristics.

**SOLUTION**

***Given:***

$R = 30$ cm, so $f = R/2 = 15$ cm
**(a)** $d_o = 45$ cm
**(b)** $d_o = 20$ cm
**(c)** $d_o = 10$ cm

***Find:*** $d_i$, $M$, and image characteristics

Note that the given object distances correspond to the regions shown in Figure 23.9a. There is no need to convert the distances to meters as long as all distances are expressed in the same unit (centimeters in this case). Ray diagrams can be drawn for each of these cases in order to find the characteristics of each image.

**(a)** In this case, $d_o > 2f = R$ and the object distance is greater than the radius of curvature:

$$\frac{1}{d_o}+\frac{1}{d_i}=\frac{1}{f} \quad \text{or} \quad \frac{1}{d_i}=\frac{1}{f}-\frac{1}{d_o}=\frac{1}{15\text{ cm}}-\frac{1}{45\text{ cm}}=\frac{2}{45\text{ cm}}$$

Then

$$d_i=\frac{45\text{ cm}}{2}=+22.5\text{ cm} \quad \text{and} \quad M=-\frac{d_i}{d_o}=-\frac{22.5\text{ cm}}{45\text{ cm}}=-\frac{1}{2}$$

Thus, the image is real (positive $d_i$), inverted (negative $M$), and reduced ($|M|=1/2$).

**(b)** Here, $R = 2f > d_o > f$ and the object is between the focal point and the center of curvature:

$$\frac{1}{d_i}=\frac{1}{15\text{cm}}-\frac{1}{20\text{cm}}=\frac{1}{60\text{cm}}$$

Thus,

$$d_i=+60\text{cm} \quad \text{and} \quad M=-\frac{60\text{cm}}{20\text{cm}}=-3.0$$

In this case, the image is real (positive $d_i$), inverted (negative $M$), and magnified ($|M|=3.0$).

**(c)** For this case, $d_o < f$, and the object is inside the focal point.

Using the alternate form of Equation. 23.3 for illustration:

$$d_i=\frac{d_o f}{d_o-f}=\frac{(10\text{ cm})(15\text{ cm})}{10\text{ cm}-15\text{ cm}}=-30\text{ cm}$$

Then

$$M=-\frac{d_i}{d_o}=-\frac{(-30\text{ cm})}{10\text{ cm}}=+3.0$$

In this case, the image is virtual (negative $d_i$), upright (positive $M$), and magnified ($|M|=3.0$).

From the denominator of the expression for $d_i$ of this alternate form, it can be seen that $d_i$ will always be negative when $d_o$ is less than $f$. Therefore, a virtual image is always formed for an object inside the focal point of a converging mirror.

**FOLLOW-UP EXERCISE.** For the converging mirror in this Example, where is the image formed and what are its characteristics if the object is at 30 cm, or $d_o = R$?

## 23.2.2 Problem-Solving Hint

When using the spherical mirror equations to find image characteristics, it is helpful to first make a quick sketch (approximate, not necessarily to scale) of the ray diagram for the situation. This sketch shows the image characteristics and helps to avoid making mistakes when applying the sign conventions. *The ray diagram and the mathematical solution must agree.*

---

INTEGRATED EXAMPLE 23.4: BIG DIFFERENCES – CHARACTERISTICS OF A DIVERGING MIRROR

An object (in this case, a candle) is 20 cm in front of a diverging mirror that has a focal length of −15 cm (see the sign conventions in Table 23.2). (a) Use a ray diagram to determine whether the image formed is (1) real, upright, magnified, (2) virtual, upright, magnified, (3) real, upright, reduced, (4) virtual, upright, reduced, (5) real, inverted, magnified, or (6) virtual, inverted, reduced. (b) Find the location and characteristics of the image by using the mirror equations.

**(a) CONCEPTUAL REASONING.** Since the object distance and the focal length of the diverging mirror are known, a ray diagram can be drawn and the image characteristics can be determined. The first thing to decide on is a scale for the ray diagram. In this example, 1 cm (on the drawing) could be used to represent 10 cm. That way, the object would be 2.0 cm in front of the mirror in the drawing. Draw the optic axis, the mirror, the object (a lighted candle), and the focal point ($F$).

Since this mirror is convex, the focal point ($F$) and the center of curvature ($C$) are behind the mirror. From Equation 23.2, $R = 2f = 2(-15\text{ cm}) = -30$ cm. So $C$ is drawn at twice the distance of $F$ from the vertex (▶ **Figure 23.11**).

Only two out of the three key rays are necessary to locate the image. The parallel ray ① starts from the tip of the flame, travels parallel to the optic axis, and then diverges from the mirror after reflection, appearing to come from $F$. The chief ray ② originates from the tip of the flame, appears to go through $C$, and then reflects straight back, but appears to come from $C$. It is clearly seen that these two rays, after reflection, diverge from each other, and there is no chance for them to intersect. However, they appear to start from a common point behind the mirror: the image point of the tip of the flame. The focal ray ③ can also be drawn to verify that all three rays appear to emanate from the same image point.

The image is virtual (the reflected rays don't actually come from the image point), upright, and reduced (the image is smaller than the object). Therefore, the answer is (4) virtual, upright, reduced. Measuring from the diagram (keep in mind the drawing scale used), $d_i = -9.0$ cm, and

$$M=\frac{h_i}{h_o}\approx\frac{0.5\text{ cm}}{1.2\text{ cm}}=+0.4.$$

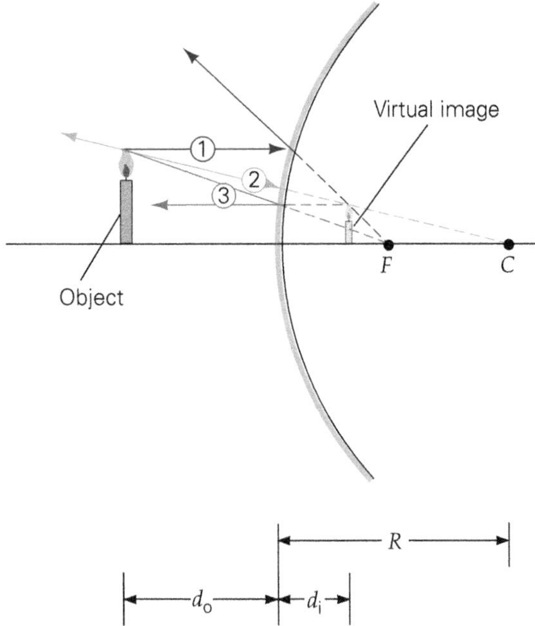


▲ **FIGURE 23.11** **Diverging mirror** Ray diagram of a diverging mirror. See Integrated Example 23.4 for a description.

**(b) QUANTITATIVE REASONING AND SOLUTION.** The image position and characteristics can be calculated with the mirror equations by using the given object distance and focal length.

***Given:***

$d_o = 20$ cm
$f = -15$ cm

***Find:*** $d_i$, $M$, and image characteristics

Note that the focal length is negative for a convex mirror. (See Table 23.2.) Using Equation 23.3,

$$\frac{1}{20\ \text{cm}} + \frac{1}{d_i} = \frac{1}{-15\ \text{cm}}$$

or

$$\frac{1}{d_i} = \frac{1}{-15\ \text{cm}} - \frac{1}{20\ \text{cm}} = -\frac{7}{60\ \text{cm}}$$

so

$$d_i = -\frac{60\,\text{cm}}{7} = -8.6\,\text{cm}$$

Then

$$M = -\frac{d_i}{d_o} = -\frac{-8.6\,\text{cm}}{20\,\text{cm}} = +0.43$$

Thus, the image is virtual ($d_i$ is negative), upright ($M$ is positive), and reduced ($|M| = 0.43$). These results agree well with those from the ray diagram. The image of an object is *always* virtual for a diverging (convex) mirror for a real object. (Can you prove this using either a ray diagram or the mirror equation?)

**FOLLOW-UP EXERCISE.** As has been pointed out, a diverging mirror always forms a virtual image of a real object. What about the other characteristics of the image – its orientation and magnification? Can any general statements be made about them?

### 23.2.3 Spherical Mirror Aberrations

Technically, our descriptions of image characteristics for spherical mirrors are true only for objects near the optic axis – that is, only for small angles of incidence and reflection. If these conditions do not hold, the images will be blurred (out of focus) or distorted, because not all of the parallel rays will converge in the same plane. As illustrated in ▼ **Figure 23.12**, incident parallel rays far from the optic axis do not converge at the focal point. The farther the incident ray from the axis, the farther the reflected ray from the focal point. This effect is called **spherical aberration**.

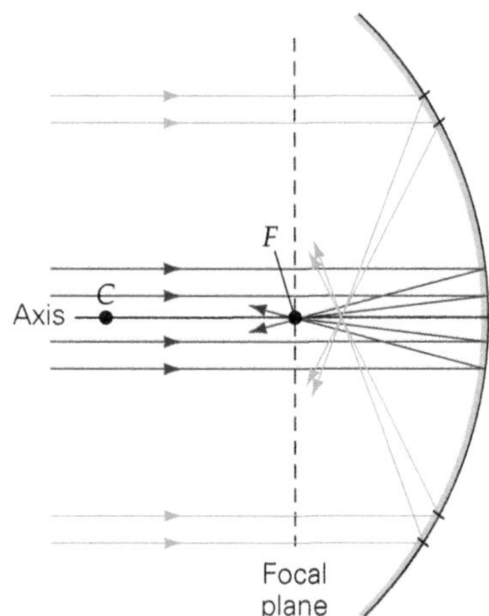


▲ **FIGURE 23.12** **Spherical aberration for a mirror** According to the small-angle approximation, rays parallel to and near the mirror's axis converge at the focal point. However, when parallel rays not near the axis are reflected, they converge in front of the focal point. This effect, called *spherical aberration*, gives rise to blurred images.

Spherical aberration does not occur with a parabolic mirror. (As the name parabolic mirror implies, a *parabolic mirror* has the form of a paraboloid.) *All* of the incident rays parallel to the optic axis of such a mirror have a common focal point. For this reason, parabolic mirrors are used in most astronomical telescopes (Chapter 25). However, these mirrors are more difficult to make than spherical mirrors and are therefore more expensive.

## 23.3 Lenses

The word lens is from the Latin *lentil*, which is a round, flattened, edible seed of a pea-like plant. Its shape is similar to that of a lens. An optical **lens** is made from transparent material such as glass or plastic. One or both surfaces usually have a spherical contour. *Biconvex spherical lenses* (with both surfaces convex) and *biconcave spherical lenses* (with both surfaces concave) are illustrated in ▼ **Figure 23.13**. Lenses can form images by refracting the light that passes through them.

A biconvex lens is an example of a **converging lens**. Incident light rays parallel to the axis of the lens converge at a focal point ($F$) on the opposite side of the lens (▼ **Figure 23.14a**). This fact provides a way to experimentally determine the focal length of a converging lens. You may have focused the Sun's rays with a magnifying glass (a biconvex, or converging, lens) and thereby witnessed the concentration of radiation energy that results (Figure 23.14b).

A biconcave lens is an example of a **diverging lens**. Incident parallel rays emerge from the lens as though they emanated from a focal point on the incident side of the lens (▶ **Figure 23.15**).

There are several types of converging and diverging lenses (▶ **Figure 23.16**). Convex and concave meniscus lenses are the type most commonly used for corrective eyeglasses. In general, a converging lens is thicker at its center than at its periphery, and a diverging lens is thinner at its center than at its periphery. Most of our discussions will be limited to spherically symmetric biconvex and biconcave lenses, for which both surfaces have the same radius of curvature.

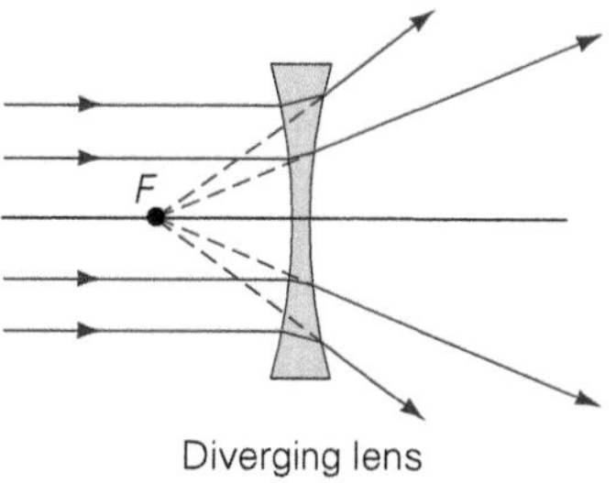


▲ **FIGURE 23.15 Diverging lens** Rays parallel to the axis of a diverging (concave) lens appear to diverge from a focal point on the incident side of the lens.

When light passes through a lens, it is refracted and displaced laterally (see Example 22.4 and Figure 22.13). If a lens is thick, this displacement may be fairly large and can complicate analysis of the lens's characteristics. This problem does not arise as much with thin lenses, for which the refractive displacement of transmitted light is negligible. Our discussion will be limited to thin lenses. A thin lens is a lens whose thickness is assumed to be negligible compared with the lens's focal length.

A lens with spherical geometry has, *for each lens surface*, a center of curvature ($C$), a radius of curvature ($R$), a focal point ($F$), and a focal length ($f$). The focal points are at equal distances on

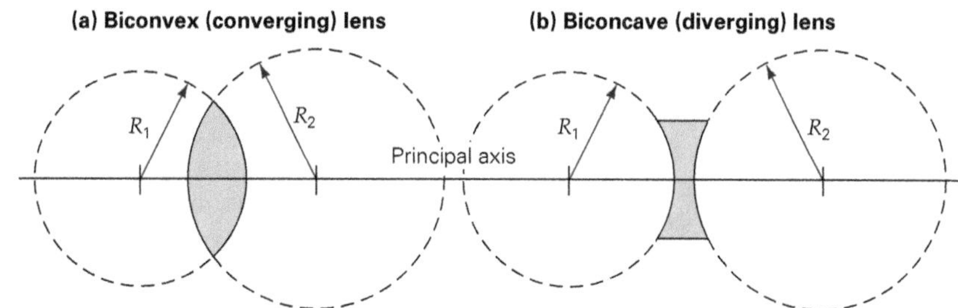


▲ **FIGURE 23.13 Spherical lenses** Spherical lenses have surfaces defined by two spheres, and the surfaces are either convex or concave. **(a)** Biconvex and **(b)** biconcave lenses are shown here. If $R_1 = R_2$, a lens is spherically symmetric.

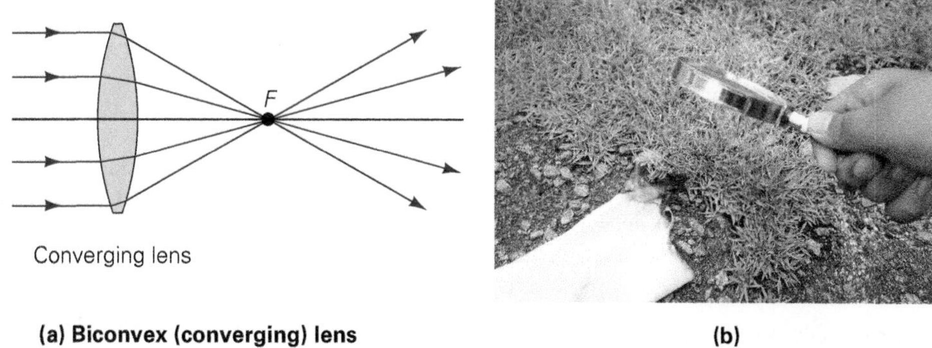


▲ **FIGURE 23.14 Converging lens (a)** For a thin converging (convex) lens, rays parallel to the axis converge at the focal point $F$. **(b)** A magnifying glass (converging lens) can be used to focus the Sun's rays to a spot – with incendiary results. Do not try this at home!

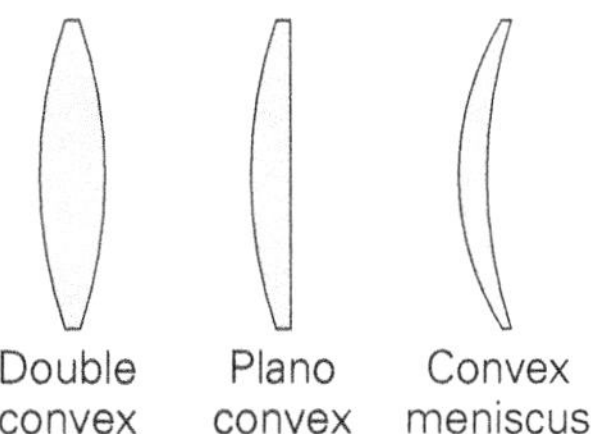


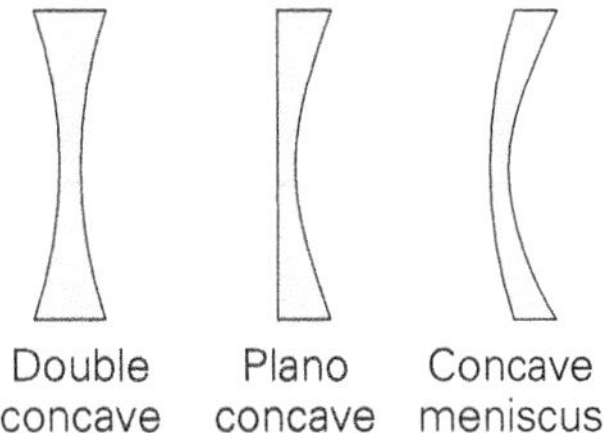


▲ FIGURE 23.16 **Lens shapes** Lens shapes vary widely and are normally categorized as converging or diverging. In general, a converging lens is thicker at its center than at the periphery, and a diverging lens is thinner at its center than at the periphery.

either side of a thin lens. However, for a spherical lens, the focal length is *not* simply related to $R$ by $f = R/2$, as it is for spherical mirrors. Because the focal length also depends on the lens's index of refraction, the focal length of a lens is usually specified, rather than its radius of curvature. This will be discussed in Section 23.4.

The general rules for drawing ray diagrams for lenses are similar to those for spherical mirrors. But some modifications are necessary, since light passes through a lens. Opposite sides of a lens are generally distinguished as the object side and the image side. The *object side* is the side on which an object is positioned, and the *image side* is the *opposite* side of the lens (where a real image would be formed). The three key rays from a point on an object are drawn as follows (see ▶ **Figure 23.17** and Example 23.5):

① A **parallel ray** is a ray that is parallel to the lens's optic axis on incidence and, after refraction, passes through the focal point on the image side of a converging lens (or *appears* to diverge from the focal point on the object side of a diverging lens).

② A **chief ray**, or **central ray**, is a ray that passes through the center of the lens and is undeviated because the lens is "thin."

③ A **focal ray** is a ray that passes through the focal point on the object side of a converging lens (or *appears* to pass through the focal point on the image side of a diverging lens) and, after refraction, is parallel to the lens's optic axis.

As with spherical mirrors, only two rays are needed to determine the image. (However, it is also generally a good idea to include the third ray as a check.)

### EXAMPLE 23.5: A LENS DIAGRAM

An object is placed 30 cm in front of a thin converging lens of focal length 20 cm. (a) Use a ray diagram to locate the image. (b) Discuss the characteristics of the image.

**THINKING IT THROUGH.** Follow the steps for lens ray diagrams, as given previously.

**SOLUTION**

***Given:***

$d_o = 30$ cm
$f = 20$ cm

***Find:***

**(a)** $d_i$ (location of the image, using a ray diagram)
**(b)** the image's characteristics

**(a)** Since a ray diagram is to be used to locate the image (see Figure 23.17), the first thing to decide on is a scale for the drawing. In this example, a scale of 1 cm to represent 10 cm is used. That way, the object would be 3.0 cm in front of the mirror in the drawing.

First the optic axis, the lens, the object (a lighted candle), and the focal points ($F$) are drawn. A vertical dashed line through the center of the lens is drawn because, for simplicity, the refraction is depicted as if it occurs at the center of the lens. In reality, it would occur at the air-glass and glass–air surfaces of the lens.

Follow steps 1–4 in Figure 23.17, *A lens ray diagram.*

1. The first ray drawn is the parallel ray (① in the drawing). From the tip of the flame, draw a horizontal ray (parallel to the optic axis). After passing through the lens, this ray goes through the focal point $F$ on the image side.
2. Then draw the chief ray (② in the drawing). From the tip of the flame, draw a ray passing through the center of the lens. This ray will go undeviated through the thin lens to the image side.
3. It can be clearly seen that these two rays intersect on the image side. The point of intersection is the image point of the tip of the candle. From this point, draw the image by extending the tip of the flame to the optic axis.
4. Only two rays are needed to locate the image. However, if the third ray, in this case the focal ray (③ in the drawing), is drawn, it must go through the same point on the image at which the other two rays intersect (if drawn carefully). The ray from the tip of the flame passing through the focal point $F$ on the object side will travel parallel to the optic axis on the image side.

**(b)** From the ray diagram in part (a), the image is real (because the rays intersect). As a result, this real image could be seen on a screen (e.g., a piece of white paper) that is positioned at the image point. The image is also inverted (the image of the candle points downward) and is magnified (the image is larger than the object).

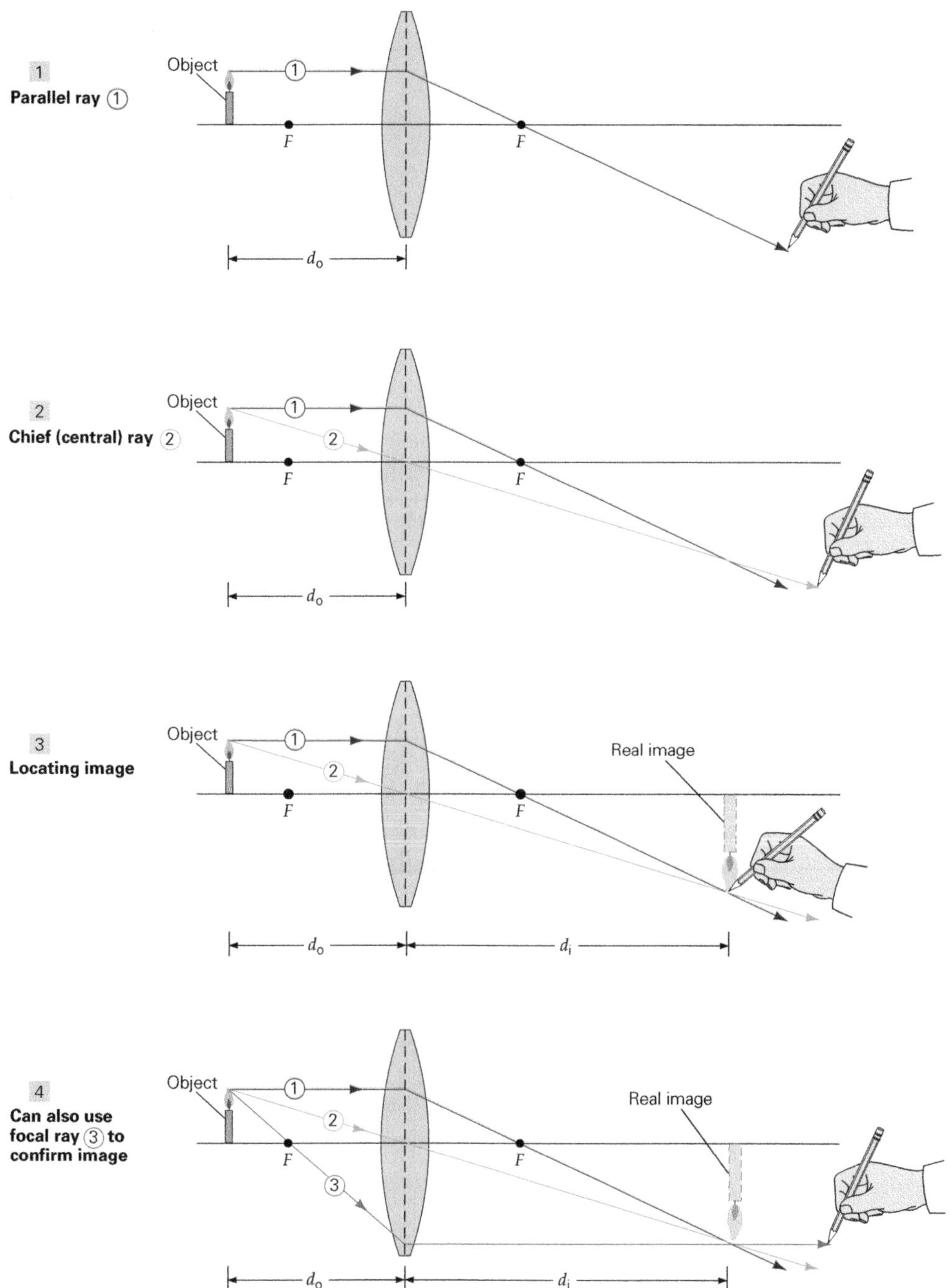


▲ FIGURE 23.17 **A lens ray diagram** Rays can be used to determine image characteristics. See Example 23.5 for a description.

In this case, $d_o = 30$ cm and $f = 20$ cm, so $2f > d_o > f$. Using similar ray diagrams, it can be proven that for any $d_o$ in this range, the image is always real, inverted, and magnified. Actually, the overhead projector or document camera in your classroom uses this type of arrangement.

**FOLLOW-UP EXERCISE.** In this Example, what does the image look like if the object is 10 cm in front of the lens? Locate the image graphically and discuss the characteristics of the image.

▶ **Figure 23.18** shows other ray diagrams with different object distances for a converging lens. The image of an object is real when it is formed on the side of the lens opposite the object's side (see Figure 23.18a). A virtual image is said to be formed on the same side of the lens as the object (see Figure 23.18b).

Regions could be similarly divided for the object distance for a converging lens as was done for a converging mirror in Figure 23.9a. Here, an object distance of $d_o = 2f$ for a converging lens has significance similar to that of $d_o = R = 2f$ for a converging mirror (▶ **Figure 23.19**).

The ray diagram for a diverging lens will be discussed shortly. Like diverging mirrors, diverging lenses can form only virtual images of real objects.

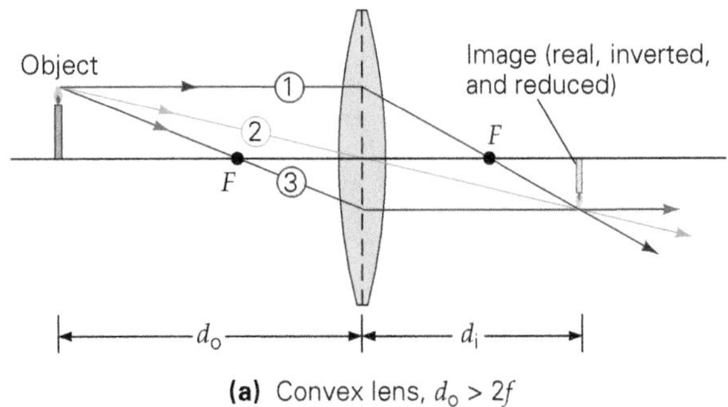

**(a)** Convex lens, $d_o > 2f$

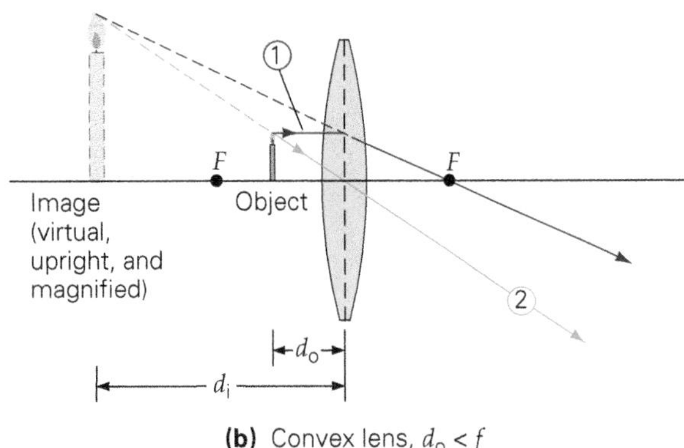

**(b)** Convex lens, $d_o < f$

▲ **FIGURE 23.18 Ray diagrams for lenses (a)** A converging biconvex lens forms a real image when $d_o > 2f$. The image is real, inverted, and reduced. **(b)** Ray diagram for a converging lens with $2f > d_o > f$. The image is virtual, upright, and magnified.

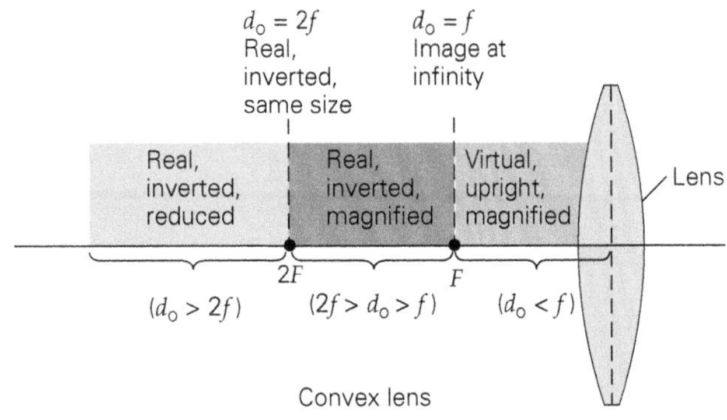


▲ **FIGURE 23.19 Converging lens** For a converging lens, the object is located within one of three regions defined by the focal length ($f$) and twice the focal length ($2f$) or at one of these two points. For $d_o > 2f$, the image is real, inverted, and reduced (Figure 23.18a). For $2f > d_o > f$, the image will also be real and inverted, but magnified (Figure 23.17 in Example 23.5). For $d_o < f$, the image will be virtual, upright, and magnified (Figure 23.18b).

The image distances and characteristics for a thin lens can also be found analytically. The equations for thin lenses are identical to those for spherical mirrors. The **thin lens equation** is

$$\frac{1}{d_o} + \frac{1}{d_i} = \frac{1}{f} \quad \text{(thin lens equation)} \tag{23.5}$$

As in the case for spherical mirrors, an alternative form of the thin lens equation is

$$d_i = \frac{d_o f}{d_o - f} \tag{23.5a}$$

which gives a quick and easy way to find $d_i$.

The **lateral magnification factor**, like that for spherical mirrors, is given by

$$M = \frac{h_i}{h_o} = -\frac{d_i}{d_o} \quad \text{(magnification factor)} \tag{23.6}$$

The sign conventions for these thin lens equations are given in ▼ **Table 23.3.**

**TABLE 23.3** Sign Conventions for Thin Lenses

| | |
|---|---|
| *Focal length* ($f$) | |
| Converging (convex) lens | $f$ is positive |
| Diverging (concave) lens | $f$ is negative |
| *Object distance* ($d_o$) | |
| Object is in front of the lens (real object) | $d_o$ is positive |
| Object is behind the lens (virtual object)[a] | $d_o$ is negative |
| *Image distance* ($d_i$) *and image type* | |
| Image is formed on the image side of the lens – opposite to the object (real image) | $d_i$ is positive |
| Image is formed on the object side of the lens – same side as the object (virtual image) | $d_i$ is negative |
| *Image orientation* ($M$) | |
| Image is upright with respect to the object | $M$ is positive |
| Image is inverted with respect to the object | $M$ is negative |

[a] In a combination of two (or more) lenses, the images formed by the first lens is taken as the object of the second lens (and so on). If this image–object falls behind the second lens, it is referred to as a virtual object distance is taken to be negative (–).

Just as when you are working with mirrors, it is helpful to sketch a ray diagram before working a lens problem analytically.

### EXAMPLE 23.6: THREE IMAGES – BEHAVIOR OF A CONVERGING LENS

A converging lens has a focal length of 12 cm. For an object (a) 60 cm, (b) 15 cm, and (c) 8.0 cm from the lens, where is the image formed, and what are its characteristics?

**THINKING IT THROUGH.** With the focal length ($f$) and the object distances ($d_o$) given, Equation 23.5 can be applied to find the image distances ($d_i$) and Equation 23.6 can be used to determine the image characteristics. Sketch ray diagrams first to get an idea of the image characteristics. The diagrams should be in good agreement with the calculations.

**SOLUTION**

***Given:***

$f = 12$ cm
**(a)** $d_o = 60$ cm
**(b)** $d_o = 15$ cm
**(c)** $d_o = 8.0$ cm

***Find:*** $d_i$, $M$, and the image characteristics for all three cases

**(a)** The object distance is greater than twice the focal length ($d_o > 2f$). Using Equation 23.5,

$$\frac{1}{d_o} + \frac{1}{d_i} = \frac{1}{f}$$

or

$$\frac{1}{d_i} = \frac{1}{f} - \frac{1}{d_o} = \frac{1}{12\,\text{cm}} - \frac{1}{60\,\text{cm}}$$

$$= \frac{5}{60\,\text{cm}} - \frac{1}{60\,\text{cm}} = \frac{4}{60\,\text{cm}} = \frac{1}{15\,\text{cm}}$$

Then

$$d_i = 15\,\text{cm} \quad \text{and} \quad M = -\frac{d_i}{d_o} = -\frac{15\,\text{cm}}{60\,\text{cm}} = -0.25$$

The image is real (positive $d_i$), inverted (negative $M$), and reduced ($|M| = 0.25$). A camera uses a similar arrangement when the object distance is usually much greater than $2f$ ($d_o \gg 2f$). The ray diagram of this situation is similar to Figure 23.18a.

**(b)** Here, $2f > d_o > f$. Using Equation 23.5,

$$\frac{1}{d_i} = \frac{1}{12\,\text{cm}} - \frac{1}{15\,\text{cm}} = \frac{5}{60\,\text{cm}} - \frac{4}{60\,\text{cm}} = \frac{1}{60\,\text{cm}}$$

Then

$$d_i = 60\,\text{cm} \quad \text{and} \quad M = -\frac{d_i}{d_o} = -\frac{60\text{ cm}}{15\text{ cm}} = -4.0$$

The image is real (positive $d_i$), inverted (negative $M$), and magnified ($|M| = 4.0$). A similar situation applies to overhead projectors and slide projectors ($2f > d_o > f$). The ray diagram of this situation is similar to Figure 23.17.

**(c)** For this case, $d_o < f$. Using the alternative form (Equation 23.5a),

$$d_i = \frac{d_o f}{d_o - f} = \frac{(8.0\,\text{cm})(12\,\text{cm})}{8.0\,\text{cm} - 12\,\text{cm}} = -24\,\text{cm}$$

Then

$$M = -\frac{d_i}{d_o} = -\frac{(-24\,\text{cm})}{8.0\,\text{cm}} = +3.0$$

The image is virtual (negative $d_i$), upright (positive $M$), and magnified ($|M| = 3.0$). This situation is an example of a simple microscope or magnifying glass ($d_o < f$). The ray diagram of this situation is similar to Figure 23.18b.

As you can see, a converging lens is very versatile. Depending on the object distance (relative to the focal length), the lens can be used as a camera, projector, or magnifying glass.

**FOLLOW-UP EXERCISE.** If the object distance of a converging lens is allowed to vary, at what object distance does the real image change from being reduced to being magnified?

## CONCEPTUAL EXAMPLE 23.7: HALF AN IMAGE?

A converging lens forms an image on a screen, as shown in ▼**Figure 23.20a**. Then the lower half of the lens is blocked, as shown in Figure 23.20b. As a result, (a) only the top half of the original image will be visible on the screen; (b) only the bottom half of the original image will be visible on the screen; (c) the entire image will still be visible.

**REASONING AND ANSWER.** At first thought, you might imagine that blocking off half of the lens would eliminate half of the image. However, rays from *every* point on the object pass through *all parts* of the lens. Thus, the upper half of the lens can form a total image (as could the lower half), so the answer is (c).

You might confirm this conclusion by drawing a chief ray in Figure 23.20b. Or you might use the scientific method and experiment – particularly if you wear eyeglasses. Block off the bottom part of your glasses, and you will find that you can still read through the top part (unless you wear bifocals).

**FOLLOW-UP EXERCISE.** Can you think of any property of the image that *would be* affected by blocking off half of the lens? Explain.

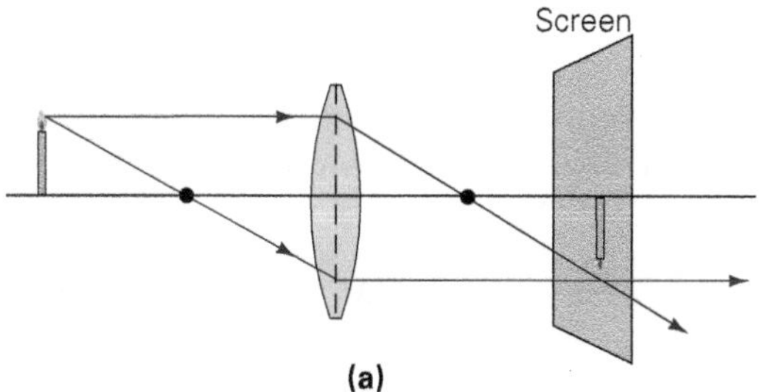


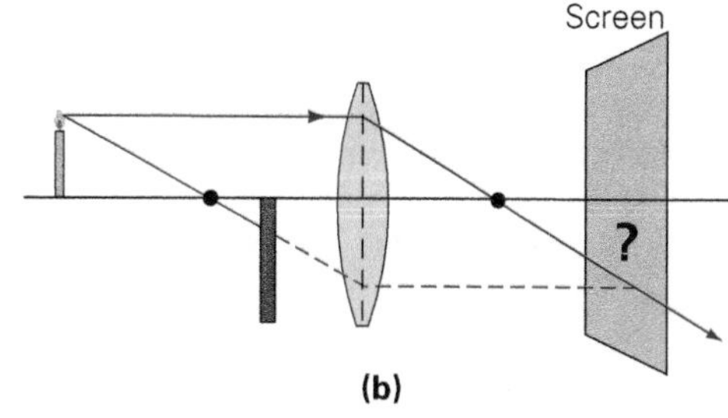


▲ FIGURE 23.20 **Half a lens, half an image? (a)** A converging lens forms an image on a screen. **(b)** The lower half of the lens is blocked. What happens to the image?

**INTEGRATED EXAMPLE 23.8: TIME FOR A CHANGE – BEHAVIOR OF A DIVERGING LENS**

An object is 24 cm in front of a diverging lens that has a focal length of −15 cm. (a) Use a ray diagram to determine whether the image is (1) real and magnified, (2) virtual and reduced, (3) real and upright, or (4) upright and magnified. (b) Find the location and characteristics of the image with the thin lens equations.

**(a) CONCEPTUAL REASONING.** (See the sign conventions in Table 23.3.) Use a scale of 1 cm (in the drawing of ▼**Figure 23.21**) to represent 10 cm. The object will be 2.4 cm in front of the lens in the drawing. Draw the optic axis, the lens, the object (in this case, a lighted candle), the focal point ($F$), and a vertical dashed line through the center of the lens.

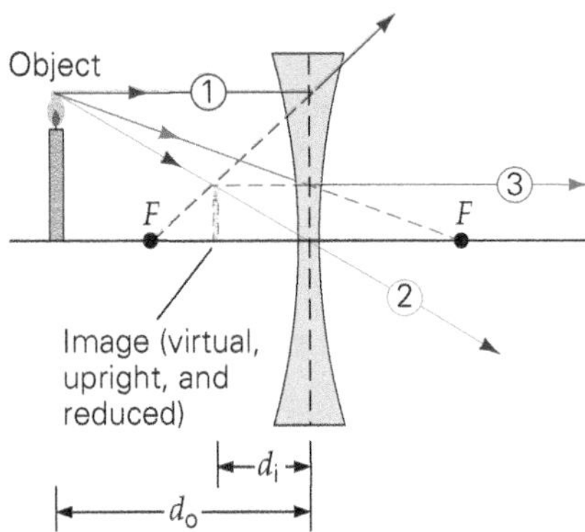


▲FIGURE 23.21 **Diverging lens** Ray diagram of a diverging lens. Here, the image is virtual and in front of the lens, upright, and reduced.

The parallel ray ① starts from the tip of the flame, travels parallel to the optic axis, diverges from the lens after refraction, and appears to diverge from $F$ on the object side. The chief ray ② originates from the tip of the flame and goes through the center of the lens, with no direction change. These two rays, after refraction, diverge and do not intersect. However, they appear to come from in front of the lens (object side), and that apparent intersection is the image point of the tip of the flame. The focal ray ③ can also be drawn to verify that these rays appear to come from the same image point. The focal ray appears to go through the focal point on the image side and travels parallel to the optic axis after refraction from the lens.

The image is virtual (why?), upright, and reduced, so the answer is (2) virtual and reduced. Measuring from the diagram (keeping in mind the drawing scale used), $d_i \approx -9$ cm (virtual image) and $M = (h_i/h_o) \approx (0.5\text{ cm}/1.4\text{ cm}) = +0.4$.

**(b) QUANTITATIVE REASONING AND SOLUTION.**

***Given:***

$d_o = 24$ cm
$f = -15$ cm (diverging lens)

***Find:*** $d_i$, $M$, and image characteristics

Note that the focal length is negative for a diverging lens. (See Table 23.3.) From Equation 23.5,

$$\frac{1}{24\,\text{cm}} + \frac{1}{d_i} = \frac{1}{-15\,\text{cm}}$$

or

$$\frac{1}{d_i} = \frac{1}{-15\,\text{cm}} - \frac{1}{24\,\text{cm}} = -\frac{13}{120\,\text{cm}}$$

so

$$d_i = -\frac{120\,\text{cm}}{13} = -9.2\,\text{cm}$$

Then

$$M = -\frac{d_i}{d_o} = -\frac{9.2\,\text{cm}}{24\,\text{cm}} = +0.38$$

Thus, the image is virtual ($d_i$ is negative) and upright ($M$ is positive), and reduced ($|M| = 0.38$). Due to the fact that $f$ is negative for a diverging lens, $d_i$ is always negative for any positive value of $d_o$, so the image of a real object is always virtual for real objects.

**FOLLOW-UP EXERCISE.** A diverging lens always forms a virtual image of a real object. What general statements can be made about the image's orientation and magnification?

## 23.3.1 Combinations of Lenses

Many optical instruments, such as microscopes and telescopes (Chapter 25), use a combination of lenses, or a compound lens system. When two or more lenses are used in combination, the overall image produced can be determined by considering the lenses individually in sequence. That is, the image formed by the first lens becomes the object for the second lens, and so on.

If the first lens produces an image in front of the second lens, that image is treated as a real object ($d_o$ is positive) for the second lens (▶**Figure 23.22a**). If, however, the lenses are close enough, the image from the first lens is not formed before the rays pass through the second lens (Figure 23.22b). In this case, the image from the first lens is *treated* as a virtual object for the second lens. The virtual object distance is taken to be *negative* in the lens equation (Table 23.3).

It can be shown that the total magnification ($M_{total}$) of a compound lens system is the product of the individual magnification factors of the component lenses. For example, for a two-lens system, as in Figure 23.19,

$$M_{total} = M_1 M_2 \tag{23.7}$$

The usual signs for $M_1$ and $M_2$ carry through to the product to indicate, from the sign of $M_{total}$, whether the final image is upright or inverted. (See Exercise 59.)

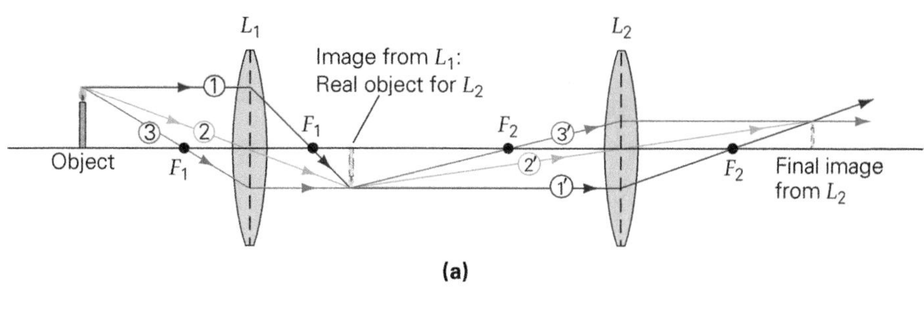


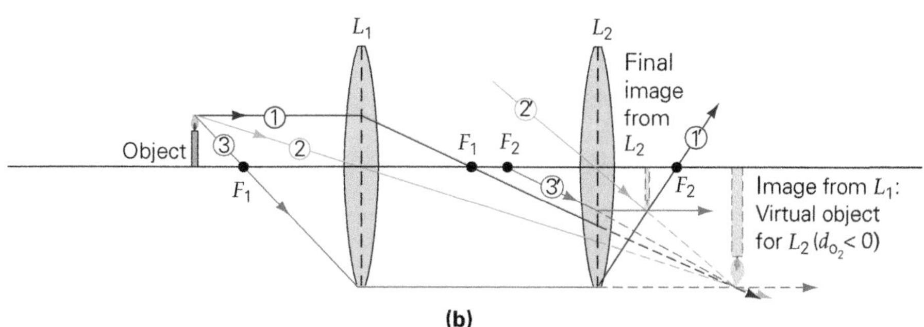


▲ **FIGURE 23.22 Lens combinations** The final image produced by a compound-lens system can be found by treating the image of one lens as the object for the adjacent lens. **(a)** If the image of the first lens ($L_1$) is formed in front of the second lens ($L_2$), the object for the second lens is said to be real. (Note that rays ①′, ②′, and ③′ are the parallel, chief, and focal rays, respectively, for $L_2$. They are *not* continuations of rays ①, ②, and ③ – the parallel, chief, and focal rays, respectively, for $L_1$.) **(b)** If the rays pass through the second lens before the image is formed, the object for the second lens is said to be virtual, and the object distance for the second lens is taken to be negative.

---

EXAMPLE 23.9: A SPECIAL OFFER – A LENS COMBO AND A VIRTUAL OBJECT

Consider two lenses similar to those illustrated in Figure 23.22b. Suppose the object is 20 cm in front of lens $L_1$, which has a focal length of 15 cm. Lens $L_2$, with a focal length of 12 cm, is 26 cm from $L_1$. What is the location of the final image, and what are its characteristics?

**THINKING IT THROUGH.** This is a double application of the thin lens equation. The lenses are treated successively: The image of lens $L_1$ becomes the object of lens $L_2$. The quantities must be distinctly labeled and the distances appropriately referenced (with signs!).

**SOLUTION**

Listing the known quantities and what is to be found:

***Given:***

$d_{o_1} = +20$ cm
$f_1 = +15$ cm
$f_2 = +12$ cm
$D = 26$ cm (distance between lenses)

***Find:*** $d_{i_2}$, $M_{\text{total}}$ and image characteristics

The first step is to apply the thin lens equation (Equation 23.5) and the magnification factor for thin leases (Equation 23.6) to $L_1$:

$$\frac{1}{d_{i_1}} = \frac{1}{f_1} - \frac{1}{d_{o_1}}$$

$$= \frac{1}{15\text{ cm}} - \frac{1}{20\text{ cm}} = \frac{4}{60\text{ cm}} - \frac{3}{60\text{ cm}} = \frac{1}{60\text{ cm}}$$

or

$$d_{i_1} = 60\text{ cm (real image from } L_1)$$

and

$$M_1 = -\frac{d_{i_1}}{d_{o_1}} = -\frac{60\text{ cm}}{20\text{ cm}} = -3.0\text{ (inverted and magnified)}$$

The image from lens $L_1$ becomes the object for lens $L_2$. This image is then 60 cm − 26 cm = 34 cm on the right, or image, side of $L_2$. Therefore, it is a virtual object (see Table 23.3), and. $d_{o_2} = -34$ cm. (Remember that $d_o$ for *virtual* objects is taken to be negative.)

Then applying the equations to the second lens, $L_2$:

$$\frac{1}{d_{i_2}} = \frac{1}{f_2} - \frac{1}{d_{o_2}} = \frac{1}{12\text{ cm}} - \frac{1}{(-34\text{ cm})} = \frac{23}{204\text{ cm}}$$

or

$$d_{i_2} = \frac{204\text{ cm}}{23} = 8.9\text{ cm (real image)}$$

and

$$M_2 = -\frac{d_{i_2}}{d_{o_2}} = -\frac{8.9\text{ cm}}{-34\text{ cm}} = 0.26\text{ (upright and reduced)}$$

(*Note:* The virtual object for $L_2$ was inverted, and thus the term upright means that the final image is also inverted.) The total magnification $M_{\text{total}}$ is then

$$M_{\text{total}} = M_1 M_2 = (-3.0)(0.26) = -0.78$$

The sign is carried through with the magnifications. We determine that the final real image is located at 8.9 cm on the right (image) side of $L_2$ and that it is inverted relative to the initial object ($M_{\text{total}} < 0$ and reduced ($|M_{\text{total}}| = 0.78$).

**FOLLOW-UP EXERCISE.** Suppose the object in Figure 23.22b were located 30 cm in front of $L_1$. Where would the final image be formed in this case, and what would be its characteristics?

## 23.4 The Lens Maker's Equation

There are a variety of other shapes of lenses, as illustrated in Figure 23.16. Lens refraction depends on the shapes of the lens's surfaces and on the index of refraction of the lens. These properties together determine the focal length of a thin lens. The thin lens focal length is given by the **lens maker's equation**, which enables us to calculate the focal length of a thin lens in air ($n_{\text{air}} = 1$) as

$$\frac{1}{f} = (n-1)\left(\frac{1}{R_1} + \frac{1}{R_2}\right) \quad \text{(for thin lens in air)} \qquad (23.8)$$

where $n$ is the index of refraction of the lens material and $R_1$ and $R_2$ are the radii of curvature of the first (front side) and second (back side) lens surfaces, respectively.

A sign convention is required for the lens maker's equation, and a common one is summarized in ▼**Table 23.4**. The signs depend only on the shape of the surface; that is, convex or concave (▶**Figure 23.23**). For the biconvex lens in Figure 23.23a, both $R_1$ and $R_2$ are positive (both surfaces are convex), and for the biconcave lens in Figure 23.23b, both $R_1$ and $R_2$ are negative (both surfaces are concave).

TABLE 23.4 Sign Conventions for the Lens Maker's Equation

| | |
|---|---|
| Convex surface | $R$ is positive |
| Concave surface | $R$ is negative |
| Plane (flat) surface | $R = \infty$ |
| Converging (convex) lens | $f$ is positive |
| Diverging (concave) lens | $f$ is negative |

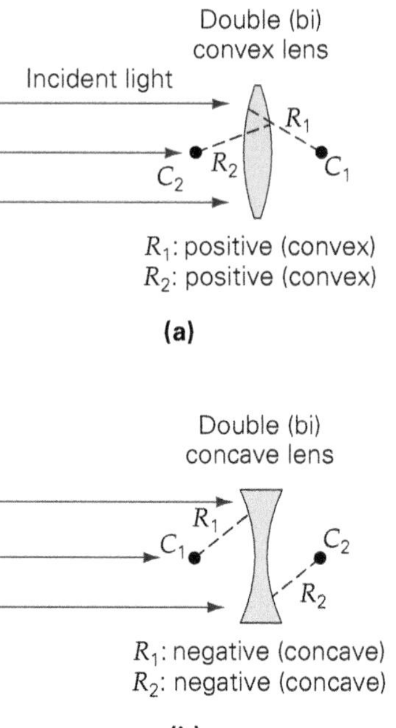


▲ FIGURE 23.23 **Centers of curvature** Lenses, such as **(a)** a biconvex lens and **(b)** a biconcave lens, have two centers of curvature, which define the signs of the radii of curvature. See Table 23.4.

If the lens is surrounded by a medium other than air, then the first term in parentheses in Equation 23.8 becomes $(n/n_{\text{m}}) - 1$, where $n$ and $n_{\text{m}}$ are the indices of refraction of the lens material and the surrounding medium, respectively. Now we can see why some converging lenses in air become diverging when submerged in water: If $n_{\text{m}} > n$, then $f$ is negative, and the lens is diverging.

### 23.4.1 Lens Power: Diopters

Notice that the lens maker's equation (Equation 23.8) gives the inverse focal length $1/f$. Optometrists use this inverse relationship to express the *lens power* ($P$) of a lens in units called **diopters** (abbreviated as D). The lens power is the reciprocal or inverse of the focal length of the lens expressed in *meters:*

$$P\text{ (expressed in diopters)} = \frac{1}{f\text{ (expressed in meters)}} \qquad (23.9)$$

So, $1\text{ D} = 1\text{ m}^{-1}$. The lens maker's equation gives a lens's power ($1/f$) in diopters if the radii of curvature are expressed in meters.

If you wear glasses, you may have noticed that the prescription the optometrist gave you for your eyeglass lenses was written in terms of diopters. Converging and diverging lenses are referred to as positive (+) and negative (−) lenses, respectively. Thus, if an optometrist prescribes a corrective lens with a power of +2.0 diopters, it is a converging lens with a focal length of

$$f = \frac{1}{P} = \frac{1}{+2.0\text{ D}} = \frac{1}{+2.0\text{ m}^{-1}} = +0.50\text{ m} = +50\text{ cm}$$

The greater the power of the lens in diopters, the shorter its focal length and the more strongly converging or diverging it is. Thus, a "stronger" prescription lens (greater lens power) has a shorter $f$ than does a "weaker" prescription lens (lesser lens power).

---

**INTEGRATED EXAMPLE 23.10: A CONVEX MENISCUS LENS - CONVERGING OR DIVERGING**

The convex meniscus lens shown in Figure 23.16 has a 15-cm radius for the convex surface and a 20-cm radius for the concave surface. The lens is made of crown glass and is surrounded by air. (a) Is this lens a (1) converging or (2) diverging lens? Explain. (b) What are the focal length and the power of the lens?

**(a) CONCEPTUAL REASONING.** The index of refraction of crown glass can be obtained from Table 22.1: $n=1.52$. For a convex meniscus lens, the first surface is convex, so $R_1$ is positive; the second surface is concave, so $R_2$ is negative. Since $R_1=15$ cm $<|R_2|=20$ cm, $1/R_1+1/R_2$ will be positive. Therefore, the lens is a converging (positive) lens, according to Equation 23.8. Thus the answer is (1) converging.

**(b) QUANTITATIVE REASONING AND SOLUTION.**

***Given:***

$R_1=15\text{ cm}=0.15\text{ m}$
$R_2=-20\text{ cm}=-0.20\text{ m}$
$n=1.52$ (from Table 22.1 for crown glass)

***Find:*** $f$ (focal length) and $P$ (lens power)

**(b)** From Equation 23.8

$$\frac{1}{f}=(n-1)\left(\frac{1}{R_1}+\frac{1}{R_2}\right)$$
$$=(1.52-1)\left(\frac{1}{0.15\text{m}}+\frac{1}{-0.20\text{m}}\right)=0.867\text{m}^{-1}$$

Hence,

$$f=\frac{1}{0.867\text{ m}^{-1}}=+1.15\text{ m}$$

and the power of lens therefore is

$$P=\frac{1}{f}=+0.867\text{ D}.$$

**FOLLOW-UP EXERCISE.** In this Example, if this lens were immersed in water, what would your answers be?

---

## 23.5 Lens Aberrations (Optional)

### 23.5.1 Spherical Aberration

The discussion of lenses thus far has concentrated on rays that are near the optic axis. Like spherical mirrors, however, lenses may show **spherical aberration**, an effect that occurs when parallel rays passing through different regions of a lens do not come together on a common focal plane. In general, rays close to the axis of a converging lens are refracted less and come together at a point farther from the lens than do rays passing through the periphery of the lens (▼ **Figure 23.24a**). Spherical aberration can be minimized by using an aperture to reduce the effective area of the lens, so that only light rays near the axis are transmitted. Also, combinations of converging and diverging lenses can be used, as the aberration of one lens can be compensated for by the optical properties of another lens.

### 23.5.2 Chromatic Aberration

**Chromatic aberration** is an effect that occurs because the index of refraction of the lens material is not the same for all wavelengths of light (i.e., the material is dispersive). When white light

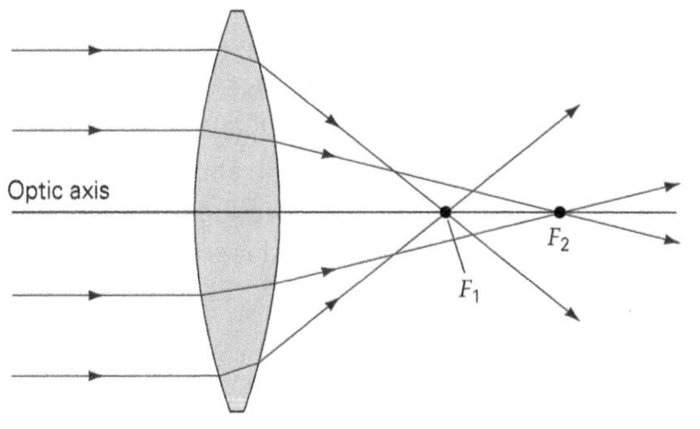


**(a) Spherical aberration**

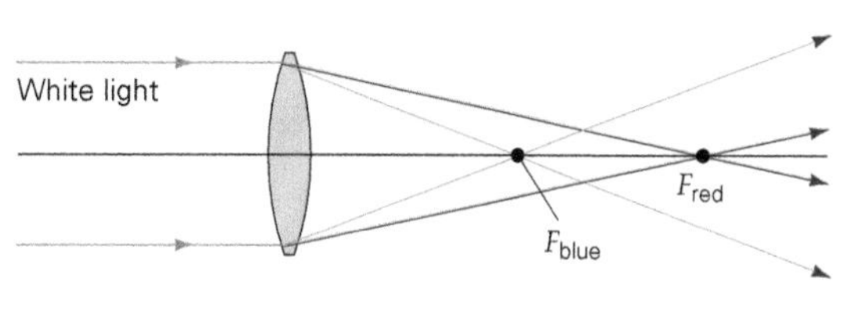


**(b) Chromatic aberration**

▲ FIGURE 23.24 **Lens aberrations (a)** Spherical aberration. In general, rays closer to the axis of a lens are refracted less and come together at a point farther from the lens than do rays passing through the periphery of the lens. **(b)** Chromatic aberration. Because of dispersion, different wavelengths (colors) of light are focused in different planes, which results in distortion of the overall image.

is incident on a lens, the refracted rays of different wavelengths (colors) do not have a common focal point, and images of different colors are produced at different locations (Figure 23.24b).

This dispersive aberration can be minimized, but not eliminated, by using a compound lens system consisting of lenses of different materials, such as crown glass and flint glass. The lenses are chosen so that the dispersion produced by one is approximately compensated for by the opposite dispersion produced by the other. With a properly constructed two-component lens system, called achromatic doublet (achromatic means "without color"), the images of any two selected colors can be made to coincide.

### 23.5.3 Astigmatism

A circular beam of light along the lens axis forms a circular illuminated area on the lens. When incident on a converging lens, the parallel beam converges at the focal point. However, when a circular beam of light from an off-axis source falls on the lens, the light forms an elliptical illuminated area on the lens. The rays entering along the major and minor axes of the ellipse then focus at different points after passing through the lens. This condition is called astigmatism.

With different focal points in different planes, the images in both planes are blurred. For example, the image of a point is no longer a point, but rather two separated short-line images (blurred points). Astigmatism can be reduced by decreasing the effective area of the lens with an aperture or by adding a cylindrical lens to compensate.

## Chapter 23 Review

- **Plane mirrors** form virtual, upright, and unmagnified images. The object distance is equal to the absolute value of the image distance ($d_o = |d_i|$).
- The **lateral magnification factor** for all mirrors and lenses is

$$M = \frac{h_i}{h_o} = -\frac{d_i}{d_o} \tag{23.4, 23.6}$$

- **Spherical mirrors** are either converging (concave) or diverging (convex). Diverging spherical mirrors always form virtual, upright, and reduced images.

  ***Focal length of a spherical mirror:***

$$f = \frac{R}{2} \tag{23.2}$$

  ***Spherical mirror equation:***

$$\frac{1}{d_o} + \frac{1}{d_i} = \frac{1}{f} = \frac{2}{R} \tag{23.3}$$

  ***Alternative form:***

$$d_i = \frac{d_o f}{d_o - f} \tag{23.3a}$$

- Lenses are either convex (converging) or concave (diverging). Diverging lenses always form virtual, upright, and reduced images of real objects.
- The **thin lens equation** relates focal length, object distance, and image distance:

$$\frac{1}{d_o} + \frac{1}{d_i} = \frac{1}{f} \tag{23.5}$$

  ***Alternative form:***

$$d_i = \frac{d_o f}{d_o - f} \tag{23.5a}$$

- The **lens maker's equation** determines the focal length of a lens based on the radii and index of refraction of the lens.

$$\frac{1}{f} = (n-1)\left(\frac{1}{R_1} + \frac{1}{R_2}\right) \quad \text{(thin lens in air only)} \tag{23.8}$$

- **Lens power in diopters** (where $f$ is in meters) is given by

$$P = \frac{1}{f} \tag{23.9}$$

## End of Chapter Questions and Exercises

### Multiple Choice Questions

**23.1 Plane Mirrors**

1. A plane mirror has (a) a greater image distance than object distance, (b) a greater object distance than image distance, (c) the same object and image distance.
2. The image of a person formed by a plane mirror is (a) real, upright, and unmagnified, (b) virtual, upright, and magnified, (c) real, upright, and magnified, (d) virtual, upright, and unmagnified.
3. A plane mirror (a) produces both real and virtual images, (b) always produces a virtual image, (c) always produces a real image, (d) forms images by diffuse reflection.
4. The lateral magnification of a plane mirror is (a) greater than 1, (b) less than 1, (c) equal to +1, (d) equal to −1.

**23.2 Spherical Mirrors**

5. A concave spherical mirror can form (a) both real and virtual images, (b) only virtual images, (c) only real images.

6. Which of the following statements concerning spherical mirrors is correct? (a) A converging mirror can produce an inverted virtual image. (b) A diverging mirror can produce an inverted virtual image. (c) A diverging mirror can produce an inverted real image. (d) A converging mirror can produce an inverted real image.
7. The image produced by a convex mirror is always (a) virtual and upright, (b) real and upright, (c) virtual and inverted, (d) real and inverted.
8. A mirror is used to form an image that is larger than the object, so it must be a (a) concave mirror, (b) convex mirror, (c) plane mirror.

**23.3 Lenses**

9. If an object is placed at the focal point of a converging lens, the image is (a) at zero, (b) also at the focal point, (c) at a distance equal to twice the focal length, (d) at infinity.
10. A converging thin lens can form (a) both magnified and reduced images, (b) only magnified images, (c) only reduced images.
11. The image produced by a diverging lens is always (a) virtual and magnified, (b) real and magnified, (c) virtual and reduced, (d) real and reduced.
12. A converging lens (a) must have at least one convex surface, (b) cannot produce a virtual and reduced image, (c) is thicker at its center than at the periphery, (d) all of the preceding.

**23.4 The Lens Maker's Equation and 23.5 Lens Aberrations (Optional)**

13. The focal length of a rectangular glass block is (a) zero, (b) infinity, (c) not defined.
14. The focal length of a thin lens depends on (a) the radii of both surfaces, (b) the index of refraction of the lens material, (c) the index of refraction of the surrounding material, (d) all of the preceding.
15. The power of a lens is expressed in units of (a) watts, (b) joules, (c) diopters, (d) meters.
16. If the focal length of a lens increases, the lens power will (a) also increase, (b) decrease, (c) remain the same.
17. A lens aberration that is caused by dispersion is called (a) spherical aberration, (b) chromatic aberration, (c) astigmatism, (d) none of the preceding.

## Conceptual Questions

**23.1 Plane Mirrors**

1. Can a virtual image be projected onto a screen? Why or why not?
2. What is the focal length of a plane mirror? Explain.
3. Day–night rearview mirrors are common in cars. At night, when you tilt the mirror backward, the intensity and glare of headlights behind you are reduced (▶ **Figure 23.25**). The mirror is wedge-shaped and is silvered on the back. The unsilvered front surface reflects about 5% of incident light; the silvered back surface reflects about 90% of the incident light. Explain how the day–night mirror works.

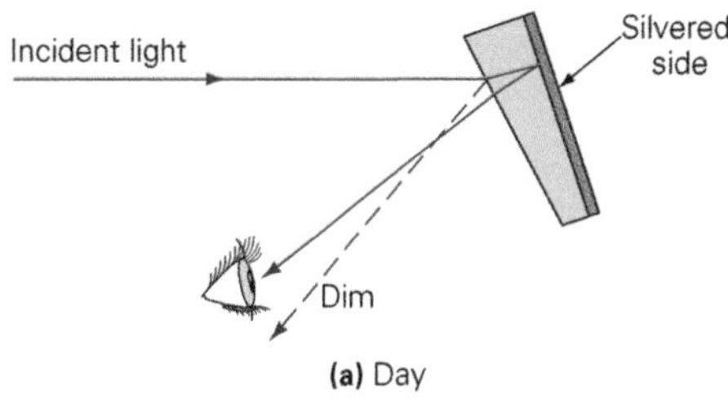


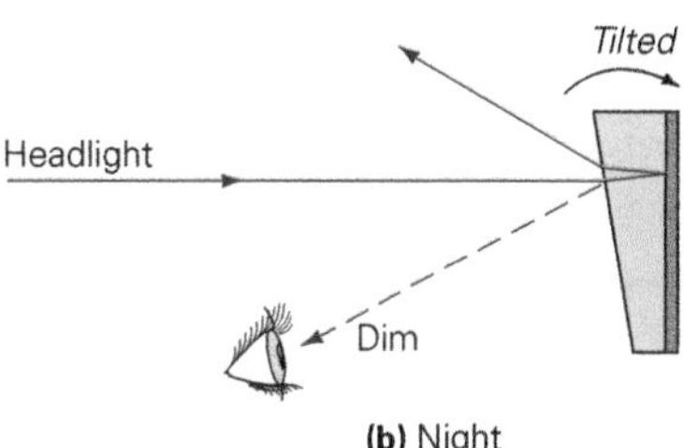


▲ FIGURE 23.25 **Automobile day–night mirror** See Conceptual Question 3.

4. When you stand in front of a plane mirror, there is a right–left reversal. (a) Why is there not a top–bottom reversal of your body? (b) Could you create an apparent top–bottom reversal by positioning your body differently?
5. Why do some fire trucks have "FIRE" printed backward and reversed on the front of the truck (▼ **Figure 23.26**)?

▲ FIGURE 23.26 **Backward and reversed** See Conceptual Question 5.

**23.2 Spherical Mirrors**

6. How can the focal length be quickly determined experimentally for a concave mirror? Can you do the same thing for a convex mirror?
7. Can a convex mirror produce an image that is taller than the object? Why or why not?
8. (a) Some side rearview mirrors have the warning "OBJECTS IN THE MIRROR ARE CLOSER THAN THEY APPEAR" as shown in ▶ **Figure 23.27**. Explain why. (b) What is the type of the mirror – concave, convex, plane?

▲ **FIGURE 23.27 Mirror applications** See Conceptual Question 8.

9. A 10-cm-tall mirror bears the following advertisement: "Full-view mini mirror. See your full body in 10 cm." How can this be?

**23.3 Lenses**

10. How can you quickly determine the focal length of a converging lens? Will the same method work for a diverging lens?
11. If you want to use a converging lens to design a simple overhead projector to project the magnified image of an object containing small writing onto a screen on a wall, how far should you place the object in front of the lens?
12. Explain why a fish in a spherical fish bowl, viewed from the side, appears larger than it really is.
13. How would you use a converging lens as a magnifying glass? Can you do the same with a diverging lens?
14. The lateral magnification of an image formed by a lens of a chair is −0.50. Discuss the image characteristics.

**23.4 The Lens Maker's Equation and 23.5 Lens Aberrations (Optional)**

15. Determine the signs of $R_1$ and $R_2$ for each lens shown in Figure 23.16.
16. When you open your eyes underwater, everything is blurry. However, when you wear goggles, you can see clearly. Explain.
17. A lens that is converging in air is submerged in a fluid whose index of refraction is greater than that of the lens. Is the lens still converging?
18. If a farsighted person is prescribed with a "stronger" or more "powerful" lens, does the lens have a longer or shorter focal length? Explain.
19. What is the cause of spherical aberration?

## Exercises*

*Integrated Exercises (IEs) are two-part exercises. The first part typically requires a conceptual answer choice based on physical thinking and basic principles. The following part is quantitative calculations associated with the conceptual choice made in the first part of the exercise.*

**23.1 Plane Mirrors**

1. • Standing 2.5 m in front of a plane mirror with your camera, you decide to take a picture of yourself. To what distance should the camera be focused to get a sharp image?
2. • A man stands 2.0 m away from a plane mirror. (a) What is the distance between the mirror and the man's image? (b) What are the image characteristics?
3. • An object 5.0 cm tall is placed 40 cm from a plane mirror. Find (a) the distance from the object to the image, (b) the height of the image, and (c) the image's magnification.
4. •• If you hold a 900-$cm^2$ square plane mirror 45 cm from your eyes and can just see the full length of an 8.5-m flagpole behind you, how far are you from the pole? [*Hint:* A diagram is helpful.]
5. •• A small dog sits 3.0 m in front of a plane mirror. (a) Where is the dog's image in relation to the mirror? (b) If the dog jumps at the mirror at a speed of 1.0 m/s, how fast does the dog approach its image?
6. IE •• A woman fixing the hair on the back of her head holds a plane mirror 30 cm in front of her face so as to look into a plane mirror on the bathroom wall behind her. She is 90 cm from the wall mirror. (a) The image of the back of her head will be from (1) only the front mirror, (2) only the wall mirror, or (3) both mirrors. (b) Approximately how far does the image of the back of her head appear in front of her?
7. IE •• (a) When you stand between two plane mirrors on opposite walls in a dance studio, you observe (1) one, (2) two, or (3) multiple images. Explain. (b) If you stand 3.0 m from the mirror on the north wall and 5.0 m from the mirror on the south wall, what are the image distances for the first two images in both mirrors?
8. •• A woman 1.7 m tall stands 3.0 m in front of a plane mirror. (a) What is the minimum height the mirror must be to allow the woman to view her complete image from head to foot? Assume that her eyes are 10 cm below the top of her head. (b) What would be the required minimum height of the mirror if she were to stand 5.0 m away?

* The bullets denote the degree of difficulty of the exercises: •, simple; ••, medium; and •••, more difficult.

9. •• Prove that $d_o = |d_i|$ (equal magnitude) for a plane mirror. [*Hint:* Refer to Figures 23.1 and 23.2 and use similar and identical triangles.]
10. ••• Draw ray diagrams that show how three images of an object are formed in two plane mirrors at right angles, as shown in ▼**Figure 23.28.** [*Hint:* Consider rays from both ends of the object in the drawing for each image.]

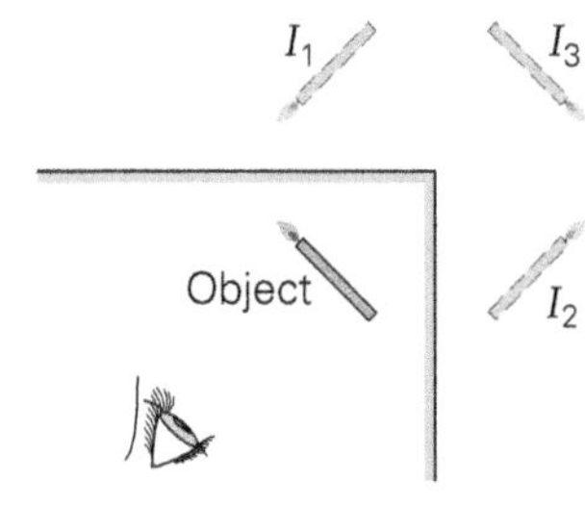


**(a)**

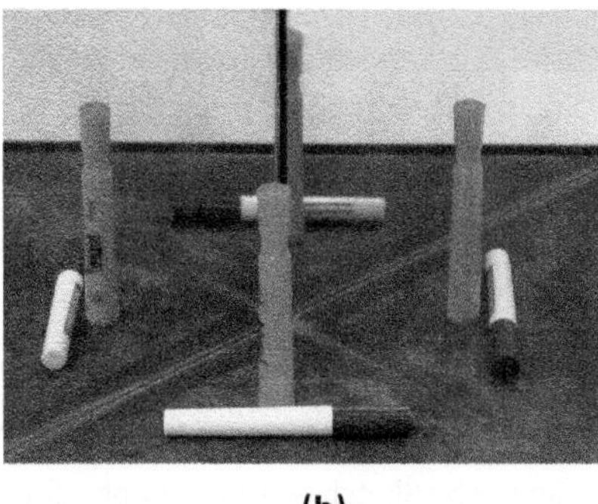

**(b)**

▲**FIGURE 23.28 Two mirrors – Multiple images** (**a**) Three images are formed by two perpendicular mirrors. Note the orientations of the candle flame. (**b**) A photograph showing three images of a white marker formed by two perpendicular mirrors. See Exercise 10.

## 23.2 Spherical Mirrors

11. IE • An object is 100 cm in front of a concave mirror that has a radius of 80 cm. (a) Use a ray diagram to determine whether the image is (1) real or virtual, (2) upright or inverted, and (3) magnified or reduced. (b) Calculate the image distance and lateral magnification.
12. • A candle with a flame 1.5 cm tall is placed 5.0 cm from the front of a concave mirror. A virtual image is formed 10 cm behind the mirror. (a) Find the focal length and radius of curvature of the mirror. (b) How tall is the image of the flame?
13. • An object is placed 50 cm in front of a convex mirror and its image is found to be 20 cm behind the mirror. What is the focal length of the mirror? What is the lateral magnification?
14. • An object 3.0 cm tall is placed 20 cm from the front of a concave mirror with a radius of curvature of 30 cm. Where is the image formed, and how tall is it?
15. • If the object in Exercise 14 is moved to a position 10 cm from the front of the mirror, what will be the characteristics of the image?
16. •• An object 3.0 cm tall is placed at different locations in front of a concave mirror whose radius of curvature is 30 cm. Determine the location of the image and its characteristics when the object distance is 40, 30, 15, and 5.0 cm, using (a) a ray diagram and (b) the mirror equation.
17. •• Use the mirror equation and the magnification factor to show that when $d_o = R = 2f$ for a concave mirror, the image is real, inverted, and the same size as the object.
18. IE •• An object is 120 cm in front of a convex mirror that has a focal length of 50 cm. (a) Use a ray diagram to determine whether the image is (1) real or virtual, (2) upright or inverted, and (3) magnified or reduced. (b) Calculate the image distance and image height.
19. •• A bottle 6.0 cm tall is located 75 cm from the concave surface of a mirror with a radius of curvature of 50 cm. Where is the image located, and what are its characteristics?
20. IE •• A virtual image of magnification +2.0 is produced when an object is placed 7.0 cm in front of a spherical mirror. (a) The mirror is (1) convex, (2) concave, (3) flat. Explain. (b) Find the radius of curvature of the mirror.
21. IE •• A virtual image of magnification +0.50 is produced when an object is placed 5.0 cm in front of a spherical mirror. (a) The mirror is (1) convex, (2) concave, (3) flat. Explain. (b) Find the radius of curvature of the mirror.
22. •• Using the spherical mirror equation and the magnification factor, show that for a convex mirror, the image of an object is always virtual, upright, and reduced.
23. IE •• When a man's face is in front of a concave mirror of radius 100 cm, the lateral magnification of the image is +1.5. What is the image distance?
24. •• A convex mirror in a department store produces an upright image 0.25 times the size of a person who is standing 200 cm from the mirror. What is the focal length of the mirror?
25. IE •• The image of an object located 30 cm from a mirror is formed on a screen located 20 cm from the mirror. (a) The mirror is (1) convex, (2) concave, (3) flat. Explain. (b) What is the mirror's radius of curvature?
26. IE •• The upright image of an object 18 cm in front of a mirror is half the size of the object. (a) The mirror is (1) convex, (2) concave, (3) flat. Explain. (b) What is the focal length of the mirror?
27. IE •• A concave mirror has a magnification of +3.0 for an object placed 50 cm in front of it. (a) The type of image produced is (1) virtual and upright, (2) real and upright, (3) virtual and inverted, (4) real and inverted. Explain. (b) Find the radius of curvature of the mirror.

28. •• A concave mirror is constructed so that a man at a distance of 20 cm from the mirror sees his image magnified 2.5 times. What is the radius of curvature of the mirror?
29. •• A child looks at a reflective Christmas tree ball ornament that has a diameter of 9.0 cm and sees an image of her face that is half the real size. How far is the child's face from the ball?
30. **IE** •• A dentist uses a spherical mirror that produces an upright image of a tooth that is magnified four times. (a) The mirror is (1) converging, (2) diverging, (3) flat. Explain. (b) What is the mirror's focal length in terms of the object distance?
31. •• A 15-cm-long pencil is placed with its eraser on the optic axis of a concave mirror and its point directed upward at a distance of 20 cm in front of the mirror. The radius of curvature of the mirror is 30 cm. Use (a) a ray diagram and (b) the mirror equation to locate the image and determine the image characteristics.
32. •• A spherical mirror at an amusement park has a radius of 10 m. If it forms an image that has a lateral magnification of +2.0, what are the object and image distances?
33. **IE** •• A pill bottle 3.0 cm tall is placed 12 cm in front of a mirror. A 9.0-cm-tall upright image is formed. (a) The mirror is (1) convex, (2) concave, (3) flat. Explain. (b) What is its radius of curvature?
34. •• A convex mirror is on the exterior of the passenger side of many trucks (see Conceptional Question 8). If the focal length of such a mirror is −40.0 cm, what will be the location and height of the image of a car that is 2.0 m high and (a) 100 m and (b) 10.0 m behind the truck mirror?
35. ••• For values of $d_o$ from 0 to $\infty$, (a) sketch graphs of (1) $d_i$ versus $d_o$ and (2) $M$ versus $d_o$ for a converging mirror. (b) Sketch similar graphs for a diverging mirror.
36. **IE** ••• A concave mirror of radius of curvature of 20 cm forms an image of an object that is twice the height of the object. (a) There could be (1) one, (2) two, (3) three object distance(s) that satisfy the image characteristics. Explain. (b) What are the object distances?

### 23.3 Lenses

37. • An object is placed 50.0 cm in front of a converging lens of focal length 10.0 cm. What are the image distance and the lateral magnification?
38. • An object placed 30 cm in front of a converging lens forms an image 15 cm behind the lens. What are the focal length of the lens and the lateral magnification of the image?
39. • A converging lens with a focal length of 20 cm is used to produce an image on a screen that is 2.0 m from the lens. What are the object distance and the lateral magnification of the image?
40. •• When an object is placed at 2.0 m in front of a diverging lens, a virtual image is formed at 30 cm in front of the lens. What are the focal length of the lens and the lateral magnification of the image?
41. **IE** •• An object 4.0 cm tall is in front of a converging lens of focal length 22 cm. The object is 15 cm away from the lens. (a) Use a ray diagram to determine whether the image is (1) real or virtual, (2) upright or inverted, and (3) magnified or reduced. (b) Calculate the image distance and lateral magnification.
42. •• (a) Design the lens in a single-lens slide projector that will form a sharp image on a screen 4.0 m away with the transparent slides 6.0 cm from the lens. (b) If the object on a slide is 1.0 cm tall, how tall will the image on the screen be?
43. •• An object is placed in front of a concave lens whose focal length is −18 cm. Where is the image located and what are its characteristics, if the object distance is (a) 10 cm and (b) 25 cm? Sketch ray diagrams for each case.
44. •• A convex lens produces a real, inverted image of an object that is magnified 2.5 times when the object is 20 cm from the lens. What are the image distance and the focal length of the lens?
45. •• A convex lens has a focal length of 0.12 m. Where on the lens axis should an object be placed in order to get (a) a real, magnified image with a magnification of 2.0 and (b) a virtual, magnified image with a magnification of 2.0?
46. •• Using the thin lens equation and the magnification factor, show that for a spherical diverging lens, the image of a real object is always virtual, upright, and reduced.
47. •• (a) For values of $d_o$ from 0 to $\infty$, sketch graphs of (1) $d_i$ versus $d_o$ and (2) $M$ versus $d_o$ for a converging lens. (b) Sketch similar graphs for a diverging lens. (Compare to Exercise 35.)
48. •• A simple single-lens camera (convex lens) is used to photograph a man 1.7 m tall who is standing 4.0 m from the camera. If the man's image fills the height of a full frame digital sensor (24 mm), what is the focal length of the lens?
49. •• To photograph a full moon, a photographer uses a single-lens camera with a focal length of 60 mm. What will be the diameter of the Moon's image on the sensor? (*Note:* Data about the Moon are given inside the back cover of this text.)
50. •• An object 5.0 cm tall is 10 cm from a concave lens. The resulting virtual image is one-fifth as large as the object. What is the focal length of the lens and the image distance?
51. •• An object is placed 80 cm from a screen. (a) At what point from the object should a converging lens with a focal length of 20 cm be placed so that it will produce a sharp image on the screen? (b) What is the image's magnification?
52. •• (a) For a convex lens, what is the minimum distance between an object and its image if the image is real? (b) What is the minimum distance if the image is virtual?
53. •• Using ▶ **Figure 23.29**, derive (a) the thin lens equation and (b) the magnification equation for a thin lens. [*Hint:* Use similar triangles.]

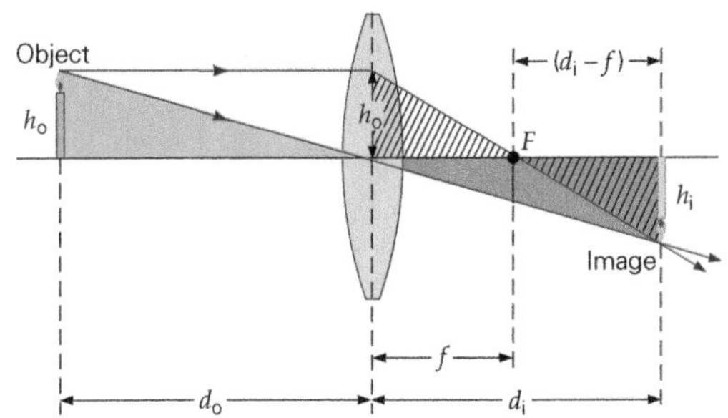


▲ FIGURE 23.29 **The thin lens equation** The geometry for deriving the thin lens equation and magnification factor. Note the two sets of similar triangles. See Exercise 53.

54. •• (a) If a book is held 30 cm from an eyeglass lens with a focal length of −45 cm where is the image of the print formed? (b) If an eyeglass lens with a focal length of +57 cm is used, where is the image formed?
55. •• To correct myopia (nearsightedness), concave lenses are prescribed. If a student can read her physics book only when she holds it no farther than 18 cm away, what focal length of lens should be prescribed so she can read when she holds the book 35 cm away?
56. •• To correct hyperopia (farsightedness), convex lenses are prescribed. If a senior citizen can read a newspaper only when he holds it no closer than 50 cm away, what focal length of lens should be prescribed so he can read when he holds the newspaper 25 cm away?
57. IE •• A biology student wants to examine a bug at a magnification of +5.00. (a) The lens should be (1) convex, (2) concave, (3) flat. Explain. (b) If the bug is 5.00 cm from the lens, what is the focal length of the lens?
58. •• The human eye is a complex multiple-lens system. However, it can be approximated to an equivalent single converging lens with an average focal length about 1.7 cm when the eye is relaxed. If an eye is viewing a 2.0-m-tall tree located 15 m in front of the eye, what are the height and orientation of the image of the tree on the retina?
59. ••• For a lens combination, show that the total magnification $M_{total} = M_1 M_2$. [*Hint:* Think about the definition of magnification.]
60. ••• Two converging lenses $L_1$ and $L_2$ have focal lengths of 30 cm and 20 cm, respectively. The lenses are placed 60 cm apart along the same axis, and an object is placed 50 cm from $L_1$ (110 cm from $L_2$). Where is the image formed relative to $L_2$, and what are its characteristics?

**23.4 The Lens Maker's Equation**
**and**
**23.5 Lens Aberrations (Optional)**

61. • An optometrist prescribes glasses with a power of −4.0 D for a nearsighted student. What is the focal length of the glass lenses?
62. • A farsighted senior citizen needs glasses with a focal length of 45 cm. What is the power of the lens?
63. IE •• A plastic convex meniscus (Figure 23.16) contact lens is made of plastic with an index of refraction of 1.55. The lens has a front radius of 2.50 cm and a back radius of 3.00 cm. (a) The signs of $R_1$ and $R_2$ are (1) +, +, (2) +, −, (3) −, +, (4) −, −. Explain. (b) What is the focal length of the lens?
64. •• A plastic plano-concave lens has a radius of curvature of 50 cm for its concave surface. If the index of refraction of the plastic is 1.35, what is the power of the lens?
65. •• An optometrist prescribes a corrective lens with a power of +1.5 D. The lens maker starts with a glass blank that has an index of refraction of 1.6 and a convex front surface whose radius of curvature is 20 cm. To what radius of curvature should the other surface be ground? Is the surface convex or concave?

# 24

# Physical Optics: The Wave Nature of Light

**The beautiful colors of the peacock's tail can be explained by the wave nature of light.**

It's always intriguing to see brilliant colors produced by objects that don't have any colors of their own. A glass prism, for example, which is clear and transparent by itself, nevertheless gives rise to a whole array of colors when white light passes through it. Prisms, like the water droplets that produce rainbows, don't create color. They merely separate the different colors that make up white light.

The phenomena of reflection and refraction are conveniently analyzed by using **geometrical optics** (Chapter 22). Ray diagrams (Chapter 23) show what happens when light is reflected from a mirror or refracted through a lens. However, other phenomena involving light, such as the colorful patterns of the peacock feathers, cannot be adequately explained or described using the ray concept. They can only be explained with the wave theory of light. The prominent wave phenomena are interference, diffraction, and polarization.

**Physical optics**, or **wave optics**, takes into account wave properties that geometrical optics ignores. The wave theory of light leads to satisfactory explanations of those phenomena that cannot be analyzed with rays. Thus, in this chapter, the wave nature of light must be used to analyze phenomena such as interference and diffraction.

Wave optics must be used to explain how light propagates around small objects or through small openings, such as the narrow track-to-track distances in CDs and DVDs. An object or opening is considered small if it is on the order of magnitude of the wavelength of light.

## 24.1 Young's Double-Slit Experiment

It has been stated that light behaves like a wave, but no proof of this assertion has been discussed. How would you go about demonstrating the wave nature of light? One method that involves the use of interference was first devised in 1801 by the English scientist Thomas Young (1773–1829). **Young's double-slit experiment** not only demonstrates the wave nature of light, but also allows the measurement of its wavelengths. Essentially, light can be shown to be a wave if it exhibits wave properties such as interference and diffraction.

Recall from the discussion of wave interference in Sections 13.4 and 14.4 that superimposed waves may interfere constructively or destructively. Total constructive interference occurs when two crests are superimposed. If a crest and a trough are superimposed, then destructive interference occurs. Interference can be easily observed with water waves, for which constructive and destructive interference produce obvious interference patterns (▼ **Figure 24.1**).

The interference of (visible) light waves is not as easily observed, because of their relatively short wavelengths ($\approx 10^{-7}$ m) and the fact that they usually are not monochromatic (single color or frequency). Also, stationary interference patterns are produced only with *coherent sources* – sources that produce light waves that have a constant phase relationship to one another. For example, for constructive interference to occur at some point, the waves meeting at that point must be in phase. As the waves meet, a crest must *always* overlap a crest, and a trough must *always* overlap a trough. If a phase difference develops between the waves over time, the interference pattern changes, and no stable or stationary pattern will be established.

In an ordinary light source, the atoms are excited randomly, and the emitted light waves fluctuate in amplitude and frequency. Thus, light from two such sources is *incoherent* and cannot produce a stationary interference pattern. Interference does occur, but the phase difference between the interfering waves changes so fast that the interference effects are not discernible.

To obtain the equivalency of two coherent sources, a barrier with one narrow slit is placed in front of a single light source, and a barrier with two very narrow slits is positioned symmetrically in front of the first barrier (▼ **Figure 24.2a**).

▲ FIGURE 24.1 **Water wave interference** The interference of water waves from two coherent sources in a ripple tank produces patterns of constructive and destructive interferences (bright and dark regions).

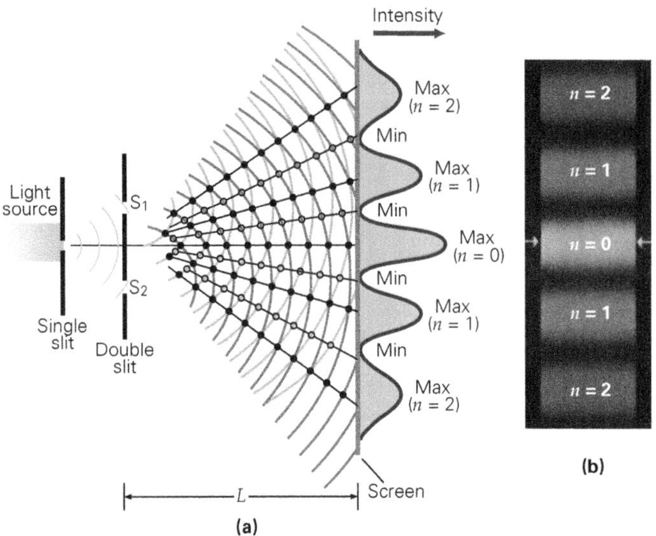


▲ FIGURE 24.2 **Double-slit interference (a)** The coherent waves from two slits are shown in light gray (top slit) and dark gray (bottom slit). The waves spread out from narrow slits. The waves interfere, producing alternating maxima and minima, on the screen. **(b)** An interference pattern. Note the symmetry of the pattern about the central maximum ($n = 0$).

Waves propagating out from the single slit are in phase, and the double slits then act as two coherent sources by separating one single wave into two parts. Any random changes in the light from the original source will thus occur for both sources passing through the two slits, and the phase difference will be constant. The modern laser beam, a coherent light source, makes the observation of a stable interference pattern much easier. A series

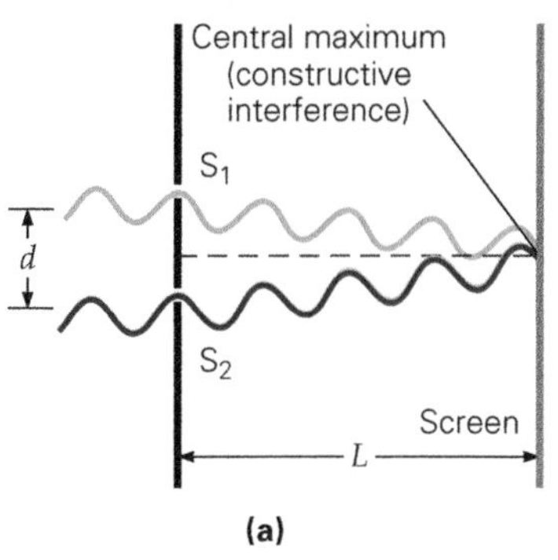


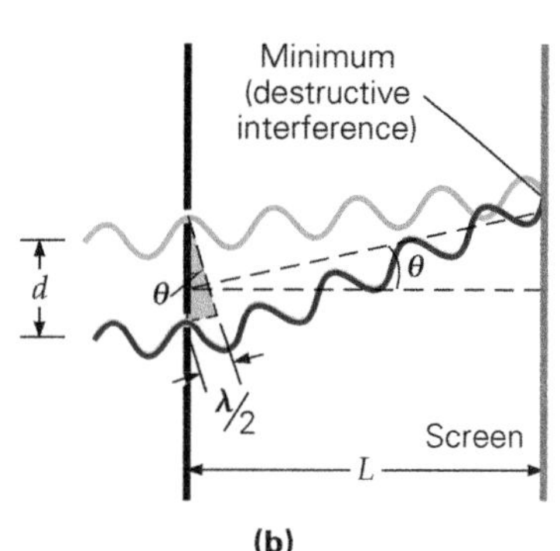


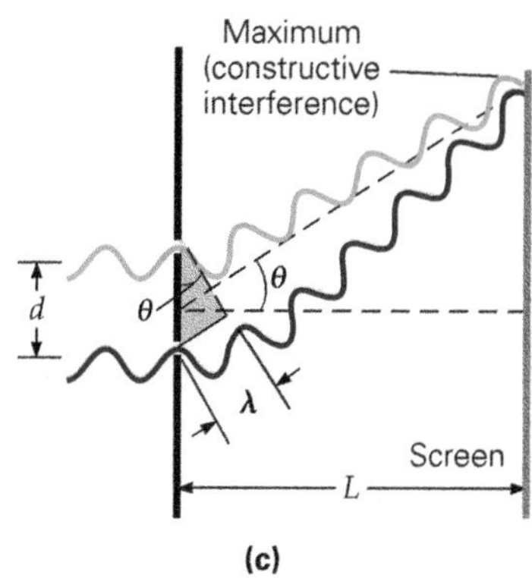


▲ FIGURE 24.3 **Interference** The interference that produces a maximum or minimum depends on the difference in the path lengths of the light from the two slits. **(a)** The path length difference at the position of the central maximum is zero, so the waves arrive in phase and interfere constructively. **(b)** At the position of the first side minimum, the path length difference is $\lambda/2$, and the waves interfere destructively. **(c)** At the position of the first side maximum, the path length difference is $\lambda$, and the interference is constructive.

of maxima or bright positions can be observed on a screen placed relatively far from the slits (Figure 24.2b).

To help analyze Young's experiment, imagine that light with a single wavelength (monochromatic light) is used. Because of diffraction (see Sections 13.4 and 14.4 and, in this chapter, Section 24.3), or the spreading of light as it passes through a slit, the waves spread out and interfere as illustrated in Figure 24.2a. Coming from two coherent "sources," the interfering waves produce a stable interference pattern on the screen. The pattern consists of a bright central maximum (▲ **Figure 24.3a**) and a series of symmetrical side minima (Figure 24.3b) and maxima (Figure 24.3c), which mark the positions at which destructive and constructive interferences occur. The existence of this interference pattern clearly demonstrates the wave nature of light. The intensities of each side maximum decrease with distance from the central maximum.

Measuring the wavelength of light requires the use of geometry in Young's experiment, as shown in ▶ **Figure 24.4**. Let the screen be a distance $L$ from the slits and $P$ be an arbitrary point on the screen. $P$ is located a distance $y$ from the center of the central maximum and at an angle $\theta$ relative to a normal line between the slits. The slits $S_1$ and $S_2$ are separated by a distance $d$. Note that the light path from slit $S_2$ to $P$ is longer than the path from slit $S_1$ to $P$. As the figure shows, the path length difference is approximately

$$\Delta L = d \sin\theta$$

The fact that the angle in the small shaded triangle is almost equal to $\theta$ can be shown by a simple geometrical argument involving similar triangles when $d \ll L$, as described in the caption of Figure 24.4.

The relationship of the phase difference of two waves to their path length difference was discussed in Section 14.4 for sound waves. These conditions hold for any wave, including light. Constructive interference occurs at any point where the path

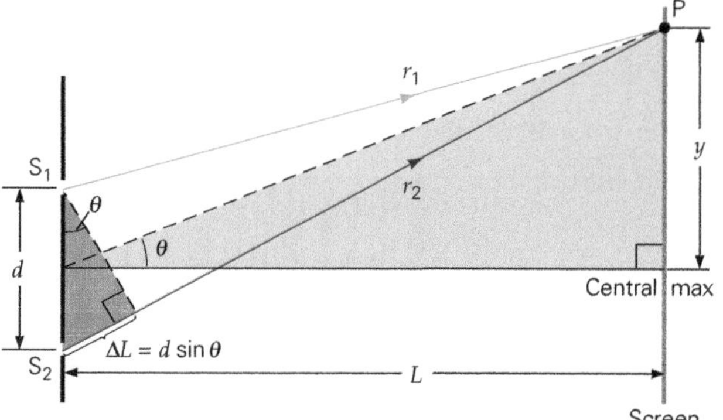


▲ FIGURE 24.4 **Geometry of Young's double-slit experiment** The difference in the path lengths for light traveling from the two slits to a point $P$ is $\Delta L = r_2 - r_1$, which forms a side of the small shaded triangle. Because the barrier with the slits is parallel to the screen, the angle between $r_2$ and the barrier (at $S_2$, in the small shaded triangle) is equal to the angle between $r_2$ and the screen. When $L$ is much greater than $y$, that angle is almost identical to the angle between the screen and the dashed line, which is an angle $\theta$ in the large shaded triangle. The two shaded triangles are then almost similar, and the angle at $S_1$ in the small triangle is almost exactly equal to $\theta$. Thus, $\Delta L = d \sin\theta$. (Not drawn to scale. Assume that $d \ll L$.)

length difference between the two waves is an integral number of wavelengths:

$$\Delta L = n\lambda \quad \text{for } n = 0, 1, 2, 3, \ldots \quad \text{(condition for constructive interference)} \tag{24.1}$$

Similarly, for destructive interference, the path length difference is an odd number of half-wavelengths:

$$\Delta L = \frac{m\lambda}{2} \quad \text{for } m = 1, 3, 5, \ldots \quad \text{(condition for destructive interference)} \tag{24.2}$$

Thus, in Figure 24.4, the maxima (constructive interference) satisfy

$$d \sin\theta = n\lambda \quad \text{for } n = 0, 1, 2, 3, \ldots \quad \text{(condition for interference maxima)} \tag{24.3}$$

where $n$ is called the *order number*. The zeroth order ($n = 0$) corresponds to the central maximum; the first order ($n = 1$) is the first maximum on either side of the central maximum, and so on. As the path length difference varies from point to point, so does the phase difference and the resulting type of interference (constructive or destructive).

The wavelength can therefore be determined by measuring $d$ and $\theta$ for a maximum of a particular order (other than the central maximum), because Equation 24.3 can be solved as $\lambda = (d \sin\theta)/n$.

The angle $\theta$ locates a side maximum relative to the central maximum. This can be measured from a photograph of the interference pattern, such as shown in Figure 24.2b. If $\theta$ is small ($y \ll L$), $\sin\theta \approx \tan\theta = y/L$.* Substituting this $y/L$ for $\sin\theta$ into Equation 24.3 and solving for ($y_n$) gives a good approximation of the distance of the $n$th maximum ($y_n$) from the central maximum on either side:

$$y_n \approx \frac{nL\lambda}{d} \quad \text{for } n = 0, 1, 2, 3, \ldots \quad \text{(lateral distance to maxima for small } \theta \text{ only)} \tag{24.4}$$

A similar analysis gives the locations of the minima. (See Exercise 6a.)

From Equation 24.3, it can be seen that, except for the zeroth order (the central maximum), the positions of the maxima depend on wavelength – that is, different wavelengths ($\lambda$) give different values of $\theta$ and $y$. Hence, when white light is used, the central maximum is white because all wavelengths are at the same location, but the other orders become a "spread out" spectrum of colors. Because $y$ is proportional to $\lambda$ ($y \propto \lambda$), in a given order, red is farther out than blue (red has a longer wavelength than blue).

By measuring the positions of the color maxima within a particular order, Young was able to determine the wavelengths of the colors of visible light. Note also that the size or "spread" of the interference pattern, $y_n$, depends inversely on the slit separation $d$. The smaller the slit separation $d$, the more spread out the pattern. For large $d$, the interference pattern is so compressed that it appears to us as a single white spot (all maxima together at the center).

In this analysis, the word *destructive* does *not* imply that energy is destroyed. Destructive interference means that light energy is not present at a particular location (minima). By energy conservation, the light energy *must* be somewhere else (maxima). This is observed with sound waves as well.

* For $\theta = 10°$, the percentage difference between $\sin\theta$ and $\tan\theta$ is only 1.5%.

### INTEGRATED EXAMPLE 24.1: MEASURING THE WAVELENGTH OF LIGHT – YOUNG'S DOUBLE-SLIT EXPERIMENT

In a lab experiment similar to the one shown in Figure 24.4, monochromatic light (single wavelength or frequency) passes through two narrow slits that are 0.050 mm apart. The interference pattern is observed on a white wall 1.0 m from the slits, and the second-order maximum is at an angle of $\theta_2 = 1.5°$. (a) If the slit separation decreases, the second-order maximum will be seen at an angle of (1) greater than 1.5°, (2) 1.5°, (3) less than 1.5°. Explain. (b) What is the wavelength of the light and what is the distance between the second-order and third-order maxima? (c) If $d = 0.040$ mm, what is $\theta_2$?

**(a) CONCEPTUAL REASONING.** According to the condition for constructive interference, $d \sin\theta = n\lambda$, the product of $d$ and $\sin\theta$ is a constant, for a given wavelength $\lambda$ and order number $n$. Therefore, if $d$ decreases, $\sin\theta$ increases, as does $\theta$. Thus the answer is (1).

**(b,c) QUANTITATIVE REASONING AND SOLUTION.** Equation 24.3 can be used to find the wavelength. Since $L \gg d$ (that is, $1.0 \text{ m} \gg 0.050 \text{ mm}$) $\theta$ is very small. We could compute $y_2$ and $y_3$ from Equation 24.4 and determine the distance between the second-order and third-order maxima ($y_3 - y_2$). However, the maxima for a given wavelength of light are evenly spaced (for a small $\theta$). That is, the distance between adjacent maxima is a constant.

***Given:***

$L = 1.0$ m
$n = 2$
$d = 0.050 \text{ mm} = 5.0 \times 10^{-5}$ m
(b) $\theta_2 = 1.5°$
$d = 5.0 \times 10^{-5}$ m
(c) $d = 4.0 \times 10^{-5}$ m

***Find:***

**(b)** $\lambda$ and ($y_3 - y_2$)
**(c)** $\theta_2$ (for $d = 0.040$ mm)

**(b)** Using Equation 24.3,

$$\lambda = \frac{d \sin\theta}{n} = \frac{(5.0 \times 10^{-5} \text{ m}) \sin 1.5°}{2} = 6.5 \times 10^{-7} \text{ m} = 650 \text{ nm}$$

This value is 650 nm, which is the wavelength of orange-red light (see Figure 20.18). From Equation 24.4, and a general approach from $n$ and $n + 1$,

$$y_{n+1} - y_n = \frac{(n+1)L\lambda}{d} - \frac{nL\lambda}{d} = \frac{L\lambda}{d}$$

In this case, the distance between successive maxima is

$$y_3 - y_2 = \frac{L\lambda}{d} = \frac{(1.0 \text{ m})(6.5 \times 10^{-7} \text{ m})}{5.0 \times 10^{-5} \text{ m}} = 1.3 \times 10^{-2} \text{ m} = 1.3 \text{ cm}$$

(c) $\sin\theta_2 = \frac{n\lambda}{d} = \frac{(2)(650\times10^{-9}\,\text{m})}{(4.0\times10^{-5}\,\text{m})} = 0.0325$

so $\theta_2 = \sin^{-1}(0.0325) = 1.9° > 1.5°$

As is reasoned in (a), $\theta_2$ is indeed greater than 1.5° for a decreased *d*.

**FOLLOW-UP EXERCISE.** Suppose white light were used instead of monochromatic light in this Example. What would be the separation distance of the red ($\lambda = 700$ nm) and blue ($\lambda = 400$ nm) components in the second-order maximum? *(Answers to all Follow-Up Exercises are given in Appendix V at the back of the book.)*

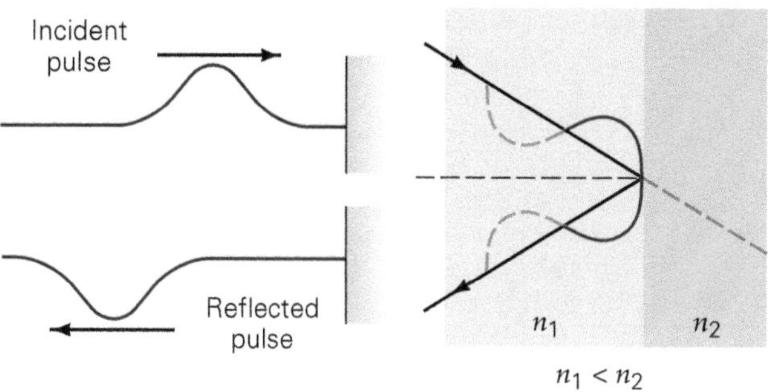

**(a)** Fixed end: 180° phase shift

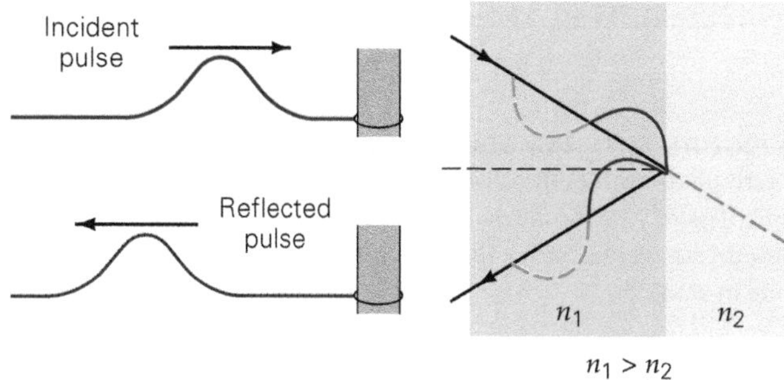

**(b)** Free end: zero phase shift

▲ **FIGURE 24.6 Reflection and phase shifts** The phase changes that light waves undergo on reflection are analogous to those for pulses in strings. **(a)** The phase of a pulse in a string is shifted by 180° or $\lambda/2$ on reflection from a fixed end, and so is the phase of a light wave when it is reflected from a medium of higher index of refraction. **(b)** A pulse in a string has a phase shift of zero (it is not shifted) when reflected from a free end. Analogously, a light wave is not phase shifted when reflected from a medium of lower index of refraction.

## 24.2 Thin Film Interference

Have you ever wondered what causes the rainbowlike colors that occur when white light is reflected from a thin film of oil or a soap bubble? (▼ **Figure 24.5**) This effect – known as *thin film interference* – is a result of the interference of light reflected from opposite surfaces of the film and may be understood in terms of wave interference.

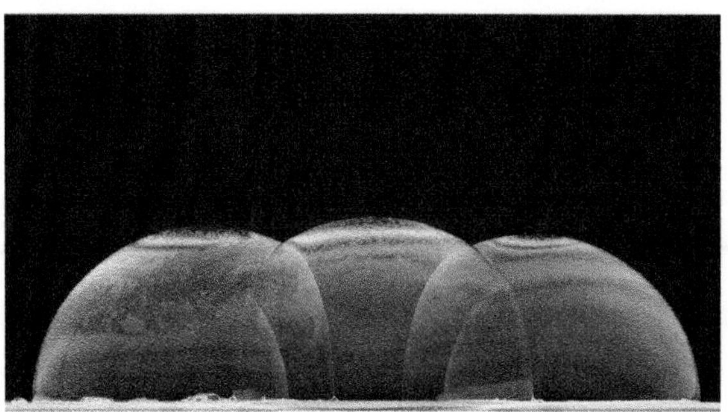

▲ **FIGURE 24.5 The rainbow colors can be seen from soap bubbles** Thin film interference displays multiple colors can be seen from soap bubbles. (Unfortunately here in black and white.)

First, however, you need to know how the phase of a light wave is affected by reflection. Recall from Section 13.4 that a wave pulse on a rope undergoes a phase change [or a *half wave shift* ($\lambda/2$)] when reflected from a rigid support and no phase shift when reflected from a free support (▶ **Figure 24.6**). Similarly, as the figure shows, the phase change for the reflection of light waves at a boundary depends on the indices of refraction (*n*) of the two materials:*

- A light wave undergoes a 180° phase change on reflection if $n_1 < n_2$.
- There is no phase change on reflection if $n_1 > n_2$.

To understand why you see colors from a soap bubble or an oil film (for example, floating on water or on a wet road), consider the reflection of monochromatic light from a thin film in ▶ **Figure 24.7**. The path length of the wave in the film depends on the angle of incidence (why?), but for simplicity, normal (perpendicular) incidence is assumed, even though the rays are drawn at an angle in the figure for clarity.

The oil film has a greater index of refraction than that of air, and the light reflected from the air–oil interface (wave 1 in the figure) undergoes a 180° phase shift. The transmitted waves pass through the oil film and are reflected at the oil–water interface. In general, the index of refraction of oil is greater than that of water (see Table 22.1) – that is, $n_1 > n_2$ – so a reflected wave in this instance (wave 2) does not undergo a phase shift.

You might think that if the path length difference of the waves in the oil film (2*t*, twice the thickness – down and back up) was an integral number of wavelengths – for example, if $2t = 2(\lambda'/2) = \lambda'$ in Figure 24.7a, where $\lambda' = \lambda/n$ is the wavelength in the oil – then the waves reflected from the two surfaces would interfere constructively. But keep in mind that the wave reflected from the top surface (wave 1) undergoes a 180° phase shift. The reflected waves from the two surfaces are therefore actually *out of phase* and would interfere destructively for this condition. This means that no reflected light for this wavelength would be observed. (The light would be transmitted.)

* The refracted wave does not shift in phase.

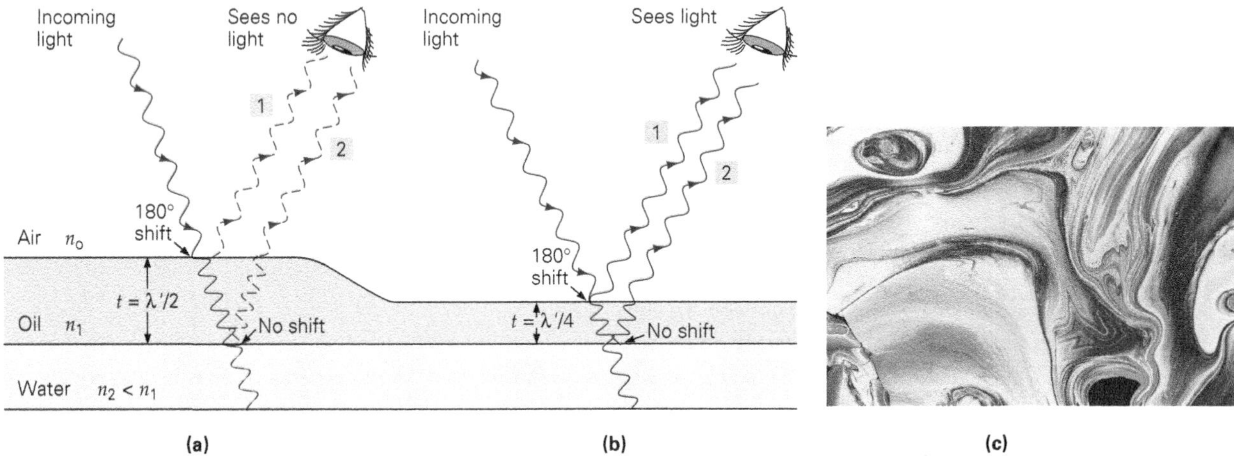


▲ **FIGURE 24.7** **Thin film interference** For an oil film on water, there is a 180° phase shift for light reflected from the air–oil interface and a zero phase shift at the oil–water interface. $\lambda'$ is the wavelength in the oil. **(a)** Destructive interference occurs if the oil film has a minimum thickness of $\lambda'/2$ for normal incidence. (Waves are displaced and angled for clarity.) **(b)** Constructive interference occurs with a minimum film thickness of $\lambda'/4$. **(c)** Thin film interference in an oil slick. Different film thicknesses give rise to the reflections of different colors (shown here in shades).

Similarly, if the path length difference of the waves in the film was an odd number of half-wavelengths [$2t = 2(\lambda'/4) = \lambda'/2$] in Figure 24.7b, again where $\lambda'$ is the wavelength in the oil, then the reflected waves would actually be *in phase* (as a result of the 180° phase shift of wave 1) and would interfere constructively. Reflected light for this wavelength would be observed from above the oil film.

Because oil films generally have different thicknesses in different locations, particular wavelengths (colors) of white light interfere constructively in different locations after reflection. As a result, a vivid display of various colors appears (Figure 24.7c). The bright colors of a peacock's tail, an example of colorful interference in nature, are a result of layers of fibers in its feathers, which mostly lack pigments. Light reflected from successive layers interferes constructively, giving bright colors. Since the condition for constructive interference depends on the angle of incidence, the color pattern changes somewhat with the viewing angle and motion of the bird (▶ **Figure 24.8**).

▲ **FIGURE 24.8** **Thin film interference** Multilayer interference in a peacock's feathers gives rise to bright colors (indicated here by shades). The brilliant throat colors of hummingbirds are produced in the same way.

A soap bubble in the air is slightly thicker near the bottom than on the top, due to the Earth's gravitational force. The gradual increase in thickness from the top to the bottom of the bubble causes the constructive interferences of different colors.

A practical application of thin film interference is nonreflective coatings for lenses. You may have noticed the green-blue tint of the coated optical lenses used in cameras and binoculars (▶ **Figure 24.9a**). The coating makes the lenses almost "nonreflecting" for all the colors you do not see in the reflected light. The incident light is mostly transmitted through the lens. Maximum transmission of light is desirable for all optical instruments, especially under low light conditions.

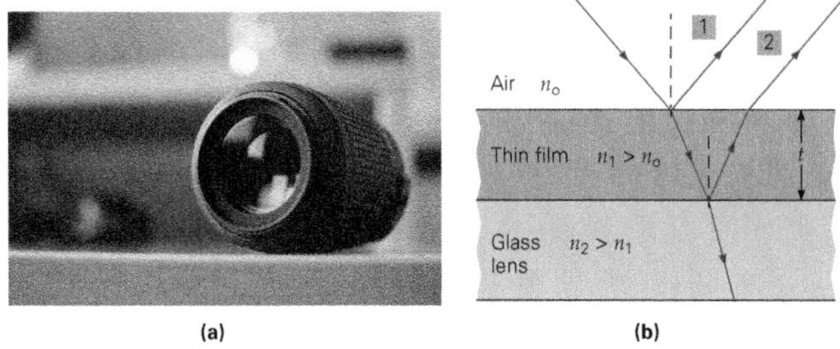


▲ **FIGURE 24.9** **Nonreflecting coating** **(a)** The green-bluish tint on the lens is a nonreflecting coating. **(b)** For a thin film on a glass lens, there is a 180° phase shift at each interface when the index of refraction of the film is less than that of the glass. The waves reflected off the top and bottom surfaces of the film interfere. For clarity, the angle of incidence is drawn to be large, but, in reality, it is almost zero.

In this situation, a film coating is used to create destructive interference between the reflected waves so as to increase the light transmission into the glass lens (Figure 24.9b). The index of refraction of the film has a value between that of air and glass ($n_o < n_1 < n_2$). Consequently, phase shifts of incident light take place at both surfaces of the film.

In such a case, the condition for constructive interference of the reflected light is

$$\Delta L = 2t = m\lambda' \quad \text{or} \quad t = \frac{m\lambda'}{2} = \frac{m\lambda}{2n_1}, \quad m = 1, 2, \ldots \tag{24.5}$$

(condition for constructive interference when $n_o < n_1 < n_2$)

and the condition for destructive interference is

$$\Delta L = 2t = \frac{m\lambda'}{2} \quad \text{or} \quad t = \frac{m\lambda'}{4} = \frac{m\lambda}{4n_1}, \quad m = 1, 3, 5, \ldots \tag{24.6}$$

(condition for destructive interference when $n_o < n_1 < n_2$)

The *minimum* film thickness for destructive interference occurs when $m = 1$, so

$$t_{\min} = \frac{\lambda}{4n_1} \quad \text{(minimum film thickness for } n_o < n_1 < n_2) \tag{24.7}$$

Nonreflective coatings are also applied to the surfaces of solar cells, which convert light into electrical energy (Section 27.2). Because the thickness of such a coating is wavelength dependent, the overall losses due to reflection can be decreased from around 30% to only 10%. This reduction of reflection improves the cell's efficiency in collecting sunlight.

If the index of refraction of the film is greater than that of air and glass, then only the reflection at the air–film interface has the 180° phase shift. Therefore $2t = m\lambda'$ will actually create destructive interference, and $2t = m\lambda'/2$ will create constructive interference. (Why?)

---

EXAMPLE 24.2: NONREFLECTIVE COATINGS – THIN FILM INTERFERENCE

A glass lens ($n = 1.60$) is coated with a thin, transparent film of magnesium fluoride ($n = 1.38$) to make the lens nonreflecting. (a) What is the minimum film thickness so that the lens will not reflect normally incident light of wavelength 550 nm? (b) Will a film thickness of 996 nm make the lens nonreflecting?

**THINKING IT THROUGH.** (a) Equation 24.7 can be used directly to get an idea of the minimum film thickness needed for a nonreflective coating. (b) We need to determine whether 996 nm satisfies the condition in Equation 24.6.

SOLUTION

***Given:***

$n_o = 1.00$ (air)
$n_1 = 1.38$ (film)
$n_2 = 1.60$ (lens)
$\lambda = 550$ nm

***Find:***

**(a)** $t_{\min}$ (minimum film thickness)
**(b)** whether $t = 996$ nm gives a nonreflecting lens

**(a)** Because $n_2 > n_1 > n_o$,

$$t_{\min} = \frac{\lambda}{4n_1} = \frac{550 \text{ nm}}{4(1.38)} = 99.6 \text{ nm}$$

which is quite thin ($\approx 10^{-5}$ cm). In terms of atoms, which have diameters on the order of $10^{-10}$ nm, or $10^{-1}$ nm, the film is about 1000 atoms thick.

**(b)** $t = 996 \text{ nm} = 10(99.6 \text{ nm}) = 10t_{\min} = 10\left(\frac{\lambda}{4n_1}\right) = 5\left(\frac{\lambda}{2n_1}\right)$

This means that this film thickness does *not* satisfy the nonreflective condition (destructive interference). Actually, it satisfies the requirement for constructive interference (Equation 24.5) with $m = 5$. Such a coating specific for infrared radiation could be useful in hot climates on car and house windows, because it maximizes reflection and minimizes transmission.

**FOLLOW-UP EXERCISE.** What would be the minimum film thickness for the glass lens in this Example to reflect, rather than transmit, the incident light through the lens?

---

## 24.2.1 Optical Flats and Newton's Rings

The phenomenon of thin film interference can be used to check the smoothness and uniformity of optical components such as mirrors and lenses. *Optical flats* are made by grinding and polishing glass plates until they are as flat and smooth as possible. The degree of flatness can be checked by putting two such plates together at a slight angle so that a very thin air wedge is between them (▼ **Figure 24.10a**).

The reflected waves off the bottom of the top plate (wave 1) and top of the bottom plate (wave 2) interfere. Note that

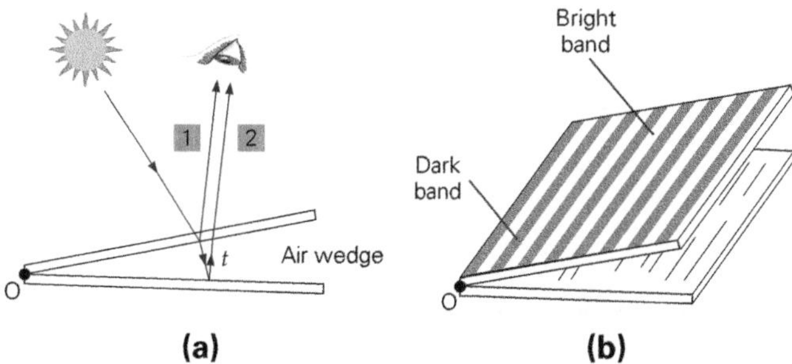


▲ FIGURE 24.10 **Optical flatness (a)** An optical flat is used to check the smoothness of a reflecting surface. The flat is placed so that there is a small air wedge between it and the surface. The waves reflected from the two plates interfere, and the thickness of the air wedge at certain points determines whether bright or dark bands are seen. **(b)** If the surfaces are smooth, a regular or symmetrical interference pattern is seen. Note that a dark band is at point O where $t = 0$.

wave 2 has a phase shift as it is reflected from an air–plate interface, whereas wave 1 does not. Therefore, at certain points from where the plates touch (point O), the condition for constructive interference is $2t = m\lambda/2$ ($m = 1, 3, 5, \ldots$), and the condition for destructive interference is $2t = m\lambda$ ($m = 0, 1, 2, \ldots$). The thickness $t$ determines the type of interference (constructive or destructive). If the plates are smooth and flat, a regular interference pattern of bright and dark bands appears (Figure 24.10b).

This pattern is a result of the uniformly varying differences in path lengths between the plates. Any irregularity in the pattern indicates an irregularity in at least one plate. Once a good optical flat is verified, it can be used to check the flatness of a reflecting surface, such as that of a precision mirror.

Direct evidence of the 180° phase shift can be clearly seen in Figure 24.10b. At the point where the two plates touch ($t = 0$), we see a *dark* band. If there were no phase shift, ($t = 0$) would correspond to $\Delta L = 0$, and a bright band would appear. The fact that it is a dark band proves that there is a phase shift in reflection from a material of higher index of refraction.

A similar technique is used to check the smoothness and symmetry of lenses. When a curved lens is placed on an optical flat, a radially symmetric air wedge is formed between the lens and the optical flat (▼**Figure 24.11a**). Since the thickness of the air wedge again determines the condition for constructive and destructive interference, the regular interference pattern in this case is a set of concentric circular bright and dark rings (Figure 24.11b). They are called *Newton's rings*, after Isaac Newton, who first described this interference effect. Note that at the point where the lens and the optical flat touch ($t = 0$), there is, once again, a dark spot. (Why?) Lens irregularities give rise to a distorted fringe pattern, and the radii of these rings can be used to calculate the radius of curvature of the lens.

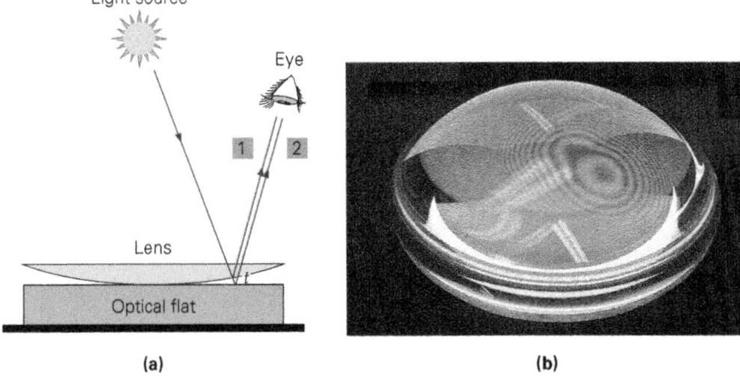


▲**FIGURE 24.11 Newton's rings (a)** Lens placed on an optical flat forms a ring-shaped air wedge, which gives rise to interference of the waves reflected from the top (wave 1) and the bottom (wave 2) of the air wedge. **(b)** The resulting interference pattern is a set of concentric rings called *Newton's rings*. Note that at the center of the pattern is a dark spot. Lens irregularities produce a distorted pattern.

## 24.3 Diffraction

In geometrical optics, light is represented by rays and pictured as traveling in straight lines. If this model were to represent the real nature of light, however, there would be no interference effects in Young's double-slit experiment. Instead, there would be only two bright images of slits on the screen, with a well-defined shadow area where no light enters. But we *do* see interference patterns, which means that the light must deviate from a straight-line path and enter the regions that would otherwise be in shadow. The waves actually "spread out" as they pass through the slits. This spreading is called **diffraction**. Diffraction generally occurs when waves pass through small openings or around sharp edges or corners. The diffraction of water waves is shown in ▼**Figure 24.12**. (See also Figure 13.16.)

▲**FIGURE 24.12 Water wave diffraction** This photograph of a beach dramatically shows single-slit diffraction of ocean waves through the barrier opening.

As Figure 13.16 shows, the amount of diffraction depends on the wavelength in relation to the size of the opening or object. In general, *the longer the wavelength compared to the width of the opening or object, the greater the diffraction.* With an opening much larger than the wavelength of the waves ($w \gg \lambda$), there is little diffraction – the wave keeps traveling without much spreading (Figure 13.16a). (There is some degree of diffraction around the edges of the opening.) With an opening about the same size as the wavelength ($w \approx \lambda$), there is noticeable diffraction – the wave spreads out and deviates from its original direction. Part of the wave keeps traveling in its original direction, but the rest bends *around* the opening and clearly spreads out (Figure 13.16b).

The diffraction of sound is quite evident (Section 14.4). Someone can talk to you from another room or around the corner of a building, and even in the absence of reflections, you can easily hear the person. Recall that audible sound wavelengths are on the order of centimeters to meters. Thus, the widths of ordinary objects and openings are about the same as or narrower than the wavelengths of sound, and diffraction will readily occur under these conditions.

Visible light waves, however, have wavelengths on the order of $10^{-7}$ m. Therefore, diffraction phenomena for these waves often go unnoticed, especially through large openings such as doors where sound readily diffracts. However, diffraction patterns produced by Sun rays coming through pine trees can be easily observed (▼ **Figure 24.13**).

▲ **FIGURE 24.13 Diffraction in action** Diffraction patterns produced by Sun rays coming through pine trees.

As an illustration of "single-slit" diffraction, consider a slit in a barrier (▼ **Figure 24.14**). Suppose that the slit (width $w$) is illuminated with monochromatic light. A diffraction pattern consisting of a bright central maximum and a symmetrical array of side maxima (regions of constructive interference) on both sides is observed on a screen at a distance $L$ from the slit ($L \gg w$).

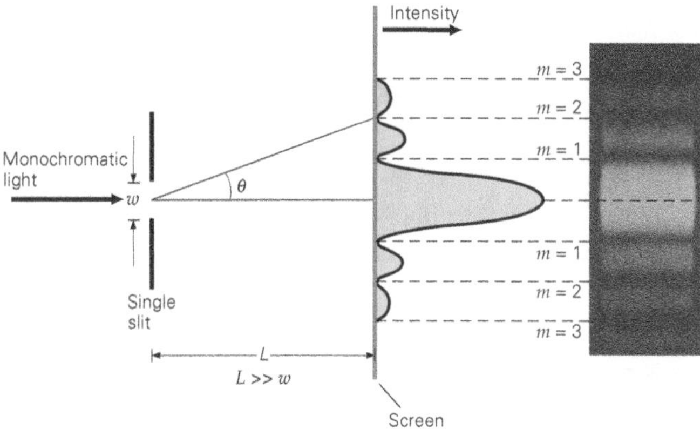


▲ **FIGURE 24.14 Single-slit diffraction** The diffraction of light by a single slit gives rise to a diffraction pattern consisting of a wide and bright central maximum and a symmetric array of side maxima. The order number $m$ corresponds to the minima or dark positions. (See text for a description.)

Thus a diffraction pattern results from the fact that various points on the wave front passing through the slit can be considered to be small point sources of light. The interference of those waves gives rise to the *diffraction* maxima and minima. The fairly complex analysis is not done here; however, from geometry, it can be proven that the *minima* (regions of destructive interference) satisfy the relationship

$$w \sin\theta = m\lambda \quad \text{for } m = 1, 2, 3, \ldots \quad \text{(condition for diffraction minima)} \tag{24.8}$$

where $\theta$ is the angle of a particular minimum, designated by $m = 1, 2, 3, \ldots$, on either side of the central maximum and $m$ is called the order number. (There is no $m = 0$. Why?)

Although this result is similar in form to that for Young's double-slit experiment (Equation 24.3), it is extremely important to realize that for the single-slit experiment, diffraction minima, rather than interference maxima, are analyzed. Also, note that the width of the slit ($w$) is used in diffraction.

The small-angle approximation, $\sin\theta \approx \tan\theta = y/L$, can be used when $y \ll L$. In this case, the distances of the minima relative to the center of the central maximum are given by

$$y_m = m\left(\frac{L\lambda}{w}\right) \quad \text{for } m = 1, 2, 3, \ldots \quad \text{(location for diffraction minima)} \tag{24.9}$$

The qualitative predictions from Equation 24.9 are interesting and instructive:

- For a given slit width ($w$), the longer the wavelength ($\lambda$), the wider (more "spread out") the diffraction pattern.
- For a given wavelength ($\lambda$) the narrower the slit width ($w$), the wider the diffraction pattern.
- The width of the central maximum is twice the width of any side maximum.

Let's look in detail at these results. As the slit is made narrower, the central maximum and the side maxima spread out and become wider. Equation 24.9 is not applicable to very small slit widths (because of the small-angle approximation). If the slit width is decreased until it is of the same order of magnitude as the wavelength of the light, then the central maximum spreads out over the whole screen. That is, diffraction becomes dramatically evident when the width of the slit is about the same as the wavelength of the light used. Diffraction effects are most easily observed when $\lambda/w \approx 1$, or $w \approx \lambda$.

Conversely, if the slit is made wider for a given wavelength, then the diffraction pattern becomes less spread out. The maxima move closer together and eventually become difficult to distinguish when $w$ is much wider than $\lambda$ ($w \gg \lambda$). The pattern then appears as a fuzzy shadow around the central maximum, which is the illuminated image of the slit. This type of pattern is observed for the image produced by sunlight entering a dark room through a hole in a curtain. Such an observation led early experimenters to investigate the wave nature of light. The acceptance of this concept was due, in large part, to the explanation of diffraction offered by physical optics.

The central maximum is twice as wide as any of the side maxima. The width of the central maximum is simply the distance between the first minima on each side ($m = 1$), or a value of $2y_1$ from Equation 24.9, $y_1 = L\lambda/w$, so

$$2y_1 = \frac{2L\lambda}{w} \quad \text{(width of central maximum)} \tag{24.10}$$

Similarly, the width of the side maxima is given by

$$y_{m+1} - y_m = (m+1)\left(\frac{L\lambda}{w}\right) - m\left(\frac{L\lambda}{w}\right) = \frac{L\lambda}{w} = y_1 \quad (24.11)$$

Thus, we have proved the width of the central maximum is twice that of the side maxima.

---

CONCEPTUAL EXAMPLE 24.3: DIFFRACTION AND RADIO RECEPTION

When you drive through a city or mountainous area, the quality of your radio reception may vary sharply from place to place, with stations seeming to fade out and reappear. Could diffraction be a cause of this phenomenon? If so, which of the following frequency bands would you expect to be least affected: (a) Satellite radio (2.3 GHz); (b) weather (162 MHz); (c) FM (88–108 MHz); or (d) AM (525–1610 kHz)?

**REASONING AND ANSWER.** Radio waves, like visible light, are electromagnetic waves and so tend to travel in straight lines when they are long distances from their sources. They can be blocked by objects in their path - especially if the objects are massive (such as buildings and hills).

However, because of diffraction, radio waves can also "wrap around" obstacles or "fan out" as they pass by obstacles and through openings, *provided* that their wavelength is at least roughly the size of the obstacle or opening. The longer the wavelength, the greater the amount of diffraction, and so the *less likely* the radio waves are to be obstructed.

To determine which band benefits most by such diffraction, we need the wavelengths that correspond to the given frequencies, as given by $c = \lambda f$. AM radio waves, with $\lambda = 186\text{-}571$ m, are the longest of the four bands (by a factor of at least 50). Thus AM broadcasts are more likely to be diffracted around such objects as buildings or mountains or through the openings between them, and the answer is (c).

**FOLLOW-UP EXERCISE.** Woodwind instruments, such as the clarinet and the flute, usually have smaller openings than brass instruments, such as the trumpet and trombone. During halftime at a football game, when a marching band faces you, you can easily hear both the woodwind instruments and the brass instruments. Yet when the band marches away from you, the brass instruments sound muted, but you can hear the woodwinds quite well. Why?

---

INTEGRATED EXAMPLE 24.4: SINGLE-SLIT DIFFRACTION - WAVELENGTH AND CENTRAL MAXIMUM

Monochromatic light passes through a slit whose width is 0.050 mm. (a) The resulting diffraction pattern is generally (1) wider for longer wavelengths, (2) wider for shorter wavelengths, (3) the same width for all wavelengths. Explain. (b) At what angle will the third order minimum be seen and what is the width of the central maximum on a screen located 1.0 m from the slit, for $\lambda = 400$ nm and 550 nm, respectively?

(a) **CONCEPTUAL REASONING.** The general size of the diffraction pattern can be characterized by the position and width of a particular maximum or minimum. From Equation 24.8, it can be seen that for a given width $w$ and order number $m$, the position of a minimum is directly proportional to the wavelength $\lambda$. Therefore, a longer wavelength will correspond to a greater $\sin\theta$ or a greater $\theta$, and the answer is (1).

(b) **QUANTITATIVE REASONING AND SOLUTION.** This part is a direct application of Equations 24.8 and 24.10.

***Given:***

$\lambda_1 = 400\text{ nm} = 4.00 \times 10^{-7}$ m
$\lambda_2 = 550\text{ nm} = 5.50 \times 10^{-7}$ m
$w = 0.050\text{ mm} = 5.0 \times 10^{-5}$ m
$m = 3$ and $L = 1.0$ m

***Find:*** $\theta_3$ and $2y_1$ (width of central maximum)

For $\lambda = 400$ nm:
From Equation 24.8,

$$\sin\theta_3 = \frac{m\lambda}{w} = \frac{3(4.00\times10^{-7}\text{ m})}{5.0\times10^{-5}\text{ m}} = 0.024$$

$$\text{so}\quad \theta_3 = \sin^{-1}0.024 = 1.4^\circ$$

Equation 24.10 gives

$$2y_1 = \frac{2L\lambda}{w} = \frac{2(1.0\text{ m})(4.00\times10^{-7}\text{ m})}{5.0\times10^{-5}\text{ m}} = 1.6\times10^{-2}\text{ m} = 1.6\text{ cm}$$

For $\lambda = 550$ nm

$$\sin\theta_3 = \frac{m\lambda}{w} = \frac{3(5.50\times10^{-7}\text{ m})}{5.0\times10^{-5}\text{ m}} = 0.033$$

$$\text{so}\quad \theta_3 = \sin^{-1}0.033 = 1.9^\circ$$

$$2y_1 = \frac{2L\lambda}{w} = \frac{2(1.0\text{m})(5.50\times10^{-7}\text{ m})}{5.0\times10^{-5}\text{ m}} = 2.2\times10^{-2}\text{ m} = 2.2\text{ cm}$$

As is reasoned in (a), the diffraction pattern for 550 nm is wider (larger $\theta_3$ and $2y_1$).

**FOLLOW-UP EXERCISE.** By what factor would the width of the central maximum change if red light ($\lambda = 700$ nm) were used instead of light with $\lambda = 550$ nm?

---

## 24.3.1 Diffraction Gratings

We have seen that maxima and minima result from diffraction followed by interference when monochromatic light passes through a set of double slits. As the number of slits is increased, the maxima become sharper (narrower) and the minima become wider. The sharp maxima are very useful in optical analysis of light sources and other applications. ▶**Figure 24.15** shows

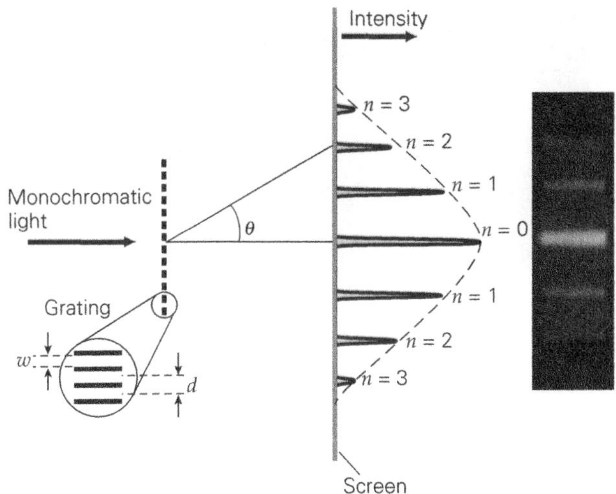


▲ **FIGURE 24.15 Diffraction grating** A diffraction grating produces a sharply defined interference/diffraction pattern. Two parameters define a grating: the slit separation $d$ and the slit width $w$. The combination of multiple-slit interference and single-slit diffraction determines the intensity distribution of the various orders of maxima.

a typical experiment with monochromatic light incident on a **diffraction grating**, which consists of large numbers of parallel, closely spaced slits. Two parameters define a diffraction grating: the slit separation between successive slits, the **grating constant**, $d$, and the individual slit width, $w$. The resulting pattern of interference and diffraction is shown in ▼ **Figure 24.16.**

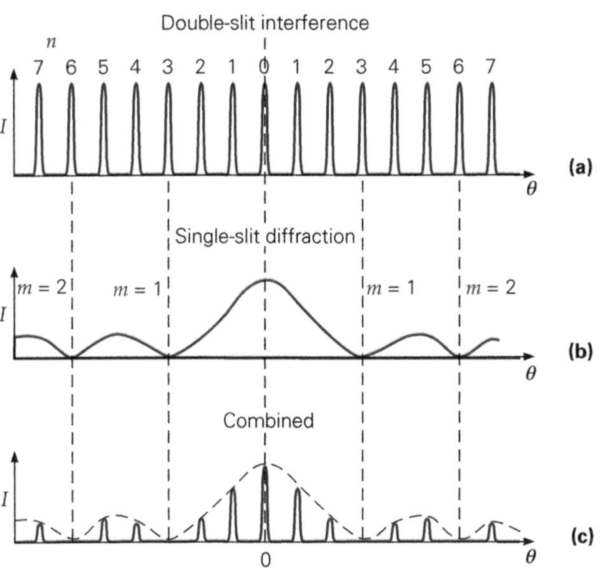


▲ **FIGURE 24.16 Intensity distribution of interference and diffraction (a)** Interference determines the positions of the interference maxima: $d \sin\theta = n\lambda$, $n = 0, 1, 2, 3, \ldots$ **(b)** Diffraction locates the positions of the diffraction minima: $w \sin\theta = m\lambda$, $m = 1, 2, 3, \ldots$, and the relative intensity of the maxima. **(c)** The combination of interference and diffraction determine the overall intensity distribution.

If light is transmitted through a grating, it is called a *transmission grating.* However, *reflection gratings* are also common. The closely spaced tracks of a compact disc (CD) act as a reflection grating, giving rise to their familiar iridescent sheen (▼ **Figure 24.17**). Commercial master gratings are made by depositing a thin film of aluminum on an optically flat surface and then removing some of the reflecting metal by cutting regularly spaced, parallel lines. Precision diffraction gratings are made using laser beams that expose a layer of photosensitive material, which is then etched. Precision gratings may have 30 000 or more lines per centimeter and are therefore expensive and difficult to fabricate. Most gratings used in laboratory instruments are *replica gratings*, which are plastic castings of high-precision master gratings.

▲ **FIGURE 24.17 Diffraction effects** The narrow tracks of a compact disc (CD) act as reflection diffraction gratings, producing colorful displays.

It can be shown that the condition for interference maxima for a grating illuminated with monochromatic light is identical to that for a double slit:

$$d \sin\theta = n\lambda \quad \text{for } n = 0, 1, 2, 3, \ldots \quad \text{(grating interference maxima)} \tag{24.12}$$

where $n$ is called the *order number* and $\theta$ is the angle at which that maximum occurs for a particular wavelength. The zeroth-order maximum is coincident with the central maximum. The grating constant $d$ is obtained from the number of lines or slits per unit length of the grating: $d = 1/N$. For example, if $N = 5000$ lines/cm, then

$$d \approx \frac{1}{N} = \frac{1}{5000/\text{cm}} = 2.0 \times 10^{-4}\ \text{cm}$$

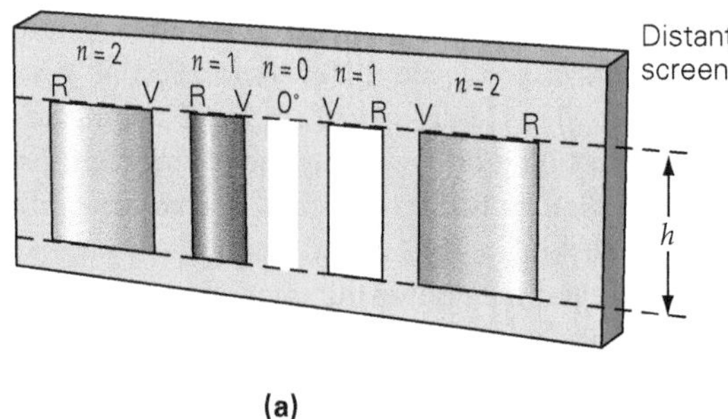


(a)

(b)

▲ **FIGURE 24.18** **Spectroscopy (a)** In each side maximum, components of different wavelengths (R = red and V = violet) are separated, because the angle depends on wavelength: $\theta = \sin^{-1}(n\lambda/d)$. **(b)** Diffraction pattern observed using a candle light.

If the light incident on a grating is white light (polychromatic), then the maxima are multicolored (▲**Figure 24.18a**). For the zeroth order, all color components are at the same location ($\sin\theta = 0$ for all wavelengths), so the central maximum is white. However, the colors separate for higher orders, since the position of the maximum depends on wavelength (Equation 24.12). Because longer wavelengths have a larger $\theta$, this produces a spectrum, Figure 24.18b.

Only a limited number of spectral orders can be obtained using a diffraction grating. The number depends on the wavelength of the light and on the grating constant ($d$). From Equation 24.12, $\theta$ cannot exceed 90° (that is, $\sin\theta \leq 1$), so

$$\sin\theta = \frac{n\lambda}{d} \leq 1 \quad \text{or} \quad n_{\text{max}} \leq \frac{d}{\lambda}$$

The sharp spectra produced by gratings are used in instruments called *spectrometers* (▼**Figure 24.19**). With a spectrometer, materials can be illuminated with light of various wavelengths to find which wavelengths are strongly transmitted or reflected. Their absorption can then be measured and material characteristics determined.

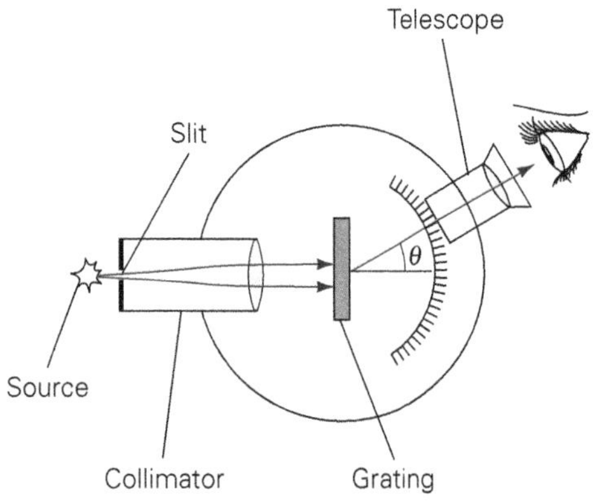


▲ **FIGURE 24.19** **Spectrometer** Gratings are used in spectrometers to determine the wavelengths present in a beam of light by measuring their angles and to separate the various wavelengths for further analysis.

---

### EXAMPLE 24.5: A DIFFRACTION GRATING – GRATING CONSTANT AND SPECTRAL ORDERS

A particular diffraction grating produces an $n = 2$ spectral order at an angle 32° for light with a wavelength of 500 nm.

**(a)** How many lines per centimeter does the grating have?
**(b)** At what angle can the $n = 3$ spectral order be seen?
**(c)** What is the highest order maximum that can be observed?

**THINKING IT THROUGH.** Equation 24.12 can be used for all three questions. (a) To find the number of lines per centimeter ($N$) the grating has, the grating constant ($d$) needs to be found, since $N = 1/d$. (b) The angle $\theta$ can be computed for $n = 3$. (c) The maximum angle is 90°, which corresponds to the highest spectral order.

**SOLUTION**

***Given:***

$\lambda = 500 \text{ nm} = 5.00 \times 10^{-7} \text{ m}$
$\theta = 32°$ for $n = 2$
(b) $n = 3$
(c) $\theta_{\text{max}} = 90°$

***Find:***

**(a)** $N$ (lines/cm)
**(b)** $\theta$ for $n = 3$
**(c)** $n_{\text{max}}$

**(a)** Using Equation 24.12, the grating constant is

$$d = \frac{n\lambda}{\sin\theta} = \frac{2(5.00 \times 10^{-7}\text{ m})}{\sin 32°} = 1.887 \times 10^{-6}\text{ m}$$
$$= 1.89 \times 10^{-4}\text{ cm}$$

Then

$$N = \frac{1}{d} = \frac{1}{1.89 \times 10^{-4}\text{ cm}} = 5300 \text{ lines/cm}$$

**(b)**

$$\sin\theta = \frac{n\lambda}{d} = \frac{3(5.00 \times 10^{-7}\text{ m})}{1.89 \times 10^{-6}\text{ m}} = 0.794$$

So

$$\theta = \sin^{-1} 0.794 = 52.6°$$

(c)

$$n_{max} = \frac{d \sin\theta_{max}}{\lambda} = \frac{(1.89\times10^{-6}\,\text{m})\sin 90°}{5.00\times10^{-7}\,\text{m}} = 3.8$$

This means the $n = 3$ order is seen, but the $n = 4$ order is not seen.

**FOLLOW-UP EXERCISE.** If white light of wavelengths ranging from 400 to 700 nm were used, what would be the angular width of the spectrum for the second order?

### 24.3.2 X-Ray Diffraction

In principle, the wavelength of any electromagnetic wave can be determined by using a diffraction grating with the appropriate grating constant. Diffraction was used to determine the wavelengths of X-rays early in the twentieth century. Experimental evidence indicated that the wavelengths of X-rays were probably around $10^{-10}$ m or 0.1 nm (much shorter than visible light wavelengths), but it is impossible to construct a diffraction grating with line spacing this close. Around 1913, Max von Laue (1879–1960), a German physicist, suggested that the regular spacing of the atoms in a crystalline solid might make the crystal act as a diffraction grating for X-rays, since the atomic spacing is on the order of 0.1 nm (▼**Figure 24.20**). When X-rays were directed at crystals, diffraction patterns were indeed observed. (See Figure 24.20b.)

Figure 24.20a illustrates diffraction by the planes of atoms in a crystal such as sodium chloride. The path length difference is $2d \sin\theta$, where $d$ is the distance between the crystal's atomic planes. Thus, the condition for constructive interference is

$$2d \sin\theta = n\lambda \quad \text{for } n = 1, 2, 3, \ldots \quad \text{(constructive interference, X-ray diffraction)} \tag{24.13}$$

This relationship is known as Bragg's law, after W. L. Bragg (1890–1971), the British physicist who first derived it. X-ray diffraction is now routinely used to investigate the internal structure not only of simple crystals, but also of large, complex biological molecules such as proteins and DNA. Because of their short wavelengths, which are comparable with interatomic distances within a molecule, X-rays provide a method for investigating atomic structures within molecules.

## 24.4 Polarization

When you think of polarized light, you may visualize polarizing (or Polaroid) sunglasses, since this is one of the more common applications of polarization. When something is polarized, it has a preferential direction, or orientation. In terms of the transverse light waves, polarization refers to the orientation of electric field oscillations.

Recall from Section 20.4 that light is an electromagnetic wave with oscillating electric and magnetic field vectors ($\vec{E}$ and, $\vec{E}$, respectively) perpendicular (transverse) to the direction of propagation. Light from most sources consists of a very large number of electromagnetic waves emitted by the atoms of the source. Each atom produces a wave with a particular $\vec{E}$ orientation, corresponding to the direction of the atomic vibration. However, since electromagnetic waves from a typical source are produced by many atoms, many random orientations of the $\vec{E}$ fields are in the emitted composite light. When the $\vec{E}$ vectors are randomly oriented, the light is said to be unpolarized. This situation is commonly represented schematically in terms of the electric field vector as shown in ▶**Figure 24.21a**.

As viewed along the direction of propagation, the $\vec{E}$ is equally distributed in all directions. However, as viewed parallel to the direction of propagation, this random or equal distribution can be represented by two directions (such as the $x$- and $y$-directions in a two-dimensional coordinate system). Here, the vertical arrows denote the $\vec{E}$ components in that direction, and the dots represent the $\vec{E}$ components going in and out of the paper. This notation will be used throughout this section.

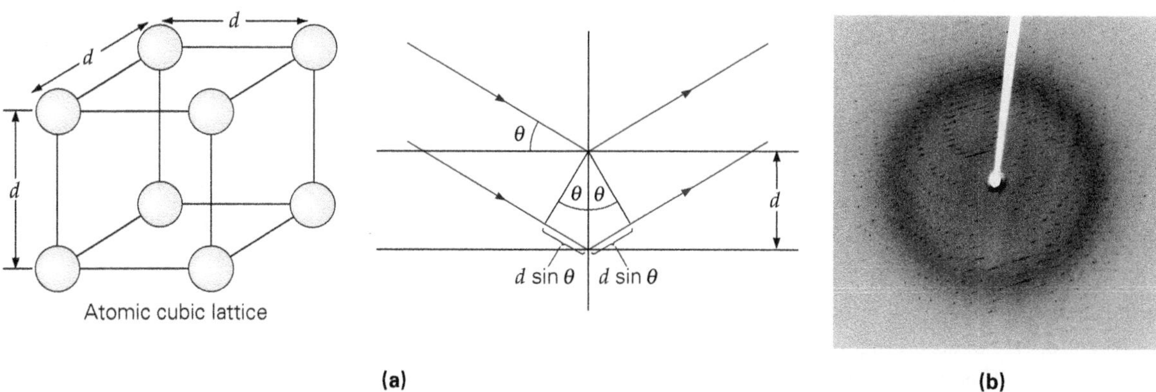


▲ FIGURE 24.20 **Crystal diffraction (a)** The array of atoms in a crystal lattice structure acts as a diffraction grating, and X-rays are diffracted from the planes of atoms. With a lattice spacing of $d$, the path length difference for the X-rays diffracted from adjacent planes is $2d \sin\theta$. **(b)** X-ray diffraction pattern of crystalline material (3Clpro, a SARS protease).

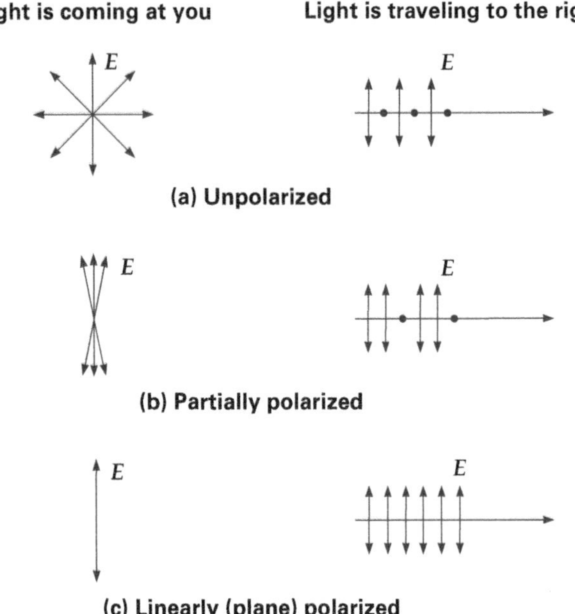


▲ **FIGURE 24.21** **Polarization** Polarization is represented by the orientation of the plane of vibration of the electric field vectors. **(a)** When the $\vec{E}$ vectors are randomly oriented, the light is unpolarized. The dots represent an electric field direction perpendicular to the paper, and the vertical arrows denote the up-and-down direction of the electric field. Equal numbers of dots and arrows are used to represent unpolarized light. **(b)** With preferential orientation of the $\vec{E}$ vectors, the light is partially polarized. Here, there are fewer dots than arrows. **(c)** When the $\vec{E}$ vectors are in one direction, the light is linearly polarized, or plane polarized. No dots are seen here.

If there is some preferential orientation of the $\vec{E}$ vectors, the light is said to be *partially polarized*. Both representations in Figure 24.21b show that there are more $\vec{E}$ vectors in the vertical direction than in the horizontal direction. If the $\vec{E}$ vectors oscillate in only *one* plane, the light is *linearly polarized* or *plane polarized*. In Figure 24.21c, the $\vec{E}$ is entirely in the vertical direction and there is no horizontal component. Note that polarization is evidence that light is a transverse wave. True longitudinal waves, such as sound waves, cannot be polarized, because the molecules of the media do not vibrate perpendicular to the direction of propagation.

Light can be polarized in many ways. Polarization by selective absorption, reflection, and double refraction will be discussed here. Polarization by scattering will be considered in Section 24.5.

### 24.4.1 Polarization by Selective Absorption (Dichroism)

Some crystals, such as those of the mineral tourmaline, exhibit the interesting property of absorbing one of the $\vec{E}$ components more than the other. This property is called **dichroism**. If a dichroic crystal is sufficiently thick, the more strongly absorbed component may be completely absorbed. In that case, the emerging beam is linearly polarized (▶ **Figure 24.22**).

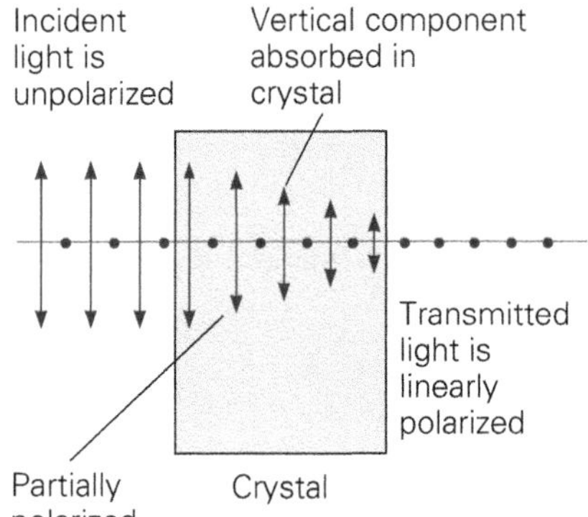


▲ **FIGURE 24.22** **Selective absorption (dichroism)** Dichroic crystals selectively absorb one polarized component (the vertical component) more than the other. If the crystal is thick enough, the emerging beam is linearly polarized.

Another dichroic crystal is quinine sulfide periodide (commonly called *herapathite*, after W. Herapath, an English physician who discovered its polarizing properties in 1852). This crystal was of great practical importance in the development of modern polarizers. Around 1930, Edwin H. Land (1909–1991), an American scientist, found a way to align tiny, needle-shaped dichroic crystals in sheets of transparent celluloid. The result was a thin sheet of polarizing material that was given the commercial name *Polaroid*™.

Better polarizing films have been developed that use synthetic polymer materials instead of celluloid. During the manufacturing process, this kind of film is stretched to align the long molecular chains of the polymer. With proper treatment, the outer (valence) electrons of the molecules can move along the oriented chains. As a result, light with $\vec{E}$ vectors parallel to the oriented chains is readily absorbed, but light with $\vec{E}$ vectors perpendicular to the chains is transmitted. The direction perpendicular to the orientation of the molecular chains is called the **transmission axis**, or **the polarization direction**. Thus, when unpolarized light falls on a polarizing sheet, the sheet acts as a polarizer and transmits polarized light (▶ **Figure 24.23**).

Since one of the two $\vec{E}$ components is absorbed, the light intensity after the polarizer is half of the intensity incident on it ($I_o/2$). The human eye cannot distinguish between polarized and unpolarized light. To tell whether light is polarized, we must use a second polarizer, or an analyzer. As shown in Figure 24.23a, if the transmission axis of an analyzer is parallel to the direction of polarization of polarized light, there is maximum transmission. If the transmission axis of the analyzer is perpendicular to the direction of polarization, little light (ideally, none) will be transmitted.

In general, the light intensity through the analyzer is given by

$$I = I_o \cos^2\theta \quad \text{(Malus' law)} \tag{24.14}$$

where $I_o$ is the light intensity after the first polarizer and $\theta$ is the angle between the transmission axes of the polarizer and analyzer. This expression is known as **Malus' law**, after its discoverer, French physicist E. L. Malus (1775–1812).

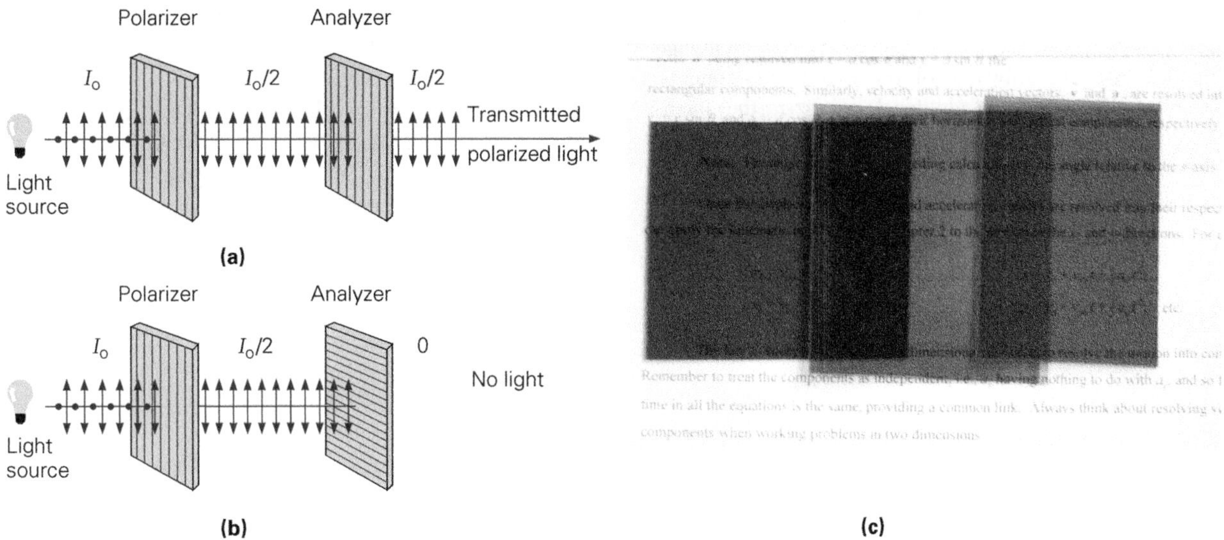


▲ FIGURE 24.23 **Polarizing sheets (a)** When polarizing sheets are oriented so that their transmission axes are in the same direction, the emerging light is polarized. The first sheet acts as a polarizer, and the second acts as an analyzer. **(b)** When one of the sheets is rotated and the transmission axes are perpendicular (crossed polarizers), little light (ideally, none) is transmitted. **(c)** The transmission axes of polarizers 1 and 2 are perpendicular and that of polarizers 2 and 3 are parallel.

Polarizing glasses whose lenses have different transmission axes are used to view some 3D movies. The pictures are projected on the screen by two projectors that transmit slightly different images, photographed by two cameras a short distance apart. The projected light from each projector is linearly polarized, but in mutually perpendicular directions. The lenses of the 3D glasses also have transmission axes that are perpendicular. Thus, one eye sees the image from one projector, and the other eye sees the image from the other projector. The brain receives a slight difference in perspective (or "viewing angle") from the two images, and interprets the image as having depth, or a third dimension, just as in normal vision.

## INTEGRATED EXAMPLE 24.6: MAKE SOMETHING OUT OF NOTHING – THREE POLARIZERS

In Figure 24.23b, no light is transmitted through the analyzer, because the transmission axes of the polarizer and analyzer are perpendicular. Assume that the unpolarized light incident on the first polarizer has an intensity of $I_o$. Another polarizer is then inserted between the first polarizer and analyzer, and the transmission axis of this additional polarizer makes an angle of $\theta$ with the first polarizer. (a) Is it possible for some light to go through this arrangement? If yes, does it occur at (1) $\theta = 0°$, (2) $\theta = 30°$, (3) $\theta = 90°$? Explain. (b) When $\theta = 30°$, what is the light intensity through the analyzer in terms of the incident light intensity?

**(a) CONCEPTUAL REASONING.** Yes, it is possible for some light to go through this arrangement at any angle other than 0° or 90°. ▼ **Figure 24.24** can help explain this situation.

With just the first polarizer and the analyzer, no light is transmitted, according to Malus' law (Equation 24.14), because the angle between the transmission axes is 90°. However, when an additional polarizer is inserted in between the first polarizer and the analyzer, some light can actually pass through the system. For example, if the transmission axis of the additional polarizer makes an angle of $\theta$ with that of the first polarizer, then the angle between the transmission axes of the additional polarizer and the analyzer will be $90° - \theta$. (Wy?)

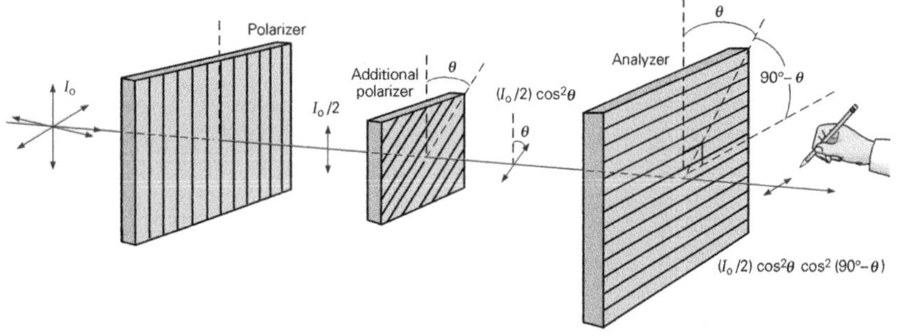


▲ FIGURE 24.24 **Three polarizers** See Integrated Example 24.6.

When unpolarized light of intensity $I_o$ is incident on the first polarizer, the light intensity after the first polarizer is $I_o/2$, because only one of the two $\vec{E}$ components is transmitted. After the additional polarizer, the intensity is decreased by a factor of $\cos^2 \theta$. After the analyzer, the intensity is decreased further by a factor of $\cos^2(90° - \theta) = \sin^2\theta$. So the overall transmitted intensity is $I = (I_o/2)(\cos^2 \theta)(\sin^2\theta)$. Therefore, as long as $\theta$ is not 0° or 90°, some light will be transmitted through the system The answer is then $2\theta = 30°$.

(b) **QUANTITATIVE REASONING AND SOLUTION.** Once this situation is understood, part (b) is a straightforward calculation.

***Given:*** $\theta = 30°$

***Find:*** (b) $I$ (after analyzer in terms of $I_o$)

When $\theta = 30°$

$$I = \frac{I_o}{2}(\cos^2 30°)(\sin^2 30°) = \frac{I_o}{2}\left(\frac{\sqrt{3}}{2}\right)^2\left(\frac{1}{2}\right)^2 = \frac{3I_o}{32}$$

**FOLLOW-UP EXERCISE.** For what value of $\theta$ will the transmitted intensity be a maximum in this Example?

## 24.4.2 Polarization by Reflection

When a beam of unpolarized light strikes a smooth, transparent medium such as glass, the beam is partially reflected and partially transmitted. The reflected light may be linearly polarized, partially polarized, or unpolarized, depending on the angle of incidence. The unpolarized case occurs for 0°, or normal incidence. As the angle of incidence is changed from 0°, both the reflected and refracted light become partially polarized. For example, the $\vec{E}$ components perpendicular to the plane of incidence (the plane containing the incident, reflected, and refracted rays) are reflected more strongly, producing partial polarization (▼ **Figure 24.25a**).

However, at one particular angle of incidence, the reflected light is linearly polarized (Figure 24.25b). (At this angle, though, the refracted light is still only partially polarized.)

David Brewster (1781–1868), a Scottish physicist, found that the linear polarization of the reflected light occurs when the reflected and refracted rays are perpendicular. The angle of incidence at which linear polarization occurs is the **polarizing angle ($\theta_p$)**, or the **Brewster angle**, and it depends on the indices of refraction of the two media. In Figure 24.25b, the reflected and refracted rays are at 90° and the angle of incidence is thus the polarizing angle. By the law of refraction (Section 22.3),

$$n_1 \sin\theta_1 = n_2 \sin\theta_2$$

Since $\theta_1 + 90° + \theta_2 = 180°$ or $\theta_2 = 90° - \theta_1$, $\sin\theta_2 = \sin(90° - \theta_1) = \cos\theta_1$.

Therefore,

$$\frac{\sin\theta_1}{\sin\theta_2} = \frac{\sin\theta_1}{\cos\theta_1} = \tan\theta_1 = \frac{n_2}{n_1}$$

With $\theta_1 = \theta_p$,

$$\tan\theta_p = \frac{n_2}{n_1} \quad \text{or} \quad \theta_p = \tan^{-1}\left(\frac{n_2}{n_1}\right) \tag{24.15}$$

If the first medium is air ($n_1 = 1$), then $\tan\theta_p = n_2/1 = n_2 = n$, where $n$ is the index of refraction of the second medium.

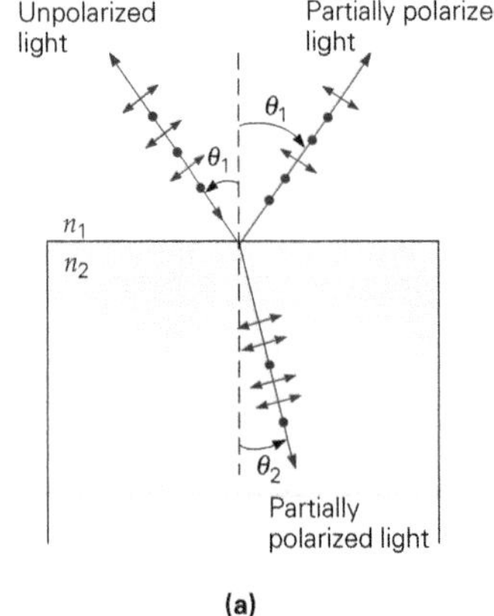


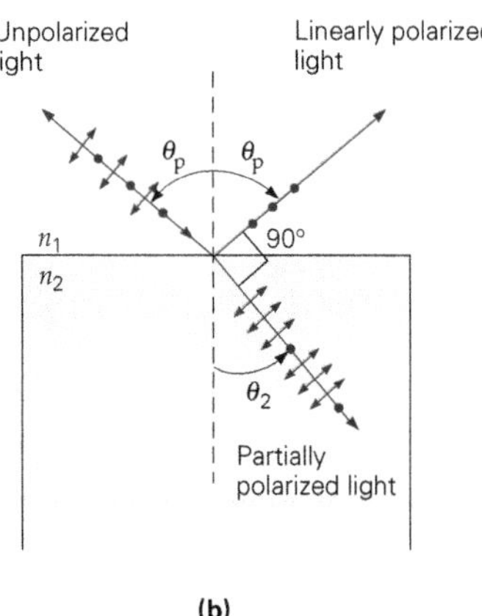


▲ **FIGURE 24.25 Polarization by reflection (a)** When light is incident on boundary, the reflected and refracted lights are normally partially polarized. **(b)** When the reflected and refracted rays are 90° apart, the reflected light is linearly polarized, and the refracted light is partially polarized. This situation occurs when $\theta_1 = \theta_p = \tan^{-1}(n_2/n_1)$.

Now you can understand the reason for inventing polarizing glasses. Light reflected from a smooth surface is partially polarized. The direction of polarization is mostly perpendicular to the plane of incidence (horizontal direction in Figure 24.25b). Reflected light can be so intense that it gives rise to visual glare (▼ **Figure 24.26a**). To reduce this effect, polarizing glasses are oriented with their transmission axes vertical so that some of the partially polarized light from reflection is absorbed. Polarizing filters also enable cameras to take "clean" pictures without interference from glare (Figure 24.26b).

**(a)**

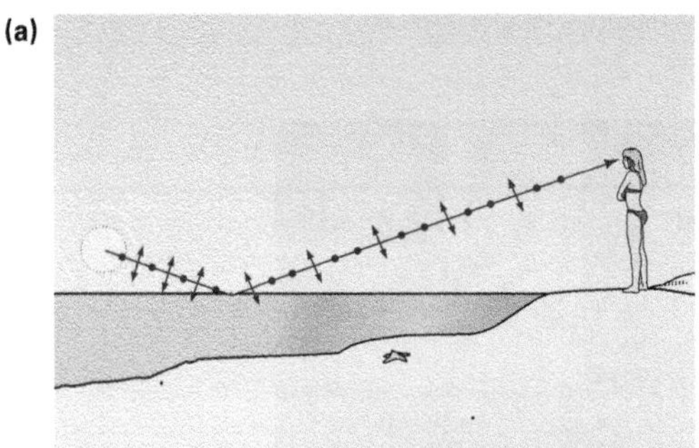

**(b)**

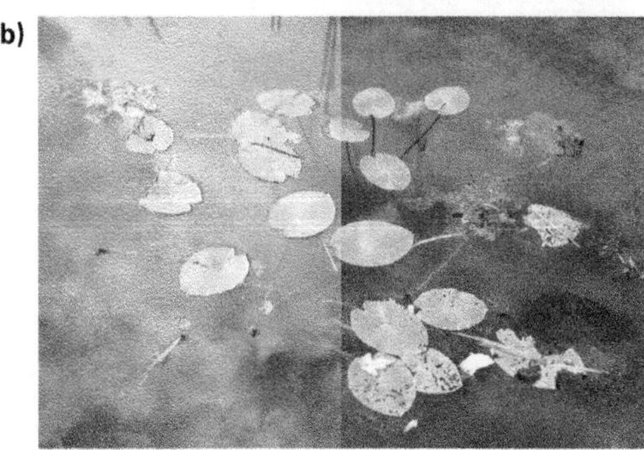

▲ FIGURE 24.26 **Glare reduction (a)** Light reflected from a horizontal surface is partially polarized in the horizontal direction. When glasses are oriented so that their transmission axis is vertical, the horizontally polarized component of such light is not transmitted, so glare is reduced. **(b)** Polarizing filters for cameras use the same principle. The photo at right was taken with such a filter. Note the reduction in reflections from the water surface.

---

EXAMPLE 24.7: CALLING BREWSTER – POLARIZATION BY REFLECTION

ZnSe (zinc selenide) is a material used for the windows of high-power $CO_2$ lasers. The index of refraction of ZnSe is 2.40 at a wavelength of 10.6 $\mu$m, and the index of refraction of $CO_2$ is 1.00. What should be the angle of incidence when the polarization of the reflected laser light is greatest?

**THINKING IT THROUGH.** Incident light at the Brewster (polarization) angle has the greatest polarization upon reflection, so it is the simple application of Equation 24.15.

**SOLUTION**

***Given:***

$n_1 = 1.00$
$n_2 = 2.40$

***Find:*** $\theta_1 = \theta_p$ (angle of incident for greatest polarization)

$\theta_p$ is the Brewster (polarization) angle. Using Equation 24.15, we find that

$$\theta_p = \tan^{-1}\left(\frac{n_2}{n_1}\right) = \tan^{-1}\left(\frac{2.40}{1.00}\right) = 67.4°$$

The ZnSe window in $CO_2$ lasers is therefore installed at this angle of incidence so as to maximize the polarization of the laser beam.

**FOLLOW-UP EXERCISE.** Light is incident on a flat, transparent material with an index of refraction of 1.52. At what angle of refraction would the transmitted light have the greatest polarization if the transparent material is in water?

---

### 24.4.3 Polarization by Double Refraction (Birefringence)

When monochromatic light travels through glass, its speed is the same in all directions and is characterized by a single index of refraction. Any material that has such a property is said to be *isotropic*, meaning that it has the same optical characteristics in all directions. Some crystalline materials, such as quartz, calcite, and ice, are *anisotropic*; that is, the speed of light, and therefore the index of refraction, are different for different directions within the material. Anisotropy gives rise to some interesting optical properties. Anisotropic materials are said to be doubly refracting, or to exhibit **birefringence**, and polarization is involved.

For example, unpolarized light incident on a birefringent crystal of calcite ($CaCO_3$, calcium carbonate) is illustrated in ▶ **Figure 24.27**. When the light propagates at an angle to a particular crystal axis, the beam is doubly refracted and separated into two components, or rays, upon refraction. These two rays are linearly polarized in mutually perpendicular directions. One ray, called the *ordinary* (o) *ray*, passes straight through the crystal and is characterized by an index of refraction $n_o$. The second ray, called the *extraordinary* (e) *ray*, is refracted and is characterized by an index of refraction $n_e$. The particular axis direction indicated by dashed lines in Figure 24.27a is called the optic axis. Along this direction, and nothing extraordinary is noted about the transmitted light.

Some transparent materials have the ability to rotate the plane of polarization of linearly polarized light. This property, called **optical activity**, is due to the molecular structure of the material (▶ **Figure 24.28a**). Optically active molecules include those of certain proteins, amino acids, and sugars.

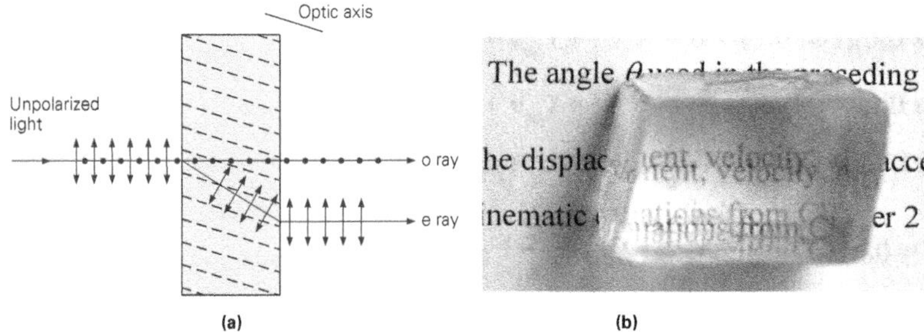


▲ **FIGURE 24.27 Double refraction or birefringence (a)** Unpolarized light incident normal to the surface of a birefringent crystal and at an angle to a particular direction in the crystal (dashed lines) is separated into two components. The ordinary (o) ray and the extraordinary (e) ray are linearly polarized in mutually perpendicular directions. **(b)** Double refraction seen through a calcite crystal.

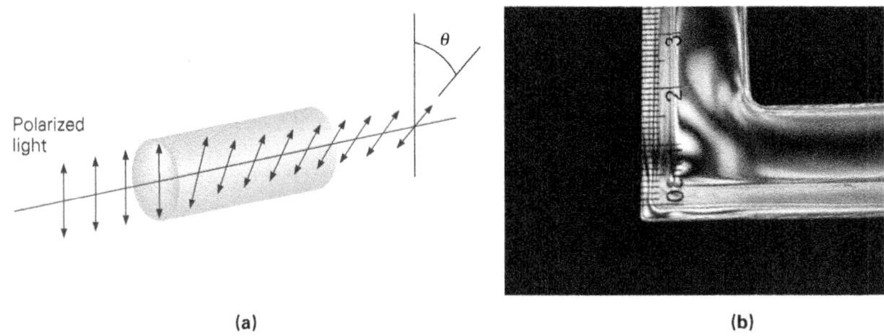


▲ **FIGURE 24.28 Optical activity and stress detection (a)** Some substances have the property of rotating the polarization direction of linearly polarized light. This ability, which depends on the molecular structure of the substance, is called *optical activity.* **(b)** Glasses and plastics become optically active under stress, and the points of greatest stress are apparent when the material is viewed through crossed polarizers. Engineers can thus test plastic models of structural elements to see where the greatest stresses will occur when the models are "loaded." Here, a model of a suspension bridge strut is being analyzed.

Glasses and plastics become optically active under stress. The greatest rotation of the direction of polarization occurs in the regions where the stress is the greatest. Viewing the stressed piece of material through crossed polarizers allows the points of greatest stress to be identified. This determination is called *optical stress* analysis (Figure 24.28b).

Today, *liquid crystal displays* (LCDs) are commonplace in items such as watches, calculators, televisions, and computer screens. A common type of LCD, called a *twisted nematic display,* makes use of polarized light. A polarizer can be used to easily check this property (▼ **Figure 24.29**).

## 24.5 Atmospheric Scattering of Light

When light is incident on a suspension of particles, such as the molecules in air, some of the light may be absorbed and reradiated in all directions. This process is called *scattering.* The scattering of sunlight in the atmosphere produces some interesting effects, including the polarization of skylight (that is, sunlight that has been scattered by the atmosphere), the blueness of the sky, and the redness of sunsets and sunrises.

Atmospheric scattering causes skylight to be polarized. When unpolarized sunlight is incident on air molecules, the electric

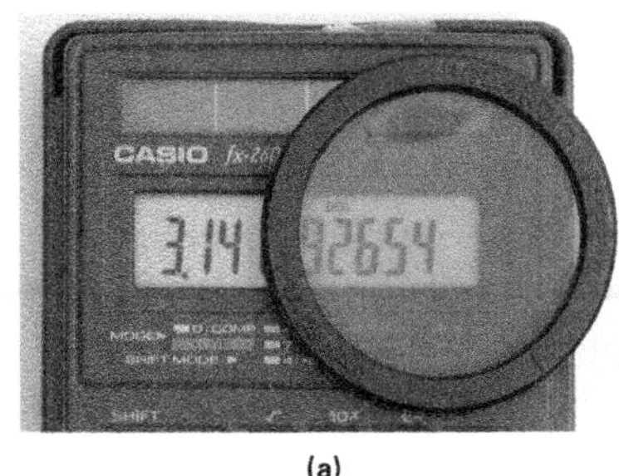


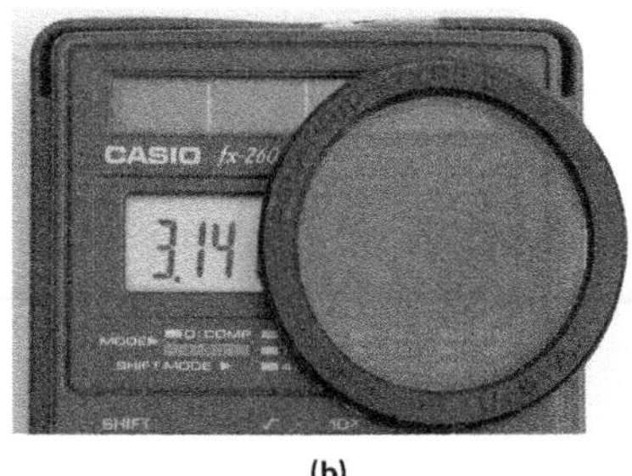


▲ **FIGURE 24.29 Liquid crystal display** (LCD) **(a)** The transmission axes of the circular polarizer and the calculator polarizer are parallel. **(b)** The circular polarizer is rotated by 90°.

field of the light wave sets electrons of the molecules into vibration. The vibrations are complex, but these accelerated charges emit radiation, like the vibrating electrons in the antenna of a radio broadcast station (see Section 20.4). As illustrated in ▼ **Figure 24.30**, an observer viewing from an angle of 90° with respect to the direction of the sunlight will receive linearly polarized light, because the electric field component in the direction of travel cannot exist in a transverse wave. At other viewing angles, both components are present, and skylight seen through a polarizing filter appears partially polarized.

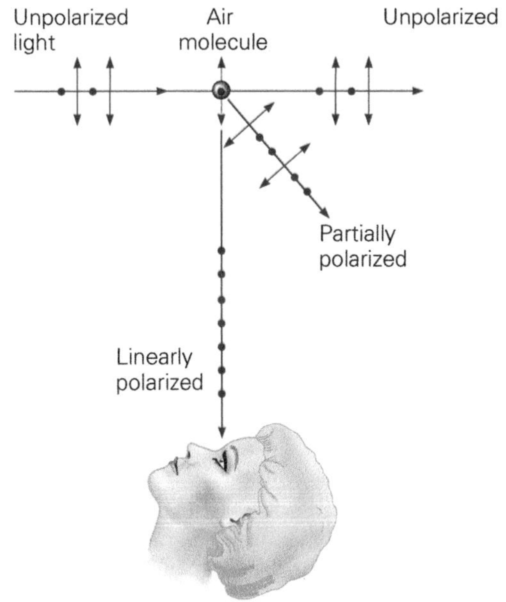


▲ FIGURE 24.30 **Polarization by scattering** When incident unpolarized sunlight is scattered by a gas molecule in the air, the light perpendicular to the direction of the incident ray is linearly polarized. Light scattered at some arbitrary angle is partially polarized. An observer at a right angle to the direction of the incident sunlight receives linearly polarized light.

Since the scattering of light with the greatest degree of polarization occurs at a right angle to the direction of the Sun, at sunrise and sunset the scattered light from directly overhead has the greatest degree of polarization. The polarization of skylight can be observed by viewing the sky through a polarizing filter (or a polarizing sunglass lens) and rotating the filter. Light from different regions of the sky will be transmitted in different degrees, depending on its degree of polarization. It is believed that some insects, such as bees, use polarized skylight to determine navigational directions relative to the Sun.

## 24.5.1 Why the Sky Is Blue

The scattering of sunlight by air molecules is the reason why the sky looks blue. This effect is not due to polarization, but to the selective absorption of light. As oscillators, air molecules have resonant frequencies (at which they scatter most efficiently) in the blue-violet region. Consequently, when sunlight is scattered, the blue end of the visible spectrum is scattered more than the red end.

For particles such as air molecules, which are much smaller than the wavelength of light, the intensity of the scattered light is inversely proportional to the wavelength to the fourth power ($1/\lambda^4$). This relationship between wavelength and scattering intensity is called **Rayleigh scattering**, after Lord Rayleigh (1842–1919), a British physicist who derived it. This inverse relationship predicts that light of the shorter wavelength, or blue, end of the spectrum will be scattered much more than light of the longer-wavelength, or red, end. The scattered blue light is rescattered in the atmosphere and eventually is directed toward the ground. This is why the sky appears blue.

---

**EXAMPLE 24.8: RED AND BLUE – RAYLEIGH SCATTERING**

How much more is light at the blue end (400 nm) of the visible spectrum scattered by air molecules than light at the red end (700 nm)?

**THINKING IT THROUGH.** We know that Rayleigh scattering is proportional to $1/\lambda^4$ and that light from the blue end of the spectrum (shorter wavelength) is scattered more than light from the red end. The wording "how much more" implies a factor or ratio.

**SOLUTION**

The Rayleigh scattering relationship is $I \propto 1/\lambda^4$, where $I$ is the intensity of scattering for a particular wavelength. Thus, forming a ratio:

$$\frac{I_{\text{blue}}}{I_{\text{red}}} = \left(\frac{\lambda_{\text{red}}}{\lambda_{\text{blue}}}\right)^4$$

Inserting the wavelengths:

$$\frac{I_{\text{blue}}}{I_{\text{red}}} = \left(\frac{\lambda_{\text{red}}}{\lambda_{\text{blue}}}\right)^4 = \left(\frac{700 \text{ nm}}{400 \text{ nm}}\right)^4 = 9.4 \quad \text{or} \quad I_{\text{blue}} = 9.4 I_{\text{red}}$$

And so, blue light is scattered almost ten times as much as red light.

**FOLLOW-UP EXERCISE.** What wavelength of light is scattered twice as much as red light? What color light is this?

---

## 24.5.2 Why Sunsets and Sunrises Are Red

Beautiful red sunsets and sunrises are sometimes observed. When the Sun is near the horizon, sunlight travels a greater distance through the denser air near the Earth's surface. Since the light therefore undergoes a great deal of scattering, you might think that only the least scattered light, the red light, would reach observers on the Earth's surface. This would explain red sunsets. However, it has been shown that the dominant color of white light after only molecular scattering is orange. Thus, other types of scattering must shift the light from the setting (or rising) Sun toward the red end of the spectrum (▼ **Figure 24.31**).

▲ FIGURE 24.31 **Red sky at night** A black and white photo of a spectacular red sunset over water. The red sky results from the scattering of sunlight by atmospheric gases and small solid particles. A directly observed reddening Sun is due to the scattering of light in the blue end of the spectrum out of the direct line of sight.

Red sunsets have been found to result from the scattering of sunlight by atmospheric gases and by small dust particles. These particles are not necessary for the blueness of the sky, but are compulsory for deep-red sunsets and sunrises. (This is why spectacular red sunsets are observed in the months after large volcanic eruptions that put tons of particulate matter into the atmosphere.) Red sunsets occur most often when there is a high-pressure air mass to the west, since the concentration of dust particles is generally greater in high-pressure air masses than in low pressure air masses. Similarly, red sunrises occur most often when there is a high-pressure air mass to the east.

## Chapter 24 Review

- **Young's double-slit experiment** provides evidence of the wave nature of light and a way to measure the wavelength of light ($\approx 10^{-7}$ m)

  The angular position ($\theta$) of the **maxima** satisfies the condition

$$d \sin\theta = n\lambda \quad \text{for } n = 0, 1, 2, 3, \ldots \qquad (24.3)$$

  where $d$ is the slit separation.

  For small $\theta$, the distance between the $n$th maximum and the central maximum is

$$y_n \approx \frac{nL\lambda}{d} \quad \text{for } n = 0, 1, 2, 3, \ldots \qquad (24.4)$$

- Light reflected at a media boundary for which $n_2 > n_1$ undergoes a 180° **phase change**. If $n_2 > n_1$, there is no phase change on reflection. The phase changes affect thin film interference, which also depends on film thickness and index of refraction.

  The **minimum thickness for a nonreflecting film** is

$$t_{\min} = \frac{\lambda}{4n_1} \quad (\text{for } n_2 > n_1 > n_o) \qquad (24.7)$$

- In a **single-slit diffraction** experiment, the **minima** at angle $\theta$ satisfy

$$w \sin\theta = m\lambda \quad \text{for } m = 1, 2, 3, \ldots \qquad (24.8)$$

  where $w$ is the slit width. In general, the longer the wavelength as compared with the width of an opening or object, the greater the diffraction.

- For a **diffraction grating**, the maxima satisfy

$$d \sin\theta = n\lambda \quad \text{for } n = 0, 1, 2, \ldots \qquad (24.12)$$

  where $d = 1/N$ and $N$ is the number of lines per unit length.

- **Polarization** is the preferential orientation of the electric field vectors that make up a light wave and is evidence that light is a transverse wave. Light can be polarized by selective absorption, reflection, double refraction (**birefringence**), and scattering.

  When the transmission axes of a polarizer and an analyzer make an angle of $\theta$, the light intensity through the analyzer is given by **Malus' law**:

$$I = I_o \cos^2\theta \qquad (24.14)$$

  In reflection, if the angle of incidence is equal to the **polarizing (Brewster) angle** $\theta_p$, then the reflected light is linearly polarized:

$$\tan\theta_p = \frac{n_2}{n_1} \quad \text{or} \quad \theta_p = \tan^{-1}\left(\frac{n_2}{n_1}\right) \qquad (24.15)$$

- The intensity of **Rayleigh scattering** is inversely proportional to the fourth power of the wavelength of the light. The blueness of the Earth's sky results from the preferential scattering of sunlight by air molecules.

## End of Chapter Questions and Exercises

### Multiple Choice Questions

**24.1 Young's Double-Slit Experiment**

1. If the path length difference between two identical and coherent beams is 2.5λ when they arrive at a point on a screen, the point will be (a) bright, (b) dark, (c) multicolored, (d) gray.

2. In a Young's double-slit experiment using monochromatic light, if the slit spacing *d* increases, the interference maxima spacing will (a) decrease, (b) increase, (c) remain unchanged, (d) disappear.
3. When white light is used in Young's double-slit experiment, many maxima with a spectrum of colors are seen. In a given maximum, the color farthest from the central maximum is (a) red, (b) blue, (c) other colors.

**24.2 Thin Film Interference**

4. When a thin film of kerosene spreads out on water, the thinnest part looks bright. The index of refraction of kerosene is (a) greater than, (b) less than, (c) the same as that of water.
5. For a thin film with $n_o < n_1 < n_2$, where $n_1$ is the index of refraction of the film, the minimum film thickness for destructive interference of the reflected light is (a)$\lambda'/4$, (b) $\lambda'/2$, (c) $\lambda'$.
6. For a thin film with $n_o < n_1 > n_2$, where $n_1$ is the index of refraction of the film, a film thickness for destructive interference of the reflected light is (a) $\lambda'/4$, (b) $\lambda'/2$, (c) $\lambda'$, (d) both (b) and (c).

**24.3 Diffraction**

7. In a single-slit diffraction pattern, (a) all maxima have the same width, (b) the central maximum is twice as wide as the side maxima, (c) the side maxima are twice as wide as the central maximum, (d) none of the preceding.
8. In a single-slit diffraction pattern, if the wavelength of light decreases, the width of the central maximum will (a) increase, (b) decrease, (c) remain the same.
9. As the number of lines per unit length of a diffraction grating increases, the spacing between the maxima (a) increases, (b) decreases, (c) remains unchanged.

**24.4 Polarization**

10. A sound wave cannot be polarized. This is because sound is (a) a transverse wave, (b) a longitudinal wave, (c) none of the preceding.
11. Light can be polarized by (a) reflection, (b) refraction, (c) selective absorption, (d) all of the preceding.
12. The polarizing (Brewster) angle depends on (a) the indices of refraction of materials, (b) Bragg's law, (c) internal reflection, (d) interference.
13. The percentage of unpolarized light that will pass through two polarizing sheets with their transmission axes parallel to each other ideally is (a) 100%, (b) 50%, (c) 25%, (d) 0%.

**24.5 Atmospheric Scattering of Light (Optional)**

14. Scattering involves (a) the reflection of light off particles, (b) the refraction of light off particles, (c) the absorption and reradiation of light by particles, (d) the interference of light with particles.
15. Which of the following colors is scattered the least in the atmosphere: (a) blue, (b) yellow, (c) red, or (d) color makes no difference?

## Conceptual Questions

**24.1 Young's Double-Slit Experiment**

1. Discuss how the interference pattern in Young's double-slit experiment would change if the distance between the double slits decreases.
2. Describe what would happen to the interference pattern in Young's double-slit experiment if the wavelength of the monochromatic light were to decrease.
3. The intensity of the central maximum in the interference pattern of a Young's double-slit experiment is about four times that of either light wave. Is this a violation of the law of conservation of energy?

**24.2 Thin Film Interference**

4. When destructive interference of two waves occurs at a certain location, there is no energy at that location. Is this situation a violation of the law of conservation of energy? Explain.
5. At the center of a Newton's ring arrangement (Figure 24.11a), the air wedge has a thickness of zero. Why is this area always dark (Figure 24.11b)?
6. Most lenses used in cameras are coated with thin films and appear bluish-green when viewed in the reflected light (Figure 24.9a). What wavelengths are refracting through the lens?

**24.3 Diffraction**

7. In our discussion of single-slit diffraction, the length of the slit was assumed to be much greater than the width. What changes would be observed in the diffraction pattern if the length were comparable with the width of the slit?
8. From Equation 24.8, can the $m = 2$ minimum be seen if $w = \lambda$? How about the $m = 1$ minimum?
9. In a diffraction grating, the slits are very closely spaced. What is the advantage of this design?

**24.4 Polarization**

10. Given two pairs of sunglasses, could you tell whether one or both were polarizing?
11. Suppose that you held two polarizing sheets in front of you and looked through both of them. How many times would you see the sheets lighten and darken (a) if one were rotated through one complete rotation, (b) if both were rotated through one complete rotation at the same speed in opposite directions, (c) if both were rotated through one complete rotation at the same speed in the same direction, and (d) if one were rotated twice as fast as the other and the slower one were rotated through one complete rotation?
12. How does selective absorption produce polarized light?
13. If you place a pair of polarizing sunglasses in front of your calculator's LCD display and rotate them, what would you observe?

**24.5 Atmospheric Scattering of Light (Optional)**

14. Explain why the sky may be red in the morning and evening and blue during the day.
15. What color would an astronaut on the Moon see when looking at the sky or into space?

## Exercises*

*Integrated Exercises (IEs) are two-part exercises. The first part typically requires a conceptual answer choice based on physical thinking and basic principles. The following part is quantitative calculations associated with the conceptual choice made in the first part of the exercise.*

**24.1 Young's Double-Slit Experiment**

1. • To study wave interference, a student uses two speakers driven by the same sound wave of wavelength 0.50 m. If the distances from a point to the speakers differ by 0.75 m, will the waves interfere constructively or destructively at that point? What if the distances differ by 1.0 m?
2. • In the development of Young's double-slit experiment, a small-angle approximation ($\tan\theta \approx \sin\theta$) was used to find the lateral displacements of the maxima (bright) and minima (dark) positions. How good is this approximation? For example, what is the percentage error for $\theta = 10°$?
3. • Two parallel slits 0.075 mm apart are illuminated with monochromatic light of wavelength 480 nm. Find the angle between the center of the central maximum and the center of the first side maximum.
4. • When two parallel slits are illuminated with monochromatic light of wavelength 632.8 nm, the angle between the center of the central maximum and the center of the second side maximum is 0.45°. What is the distance between the parallel slits?
5. • In a double-slit experiment that uses monochromatic light, the angular separation between the central maximum and the second-order maximum is 0.160°. What is the wavelength of the light if the distance between the slits is 0.350 mm?
6. IE •• Monochromatic light passes through two narrow slits and forms an interference pattern on a screen. (a) If the wavelength of light used increases, will the distance between the maxima (1) increase, (2) remain the same, or (3) decrease? Explain. (b) If the slit separation is 0.25 mm, the screen is 1.5 m away from the slits, and light of wavelength 550 nm is used, what is the distance from the center of the central maximum to the center of the third-order maximum? (c) What if the wavelength is 680 nm?
7. •• (a) If the wavelength used in a double-slit experiment is decreased, the distance between adjacent maxima will (1) increase, (2) decrease, (3) remain the same. Explain. (b) If the separation between the two slits is 0.20 mm and the adjacent maxima of the interference pattern on a screen 1.5 m away from the slits are 0.45 cm apart, what is the wavelength and color of the light? (c) If the wavelength is 550 nm, what is the distance between adjacent maxima?
8. •• In a double-slit experiment using monochromatic light, a screen is placed 1.25 m away from the slits, which have a separation distance of 0.0250 mm. The position of the third-order maximum is 6.60 cm from the center of the central maximum. Find (a) the wavelength of the light and (b) the position of the second-order maximum.
9. •• In a double-slit experiment with monochromatic light and a screen at a distance of 1.50 m from the slits, the angle between the second-order maximum and the central maximum is 0.0230 rad. If the separation distance of the slits is 0.0350 mm, what are (a) the wavelength and color of the light and (b) the lateral displacement of this maximum?
10. IE •• Two parallel slits are illuminated with monochromatic light, and an interference pattern is observed on a screen. (a) If the distance between the slits were decreased, would the distance between the maxima (1) increase, (2) remain the same, or (3) decrease? Explain. (b) If the slit separation is 1.0 mm, the wavelength is 640 nm, and the distance from the slits to the screen is 3.00 m, what is the separation between adjacent interference maxima? (c) What if the slit separation is 0.80 mm?
11. IE •• (a) In a double-slit experiment, if the distance from the double slits to the screen is increased, the separation between the adjacent maxima will (1) increase, (2) decrease, (3) remain the same. Explain. (b) Yellow-green light ($\lambda = 550$ nm) illuminates a double-slit separated by $1.75 \times 10^{-4}$ m. If the screen is located 2.00 m from the slits, determine the separation between the adjacent maxima. (c) What if the screen is located 3.00 m from the slits?
12. •• (a) Derive a relationship that gives the locations of the minima in a Young's double-slit experiment. What is the distance between adjacent minima? (b) For a third-order minimum (the third side dark position from the central maximum), what is the path length difference between that location and the two slits?
13. •• When a double-slit setup is illuminated with light of wavelength 632.8 nm, the distance between the center of the central bright position and the second side dark position is 4.5 cm on a screen that is 2.0 m from the slits. What is the distance between the slits?

---

* The bullets denote the degree of difficulty of the exercises: •, simple; ••, medium; and •••, more difficult.

14. IE ••• (a) If the apparatus for a Young's double-slit experiment were completely immersed in water, would the spacing of the interference maxima (1) increase, (2) remain the same, or (3) decrease? Explain. (b) What would the lateral displacements in Exercise 6 be if the entire system were immersed in still water?
15. ••• Light of two different wavelengths is used in a double-slit experiment. The location of the third-order maximum for the first light, yellow-orange light ($\lambda = 600$ nm), coincides with the location of the fourth-order maximum for the other color's light. What is the wavelength of the other light?

## 24.2 Thin Film Interference

16. • A film on a lens with an index of refraction of 1.5 is $1.0 \times 10^{-7}$ m is thick and is illuminated with white light. The index of refraction of the film is 1.4. (a) The number of waves that experience the 180° phase shift is (1) zero, (2) one, (3) two. Explain. (b) For what wavelength of visible light will the lens be nonreflecting?
17. •• A solar cell is designed to have a nonreflective film of a transparent material for a wavelength of 550 nm. (a) Will the thickness of the film depend on the index of refraction of the underlying material in the solar cell? Discuss the possible scenarios. (b) If $n_{solar} > n_{film}$ and $n_{film} = 1.22$, what is the minimum thickness of the film? (c) Repeat the calculation in (b) if $n_{solar} < n_{film}$ and $n_{film} = 1.40$.
18. IE •• A thin layer of oil ($n = 1.50$) floats on water. Destructive interference is observed for reflected light of wavelengths 480 and 600 nm, each at a different location. (a) If the order number is the same for both wavelengths, which wavelength is at a greater thickness: (1) 480 nm, or (2) 600 nm? Explain. (b) Write the general condition of destructive interference for reflected light. (c) Find the two minimum thicknesses of the oil film, assuming normal incidence.
19. •• Two parallel plates are separated by a small distance as illustrated in ▼**Figure 24.32**. If the top plate is illuminated with light from a He–Ne laser ($\lambda = 632.8$ nm), for what minimum separation distances will the light be (a) constructively reflected and (b) destructively reflected? [*Note*: $t = 0$ is *not* an answer for part (b).]

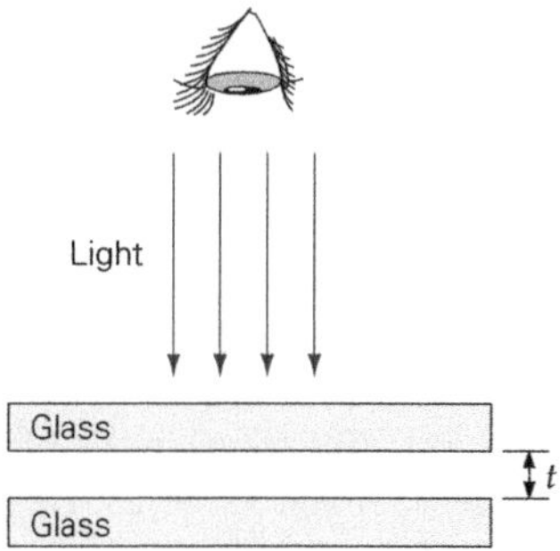


▲FIGURE 24.32 **Reflection or transmission?** See Exercise 19.

## 24.3 Diffraction

20. • A slit of width 0.15 mm is illuminated with monochromatic light of wavelength 632.8 nm. At what angle will the first minimum occur?
21. • In a single-slit diffraction pattern using light of wavelength 550 nm, the second-order minimum is measured to be at 0.32°. What is the slit width?
22. • A slit of width 0.20 mm is illuminated with monochromatic light of wavelength 480 nm, and a diffraction pattern is formed on a screen 1.0 m from the slit. (a) What is the width of the central maximum? (b) What are the widths of the second- and third-order maxima?
23. • A slit 0.025 mm wide is illuminated with red light ($\lambda = 680$ nm). How wide are (a) the central maximum and (b) the side maxima of the diffraction pattern formed on a screen 1.0 m from the slit?
24. • At what angle will the second-order maximum be seen from a diffraction grating of spacing 1.25 $\mu$m when illuminated by light of wavelength 550 nm?
25. • A venetian blind is essentially a diffraction grating – not for visible light, but for waves with much longer wavelengths. If the spacing between the slats of a blind is 2.5 cm, (a) for what wavelength would there be a first-order maximum at an angle of 10°, and (b) what type of radiation is this?
26. IE •• A single slit is illuminated with monochromatic light, and a screen is placed behind the slit to observe the diffraction pattern. (a) If the width of the slit is increased, will the width of the central maximum (1) increase, (2) remain the same, or (3) decrease? Why? (b) If the width of the slit is 0.50 mm, the wavelength is 680 nm, and the screen is 1.80 m from the slit, what would be the width of the central maximum? (c) What if the width of the slit is 0.60 mm?
27. IE •• (a) If the wavelength used in a single-slit diffraction experiment increases, will the width of the central maximum (1) increase, (2) remain the same, or (3) decrease? Why? (b) If the width of the slit is 0.45 mm, the wavelength is 400 nm, and the screen is 2.0 m from the slit, what would be the width of the central maximum? (c) What if the wavelength is 700 nm?
28. •• Find the angles of the blue ($\lambda = 420$ nm) and red components of the first- and second-order maxima in a pattern produced by a diffraction grating with 7500 lines/cm.
29. •• A certain crystal gives a deflection angle of 25° for the first-order maximum of monochromatic X-rays with a frequency of $5.0 \times 10^{17}$ Hz. What is the lattice spacing of the crystal?
30. IE •• (a) Only a limited number of maxima can be observed with a diffraction grating. The factor(s) that limit(s) the number of maxima seen is (are) (a) (1) the wavelength, (2) the grating spacing, (3) both. Explain. (b) How many maxima appear when monochromatic

light of wavelength 560 nm illuminates a diffraction grating that has 10 000 lines/cm, and what are their order numbers?

31. •• A diffraction grating with 6000 lines/cm is illuminated with a red light from a He–Ne laser ($\lambda = 632.8$ nm). How many side maxima are formed in the diffraction pattern, and at what angles are they observed?
32. •• In a particular diffraction grating pattern, the red component (700 nm) in the second-order maximum is deviated at an angle of 20°. (a) How many lines per centimeter does the grating have? (b) If the grating is illuminated with white light, how many maxima of the complete visible spectrum would be produced?
33. •• The commonly used CD (compact disc) consists of many closely spaced tracks that can be used as reflecting gratings. The industry standard for the track-to-track distance is 1.6 $\mu$m. If a He–Ne laser with a wavelength of 632.8 nm is incident normally onto a CD, calculate the angles for all the visible maxima.
34. **IE** •• White light of wavelength ranging from 400 to 700 nm is used for a diffraction grating with 6500 lines per centimeter. (a) In a particular order of maximum, red color will have (1) a larger, (2) the same, or (3) a smaller angle than blue color. Explain. (b) Calculate the angles for 400 and 700 nm in the second-order maximum. (c) What is the angular width of the whole spectrum in the second order?
35. **IE** •• White light ranging from blue (400 nm) to red (700 nm) illuminates a diffraction grating with 8000 lines/cm. (a) For the first maxima measured from the central maximum, the blue color is (1) closer to, (2) farther from, or (3) at the same location as the red color. Explain. (b) What are the angles of the first-order maximum for blue and red?

### 24.4 Polarization

36. **IE** • Unpolarized light is incident on a polarizer–analyzer pair that can have their transmission axes at an angle of either 30° or 45°. (a) The 30° angle will allow (1) more, (2) the same, or (3) less light to go through. (b) Calculate the percentage of light that goes through the polarizer–analyzer pair in terms of the incident light intensity.
37. **IE** • When unpolarized light is incident on a polarizer–analyzer pair, 30% of the original light intensity passes the analyzer. What is the angle between the transmission axes of the polarizer and analyzer?
38. • Some types of glass have a range of indices of refraction of about 1.4 to 1.7. What is the range of the polarizing (Brewster) angle for these glasses when light is incident on them from air?
39. **IE** • Light is incident on a certain material in air. (a) If the index of refraction of the material increases, the polarizing (Brewster) angle will (1) also increase, (2) decrease, (3) remain the same. Explain. (b) What are the polarizing angles if the index of refraction is 1.6 and 1.8?
40. **IE** •• Unpolarized light of intensity $I_0$ is incident on a polarizer–analyzer pair. (a) If the angle between the polarizer and analyzer increases in the range of 0°–90°, the transmitted light intensity will (1) also increase, (2) decrease, (3) remain the same. Explain. (b) If the angle between the polarizer and analyzer is 30°, what light intensity would be transmitted through the polarizer and the analyzer, respectively? (c) What if the angle is 60°?
41. •• A beam of light is incident on a glass plate ($n = 1.62$) in air and the reflected ray is completely polarized. What is the angle of refraction for the beam?
42. •• The critical angle for total internal reflection in a certain media boundary is 45°. What is the polarizing (Brewster) angle for light externally incident on the same boundary?
43. •• The polarizing (Brewster) angle for a certain media boundary is 33°. What is the critical angle for total internal reflection for the same boundary?
44. **IE** •• The angle of incidence is adjusted so there is maximum linear polarization for the reflected light from a transparent piece of plastic in air. (a) There is (1) no, (2) maximum, or (3) some light transmitted through the plastic. Explain. (b) If the index of refraction of the plastic is 1.40, what would be the angle of refraction in the plastic?
45. **IE** •• (a) The polarizing (Brewster) angle of a piece of flint glass ($n = 1.66$) in water is (1) greater than, (2) less than, (3) the same as that of the glass in air. Explain. (b) What are the polarizing angles when it is in air and submerged in water, respectively?
46. •• Sunlight is reflected off a vertical plate-glass window ($n = 1.55$). What would the Sun's altitude (angle above the horizon) have to be for the reflected light to be completely polarized?
47. ••• A plate of crown glass ($n = 1.52$) is covered with a layer of water. A beam of light traveling in air is incident on the water and partially transmitted. Is there any angle of incidence for which the light reflected from the water–glass interface will have maximum linear polarization? Justify your answer mathematically.

### 24.5 Atmospheric Scattering of Light (Optional)

48. **IE** •• Sunlight is scattered by air molecules. (a) The intensity of the scattered blue light is (1) greater, (2) the same as, or (3) less than that of the scattered red light. Explain. (b) Calculate the ratio of the scattered light intensity of blue color (400 nm) to that of red color (700 nm).
49. **IE** •• When sunlight is scattered by air molecules, the intensity of scattered light for a wavelength of 550 nm is greater than another color by a factor of 5.0. (a) The wavelength of the other color is (1) longer, (2) the same as, or (3) shorter than 550 nm. Explain. (b) What is the wavelength of the other color?

# 25

# Vision and Optical Instruments



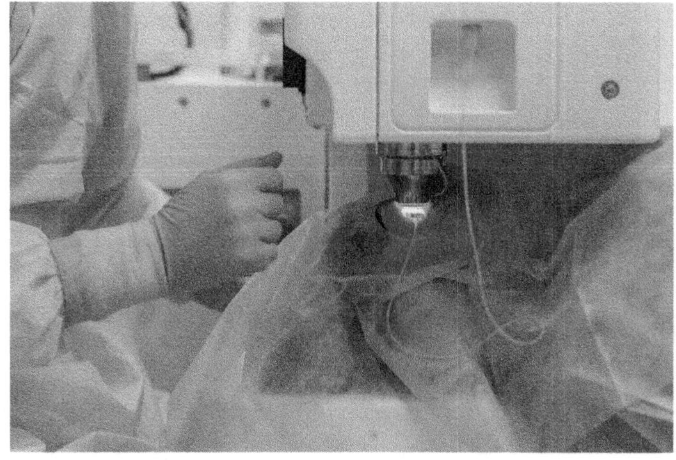

**An ophthalmology laser operation to correct a vision defect.**

Vision is one of our chief means of acquiring information about the world around us. However, the images seen by many eyes are not always clear or in focus, so glasses or some other remedy are needed. Great progress has been made in the last several decades in contact lens therapy and surgical correction of vision defects. A popular procedure is laser surgery, as shown in the chapter-opening photograph. Laser surgery can be used for such procedures as repairing torn retinas, destroying eye tumors, and stopping abnormal growth of blood vessels that can endanger vision.

Optical instruments, whose basic function is to improve and extend the power of observation beyond that of the human eye, augment our vision. Mirrors and lenses are used in a variety of optical instruments, including microscopes and telescopes.

The earliest magnifying glasses were drops of water captured in small holes. By the seventeenth century, artisans were able to grind fair-quality lenses for spectacles, simple microscopes, or magnifying glasses, which were used primarily for botanical studies. Soon thereafter, the basic compound microscope, which uses two lenses, was developed. Modern compound microscopes, which can magnify an object up to 2000 times, extended our vision into the microscopic world.

Around 1609, Galileo used lenses to construct an astronomical telescope that allowed him to observe valleys and mountains on the Moon, sunspots, and the four largest moons of Jupiter. Today, huge telescopes that use multiple lenses and mirrors (to form very large equivalent lenses and mirrors) have extended our vision far into the past as we look at more distant, and therefore younger, galaxies from which the light takes years to reach us.

How much knowledge would we lack if these instruments had never been invented? Bacteria would still be unknown, and planets, stars, and galaxies would have remained nothing but mysterious points of light.

Mirrors and lenses were discussed in terms of geometrical optics in Chapter 23, and the wave nature of light was investigated in Chapter 24. These principles can be applied to the study of vision and optical instruments. In this chapter, you will learn about our fundamental optical instrument - the human eye, without which all others would be of little use. Also, microscopes and telescopes will be discussed, along with the factors that limit their resolutions.

## 25.1 The Human Eye

The human eye is the most important of all optical instruments. Without it we would know little about our world and the study of optics would not exist. The human eye is analogous to a camera in several respects (▼**Figure 25.1**). A camera consists of a converging lens, which is used to focus images on light-sensitive film (traditional camera) or a *digital sensor* (digital cameras) at the back of the camera's interior chamber. (Recall from Chapter 23 that for relatively distant objects, a converging lens produces a real, inverted, and reduced image.) There is an adjustable diaphragm opening, or aperture, and a shutter to control the amount of light entering the camera.

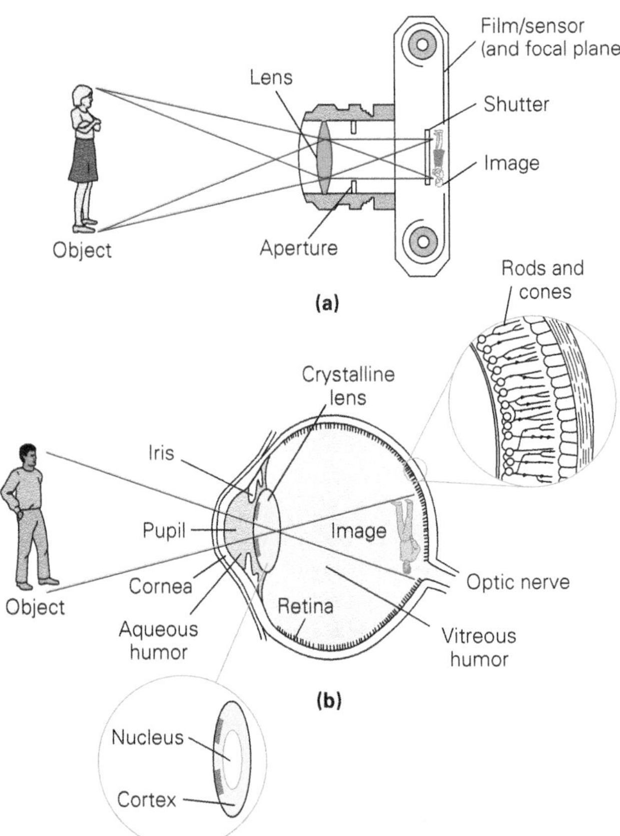


▲FIGURE 25.1 **Camera and eye analogy** In some respects, **(a)** a camera is similar to **(b)** the human eye. An image is formed on the film or digital sensor in a camera and on the retina of the eye. (The complex refractive properties of the eye are not shown here, because multiple refractive media are involved.) See text for a comparative description.

The eye, too, focuses images onto a light-sensitive lining (the retina) on the rear surface of the eyeball. The eyelid might be thought of as a shutter; however, the shutter of a camera, which controls the exposure time, is generally opened only for a fraction of a second, while the eyelid normally remains open for continuous exposure. The human nervous system actually performs a function analogous to a shutter by analyzing image signals from the eye at a rate of 20 to 30 times per second. The eye might therefore be better likened to a movie or video camera, which exposes a similar number of frames (images) per second.

Although the optical functions of the eye are relatively simple, its physiological functions are quite complex. As Figure 25.1b shows, the eyeball is a nearly spherical chamber. It has an internal diameter of about 1.5 cm and is filled with a transparent jellylike substance called the *vitreous humor*. The eyeball has a white outer covering called the *sclera*, part of which is visible as the "white" of the eye. Light enters the eye through a curved, transparent tissue called the *cornea* and passes into a clear fluid known as the *aqueous humor*. Behind the cornea is a circular diaphragm, the *iris*, whose central opening is the *pupil*. The iris contains the pigment that determines eye color. Through muscle action, the area of the pupil can change (from 2–8 mm in diameter), thereby controlling the amount of light entering the eye.

Behind the iris is a *crystalline lens*, a converging lens composed of microscopic glassy fibers. (A cortex surrounding a nucleus, inset of Figure 25.1.) When tension is exerted on the lens by attached ciliary muscles, the glassy fibers slide over each other, causing the shape, and the focal length, of the lens to change, this allows an image to focus on the retina properly. Notice that this is an *inverted* image (Figure 25.1b). We do not "see" an inverted image, however, because the brain reinterprets this image as being upright.

On the back interior wall of the eyeball is a light-sensitive surface called the *retina*. From the retina, the optic nerve relays retinal signals to the brain. The retina is composed of nerves and two types of light receptors, or photosensitive cells, called rods and cones because of their shapes. The rods are more sensitive to light than the cones and distinguish light from dark in low light intensities (twilight vision). The cones can distinguish frequency ranges but require brighter light. The brain interprets these different frequencies as colors (color vision). Most of the cones are clustered in a central region of the retina called the *macula*. The rods, which are more numerous than the cones, are outside this region and are distributed nonuniformly over the retina.

The focusing mechanism of the eye differs from that of a simple camera. A non-zoom camera lens has a fixed focal length, and the image distance is varied by moving the lens relative to the film to produce sharp images for different object distances. In the eye, the image distance is constant, and the focal length of the lens is varied (as the attached ciliary muscles change the lens's shape) to produce sharp images on the retina regardless of object distance. When the eye is focused on distant objects, the muscles are relaxed and the crystalline lens is thinnest with a power of about 20 D (diopters). Recall from Chapter 23 that the power ($P$) of a lens in diopters (D) is the reciprocal of its focal length *in meters*. So 20 D corresponds to a focal length of $f = 1/(20\text{ D}) = 0.050\text{ m} = 50\text{ mm}$.

When the eye is focused on closer objects, the lens becomes thicker. Then its radius of curvature and hence its focal length are decreased. For close-up vision, the lens power may increase to 30 D ($f = 33$ mm), or even more in young children. The adjustment of the focal length of the crystalline lens is called *accommodation*. (Look at a nearby object and then at an object in the distance, and notice how fast accommodation takes place. It's practically instantaneous!)

The distance extremes over which sharp focus is possible are known as the *far point* and the *near point*. The *far point* is the greatest distance at which the eye can see objects clearly, and is infinity for a normal eye. The *near point* is the position closest to the eye at which objects can be seen clearly. This position depends on the extent to which the crystalline lens can be deformed (thickened) by accommodation. The range of accommodation gradually diminishes with age as the crystalline lens loses its elasticity. Generally, in the normal eye the near point gradually recedes (increases) with age. The approximate positions of the near point at various ages are listed in ▼ **Table 25.1**.

TABLE 25.1 Approximate Near Points of the Normal Eye at Different Ages

| Age (years) | Near Point (cm) |
|---|---|
| 10 | 10 |
| 20 | 12 |
| 30 | 15 |
| 40 | 25 |
| 50 | 40 |
| 60 | 100 |

Children can see sharp images of objects that are within 10 cm of their eyes, and the crystalline lens of the eye of a normal young adult can do the same for objects as close as 12 to 15 cm. However, adults at age 40 normally experience a shift in the near point to about 25 cm. You may have noticed middle-aged people holding reading material fairly far from their eyes so as to keep it within their range of accommodation. When the print becomes too small (or the arms too short), corrective reading glasses are one solution. The recession of the near point with age is not considered an abnormal defect. Since it proceeds at about the same rate in most normal eyes, it is considered a part of the natural aging process.

### 25.1.2 Vision Defects

The existence of a "normal" eye implies that some eyes must have defects. This is indeed the case, as is quite apparent from the number of people who wear corrective glasses or contact lenses. Many people have eyes that cannot accommodate within the normal range (25 cm to infinity).* These people usually have one or both of the two most common visual defects: nearsightedness (myopia) or farsightedness (hyperopia). Both of these conditions can usually be corrected with glasses, contact lenses, or surgery (▲ **Figure 25.2a**).

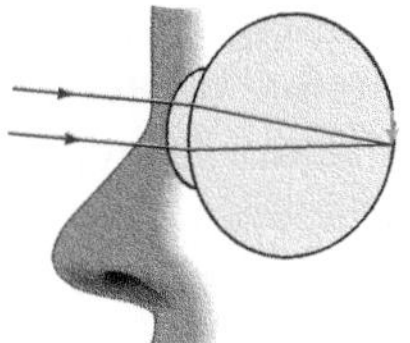

(a) Normal

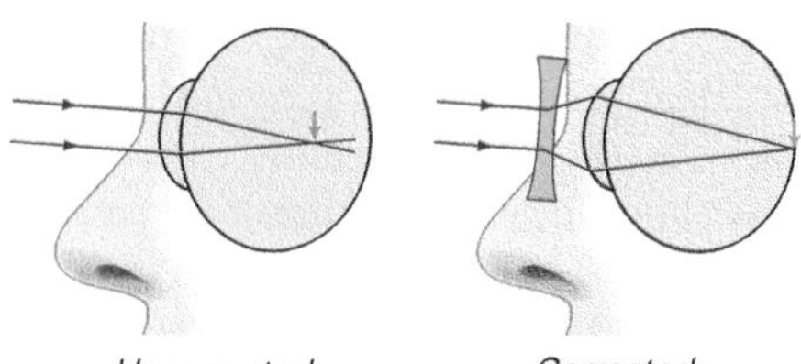



(b) Nearsightedness (myopia)

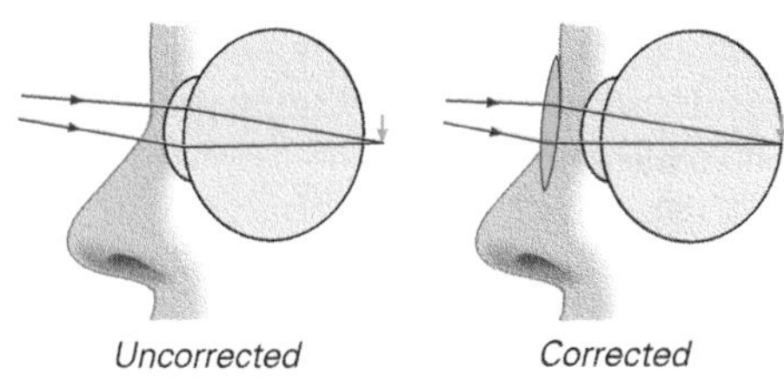



(c) Farsightedness (hyperopia)

▲ FIGURE 25.2 **Nearsightedness and farsightedness (a)** The normal eye produces sharp images on the retina for objects located between its near point and its far point. The image is real, inverted, and always smaller than the object. (Why?) Here, the object is a distant, upward-pointing arrow (not shown) and the light rays come from its tip. **(b)** In a nearsighted eye, the image of a *distant* object is focused *in front of* the retina. This defect is corrected with a diverging lens. **(c)** In a farsighted eye, the image of a *nearby* object is focused *behind* the retina. This defect is corrected with a converging lens. (Not drawn to scale.)

**Nearsightedness** (or *myopia*) is the ability to see nearby objects clearly, but not distant objects. That is, the far point is less than infinity. When an object is beyond the far point, the rays focus in *front* of the retina (Figure 25.2b). As a result, the image on the retina is blurred, or out of focus. As the object is moved closer, its image moves back toward the retina. When the object reaches the far point for that eye, a sharp image is formed on the retina.

Nearsightedness usually arises because the eyeball is too long (elongated in the horizontal direction) or the curvature of the cornea is too great (bulging cornea). In the former, the retina is placed at a farther distance from the cornea–lens system, and in

* A standard near point distance of 25 cm is typically assumed in the design of optical instruments.

the latter, the eyeball converges the light from distant objects to a spot in front of the retina. Whatever the reason, the image is focused in front of the retina. Appropriate diverging lenses correct this condition. Such lenses cause the rays to diverge before reaching the cornea. The image is thus focused on the retina. The optical purpose of the corrective lens is to form an image of a distant object (at infinity) at the patient's far point.

**Farsightedness** (or *hyperopia*) is the ability to see distant objects clearly, but not nearby ones. That is, the near point is farther from the eye than normal. The image of an object that is closer than the near point is formed behind the retina (Figure 25.2c). Farsightedness arises because the eyeball is too short, because of insufficient curvature of the cornea, or because of the weakening of the ciliary muscles and insufficient elasticity of the crystalline lens. Appropriate converging lenses correct this condition. Such lenses cause the rays to converge before reaching the cornea. The image is thus focused on the retina. The optical purpose of the corrective lens is to form an image of an object at 25 cm (the normal near point) at the patient's near point. If this occurs as part of the aging process as previously discussed, it is called *presbyopia*.

### INTEGRATED EXAMPLE 25.1: CORRECTING NEARSIGHTEDNESS - USE OF DIVERGING LENSES

An optometrist can give a patient either regular glasses or contact lenses to correct nearsightedness (▼**Figure 25.3**). Usually, regular glasses sit a few centimeters in front of the eye and contact lenses sit right on the eye. (a) Should the power of the contact lenses prescribed be (1) the same as, (2) greater than, or (3) less than that of the regular glasses? Explain. (b) A certain nearsighted person cannot see objects clearly when they are more than 78.0 cm from either eye. What power must corrective lenses have, for both regular glasses and contact lenses, if this person is to see distant objects clearly? Assume that the glasses are 2.00 cm in front of the eye.

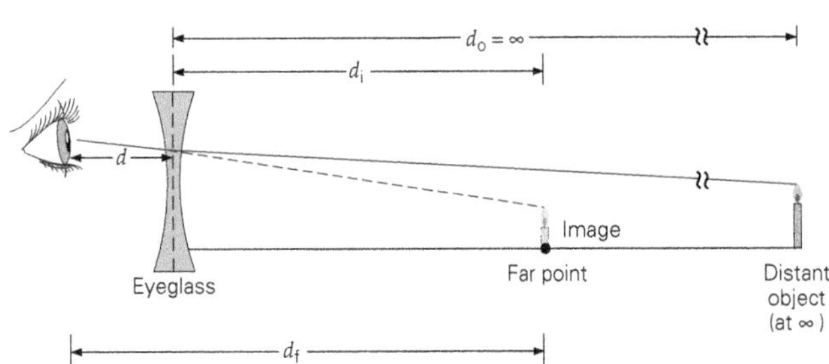


▲**FIGURE 25.3 Correcting nearsightedness** A diverging lens is used. Only regular glasses are shown. For contact lenses, the lens is immediately in front of the eye ($d=0$).

(a) **CONCEPTUAL REASONING.** For nearsightedness, the corrective lens is a diverging one (Figure 25.3). The lens must effectively put the virtual image of a distant object ($d_o=\infty$) at the patient's far point of the eye; that is, $d_f$ from the eye. The image, which acts as an object for the eye, is then within the range of accommodation. Because the image distance is *measured from the lens*, a contact lens will have a *longer* image distance. For a contact lens, $d_i=-(d_f)$. For regular glasses, $d_i=-(d_f-d)$, where $d$ is the distance between the regular glasses and the eye.

Note that $d_i$ is negative. Recall that the power of a lens is $P=1/f$ (Equation 23.9). We can use the thin lens equation (Equation 23.5) to find $P$ if we can determine the object and image distances, $d_o$ and $d_i$:

$$P=\frac{1}{f}=\frac{1}{d_o}+\frac{1}{d_i}=\frac{1}{\infty}+\frac{1}{d_i}=\frac{1}{d_i}=-\frac{1}{|d_i|}$$

That is, a longer $|d_i|$ will yield a smaller $P$, so the contact lenses should have a lower power than the regular glasses. Thus, the answer is (3).

(b) **QUANTITATIVE REASONING AND SOLUTION.** Once it is understood how corrective lenses work, the calculation for part (b) is straightforward.

***Given:***

$d_f = 78$ cm $= 0.780$ m (far point)
$d = 2.0$ cm $= 0.0200$ m (glass-eye distance)

***Find:***

$P$ (in diopters for regular glasses)
$P$ (in diopters for contact lenses)

For regular glasses (see Figure 25.3, which is not drawn to scale),

$$d_i=-(d_f-d)=-(0.780\,\text{m}-0.0200\,\text{m})=-0.760\,\text{m}$$

Then, using the thin lens equation,

$$P=\frac{1}{f}=\frac{1}{d_o}+\frac{1}{d_i}=\frac{1}{\infty}+\frac{1}{-0.760\,\text{m}}=-\frac{1}{0.760\,\text{m}}=-1.32\,\text{D}$$

A negative, or diverging, lens with a power of 1.32 D is needed. For contact lenses, $d=0$,

$$d_i=-0.780\,\text{m}$$

And using the thin lens equation,

$$P=\frac{1}{\infty}+\frac{1}{-0.780\,\text{m}}=-\frac{1}{0.780\,\text{m}}=-1.28\,\text{D}$$

The contact lenses have lower power, which is in agreement with the result of (a).

**FOLLOW-UP EXERCISE.** Suppose a mistake was made for the regular glasses in this Example such that a "corrective" lens of +1.32 D were used. What would happen to the image of objects at infinity? *(Answers to all Follow-Up Exercises are given in Appendix V at the back of the book.)*

If the far point for a nearsighted person is changed using diverging lenses (see Integrated Example 25.1), the near point will be affected as well. This causes the close-up vision to worsen, but *bifocal lenses* (▼ **Figure 25.4**) can be used in this situation to address the problem. Bifocals were invented by Benjamin Franklin, who glued two lenses together. They are now made by grinding or molding lenses with different curvatures in two different regions. Both nearsightedness and farsightedness can be treated at the same time with bifocals. Trifocals, or lenses having three different curvatures, are also available. The top lens is for far vision and the bottom lens for near vision. The middle lens is for intermediate vision and is sometimes referred to as a lens for "computer" vision.

▲ FIGURE 25.4 **Bifocal glasses** Both nearsightedness and farsightedness can be treated at the same time with bifocals (top for nearsightedness and bottom for farsightedness).

More modern techniques involve contact lens therapy or the use of a laser to correct nearsightedness. The purpose of either technique is to change the shape of the exposed surface of the cornea, which changes its refractive characteristics.

## INTEGRATED EXAMPLE 25.2: CORRECTING FARSIGHTEDNESS - USE OF CONVERGING LENS

A farsighted person has a near point of 75 cm for the left eye and a near point of 100 cm for the right one. (a) If the person is prescribed contact lenses, the power of the left lens should be (1) greater than, (2) the same as, (3) less than the power of the right lens. Explain. (b) What powers should contact lenses have to allow the person to see an object clearly at a distance of 25 cm?

**(a) CONCEPTUAL REASONING.** The normal eye's near point is 25 cm. For farsightedness, the corrective lens must be converging and form an image of an object at 25 cm (the normal near point) at the patient's near point. Since the near point of the left eye (75 cm) is closer to the 25-cm normal position than the right eye, the left lens should have lower power, so the answer is (3).

**(b) QUANTITATIVE REASONING AND SOLUTION.** The two different eyes are labeled as L (left) and R (right). The image distances are negative. (Why?)

***Given:***

$d_{iL} = -75\text{ cm} = -0.75\text{ m}$ (left image distance)
$d_{iR} = -100\text{ cm} = -1.0\text{ m}$ (right image distance)
$d_o = 25\text{ cm} = 0.25\text{ m}$ (object distance)

***Find:*** $P_L$ and $P_R$ (lens power)

A different lens prescription is usually required for each eye. In this case, each lens is to form an image of an object that is at a distance ($d_o$) of 0.25 m at its eye's near point. The image will then act as an object within the eye's range of accommodation. This situation is similar to a person wearing reading glasses (▼ **Figure 25.5**).

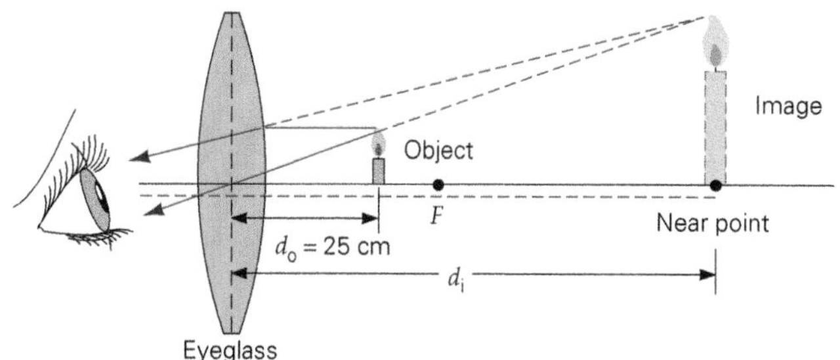


▲ FIGURE 25.5 **Reading glasses and correcting farsightedness** When an object at the normal near point (25 cm) is viewed through reading glasses with converging lenses, the image is formed farther away, but within the eye's range of accommodation (beyond the receded near point).

The image distances are negative, because the images are virtual (that is, the image is on the same side as the object). With contact lenses, the distance from the eye to the object and the distance from the lens to the object are the same. Then

$$P_L = \frac{1}{f_L} = \frac{1}{d_o} + \frac{1}{d_{iL}} = \frac{1}{0.25\text{m}} + \frac{1}{-0.75\text{m}} = \frac{2}{0.75\text{m}} = +2.7\text{ D}$$

and

$$P_R = \frac{1}{f_R} = \frac{1}{d_o} + \frac{1}{d_{iR}} = \frac{1}{0.25\text{m}} + \frac{1}{-1.0\text{m}} = \frac{3}{1.0\text{m}} = +3.0\text{ D}$$

Note that the left lens has lower power than the right lens, as expected.

**FOLLOW-UP EXERCISE.** A mistake is made in grinding or molding the corrective lenses in this Example such that the left lens is made to the prescription intended for the right eye, and vice versa. Discuss what happens to the images of an object at a distance of 25 cm.

Another common defect of vision is **astigmatism**, which is usually due to a refractive surface, usually the cornea or crystalline lens, being "out of round" (nonspherical). As a result, the eye has different focal lengths in different planes (▶ **Figure 25.6a**). Points may appear as lines, and the image of a line may be distinct in one direction and blurred in another or blurred in both directions. A test for astigmatism is given in Figure 25.6b.

Astigmatism can be corrected with lenses that have greater curvature in the plane in which the cornea or crystalline lens has deficient curvature (Figure 25.6c). Astigmatism is lessened in bright light, because the pupil of the eye becomes smaller, so only rays near the axis are entering the eye, thus avoiding the outer edges of the cornea.

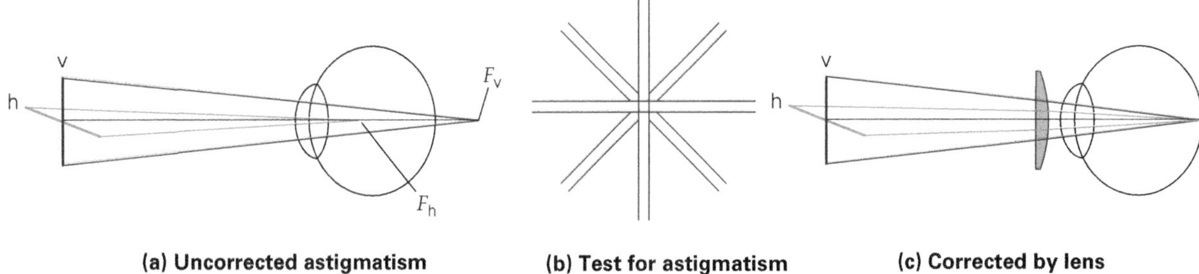


▲ **FIGURE 25.6** **Astigmatism** When one of the eye's refracting components is not spherical, the eye has different focal lengths in different planes. **(a)** The effect occurs because rays in the vertical plane (dark red) and horizontal plane (light red) are focused at different points: $F_v$ and $F_h$, respectively. **(b)** To someone with eyes that are astigmatic, some or all of the lines in this diagram will appear blurred. **(c)** Nonspherical lenses, such as plano-convex cylindrical lenses, are used to correct astigmatism.

You have probably heard of 20/20 *vision*. But what is it? *Visual acuity* is a measure of how vision is affected by object distance. This quantity is commonly determined by using a chart of letters placed at a given distance (usually 20 ft) from the eyes. The result is usually expressed as a fraction: The *numerator* is the distance at which the test eye sees a standard symbol, such as the letter E, clearly; the *denominator* is the distance at which the letter is seen clearly by a *normal* eye. A 20/20 (test/normal) rating, which is sometimes called "normal" vision, means that at a distance of 20 ft, the eye being tested can see standard-sized letters as clearly as can a normal eye. It is possible (and, in fact, very common) to see better than that. For example, a person with 20/15 acuity can see objects as clearly at 20 feet away as a person with normal vision at 15 feet away from the object. However, a person with 20/30 acuity indicates that at 20 feet, the person can see what the normal eye can see at 30 feet.

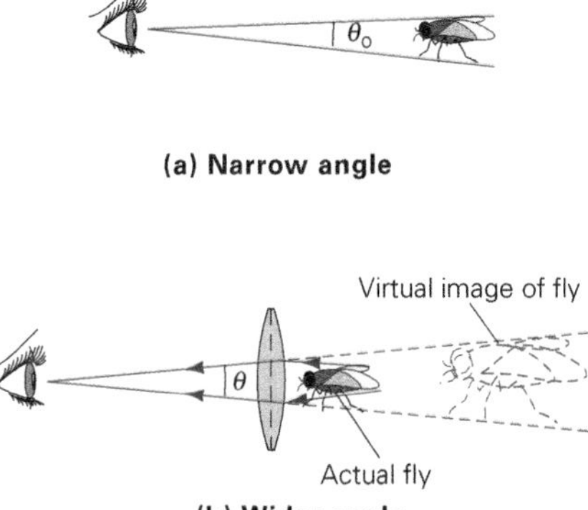


▲ **FIGURE 25.7** **Magnification and angle** **(a)** How large an object appears is related to the angle subtended by the object. **(b)** The angle and the size of the virtual image of an object are increased with a converging lens.

## 25.2 Microscopes

Microscopes are used to magnify objects so that we can see more detail or see features that are normally indiscernible. Two basic types of microscopes will be considered here.

### 25.2.1 The Magnifying Glass (A Simple Microscope)

When we look at a distant object, it appears very small. As it moves closer to our eyes, it appears larger. How large an object appears depends on the size of the image on the retina. This size is related to the angle subtended by the object – the greater the angle, the larger the image (▶ **Figure 25.7**).

When wanting to look at something closely, we bring it close to our eyes so that it subtends a greater angle. For example, you may examine the detail of a figure in this book by bringing it closer to your eyes. You'll see the greatest amount of detail when the book is at your near point. If your eyes were able to accommodate to shorter distances, an object brought closer would appear even larger. However, as you can easily prove by bringing this book very close to your eyes, images are blurred when objects are inside the near point.

A **magnifying glass** (sometimes called a *simple microscope*), which is just a single converging lens, forms a magnified image of an object when it is closer than the focal point (Figure 23.18b). In such a position, the image of an object subtends a greater angle than the object itself and therefore appears larger, or magnified (▶ **Figure 25.8**). The lens produces a virtual image beyond the near point on which the eye focuses. If a handheld magnifying glass is used, its position is usually adjusted until this image is seen clearly.

As illustrated in Figure 25.8, the angle subtended by the virtual image of an object is much greater when a magnifying glass is used. The angular magnification, or *magnifying power*, of an object *viewed through a magnifying glass* is expressed in terms of this angle. The **angular magnification**, *m*, is defined as the ratio of the angular size of the object as viewed through the magnifying glass ($\theta$) to the angular size of the object as viewed without the magnifying glass ($\theta_o$):

$$m = \frac{\theta}{\theta_o} \quad \text{(angular magnification)} \tag{25.1}$$

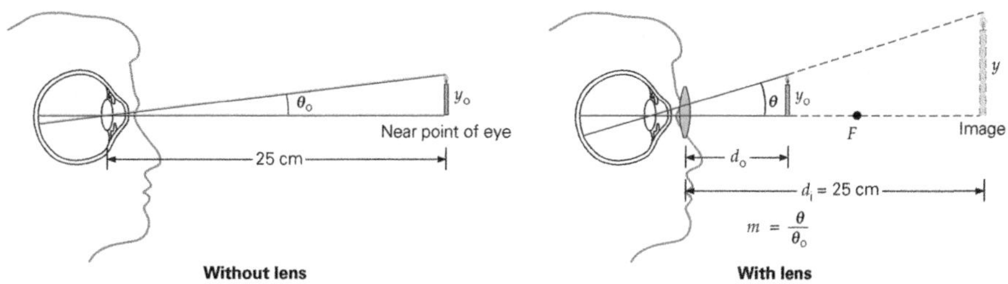


▲ **FIGURE 25.8 Angular magnification** The angular magnification ($m$) of a lens is defined as the ratio of the angular size of an object viewed through the lens to the angular size of the object viewed without the lens: $m = \theta/\theta_o$. Both the object (without lens) and the image (with lens) are at the near point (25 cm) for maximum magnification.

This $m$ is not defined the same as $M$, the lateral magnification, which is a ratio of heights: $M = h_i/h_o$ (Section 23.1).

The maximum angular magnification occurs when the image is at the eye's near point, $d_i = -25$ cm, since this position is as close as it can be seen clearly. (A value of 25 cm will be assumed to be typical for the near point of the normal eye. The negative sign is used because the image is virtual; see Chapter 23.) The corresponding object distance can be calculated from the thin lens equation, Equation 23.5, as

$$d_o = \frac{d_i f}{d_i - f} = \frac{(-25\,\text{cm})f}{-25\,\text{cm} - f}$$

or

$$d_o = \frac{(25\,\text{cm})f}{25\,\text{cm} + f} \qquad (25.2)$$

where $f$ must be in centimeters as well.

The angular size of the object is related to its height by

$$\tan\theta_o = \frac{y_o}{25} \quad \text{and} \quad \tan\theta = \frac{y_o}{d_o}$$

(See Figure 25.8.) Assuming that a small-angle approximation ($\tan\theta \approx \theta$) is valid,

$$\theta_o \approx \frac{y_o}{25} \quad \text{and} \quad \theta \approx \frac{y_o}{d_o}$$

Then the maximum angular magnification can be expressed as

$$m = \frac{\theta}{\theta_o} = \frac{y_o / d_o}{y_o / 25} = \frac{25}{d_o}$$

Substituting for $d_o$ from Equation 25.2 gives

$$m = \frac{25}{25f/(25+f)}$$

which simplifies to

$$m = 1 + \frac{25\,\text{cm}}{f} \quad \begin{pmatrix}\text{angular magnification} \\ \text{for image at 25cm}\end{pmatrix} \qquad (25.3)$$

where $f$ is in centimeters. Lenses with shorter focal lengths give greater angular magnifications.

In the derivation of Equation 25.3, the object being viewed by the unaided eye was taken to be at the near point, as was the image viewed through the lens. Actually, the normal eye can focus on an image located anywhere between the near point and infinity. When the image is at infinity, the eye is more relaxed – the muscles attached to the crystalline lens are relaxed, and the lens is thin. For the image to be at infinity, the object must be at the focal point of the lens. In this case,

$$\theta \approx \frac{y_o}{f}$$

and the angular magnification is

$$m = \frac{25\,\text{cm}}{f} \quad \begin{pmatrix}\text{angular magnification} \\ \text{for image at infinity}\end{pmatrix} \qquad (25.4)$$

Mathematically, it seems that the magnifying power can be increased to any desired value by using lenses that have sufficiently short focal lengths. However, lens aberrations limit the practical range of a single magnifying glass to about three or four times the size of the object (3× or 4×).

---

**EXAMPLE 25.3: ELEMENTARY HOLMES – ANGULAR MAGNIFICATION OF A MAGNIFYING GLASS**

Sherlock Holmes uses a converging lens with a focal length of 12 cm to examine the fine detail of some cloth fibers found at the scene of a crime. (a) What is the maximum magnification given by the lens? (b) What is the magnification for relaxed eye viewing?

**THINKING IT THROUGH.** Equations 25.3 and 25.4 apply here. Part (a) asks for the maximum magnification, which is discussed in the derivation of Equation 25.3 and occurs when the image formed by the lens is at the near point of the eye. For part (b), note that the eye is most relaxed when viewing distant objects.

**SOLUTION**

***Given:***

$f = 12$ cm (focal length)

***Find:***

**(a)** $m$ ($d_i$ = near point)
**(b)** $m$ ($d_i = \infty$)

**(a)** For Equation 25.3, the near point was taken to be 25 cm:

$$m = 1 + \frac{25\,\text{cm}}{f} = 1 + \frac{25\,\text{cm}}{12\,\text{cm}} = 3.1\times$$

**(b)** Equation 25.4 gives the magnification for the image formed by the lens at infinity:

$$m = \frac{25\,\text{cm}}{f} = \frac{25\,\text{cm}}{12\,\text{cm}} = 2.1\times$$

**FOLLOW-UP EXERCISE.** Taking the maximum practical magnification of a magnifying glass to be 4×, which would have the longer focal length, a lens for near-point viewing or one for distant viewing, and how much longer would its focal length be?

## 25.2.2 The Compound Microscope

A compound microscope provides greater magnification than that attained with a single lens, or a simple microscope. A basic compound microscope consists of a pair of converging lenses, each of which contributes to the overall magnification (▼ **Figure 25.9a**). The converging lens with a relatively short focal length ($f_o < 1$ cm) is known as the *objective*. It produces a real, inverted, and magnified image of an object positioned slightly beyond its focal point. The other lens, called the *eyepiece*, or *ocular*, has a longer focal length ($f_e$ is a few centimeters) and is positioned so that the image formed by the objective falls just *inside* its focal point. The eyepiece forms a virtual, upright, and magnified image of the image of the objective. Therefore, the final image observed is virtual, inverted, and magnified. In essence, the objective gives a magnified real image, and the eyepiece is a simple magnifying glass.

The **total magnification** ($m_{total}$) of a lens combination is the product of the magnifications produced by the two lenses. The image formed by the objective is larger than its object by a factor $M_o$ that is equal to the lateral magnification ($M_o = -d_i/d_o$). In Figure 25.9a, note that the image distance for the objective lens is approximately equal to the distance between the lenses, $L$ - that is, $d_i \approx L$. This is because the focal length of the eyepiece is usually much shorter than the distance between the two lenses and the image $I_o$ is formed by the objective just inside the focal point of the eyepiece. Also, the object is very close to the focal point of the objective, $d_o \approx f_o$. With these approximations,

$$M_o \approx -\frac{L}{f_o}$$

Equation 25.4 gives the angular magnification of the eyepiece for an image at infinity.

$$m_e = \frac{25\,\text{cm}}{f_e}$$

The total magnification is then equal to

$$m_{total} = -\frac{(25\,\text{cm})L}{f_o f_e} \quad \text{(angular magnification of compound microscope)} \tag{25.5}$$

where $f_o$, $f_e$, and $L$ are in centimeters.

The angular magnification of a compound microscope is negative, indicating that the final image is inverted compared to the initial orientation of the object. However, we usually state only the magnification (100×, not −100×).

**EXAMPLE 25.4: A COMPOUND MICROSCOPE - FINDING THE MAGNIFICATION**

A compound microscope has an objective with a focal length of 10 mm and an eyepiece with a focal length of 4.0 cm. The lenses are positioned 20 cm apart in the barrel. Determine the approximate total magnification of the microscope.

**THINKING IT THROUGH.** This is a direct application of Equation 25.5.

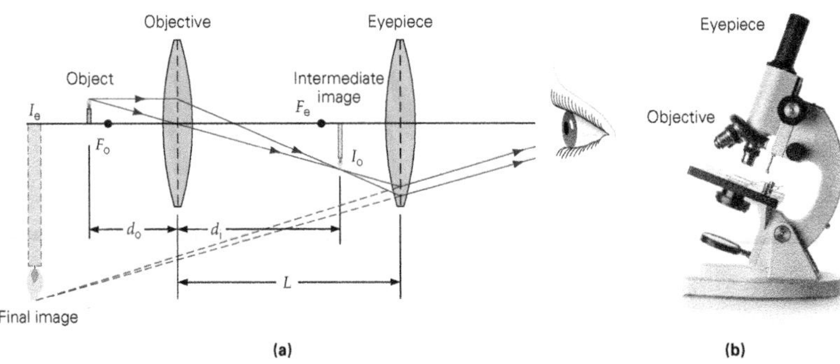


▲ **FIGURE 25.9 The compound microscope (a)** In the optical system of a compound microscope, the real image formed by the objective falls just within the focal point of the eyepiece ($f_e$) and acts as an object for the eyepiece. An observer looking through the eyepiece sees an enlarged image. **(b)** A compound microscope.

**SOLUTION**

***Given:***

$f_o = 10$ mm $= 1.0$ cm (objective focal length)
$f_e = 4.0$ cm (eyepiece focal length)
$L = 20$ cm (distance between the lenses)

***Find:*** $m_{total}$ (total magnification)

Using Equation 25.5, we get

$$m_{total} = -\frac{(25\text{cm})L}{f_o f_e} = -\frac{(25\text{cm})(20\text{cm})}{(1.0\text{cm})(4.0\text{cm})} = -125\times$$

Note the relatively short focal length of the objective. The negative sign indicates that the final image is inverted.

**FOLLOW-UP EXERCISE.** If the focal length of the eyepiece in this Example were doubled, how would the length of the microscope change if the same magnification were still desired?

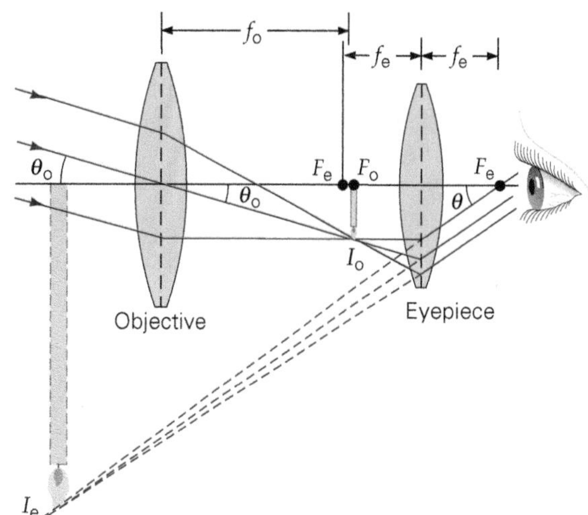


▲ **FIGURE 25.10 The refracting astronomical telescope** In an astronomical telescope, rays from a distant object form an intermediate image ($I_o$) at the focal point of the objective ($f_o$). The eyepiece ($f_e$) is moved so that the image is at or slightly inside its focal point. An observer sees an enlarged image at infinity ($I_e$, shown at a finite distance here for illustration).

A modern compound microscope is shown in Figure 25.9b. Interchangeable eyepieces with magnifications from about 5× to more than 100× are available. For standard microscopic work in biology or medical laboratories, 5× and 10× eyepieces are normally used. Microscopes are often equipped with rotating turrets, which usually contain three objectives for different magnifications, such as 10×, 43×, and 97×. These objectives and the 5× and 10× eyepieces can be used in various combinations to provide magnifying powers from 50× to 970×. The maximum magnification obtained from a compound optical microscope is about 2000×.

Opaque objects are usually illuminated with a light source placed above them. Specimens that are transparent, such as cells or thin sections of tissues on glass slides, are illuminated with a light source beneath the microscope stage so that light passes through the specimen.

Chapter 28.1 discusses electron microscope which has resolution about 500 000 times greater than that of a human eye.

## 25.3 Telescopes

Telescopes apply the optical principles of mirrors and lenses to allow some objects to be viewed in greater detail and other fainter or more distant objects simply to be seen. Basically, there are two types of telescopes – refracting and reflecting – which are characterized by the gathering and converging of light by lenses or mirrors, respectively.

### 25.3.1 Refracting Telescope

The principle underlying a **refracting telescope** is similar to that behind a compound microscope. The major components of a refracting telescope are the objective and eyepiece lenses, as illustrated in ▶ **Figure 25.10**. The objective is a large converging lens with a long focal length, and the movable eyepiece has a relatively short focal length. Rays from a distant object are essentially parallel and form an image ($I_o$) at the focal point ($f_o$) of the objective. This image acts as an object for the eyepiece ($f_e$), which is moved until the image lies just inside its focal point. The final image seen by the observer is virtual, inverted, and magnified.

For relaxed viewing, the eyepiece is adjusted so that its image ($I_e$) is at infinity, which means that the image of the objective ($I_o$) is at the focal point of the eyepiece ($f_e$). As Figure 25.10 shows, the distance between the lenses is then the sum of the focal lengths ($f_o + f_e$), which is the length of the telescope tube. The **magnification of a refracting telescope** focused for the final image at infinity can be shown to be

$$m = -\frac{f_o}{f_e} \quad \text{(angular magnification of refracting telescope)} \quad (25.6)$$

where the negative sign indicates that the image is inverted, as in our lens sign convention described in Section 23.3. Thus, to achieve the greatest magnification, the focal length of the objective should be made as long as possible and the focal length of the eyepiece as short as possible.

The telescope illustrated in Figure 25.10 is called an *astronomical telescope*. The final image produced by an astronomical telescope is inverted, but this condition poses little problem to astronomers. (Why?) Someone viewing an object on Earth through a telescope finds it more convenient to have an upright image. A telescope in which the final image is upright is called a *terrestrial telescope*. An upright final image can be obtained in several ways; two are illustrated in ▶ **Figure 25.11**.

In the telescope diagrammed in Figure 25.11a, a diverging lens is used as an eyepiece. This type of terrestrial telescope is referred to as a *Galilean telescope*, because Galileo built one in 1609. A real and inverted image is formed by the objective to the left of the eyepiece, and this image acts as a "virtual" object for the

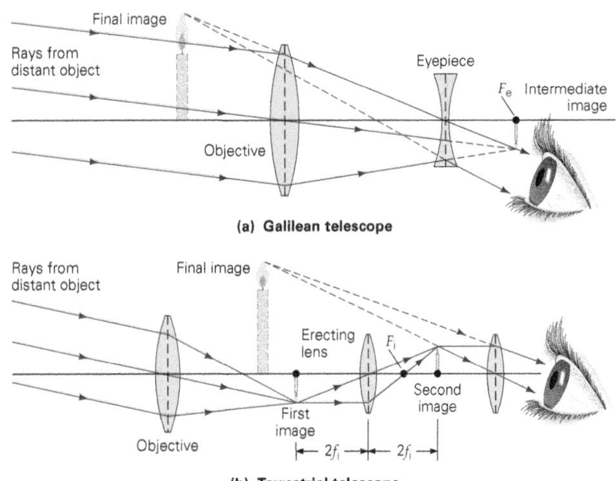


▲ **FIGURE 25.11** **Terrestrial telescopes (a)** A Galilean telescope uses a diverging lens as an eyepiece, producing upright, virtual images. **(b)** Another way to produce upright images is to use a converging "erecting" lens (focal length $f_i$) between the objective and eyepiece in an astronomical telescope. This addition elongates the telescope, but the length can be shortened by using internally reflecting prisms.

eyepiece. (See Section 23.3.) The diverging lens then forms a virtual, inverted, and magnified image of the image of the objective. Therefore, the final image observed is virtual, upright, and magnified. (The image is inverted twice so the final image is upright.) Also note that with a diverging lens and negative focal length, Equation 25.6 gives a positive $m$, indicating an upright image.

Another type of terrestrial telescope, illustrated in Figure 25.11b, uses a third lens, called the *erecting lens*, or *inverting lens*, between the converging objective and eyepiece lenses. If the image is formed by the objective at a distance that is twice the focal length of the intermediate erecting lens ($2f_i$), then the lens merely inverts the image without magnification, and the telescope magnification is still given by Equation 25.6.

However, achieving the upright image in this way increases the length of the telescope by four times the focal length of the erecting lens ($2f_i$ on each side). The inconvenient extra length can be decreased by using internally reflecting prisms. This is the principle behind prism binoculars, which are really double telescopes – one for each eye (▶ **Figure 25.12**).

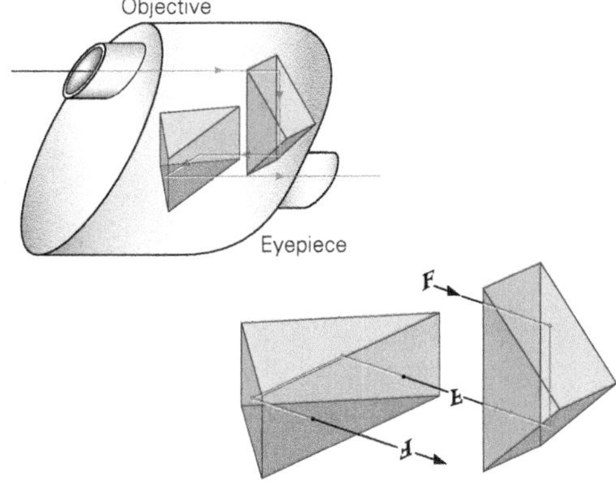


▲ **FIGURE 25.12** **Prism binoculars** A schematic cutaway view of one ocular (one half of a pair of prism binoculars), showing the internal reflections in the prisms, which reduce the overall physical length. The prisms also erect the image.

### EXAMPLE 25. 5: AN ASTRONOMICAL TELESCOPE – AND A LONGER TERRESTRIAL TELESCOPE

An astronomical telescope has an objective lens with a focal length of 30 cm and an eyepiece with a focal length of 9.0 cm. (a) What is the magnification of the telescope? (b) If an erecting lens with a focal length of 7.5 cm is used to convert the telescope to a terrestrial type, what is the overall length of the telescope tube?

**THINKING IT THROUGH.** Equation 25.6 applies directly to part (a). In part (b), the erecting lens elongates the telescope by four times the focal length of the erecting lens ($4f_i$) (Figure 25.11b).

**SOLUTION**

Listing the data:

***Given:***

$f_o = 30$ cm (objective focal length)
$f_e = 9.0$ cm (eyepiece focal length)
$f_i = 7.5$ cm (intermediate lens focal length)

***Find:***

**(a)** $m$ (magnification)
**(b)** $L$ (length of telescope tube)

**(a)** The magnification is given by Equation 25.6 as

$$m = -\frac{f_o}{f_e} = -\frac{30\,\text{cm}}{9.0\,\text{cm}} = -3.3\times$$

where the negative sign indicates that the final image is inverted.

**(b)** Taking the length of the astronomical tube to be the distance between the lenses, we find that this length is just the sum of the lenses' focal lengths (without the erecting lens):

$$L_1 = f_o + f_e = 30\,\text{cm} + 9.0\,\text{cm} = 39\,\text{cm}$$

With the erecting lens, the overall length is then

$$L = L_1 + L_2 = 39\,\text{cm} + 4f_i = 39\,\text{cm} + 4(7.5\,\text{cm}) = 69\,\text{cm}$$

Hence, the telescope length is longer, with an upright image, but the same magnification, $-3.3\times$ (why?).

**FOLLOW-UP EXERCISE.** A terrestrial telescope 66 cm in length has an intermediate erecting lens with a focal length of 12 cm. What is the focal length of an erecting lens that would reduce the telescope length to a more manageable 50 cm?

CONCEPTUAL EXAMPLE 25.6: CONSTRUCTING A TELESCOPE

A student is given two converging lenses, one with a focal length of 5.0 cm and the other with a focal length of 20 cm. To construct a telescope to best view distant objects with these lenses, the student should hold the lenses (a) more than 25 cm apart, (b) less than 25 cm but more than 20 cm apart, (c) less than 20 cm but more than 5.0 cm apart, (d) less than 5.0 cm apart. Specify which lens should be used as the eyepiece.

**REASONING AND ANSWER.** First, let's see which lens should be used as the eyepiece. The only type of telescope that can be constructed with two converging lenses is an astronomical telescope. In this type of telescope, the lens with the longer focal length is used as an objective lens to produce a real image of a distant object. That image is then viewed with the lens with the shorter focal length, the eyepiece, used as a simple magnifier.

If the object is at a great distance, a real image is formed by the objective lens in the focal plane of the lens (Figure 25.10). This image acts as the object for the eyepiece, which is positioned so that the by the objective lies just inside its focal point so as to produce a second inverted and magnified image.

The two lenses must be *slightly* less than 25 cm apart, so answer (a) is not correct. Answers (c) and (d) are also not correct, because the eyepiece would be too close to the objective to produce the second magnified image needed for optimal viewing of a distant object. In these cases, the rays would pass through the second lens before the image was formed, and a *reduced* image might be produced. (See Section 23.3.) Thus, answer (b), with the image by the objective just inside the eyepiece's focal point, is the correct answer.

**FOLLOW-UP EXERCISE.** A third converging lens with a focal length of 4.0 cm is used with the aforementioned two lenses to produce a terrestrial telescope in which the third lens does nothing more than invert the image. How should the lenses be positioned and how far apart should they be for the final image to be of maximum size and upright?

## 25.3.2 Reflecting Telescope

For viewing the Sun, Moon, and nearby planets, large magnifications are important to see details. However, even with the highest feasible magnification, stars appear only as faint points of light. For distant stars and galaxies, it is more important to gather more light than to increase the magnification, so that the object can be seen and its spectrum analyzed. The intensity of light from a distant source is sometimes very low. In many instances, such a source can be detected only when the light is gathered and focused on a photographic sensor over a long period of time.

Recall from Section 14.3 that intensity is energy per unit time per unit *area*. Thus, more light energy can be gathered if the size of the objective is increased. However, producing a large lens involves difficulties associated with glass quality, grinding, and polishing. Compound lens systems are required to reduce aberrations, and a very large lens may sag under its own weight, producing further aberrations. Currently, the largest objective lens in operation has a diameter of 102 cm (40 in.) and is part of the refracting telescope of the Yerkes Observatory at Williams Bay, Wisconsin.

These problems can be reduced by using a reflecting telescope, which uses a large, concave, parabolic mirror (▼ **Figure 25.13**). A parabolic mirror does not exhibit spherical aberration, and a mirror has no inherent chromatic aberration. (Section 23.5; why?) High-quality glass is not needed, because the light is reflected by a mirrored surface. Only one surface has to be ground, polished, and silvered.

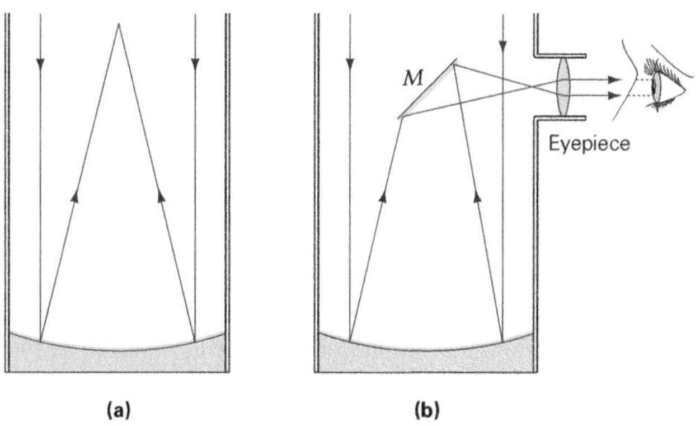


▲ FIGURE 25.13 **Reflecting telescopes** A concave parabolic mirror can be used in a telescope to converge light to form an image of a distant object. **(a)** The image may be at the prime focus, or **(b)** a small mirror and lens can be used to focus the image outside the telescope, a configuration called a *Newtonian focus.*

The largest optical reflecting telescope, with two mirrors, each 8.4 m (27 ft) in diameter, is the Large Binocular Telescope (LBT) in Mount Graham International Observatory, Arizona, USA (▼ **Figure 25.14a**).

(a) (b)

▲ FIGURE 25.14 **Large reflecting telescopes (a)** Two 8.2-m-diameter mirror for the Large Binocular Telescope, in Mount Graham, Arizona. **(b)** Seven 8.4-m diameter mirrors forming a single primary mirror with an effective mirror diameter of 24.5 m (80 ft) is planned for the Giant Magellan Telescope (GMT).

Even though reflecting telescopes have advantages over refracting telescopes, they do have their own problems. Like a large lens, a large mirror may sag under its own weight. The weight factor also increases the costs of construction, because the supporting elements for a heavier mirror must be more massive.

These problems are being addressed by new technologies. One approach is to use an array of mirrors, coordinated to function as a single large mirror. Examples include the European Southern Observatory's four 8.2-m-diameter mirrors linked to form a VLT (Very Large Telescope) with an equivalent diameter of 16 m. The Giant Magellan Telescope (GMT) is planned to be completed in 2025. Seven 8.4-m diameter mirrors will form a single primary mirror with an effective mirror diameter of 24.5 m or 80 ft (Figure 25.14b).

Another way of extending our view into space is to put telescopes into orbit around the Earth. Above the atmosphere, the view is unaffected by the twinkling effect of atmospheric turbulence and refraction, and there is no background problem from city lights. In 1990, the optical Hubble Space Telescope (HST) was launched into orbit. Even with a mirror diameter of only 2.4 m, its privileged position has allowed the HST to produce images seven times as clear as those formed by similarly sized Earthbound telescopes. A flawed optics design of the HST was repaired in 1993, and more repairs were performed in 2009 (▼**Figure 25.15**). The repairs, along with the addition of new instruments, made the HST 90 times as powerful as it was in 1993.

▲FIGURE 25.15 **Hubble Space Telescope (HST)** The HST in the cargo bay of the Space Shuttle Atlantis during its last servicing mission in May, 2009.

Last, you may know that some telescopes use nonvisible radiation such as radio waves, infrared radiation, ultraviolet radiation, X-rays, and gamma rays.

## 25.4 Diffraction and Resolution

The diffraction of light places a limitation on our ability to distinguish objects that are close together when microscopes or telescopes are used. This effect can be understood by considering two point sources located far from a narrow slit of width $w$ (▶**Figure 25.16**). The sources could represent distant stars, for example. In the absence of diffraction, two bright spots, or images, would be observed on a screen. As you know from Section 24.3, however, the slit diffracts the light, and each image is a diffraction pattern that consists of a central maximum with a pattern of weaker bright and dark positions on either side. If the sources are close together, the two central maxima may overlap. In this case, the images cannot be distinguished, or are said to be *unresolved*. For the images to be *resolved*, the central maxima must not overlap appreciably.

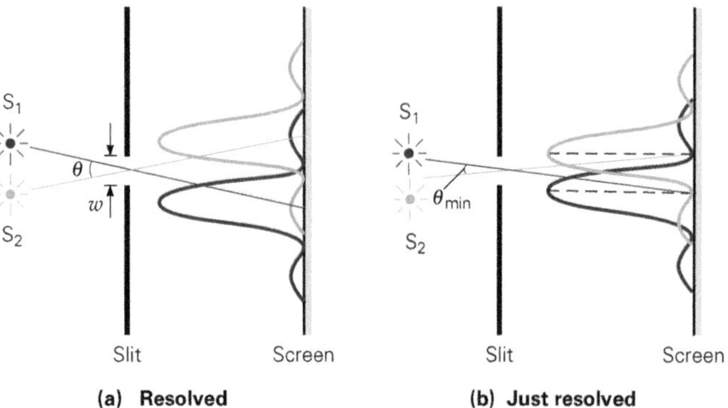


▲FIGURE 25.16 **Resolution** Two light sources in front of a slit produce diffraction patterns. **(a)** When the angle subtended by the sources at the slit is large enough for the images to be distinguishable, the images are said to be resolved. **(b)** At smaller angles, the central maxima are closer together. At $\theta_{min}$, the center of the central maximum of one image falls on the first minimum of the other image, and the images are said to be just resolved. For smaller angles, the patterns are unresolved.

In general, images of two sources can be resolved if the center of the central maximum of one falls at or beyond the first minimum of the other. This limiting condition for the resolution of two images – that is, the ability to distinguish them as separate – was first proposed by Lord Rayleigh (1842–1919), a British physicist. The condition is known as the **Rayleigh criterion**:

> Two images are said to be just resolved when the center of the central maximum of one image falls on the first minimum of the diffraction pattern of the other image.

The Rayleigh criterion can be expressed in terms of the angular separation of the sources. (See Figure 25.16.) The first minimum ($m=1$) for a single-slit diffraction pattern satisfies this relationship:

$$w\sin\theta = m\lambda = (1)\lambda \quad \text{or} \quad \sin\theta = \frac{\lambda}{w}$$

According to Figure 25.16, this is the minimum angular separation for two images to be just resolved according to the Rayleigh criterion. In general, for visible light, the wavelength is much smaller than the slit width ($\lambda < w$), so $\theta$ is small and $\sin\theta \approx \theta$. In this case, the limiting, or **minimum angle of resolution ($\theta_{min}$)** for a slit of width $w$ is

$$\theta_{min} = \frac{\lambda}{w} \quad \text{(minimum angle of resolution for a slit)} \qquad (25.7)$$

Note that $\theta_{min}$ is dimensionless (a pure number) and is therefore in radians. Thus, the images of two sources will be *distinctly* resolved if the angular separation of the sources is greater than $\lambda/w$.

The apertures (openings) of cameras, microscopes, and telescopes are generally circular. Thus, there is a circular diffraction pattern around the central maximum, in the form of a bright circular disk (▼ **Figure 25.17**). Detailed analysis shows that the **minimum angle of resolution for a circular aperture** for the images of two objects to be just resolved is similar to, but slightly different from, Equation 25.7. It is

$$\theta_{\min} = \frac{1.22\lambda}{D} \quad \text{(minimum angle of resolution for a circular aperture)} \tag{25.8}$$

where $D$ is the diameter of the aperture and $\theta_{\min}$ is again in radians.

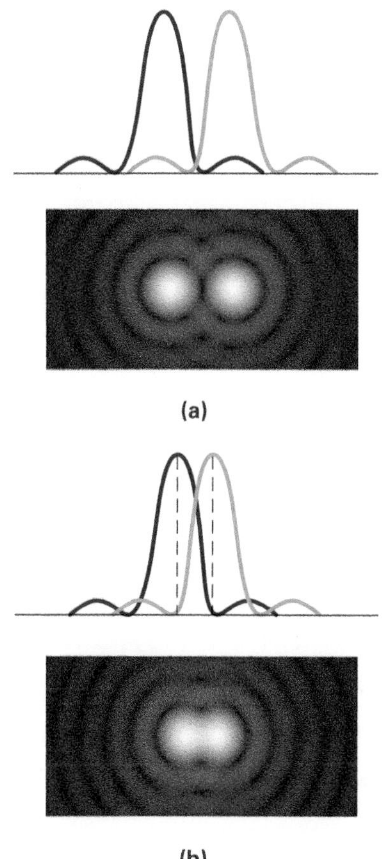

▲ FIGURE 25.17 **Circular aperture resolution (a)** When the angular separation of two objects is large enough, the images are well resolved. (Compare with Figure 25.16a.) **(b)** Rayleigh criterion: The center of the central maximum of one image falls on the first minimum of the other image. (Compare with Figure 25.16b.) The images of objects with smaller angular separations cannot be clearly distinguished as individual images.

Equation 25.8 applies to the objective lens of a microscope or telescope, or the iris of the eye, all of which may be considered to be circular apertures for light. According to Equations 25.7 and 25.8, the smaller $\theta_{\min}$, the better the resolution. The minimum angle of resolution $\theta_{\min}$ should be small so that objects close together can be resolved; therefore, the aperture should be as *large* as possible. This is yet another reason for using large lenses (and mirrors) in telescopes.

For a microscope, it is more convenient to specify the actual separation ($s$) between two point sources. Since the objects are usually near the focal point of the objective, to a good approximation,

$$\theta_{\min} = \frac{s}{f} \quad \text{or} \quad s = f\theta_{\min}$$

where $f$ is the focal length of the lens and $\theta_{\min}$ is expressed in radians. (Here, $s$ is taken as the arc length subtended by $\theta_{\min}$, and $s = r\theta_{\min} = f\theta_{\min}$.) Then, using Equation 25.8, we get

$$s = f\theta_{\min} = \frac{1.22\lambda f}{D} \quad \text{(resolving power of a microscope)} \tag{25.9}$$

This minimum distance between two points whose images can be just resolved is called the resolving power of the microscope. Note that $s$ is directly proportional to $\lambda$, so shorter wavelengths give better resolution. In practice, the resolving power of a microscope indicates the ability of the objective to distinguish fine detail in specimens' structures. For another real-life example of resolution, see Figure 25.17.

## EXAMPLE 25.7: VIEWING FROM SPACE – THE GREAT WALL OF CHINA

The Great Wall of China was originally about 2400 km (1500 mi) long, with a base width of about 6.0 m and a top width of about 3.7 m. Several hundred kilometers of the wall remain intact (▼ **Figure 25.18**). It is sometimes said that the wall is the only human construction that can be seen with the unaided eye by an astronaut in an orbit that is 200 km (125 mi) above the Earth. If the pupil of the eye is 4.0 mm in diameter for daytime visible light with a wavelength of 550 nm, determine whether the wall is visible according to the Rayleigh criterion. (Neglect any atmospheric effects.)

▲ FIGURE 25.18 **The Great Wall** The walkway of the Great Wall of China, which was built as a fortification along China's northern border.

**THINKING IT THROUGH.** Despite the great length of the wall, it would not be visible from space unless its *width* subtends an angle that is greater than the minimum angle of resolution for the eye of an observing astronaut. The angle subtended by the width of the wall is calculated by $s = r\theta$ (Equation 7.3), where $s$ is the maximum observable width of the wall and $r$ is the distance from the wall to the astronaut. This is then compared with for the eye - a direct application of Equation 25.8.

**SOLUTION**

***Given:***

$D = 4.0\ \text{mm} = 4.0 \times 10^{-3}\ \text{m}$
$\lambda = 550\ \text{nm} = 5.50 \times 10^{-7}\ \text{m}$
$s = 6.0\ \text{m}$ (width of wall)
$r = 200\ \text{km} = 2.0 \times 10^{5}\ \text{m}$ (distance)

***Find:*** $\theta$ (for wall) and $\theta_{min}$ (for eye)

The angle subtended by the width of the wall to the astronaut is

$$\theta = \frac{s}{r} = \frac{6.0\,\text{m}}{2.0\times10^{5}\,\text{m}} = 3.0\times10^{-5}\,\text{rad}$$

The minimum angle of resolution for the eye is

$$\theta_{min} = \frac{1.22\lambda}{D} = \frac{1.22(5.50\times10^{-7}\,\text{m})}{4.0\times10^{-3}\,\text{m}} = 1.7\times10^{-4}\,\text{rad}$$

Since $\theta \ll \theta_{min}$, the wall would not be able to be seen with the unaided eye.

Now, here is the living proof. In 2003, Chinese astronaut Yang Liwei went to space in a historic mission. (He was the very first Chinese to go into space.) After he returned to the Earth, he was asked if he was able to see the Great Wall in space. "I did not see our Great Wall from space," Yang said in an interview with China Central Television. This was also confirmed by NASA in 2004.

**FOLLOW-UP EXERCISE.** What would be the minimum diameter of the objective of a telescope that would allow an astronaut orbiting the Earth at an altitude of 300 km to actually see the Great Wall?

---

Note from Equation 25.8 that higher resolution can be gained by using radiation of a shorter wavelength. Thus, a telescope with an objective of a given size will have greater resolution with violet light than with red light. For microscopes, it is possible to increase resolving power by shortening the wavelengths of the light used to create the image. This can be done with a specialized objective called an *oil immersion* lens. When such a lens is used, a drop of transparent oil fills the space between the objective and the specimen. Recall that the wavelength of light in oil is $\lambda_m = \lambda/n$, where $n$ is the index of refraction of the oil and $\lambda$ is the wavelength of light in air. For values of $n$ about 1.50 or higher, the wavelength is significantly reduced, and the resolution is increased proportionally.

# Chapter 25 Review

- Nearsighted people cannot see distant objects clearly. Farsighted people cannot see nearby objects clearly. These conditions may be corrected by diverging and converging lenses, respectively.
- The magnification of a magnifying glass (or simple microscope) is expressed in terms of **angular magnification (*m*)**, as distinguished from the lateral magnification ($M$; see Chapter 23):

$$m = \frac{\theta}{\theta_o} \quad (25.1)$$

- The magnification of a magnifying glass with the image at the near point (25 cm) is expressed as

$$m = 1 + \frac{25\,\text{cm}}{f} \quad (25.3)$$

- The magnification of a magnifying glass with the image at infinity (relaxed viewing) is expressed as

$$m = \frac{25\,\text{cm}}{f} \quad (25.4)$$

- The objective of a compound microscope has a relatively short focal length, and the eyepiece, or ocular, has a longer focal length. Both contribute to the **total magnification $m_{total}$**, given by

$$m_{total} = M_o m_e = -\frac{(25\,\text{cm})L}{f_o f_e} \quad (25.5)$$

where $L$, $f_o$, and $f_e$ are in centimeters.
- A refracting telescope uses a converging lens to gather light, and a reflecting telescope uses a converging mirror. The image created by either one is magnified by the eyepiece. The **magnification of a refracting telescope** is

$$m = -\frac{f_o}{f_e} \quad (25.6)$$

- Diffraction places a limit on **resolution** - the ability to resolve, or distinguish, objects that are close together. Two images are said to be just resolved when the center of the central maximum of one image falls on the first minimum of the other image (the **Rayleigh criterion**).

  For a rectangular slit, the **minimum angle of resolution** is

$$\theta_{min} = \frac{\lambda}{w} \quad (25.7)$$

For a circular aperture of diameter $D$, the **minimum angle of resolution** is

$$\theta_{min} = \frac{1.22\lambda}{D} \quad (25.8)$$

The resolving power of a microscope is

$$s = f\theta_{min} = \frac{1.22\lambda f}{D} \quad (25.9)$$

# End of Chapter Questions and Exercises

## Multiple Choice Questions

**25.1 The Human Eye**

1. The cones of the retina are responsible for (a) vision, (b) black-and-white twilight vision, (c) color vision, (d) close-up vision.
2. An imperfect cornea can cause (a) astigmatism, (b) nearsightedness, (c) farsightedness, (d) all of the preceding.
3. The image of an object formed on the retina is (a) inverted, (b) upright, (c) the same size as the object, (d) all of the preceding.
4. Nearsightedness can be corrected by using (a) converging lens, (b) diverging lens, (c) flat lens, (d) none of the preceding.

**25.2 Microscopes**

5. A magnifying glass (a) is a converging lens, (b) forms virtual images, (c) magnifies by effectively increasing the angle the object subtends, (d) all of the preceding.
6. When using a magnifying glass, the magnification is greater when the magnified image is at (a) the near point, (b) the far point, (c) infinity.
7. Compared with the focal length of the eyepiece in a compound microscope, the objective has (a) a longer focal length, (b) a shorter focal length, (c) the same focal length.

**25.3 Telescopes**

8. An astronomical telescope has (a) unlimited magnification, (b) two lenses of the same focal length, (c) an objective of relatively long focal length, (d) an objective of relatively short focal length.
9. An inverted image is produced by (a) a terrestrial telescope, (b) an astronomical telescope, (c) a Galilean telescope, (d) all of the preceding.
10. The image formed by a terrestrial telescope is (a) inverted, (b) upright, (c) real, (d) none of the preceding.

**25.4 Diffraction and Resolution**

11. The images of two sources are said to be just resolved when (a) the central maxima of the diffraction patterns fall on each other, (b) the first maxima of the diffraction patterns fall on each other, (c) the central maximum of one diffraction pattern falls on the first minimum of the other, (d) none of the preceding.
12. For a particular wavelength, the minimum angle of resolution is (a) smaller for a lens of a larger radius, (b) smaller for a lens of a smaller lens, (c) the same for lenses of all radii.
13. The purpose of using oil immersion lenses on microscopes is to (a) reduce the size of the microscope, (b) increase the magnification, (c) increase the wavelength of light so as to increase resolution, (d) reduce the wavelength of light so as to increase resolution.

## Conceptual Questions

**25.1 The Human Eye**

1. Which parts of a camera correspond to the iris, crystalline lens, and retina of the eye?
2. (a) If an eye has a far point of 15 m and a near point of 25 cm, is that eye nearsighted or farsighted? (b) How about an eye with a far point at infinity and a near point at 50 cm? (c) What type of corrective lenses (converging or diverging) would you use to correct the vision defects in parts (a) and (b)?
3. Will wearing glasses to correct nearsightedness and farsightedness, respectively, affect the size of the image on the retina? Explain.
4. A fifty-year-old person has a far point of 20 m and near point of 45 cm. What type of corrective glasses would be necessary to correct this person's vision?
5. A person with nearsightedness wishes to switch from regular glasses to contact lenses. Should the contact lenses have a stronger or a weaker prescription than the glasses? Explain.

**25.2 Microscopes**

6. When you use a simple convex lens as a magnifying glass to view an object, where should you put the object, farther away than the focal length or closer than the focal length? Explain.
7. With an object at the focal point of a magnifying glass, the magnification is given by (Equation 25.4). According to this equation, the magnification could be increased indefinitely by using lenses with shorter focal lengths. Why, then, are compound microscopes needed?
8. In a compound microscope, which lens, the objective or the eyepiece, plays the same role as a simple magnifying glass?

**25.3 Telescopes**

9. If you are given two lenses with different focal lengths, how would you decide which should be used as the objective and which should be used as the eyepiece for a telescope? Explain.
10. What are the main differences among the following refracting telescopes: an astronomical telescope, a Galilean telescope, and a terrestrial telescope?

11. Why are chromatic and spherical aberrations important factors in refracting telescopes, but not in reflecting telescopes?
12. In Figure 25.13b, part of the light entering the concave mirror is obstructed by a small plane mirror that is used to redirect the rays to a viewer. Does this mean that only a portion of an object can be seen? How does the size of the obstruction affect the image?

**25.4 Diffraction and Resolution**

13. When an optical instrument is designed, a high resolution is often desired so that the instrument may be used to observe fine details. Does a higher resolution mean a smaller or larger minimum angle of resolution? Explain.
14. A reflecting telescope with a large objective mirror can collect more light from stars than a reflecting telescope with a smaller objective mirror. What other advantage is gained with a large mirror? Explain.
15. Modern digital cameras are getting smaller and smaller. Discuss the image resolution of these small cameras.
16. In order to observe fine details of small objects in a microscope, should you use blue light or red light? Explain.

## Exercises*

*Integrated Exercises (IEs) are two-part exercises. The first part typically requires a conceptual answer choice based on physical thinking and basic principles. The following part is quantitative calculations associated with the conceptual choice made in the first part of the exercise.*

**25.1 The Human Eye†**

1. • What are the powers of (a) a converging lens of focal length 20 cm and (b) a diverging lens of focal length −50 cm?
2. • A person is prescribed with contact lenses that have powers of −3.0 D. What type of lenses are these? What is the lenses' focal length?
3. **IE** • The far point of a certain nearsighted person is 90 cm. (a) Which type of contact lenses, (1) converging, (2) diverging, or (3) bifocal, should an optometrist prescribe to enable the person to see more distant objects clearly? Explain. (b) What would the power of the lenses be, in diopters?
4. **IE** • A certain farsighted person has a near point of 50 cm. (a) Which type of contact lenses, (1) converging, (2) diverging, or (3) bifocal, should an optometrist prescribe to enable the person to see clearly objects as close as 25 cm? Explain. (b) What is the power of the lenses, in diopters?
5. •• A nearsighted person has an uncorrected far point of 200 cm. Which type of contact lenses would correct this condition, and of what focal length should it be?
6. •• A person can just see the print in a book clearly when she holds the book no closer than at arm's length (0.45 m from the eyes). (a) Does she have (1) nearsightedness, (2) farsightedness, or (3) astigmatism? Explain. (b) Which type of lens will allow her to read the text at the normal near point (0.25 m), and what is that lens's power?
7. •• To correct a case of farsightedness, an optometrist prescribes converging contact lenses that effectively move the patient's near point from 85 cm to 25 cm. (a) What is the power of the lenses? (b) To see distant objects clearly, should the patient wear the contact lenses or take them out? Explain.
8. **IE** •• A woman cannot see objects clearly when they are farther than 12.5 m away. (a) Does she have (1) nearsightedness, (2) farsightedness, or (3) astigmatism? Explain. (b) Which type of lens will allow her to see distant objects clearly, and of what power should the lens be?
9. **IE** •• A man is unable to focus on objects nearer than 1.5 m. (a) Does he have (1) nearsightedness, (2) farsightedness, or (3) astigmatism? Explain. (b) The type of contact lenses that allows him to focus on the print of a book held 25 cm from his eyes should be (1) converging, (2) diverging, (3) flat. Explain. (c) What should be the power of the lenses?
10. •• A nearsighted student wears contact lenses to correct for a far point that is 4.00 m from her eyes. When she is not wearing her contact lenses, her near point is 20 cm. What is her near point when she is wearing her contacts?
11. •• A nearsighted woman has a far point located 2.00 m from one eye. (a) If a corrective lens is worn 2.00 cm from the eye, what would be the necessary power of the lens for her to see distant objects? (b) What would be the necessary power if a contact lens were used?
12. •• A nearsighted man wears eyeglasses whose lenses have a focal length of −0.25 m. How far away is his far point?
13. •• An eyeglass lens with a power of +2.8 D allows a farsighted person to read a book held at a distance of 25 cm from her eyes. At what distance must she hold the book to read it without glasses?
14. •• A college professor can see objects clearly only if they are between 70 and 500 cm from her eyes. Her optometrist prescribes bifocals (Figure 25.4) that enable her to see distant objects through the top half of the lenses and read students' papers at a distance of 25 cm through the lower half. What are the respective powers of the top and bottom lenses?

* The bullets denote the degree of difficulty of the exercises: •, simple; ••, medium; and •••, more difficult.

† Assume that corrective lenses are in contact with the eye (contact lenses) unless otherwise stated.

15. •• A senior citizen wears bifocals (Figure 25.4) in which the top half of the lens has a focal length of −0.850 m and the bottom half of the lens has a focal length of +0.500 m. What are this person's near point and far point?

**25.2 Microscopes***

16. • An object is placed 10 cm in front of a converging lens with a focal length of 18 cm. What are (a) the lateral magnification and (b) the angular magnification?
17. • A biology student uses a converging lens to examine the details of a small insect. If the focal length of the lens is 12 cm, what is the maximum angular magnification?
18. • A converging lens can give a maximum angular magnification of 1.5×. What is the focal length of the lens?
19. • When viewing an object with a magnifying glass whose focal length is 10 cm, a student positions the lens so that there is minimum eyestrain. What is the observed magnification?
20. IE • A physics student uses a converging lens with a focal length of 14 cm to read a small measurement scale. (a) Maximum magnification is achieved if the image is at (1) the near point, (2) infinity, (3) the far point. Explain. (b) What are the magnifications when the image is at the near point and infinity, respectively?
21. IE • A detective wants to achieve maximum magnification when looking at a fingerprint with a magnifying glass. (a) He should use a lens with (1) a long focal length, (2) a short focal length, (3) a larger size. Explain. (b) If he uses lenses of focal length +28 cm and +40 cm, what are the maximum magnifications of the print?
22. •• What is the maximum magnification of a magnifying glass with a power of +3.0 D for (a) a person with a near point of 25 cm and (b) a person with a near point of 10 cm?
23. •• If a magnifying glass gives an angular magnification of 1.5× when viewed with relaxed eyes, what is the power of the lens?
24. •• A compound microscope has an objective with a focal length of 4.00 mm and an eyepiece with a magnification of 10.0×. If the objective and eyepiece are 15.0 cm apart, what is the total magnification of the microscope?
25. •• A compound microscope has a distance of 15 cm between lenses and an objective with a focal length of 8.0 mm. What power should the eyepiece have to give a total magnification of −360×?
26. •• The focal length of the objective lens of a compound microscope is 4.5 mm. The eyepiece has a focal length of 3.0 cm. If the distance between the lenses is 18 cm, what is the magnification of a viewed image?

* The normal near point should be taken as 25 cm unless otherwise specified.

27. •• A compound microscope has an objective lens with a focal length of 0.50 cm and an eyepiece with a focal length of 3.25 cm. The separation distance between the lenses is 22 cm. (a) What is the total magnification? (b) Compare (as a percentage) the total magnification with the magnification of the eyepiece alone as a simple magnifying glass.
28. •• The lenses used in a compound microscope have powers of +100 D and +50 D. If a total magnification of −200× is desired, what should be the distance between the two lenses?
29. IE •• Two lenses of focal length 0.45 and 0.35 cm are available for a compound microscope using an eyepiece of focal length of 3.0 cm, and the distance between the lenses has to be 15 cm. (a) Which lens should be used as the objective: (1) the one with the longer focal length, (2) the one with the shorter focal length, or (3) either? (b) What are the two possible total magnifications of the microscope?
30. •• A −150× microscope has an eyepiece whose focal length is 4.4 cm. If the distance between the lenses is 20 cm, find the focal length of the objective.
31. •• A specimen is 5.0 mm from the objective of a compound microscope that has a lens power of +250 D. What must be the magnifying power of the eyepiece if the total magnification of the specimen is −100×?

**25.3 Telescopes**

32. • Find the magnification and length of a telescope whose objective has a focal length of 50 cm and whose eyepiece has a focal length of 2.0 cm.
33. • An astronomical telescope has an objective and an eyepiece whose focal lengths are 60 and 15 cm, respectively. What are the telescope's (a) magnifying power and (b) length?
34. •• An astronomical telescope has an eyepiece with a focal length of 10.0 mm. If the length of the tube is 1.50 m, (a) what is the focal length of the objective? (b) What is the angular magnification of the telescope when it is focused for an object at infinity?
35. •• A telescope has an angular magnification of −50× and a barrel 1.02 m long. What are the focal lengths of the objective and the eyepiece?
36. IE •• A terrestrial telescope has three lenses: an objective, an erecting lens, and an eyepiece. (a) Does the erecting lens (1) increase the magnification, (2) increase the physical length of the telescope, (3) decrease the magnification, or (4) decrease the physical length of the telescope? Explain. (b) This terrestrial telescope has focal lengths of 40, 10, and 5.0 cm for the objective, erecting lens, and eyepiece, respectively. What is the magnification of the telescope for an object at infinity? (c) What is the length of the telescope barrel?

37. •• A terrestrial telescope uses an objective and eyepiece with focal lengths of 42 and 6.0 cm, respectively. (a) What should the focal length of the erecting lens be if the overall length of the telescope is to be 1.0 m? (b) What is the magnification of the telescope for an object at infinity?
38. •• An astronomical telescope uses an objective of power +2.0 D. If the length of the telescope is 52 cm, (a) what is the focal length of the eyepiece? (b) What is the angular magnification of the telescope?
39. IE •• You are given two objectives and two eyepieces and are instructed to make a telescope with them. The focal lengths of the objectives are 60.0 and 40.0 cm, and the focal lengths of the eyepieces are 0.90 and 0.80 cm. (a) Which lens combination would you pick if you want to have maximum magnification? How about minimum magnification? Explain. (b) Calculate the maximum and minimum magnifications.

### 25.4 Diffraction and Resolution*

40. IE • (a) For a given wavelength, a wider single slit will give (1) a greater, (2) a smaller, (3) the same minimum angle of resolution as a narrower slit, according to the Rayleigh criterion. (b) What are the minimum angles of resolution for two point sources of red light in the diffraction pattern produced by single slits with widths of 0.55 and 0.45 mm, respectively?
41. • The minimum angle of resolution of the diffraction patterns of two identical monochromatic point sources in a single-slit diffraction pattern is 0.0065 rad. If a slit width of 0.10 mm is used, what is the wavelength of the sources?
42. • What is the resolution limit due to diffraction for the European Southern Observatory reflecting telescope (which has an 8.20-m, or 323-in., diameter) for light with a wavelength of 550 nm?
43. • What is the resolution limit due to diffraction for the Hale telescope at Mount Palomar, with its 200-in.-diameter mirror, for light with a wavelength of 550 nm? Compare this value with the resolution limit for the European Southern Observatory telescope found in Exercise 42.
44. •• From a spacecraft in orbit 150 km above the Earth's surface, an astronaut wishes to observe her hometown as she passes over it. What size features will she be able to identify with the unaided eye, neglecting atmospheric effects? [*Hint:* Estimate the diameter of the human iris.]
45. IE •• A human eye views small objects of different colors, and the eye's resolution is measured. (a) The eye sees the finest details for objects of which color: (1) red, (2) yellow, (3) blue, or (4) any color? Explain. (b) The maximum diameter of the eye's pupil at night is about 7.0 mm. What are the minimum angles of resolution for sources with wavelengths of 400 and 700 nm, respectively?
46. •• Some African tribes people claim to be able to see the moons of Jupiter with the unaided eye. If two moons of Jupiter are at a minimum distance of $3.1 \times 10^8$ km away from Earth and at a maximum separation distance of $3.0 \times 10^6$ km, is this possible in theory? Explain. Assume that the moons reflect sufficient light and that their observation is not restricted by Jupiter. [*Hint:* See Exercise 45b.]
47. •• Assuming that the headlights of a car are point sources 1.7 m apart, what is the maximum distance from an observer to the car at which the headlights are distinguishable from each other? [*Hint:* See Exercise 45b.]
48. •• If a camera with a 50-mm lens is to resolve two objects that are 4.0 mm from each other and both objects are 3.5 m from the camera lens, (a) what is the minimum diameter of the camera lens? (b) What is the resolving power? (Assume the wavelength of light is 550 nm.)
49. •• The objective of a microscope is 2.50 cm in diameter and has a focal length of 0.80 mm. (a) If blue light with a wavelength of 450 nm is used to illuminate a specimen, what is the minimum angular separation of two fine details of the specimen for them to be just resolved? (b) What is the resolving power of the lens?
50. •• A refracting telescope with a lens whose diameter is 30.0 cm is used to view a binary star system that emits light in the visible region. (a) What is the minimum angular separation of the two stars for them to be barely resolved? (b) If the binary star is a distance of $6.00 \times 10^{20}$ km from the Earth, what is the distance between the two stars? (Assume that a line joining the stars is perpendicular to our line of sight.)

* Ignore atmospheric blurring unless otherwise stated.

# 26

# Relativity



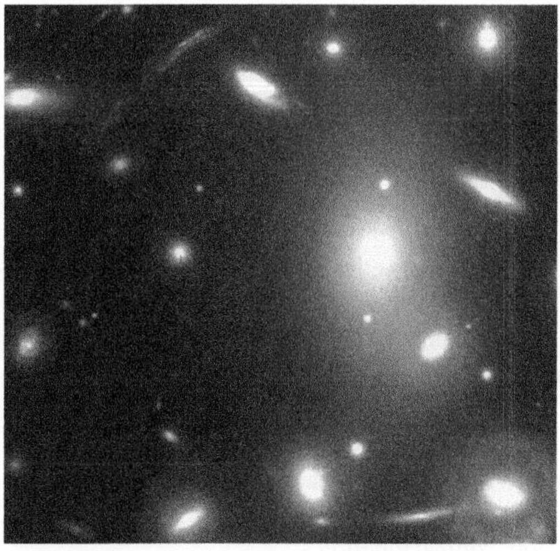

**The light from these distant galaxies has taken billions of years to reach us and in the travel has suffered directional changes and distortions due to encounters with interstellar gravitational fields.**

You might not think so, but the chapter-opening photograph tells us something very remarkable about our universe. The bright shapes are galaxies, each consisting of billions of stars. They are very far from us – some billions of light-years away. The faint arcs that make the photograph resemble a spider's web are from galaxies even more distant. What is remarkable about these arcs of light, however, is not the billions of years they took to reach us, but their paths to us. We usually think of light traveling in straight lines. Yet the light from these distant galaxies has had its direction changed by the gravitational fields of the galaxies in the foreground, creating the arcs in the photo.

The idea that light can be affected by gravity was first predicted by Albert Einstein as a consequence of his theory of general relativity. Relativity originated from the analysis of physical phenomena involving speeds approaching that of light. Indeed, modern relativity caused us to rethink our understanding of space, time, and gravitation. It successfully challenged Newtonian concepts that had dominated science for nearly 300 years.

The impact of relativity has been especially significant in the branches of science concerned with two extremes of physical reality: the subatomic realm of nuclear and particle physics, in which time intervals and distances are inconceivably small (Chapters 29 and 30), and the cosmic realm, in which time intervals and distances are unimaginably large. All modern theories about the birth, evolution, and ultimate fate of our universe are inextricably linked to our understanding of relativity.

In this chapter, you will learn how Einstein's relativity explains the changes in length and time that are observed for rapidly moving objects, the equivalence of energy and mass, and the bending of light by gravitational fields - phenomena that all seem strange from the classical Newtonian view.

## 26.1 Classical Relativity and the Michelson–Morley Experiment

Physics is concerned with the description of the world around us and depends on observations and measurements. Some aspects of nature are expected to be unvarying; that is, the ground rules by which nature plays should be consistent, and physical principles should not change from observation to observation. This consistency is emphasized by referring to such principles as *laws* - for example, the laws of motion. Not only have physical laws proved valid over time, but they also are the same for all observers.

The last sentence implies that a physical principle or law should not depend on an observer's frame of reference. When a measurement is made or experiment performed, reference is usually made to a particular frame or coordinate system - most often the laboratory, which is considered to be "at rest." Now envision the same experiment observed by a passerby (moving relative to the laboratory). Upon comparing notes, the experimenter and observer should find the results of the experiment should be explained by the same physical principles. That is, physicists believe that the laws of nature are the same regardless of the observer. Measured quantities may vary and descriptions may be different, but the *laws* that these quantities obey must be the same for all observers.

Suppose that you are at rest and observe two cars traveling in the same direction on a straight road at speeds of 60 and 90 km/h, respectively. Even though we rarely say it, it is assumed that these speeds are measured relative to your reference frame - the ground. However, a woman in the car traveling at 60 km/h will note the other car traveling at 30 km/h down the road *relative to her reference frame* - the car in which she is riding. (What does someone in the car traveling at 90 km/h observe?) That is, each person observes a *relative* velocity - the one relative to his or her own reference frame.

In measuring relative velocities, there seems to be no "true" rest frame. Any reference frame can be considered at rest if the observer moves with it. A distinction can be made, however, between *inertial* and *noninertial* reference frames. An **inertial reference frame** is a reference frame in which Newton's first law of motion holds. That is, in an inertial frame, an object on which there is no net force does not accelerate. Since Newton's first law holds in this frame, the second law of motion, $\vec{\mathbf{F}}_{\text{net}} = m\vec{\mathbf{a}}$, also applies.

Conversely, in a **noninertial reference frame** (one that is accelerating relative to an inertial frame), an object with no net force acting on it would *appear* to accelerate. Note, however, that it is the frame and its observer that are accelerating, not the object. Thus for observations made from a noninertial frame, Newton's second law will not correctly describe motion. One example of a noninertial reference frame is an automobile accelerating forward from rest. A cup placed on its dashboard, when viewed from the car's (noninertial) frame, would appear to accelerate backward while having no net force acting on it. In fact, noninertial observers would be forced to invoke a *fictitious* backward force to explain the cup's *apparent* acceleration. From the inertial frame of a sidewalk observer, however, the cup stays put (in accordance with the first law, with no friction, no net force acts on it) and it is the car that accelerates away from the cup.

Any reference frame moving with a constant velocity relative to an inertial reference frame is itself an inertial frame. Given a constant relative velocity, no acceleration effects are introduced in comparing one frame with another. In such cases, $\vec{\mathbf{F}}_{\text{net}} = m\vec{\mathbf{a}}$ can be used by observers in *either* frame to analyze a situation, and both observers will come to the same conclusions. That is, Newton's second law holds in both inertial frames. Thus, at least with respect to the laws of mechanics, no inertial frame is preferred over another. This is called the **principle of Newtonian (or classical) relativity**:

> The laws of *mechanics* are the same in all inertial reference frames.

### 26.1.1 The "Absolute" Reference Frame: The Ether

With the development of the theories of electricity and magnetism in the 1800s, some serious questions arose. Maxwell's equations (Section 20.4) predicted light to be an electromagnetic wave that travels with a speed of $c = 3.00 \times 10^8$ m/s in a vacuum. But relative to what reference frame does light have this speed? Classically, this speed would be expected to be different when measured from different reference frames. For example, it might be expected to be greater than $c$ if you were approaching the beam of light, as in the relative case of another car approaching your car on a highway.

Consider the situation in ▶**Figure 26.1a**: A person in a reference frame (truck) moving relative to another frame (ground) with a constant velocity $\vec{\mathbf{v}}'$ throws a ball with a velocity

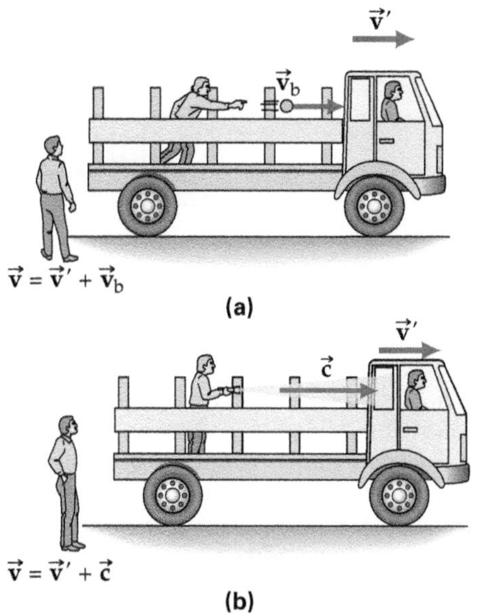


▲FIGURE 26.1 **Relative velocity (a)** According to a stationary observer on the ground, the ball velocity would be $\vec{v} = \vec{v}' + \vec{v}_b$. **(b)** If the same were true for light, then its speed would be $\vec{v} = \vec{v}' + \vec{c}$ (i.e., a magnitude greater than $c$).

$\vec{v}_b$ relative to the truck. Then the so-called stationary observer (ground) would say the ball had a velocity of $\vec{v} = \vec{v}' + \vec{v}_b$ relative to the ground. Suppose the truck were moving at 20 m/s east relative to the ground and a ball were thrown at 10 m/s (relative to the truck), also easterly. The ball would then have a speed of 20 m/s + 10 m/s = 30 m/s to the east as observed by someone on the ground.

Now suppose that the person on the truck turned on a flashlight, projecting a beam of light to the east as in Figure 26.1b. According to Newtonian relativity, $\vec{v} = \vec{v}' + \vec{c}$, thus the speed of light measured by the ground observer would be greater than $3.00 \times 10^8$ m/s. So in general, according to classical relativity, the speed of light can have *any* value, depending on the observer's reference frame.

Assuming Newtonian relativity, it followed that the particular speed of light value of $3.00 \times 10^8$ m/s must be specific to some unique inertial frame. This certainly is true for other types of waves; for example, sound wave speeds are relative to still air, etc. Thus, in analogy to all other known waves, the assumption of a unique reference frame for light seemed quite natural. Since the Earth receives light from the Sun and distant stars, it was assumed that a light-transporting medium must permeate all space. This medium was called the *luminiferous ether*, or simply, **ether**. The idea of this (as yet undetected) ether became popular in the latter part of the nineteenth century. Maxwell himself believed in the existence of the ether, as evidenced by a quote from his writings:

> Whatever difficulties we may have in forming a consistent idea of the constitution of the ether, there can be no doubt that the interplanetary and interstellar spaces are not empty, but are occupied by a material substance or body which is certainly the largest, and probably the most uniform body of which we have any knowledge.

It seemed, then, that Maxwell's equations, which describe the propagation of light, did *not* satisfy the Newtonian relativity principle, as did the laws of mechanics. On the basis of the preceding discussion, a preferential reference frame would appear to exist – one that could be considered absolutely at rest – the ether frame.

Thus at the end of the nineteenth century, scientists set out to determine if the ether existed. If so, then presumably a truly absolute rest frame would finally be identified. This was the purpose of the famous *Michelson–Morley experiment.*

During the 1880s, two American scientists, A. A. Michelson and E. W. Morley, carried out a series of experiments designed to measure the Earth's velocity relative to this ether.* They sought to do this by measuring differences in the speed of light due to the Earth's orbital velocity. According to the ether theory, if an observer were moving relative to the ether (the absolute frame), then he or she would measure the speed of light different from $c$. Their experimental apparatus, while crude by today's standards, was capable of making such a measurement, but always yielded a speed of $c$, regardless of the Earth's velocity – a *null* result for the existence of the ether!

## 26.2 The Special Relativity Postulate and the Relativity of Simultaneity

The failure of the Michelson–Morley experiment to detect an absolute ether frame left the scientific community in a quandary. The inconsistency between the classical relativity results of Newtonian mechanics (as in Figure 26.1a where $v > v_b$) and the Michelson–Morley experiment (as in Figure 26.1b, $v = c$) remained unexplained. Many physicists simply could not believe that light didn't need a medium through which to propagate. Albert Einstein finally resolved these issues in 1905. Interestingly, he was apparently not motivated by the results of the Michelson–Morley experiment. When asked later, Einstein could not recall whether or not he knew about the experiment when formulating his relativity theory.

In fact, Einstein's insight was based on an intuitive feeling that the laws of mechanics should not be the only ones to obey the relativity principle. He reasoned that nature should be symmetrical and that *all* physical laws should obey the relativity principle. In Einstein's view, the inconsistencies in electromagnetic theory were due to the assumption that an

* Albert Abraham Michelson (1852–1931) was a German-born American physicist who devised the Michelson interferometer in an attempt to detect the motion of the Earth through the ether. Edward W. Morley (1838–1923) was an American chemist who collaborated with Michelson. The interferometer is a sensitive instrument that can measure tiny distance and speed changes using the superposition of light. Its details are beyond the scope of this text.

absolute rest frame (the ether frame) existed. His theory did away with the need for such a frame (and the "ether") by placing all laws of physics on an equal footing, thus eliminating any way of measuring the absolute speed of an inertial reference frame.

The postulate on which relativity is based is thus a generalization of Newtonian relativity. Einstein's **principle of relativity** (sometimes called the special relativity postulate) applies to all the laws of physics, including those of electricity and magnetism:*

*All the laws of physics are the same in all inertial reference frames.*

This reasoning means that all inertial frames must be physically equivalent. That is, all physical laws, not just those of mechanics, must be the same in all inertial frames. As a consequence, no experiment - mechanical, electromagnetic or any other type - performed entirely within an inertial frame could enable an observer in that frame to detect its motion. That is, *there is no absolute reference frame.* In hindsight, this idea seems reasonable: There is no reason to think that nature would play favorites by picking the laws of mechanics over other fundamental laws.

Pulling the laws of electromagnetic waves (light) into this principle to ensure that different inertial observers would not interpret the laws of electromagnetism differently led Einstein to a result called the constancy of the speed of light:

*The speed of light in a vacuum has the same value in all inertial systems.*

If this were not the case, then the speed of electromagnetic waves (light) would be measurably different from $c$ in different inertial frames, thus violating the special relativity postulate.

The constancy of the speed of light means that *the speed of light is independent of the speed of source or observer.* For example, if a person moving toward you at a constant velocity turned on a flashlight, you both would measure the speed of the emitted light to be $c$, regardless of the relative velocity (▶ **Figure 26.2**). This conclusion is consistent with the null result of the Michelson–Morley experiment. By doing away with an absolute reference frame, Einstein could reconcile the apparently fundamental differences between mechanics and electromagnetism.

The ultimate test of any theory is provided by the scientific method. What does Einstein's theory predict, and can it be experimentally verified? The answer to the latter is "yes," as will be seen and discussed in the following sections.

With the relativity postulate in place, some of its implications can be explored. Because many of the results are discernible only when speeds are near to that of ight, many times special relativity can be better understood by *imagining* simple situations. These can often tell us what phenomena are predicted. Einstein used this idea by employing what he called *gedanken,* or "thought," experiments; that is, "experiments" done completely in the mind. Let us begin with a series of famous such experiments related to simultaneity and how length and time are measured.

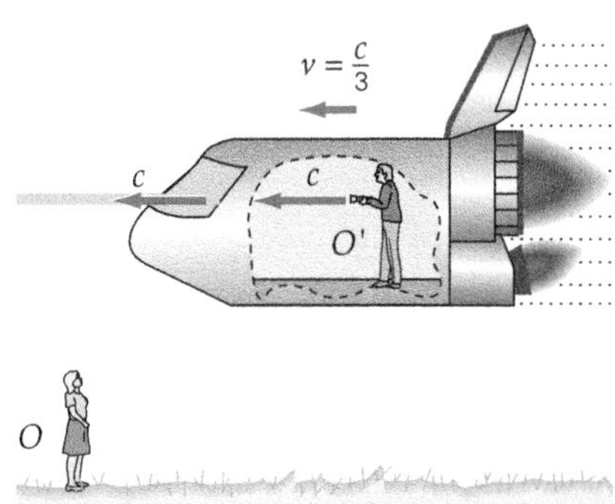


▲ FIGURE 26.2 **Constancy of the speed of light** Two observers in different inertial frames measure the speed of the same beam of light. The observer in frame $O'$ measures a speed of $c$ inside the ship. According to Newtonian ideas, the observer in frame $O$ would measure a speed of $c + (c/3) = (4c/3) > c$ as the beam passes her, but instead, according to Einstein, she measures $c$.

### 26.2.1 The Relativity of Simultaneity

In everyday life, two events that are simultaneous to one person are thought to be simultaneous to everyone. That is, the concept of simultaneity is assumed absolute - the same for everyone. What could be more obvious? Simultaneous events occur at the same time; isn't that the same for all observers? The answer is no - but this is obvious only for observers moving at relative velocities near that of light. In other words, as will be shown, at everyday speeds this lack of agreement about simultaneity is too small to be observed.

Think of an inertial reference frame (called $O$) in which two events are *designed* to be simultaneous. For example, suppose that two firecrackers (at A and B on the $x$-axis) are arranged to explode when a switch, placed midway between them, is flipped to the "on" position (▶ **Figure 26.3a**). Now *imagine* equipping the observer in the $O$ frame with a light receptor at point R (for receptor), exactly midway between the them. This detector is capable of detecting whether the light flashes from the firecrackers arrive simultaneously (i.e., are detected "in coincidence") or not. (In actuality, the receptor could be anywhere, but the results would have to be corrected for unequal travel distances. To avoid this complication, it will be assumed that all simultaneity detectors are midway between the two events.)

After detonation, the light receptor determines that the two explosions went off simultaneously *in the O frame.* But now consider the same explosions as measured by an observer in a different inertial frame, $O'$, moving to the right at a speed $v$ relative to $O$. Assume that observer in $O'$ has equipped himself with a *series* of light receptors on his $x'$-axis, because he is not sure which one will be midway between the explosions.

* The *special* designation indicates that the theory deals only with inertial reference frames. The *general* theory of relativity, discussed later in this chapter, deals with the more general case of noninertial, or accelerating, frames.

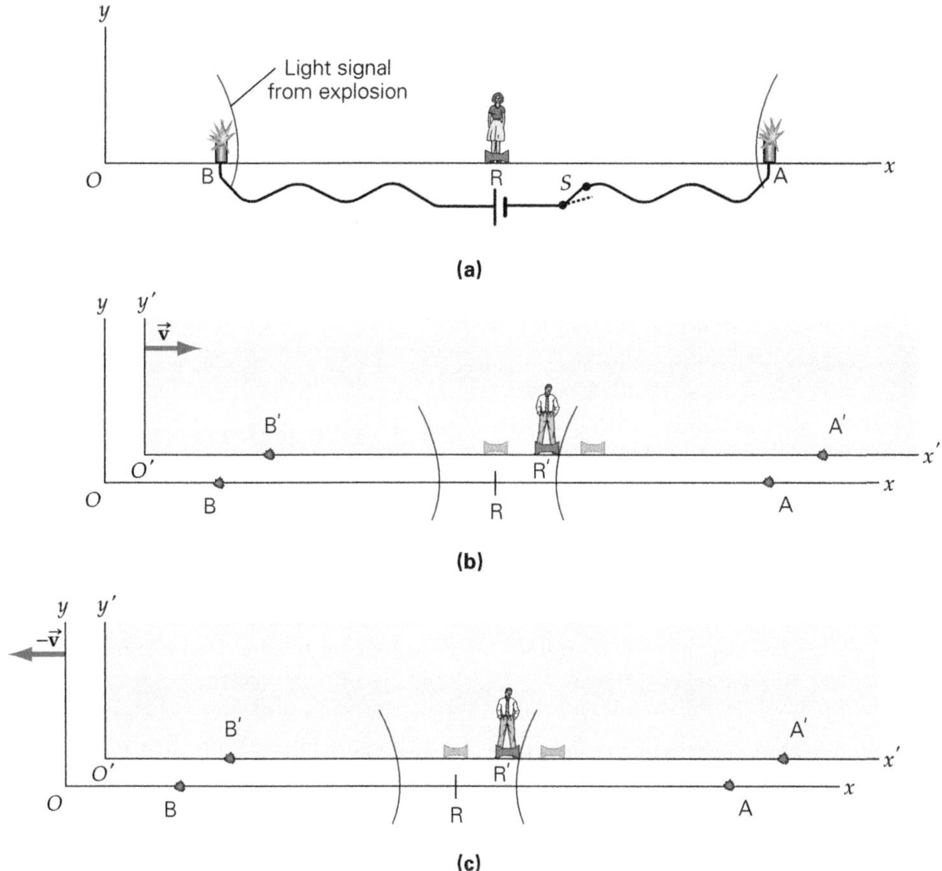


▲ FIGURE 26.3 **The relativity of simultaneity (a)** An observer in reference frame $O$ triggers two explosions (at A and B) that occur simultaneously. A light receptor R, located midway between them, records the two light signals as arriving at the same time. **(b)** An observer midway between the two explosions, but in frame $O'$, moving with respect to $O$, sees the burn marks made by the two explosions on the $x'$-axis, but sees the explosion at A happen before that at B. **(c)** The situation as viewed from the observer in $O'$ who sees the explosion at A before that at B. To him, $O$ is moving to the left.

After the explosive events, there are burn marks on both the $x$-axis (at A and B) and the $x'$-axis (at $A'$ and $B'$), as in Figure 26.3b. These marks can be used to identify the particular $O'$ receptor (call it $R'$) that was, in fact, midway between $A'$ and $B'$. But when the observer in $O'$ reviews the data from this receptor, he finds that it did *not* record the explosions simultaneously. This result, however, does not cause the observer in $O$ to doubt her conclusion. She has an explanation of what happened. As she sees the situation, during the time it took for the light to get to $R'$, that receptor had moved toward A and away from B. This is the reason, then, according to the observer in $O$, that the receptor in $O'$ received the flash from A before that from B.

The question then appears to be: Which observer is correct? It's hard to find any objection to the conclusion reached by the observer in $O$ – so isn't the observer in $O'$ mistaken? It should be obvious to the $O'$ observer that he is moving with respect to the firecrackers. Why doesn't he realize this and take his motion into account? After all, wasn't his light receptor moving toward A and away from B? If so, then it should not surprise him that it recorded the flash from A before the flash from B. All he has to do is to allow for this motion in his calculations, and he will conclude that the flashes "really" were simultaneous.

But this type of reasoning ignores the postulate of relativity. The "logical" argument laid out previously assumes that when the situation is viewed from the $O$ frame (Figure 26.3a and b), the observer is looking at what "really" happened from the vantage point of the frame that is "really" at rest. But according to the special relativity postulate, no inertial reference frame is more valid than any other, and none can be considered absolutely at rest. To the observer in $O$, *it is the O frame that is moving, and that his, the O′ frame, is the rest frame*. From the $O'$ viewpoint, the firecrackers are moving at a speed $v$ to the left (Figure 26.3c), but this motion would not affect his conclusions. To him, the explosions, equally distant from $R'$, arrived at $R'$ at different times; therefore, they were *not* simultaneous.

You might wonder whether it could be arranged so that the observer in $O'$ would agree that the explosions were simultaneous. The answer is yes. However, to accomplish this, the observer in $O$ would have to *delay* the firing of firecracker A relative to B so that $R'$ would receive the two signals at the same time. This does not change the lack of agreement on simultaneity, because now the explosions would no longer be simultaneous in $O$!

What is to be made of this curious situation? In a nonrelativistic ($v \ll c$) world, one of the observers would have to be wrong. But as this example shows, both observers performed the measurements correctly. Neither one used faulty instruments or made any errors in logic. So the conclusion must be that *both* are correct.

Furthermore, there is nothing special about firecracker explosions. Any "happening" at a particular point in space at a particular time – a karate kick, a soap bubble bursting, a heartbeat – would have done just as well. Such a happening is called an *event* (an occurrence specified by a location in space *and* a time) in the language of relativity. Based on the postulate of relativity is the following result:

> Events that are simultaneous in one inertial reference frame may not be simultaneous in a different inertial frame.

This *gedanken* experiment convinced Einstein to give up on the concept of absolute simultaneity.

Note that if the relative speed of the reference frames is much, much less than that of light (as in everyday speeds), this lack of simultaneity is completely undetectable. *Most relativistic effects have this property.* That is, their departure from familiar experience is not apparent when the speeds involved are much less than the speed of light. Since most of us have no experience with such high speeds, it is hardly surprising that the predictions of special relativity seem "strange." In fact, it is our everyday assumptions of absolute simultaneity, lengths, time intervals, etc., observed on our low-speed world, that are not correct!

---

**CONCEPTUAL EXAMPLE 26.1: AGREEING TO DISAGREE – THE RELATIVITY OF SIMULTANEITY**

(a) In Figure 26.3, estimate the relative speed of the two observers.
(b) If the relative speed were only 10 m/s, would there be better agreement on simultaneity? Why?

**REASONING AND ANSWER**

**(a)** To estimate the relative speed of the observers, compare the distance between B and B′ in Figure 26.3b with the distance the light has traveled from B.
The figure indicates that the $O'$ frame has moved about 25% as far as the light. Therefore, the relative speed between the two reference frames, $v$, is approximately 25% of the speed of light, or $v \approx 0.25c$.

**(b)** At a relative speed of 10 m/s, the two frames would not have moved a noticeable distance; thus, both observers would agree on simultaneity, within the limits of measurability.

**FOLLOW-UP EXERCISE.** Show that two events that occur simultaneously on the y-axis of $O$ are perceived as simultaneous by an observer in $O'$, regardless of the relative speed, as long as the relative motion is along their common $x$–$x'$-axes.

---

To grasp the importance of the relativity of simultaneity, imagine trying to measure the length of a moving object. To do so properly, the positions of both ends of the object must be marked *simultaneously*. However, as has been seen, two different inertial observers will, in general, disagree on simultaneity. Thus, they will also disagree on the object's length, as will be shown in the next section.

## 26.3 The Relativity of Time and Length: Time Dilation and Length Contraction

### 26.3.1 Time Dilation

Another of Einstein's *gedanken* experiments pertained to the measurement of time intervals in different inertial frames. To compare time intervals as measured in different inertial frames, he envisioned a *light pulse clock*, as illustrated in ▶ **Figure 26.4a**. A tick (time interval) on the clock corresponds to the time a light pulse takes to make a round trip between the source and the mirror. Let's assume that the observers in the two frames, $O$ and $O'$, have identical light clocks, and the clocks run at the same rate when at rest relative to one another. For an observer at rest with respect to one of these clocks, the time interval ($\Delta t_o$) for a round trip of a light pulse is the total distance traveled divided by the speed of light, or

$$\Delta t_o = \frac{2L}{c} \tag{26.1}$$

Now, suppose $O'$ is moving relative to $O$ with a constant velocity $\vec{\mathbf{v}}$ to the right. With his clock (at rest in $O'$), the observer in $O'$ measures the same time interval for his clock, $\Delta t_o$. However, according to the observer in $O$, the clock in the $O'$ system is moving, and the path of its light pulse forms the sides of two right triangles (Figure 26.4b). Hence the observer in $O$ sees the light pulse from the clock in $O'$ take a longer path than the light in her own clock. From the $O$ frame, the Pythagorean theorem gives:

$$\left(\frac{c\Delta t}{2}\right)^2 = \left(\frac{v\Delta t}{2}\right)^2 + L^2$$

(*Note:* $\Delta t$ is the time interval of the $O'$ clock but as measured by the observer in $O$.) Since the speed of light is the same for all observers, the light in the "moving" clock's reference frame takes a longer time to cover the path, *according to the observer in O*. Therefore for the observer in the $O$ frame, the "moving" clock runs slowly since the ticks occur at a lower rate. To determine

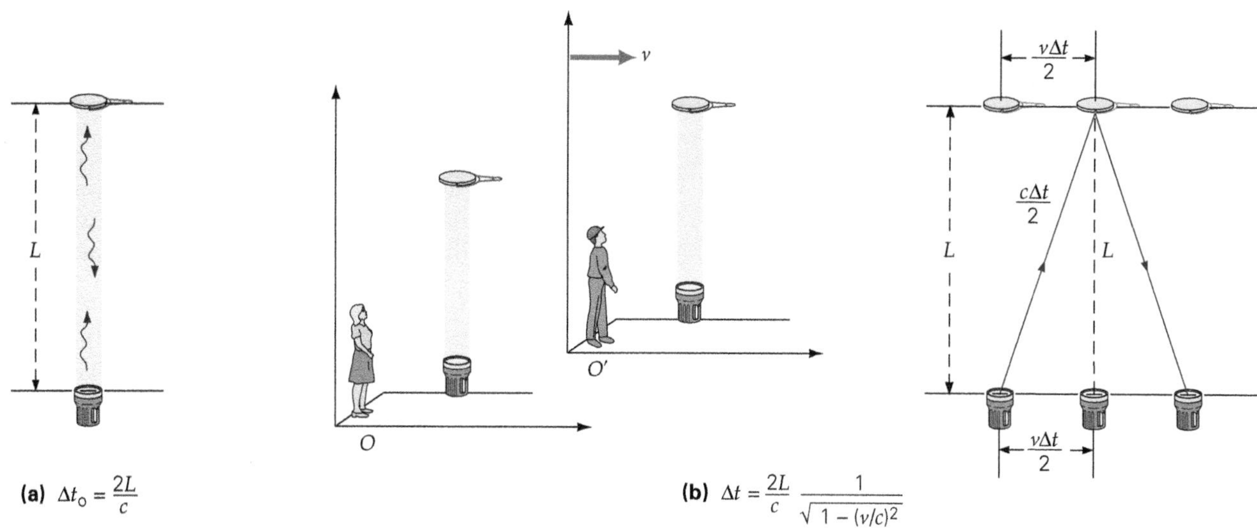


▲ FIGURE 26.4 **Time dilation (a)** A light clock that measures time in units of round-trip reflections of light pulses. The time for light to travel up and back is $\Delta t_o = 2L/c$. **(b)** An observer in $O$ measures a time interval of $\Delta t = (2L/c)\left[1/\sqrt{1-(v/c)^2}\right]$ on the clock in the $O'$ frame. Thus, the moving clock appears to run slowly to the observer in $O$.

the relationship between the times, the preceding equation can be solved for $\Delta t$

$$\Delta t = \frac{2L}{c}\left[\frac{1}{\sqrt{1-(v/c)^2}}\right] \quad (26.2)$$

But the time interval measured by an observer at rest with respect to a clock is $\Delta t_o = 2L/c$ (Equation 26.1). Thus,

$$\Delta t = \frac{\Delta t_o}{\sqrt{1-(v/c)^2}} \quad \text{(relativistic time dilation)} \quad (26.3)$$

Since $\frac{1}{\sqrt{1-(v/c)^2}} > 1$, then $\Delta t > \Delta t_o$. Therefore, an observer in $O$ measures a longer time interval ($\Delta t$) on the $O'$ clock than does the observer in $O'$ on the same clock ($\Delta t_o$). This effect is called **time dilation** (dilation in the sense of "stretching out" or "elongating"). With a longer time between ticks, the $O'$ clock appears, to an observer in $O$, to run more slowly than the $O$ clock. The situation is symmetric and relative: The observer in $O'$ would observe the clock in the $O$ frame running slowly relative to the $O'$ clock, thus leading to the following conclusion:

> Moving clocks are observed to run more slowly than clocks at rest in an observer's own reference frame.

This effect, like all relativistic effects, is significant only if the relative speeds are close to the speed of light. Under these conditions, can you explain why, using Figure 26.4b, the time intervals would appear to be the same?

To distinguish between these two time intervals, the term **proper time interval** is used. As with most relative measurements (e.g., see length measurements), here the term "proper" refers to the time interval measured by an observer using a clock at rest with respect to that observer. In the preceding development, the proper time interval is $\Delta t_o$. Stated more formally:

> The *proper time interval* between two events is that interval as measured by an observer at rest relative to those events and who observes them occur *at the same location.*

To understand the meaning of this, ask yourself: What are the two events for the light-pulse clock? Are they at the same location in $O'$ for the $O'$ clock? In Figure 26.4, the observer in $O$ sees the events by which the time interval of the $O'$ clock is measured at *different* locations. Because that clock is moving, the starting event (light pulse leaving) occurs at a different location in $O$ from that of the ending event (light pulse returning). Thus, $\Delta t$, the time interval measured by the observer in $O$, is *not* the proper time interval.

Many of the equations of relativity can be written more compactly if a single symbol $\gamma$ (Greek letter "gamma") is defined as

$$\gamma = \frac{1}{\sqrt{1-(v/c)^2}} \quad (26.4)$$

Note that $\gamma \geq 1$. (When would it be equal to 1?) Also notice that as $v$ approaches $c$, $\gamma$ approaches infinity. Since an infinite time interval is not physically possible, *relative speeds equal to or greater than that of light are not possible.* The values of $\gamma$ for several values of $v$ (expressed as fractions of $c$) are listed in ▶ **Table 26.1**. This table clearly indicates that speeds must be an appreciable fraction of $c$ before relativistic effects can be readily observed. For example, at $v = 0.10c$, $\gamma$ differs from 1.00 by only 1%.

TABLE 26.1 Some Values of $\gamma = \frac{1}{\sqrt{1-(v/c)^2}}$

| $v$ | $\gamma$ | $v$ | $\gamma$ |
|---|---|---|---|
| 0 | 1.00 | 0.800$c$ | 1.67 |
| 0.100$c$ | 1.01 | 0.900$c$ | 2.29 |
| 0.200$c$ | 1.02 | 0.950$c$ | 3.20 |
| 0.300$c$ | 1.05 | 0.990$c$ | 7.09 |
| 0.400$c$ | 1.09 | 0.995$c$ | 10.0 |
| 0.500$c$ | 1.15 | 0.999$c$ | 22.4 |
| 0.600$c$ | 1.25 | $c$ | $\infty$ |
| 0.700$c$ | 1.40 | | |

Using Equation 26.4, the time dilation relationship (Equation 26.3) can be written more compactly as

$$\Delta t = \gamma \Delta t_o \quad \text{(relativistic time dilation)} \qquad (26.5)$$

Suppose you observed a clock at rest in a system moving relative to you at a constant speed of $v = 0.60c$. For that speed, $\gamma = 1.25$. Thus if, for example, 20 min have elapsed on that clock, an interval of $\Delta t = \gamma \Delta t_o = (1.25)(20 \text{ min}) = 25$ min would be measured by you. The 20-min interval is the proper interval, because the events defining this interval took place at the same location. Thus, the "moving" clock runs more slowly (20 min elapsed, as opposed to 25 min on your clock) when viewed by an observer (you) moving relative to it.

Finally, note that the time dilation effect cannot apply just to our admittedly artificial (*gedanken*) light pulse clock. *It must be true for all clocks and hence all time intervals* (i.e., anything that keeps a rhythm or frequency, including the heart). If this were not the case - if a mechanical watch, for example, did not exhibit time dilation - then that watch and a light-pulse clock would run at different rates in the same frame. This would mean that observers in that frame would be able to tell their motion by making a comparison between the two different types of clock *solely within their frame.* Since this violates the postulate of special relativity, it must be that *all* moving clocks, regardless of their nature, exhibit time dilation. Example 26.2 illustrates an actual situation of time dilation occurring in nature.

### EXAMPLE 26.2: MUON DECAY VIEWED FROM THE GROUND - TIME DILATION VERIFIED BY EXPERIMENT

Subatomic particles, called *muons,* can be created in the Earth's atmosphere when cosmic rays (mostly protons) collide with the nuclei of the atoms that compose air molecules. Once created, they approach the Earth's surface with speeds near $c$ (typically, about 0.998$c$). However, muons are unstable and decay into other particles. The average lifetime of a muon at rest has been measured in the laboratory to be $2.20 \times 10^{-6}$ s. During this time, the muon travels a distance

$$\begin{aligned} d &= v\Delta t = (0.998c)(2.20 \times 10^{-6} \text{ s}) \\ &= 0.998(3.00 \times 10^8 \text{ m/s})(2.20 \times 10^{-6} \text{ s}) \\ &= 660 \text{ m} = 0.660 \text{ km} \end{aligned}$$

Since muons are created at altitudes of 5 to 15 km, it would be reasonable to expect that very few of them would reach the Earth's surface. However, experimentally, an appreciable number actually do reach the surface. Using time dilation, explain this apparent paradox.

**THINKING IT THROUGH.** The paradox arises because the preceding calculation does *not* take time dilation into account. That is, a muon decays by its own "internal clock," as measured in its own reference frame. Its "lifetime" of $2.20 \times 10^{-6}$ s is a proper time interval, and so is its "proper" lifetime. This is because in the muon's rest frame and "birth" and "death" events take place at the same location. Thus, to an observer on the Earth, any "clock" in the muon's reference frame would appear to run more slowly than a clock on the Earth (▼ **Figure 26.5**). If the muon's lifetime is dilated, as observed from the surface of the Earth, then that could explain their presence at the surface.

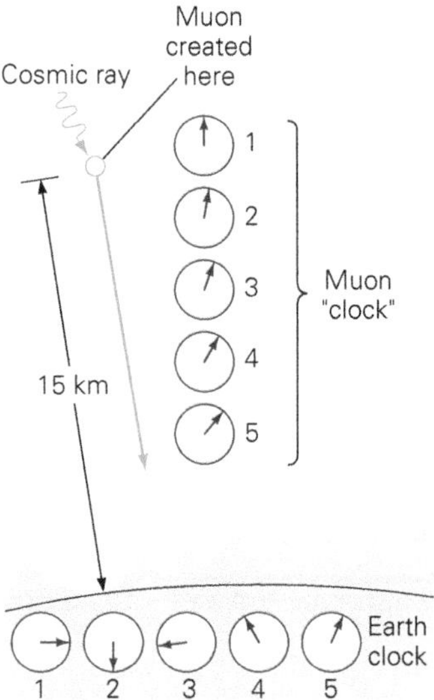


▲ FIGURE 26.5 **Experimental evidence of time dilation** Muons are observed at the surface of the Earth, as predicted by special relativity. The sequence of integers (1 through 5 on each clock) refers to the order of events. From the Earth's viewpoint, the muon's "clock" runs more slowly than Earth clocks. See Example 26.2.

**SOLUTION**

Listing the given quantities:

***Given:***

$v = 0.998c$

$\Delta t_o = 2.20 \times 10^{-6}$ s (muon proper lifetime)

***Find:*** Use time dilation to explain the larger-than-expected number of muons that reach the Earth's surface.

Instead of the proper lifetime, an observer on the Earth would observe one that is longer by a factor of $\gamma$. From Equation 26.4

$$\gamma = \frac{1}{\sqrt{1-(v/c)^2}} = \frac{1}{\sqrt{1-(0.998c/c)^2}} = 15.8$$

From Equation 26.5, the lifetime of the muon, according to an observer on the Earth, is

$$\Delta t = \gamma \Delta t_o = (15.8)(2.20 \times 10^{-6}\,\text{s}) = 3.48 \times 10^{-5}\,\text{s}$$

The distance the muon travels, according to an observer on the Earth (using the dilated time interval), is

$$\begin{aligned} d = v\Delta t &= (0.998c)\Delta t \\ &= 0.998(3.0 \times 10^8\,\text{m/s})(3.48 \times 10^{-5}\,\text{s}) \\ &= 1.04 \times 10^4\,\text{m} = 10.4\,\text{km} \end{aligned}$$

This distance is in the altitude range at which muons are created. Hence, the detection of more muons than expected is explained by relativistic time dilation.

**FOLLOW-UP EXERCISE.** In this Example, what speed would enable an average muon to travel 20.8 km relative to the Earth (i.e., twice as far as the distance in the Example)? Would the muon have to travel twice as fast? Why or why not? Explain.

### 26.3.2 Length Contraction

When measuring the length of a linear (one-dimensional) object not at rest in our reference frame, care must be taken to mark both ends simultaneously. Consider again the two inertial reference frames used in the discussion of simultaneity and imagine a measuring stick lying on the $x$-axis at rest in $O$ (▼ **Figure 26.6a**).

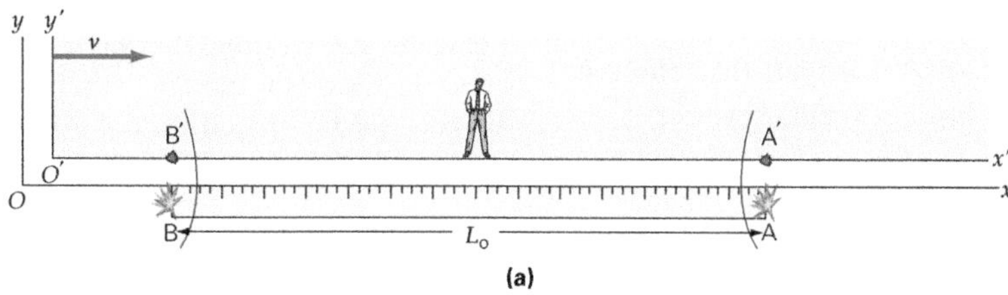


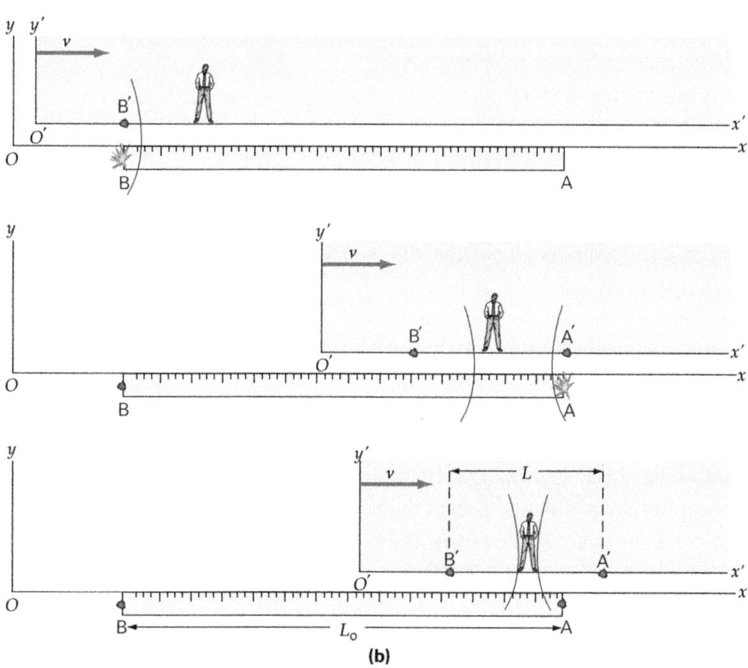


▲ FIGURE 26.6 **Measuring lengths "correctly" and length contraction** To measure the length of a moving object correctly, the ends must be marked simultaneously. **(a)** When the observer in frame $O$ marks the ends simultaneously (thus measuring the "correct" length according to $O$), the observer in $O'$ observes A marked before B (i.e., an incorrect method according to $O'$), resulting in too long a length from the viewpoint of $O'$. **(b)** If the observer in $O$, instead, delays marking A relative to B by just the correct amount (how do we know this was done from the sketch?), the observer in $O'$ measures the correct length *from his point of view*. Thus the length measured by the observer in $O'$ is less than the rest (proper) length measured by the observer in $O$ – length contraction of the moving object observed by $O'$.

If the observer in $O$ marks the ends simultaneously, the observer in $O'$ will observe A marked before B. Thus, for the observer in $O'$ to make a correct length measurement, the observer in $O$ should delay marking A relative to B (Figure 26.6b). Imagine the observer in $O$ setting off explosions that create burn marks in both reference frames (on both the $x$- and $x'$-axes). When the ends are marked so that the observer in $O'$ agrees that they were done simultaneously, all that needs to be done is to subtract the two positions to get the length of the stick *as measured* in $O'$. Note that it will be *less* than the length measured by $O$.

Again, it might be asked, "Which observer makes the correct measurement?" By now you know the answer: Both are correct. *Both have measured the length correctly* in their own frames. Neither thinks that the other has done things correctly, but each is satisfied with their own measurement. Fundamentally, then, it is the lack of agreement on simultaneity that leads to the following qualitative statement about length contraction:

> An object's length is largest when measured by an observer at rest with respect to it (the "proper" observer). If the object is moving relative to an inertial observer, that observer measures a smaller length than the proper observer.*

As usual, this effect is entirely negligible at speeds that are slow compared with $c$.

The distance between two points as measured by the observer at rest with respect to them is designated by $L_o$, the **proper length**. The proper length (or *rest length*) is the largest possible length. Note that the term "proper" has nothing to do with the correctness of the measurement, because each observer is measuring correctly from his or her point of view.

A *gedanken* experiment can help develop a mathematical expression for length contraction. Consider a rod at rest in frame $O$. This means that the observer in $O$ is the proper length measurer and he measures $L_o$. An observer in $O'$, traveling at a constant speed $v$ parallel to the rod, also measures the length of the rod (▶ **Figure 26.7**). She does this by using the clock she is holding to measure the time required for the two ends of the rod to pass her. Since she measures the proper time interval (how do we know this?), she measures is $\Delta t_o$. In her reference frame, the rod is moving to the left at speed $v$. Hence from her viewpoint the length is $L = v\Delta t_o$.

Suppose the observer in $O$ measures the rod's length using the same method. To him, the observer in $O'$ is moving relative to the rod. If he notes the times on his clock when $O'$ passes the ends of the rod, he measures a time interval $\Delta t$. For him, the length of the stick is the proper length; therefore, $L_o = v\Delta t$. Then, dividing one length by the other yields

$$\frac{L}{L_o} = \frac{v\Delta t_o}{v\Delta t} = \frac{\Delta t_o}{\Delta t} \qquad (26.6)$$

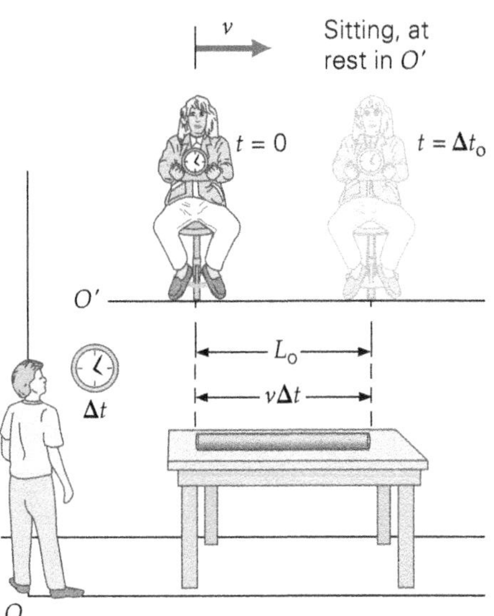


▲ **FIGURE 26.7 Derivation of length contraction expression** The observer in $O$ measures the time it takes for the observer in $O'$ to move past the ends of the rod. Similarly, the observer in $O'$ measures the time it takes for the ends of the rod to pass her. The observer in $O'$ is the *proper time measurer*; she measures the shortest possible time between these two events. The measured lengths of the rod are not the same. The observer in $O$ is the *proper length measurer;* he measures the longest possible length.

But $\Delta t = \gamma\Delta t_o$ (Equation 26.5), or $\Delta t_o/\Delta t = 1/\gamma$. Thus Equation 26.6 can be rewritten as

$$L = \frac{L_o}{\gamma} = L_o\sqrt{1-(v/c)^2} \quad \text{(relativistic length contraction)} \qquad (26.7)$$

Since $\gamma > 1$, then, as expected, $L < L_o$. This is called **relativistic length contraction**.

To see the effects of length contraction and time dilation, consider the following two high-speed Examples.

### EXAMPLE 26.3: WARP SPEED? LENGTH CONTRACTION AND TIME DILATION

An observer sees a spaceship, 100 m long when at rest with respect to her, pass completely by (i.e., nose to tail) at a speed of $0.500c$ (▶ **Figure 26.8**). While this observer is watching the ship, a time of 2.00 s elapses on the ship's clock. (a) What is the length of the ship as measured by the observer? (b) What time interval elapses on the observer's clock during the 2.00 s interval on the ship's clock?

**THINKING IT THROUGH.** The 100 m is the ship's proper length. (Why?) The time interval of 2.00 s is the proper time interval, because the same clock, in the same location, measures the beginning and end of the interval. In (a), Equation 26.7 can be used to find the contracted length, and in (b), Equation 26.5 will allow the determination of the dilated time interval.

* Here, "length" refers to the dimension of the object in the direction of its relative velocity. That is, if a cylindrical object is moving parallel to its long axis, then only its long axis, and not its diameter, would exhibit length contraction.

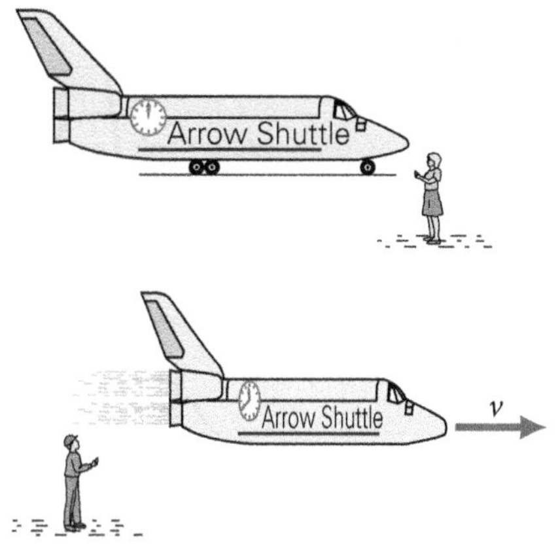


▲ FIGURE 26.8 **Length contraction and time dilation** As a result of length contraction, moving objects are observed to be shorter, or contracted, in the direction of motion, and moving clocks are observed to run more slowly because of time dilation.

**SOLUTION**

Listing the given quantities:

***Given:***

$L_o = 100$ m (proper length)
$v = 0.500$
$\Delta t_o = 2.00$ s (proper time interval)

***Find:***

**(a)** $L$ (contracted length)
**(b)** $\Delta t$ (dilated time interval)

**(a)** By calculation (or) Table 26.1, $\gamma = 1.15$ for $v = 0.500c$, and the contracted length is

$$L = \frac{L_o}{\gamma} = \frac{100\,\text{m}}{1.15} = 87.0\,\text{m}$$

**(b)** The time interval $\Delta t$ measured by the observer is longer than the proper time interval $\Delta t_o$:

$$\Delta t = \gamma \Delta t_o = 1.15(2.00\,\text{s}) = 2.30\,\text{s}$$

**FOLLOW-UP EXERCISE.** In this Example, find the time it takes the spaceship to pass a given point in the observer's reference frame, as seen by (a) a person in the spaceship and (b) the observer watching the ship move by. Explain clearly why these time intervals are *not* the same.

---

**EXAMPLE 26.4: MUON DECAY REVISITED – ALTERNATIVE EXPLANATIONS**

Example 26.2 showed, using relativistic time dilation, how more muons than expected classically could reach the Earth's surface. Since a hypothetical observer *on the muon* could *not* use this argument (why not?), how would he or she explain the fact that the average muon does make it to the surface? Which explanation is "correct?"

**THINKING IT THROUGH.** A muon traveling at $v = 0.998c$ decays, in its rest frame in $\Delta t_o = 2.20 \times 10^{-6}$ s. In that time, it would travel only 659 m, not nearly enough to reach the surface. In Example 26.2, this apparent paradox was explained (at least for the Earth observer) by invoking relativistic time dilation. But how is this "paradox" explained by a hypothetical observer on the muon? For such an observer, the muon "clock" is correct – as a result, the proper lifetime is *not* sufficiently long to enable the muon to reach the Earth's surface! How can this be reconciled with the experimental observation of the Earth observer, which finds that the average muon does make it to the surface? After all, two observers cannot disagree on this experimental result. To the observer on the muon, the distance from the spot of its "creation" to the surface moves by quickly, so the explanation from the muon viewpoint involves length contraction.

**SOLUTION**

***Given:***

See Example 26.2.

***Find:*** the explanation for muons reaching the surface according to an observer in the muon reference frame.

The apparent paradox disappears when length contraction is taken into account. For the observer on the muon, the muon's "clock" reads correctly ($\Delta t_o = 2.20 \times 10^{-6}$ s), but the travel distance is shorter because of length contraction. Since $\gamma = 15.8$ here, a length of 10.0 km in the Earth frame, the proper length $L_o$, is measured from the muon frame to be considerably shorter:

$$L = \frac{L_o}{\gamma} = \frac{10.0\,\text{km}}{15.8} = 0.633\,\text{km} = 633\,\text{m}$$

According to an observer in the muon frame, traveling this distance would take a time

$$\Delta t = \frac{L}{v} = \frac{L}{0.998c} = \frac{633\,\text{m}}{0.998(3.00 \times 10^8\,\text{m/s})} = 2.11 \times 10^{-6}\,\text{s}$$

This is about equal to the muon lifetime in the muon's reference frame. Thus, through relativistic considerations, the two observers do agree that many more muons than classically expected reach the Earth (the experimental result). The Earth observer explains by saying, "The muon clock is running slow" (time dilation). The observer on the muon says, "No, the clock is fine, but the distance the muons must travel is considerably less than you claim" (length contraction).

Who is correct? It must be that both are. The reasoning may be different for the two observers, but the experimental result (the number of muons reaching the Earth's surface) *is* the same. This is all that matters – what is measured!

**FOLLOW-UP EXERCISE.** Muons are actually created with a range of speeds. What is the speed of a muon if it decays 5.00 km from its creation point as measured by an Earth observer?

## 26.4 Relativistic Kinetic Energy, Momentum, Total Energy, and Mass–Energy Equivalence

The ramifications of special relativity are very important in particle physics, in which speeds routinely approach $c$. Most of the expressions from classical (low-speed) mechanics are incorrect at such speeds. Relativistic kinetic energy, for example, is not the familiar $K = (1/2)mv^2$ that is routinely used at low speeds. Einstein showed that kinetic energy does increase with speed, but in a different way. His derivation is beyond the mathematical scope of this book, but the result for the **relativistic kinetic energy** of a particle of mass $m$ moving at speed $v$ is

$$K = \left[\frac{1}{\sqrt{1-(v/c)^2}} - 1\right] mc^2 = (\gamma - 1)mc^2 \qquad (26.8)$$

(relativistic kinetic energy)

[It can be shown that this expression does reduce, as it should, to the more familiar $K = (1/2)mv^2$ when $v \ll c$.]

According to Equation 26.8, as $v$ approaches $c$, the kinetic energy of an object becomes infinite. In other words, to get an object to move at the speed of light would require an infinite amount work, which is not possible. Thus, no object can travel as fast as, or faster than, the speed of light.

Particle accelerators can accelerate charged particles to very high speeds. There is complete agreement between experimentally measured kinetic energies of these particles and Equation 26.8. A graphical comparison of the relativistic expression for kinetic energy and the classical (low-speed) expression is shown in ▶ **Figure 26.9**. As can be seen, they agree at low speeds.

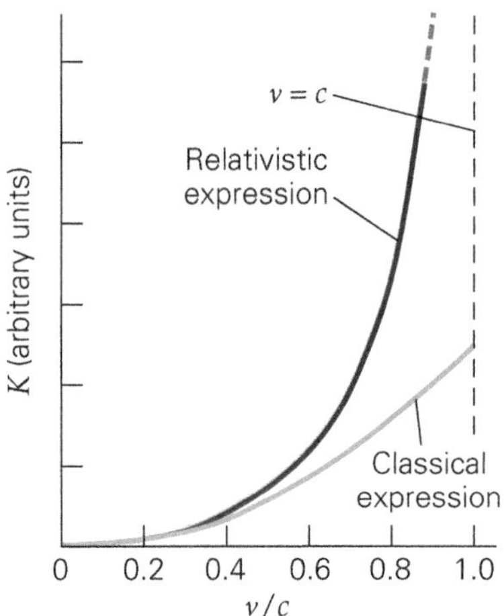


▲ **FIGURE 26.9 Relativistic versus classical kinetic energy** The variation in kinetic energy with particle speed, expressed as a fraction of $c$, is shown for the relativistically correct expression and for the classical expression. The classical expression is not noticeably different from the relativistic one for speeds less than about $0.2c$. Objects cannot have a speed of $c$, because that would require them to have an infinite kinetic energy.

### 26.4.1 Relativistic Momentum

Just as for kinetic energy, the relativistic expression for the momentum of an object is different from the low-speed expression ($\vec{\mathbf{p}} = m\vec{\mathbf{v}}$). The expression for **relativistic momentum** is

$$\vec{\mathbf{p}} = \frac{m\vec{\mathbf{v}}}{\sqrt{1-(v/c)^2}} = \gamma m\vec{\mathbf{v}} \quad \text{(relativistic momentum)} \qquad (26.9)$$

Relativistic momentum is still a vector, and total (vector) momentum is still a conserved quantity under relativistic conditions.

### 26.4.2 Relativistic Total Energy and Rest Energy: The Equivalence of Mass and Energy

In classical mechanics, the total mechanical energy of an object is the sum of its kinetic and potential energies. When there is no potential energy (i.e., for a free object), then the total energy is purely kinetic, or $E = K = (1/2)mv^2$. However, Einstein was able to show that the **relativistic total energy** of such a free object is correctly given by

$$E = \frac{mc^2}{\sqrt{1-(v/c)^2}} = \gamma mc^2 \quad \text{(relativistic total energy)} \qquad (26.10)$$

Thus, according to relativity, an object that is at rest and still has no kinetic energy ($K = 0$, $v = 0$), however does possess an energy of $mc^2$, *not zero*. This minimum energy that an object always has is its **rest energy** $E_o$ and is given by Equation 26.10 with $v = 0$:

$$E_o = mc^2 \quad \text{(rest energy)} \qquad (26.11)$$

Thus even when $v = 0$, the total energy of the particle is not zero, but $E_o$. Since $K = (\gamma - 1)mc^2$, the total energy of such a free particle can be expressed as the sum of kinetic and rest energies.

To see this, note that the relativistic kinetic energy expression (Equation 26.8) can be rewritten as

$$K=(\gamma-1)mc^2=\gamma mc^2-mc^2=E-E_o$$

Thus, an alternative to Equation 26.10 is

$$E=K+E_o=K+mc^2 \quad \text{(relativistic total energy – free particle)} \quad (26.12)$$

A relationship between a particle's total energy and its rest energy can be obtained by replacing $mc^2$ with $E_o$ in Equation 26.10:

$$E=\gamma E_o \quad \text{(relativistic total energy – free particle)} \quad (26.13)$$

As a double check, note that if $v=0$ (i.e., $K=0$) then $\gamma=1$, yielding $E=E_o$ as expected.

Equation 26.11 expresses Einstein's famous **mass–energy equivalence**. *An object has energy even at rest* – its rest energy. As a consequence, mass is now viewed as a form of energy. In nuclear and particle physics, for total energy to be conserved mass must be treated as a form of energy.

Mass–energy equivalence does not mean that mass can be converted into useful energy at will. If so, our energy problems would be solved, since there is a lot of mass on the Earth! However, practical conversion of mass into other forms of energy, such as heat, does take place in applications such as nuclear reactors, which generate electric energy (see Chapter 30).

The (rest) energy of an object obviously depends on its mass. For example, the mass of an electron is $9.109\times10^{-31}$ kg, and therefore its rest energy is

$$E_o=mc^2=(9.109\times10^{-31}\,\text{kg})(2.998\times10^8\,\text{m/s})^2=8.187\times10^{-14}\,\text{J}$$

In particle and nuclear physics, it is common to express rest energies in terms of the electron-volt (eV) (see Chapter 16) or multiples of it, such as the keV and the MeV. Thus for the electron, the rest energies would be

$$E_o=(8.187\times10^{-14}\,\text{J})\left(\frac{1\,\text{eV}}{1.602\times10^{-19}\,\text{J}}\right)=5.11\times10^5\,\text{eV}$$
$$=511\,\text{keV}=0.511\,\text{MeV}$$

How do you know whether the relativistic expressions are needed or whether you can "get away" with the classical ones? The rule of thumb is that if an object's speed is 10% of the speed of light or less, then the error in using the nonrelativistic expressions is generally less than 1%. (The classical expression always yields a lower result than the relativistic one.) At $v<0.1c$, an object's kinetic energy is less than 0.5% of its rest energy. Thus, a commonly accepted practice is to use this speed and kinetic energy region as a dividing line:

> For speeds below 0.1c, or kinetic energies less than 0.5% of an object's rest energy, the error in using the nonrelativistic formulas is less than 1%, and it is then usually acceptable to employ nonrelativistic expressions.

For example, an electron with a kinetic energy of 50 eV would qualify as a "nonrelativistic electron" because 50 eV is only about 0.01% of its rest energy. An electron with a kinetic energy of 0.511 MeV would, however, be *highly* relativistic, because its kinetic energy is the same as its rest energy. Thus, what counts in this determination is a particle's kinetic energy relative to its rest energy, or its speed relative to light – even the dividing line between "relativistic" and "nonrelativistic" is relative! Example 26.5 shows how particle energies are calculated using the correct relativistic expressions.

---

### EXAMPLE 26.5: A SPEEDY ELECTRON – ENERGY REQUIRED FOR ACCELERATION

(a) How much work is required to accelerate an electron from rest to a speed of $0.900c$? (b) How much different is the nonrelativistic answer from the relativistically correct one in (a)?

**THINKING IT THROUGH.** (a) By the work-energy theorem, the work needed is equal to the electron's gain in kinetic energy. The gain in kinetic energy is the same as the final kinetic energy, since the initial kinetic energy is zero. The rest energy of the electron is 0.511 MeV. From its speed, its kinetic energy can be determined with Equation 26.8. (b) Here the nonrelativistic kinetic energy is used and its result compared to the answer in (a).

**SOLUTION**

The data are as follows:

***Given:***

$v=0.900c$
$E_o=0.511$ MeV

***Find:***

**(a)** $W$ (work required)
**(b)** $\Delta W$ (difference between the nonrelativistic work and relativistic work)

**(a)** The relativistic kinetic energy is given by Equation 26.8. By calculation (or Table 26.1) $\gamma=2.29$; thus,

$$K=(\gamma-1)E_o=(2.29-1)(0.511\,\text{MeV})=0.659\,\text{MeV}$$

Thus 0.659 MeV of work is required to accelerate an electron to a speed of $0.900c$.

**(b)** Many times, even with nonrelativistic expressions, a shortcut can be used by working with mass expressed in energy units. The technique involves first multiplying and then dividing by $c^2$, which enables the use of $E_o$. As an example, let's try it here:

$$K_{\text{nonrel}} = \tfrac{1}{2}mv^2 = \tfrac{1}{2}(mc^2)\left(\frac{v}{c}\right)^2 = \tfrac{1}{2}(0.511\,\text{MeV})(0.900)^2$$
$$= 0.207\,\text{MeV}$$

Thus the nonrelativistic answer is low by an amount $\Delta W = -0.452$ MeV, or about 70%.

**FOLLOW-UP EXERCISE.** In this Example, (a) what is the relativistic total energy of the electron? (b) What would be the electron's total energy if it were treated nonrelativistically?

To see how conservation of relativistic momentum and energy is applied, consider Integrated Example 26.6.

**INTEGRATED EXAMPLE 26.6: WHEN 1 + 1 DOESN'T EQUAL 2 - CONSERVATION OF RELATIVISTIC MOMENTUM AND ENERGY**

A particle of mass $m$, initially moving, collides head-on with an identical particle initially at rest. The two stick together, forming a single particle of mass $m'$. (a) Do you expect (1) $m' > 2m$, (2) $m' < 2m$, or (3) $m' = 2m$? Explain. (b) If the incoming particle is initially moving at a speed $v = 0.800c$ to the right, what is $m'$ in terms of $m$?

**(a) CONCEPTUAL REASONING.** This is an example of an inelastic collision (Section 6.4). In such a collision, kinetic energy is not conserved. To conserve linear momentum, some, but not all, of the initial kinetic energy is lost. Since total energy is also conserved, any loss of kinetic energy, according to relativity, must be converted into mass (energy). If this weren't true, then (3) would be the correct answer. However, the combined mass must include the mass equivalent of the "lost" kinetic energy, or $m' = 2m$ + "some lost kinetic energy in the form of mass–energy." Therefore, the correct answer is (1): $m' > 2m$.

**(b) QUANTITATIVE REASONING AND SOLUTION.** Collisions are analyzed employing both conservation of momentum and total energy. After the collision, the combined particle must be moving to the right to conserve the direction of the total system momentum. Also, the magnitude of the moving particle's momentum before the collision must equal the magnitude of the single combined particle's momentum afterward. Total relativistic energy must also be conserved. From these considerations, the combined particle's mass can be determined.

***Given:***

$v = 0.800c$

$m$ = mass of one particle

***Find:*** $m'$ (mass of combined particle in terms of m)

For the incoming particle,

$$\gamma = 1/\sqrt{1-(v/c)^2} = 1/\sqrt{1-(0.800)^2} = 1.67$$

The total system momentum $\vec{\mathbf{P}}$ conserved, that is, $\vec{\mathbf{P}}_i = \vec{\mathbf{P}}_f$. Using the expression for relativistic momentum (Equation 26.9) and equating the magnitude of the momentum of the incoming particle to that of the combined particle gives

$$\gamma m v = \gamma' m' v'$$

where $\gamma'$ refers to the combined particle after the collision. Putting in the numbers,

$$1.67m(0.800c) = \gamma' m' v'$$

Next, conserve total relativistic energy remembering that the total energy of a particle is related to its rest energy by $E = \gamma E_o$. The initial total energy is the sum of the energy due to the incoming particle's energy and the rest energy of the "target," or $E_i = 1.67mc^2 + mc^2$. The final total energy of the combined particle is given by $E_f = \gamma' m' c^2$, thus energy conservation gives

$$1.67mc^2 + mc^2 = \gamma' m' c^2 \quad \text{or} \quad 2.67m = \gamma' m'$$

Dividing this result into the momentum result yields:

$$\frac{1.67m(0.800c)}{2.67m} = \frac{\gamma' m' v'}{\gamma' m'} = v'$$

Finally, solving for the final speed of the combined particle gives $v' = 0.500c$.

Using this result, $\gamma' = 1/\sqrt{1-(v/c)^2} = 1/\sqrt{1-(0.500)^2} = 1.15$. Now the energy equation can be solved for $m'$ as follows:

$$2.67mc^2 = 1.15m'c^2 \quad \text{or} \quad m' = 2.32m$$

As expected, the mass of the combined particle is greater than $2m$ because some kinetic energy is converted into mass (energy).

**FOLLOW-UP EXERCISE.** (a) In this Example, how much kinetic energy is lost? (b) What would be the mass of the combined particle if the two particles initially approached head-on, each with a speed of $0.800c$, and stuck? (Your answers should be in terms of $m$ and $c$.)

## 26.5 The General Theory of Relativity

Special relativity applies to inertial systems, not to accelerating systems. Accelerating systems require a different approach, first described by Einstein around 1915. Called the **general theory of relativity**, it contains many implications about the theory of gravity.

### 26.5.1 The Principle of Equivalence

A fundamental principle of general relativity was first stated by Einstein. He called it the **principle of equivalence**, which can be stated as follows:

> An inertial reference frame in a uniform gravitational field is physically equivalent to a reference frame that is not in a gravitational field, but one that is in uniform linear acceleration.

In essence, what this means is that *no experiment performed in a closed system can distinguish between the effects of a gravitational field and the effects of an acceleration*. In other words, an observer in an accelerating system would find the effects of a gravitational field and those of the acceleration to be indistinguishable. For simplicity, only linearly accelerating systems are considered here (i.e., those that have no rotational acceleration).

To understand this principle, consider the situations in ▶ **Figure 26.10**. Imagine yourself as an astronaut in a closed spaceship. Suppose when you drop a pencil on Earth, you observe that it accelerates to the floor. What does this mean? According to the principle of equivalence, it could mean (a) that you are in a downward gravitational field or (b) that you are in an upward accelerating system (Figure 26.10a). Any experiment performed entirely inside the ship can't determine which of these interpretations is correct.

Whether the spaceship is in free space and accelerating with an acceleration $a = g$ or it is in a gravitational field with $a = -g$, the pencil has the same observed acceleration. In your closed system (you can't look outside), there is *no experiment* that could distinguish between these two choices – they are physically indistinguishable.

As another example, suppose the pencil does not accelerate, but instead remains suspended next to you. This could mean that (a) you are in an inertial frame with no gravitational field present, or (b) you are in free fall in a gravitational field. Once again, in the closed spaceship (Figure 26.10b), there is no way of distinguishing between these – they are *physically equivalent*.

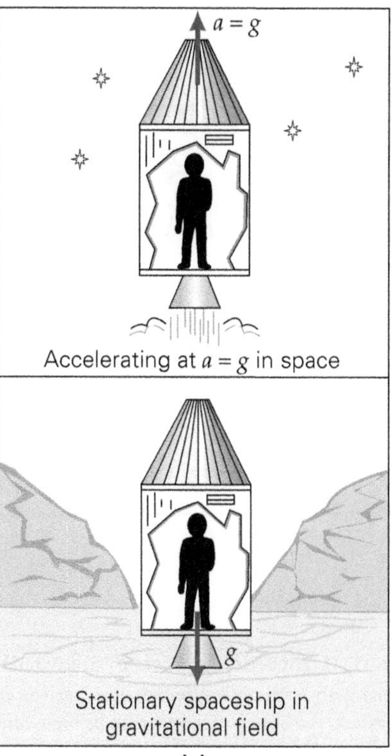


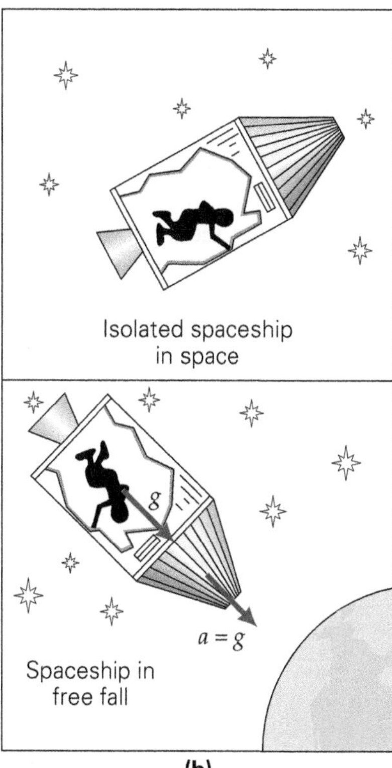


▲ FIGURE 26.10 **The principle of equivalence (a)** In a closed spaceship, the astronaut can perform no experiment that would determine whether he was in a gravitational field or an accelerating system. **(b)** Similarly, an inertial frame without gravity cannot be distinguished from free fall in a gravitational field.

## 26.5.2 Light and Gravitation

The principle of equivalence leads to an important prediction: that a gravitational field bends light. To see how this prediction arises, let's use another *gedanken* experiment. Suppose a beam of light traverses a spaceship that is accelerating upward. If the spaceship were stationary, light entering the ship at A would arrive at B on the far wall; however, because the spaceship is accelerating, the light actually lands at *C* (▶ **Figure 26.11a**).

From the point of view of the astronaut on the ship (Figure 26.11b), the light's path is a downward-curving one. To her, it seems that the gravitational field she is in has bent the light path, much like a baseball's path, since, according to the equivalence principle, the acceleration is indistinguishable from a gravitational field.

Under most conditions, this effect must be very small, since light bending in the Earth's gravitational field is never observed. However, this prediction was experimentally verified in 1919 during a solar eclipse. During times of the year when they aren't near the Sun and can be seen at night, distant stars have a constant angular separation between them (▶ **Figure 26.12a**). During other times of the year, light from some of these stars may pass near the Sun, which has a relatively strong gravitational field. However, any evidence of bending would not be observable because the starlight is masked by the glare of the Sun.

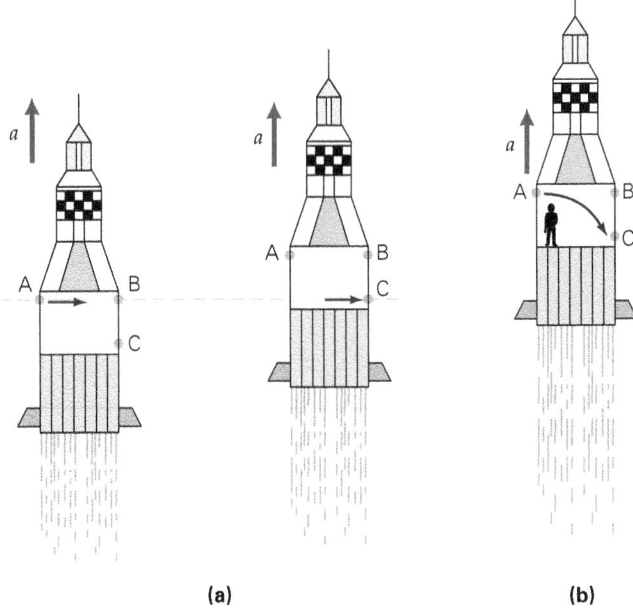


▲ **FIGURE 26.11 Light bending by gravity predicted by equivalence** (**a**) Light entering an accelerating rocket at A arrives at C. (**b**) In the accelerating system, the light path appears to be bent. Since an acceleration produces this effect, by the principle of equivalence, it is predicted that light should be bent by a gravitational field.

However, these stars may be seen during a total solar eclipse. When the Moon comes between the Earth and the Sun, an observer in the Moon's shadow can see stars not normally visible during the day (Figure 26.12b). If the light from a star passing near the Sun is bent, then the star will be in apparent location that is different from its normal one. In particular, the angular distance between pairs of stars should be slightly larger than normal (see Figure 26.12b). Einstein's theory predicted this angular difference should be tiny, only about 1.75 s of arc ($\approx 0.00005°$) but not zero. The experimental result was $1.61 \pm 0.30$ s of arc, in good agreement with Einstein's calculations, within the uncertainty of the experiment!

General relativity conceptually pictures a gravitational field as "warping" space and time, as illustrated in Figure 26.12c. A light beam follows the curvature of space–time like a ball rolling on a curved surface.

### 26.5.3 Relativity in Everyday Applications

From the previous discussions, one might believe that relativistic effects are not relevant to everyday life. However, this is not true in our modern technological world. Relativistic effects, in fact, are crucial in global positioning systems (GPS), used for locating objects to an accuracy of several meters. GPS is crucial, for example, to modern airplane navigation and military operations, and GPS capability is now included in automobiles. What is not commonly known is that GPS systems work properly only because of crucial corrections made for both special and general relativistic effects.

The GPS consists of an array of satellites, each with a very accurate atomic "clock" on board. To determine the position of an object, that object must have a GPS receiver. The receiver detects light (radio) signals from several satellites at once. Knowing the speed of light and measuring travel times, the receiver's computer determines distances and directions to several satellites. Through triangulation, the GPS receiver's computer can then rapidly calculate its location. Finally, as the object's location changes, its velocity can be calculated.

However, for a GPS to work, the clocks in orbit must be "in sync" with the corresponding clocks on Earth. If they are not in sync, then travel times and thus distances will be inaccurate and useless. An error of just 100 ns ($10^{-7}$ s) can lead to an error in location of several tens of meters – clearly unacceptable when trying to land a plane in bad weather.

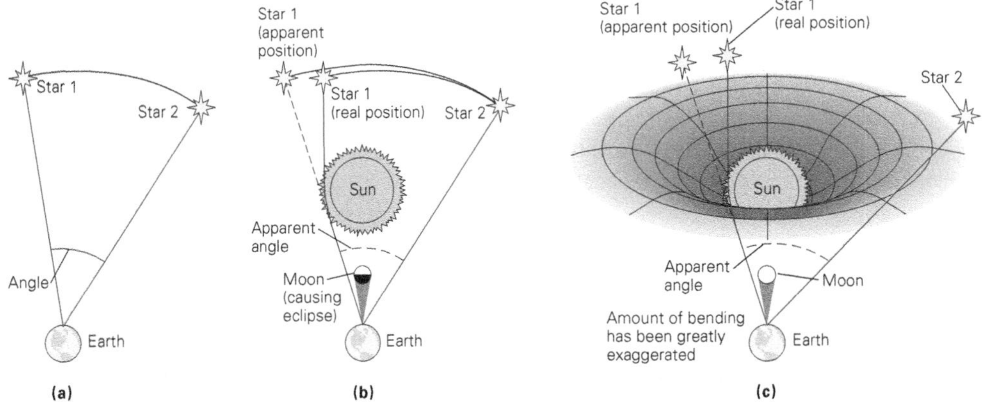


▲ **FIGURE 26.12 Gravitational attraction of light** (**a**) Normally, two distant stars are observed at a certain angular separation. (**b**) During a solar eclipse, stars near the Sun in the sky will have a slightly different position, due to the effect of solar gravity on their light. In this case, the two stars shown would exhibit a larger angular separation than in the absence an eclipse. (**c**) General relativity views a gravitational field as a warping of space and time.

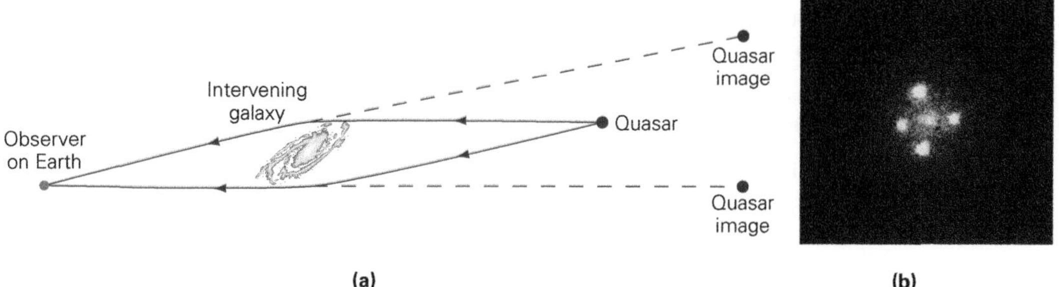


▲ FIGURE 26.13 **Gravitational lensing (a)** The bending of light by a massive object such as a galaxy theoretically should be able to produce multiple images of a distant object. **(b)** The European Space Agency's Faint Object Camera on NASA's Hubble Space Telescope provided astronomers with one of the most detailed images taken of this "gravitational lens" effect – referred to as the Einstein Cross. The photograph shows four images of a very distant quasar that has been multiple-imaged by a relatively nearby galaxy acting as a gravitational lens.

Due to the satellites' orbital speeds (several km/s) there are time dilation effects to account for. In addition, according to general relativity, satellite clocks run at a faster rate because they are in a weaker gravitational field than the Earth-bound clocks. The net result of these two effects is that the orbiting clocks run at a faster rate than the Earth-bound ones. Therefore, to keep them in synchronization, the orbiting atomic clocks must be set to run at a slower rate *before* they are launched. When they reach the proper orbit, their rate increases to the same rate as that of the surface clocks.

### 26.5.4 Gravitational Lensing

Another effect of gravity on light is **gravitational lensing**. In the late 1970s, a double quasar was discovered. (A quasar is a powerful astronomical radio source.) The fact that it was a double quasar was not unusual, but everything about the two quasars seemed to be exactly the same. It was suggested that perhaps there was really only one quasar and that, somewhere between it and the Earth, a massive, but optically faint, object had bent its light, producing multiple images (**▲ Figure 26.13a**). The subsequent detection of such a faint galaxy confirmed this hypothesis, and other examples have since been discovered, such as the "Einstein Cross" (Figure 26.13b).

The discovery of gravitational lenses gave general relativity a new role in modern astronomy. By examining such multiple images of distant objects, astronomers can gain information about an intervening "lensing" galaxy.

### 26.5.5 Black Holes

The idea that gravity can affect light finds its most extreme application in the concept of a black hole. One mechanism by which a **black hole** is thought to form is when a massive star, typically many times the mass of our Sun, collapses near the end of its light-emitting lifetime. Such an object possesses a gravitational field so intense that nothing can escape it, hence the name. In terms of the space–time warp analogy, a black hole is graphically represented as a bottomless pit in the fabric of space–time. It is speculated that black holes also may originate in other ways, such as from the collapse of entire star clusters in the center of a galaxy or at the beginning of the universe during the big bang. The discussion here will be limited to stellar collapse.

An estimate of the radius of a black hole (assumed to be spherical) can be obtained by using the concept of escape speed. Recall from Section 7.6 that the escape speed from the surface of a spherical body of mass $M$ and radius $R$ is given by

$$v_{esc} = \sqrt{\frac{2GM}{R}} \tag{26.14}$$

If light can't escape from a black hole, its escape speed must exceed the speed of light. The critical radius, $R_s$, of the region from which light can't escape is the **Schwarzschild radius**, named after Karl Schwarzschild (1873–1916), a German astronomer who first developed the concept. The expression for it is obtained by substituting $v_{esc} = c$ into Equation 26.14. Solving for $R$,

$$R_s = \frac{2GM}{c^2} \quad \text{(Schwarzschild radius)} \tag{26.15}$$

The "surface" of a sphere of radius $R_s$ defines the black hole's **event horizon**. Any happening (event) occurring within this horizon cannot be known to an outside observer, since even light from the event cannot escape. The event horizon represents the limiting distance within which light cannot escape from a black hole. Information about what goes on inside the event horizon can never reach us, so questions such as, "What does it look like inside the event horizon?" are unanswerable (at least, given the current state of knowledge in physics*).

* There is a discrepancy here that troubles physicists. According to quantum theory, information cannot be "lost." Theorists have proposed that black holes can actually radiate energy via a quantum phenomenon called "tunneling" (Section 28.2). If true, this would bring relativity and quantum theory into better agreement, and thus it is currently an active area of theoretical and experimental research.

### EXAMPLE 26.7: IF THE SUN WERE A BLACK HOLE – SCHWARZSCHILD RADIUS

What would be the Schwarzschild radius if our Sun collapsed to a black hole? (Take the mass of the Sun to be $2.0 \times 10^{30}$ kg.)

**THINKING IT THROUGH.** This is a straightforward calculation using Equation 26.15. The gravitational constant and speed of light will be needed.

**SOLUTION**

***Given:***

$M_s = 2.0 \times 10^{30}$ kg
$G = 6.67 \times 10^{-11}$ N·m²/kg²
$c = 3.00 \times 10^8$ m/s

***Find:*** $R_s$ (Schwarzschild radius)

Thus

$$R_s = \frac{2GM_s}{c^2} = \frac{2(6.67\times10^{-11}\ \text{N·m}^2/\text{kg}^2)(2.0\times10^{30}\ \text{kg})}{(3.00\times10^8\ \text{m/s})^2} = 3.0\times10^3\ \text{m} = 3.0\ \text{km}$$

This is less than 2 miles. The Sun's actual radius is about $7 \times 10^5$ km. Note however that the Sun will *not* become a black hole. This fate befalls only stars much more massive than the Sun.

**FOLLOW-UP EXERCISE.** Once black holes form, they continuously draw in matter, increasing their Schwarzschild radius. How many times more massive would our Sun have to be for its Schwarzschild radius to extend to, and swallow up, Mercury, the innermost planet? (The average distance from the Sun to Mercury is $5.79 \times 10^{10}$ m.)

If nothing, including radiation, escapes a black hole from inside its event horizon, how, then, might we observe or even merely locate a black hole? One possibility comes from a *binary star system*; that is, a system that consists of two stars orbiting their common center of mass. Cygnus X-1, the first X-ray source discovered in the constellation Cygnus, provided the first evidence of the existence of a black hole in our galaxy. Where Cygnus X-1 is in the sky, an (apparently single) giant star is observed. However, this star's spectrum shows periodic Doppler red and then blue shifts, indicating a periodic orbital motion away from and the toward us, respectively. This indicates the presence of a binary system with an "invisible" companion to the giant star. The orbital data of both stars allows us to compute their masses. In Cygnus X-1, the unseen companion to the giant star *seems* to have enough mass (i.e., be above the minimum required mass according to theory) to qualify as a black hole.

Astronomers speculate that, in such binary star systems, one member has evolved to become a black hole. The matter drawn toward the black hole is accelerated and heated. Collisions and deceleration then produce a characteristic spectrum of X-rays (which are emitted before the matter is drawn into the black hole). This type of mechanism is believed to be the source of the observed X-rays (▶ **Figure 26.14**).

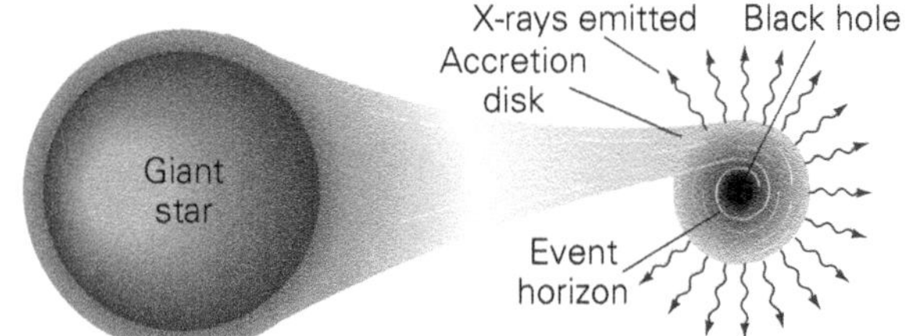


▲ FIGURE 26.14 **X-rays and black holes** Matter drawn from the "normal" member of a binary star system is accelerated, and collisions give rise to the emission of X-rays before the matter enters the black hole.

A similar mechanism may explain the enormous energy output of *active galaxies* and *quasars*. It has been proposed that the centers of these brilliant objects, which produce vast amounts of radiation, might contain black holes with masses millions or even billions of times that of our Sun. In fact, recent observations suggest that even many "normal" galaxies, including our own Milky Way galaxy, may harbor enormously massive black holes in their centers.

According to general relativity, a gravitationally violent event, such as black holes coalescing, should emit gravitational waves traveling at the speed of light. These waves should cause tiny displacements (about $10^{-14}$ m – not much larger than the diameter of a nucleus!) of the region through which they are traveling. To detect this distortion, a project called LIGO (*L*aser *I*nterferometer *G*ravitational-Wave *O*bservatory) was initiated several decades ago. The interferometer consists of two very long laser light paths arranged to produce an interference pattern when the two laser lights recombine. If a gravitational wave passes, the path distances, and thus the interference pattern, should change.

There are two identical installations, one in Washington and one in Louisiana (▼ **Figure 26.15**). If a gravitational wave does pass through the Earth, the installations should detect the passage simultaneously. (The detection of two simultaneous signals reduces the likelihood that the event resulted from a local disturbance.) In 2015, these waves were actually detected – from

▲ FIGURE 26.15 **LIGO** LIGO consists of two (interferometer) detector sites, one near Hanford in eastern Washington, and another near Livingston, Louisiana. This photo shows the Livingston site. In 2015, minute changes in the lengths of the two interferometer arms were identified as the first detection of gravitational waves – from a cataclysmic event 1.3 billion light-years away. This (and subsequent) observations confirmed general relativity's prediction made a century before.

the merger of two black holes. Then again in 2017 LIGO detected the merger of two neutron stars. The details of the waves provide information about the involved objects as well as confirm the existence of gravitational waves as predicted by general relativity. LIGO, along with several new detector designs, should provide astronomers with a new window into the universe – gravitational wave astronomy.

## Chapter 26 Review

- **Newtonian** or **classical relativity** is founded on the existence of an absolute **inertial reference frame** somewhere in the universe that is "at rest." In this frame is the **ether**, the medium through which light could propagate.
- The **Michelson–Morley experiment** attempted to measure the speed of the Earth relative to the ether frame. The results were always zero, leading physicists to give up the idea of an absolute reference frame.
- The **special theory of relativity** involves inertial reference frames moving relative to one another and is based on the principle of relativity:

  All the laws of physics are the same in all inertial reference frames.

  This results in the constancy of the speed of light:

  The speed of light in a vacuum has the same value in all inertial systems.

- A time interval measured by a clock present at both the starting and stopping events of the interval is a **proper time interval** $\Delta t_o$. The time interval $\Delta t$ measured by an observer in any other inertial frame is larger than the proper time interval. This effect is **time dilation**, and the two intervals are related by

$$\Delta t = \gamma \Delta t_o \tag{26.4}$$

  where

$$\gamma = \frac{1}{\sqrt{1-(v/c)^2}} \tag{26.5a}$$

- The length of an object, measured by an observer at rest with respect to it, is its **proper length** $L_o$. To an observer in any other inertial frame, the length $L$ is smaller than the proper length

$$L = \frac{L_o}{\gamma} = L_o\sqrt{1-(v/c)^2} \tag{26.7a}$$

  This phenomenon is known as **length contraction**.
- An object's **(relativistic) kinetic energy** is

$$K = \left[\frac{1}{\sqrt{1-(v/c)^2}} - 1\right]mc^2 = (\gamma - 1)mc^2 \tag{26.8a}$$

- An object's **(relativistic) momentum** is

$$\vec{\mathbf{p}} = \frac{m\vec{\mathbf{v}}}{\sqrt{1-(v/c)^2}} = \gamma m\vec{\mathbf{v}} \tag{26.9a}$$

- The **(relativistic) total energy** of an object is

$$E = \frac{mc^2}{\sqrt{1-(v/c)^2}} = \gamma mc^2 \tag{26.10a}$$

- When an object has no kinetic energy, it has a minimum (non-zero) energy, called its **rest energy**, which is

$$E_o = mc^2 \tag{26.11a}$$

  An object's total energy can be written in terms of its rest and kinetic energies as

$$E = K + E_o = K + mc^2 \tag{26.12a}$$

- The **general theory of relativity** is concerned with accelerating reference frames and gravitation.
- The **principle of equivalence** states that an inertial reference frame in a uniform gravitational field is physically equivalent to a reference frame that is not in a gravitational field but is, instead, accelerating. Thus no experiment performed completely within a reference frame can distinguish between the effects of a gravitational field and the effects of acceleration.

## End of Chapter Questions and Exercises

### Multiple Choice Questions

**26.1 Classical Relativity and the Michelson–Morley Experiment**

1. An object free of all forces exhibits a changing velocity in a certain reference frame. It follows that (a) the frame is inertial, (b) Newton's second law applies in this frame, (c) the laws of mechanics are the same in this reference frame as in all inertial frames, (d) none of the preceding.
2. Car A is traveling eastward at 85 km/h. Car B is traveling westward with a speed of 65 km/h. According to classical relativity, the velocity of B as measured by the driver of A would be (a) 150 km/h eastward, (b) 20 km/h westward, (c) 150 km/h westward, (d) 20 km/h eastward.
3. A space probe is moving rapidly away from the Sun at 0.2*c*. As it passes the probe, the speed of the sunlight measured by the probe is: (a) 1.2*c*, (b) exactly *c*, or (c) 0.8*c*?
4. According to classical relativity, all velocities are absolute and measured relative to which reference frame: (a) Earth, (b) Sun, (c) the Milky Way galaxy, or (d) the ether?

5. According to classical relativity, an observer traveling at $0.5c$ (with respect to the ether) toward a source of light (at rest with respect to the ether) will measure what value for the speed of the emitted light: (a) $0.5c$, (b) $1.5c$, (c) $c$, or (d) none of these?

**26.2 The Special Relativity Postulate and the Relativity of Simultaneity**

6. According to special relativity, events that are simultaneous in one inertial reference frame are (a) always simultaneous in other inertial reference frames, (b) never simultaneous in other inertial reference frames, (c) sometimes simultaneous in other inertial reference frames, (d) none of the preceding.
7. An object is at rest in an inertial reference frame. According to special relativity, what will be the same about the object as measured by an inertial observer moving relative to the object: (a) its free-body diagram, (b) its velocity, (c) its kinetic energy, or (d) none of the preceding?
8. An object is moving in an inertial reference frame. According to special relativity, what will be the same about the object as measured by an inertial observer moving relative to the object: (a) its velocity, (b) its speed, (c) its kinetic energy, (d) or none of the preceding?
9. Events A and B occur simultaneously on the $x$-axis of your reference frame. A occurred at $x = +200$ m and B at $x = +1000$ m. Which observer (there might be more than one, or even none) could possibly claim that A occurs before B: (a) one moving relative to you in the negative $x$-direction, (b) one moving relative to you in the positive $x$-direction, (c) one moving relative to you in the positive $y$-direction, or (d) none of these could observe A before B?
10. Event A occurs before event B according to you. Both occurred on the $x$-axis of your reference frame. A was at $x = +200$ m and B at $x = +1000$ m. Which observer (there might be more than one, or none) could possibly claim that B occurs before A: (a) one moving relative to you in the negative $x$-direction, (b) one moving relative to you in the positive $x$-direction, (c) one moving relative to you in the positive $y$-direction, or (d) none of these could claim that B happened before A?

**26.3 The Relativity of Time and Length: Time Dilation and Length Contraction**

11. An observer sees a friend passing her in a rocket ship that has a uniform velocity of $0.90c$. The observer knows her friend to be 1.45 m tall, and he is standing such that his length is perpendicular to their relative velocity. To the observer, her friend will appear (a) taller than 1.45 m, (b) shorter than 1.45 m, (c) exactly 1.45 m tall.
12. An observer sees a friend passing her in a rocket ship that has a uniform velocity of $0.90c$. Her friend claims that exactly 10 s have elapsed on his clock. For the same time interval, the observer's identical clock will read (a) less than 10 s, (b) greater than 10 s, (c) exactly 10 s.
13. A "race rocket" completes one straight leg of a lap in an "outer space" race at a constant speed of $0.9c$. The race organizers have carefully laid out this leg to be exactly 0.100 light-years long. Which observer(s) (if any) measure the leg's proper length: (a) the organizers, (b) the driver of the rocket, (c) both the organizers and the driver, or (d) neither?
14. In Question 13, which observer(s) (if any) measure the proper time for the rocket to complete the leg: (a) the organizers, (b) the driver of the rocket, (c) both, or (d) neither?
15. In Question 13, how does the speed of the organizers, as measured by the rocket driver, compare with the speed of the rocket as measured by the organizers: (a) the organizers measure a faster speed for the rocket than the rocket driver does for them, (b) the speeds are the same, or (c) the organizers measure a slower speed for the rocket than the rocket driver does for them?

**26.4 Relativistic Kinetic Energy, Momentum, Total Energy, and Mass–Energy Equivalence**

16. How does an object's relativistically correct kinetic energy compare to its kinetic energy calculated from the Newtonian expression: (a) the relativistic result is always larger, (b) the Newtonian result is always larger, (c) they are the same, or (d) one can be larger or smaller than the other depending upon the object's speed?
17. The total energy $E$ of a free-moving particle of mass $m$ with speed $v$ is given by which of the following: (a) $mv^2$, (b) $\gamma mc^2$, (c) $(1/2)mc^2$, or (d) $K + \gamma mc^2$?
18. How does an object's relativistically correct linear momentum (magnitude) compare to its momentum calculated from the Newtonian expression: (a) the relativistic result is always less, (b) the Newtonian result is always less, (c) they are the same, or (d) one can be larger or smaller than the other, depending on the object's velocity?
19. Which particle has the most kinetic energy: (a) an electron with a total energy of 1000 MeV, (b) a proton with a total energy of 1000 MeV, or (c) a neutron with a total energy of 1500 MeV?
20. Which particle has the most rest energy: (a) an electron with a total energy of 1000 MeV, (b) a proton with a total energy of 1000 MeV, or (c) a neutron with a total energy of 1500 MeV?
21. Which particle has the greatest speed: (a) an electron with a total energy of 1000 MeV, (b) a proton with a total energy of 1000 MeV, or (c) a neutron with a total energy of 1500 MeV?

26.5 The General Theory of Relativity

22. General relativity (a) provides a theoretical basis for explaining the gravitational force, (b) applies only to rotating systems, (c) applies only to inertial systems.
23. One of the predictions of general relativity is (a) the mass–energy equivalence, (b) time dilation, (c) the twin paradox, (d) the bending of light in a gravitational field.
24. Black hole A has three times the mass of black hole B. How do their Schwarzschild radii compare: (a) $R_{S,A}=R_{S,B}$, (b) $R_{S,A}=(1/3)R_{S,B}$, (c) $R_{S,A}=3R_{S,B}$, or (d) $R_{S,A}=9R_{S,B}$?

## Conceptual Questions

26.1 Classical Relativity and the Michelson–Morley Experiment

1. A physicist is riding on a rapidly rotating carousel. Can she apply Newton's laws of motion to correctly explain the motion of a ball rolling on the carousel's floor? Explain.
2. A car accelerating from rest forms a noninertial system. Explain the "forces" (and draw the free-body diagram) that the driver uses to explain why a coffee mug on the slippery dashboard might slide "backward." How does the inertial observer on the roadside explain this motion?
3. A person rides an elevator that is moving upward at constant velocity with a package next to her on the elevator floor. How does the free-body diagram of the package as drawn by the person in the elevator differ from that drawn by an observer at rest with respect to the building? Explain.

26.2 The Special Relativity Postulate and the Relativity of Simultaneity

4. Events A and B happen on your $x$-axis. They are separated by 500 m. You observe that A occurs 1.00 $\mu$s before B. Could A have caused B? Explain.
5. In Question 4, does there exist an inertial reference frame moving along the $x$-axis whose observer might observe the two events as being simultaneous? If so, what direction would that frame's velocity have to be? Explain.
6. In Question 4, does there exist an inertial reference frame moving along the $x$-axis whose observer might observe the two events at the same location? If so, what direction would that frame's velocity have to be? Explain.
7. In the *gedanken* experiment shown in ▶**Figure 26.16**, two events in the same inertial reference frame $O$ are related by cause and effect: (1) A gun at the origin fires a bullet along the $x$-axis at a speed of 300 m/s. (Assume no gravitational or frictional forces.) (2) The bullet hits a target at $x=+300$ m. Show, using qualitative arguments, that the two events cannot be viewed simultaneously by any inertial observer. (*Note:* This question shows that special relativity preserves the time *sequence* of two events if they are related as cause and effect. In this situation, this means that all observers agree that the gun fires before the bullet hits the target.)

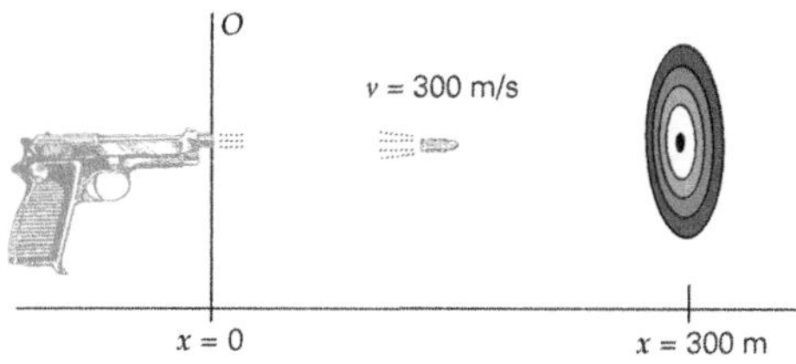


▲FIGURE 26.16 **A thought experiment** See Conceptual Question 7.

8. In a *gedanken* experiment (▼**Figure 26.17**), two events that cannot be related by cause and effect occur in the same inertial reference frame $O$: (1) strobe light A, at the origin of the $x$-axis, flashes; (2) strobe light B, located at $x=+600$ m, flashes 1.00 $\mu$s later. B's flash cannot be caused by A's flash. (The fastest signal, light, would travel only 300 m between the event locations in the time between the two events.) (a) Use qualitative arguments to show that there exists another inertial reference frame, traveling at a speed less than $c$, in which these two events would be observed to occur simultaneously. (b) What is the direction of the velocity of the reference frame in part (a)? (*Note:* This question shows that relativity does not have to preserve the time order of events if they are not related by cause and effect. Since one event cannot cause the other, no physical principles are violated if they are seen in reverse order.)

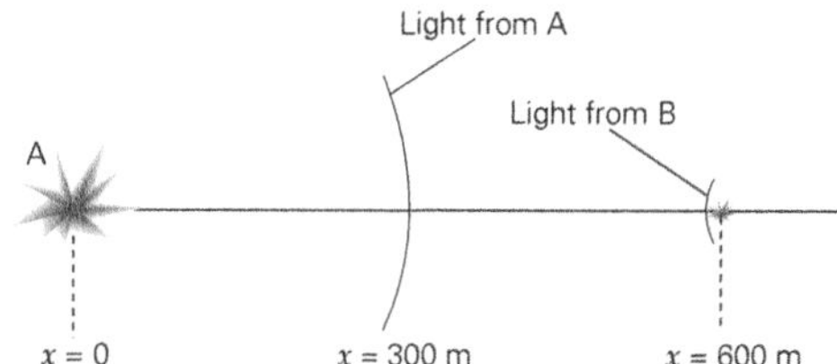


▲FIGURE 26.17 **Another thought experiment** See Conceptual Question 8.

26.3 The Relativity of Time and Length: Time Dilation and Length Contraction

9. Two identical high-speed rockets pass your (inertial) space station. Their pilots each claim that their rocket is 100 m long. You measure the length of rocket A to be 89 m and that of rocket B to be 79 m. Which one is traveling faster relative to you? Which of the two has the slower-running clock, according to you?
10. A boy wants to store a 5-m-long pole in a shed that is only 4 m long (it does have both front and rear doors). He claims that if he runs through the shed sufficiently fast, according to an observer at rest, the pole will

fit in the shed (with both doors closed at least for an instant) as a result of length contraction. Can this be true? Could it be true from the boy's reference frame? Explain any differences and why they occur.

11. You are standing on the Earth and observe a spacecraft speeding by with your professor on board. (a) If both you and your professor are observing *your* wristwatch, who is measuring the proper time? (b) Who measures the proper length of the spacecraft?

**26.4 Relativistic Kinetic Energy, Momentum, Total Energy, and Mass–Energy Equivalence**

12. The special theory of relativity places an upper limit on the speed an object can have. Are there similar limits on energy and momentum? Explain.
13. An object subject to a large constant net force approaches the speed of light. Is its acceleration constant? Explain.
14. If an electron has a kinetic energy of 2 keV, can the classical expression for kinetic energy be used to compute its speed accurately? What if its kinetic energy was 2 MeV? Explain.
15. If a proton has a kinetic energy of 2 MeV, could the classical expression for kinetic energy be used to compute its speed accurately? What if its kinetic energy is 2000 MeV? Explain.

**26.5 The General Theory of Relativity**

16. Suppose a meter stick is dropped vertically toward a black hole. Describe what happens to its shape length as it gets close to the event horizon. [*Hint:* The force of gravity will be a lot different on one end of the stick than on the other.]
17. A puzzle like that in ▼**Figure 26.18** was given to Albert Einstein on his 76th birthday by Eric M. Rogers, a physics professor at Princeton University. The goal is to get the ball into the cup without touching the ball. (Jiggling the pole up and down will not do it.) Einstein solved the puzzle immediately and then confirmed his answer with an experiment. How did Einstein get the ball into the cup? [*Hint:* He used a fundamental concept of general relativity. (This problem is adapted from R. T. Weidner, *Physics*. Boston: Allyn & Bacon, 1985)].

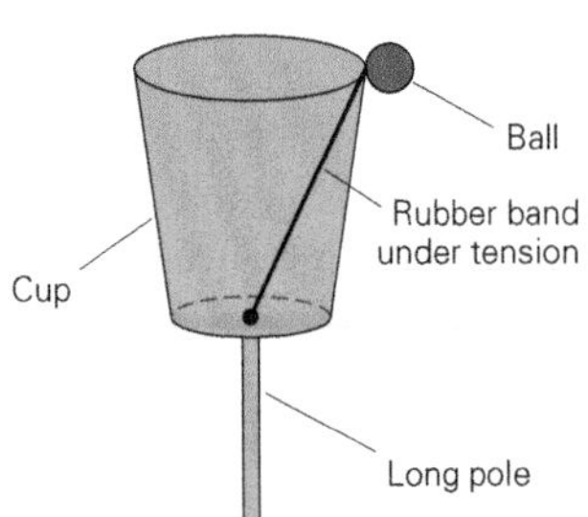


▲FIGURE 26.18 **How to get the ball into the cup** see Conceptual Question 17.

18. It is known that an enormous black hole exists at the center of our galaxy, the Milky Way. Our solar system is located near the outer edge of the galaxy, some 30 000 light-years from its center. Explain how astronomers might detect this black hole even though it is impossible to detect electromagnetic radiation from the center due to dust and gas.
19. If you are located at a certain distance above the event horizon of a black hole, you can look straight in a direction tangent to the horizon and see the back of your head. Explain how this is possible.

# Exercises*

*Integrated Exercises* (IEs) *are two-part exercises. The first part typically requires a conceptual answer choice based on physical thinking and basic principles. The following part is quantitative calculations associated with the conceptual choice made in the first part of the exercise.*

**26.1 Classical Relativity and the Michelson–Morley Experiment**

1. • A person 1.20 km away from you fires a gun. A wind is blowing at 10.0 m/s. How much difference from the "no wind" time of travel is there for the sound to reach you compared to the situation when the wind is blowing (a) toward you and (b) toward the person who fired the gun? (Take the speed of sound to be 345 m/s.)
2. • A small airplane has an airspeed (speed with respect to air) of 200 km/h. Find the time for the airplane to travel 1000 km if there is (a) no wind, (b) a headwind of 35 km/h, and (c) a tailwind of 35 km/h.
3. • A speedboat can travel with a speed of 50 m/s in still water. If the boat is in a river that has a flow speed of 5.0 m/s, (a) find the maximum and minimum values of the boat's speed relative to an observer on the riverbank. (b) What is the time difference between a downriver trip (with the current) of 1000 m and an upriver trip (against the current)?

**26.2 The Special Relativity Postulate and the Relativity of Simultaneity**

4. • Suppose, in Figure 26.4a, that the observer was 600 m to the left of event B, and A and B were separated by 900 m. If she observes that the light flash from B arrives exactly 1.00 $\mu$s after that from A, determine whether these two events are simultaneous in this reference frame.
5. • Suppose, in Figure 26.4a, that the distance between events A and B is 6000 m. The observer is stationed exactly in the middle as shown; however, she does *not* receive the signals simultaneously. Her detector receives the flash from B 25.0 $\mu$s before the flash

* The bullets denote the degree of difficulty of the exercises: •, simple; ••, medium; and •••, more difficult.

from A. (a) Could B have caused A? (b) Could B have been the cause of A if the time difference instead was 5.00 $\mu$s? Explain the difference between these two situations.

6. •• Suppose, in Figure 26.4a, that the distance between events A and B is 1200 m. The observer is stationed exactly in the middle as shown but does *not* receive the signals simultaneously. Her detector receives the flash from B 2.00 $\mu$s before the flash from A. (a) Are the events simultaneous according to her? Explain. (b) Where should she have stood, relative to A, to have received the signals simultaneously? (c) Is there another observer, moving parallel to her $x$-axis, who could observe the two events simultaneously? If not, explain why not. If so, in which direction should this observer be moving? (d) Could event B have caused event A? Explain.

**26.3 The Relativity of Time and Length: Time Dilation and Length Contraction**

7. • A spacecraft moves past a student with a relative velocity of 0.90$c$. If the pilot of the spacecraft observes 10 min to elapse on his watch, how much time has elapsed according to the student's watch?
8. **IE** • Your pulse rate is 70 beats/min, and your physics professor in a spacecraft is moving with a speed of 0.85$c$ relative to you. (a) According to your professor, your pulse rate is (1) greater than 70 beats/min, (2) equal to 70 beats/min, (3) less than 70 beats/min. Why? (b) What is your pulse rate, according to your professor?
9. • You fly your 15.0-m-long spaceship at a speed of $c/3$ relative to your friend. Your velocity is parallel to the ship's length. (a) How long is your spaceship, as observed by your friend? (b) What is the speed of your friend relative to you?
10. **IE** • An astronaut in a spacecraft moves past a field 100 m long (according to a person standing on the field) and parallel to the field's length at a speed of 0.75$c$. (a) Will the length of the field, according to the astronaut, be (1) longer than 100 m, (2) equal to 100 m, or (3) shorter than 100 m? Why? (b) What is the length as measured by the astronaut? (c) Which length is the proper length?
11. •• The proper lifetime of a muon is 2.20 $\mu$s. If the muon has a lifetime of 34.8 $\mu$s according to an observer on Earth, what is the muon's speed, expressed as a fraction of $c$, relative to the observer?
12. •• One of a pair of 25-year-old twins takes a round trip through space while the other twin remains on Earth. The traveling twin moves at a speed of 0.95$c$ for a total of 39 years, according to Earth time. Assuming that special relativity applies for the entire trip (i.e., neglect accelerations at the start, end, and turnaround), (a) what are the twins' ages when the traveling twin returns to Earth? (b) On the round trip, how far did the traveling twin go according to the traveling twin? (c) On the round trip, how far did the traveling twin go according to the Earth twin? Which of your answers to parts (b) and (c) are proper lengths, if any?
13. •• Alpha Centauri, a star close to our solar system, is about 4.3 light-years away. Suppose a spaceship traveled this distance at a constant speed of 0.90$c$ relative to Earth. (a) How long did the trip take according to an Earth-based clock? (b) How long did the trip take according to the traveler's clock? Which of your answers to parts (a) and (b) are proper time intervals, if any? (c) What is the trip distance according to an Earth-based observer? (d) What is the trip distance according to the traveler? Which of your answers to parts (b) and (c) are proper lengths, if any?
14. •• A cylindrical spaceship of length 35.0 m and diameter 8.35 m is traveling in the direction of its cylindrical axis (length). It passes by the Earth at a relative speed of $2.44 \times 10^8$ m/s. (a) What are the dimensions of the ship, as measured by an Earth observer? (b) How long does it take the spaceship to travel a distance of 10.0 km according to the ship's pilot? (c) How long does the spaceship take to travel the same 10.0 km according to an Earth-based observer?
15. •• A pole vaulter at the Relativistic Olympics sprints past you to do a vault at a speed of 0.65$c$. When he is at rest, his pole is 7.0 m long. (a) What length do you perceive the pole to be as he passes you, assuming his relative velocity is parallel to the length of the pole? (b) How long does it take the pole to pass a given location on the track according to a track-based observer? (c) Repeat part (b) from the vaulter's reference frame and explain why the two answers are different.
16. •• The distance to Planet X from Earth is 1.00 light-year. (a) How long does it take a spaceship to reach X, according to the pilot of the spaceship, if the speed of the ship is 0.700$c$ relative to X? (b) How long does it take the ship to make the trip according to an astronaut already stationed on Planet X? (c) Determine the distance between Earth and Planet X according to the pilot and according to the X-based astronaut and explain why the two answers are different.
17. •• A student notes that the length of a meter stick held by her professor (who is moving relative to her) is the same as that of her 12-in-long ruler (when oriented parallel to his ruler). (a) What is their relative speed (assuming they move in a direction parallel to their respective sticks)? (b) If the professor takes 5.00 min to write an email according to his watch, how long did it take him to write it according to the student?
18. •• Sirius is about 9.0 light-years from Earth. (a) To reach the star by spaceship in 12 years (ship time), how fast must you travel? (b) How long would the trip take according to an Earth-based observer? (c) How far is the trip according to you?

19. ••• (a) To see that length contraction is negligible at everyday speeds, determine the length contraction ($\Delta L$) of an automobile 5.00 m long when it is traveling at 100 km/h. The diameter of an atomic nucleus is on the order of $10^{-15}$ m. How does your answer compare to this? (b) Suppose that it is possible to measure a length contraction for this car of 0.0100 mm or larger. What minimum car speed would be required to detect this effect? [*Hint for (a):* Express the length contraction in terms of $x = v/c$ and recall that if $x \ll 1$, then $\sqrt{1-x^2} \approx 1-(x^2/2)$.]

**26.4 Relativistic Kinetic Energy, Momentum, Total Energy, and Mass–Energy Equivalence**

20. • An electron travels at a speed of $0.600c$. What is its kinetic and total energy?
21. • An electron is accelerated from rest through a potential difference of 2.50 MV. Find the electron's (a) speed, (b) kinetic energy, and (c) momentum.
22. • How fast must an object travel for its total energy to be (a) 1% more than its rest energy and (b) 99% more than its rest energy?
23. • An average home uses about $1.5 \times 10^4$ kWh of electricity per year. How much matter would have to be converted to energy (assuming 33% efficiency) to supply energy for 1 year to a city with 250 000 such homes?
24. • The United States uses approximately 3.0 trillion kWh of electricity annually. If 20% of this electrical energy were supplied by nuclear generating plants, how much nuclear mass would have to be converted to energy, assuming a production efficiency of 25%?
25. • To travel to a nearby star, a spaceship travels at $0.99c$ to take advantage of time dilation. If the ship has a mass of $3.0 \times 10^6$ kg, how much work must be done to get it up to speed from rest? Compare this value with the annual electricity usage of the United States. (See Exercise 24.) Does your answer make you believe that star travel is feasible or not?
26. •• An electron has a total energy of 5.6 MeV. What is its (a) kinetic energy and (b) momentum?
27. •• (a) How much work (in keV) is required to accelerate an electron from rest to $0.50c$? (b) How much kinetic energy would it have at this speed? (c) What would be its momentum?
28. •• A proton is traveling at a speed of $0.55c$. What are its (a) total energy, (b) kinetic energy, and (c) momentum?
29. •• A proton moving with a constant speed has a total energy 3.5 times its rest energy. What are the proton's (a) speed, (b) kinetic energy, and (c) momentum?
30. **IE** •• The Sun's mass is $1.989 \times 10^{30}$ kg and it radiates at a rate of $3.827 \times 10^{23}$ kW. (a) Over time, must the mass of the Sun (1) increase, (2) remain the same, or (3) decrease? (b) Estimate the lifetime of the Sun from this data, assuming it converts all its mass into energy. (c) The actual lifetime of the Sun is predicted to be much less than the answer to part (b), even though its energy emission rate will remain approximately constant. What does this tell you about the 100% conversion assumption? (d) Theoretical calculations predict the Sun's lifetime (in its current stage) to be about 5 billion years. During that time, what percentage of its mass will be lost?
31. **IE** •• Phase changes require energy in the form of latent heat (Chapter 11). (a) If 1 kg of ice at 0 °C is converted to water at 0 °C, will the water have (1) more, (2) the same, or (3) less mass compared to the ice? Why? (b) What is the difference in mass between the ice and the water? Would this difference be detectable?
32. **IE** ••• A particle of mass $m$, initially moving at speed $v$, collides head on elastically with an identical particle initially at rest. (a) Do you expect the total mass of the two particles after the collision to be (1) greater than $2m$, (2) equal to $2m$, or (3) less than $2m$? Why? (b) What are the total energy and momentum of the two particles after the collision, in terms of $m$, $v$, and $c$?
33. ••• (a) Using the relativistic expression for total energy $E$ and the magnitude $p$ of the momentum of a particle, show that the two quantities are related by $E^2 = p^2c^2 + (mc^2)^2$. (b) Use this expression to determine the linear momentum of a proton with a kinetic energy of 1000 MeV.
34. **IE** ••• The operator of a linear accelerator tells a tour group that it is used to give protons an energy of 600 MeV. (a) This 600 MeV must refer to the proton's (1) total energy, (2) kinetic energy, (3) rest energy. (b) What are the values for these three proton energies? (c) What is the protons' speed? (d) What is their momentum?

**26.5 The General Theory of Relativity**

35. • If the Sun became a black hole, what would be its average density, assuming it to be a sphere with a radius equal to the Sun's Schwarzschild radius? Compare your answer to the actual average density of the Sun.
36. • In Exercise 35, what would be the acceleration due to gravity at a distance of two Schwarzschild radii from the center of the "black hole" Sun? Compare your answer to the actual gravitational acceleration at a distance of twice the real Sun's radius from its center.
37. • In Exercise 35, what would be the escape speed at a distance of two Schwarzschild radii from the center of the "black hole" Sun? Compare your answer to the actual escape speed at a distance of twice the real Sun's radius from its center.
38. •• A black hole has an event horizon radius of $5.00 \times 10^3$ m. (a) What is its mass? (b) Determine the gravitational acceleration it produces at $5.01 \times 10^3$ m from its center. (c) Determine the escape speed at $5.01 \times 10^3$ m from its center.
39. •• Suppose two black holes meet and "coalesce" into one larger black hole. If they each have the same mass $M$ and Schwarzschild radius $R_S$, express the single black hole's Schwarzschild radius as a multiple of $R_S$.

# 27

# Quantum Physics



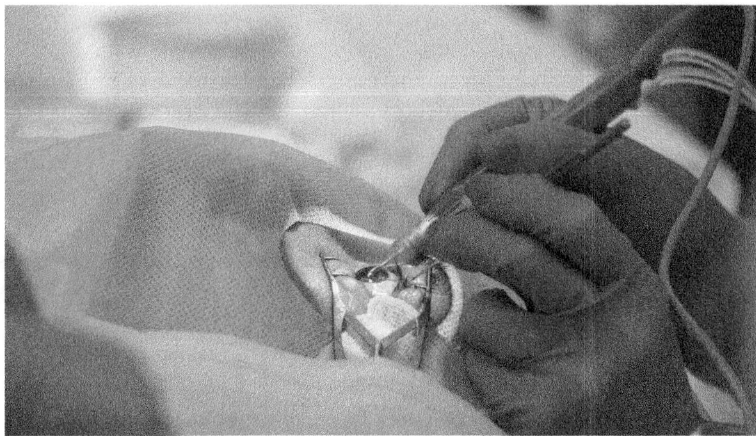

**Lasers, whose operation is based on quantum mechanics, now routinely are used to correct eye defects, as shown in this photo.**

Lasers are used in a variety of everyday applications – for example, in bar code scanners, DVD players, computer hard drives, and in various types of surgery. This chapter-opening photo shows a laser beam being used during cataract surgery on an eye lens.

The laser is a practical application based on principles that revolutionized physics. These principles were first developed in the early twentieth century, one of the most productive eras in the history of physics. For example, special relativity (Chapter 26) resolved problems faced by classical (Newtonian) theories describing objects moving at speeds near that of light. However, there were other troublesome areas in which classical theories did not agree with experimental results. To address these issues, scientists devised new hypotheses based on nontraditional approaches and ushered in a revolution in our understanding of the physical world. Chief among these new theories was the idea that light is *quantized* into discrete amounts of energy. This concept and others like it led to the formulation of a new set of principles and a new branch of physics called *quantum mechanics.*

Quantum theory demonstrated that particles often exhibit wave properties and that waves frequently behave as particles. Thus was born the *wave–particle duality* of matter. As a result of quantum theory, calculations in the realm of the very small – dimensions the sizes of atoms and smaller – deal with probabilities rather than the precisely determined values of classical theories.

A detailed treatment of quantum mechanics requires complex mathematics. However, a general overview of the important results is essential to an understanding of physics as it is known today. The important developments of "quantum physics" are presented here, and an introduction to quantum mechanics is provided in Chapter 28.

## 27.1 Quantization: Planck's Hypothesis

One of the problems scientists faced at the end of the nineteenth century was how to explain the spectra of electromagnetic radiation emitted by hot objects – solids, liquids, and dense gases. This radiation is sometimes called **thermal radiation**. It was learned in Section 11.4 that the total intensity of the emitted radiation an object is proportional to the fourth power of its absolute Kelvin temperature ($T^4$). Thus, all objects emit thermal radiation to some degree.

However, at everyday temperatures, this radiation is almost all in the infrared (IR) region and not visible to our eyes. However, at temperatures of about 1000 K, a solid object will begin to emit an appreciable amount of radiation in the long wavelength end of the visible spectrum, observed as a reddish glow. A red-hot electric stove burner is a good example of this. Still-higher temperatures cause the dominant light to shift to shorter wavelengths and the color to change to yellow-orange. Above a temperature of about 2000 K, an object glows yellowish-white, like the filament of a lightbulb, and gives off appreciable amounts of all the visible colors in varying percentages – called a *spectrum*.

Hot solids emit a *continuous spectrum*. A spectrum may be displayed as a graph of emitted energy intensity versus on wavelength. The continuous spectrum emitted by a hot solid is illustrated in ▶ **Figure 27.1**. Notice that practically all wavelengths are present, but there is a dominant color (wavelength region), which depends on the object's temperature. Two things happen to the spectrum as the temperature increase. As expected, more radiation is emitted at every wavelength, but also the wavelength of the maximum-intensity component ($\lambda_{max}$) becomes shorter. This relationship between $\lambda_{max}$ and $T$ is described experimentally by **Wien's displacement law**:

$$\lambda_{max}T = 2.90\times10^{-3}\,\text{m·K} \quad \text{(Wien's displacement law)} \quad (27.1)$$

Here $\lambda_{max}$ is the wavelength of the radiation (in meters) at which maximum intensity occurs and $T$ is the temperature of the body (in kelvins).*

Wien's displacement law can be used to determine the wavelength of the maximum-intensity component if the temperature of the object is known or, conversely, the temperature of the object if that particular wavelength is known. Hot dense gases emit the same type of spectrum as a solid, and obey Wien's law. Thus this law can be used to estimate the temperatures of stars (dense gases) from their spectrum, as Example 27.1 shows.

* Strictly speaking, the spectrum in Figure 27.1 as well as Wien's law, apply only to an object that behaves as a perfect (ideal) emitter/absorber called a **blackbody**. A blackbody is an idealized object that absorbs and emits all of the radiation that falls on it. Many everyday objects come close to behaving as a blackbody. For the discussion that follows, the assumption is that all objects exhibit blackbody behavior.

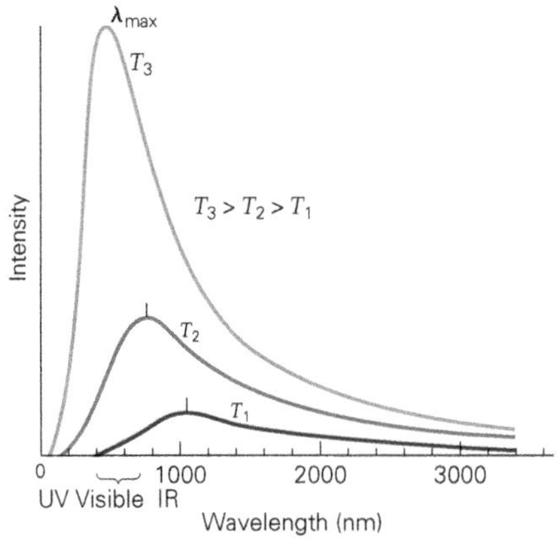


▲ **FIGURE 27.1 Thermal radiation spectrum** Intensity versus wavelength curves for the thermal radiation emitted by a hot object at different object temperatures. The wavelength associated with the maximum intensity ($\lambda_{max}$) becomes shorter with increasing temperature.

### EXAMPLE 27.1: SOLAR COLORS – USING WIEN'S DISPLACEMENT LAW FOR A HOT DENSE GAS

The visible surface of our Sun is the gaseous photosphere from which its radiation escapes. At the top of the photosphere, the temperature is about 4500 K; just below that the temperature gradually increases to about 6800 K. Assuming the Sun's photosphere radiates energy in accordance with Wien's law, (a) what are the wavelengths of the radiation of maximum intensity for these temperatures, and (b) to what colors do these wavelengths correspond?

**THINKING IT THROUGH.** Wien's displacement law (Equation 27.1) is used to determine the wavelengths.

**SOLUTION**

***Given:***

$T_1 = 4500$ K
$T_2 = 6800$ K

***Find:***

**(a)** $\lambda_{max}$ (for the two different temperatures)
**(b)** Colors corresponding to these $\lambda_{max}$ values

**(a)** At the top of the photosphere,

$$\lambda_{max} = \frac{2.90\times10^{-3}\,\text{m·K}}{4500\,\text{K}} = (6.44\times10^{-7}\,\text{m})(10^9\,\text{nm/m}) = 644\,\text{nm}$$

and at the 260-km depth,

$$\lambda_{max} = \frac{2.90\times10^{-3}\,\text{m·K}}{6800\,\text{K}} = (4.26\times10^{-7}\,\text{m})(10^9\,\text{nm/m}) = 426\,\text{nm}$$

**(b)** As the temperature increases with depth, the wavelength of the radiation of maximum intensity shifts from red (644 nm) to the blue end (426 nm) of the spectrum. This result correlates with the experimental fact that most of the Sun's radiation is emitted between 400 and 700 nm - the visible spectrum. The average photosphere temperature of about 5800 K corresponds to a maximum-energy wavelength in the yellow region, accounting for the dominant color of the Sun.

**FOLLOW-UP EXERCISE.** (a) What is the approximate surface temperature of the star Betelgeuse (that appears reddish) whose light output maximum is at a wavelength of about 1000 nm (IR)? (b) Repeat part (a) for the star Rigel (that appears bluish) and whose light output maximum is at a wavelength of about 300 nm (UV).

### 27.1.1 The Ultraviolet Catastrophe and Planck's Hypothesis

Classically, the thermal radiation emitted by an object results from the oscillations of electric charges associated with the object's atoms. Since these charges oscillate at different frequencies, a continuous spectrum of emitted radiation is expected. Classical calculations describing radiation spectra predict an intensity that is inversely related to wavelength. At long wavelengths, the classical theory agrees fairly well with experimental data. However, at short wavelengths, the agreement disappears. Contrary to experimental observations, the classical theory predicts that the radiation intensity should increase without bound as the wavelength gets shorter. This is illustrated in ▼**Figure 27.2**. This very wrong prediction is called the *ultraviolet catastrophe* - ultraviolet because the discrepancy is most obvious for the wavelengths in the UV and shorter, and the hypothesis predicts that the emitted energy grows without limits at these wavelengths.

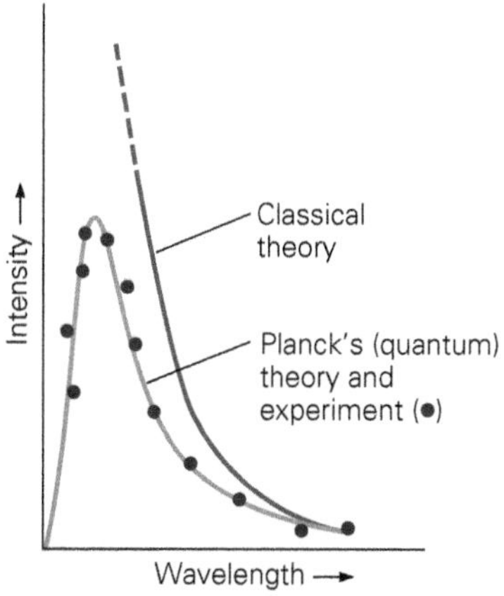


▲FIGURE 27.2 **The ultraviolet catastrophe** Classical theory predicts that the intensity of thermal radiation emitted by a blackbody should be inversely related to the wavelength of the emitted radiation. If this were true, the intensity would become infinite as the wavelength approaches zero. In contrast, Planck's quantum theory agrees with the observed radiation distribution (solid dots).

The failure of classical electromagnetic theory to explain thermal radiation spectra led German physicist Max Planck (1858–1947) to formulate a new theory. In 1900, Planck showed that his new idea did, in fact, correctly predict observed thermal radiation spectra. (Compare the solid dark red curve with the data points in Figure 27.2.) However, Planck's theory depended upon a very radical (at the time) idea. He had to assume that the thermal oscillators (the atoms emitting the radiation) have only discrete, or particular, amounts of energy rather than a continuous distribution of energies. Only with this assumption did his theory agree with experiment.

Planck found that these discrete amounts of energy were related to the frequency $f$ of the atomic oscillations by

$$E_n = n(hf) \quad \text{for } n = 1, 2, 3, \ldots \tag{27.2}$$

(Planck's quantization hypothesis)

The symbol $h$ is called **Planck's constant** and its experimental value is $6.63 \times 10^{-34}$ J·s. Thus, rather than allowing the atomic oscillator energy to take on any value, this hypothesis states that the energy is *quantized*. In other words, it occurs only in discrete amounts. The least possible oscillator energy, according to Equation 27.2 (using $n=1$), is $E_1 = hf$. All other values of the energy are integral multiples of $hf$. The quantity $hf$ is called a **quantum** of energy (from the Latin *quantus*, meaning "how much"). As a result, Planck postulated, the energy of an atom can change only by the absorption or emission of energy in discrete, or *quantum*, amounts with a value of $hf$.

Although his theoretical predictions agreed with experiment, Planck himself was not convinced of the validity of his quantum hypothesis. However, later on the concept of quantization was extended to explain other phenomena that could not be explained classically. Despite Planck's hesitation, this bold quantum hypothesis earned him a Nobel Prize in 1918.

## 27.2 Quanta of Light: Photons and the Photoelectric Effect

The concept of the quantization of light energy (as opposed to Planck's idea of atomic energy quantization) was introduced in 1905 by Albert Einstein in a paper concerning light absorption and emission by atoms, at about the same time he published his famous paper on special relativity. Einstein suggested that energy quantization should also be a fundamental property of electromagnetic waves (light). He reasoned that if the energy of the atomic oscillations is quantized, then it necessarily followed that, to conserve energy, *any emitted or absorbed radiation by an atom must also be quantized*. For example, suppose an atom initially had an energy of $5hf$ ($n=5$ in Equation 27.2) and ended up in a final (lower) energy of $4hf$ ($n=4$ in Equation 27.2). To Einstein this meant that the atom *must* emit a specific amount (or *quantum*) of light energy - in this case, $5hf - 4hf = hf$. He named this quantum, or package, of light energy the **photon**. Therefore, according to his theory, each photon possesses a definite amount of energy, $E$, that depends on the frequency $f$ of the light according to

$$E = hf \quad \text{(photon energy and light frequency)} \qquad (27.3)$$

This idea suggests that light can, under certain circumstances, behave as discrete quanta (plural of "quantum"), or "particles," of energy rather than as a wave. One way to interpret Equation 27.3 is that it represents a mathematical "connection" between the wave nature of light (wave of frequency $f$) and the particle nature of light (photons each of energy $hf$). For light of a certain frequency (or wavelength), Equation 27.3 enables us to calculate the energy of each photon, or vice versa.

Einstein used the photon concept to explain the **photoelectric effect**, another phenomenon for which classical theory was inadequate. Certain metallic materials are *photosensitive*; that is, when light strikes their surfaces, electrons may be emitted. The radiant energy supplies the energy necessary to free the electrons from the material. A schematic representation of a photoelectric effect experiment is shown in ▶ **Figure 27.3a**. A variable voltage is maintained between the anode and the cathode. When light strikes the cathode (maintained at ground or zero voltage), which is photosensitive, electrons are emitted. Because they are released by absorption of light energy, these emitted electrons are called *photoelectrons*. They collect at the anode, which is maintained initially at some positive voltage to attract the photoelectrons. Thus, in the complete circuit a current is registered on the ammeter.

When a photocell is illuminated with monochromatic (single-wavelength) light of different intensities, characteristic curves are obtained as a function of the applied voltage (Figure 27.3b). For positive voltages, the anode attracts the electrons. Under these conditions, the *photocurrent* $I_p$, which is the flow rate of photoelectrons, does not vary with voltage. With the application of a positive voltage the photoelectrons are attracted to the anode and all of them reach it. As expected classically, $I_p$ is proportional to the incident light intensity – the greater the intensity ($I_2 > I_1$ in Figure 27.3b), the more energy that is available to free more electrons.

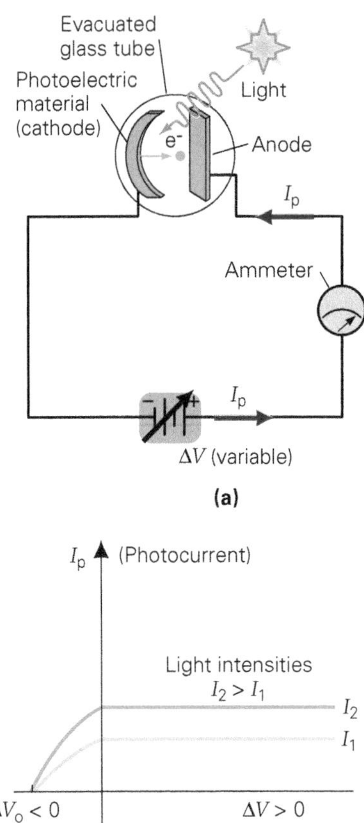


▲ **FIGURE 27.3 The photoelectric effect and characteristic curves (a)** Incident monochromatic light on the photoelectric material in a photocell (or phototube) causes the emission of electrons, which results in a current in the circuit. The applied voltage is variable. **(b)** As the plots of photocurrent versus voltage for two intensities of light show, the current stays constant as the voltage is increased. However, for negative voltages, the current becomes zero when the stopping voltage has a magnitude of $|\Delta V_o|$. For a given frequency, $|\Delta V_o|$ depends only upon the type of material and is independent of intensity.

The kinetic energy of the photoelectrons can be measured by *reversing* the voltage across the electrodes. This is accomplished by reversing the battery terminals, thus making $\Delta V < 0$. This creates a *retarding-voltage* condition. Now the electrons are repelled from the anode. The electrons are slowed as they approach the now negatively charged anode – their kinetic energy being converted into electric potential energy.

As expected, as the retarding voltage is increased (negatively), the photocurrent decreases. This is because (by energy conservation) only electrons with initial kinetic energies greater than $e|\Delta V|$ will reach the negative anode and produce the photocurrent. At some value of retarding voltage, $\Delta V_o$, called the **stopping potential**, the photocurrent drops to zero. No electrons are collected at that voltage or greater because even the fastest photoelectrons are turned around before reaching the anode. Hence, the maximum kinetic energy ($K_{max}$) of the photoelectrons is related to the magnitude of the stopping potential ($\Delta V_o$) by

$$K_{max} = e \cdot \Delta V_o \quad \text{(stopping potential)} \qquad (27.4)$$

Experimentally, if the frequency of the incident light is varied, this maximum kinetic energy increases linearly (▶ **Figure 27.4**). No emission of electrons is observed for light with a frequency below a certain *cutoff frequency* $f_o$, no matter how intense the incident light is. Another interesting observation is that even if the light intensity is very low, the current begins essentially instantaneously with no observable time delay, as long as a material is being illuminated by light with a frequency $f > f_o$.

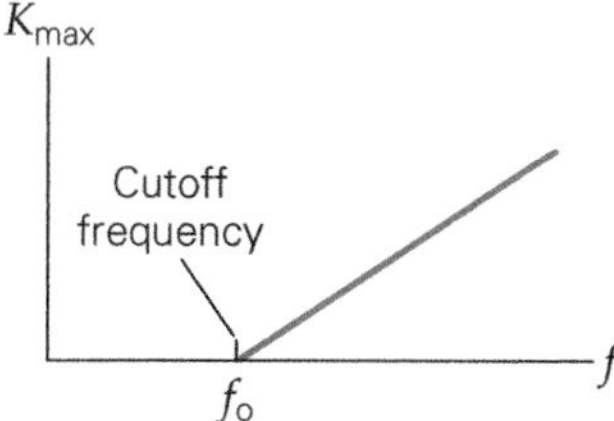


▲ FIGURE 27.4 **Maximum kinetic energy versus light frequency in the photoelectric effect** The maximum kinetic energy ($K_{max}$) of the photoelectrons is a linear function of the incident light frequency. Below a certain cutoff frequency $f_o$, no photoemission occurs, regardless of the intensity of the light.

The important characteristics of the photoelectric effect are summarized in ▼**Table 27.1**. Notice that only one of the characteristics is predicted correctly by classical wave theory, whereas Einstein's photon idea explains all of the results. When an electron absorbs a photon, some of that photon's energy goes to freeing the electron, with the remainder showing up as kinetic energy. The amount of work (or energy) required to free an electron is designated by the symbol $\phi$. Thus when a photon of energy $E$ is absorbed, conservation of energy requires that $E = K + \phi$, or, using Equation 27.3 to replace $E$,

$$hf = K + \phi \tag{27.5}$$

Since the energies involved are very small, the commonly used energy unit is the electron-volt (eV; see Section 16.2). Recall that $1.00\ \text{eV} = 1.60 \times 10^{-19}\ \text{J}$.

TABLE 27.1 Characteristics of the Photoelectric Effect

| Characteristic | Predicted by Wave Theory? |
|---|---|
| 1. The photocurrent is proportional to light intensity | Yes |
| 2. The maximum kinetic energy of the emitted electrons depends on light frequency, but not intensity | No |
| 3. No photoemission occurs for light with a frequency below a certain frequency $f_o$, regardless of the intensity | No |
| 4. A photocurrent is observed immediately when the light frequency is greater than $f_o$ even if its intensity is very low | No |

The least tightly bound electron will have the maximum kinetic energy $K_{max}$. (Why?) The *minimum* energy needed to free this electron is called the **work function** ($\phi_o$) of the material. Therefore if the situation involves the removal of the least-bound electron, Equation 27.5 becomes

$$hf = K_{max} + \phi_o \quad \text{(removal of least bound electron)} \tag{27.6}$$

Other more tightly bound electrons will require more energy than the minimum to be freed from the metal, hence their kinetic energy will be less than $K_{max}$. This concept is explored visually in ▶ **Figure 27.5**, which displays the photoelectric effect in relation to energy conservation. Some typical numerical values encountered in the photoelectric effect are shown in Example 27.2.

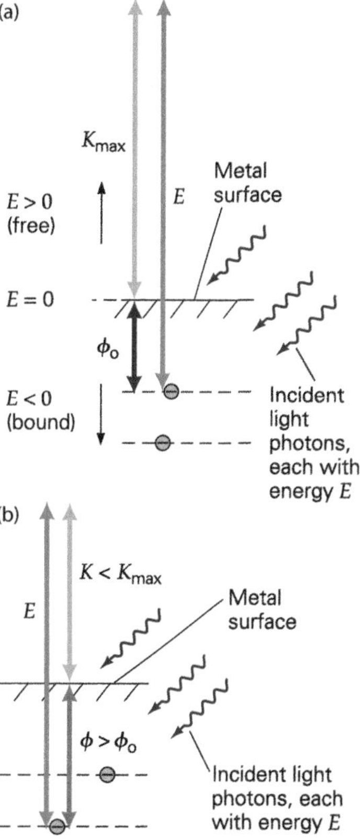


▲ FIGURE 27.5 **Visualizing energy conservation in the photoelectric effect** (a) When freeing an electron from a photosensitive material, some of the photon's energy ($hf$) goes into work done to remove it and the remainder shows up as electron kinetic energy. The least-bound electrons will have the maximum amount of kinetic energy. (b) More tightly bound electrons will have kinetic energies less than the maximum.

### EXAMPLE 27.2: THE PHOTOELECTRIC EFFECT – ELECTRON SPEED AND STOPPING POTENTIAL

The work function of a particular metal is known to be 2.00 eV. If the metal is illuminated with light of wavelength 550 nm, what will be (a) the maximum kinetic energy of the emitted electrons and (b) their maximum speed? (c) What is the stopping potential?

**THINKING IT THROUGH.** (a) By energy conservation (Equation 27.5), the maximum kinetic energy is the difference between the incoming photon energy and the work function. (b) Speed can be determined from kinetic energy, since the mass of an electron is known ($9.11 \times 10^{-31}$ kg). (c) The stopping potential is found by requiring that the maximum kinetic energy be converted to electric potential energy (Equation 27.4).

**SOLUTION**

First, convert the data into standard SI units.

***Given:***

$$\phi_o = (2.00\,\text{eV})(1.60\times10^{-19}\,\text{J/eV}) = 3.20\times10^{-19}\,\text{J}$$

$$\lambda = 550\,\text{nm} = 5.50\times10^{-7}\,\text{m}$$

***Find:***

**(a)** $K_{max}$ (maximum kinetic energy)
**(b)** $v_{max}$ (maximum speed)
**(c)** $\Delta V_o$ (stopping potential)

**(a)** Using $\lambda f = c$, the photon energy of light with the given wavelength is

$$E = hf = \frac{hc}{\lambda} = \frac{(6.63\times10^{-34}\,\text{J·s})(3.00\times10^{8}\,\text{m/s})}{5.50\times10^{-7}\,\text{m}} = 3.62\times10^{-19}\,\text{J}$$

Then from Equation 27.6

$$K_{max} = E - \phi_o = 3.62\times10^{-19}\,\text{J} - 3.20\times10^{-19}\,\text{J} = 4.20\times10^{-20}\,\text{J} = (4.20\times10^{-20}\,\text{J})\left(\frac{1\,\text{eV}}{1.60\times10^{-19}\,\text{J}}\right) = 0.263\,\text{eV}$$

**(b)** $v_{max}$ can be found, since $K_{max} = \frac{1}{2}mv_{max}^2$

$$v_{max} = \sqrt{\frac{2K_{max}}{m}} = \sqrt{\frac{2(4.20\times10^{-20}\,\text{J})}{9.11\times10^{-31}\,\text{kg}}} = 3.04\times10^{5}\,\text{m/s}$$

**(c)** The stopping potential is related to $K_{max}$ by $K_{max} = eV_o$; therefore,

$$\Delta V_o = \frac{K_{max}}{e} = \frac{4.20\times10^{-20}\,\text{J}}{1.60\times10^{-19}\,\text{C}} = 0.263\,\text{V}$$

**FOLLOW-UP EXERCISE.** In this Example, suppose that a different wavelength of light is used and the associated stopping voltage is now 0.50 V. What is the wavelength of this new light? Explain why this wavelength requires a larger stopping voltage.

---

Einstein's photon model of light is, in fact, consistent with all the experimental results of the photoelectric effect. In the photon model, an increase in light intensity means an increase in the number of photons and therefore an increase in the number of photoelectrons (i.e., the photocurrent). However, an increase in intensity would not mean a change in the energy of any one photon, since individual photon energy depends on only the light frequency. Therefore, $K_{max}$ should be independent of intensity, but linearly dependent on the frequency of the incident light – as is observed experimentally.

Einstein's photon idea also explains the existence of a cutoff frequency. In this theory, since photon energy depends on frequency, below a certain (*cutoff*) frequency (symbol $f_o$) the photons simply don't have enough energy to become free – even the most loosely bound electron. Therefore, no current is observed for those frequencies. Since, at the cutoff frequency, no electrons are emitted, the cutoff frequency can be found by setting $K_{max} = 0$ in Equation 27.6: $hf_o = K_{max} + \phi_o = 0 + \phi_o$. Solving for frequency:

$$f_o = \frac{\phi_o}{h} \quad \text{(threshold, or cutoff, frequency)} \qquad (27.7)$$

The cutoff frequency $f_o$ is also called the **threshold frequency**, below which the binding energy of even the least-bound electron exceeds the photon energy. (How would you explain this visually, using a sketch such as Figure 27.5?) To see the threshold frequency concept in action, check the following example.

---

**EXAMPLE 27.3: THE PHOTOELECTRIC EFFECT – THRESHOLD FREQUENCY AND WAVELENGTH**

The work function of a particular metal is measured to be 2.50 eV. What is the minimum frequency of light required to emit photoelectrons? What is the longest corresponding wavelength?

**THINKING IT THROUGH.** Equation 27.7 can be used to determine the threshold frequency. The threshold wavelength can then be found using $\lambda = c/f$.

**SOLUTION**

Listing the data:

***Given:***

$\phi_o = 2.50\ \text{eV} = 4.00\times10^{-19}\ \text{J}$

***Find:***

$f_o$ (threshold frequency)
$\lambda_o$ (threshold wavelength)

The threshold frequency is computed using Equation 27.7:

$$f_o = \frac{\phi_o}{h} = \frac{4.00\times10^{-19}\,\text{J}}{6.63\times10^{-34}\,\text{J·s}} = 6.03\times10^{14}\,\text{Hz}$$

The wavelength (i.e., the threshold wavelength) corresponding to this frequency is

$$\lambda_o = \frac{c}{f_o} = \frac{3.00\times10^{8}\,\text{m/s}}{6.03\times10^{14}\,\text{Hz}} = 4.98\times10^{-7}\,\text{m} = 498\,\text{nm}$$

Any frequency lower than $6.03\times10^{14}$ Hz, or, alternatively, any wavelength longer than 498 nm, would not yield photoelectrons. Since this wavelength lies in the blue-green end of the electromagnetic spectrum, blue light, for example, would dislodge electrons, but green light would not.

**FOLLOW-UP EXERCISE.** In this Example, what would be the stopping potential if the frequency of the light were twice the cutoff frequency?

---

### 27.2.1 Problem-Solving Hint: A Handy Value for *hc*

In photon calculations, the wavelength of the light is often given rather than the frequency. Typically, what is needed is the photon *energy*. Instead of first calculating the frequency ($f = c/\lambda$), then the energy in joules ($E = hf$), and finally converting to electron-volts, this can all be done in one step. To do so, combine the preceding two equations to form $E = hf = hc/\lambda$ and express the product $hc$ in electron-volts times nanometers, or eV·nm. The value of this useful product is

$$hc = (6.63 \times 10^{-34}\ \text{J·s})(3.00 \times 10^{8}\ \text{m/s}) = 1.99 \times 10^{-25}\ \text{J·m}$$
$$= \frac{(1.99 \times 10^{-25}\ \text{J·m})(10^{9}\ \text{nm/m})}{1.60 \times 10^{-19}\ \text{J/eV}}$$
$$= 1.24 \times 10^{3}\ \text{eV·nm} \approx 1240\ \text{eV·nm}$$

Besides saving calculational time, this version of $hc$ allows for a quick estimation of the photon energy associated with light of a given wavelength (or vice versa). For example, if orange light ($\lambda = 600$ nm) is used, you need only divide in your head to realize that each photon carries approximately 2 eV of energy. A more exact result could be calculated if needed, as

$$E = \frac{hc}{\lambda} = \frac{1.24 \times 10^{3}\ \text{eV·nm}}{600\text{nm}} = 2.07\,\text{eV}.$$

There are many applications of the photoelectric effect. For example, the fact that the current produced by photocells is proportional to the intensity of the light makes them ideal for use in light meters. Photosensitive materials are also used to convert sunlight to electric energy.

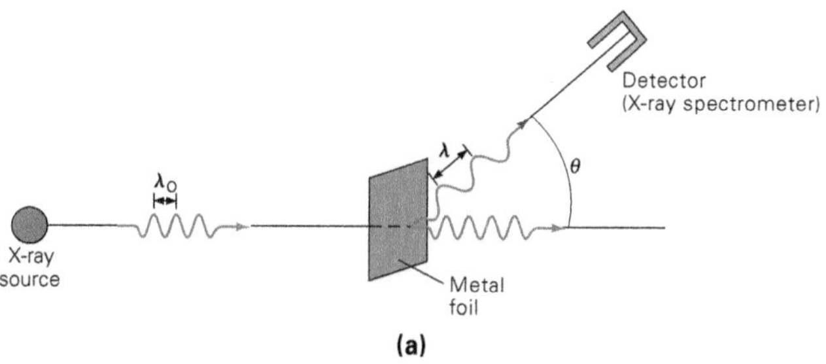


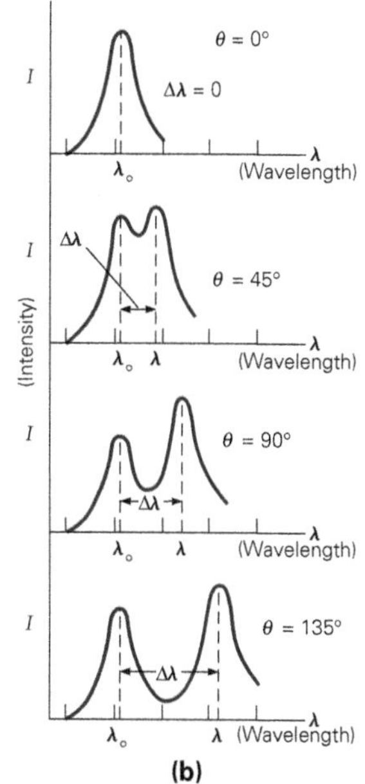


▲ **FIGURE 27.6 X-ray scattering and the Compton effect (a)** In Compton scattering, when X-rays are scattered by electrons in metal foil, the scattered wavelength ($\lambda$) is longer than the incident wavelength ($\lambda_o$). Most of the x rays pass through without interacting (and undergo no change in wavelength). **(b)** The change in wavelength increases with the scattering angle ($\theta$). Note that an unshifted peak at $\lambda_o$ remains at all angles. This corresponds to photons that scatter from electrons that are tightly bound to atoms. Because these electrons essentially gain no energy, the photon loses a negligible amount, leaving the scattered photon with the same energy as the incident one. The wavelength-shifted peak is due to scattering from the very least-bound electrons that can be treated as free electrons energy-wise.

## 27.3 Quantum "Particles": The Compton Effect

In 1923, the American physicist Arthur H. Compton (1892–1962) explained the scattering of X-rays from a graphite (carbon) block by assuming the radiation to be composed of photons. His explanation of the observed effect provided additional convincing evidence that, at least in certain types of experiments, light energy (here X-rays) is quantized into photons rather than a wave.

Compton observed that when a beam of monochromatic (single-wavelength) X-rays was scattered by various materials, the wavelength of the scattered X-rays was longer than the wavelength of the incident X-rays. In addition, he noted that the *change* in wavelength depended on the angle $\theta$ through which the X-rays were scattered, but *not* on the nature of the scattering material. This phenomenon came to be known as the **Compton effect** (▶ **Figure 27.6**).

According to classical wave theory, the electrons in the atoms of the scattering material are accelerated by the oscillating electric field of the electromagnetic wave and therefore oscillate at the same frequency as that wave. The scattered (reradiated) radiation should therefore have the same frequency (and wavelength), regardless of scattering direction.

According to Einstein's photon theory, the energy of a photon is proportional to the frequency $f$ of its associated light wave. Therefore, a change in frequency (or wavelength) would indicate a change in photon energy. Because the wavelength increased (and the frequency decreased), the scattered photons had less energy than the incident ones. Moreover, the decrease in photon energy increased with the scattering angle – which reminded Compton of a collision of two particles. Could the same principles apply in the "collisions" (scattering) of these quantum "particles" called photons?

Pursuing this idea, Compton assumed that a photon behaves as a particle when it collides with the material's electrons. It follows then that if an incident photon collides with an electron at rest, the photon will transfer some energy and momentum to that electron. Therefore one should expect that the energy of the scattered photon, and thus its frequency, should decrease (since $E = hf$). By conserving both energy and linear momentum, Compton showed, using the photon model, that the change in wavelength of the light scattered at an angle $\theta$ (from its incoming direction) by an electron is

$$\Delta\lambda = \lambda - \lambda_o = \lambda_c(1 - \cos\theta) \quad \text{(Compton scattering)} \qquad (27.8)$$

where

$$\lambda_c = \frac{h}{m_e c} = 2.43 \times 10^{-12}\,\text{m} = 2.43 \times 10^{-3}\,\text{nm}.$$

Here $\lambda_o$ is the wavelength of the incident light and $\lambda$ is that of the scattered light. The constant $\lambda_c$ is called the **Compton wavelength** of the electron. However Compton scattering can occur from any particle; hence, there is a different Compton wavelength for the proton, neutron, and so on. These values are much smaller than the electron's Compton wavelength, because the Compton wavelength is inversely related to the mass of the scattering particle.

For a given scattering object, say an electron, note that the change in wavelength $\Delta\lambda$ depends solely on the scattering angle and *not* the incident wavelength. Also note that the maximum wavelength increase occurs when $\theta = 180°$. This situation is referred to as *backscattering*, since the scattered photon's direction is completely opposite that of the incident photon's direction. For scattering from electrons, the maximum wavelength increase is $\Delta\lambda_{max} = 2\lambda_c = 4.86 \times 10^{-3}$ nm. (To see this, use Equation 27.8 and the fact that for $\theta = 180°$, $\cos\theta = -1$ and therefore $1 - \cos\theta = 2$.)

During a backscattering event, the electron must go forward in the incident photon direction. (Why?) [*Hint:* Apply conservation of linear momentum.] Since the loss of photon energy is at a maximum, it must be that the electron gains the maximum amount of kinetic energy.

Compton scattering is most readily observable for X-rays and gamma rays but is negligible for UV light or any other light of longer wavelength, such as visible light. This is because the maximum change in wavelength is extremely small compared to incident wavelengths for all types of light *except* X-rays and gamma rays which have very small wavelengths. Study the use of Compton's ideas in the next example.

**EXAMPLE 27.4: X-RAY SCATTERING – THE COMPTON EFFECT**

A beam of X-rays of wavelength $1.35 \times 10^{-10}$ m is scattered by the electrons in a metal foil. By what percentage is the wavelength shifted if the scattered X-rays are detected at an angle of 90° from the incident direction?

**SOLUTION**

Listing the data:

***Given:***

$\lambda_o = 1.35 \times 10^{-10}$ m
$\theta = 90°$

***Find:*** Percentage change in wavelength

Equation 27.8 can be used to compute the fractional change directly (here $\cos 90° = 0$):

$$\frac{\Delta\lambda}{\lambda_o} = \frac{\lambda_c}{\lambda_o}(1 - \cos\theta) = \frac{2.43 \times 10^{-12}\,\text{m}}{1.35 \times 10^{-10}\,\text{m}}(1 - \cos 90°)$$
$$= 1.80 \times 10^{-2} \text{ which is } 1.80\%$$

**FOLLOW-UP EXERCISE.** In this Example, (a) what would be the maximum percentage change if gamma rays with a much shorter wavelength of $1.50 \times 10^{-14}$ m were used instead? (b) Is this change larger or smaller than the maximum possible percentage change for the X-rays in the Example? Why?

The success in explaining experiments using photons (instead of waves) left scientists with two apparently competing models. Classically, light is a traveling wave, and the wave theory satisfactorily does explain phenomena such as interference and diffraction. However, quantum theory is needed to explain the photoelectric and Compton effects, among others. These two apparently disparate models of light forced scientists to adopt the **wave–particle duality of light**, which can be stated as:

> To explain all phenomena, light must be considered to behave as a wave at some times, and at others as a stream of photons. When it interacts with small (quantized) systems, such as atoms, nuclei, and molecules, the photon model must be used. In everyday-sized systems – for example, slit systems causing diffraction and interference – the wave model is needed.

## 27.4 The Bohr Theory of the Hydrogen Atom

In the 1800s, much experimental work was done with gas discharge tubes – for example, those containing hydrogen, neon, and mercury vapor. Common neon "lights" are actually gas discharge tubes (▶**Figure 27.7**). Recall that light from an incandescent source, such as a lightbulb's hot filament, exhibits a continuous spectrum. However, when light emissions from gas discharge tubes were analyzed, discrete spectra containing only certain wavelengths were observed. These emissions are classified as *bright-line spectra*, or **emission spectra**. The wavelengths present in an emission spectrum are characteristic of the individual atoms or molecules of the particular gas.

Atoms can absorb light as well as emit it. If white light is passed through a cool gas, the energy at only certain frequencies or wavelengths is absorbed. The result is a *dark-line spectrum*,

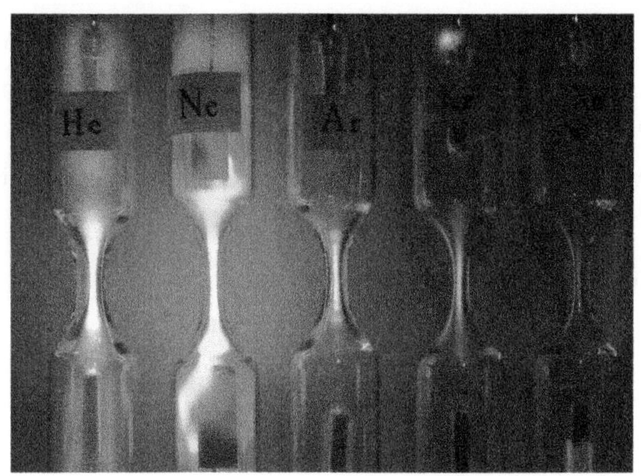


▲ **FIGURE 27.7 Gas discharge tubes** These luminous glass tubes are gas discharge tubes, in which atoms of various gases emit light when electrically excited. Each gas radiates at its own characteristic wavelengths. From left to right they are helium, neon, argon, krypton, and xenon.

or **absorption spectrum** – a series of dark lines superimposed on a continuous spectrum. Just as in emission spectra, the missing wavelengths are uniquely related to the type of atom or molecule doing the absorbing. From the pattern of emitted and/or absorbed wavelengths, the type of atoms or molecules can be identified. This method of *spectroscopic analysis* is widely used in physics, astrophysics, biology, and chemistry. For example, the element helium was first discovered on the Sun when scientists found an absorption line pattern in sunlight that did not match any known pattern. That unknown pattern belonged to helium.

Although the reasons for these line spectra were not understood in the 1800s, they provided an important clue as to the electron structure of atoms. Hydrogen, with its relatively simple visible spectrum, received much of the attention. It is also the simplest atom, consisting of only one electron and one proton. In the late nineteenth century, the Swiss physicist J. J. Balmer (1825–1898) developed an empirical formula that gives the wavelengths of the four spectral lines of hydrogen in the visible region:

$$\frac{1}{\lambda} = R\left(\frac{1}{2^2} - \frac{1}{n^2}\right) \quad \text{for } n = 3, 4, 5, \text{and } 6 \tag{27.9}$$

(visible spectrum of hydrogen)

$R$ is called the *Rydberg constant*, named after Swedish physicist Johannes Rydberg (1854–1919). It has an experimental value of $1.097 \times 10^{-2}\ \text{nm}^{-1}$. The four spectral lines of hydrogen in the visible region (note four values for $n$ in Equation 27.9), are part of the **Balmer series**. Using $R$, the formula does predict their wavelengths, but it was not understood why. Similar formulas were found to fit other spectral line series in the UV and IR regions.

An explanation of these line spectra was given in a theory of the hydrogen atom put forth in 1913 by the Danish physicist Niels Bohr (1885–1962). Bohr assumed that the electron of the hydrogen atom orbits the (stationary) proton in a circular orbit analogous to a planet orbiting the Sun. The attractive electrical force between the electron and proton supplies the centripetal force for the circular motion. Recall (Section 7.3) that the centripetal force is given by $F_c = mv^2/r$, where $v$ is the electron's orbital speed, $m$ is its mass, and $r$ is the radius of its orbit. The force between the proton and electron is given by Coulomb's law as $F_e = kq_1q_2/r^2 = ke^2/r^2$, where $e$ is the magnitude of the charge of the proton and the electron. Equating these two forces yields

$$\frac{mv^2}{r} = \frac{ke^2}{r^2} \tag{27.10}$$

The total energy of the atom is the sum of its kinetic and potential energies. Recall from Chapter 16 that the electric potential energy of two point charges is given by $U_e = kq_1q_2/r$. Since the electron and proton are oppositely charged, $U_e = -ke^2/r$. Thus, the total energy is

$$E = K + U_e = \frac{1}{2}mv^2 - \frac{ke^2}{r}$$

Using Equation 27.10, the kinetic energy can be written as $\frac{1}{2}mv^2 = ke^2/2r$. Using this relationship, the previous expression for total energy becomes

$$E = \frac{ke^2}{2r} - \frac{ke^2}{r} = -\frac{ke^2}{2r} \tag{27.11}$$

Note that $E$, the total energy of the atom, is negative, indicating that the system is bound. As the radius gets very large, $E$ becomes less negative (i.e., increases) and approaches zero. $E = 0$ indicates that the electron is no longer be bound to the proton, and the atom, having lost its electron, is said to be *ionized.*

Up to this point, only classical physics had been applied. At this step in his theory, Bohr made a radical assumption – radical in the sense that he introduced a quantum concept in his attempt to explain atomic line spectra:

> Bohr assumed that the angular momentum of the electron was quantized and could have only discrete values that were integral multiples of $h/(2\pi)$, where $h$ is Planck's constant.

Recall that in a circular orbit of radius $r$, the angular momentum $L$ of an object of mass $m$ is given by $mvr$ (Equation 8.14). Therefore, in equation terms, Bohr's assumption translates into

$$L_n = mv_nr_n = n\left(\frac{h}{2\pi}\right) \quad \text{for } n = 1, 2, 3, 4, \ldots \tag{27.12}$$

(quantization of angular momentum)

The integer $n$ is called the **principal quantum number.**[*] With this assumption, the orbital speed of the electron can be determined from Equation 27.12:

$$v_n = \frac{nh}{2\pi m r_n} \quad \text{for } n = 1, 2, 3, 4, \ldots$$

Substituting this expression for $v_n$ into Equation 27.12 and solving for $r_n$,

$$r_n = \left(\frac{h^2}{4\pi^2 k e^2 m}\right) n^2 \quad \text{for } n = 1, 2, 3, 4, \ldots \tag{27.13}$$

(hydrogen quantized orbital radii)

Thus under Bohr's angular momentum quantization assumption, it can be seen that only certain radii are possible – that is, the size of the orbit is also quantized. The energy for any of these orbits can be found by substituting this expression for $r_n$ (Equation 27.13) into Equation 27.11, which yields

$$E_n = -\left(\frac{2\pi^2 k^2 e^4 m}{h^2}\right)\frac{1}{n^2} \quad n = 1, 2, 3, \ldots \tag{27.14}$$

(hydrogen quantized energy)

thus showing that the energy of the atom is also quantized. The quantities in parentheses on the right-hand sides of both Equations 27.13 and 27.14 are constants and can be evaluated numerically. (You should be able to show how these values were attained!) Because the radii are so small, they are typically expressed in nanometers (nm). Likewise, energies are expressed in electron-volts (eV):

$$r_n = 0.0529 n^2 \text{ nm} \quad \text{for } n = 1, 2, 3, 4, \ldots \tag{27.15}$$

(orbital radii of the hydrogen atom)

$$E_n = \frac{-13.6}{n^2} \text{eV} \quad \text{for } n = 1, 2, 3, 4, \ldots \tag{27.16}$$

(total energy of the hydrogen atom)

Although these results apply only to the hydrogen atom (a one-proton nucleus), the Bohr model can be extended, with reasonable success, to any (ionized) atom that has a *single* electron in orbit around its nucleus containing any number, $Z$, of protons. For that situation, the energies $E_n$ are multiplied by $Z^2$ over their hydrogen values and the radii $r_n$ are determined by multiplying the hydrogen values by $1/Z$. For example, singly ionized helium has a value of $Z = 2$. Therefore $|(13.6\text{eV})|(2)^2$ or 54.4eV would be used in Equation 27.16 in place of the 13.6 eV. In Equation 27.15, 0.0529 nm/2 or 0.0265 nm would replace 0.0529 nm.

The use of Equations 27.15 and 27.16 is shown in Example 27.5.

[*] The principal quantum number is only one of four quantum numbers necessary to completely describe each electron in an atom. See Section 28.3.

**EXAMPLE 27.5: A BOHR ORBIT – RADIUS AND ENERGY**

Find the orbital radius and energy of an electron in a hydrogen atom characterized by the principal quantum number $n = 3$.

**THINKING IT THROUGH.** This is an application of Equations 27.15 and 27.16 with $n = 3$.

**SOLUTION**

For $n = 3$,

$$r_3 = 0.0529 n^2 \text{ nm} = 0.0529(3)^2 \text{ nm} = 0.476 \text{ nm}$$

and

$$E_3 = \frac{-13.6}{n^2} \text{ eV} = \frac{-13.6}{3^2} \text{ eV} = -1.51 \text{ eV}$$

**FOLLOW-UP EXERCISE.** In this Example, what are (a) the speed and (b) the kinetic energy of the orbiting electron?

At this point, Bohr realized there was still a problem with his quantum model for hydrogen. According to classical physics, an accelerating charge should radiate electromagnetic energy (light). For the Bohr circular orbits, the electron is accelerating centripetally, hence it should lose energy and spiral into the nucleus. Clearly, this doesn't happen in the hydrogen atom, so Bohr had to make another non-classical assumption. He postulated that:

> Orbiting atomic electrons do *not* radiate energy when in one of the "allowed" (quantized) orbits. Electromagnetic energy is only radiated when an electron makes a *downward transition* to an orbit of lower energy. It makes an *upward transition* to an orbit of higher energy by absorbing energy.

## 27.4.1 Energy Levels

The allowed orbits for the electron in a hydrogen atom are commonly expressed in terms of their energy (▶ **Figure 27.8**). In this context, the electron is referred to as being in a particular "energy level" or state. The principal quantum number is the label for each energy level. The lowest energy level ($n = 1$) is the **ground state**. Energy levels above the ground state are called **excited states**.

In the hydrogen atom, the electron is normally in the ground state and must be given energy to raise it to an excited state. Since the energy levels have specific energies, it follows that the electron can be excited only by absorbing certain discrete amounts of energy.

If enough energy is absorbed, it is possible for the electron to no longer be bound to the atom; that is, it is possible for the atom to be *ionized*. For example, ionizing a hydrogen atom initially in its ground state requires a minimum of 13.6 eV of energy. This minimum energy makes the final energy of the electron zero (meaning it is free of the atom), and it has a

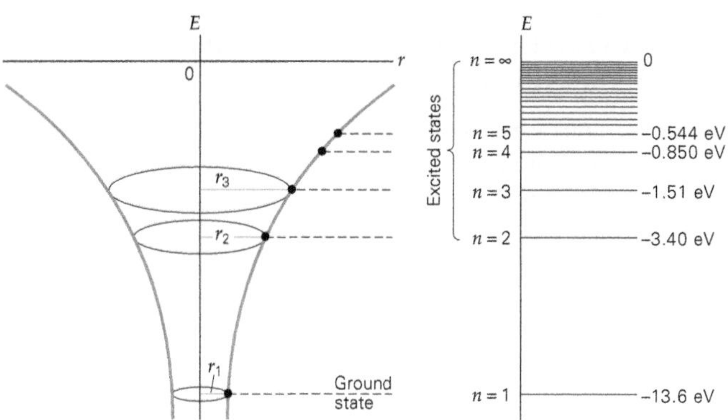


▲ FIGURE 27.8 **Orbits and energy levels of the hydrogen electron** The Bohr theory predicts that the hydrogen electron can occupy only certain orbits of discrete radii. Each allowed orbit has a corresponding total energy, conveniently displayed as an energy-level diagram. The lowest energy level ($n=1$) is the *ground* state; the levels above it ($n>1$) are *excited* states. The orbits are shown on the left, plotted in the $1/r$ energy "well" of the atom. Neither the radii nor the energy levels are drawn to scale.

principal quantum number $n=\infty$. If, on the other hand, the electron is already in an excited state, less energy than 13.6 eV would be needed to ionize the atom. Since the energy of the electron in any state is $E_n$, the energy needed to free it from the atom is $-E_n$. This energy is called the **binding energy** of the electron. Note that $-E_n$ is positive and represents the required *ionization energy* if the electron initially is in a state with a principal quantum number of $n$.

An electron generally does not remain in an excited state for long; it decays, or makes a downward transition to a lower energy level, in a very short time. The time an electron spends in an excited state is called the *lifetime* of that state and typically has a value on the order of nanoseconds or less. In making a transition to a lower state, the atom emits a quantum of light energy in the form of a photon. The energy $\Delta E$ of the photon is equal in magnitude to the energy difference of the levels. So for the hydrogen atom this is

$$\Delta E = \left|E_{n_i} - E_{n_f}\right| = \left|\left(\frac{-13.6}{n_i^2}\text{eV}\right) - \left(\frac{-13.6}{n_f^2}\text{eV}\right)\right|$$

or, more compactly,

$$\Delta E = 13.6\left[\frac{1}{n_f^2} - \frac{1}{n_i^2}\right]\text{eV} \quad \text{(hydrogen photon energies)} \qquad (27.17)$$

where the subscripts i and f refer to initial and final states, respectively. According to the Bohr theory, this energy difference is emitted as a photon with an energy $E$. This idea can be used to determine the wavelength of the emitted light: $E=\Delta E=hc/\lambda$ therefore $\lambda=hc/\Delta E$. Because photon energies are discrete, in the wave model this means that only particular wavelengths of light are emitted. These particular and unique wavelengths correspond to the various transitions between energy levels and thus explain the existence of an emission spectrum.

The final principal quantum number $n_f$ refers to the energy level at which the electron ends up after the emission process. The original Balmer series (the wavelengths in the visible region) corresponds to $n_f=2$ and $n_i=3$, 4, 5, and 6. There is only one emission series entirely in the ultraviolet range, called the *Lyman series*, in which all the transitions end in the $n_f=1$ ground state. There are many series entirely in the infrared region, most notably the *Paschen series*, which ends with the electron in the second excited state, $n_f=3$. (These series take their names from their discoverers.) See ▼ Figure 27.9.

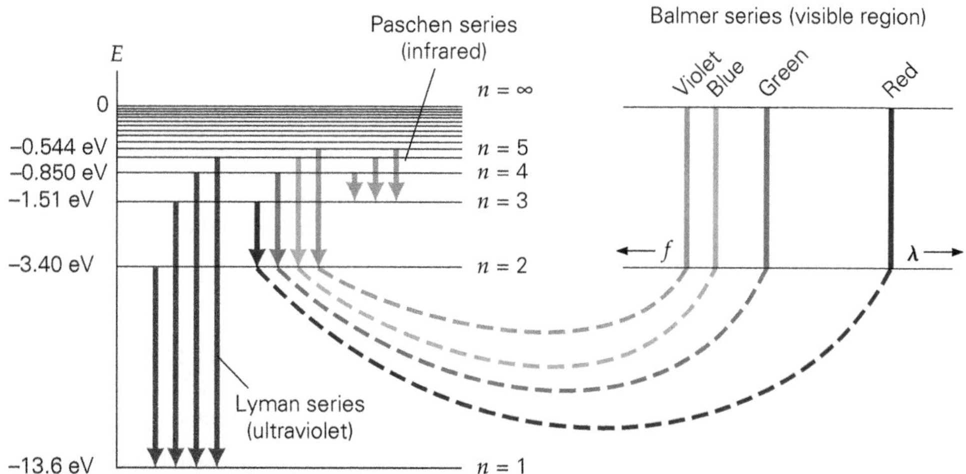


▲ FIGURE 27.9 **Hydrogen emission spectrum** Transitions may occur between two or more energy levels as the electron returns to the ground state. Transitions down to the $n=2$ state yields light with specific wavelengths in the visible region (the Balmer series). Transitions to other levels give rise to other series (not in the visible region), as shown. See Example 27.6 for the explanation of the red light emission at 656 nm.

Usually, the wavelength of the light is what is measured during the emission process. Since photon energy and light wavelength are related by Einstein's equation (Equation 27.4), the wavelength of the emitted light, $\lambda$, can be obtained from

$$\lambda\,(\text{in nm}) = \frac{hc}{\Delta E} = \frac{1.24\times10^3\,\text{eV}\cdot\text{nm}}{\Delta E\,(\text{in eV})}$$

(emitted radiation wavelength) (27.18)

(This equation uses the result for ***hc*** obtained in the Problem-Solving Hint in Section 27.2.1.) Consider its usage in Example 27.6.

### EXAMPLE 27.6: INVESTIGATING THE BALMER SERIES – VISIBLE LIGHT FROM HYDROGEN

What is the wavelength (and color) of the emitted light when an electron in a hydrogen atom undergoes a transition from the $n=3$ energy level to the $n=2$ energy level?

**THINKING IT THROUGH.** The emitted photon has an energy equal to the energy difference between the energy levels (Equation 27.17). The wavelength of the emitted light is then found from Equation 27.18.

**SOLUTION**

***Given:***

$n_i=3$
$n_f=2$

***Find:*** $\lambda$ (wavelength of emitted light)

The energy of the emitted photon is equal to the magnitude of the atom's change in energy. Thus,

$$\Delta E = 13.6\left(\frac{1}{n_f^2}-\frac{1}{n_i^2}\right)\text{eV} = 13.6\left(\frac{1}{2^2}-\frac{1}{3^2}\right)\text{eV} = 1.89\,\text{eV}$$

Using Equation 27.18,

$$\lambda = \frac{1.24\times10^{-3}\,\text{eV}\cdot\text{nm}}{\Delta E} = \frac{1.24\times10^3\,\text{eV}\cdot\text{nm}}{1.89\,\text{eV}} = 656\,\text{nm (red)}$$

See Figure 27.9 which shows this transition among others.

**FOLLOW-UP EXERCISE.** Light of what wavelength would be just barely sufficient to ionize a hydrogen atom if it was initially in its first excited state? Classify this type of light. Is it visible, UV, or IR?

In summary, since the Bohr model of hydrogen requires that the electron make transitions only between discrete energy levels, the atom emits photons of discrete energies (or light of definite wavelengths), which results in the characteristic emission spectra.

### INTEGRATED EXAMPLE 27.7: THE BALMER SERIES – ENTIRELY VISIBLE?

There are four wavelengths of the Balmer series that are in the visible range (Figure 27.9). (a) However there are more wavelengths than just these in this series. What type of light are they likely to be: (1) infrared, (2) visible, or (3) ultraviolet? (b) What is the longest wavelength of *nonvisible* light in the Balmer series?

**(a) CONCEPTUAL REASONING.** The Balmer series of emission lines is emitted when the electron ends in the first excited state, that is, $n_f=2$ (Figure 27.9). There are four distinct visible wavelengths, corresponding to $n_i=3$, 4, 5, and 6. Any other lines in this series must start with $n_i=7$ or higher and therefore represent a greater energy difference than those for the visible lines. This means that these photons carry more energy than visible light photons, and that light must have a shorter wavelength than visible light. Wavelengths shorter than visible light wavelengths are UV. Thus the answer is (3): the other Balmer lines must be in the UV region.

**(b) QUANTITATIVE REASONING AND SOLUTION.** The longest nonvisible Balmer series wavelength corresponds to the smallest photon energy above the $n=6\rightarrow2$ transition. Hence, it is the $n_i=7$ to $n_f=2$ transition. The energy difference can be computed from Equation 27.17. Then $\lambda$ can be calculated from Equation 27.18.

***Given:***

$n_i=7$
$n_f=2$

***Find:*** $\lambda$ (the longest nonvisible wavelength in the Balmer series)

From Equation 27.17,

$$\Delta E = 13.6\left(\frac{1}{n_f^2}-\frac{1}{n_i^2}\right)\text{eV} = 13.6\left(\frac{1}{2^2}-\frac{1}{7^2}\right)\text{eV} = 3.12\,\text{eV}$$

This energy difference corresponds to light of wavelength

$$\lambda = \frac{1.24\times10^3\,\text{eV}\cdot\text{nm}}{\Delta E} = \frac{1.24\times10^3\,\text{eV}\cdot\text{nm}}{3.12\,\text{eV}} = 397\,\text{nm}$$

This is UV light, just below the lower limit of the visible spectrum, which ends at about 400 nm.

**FOLLOW-UP EXERCISE.** In hydrogen, what is the longest wavelength of light emitted in the Lyman series? In what region of the spectrum is this light?

### CONCEPTUAL EXAMPLE 27.8: UP AND DOWN IN THE HYDROGEN ATOM – ABSORBED AND EMITTED PHOTONS

Assume that a hydrogen atom, initially in its ground state, absorbs a photon. In general, how many emitted photons would you expect to be associated with the de-excitation process back to the ground state? Can there be (a) more than one photon, or must there be (b) only one photon?

**REASONING AND ANSWER.** Since the difference between light emission and absorption is the direction of the transition (down for emission, up for absorption), you might be tempted to answer (b), because only one photon was required for the excitation process. In absorbing a photon's energy, a hydrogen atom will make a transition from its ground state to an excited state – say from $n=1$ to $n=3$. This process requires a single photon of a unique energy (i.e., light of a unique wavelength).

However, in returning to the ground state, the atom may take any one of *several* possible routes. For example, the atom could go from the $n=3$ state to the $n=2$ state, followed by a transition to the ground state ($n=1$). This process would involve the emission of two photons. Or the atom could go directly from the $n=3$ state to the $n=1$ state. Therefore, the answer is (a). In general, following the absorption of a single photon, several photons, of lesser energy, could be emitted.

**FOLLOW-UP EXERCISE.** (a) How many photons of different energies may a hydrogen atom emit in de-exciting from the third excited state to the ground state? (b) Starting from the ground state, which excitation transition *must* result in only one emitted photon when the hydrogen atom de-excites? Explain your reasoning.

Bohr's theory gave excellent agreement with experiments for hydrogen gas as well as for other ions with just one electron, such as singly ionized helium. However, it could not successfully describe multi-electron atoms. Bohr's theory was incomplete in the sense that it patched new quantum ideas into a basically classical framework. The theory contains some correct concepts, but a complete description of the atom did not come until the development of quantum mechanics (Chapter 28). Nevertheless, the idea of discrete energy levels in atoms enables us to qualitatively understand phenomena such as fluorescence.

In **fluorescence**, an electron in an excited state returns to the ground state in two or more steps. As seen in the previous Conceptual Example, at each step a photon is emitted. Thus each step represents a smaller energy transition than the original energy required for the upward transition. Therefore, each emitted photon must have a lower energy and (alternatively longer wavelength light) than the original exciting photon. For example, the atoms of many minerals can be excited by absorbing ultraviolet (UV) light and *fluoresce*, or glow, in the visible region when they de-excite (▶ **Figure 27.10**). A variety of living organisms, from corals to butterflies, produce fluorescent pigments that emit visible light.

▲ FIGURE 27.10 **Fluorescence** minerals can emit light in the visible spectrum when illuminated by invisible ultraviolet (UV) light (so-called black light). The visible light (the bright spots in this photo) is produced when atoms that comprise the minerals, after absorbing UV light, de-excite to lower energy levels in several steps, yielding photons of less energy and longer (visible) wavelengths.

## 27.5 A Quantum Success: The Laser

The development of the **laser**, an acronym that stands for *l*ight *a*mplification by *s*timulated *e*mission of *r*adiation, was a major technological success. Unlike the numerous inventions that have come about by trial and error or accident, including Roentgen's discovery of X-rays and Edison's electric lamp, the laser was developed on theoretical grounds. Using quantum physics, the laser was predicted and then designed, built, and, finally, applied. It has found widespread applications, some of which are discussed at the end of this section.

The existence of atomic energy levels is of prime importance in understanding laser operation. Usually, an electron remains in an excited state for only about $10^{-8}$ s and makes a transition to a lower energy level almost immediately. However, the lifetimes of some excited states are appreciably longer than this. An atomic state with a relatively long lifetime is called a **metastable state**.

In **phosphorescence**, for example, materials are composed of atoms that have such metastable states. These materials are used on luminous watch dials, toys, and other items that "glow in the dark." When a phosphorescent material is exposed to light, the atoms are excited to higher energy levels. Many of the atoms return to their normal state very quickly. However, there are also metastable states in which the atoms may remain for seconds, minutes, or even longer than an hour. Consequently, this type of material can emit light and thus glow for some time (▶ **Figure 27.11**).

A major consideration in laser operation is the emission process. ▶ **Figure 27.12** shows the absorption and spontaneous re-emission of radiation between two atomic energy levels, where a photon is absorbed re-emitted almost immediately. However, when the higher energy state is metastable, there is another possible emission process, called **stimulated emission**. Einstein first proposed this process in 1919. If a photon with an energy equal to an allowed downward transition strikes an atom already in a metastable state, it may stimulate that atom to make a transition to a lower energy level. This transition yields a second photon that is identical to the first one (the one that

▲ FIGURE 27.11 **Phosphorescence and metastable states** When atoms in a phosphorescent material are excited, some of them do not immediately return to the ground state but remain in metastable states for longer than normal periods of time. This photo shows the Doi Wang Hua Temple in Lampang Province, Thailand, which has a phosphorescent Buddha image that can glow for many seconds in the dark after being exposed to light.

strikes the atom). Thus, two photons with the same frequency and phase will go off in the same direction. Notice that stimulated emission is an amplification process - one photon in, two out. But this process is not a case of getting something for nothing, since the atom must be initially excited, and energy is required to do this.

Ordinarily, when light passes through a material, photons are more likely to be absorbed than to give rise to stimulated emission. This is because there normally are many more atoms in their ground state than in excited states. However, it is possible to prepare a material so that more of its atoms are in an excited metastable state than in the ground state. This condition is known as a **population inversion**. In this case, there may be more stimulated emission than absorption, and the net result is amplification. With the proper instrumentation, the result is a laser.

Today, there are many types of lasers capable of producing light of different wavelengths. The helium–neon (He–Ne) gas laser is probably the most familiar. For example, the characteristic reddish-pink light produced by this laser ($\lambda = 632.8$ nm) is commonly used in barcode scanners. The gas mixture is about 85% helium and 15% neon. Essentially, the helium is used for energizing and the neon is used for amplification. The gas mixture is subjected to a high-voltage discharge of electrons produced by a radio frequency power supply or direct current (dc). The helium atoms are excited by collision with these electrons (▼ **Figure 27.13a**), a process referred to as *pumping*. Here the helium atoms are pumped into an excited state that is 20.61 eV above its ground state.

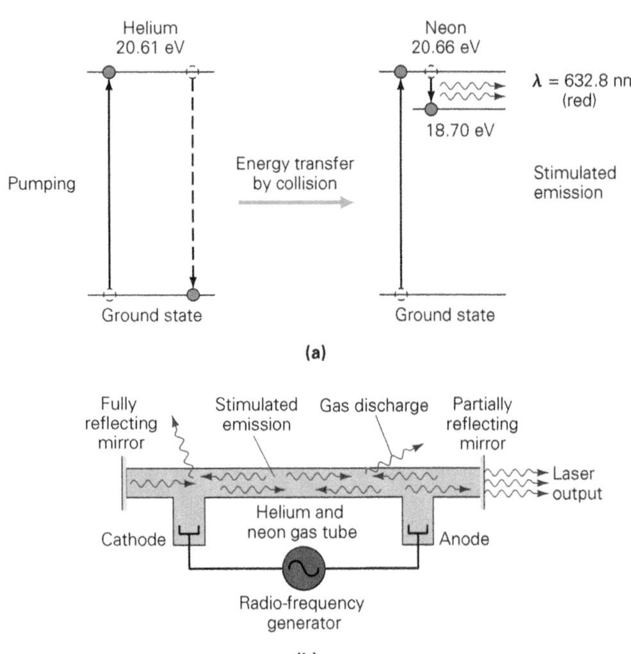


▲ FIGURE 27.13 **The helium–neon laser** (**a**) Helium atoms are first excited (or "pumped") by collision with electrons. This energy is then transferred from the helium atoms to the neon atoms in a metastable state. All that is needed is one photon from a downward transition to stimulate emission from the other excited neon atoms. (**b**) End mirrors on the laser tube are used to enhance the light beam (from stimulated emission) in a direction along the tube's axis. One of the mirrors is only partially reflecting, allowing for some of the light to come out of the tube and resulting in the beam of red light that is observed.

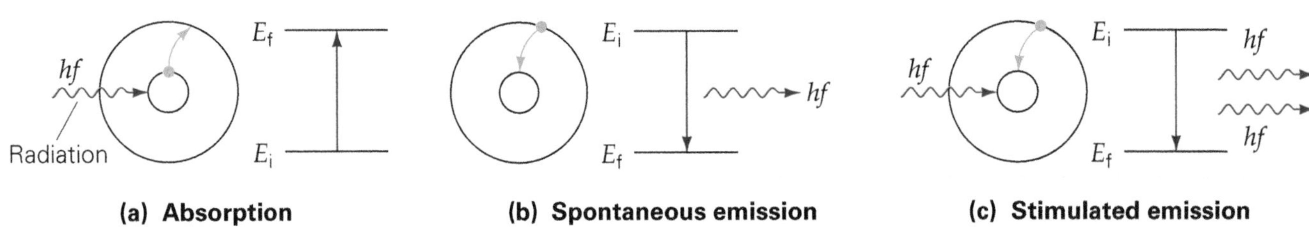


▲ FIGURE 27.12 **Photon absorption and emission** (**a**) Once a photon is absorbed, the atom is excited to a higher energy level. (**b**) After a short time, the atom spontaneously decays to a lower energy level with the emission of a photon. (**c**) If another photon with an energy equal to that of the downward transition strikes an excited atom, *stimulated* emission can occur. The result is two photons (the original plus the new one) with the same frequency. They travel in the same direction as that of the incident photon and are in phase with one another.

This excited state in helium has a relatively long lifetime of about $10^{-4}$ s and has almost the same energy as an excited state in the Ne atom at 20.66 eV. Because this lifetime is so long, there is a good chance that before an excited He atom can spontaneously emit a photon, it will collide with a Ne atom in its ground state. When such a collision occurs, energy can be transferred to the Ne atom. The lifetime of the 20.66-eV neon state is also relatively long. The delay in this metastable state of neon causes a population inversion in the neon atoms – that is, a situation with a higher percentage of neon atoms in their 20.66-eV state than in ones below it. When the neon drops to one of its states at 18.70 eV, it emits a photon of energy equal to the difference in energy of the two levels, or about 1.96 eV, which corresponds to the red light with a wavelength of 632.8 nm that is observed.

The stimulated emission of the light emitted by these neon atoms is enhanced by reflections from mirrors placed at each end of the laser tube (Figure 27.13b). Some excited Ne atoms spontaneously emit photons in all directions, and these photons, in turn, induce stimulated emissions. In stimulated emission, the two photons leave the atom in the same direction as that of the incident photon. Photons traveling in the direction of the tube axis are reflected back through the tube by the end mirrors. These photons, in reflecting back and forth, cause even more stimulated emissions. The result is an intense, highly directional, coherent (in phase), monochromatic (single-wavelength) beam of light traveling back and forth along the tube axis. Part of the beam emerges through one of the end mirrors, because it is only partially reflecting.

The monochromatic, coherent, and directional properties of laser light are responsible for its unique properties (▼ **Figure 27.14**). Light from sources such as incandescent lamps is emitted from the atoms randomly and at different frequencies, so as a result the light is out of phase, or incoherent. Such beams spread out and become less intense. The properties of laser light allow the formation of a very narrow beam, which with amplification can be very intense. An industrial laser application is shown in ▶ **Figure 27.15**.

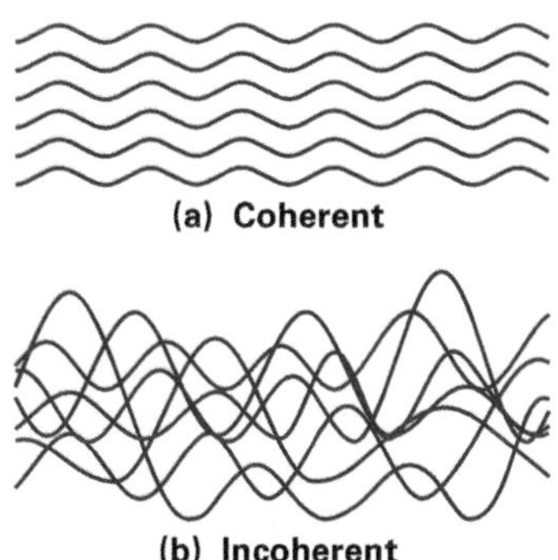


▲ FIGURE 27.14 **Coherent light** (**a**) Laser light is monochromatic (single-frequency and single wavelength or color) and coherent, meaning that all the light waves are in phase. (**b**) Light waves from sources such as a lightbulb filament are emitted randomly. They consist of many different wavelengths (colors) and are incoherent or, on average, out of phase.

▲ FIGURE 27.15 **Industrial laser applications** An industrial laser cuts through a steel plate.

The use of lasers has also had a large impact on modern medicine, in areas ranging from cosmetic surgery to cancer surgery. More and more, lasers are being chosen over other treatments, such as surgical excision, dermabrasion (skin sanding), and chemical peels. One such skin application is tattoo removal without harming surrounding cells. Laser treatment targets only the inks that make up the tattoo by adjusting the laser wavelength (color) to match the color of the ink, thus enhancing absorption of the light energy. When the ink particles absorb the light, they are heated and fragment. Then they are absorbed into the bloodstream and eliminated from the body.

## Chapter 27 Review

- **Wien's displacement law** for a continuous spectrum is the relationship between the wavelength of maximum intensity ($\lambda_{max}$) and absolute temperature of the emitting object:

$$\lambda_{max}T = 2.90\times10^{-3}\,\text{m}\cdot\text{K} \qquad (27.1)$$

- **Planck's hypothesis** is that the energy of atoms in a material is quantized into multiples of their vibrational frequency, or

$$E_n = n(hf) \quad \text{for } n = 1,2,3,\ldots \qquad (27.2)$$

where $h$, **Planck's constant**, has a numerical value of $6.63\times10^{-34}$ J·s.

- To explain the photoelectric effect, Einstein postulated that light consisted of discrete quanta of energy, called **photons**.

$$E = hf \qquad (27.3)$$

- In the **photoelectric effect**, light incident on a surface causes *photoelectrons* to be emitted from that surface. Their maximum kinetic energy ($K_{max}$) is related to the photon energy and the work function of a material ($\phi_o$) by conservation of energy

$$hf = K_{max} + \phi_o \qquad (27.6)$$

- Light can also interact with electrons via a collision process called **Compton scattering**. When light photons collide with electrons, they imparts kinetic energy to them. Thus the scattered photons have less energy (i.e., longer wavelengths) after the collision. The change in wavelength is given by

$$\Delta\lambda = \lambda - \lambda_o = \lambda_c(1-\cos\theta) \quad \text{(Compton scattering)} \quad (27.8)$$

where, for electrons,

$$\lambda_c = \frac{h}{m_e c} = 2.43\times10^{-12}\,\text{m} = 2.43\times10^{-3}\,\text{nm}$$

- The **wave-particle duality of light** states that light has both a particle and wave nature.
- The **Bohr model** of the hydrogen atom treats the electron as a classical particle held in circular orbit around the proton by the attractive electrical force. Bohr made two non-classical, quantum postulates:
  1. The angular momentum of the electron is quantized and can have only discrete values that are integral multiples of $h/(2\pi)$, where $h$ is Planck's constant.

$$L_n = mv_n r_n = n\left(\frac{h}{2\pi}\right) \quad \text{for } n = 1,2,3,4,\ldots \quad (27.12)$$

  2. The electron does *not* radiate energy when it is in a bound, discrete orbit. It radiates energy only when it makes a downward transition to an orbit of lower energy. It makes an upward transition to an orbit of higher energy by absorbing energy.
- The Bohr model of hydrogen leads directly to the quantization of the allowed orbital radii and the corresponding energies of the atom.

$$r_n = \left(\frac{h^2}{4\pi^2 ke^2 m}\right)n^2 \quad \text{for } n = 1,2,3,4,\ldots \quad (27.13)$$

$$E_n = -\left(\frac{2\pi^2 k^2 e^4 m}{h^2}\right)\frac{1}{n^2} \quad n = 1,2,3,\ldots \quad (27.14)$$

and when the numerical constants are substituted and evaluated in units of nm and eV, these can be rewritten as

$$r_n = 0.0529n^2\,\text{nm} \quad \text{for } n = 1,2,3,4,\ldots \quad (27.15)$$

$$E_n = \frac{-13.6}{n^2}\,\text{eV} \quad \text{for } n = 1,2,3,4,\ldots \quad (27.16)$$

Here $n$ is the **principal quantum number**. When the hydrogen atom is in the $n=1$ state, it has its smallest size and lowest energy and is in its **ground state**. States of larger size and higher energy are **excited states.**

- Atoms emit light in an **emission spectrum**, and absorb light in an **absorption spectrum.** The wavelengths of light emitted or absorbed are discrete and characteristic of the atom. The emitted or absorbed photon wavelength is related to the energy difference between the two atomic levels by

$$\lambda\,(\text{in nm}) = \frac{hc}{\Delta E} = \frac{1.24\times10^3\,\text{eV·nm}}{\Delta E\,(\text{in eV})} \quad (27.18)$$

- A **metastable state** is a relatively long-lived excited atomic state.
- **Stimulated emission** can occur when a photon induces a downward atomic transition, resulting in a second photon of the same energy.
- A **laser** works based on the concept of **population inversion**. This is a situation that can occur involving a metastable excited state and resulting in more electrons in a higher energy state than in a lower energy state.

# End of Chapter Questions and Exercises

## Multiple Choice Questions

### 27.1 Quantization: Planck's Hypothesis

1. Blackbody A is at a temperature of 3000 K and B is at 6000 K. What can you say about the wavelength at which they radiate the maximum intensity: (a) $\lambda_{\text{max,A}} = \frac{1}{2}\lambda_{\text{max,B}}$, (b) $\lambda_{\text{max,A}} = 2\lambda_{\text{max,B}}$, (c) $\lambda_{\text{max,A}} = \lambda_{\text{max,B}}$, or (d) can't tell from the data?
2. If the absolute temperature of a blackbody radiator is doubled, the total energy emitted by this object increases by a factor of (a) 2, (b) 4, (c) 8, (d) 16.
3. If the frequency of vibration of an atom is $f$, the atom's energy can be (a) zero, (b) $0.5hf$, (c) $hf$, (d) $1.5hf$.

### 27.2 Quanta of Light: Photons and the Photoelectric Effect

4. For the photoelectric effect, classical theory predicts that: (a) no photoemission occurs below a certain frequency, (b) the photocurrent is proportional to the light intensity, (c) the maximum kinetic energy of the emitted electrons depends on the light frequency, (d) no matter how low the light intensity, a photocurrent is observed immediately.
5. In the photoelectric effect, what happens to the stopping voltage when the light intensity is increased: (a) it increases, (b) it stays the same, or (c) it decreases?
6. In the photoelectric effect, what happens to the stopping voltage when the light frequency is decreased: (a) it increases, (b) it stays the same, or (c) it decreases?
7. The work function has what SI units: (a) joules, (b) volts, (c) coulombs, or (d) amperes?

### 27.3 Quantum "Particles": The Compton Effect

8. In the Compton effect, the scattered light (a) always has a longer, (b) always has the same, (c) always has a shorter, or (d) sometimes has a shorter wavelength than the incident light.
9. The wavelength shift for Compton scattering is a maximum when the scattering angle is (a) 0°, (b) 45°, (c) 90°, (d) 180°.

10. At what photon-scattering angle will the electron receive the least recoil energy in the Compton effect: (a) 20°, (b) 45°, (c) 60°, or (d) 80°?

**27.4 The Bohr Theory of the Hydrogen Atom**

11. In his theory of the hydrogen atom, Bohr postulated the quantization of (a) energy, (b) centripetal acceleration, (c) light, (d) angular momentum.
12. A hydrogen atom absorbs radiation when its electron (a) makes a transition to a lower energy level, (b) is excited to a higher energy level, (c) stays in the ground state.
13. A hydrogen atom in its second excited state absorbs a photon and makes a transition to a higher excited state. The longest-wavelength photon possible is absorbed. The quantum number of the final state is (a) 1, (b) 2, (c) 3, (d) 4.

**27.5 A Quantum Success: The Laser**

14. Which of the following is *not* essential for laser action: (a) population inversion, (b) phosphorescence, (c) pumping, or (d) stimulated emission?
15. The two photons involved in a stimulated emission have the same (a) frequency, (b) direction, (c) phase, (d) all of the preceding.
16. Population inversion refers to a state in which there are (a) more electrons in the ground state, (b) more electrons in an excited state, (c) the same number of electrons in the ground and an excited state, (d) no electrons in the ground state.

## Conceptual Questions

**27.1 Quantization: Planck's Hypothesis**

1. Some stars appear reddish, and others appear blue. Which of these types would have the higher surface temperature? Explain.
2. As a piece of iron is heated it first begins to glow red, then orange, then yellow, but then, instead of appearing green and then blue as its temperature rises, it appears white. Explain.
3. Make a graph showing how the wavelength of the most intense radiation component of blackbody radiation varies with the body's absolute temperature. By what ratio does $\lambda_{max}$ change (final/initial) if the body's absolute temperature is tripled?

**27.2 Quanta of Light: Photons and the Photoelectric Effect**

4. It is more dangerous to stand in front of a beam of X-ray radiation at a low intensity than a beam of red light at a much higher intensity. How does the photon model of light explain this apparent paradox?
5. Is it possible for a photon of IR radiation to contain more energy than a photon of UV radiation? Explain.
6. During the photoelectric effect, when the incident light is below the threshold frequency of the material, the energy of the light is still absorbed by that material. However, electrons are not emitted from the material. Explain where this energy goes.
7. Light of the same frequency is incident on two materials with different work functions. Discuss how the stopping voltage of the photoelectrons is affected by the work function.

**27.3 Quantum "Particles": The Compton Effect**

8. A photon undergoes Compton scattering from an electron or a neutron. How does the maximum wavelength shift of the light scattered from a neutron compare to that from an electron? Explain.
9. The Sun's energy production (near its center) is initially in the form of X-rays and gamma rays. By the time it reaches the surface, it is mostly in the visible range. Show how Compton scattering is a mechanism that can, at least partially, explain how this happens.
10. In Compton scattering, how does the maximum wavelength shift for 0.100-nm X-ray photons compare to that of visible-light photons (500-nm wavelength)?
11. In Compton scattering, if the photon is scattered at a 90° angle, in which general direction will the electron recoil? Why?

**27.4 The Bohr Theory of the Hydrogen Atom**

12. List the physical quantities of a hydrogen atom determined by the principal quantum number?
13. Does it take more or less energy to ionize a hydrogen atom if the electron is in an excited state than if it is in the ground state? Explain.
14. Very accurate measurements of the wavelengths emitted by a hydrogen atom indicate that they are all slightly longer than expected from the Bohr theory. Explain how conservation of linear momentum explains this. [*Hint:* Photons carry linear momentum and energy, and both need to be conserved.]

**27.5 A Quantum Success: The Laser**

15. "Pumping" is a necessary process in laser light production. Briefly describe what it is.
16. In what sense is a laser an "amplifier" of energy? Explain why this concept does not violate conservation of energy.
17. Explain the difference between spontaneous emission and stimulated emission.

# Exercises*

*Integrated Exercises* (*IEs*) *are two-part exercises. The first part typically requires a conceptual answer choice based on physical thinking and basic principles. The following part is quantitative calculations associated with the conceptual choice made in the first part of the exercise.*

*Note:* In all cases, assume the emitting object behaves "ideally" as a blackbody.

* The bullets denote the degree of difficulty of the exercises: •, simple; ••, medium; and •••, more difficult.

## 27.1 Quantization: Planck's Hypothesis*

1. • A warm solid is at 27°C. What is the wavelength of the radiation at maximum intensity?
2. • Find the approximate temperature of a red star that emits light with a wavelength of maximum emission of 700 nm (deep red).
3. • What are the wavelength and frequency of the most intense radiation component emitted by a thermal radiator at a temperature of 0°C?
4. **IE** • (a) If you have a fever, is the wavelength of the radiation component of maximum intensity emitted by your body (1) greater, (2) the same, or (3) smaller as compared with its value when your temperature is normal? Why? (b) Assume that human skin has a temperature of 34°C. What is the wavelength of the radiation component of maximum intensity emitted by our bodies? In what region of the EM spectrum is this wavelength?
5. •• A "red-hot" object emits thermal radiation with a maximum at frequency of $1.0 \times 10^{14}$ Hz. What is the Celsius temperature of the object?
6. •• What is the minimum energy of a atomic oscillator in an object producing the radiation at $\lambda_{max}$ if the object's temperature is 212°F?
7. •• The minimum energy of an atomic oscillator in a blackbody is $3.5 \times 10^{-19}$ J while emitting at that body's most intense wavelength. What is the Celsius temperature of the object?
8. **IE** •• The temperature of an object increases from 200°C to 400°C. (a) Will the frequency of the most intense spectral component emitted (1) increase, but not double; (2) double; (3) be reduced in half; or (4) decrease, but not by half? Why? (b) What is the change in the frequency of the most intense spectral component?
9. •• The temperature of an object is 500°C. If the intensity of the emitted radiation, 2.0 W/m$^2$, were due entirely to the most intense frequency component, how many quanta of radiation would be emitted per second per square meter?
10. ••• The wavelength at which the Sun emits its most intense light is about 550 nm. Assuming the Sun radiates ideally, estimate (a) its surface temperature and (b) its total emitted power. [*Hint*: The Sun's data can be found in the inside back cover of the book.]

## 27.2 Quanta of Light: Photons and the Photoelectric Effect

11. • Each photon in a beam of light has an energy of $6.50 \times 10^{-19}$ J. What is the light's wavelength? What type of light is this?
12. **IE** • (a) Compared with a quantum of red light ($\lambda = 700$ nm), a photon of violet light ($\lambda = 400$ nm) has (1) more, (2) the same amount of, (3) less energy. Why? (b) Determine the ratio of the photon energy associated with violet light to that of red light.
13. • A source of UV light has a wavelength of 150 nm. How much energy does one of its photons have expressed in (a) joules and (b) electron-volts?
14. • The work function of a material is $5.0 \times 10^{-19}$ J. If light of wavelength of 300 nm is incident on the surface, what is the maximum kinetic energy of the photoelectrons ejected during the photoelectric effect?
15. • When light of a wavelength of 200 nm is incident on a material's surface, the maximum kinetic energy of the photoelectrons is $6.0 \times 10^{-19}$ J. What is the work function of that material?
16. • The photoelectrons ejected from a material require a stopping voltage of 5.0 V. If the intensity of the light is tripled, what is the new stopping voltage?
17. • What is the longest wavelength light that can cause the emission of electrons from a metal that has a work function of 3.50 eV?
18. •• Assume that a 100-W lightbulb gives off 2.50% of its energy as visible light. How many photons of visible light are given off in 1.00 min? (Use an average visible wavelength of 550 nm.)
19. •• A metal with a work function of 2.40 eV is illuminated by a beam of monochromatic light. If the stopping potential is 2.50 V, what is the wavelength of that light?
20. **IE** •• The work function of metal A is less than that of metal B. (a) The threshold wavelength for metal A is (1) shorter than, (2) the same as, (3) longer than that of metal B. Why? (b) If the threshold wavelength for metal B is 620 nm and the work function of metal A is twice that of metal B, what is the threshold wavelength for metal A?
21. •• ▼**Figure 27.16** shows data of stopping potential versus frequency for a certain material. Use the graph to determine (a) Planck's constant and (b) the work function of that material.

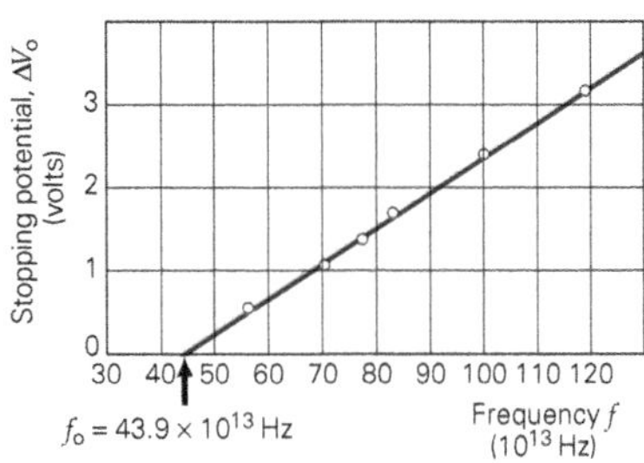


▲**FIGURE 27.16 Stopping potential versus frequency** See Exercise 21.

22. •• The photoelectric effect threshold wavelength for a metal is 400 nm. Calculate the maximum speed of photoelectrons if the light used has a wavelength of (a) 300 nm, (b) 400 nm, and (c) 500 nm.

* Take $h$ to have an exact value of $6.63 \times 10^{-34}$ J·s for significant figure purposes, and use $hc = 1.24 \times 10^3$ eV · nm (three significant figures).

23. •• When light of wavelength of 250 nm is incident on a metal, the maximum speed of the photoelectrons is $4.0 \times 10^5$ m/s. What is the work function of this metal in eV?
24. •• The work function of a material is 3.5 eV. If it is illuminated with monochromatic light ($\lambda = 300$ nm), what are (a) the stopping potential and (b) the cutoff frequency?
25. •• Blue light with a wavelength of 420 nm is incident on a certain material and causes the emission of photoelectrons with a maximum kinetic energy of $1.00 \times 10^{-19}$ J. (a) What is the stopping voltage? (b) What is the material's work function? (c) What is the stopping voltage if red light ($\lambda = 700$ nm) is used instead? Explain.
26. ••• When the surface of a particular material is illuminated with monochromatic light of various frequencies, the experimental stopping potentials as a function of light frequency are determined to be:

| Frequency (in Hz) | | | |
|---|---|---|---|
| $9.9 \times 10^{14}$ | $7.6 \times 10^{14}$ | $6.2 \times 10^{14}$ | $5.0 \times 10^{14}$ |
| Stopping potential (in V) | | | |
| 2.6 | 1.6 | 1.0 | 0.60 |

Plot stopping potential versus frequency, and from the graph determine Planck's constant and the material's work function.

27. ••• When a material is illuminated with red light ($\lambda = 700$ nm) and then blue light ($\lambda = 400$ nm), it is found that the maximum kinetic energy of the photoelectrons resulting from the blue light is twice that from the red light. What is the work function of the material?

### 27.3 Quantum "Particles": The Compton Effect

28. • What is half the maximum wavelength shift for Compton scattering from a free electron?
29. • When the wavelength shift for Compton scattering from a free electron is a maximum, what is the scattering angle?
30. • What is the change in wavelength when monochromatic X-rays are scattered by electrons through an angle of 30°?
31. • A monochromatic beam of X-rays with a wavelength of 0.280 nm is scattered by a metal foil. If the scattered beam has a wavelength of 0.281 nm, what is the scattering angle?
32. •• X-rays with a wavelength of 0.0045 nm are used in a Compton scattering experiment. If they are scattered at 45°, what is the wavelength of the scattered radiation?
33. **IE** •• A photon with an energy of 5.0 KeV is scattered by a free electron. (a) The recoiling electron could have an energy of (1) zero, (2) less than 5.0 KeV, but not zero, (3) 5.0 keV. Explain. (b) If the wavelength of the scattered photon is 0.25 nm, what is the recoiling electron's kinetic energy?
34. •• X-rays of wavelength 0.01520 nm are scattered from a carbon atom. The wavelength shift is measured to be 0.000326 nm. (a) What is the scattering angle? (b) How much energy, in eV, does each photon impart to each electron?
35. •• X-rays of frequency $1.210 \times 10^{18}$ Hz are scattered from electrons in an aluminum foil. The frequency of the scattered X-rays is $1.203 \times 10^{18}$ Hz. (a) What is the scattering angle? (b) What is the recoiling speed of the electrons?
36. **IE** ••• The Compton effect can occur for scattering from any particle – for example, from a proton. (a) Compared with the Compton wavelength for an electron, the Compton wavelength for a proton is (1) longer, (2) the same, (3) shorter. Why? (b) What is the value of the Compton wavelength for a proton? (c) Determine the ratio of the maximum Compton wavelength shift for scattering by an electron to that for scattering by a proton.

### 27.4 The Bohr Theory of the Hydrogen Atom

37. • Find the energy of a hydrogen atom whose electron is in the (a) $n = 2$ state and (b) $n = 3$ state.
38. • Find the radius of the electron orbit in a hydrogen atom for states with the following principal quantum numbers: (a) $n = 2$, (b) $n = 4$, (c) $n = 5$.
39. • Some scientists study "large" atoms; that is, atoms with orbits that are almost large enough to be measured in our everyday units of measurement. For what excited state (give an approximate principal quantum number) of a hydrogen atom would the diameter of the orbit be on the order of $10^{-5}$ m – that is, close to the diameter of a human hair?
40. • Find the binding energy of the hydrogen electron for states with the following principal quantum numbers: (a) $n = 3$, (b) $n = 5$, (c) $n = 7$.
41. •• Find the energy required to excite a hydrogen electron from (a) the ground state to the first excited state and (b) the first excited state to the second excited state. (c) Classify the type of light needed to create each of the transitions.
42. •• What is the frequency of light that would excite the electron of a hydrogen atom (a) from a state with a principal quantum number of $n = 2$ to that with a principal quantum number of $n = 5$? (b) What about from $n = 2$ to $n = \infty$?
43. **IE** •• A hydrogen atom has an ionization energy of 13.6 eV. When it absorbs a photon with an energy greater than this energy, the electron will be emitted with some kinetic energy. (a) If the energy of such a photon is doubled, the kinetic energy of the emitted electron will (1) more than double, (2) remain the same, (3) exactly double, (4) increase, but less than double. Why? (b) Photons associated with light of a frequency of $7.00 \times 10^{15}$ Hz and $1.40 \times 10^{16}$ Hz are absorbed by a hydrogen atom. What is the kinetic energy of the emitted electron?

44. •• A hydrogen atom in its ground state is excited to the $n=5$ level. It then makes a transition directly to the $n=2$ level before finally returning to the ground state. (a) What are the wavelengths of the emitted light? (b) Is any of the emitted light in the visible?
45. **IE** •• (a) For which of the following transitions in a hydrogen atom is the light of the longest wavelength emitted: (1) $n=5$ to $n=3$, (2) $n=6$ to $n=2$, or (3) $n=2$ to $n=1$? (b) Justify your answer mathematically.
46. •• The hydrogen spectrum has a series of lines called the Lyman series, which results from transitions to the ground state. What is the longest wavelength in this series, and in what region of the EM spectrum does it lie?
47. •• A hydrogen atom absorbs light of wavelength 486 nm. (a) How much energy did the atom absorb? (b) What are the values of the principal quantum numbers of the initial and final states of this transition?
48. •• If the electron in a hydrogen atom is to make a transition from the first excited state to the fourth excited state, what frequency of light is needed? What type of light is this?
49. **IE** •• (a) How many transitions in a hydrogen atom result in the absorption of red light: (1) one, (2) two, (3) three, or (4) four? (b) What are the principal quantum numbers of the initial and final states for this process? (c) What is the energy of the required photon and the wavelength of the light associated with it?
50. •• What is the binding energy for an electron in the ground state in the following hydrogen-like ions: (a) $He^+$ ($Z=2$) and (b) $Li^{2+}$ ($Z=3$)?
51. ••• Show that the orbital speeds of an electron in the Bohr orbits are given (to two significant figures) by $v_n=(2.2\times10^6\text{ m/s})/n$.
52. ••• For an electron in the ground state of a hydrogen atom, calculate its (a) potential energy, (b) kinetic energy, and (c) total energy. [*Hint:* You will need to use the orbital radius.]

**27.4 A Quantum Success: The Laser**

53. • Suppose a hypothetical atom had two metastable excited states, one 2.0 eV above the ground state and one 4.0 eV above the ground state. If used in a laser with transitions only to the ground state, (a) what would be the wavelength emitted from each excited state? (b) Which transition is in the visible range?
54. • To achieve population inversion between two atomic states separated by an energy of 3.5 eV, what wavelength of pumping light should be used?

# 28

# Quantum Mechanics and Atomic Physics



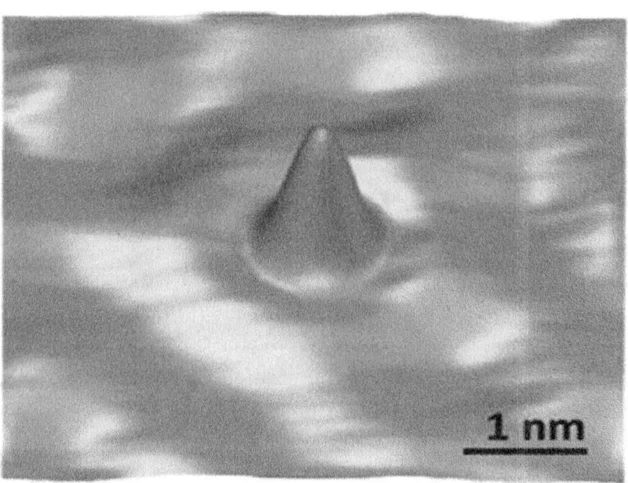


**This image of a single cobalt atom is made possible by the quantum mechanical phenomenon called tunneling.**

Just a few decades ago, if someone had claimed to have a photograph of an atom, people would have laughed. Today, a device called the scanning tunneling microscope (STM) routinely produces images such as the chapter-opening photograph, which shows a single atom. In that image, the shape is a single cobalt (Co) atom resting on a surface of copper atoms. The STM, as well as other modern instruments, operate on a quantum-mechanical phenomenon called *tunneling*. Tunneling reflects some of the fundamental features of the subatomic realm: the probabilistic character of quantum processes and the wave nature of particles.

In the 1920s, a new kind of physics, based on the synthesis of wave and quantum ideas, was introduced. This theory, called **quantum mechanics**, combined the wave–particle duality of matter (Section 27.3) into a single consistent description. It revolutionized scientific thought and today provides the basis of our understanding of phenomena at the molecular, atomic, and nuclear levels.

In this chapter, some of the basic ideas of quantum mechanics are also discussed.

## 28.1 Matter Waves: The de Broglie Hypothesis

Since a photon travels at the speed of light, it must be treated relativistically as a particle with no mass. Thus its energy $E$ and momentum $p$ are related by $p = E/c$. Recall that a photon's energy can alternatively be written in terms of its associated wave frequency and wavelength as $E = hf = hc/\lambda$. Combining these shows that the momentum of a photon is inversely related to the wavelength of the associated light wave by

$$p = \frac{E}{c} = \frac{hf}{c} = \frac{h}{\lambda} \quad \text{(photon momentum)} \tag{28.1}$$

In the early part of the twentieth century, French physicist Louis de Broglie (1892–1987) suggested that there might be symmetry between waves and particles. He conjectured that if light sometimes behaves like particles do, then perhaps particles, such as electrons, might sometimes behave like waves and exhibit wave-like properties. In 1924, de Broglie hypothesized that a moving particle has a wave associated with it. He proposed that its wavelength is related to the magnitude of the particle's momentum $p$ by an equation similar to that for a photon (Equation 28.1), *except* that the momentum is given by the expression for a particle with mass, $p = mv$. Thus the **de Broglie hypothesis** was: A particle with momentum $p$ has a wave associated with it. The wavelength is given by

$$\lambda = \frac{h}{p} = \frac{h}{mv} \quad \text{(material particles)} \tag{28.2}$$

The waves associated with moving particles were called **matter waves**, or, more commonly, **de Broglie waves**, and were thought to somehow influence, or guide, the particle's motion. The electromagnetic wave is the associated wave for a photon. However, de Broglie waves associated with particles, such as electrons, are a different type and decidedly not electromagnetic in nature.

Needless to say, de Broglie's hypothesis was met with great skepticism. The idea that the motions of photons were somehow governed by the wave properties of light seemed reasonable. But the extension of this idea to the motion of a particle was difficult to accept. Moreover, there was no evidence at the time that particles exhibited any wave properties, such as interference and diffraction.

In support of his hypothesis, de Broglie showed how it could give an interpretation of the quantization of the angular momentum postulated by Bohr in his theory of the hydrogen atom. Recall that Bohr had to hypothesize that the angular momentum of the orbiting electron was quantized in integer multiples of $h/(2\pi)$ (Section 27.4). De Broglie argued that for a free particle, the associated wave would be a traveling wave. However, the bound electron of a hydrogen atom travels repeatedly in discrete circular orbits. The associated matter wave might therefore be expected to be a standing wave (Section 13.5). For a standing wave to be produced, an integral number of wavelengths must fit into the orbital circumference, much as a standing wave in a circular string (▼ **Figure 28.1**).

The circumference of a Bohr orbit of radius $r_n$ is $2\pi r_n$, where $n$ is an integer quantum number. de Broglie equated this circumference to the $n$ electron "wavelengths" in the following manner: $2\pi r_n = n\lambda$ for $n = 1, 2, 3,\ldots$ Substituting for the wavelength $\lambda$ from Equation 28.2 yields

$$2\pi r_n = \frac{nh}{mv}$$

Recall that for a particle in a circular orbit, the angular momentum is $L = mvr$; rearranging the above equation, an expression for the angular momentum results:

$$L_n = mvr_n = n\left(\frac{h}{2\pi}\right) \quad \text{for } n = 1, 2, 3, \ldots$$

Thus, Bohr's angular momentum quantization hypothesis is equivalent to the de Broglie assumption that, in some way, the electron behaves like a wave.

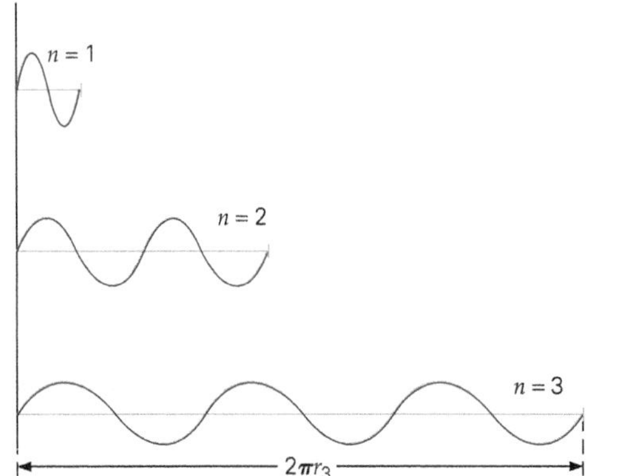


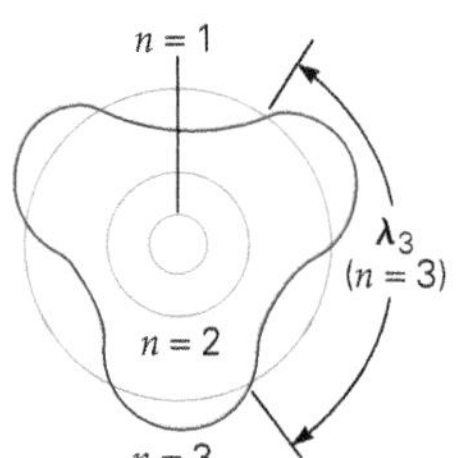


▲ FIGURE 28.1 **de Broglie waves and Bohr orbits** Similar to standing waves on a stretched string, de Broglie waves form circular standing waves on the circumferences of the Bohr orbits. The number of wavelengths in a particular orbit of radius $r$, shown here for $n = 3$, is equal to the principal quantum number of that orbit. (Not to scale.)

For orbits other than those allowed by the Bohr theory, the de Broglie wave for the orbiting electron would not close on itself. This is consistent with the Bohr postulate of the electron being only in certain allowed orbits and implies that *the amplitude of the de Broglie wave might be related to the location of the electron.* This idea is actually a fundamental cornerstone of modern quantum mechanics, as will be seen.

If particles really have wavelike properties, why isn't their wave nature observed in everyday phenomena? This is because effects such as diffraction are significant only when the wavelength $\lambda$ is on the order of the size of the object or the opening it meets. (See Section 24.3.) If $\lambda$ is much smaller than these dimensions, diffraction effects are negligible. The numbers in Examples 28.1 and 28.2 should convince you of the difference between the atomic world and our everyday world.

---

### EXAMPLE 28.1: SHOULD BALLPLAYERS WORRY ABOUT DIFFRACTION? DE BROGLIE WAVELENGTH

A baseball pitcher throws a fastball to the catcher at 40 m/s through a square opening cut in a sheet of canvas and is 50 cm on a side. If the ball's mass is 0.15 kg, (a) what is its associated de Broglie wavelength? (b) From this result, should the catcher be concerned about diffraction effects as the ball passes through the opening?

**THINKING IT THROUGH.** (a) The de Broglie relationship (Equation 28.2) can be used to calculate the wavelength of the matter wave associated with the ball. (b) The question is whether the wavelength is much larger than, much smaller than, or approximately the same size as the opening.

### SOLUTION

***Given:***

$v = 40$ m/s
$d = 50$ cm $= 0.50$ m
$m = 0.15$ kg

***Find:***

**(a)** $\lambda$ (de Broglie wavelength)
**(b)** Whether diffraction is likely

**(a)** Equation 28.2 gives the wavelength of the baseball:

$$\lambda = \frac{h}{mv} = \frac{6.63\times10^{-34}\ \text{J}\cdot\text{s}}{(0.15\ \text{kg})(40\ \text{m/s})} = 1.1\times10^{-34}\ \text{m}$$

**(b)** For significant diffraction to occur when a wave passes through an opening, that wave must have a wavelength on the order of the size of the opening. Since $1.1 \times 10^{-34}$ m $\ll 0.50$ m, the baseball travels straight into the catcher's mitt, with no noticeable wave diffraction.

**FOLLOW-UP EXERCISE.** In this Example, (a) how fast would the ball have to be thrown for diffractive effects to become important? (b) At that speed, how long would the ball take to travel the 20 m to the plate? (Compare your answer with the age of the universe – about 15 billion years. Would someone watching this ball think that it is moving?)

---

The results of Example 28.1 show that it is little wonder that the wave nature of matter isn't observable in our everyday lives. Notice, however, that a particle's de Broglie wavelength varies inversely with its mass and speed. So particles with very small masses traveling at low speeds might be another story, as Example 28.2 shows.

---

### EXAMPLE 28.2: A WHOLE DIFFERENT BALL GAME – DE BROGLIE WAVELENGTH OF AN ELECTRON

(a) What is the de Broglie wavelength of the wave associated with an electron that has been accelerated from rest through a potential of 50.0 V? (b) Compare your answer with the typical distance between atoms in a crystal, about $10^{-10}$ m. Would you expect diffraction to occur as these electrons pass between such atoms?

**THINKING IT THROUGH.** (a) The de Broglie hypothesis (Equation 28.2) can be used, but the electron's speed must first be calculated from the given accelerating voltage. This computation involves consideration of energy conservation – the conversion of electric potential energy into kinetic energy. (b) For diffractive effects to be important, $\lambda$ must be on the order of $10^{-10}$ m.

### SOLUTION

Listing the data, as well as the mass of an electron:

***Given:***

$\Delta V = 50.0$ V
$m = 9.11 \times 10^{-31}$ kg

***Find:***

**(a)** $\lambda$ (de Broglie wavelength)
**(b)** Whether diffraction is likely

**(a)** The magnitude of the potential energy lost by the electron is $|\Delta U_e| = e{\cdot}\Delta V$ and is equal to its gain in kinetic energy $\Delta K = (1/2)mv^2$, because $K_o = 0$. Equating these quantities allows the speed to be calculated:

$$\frac{1}{2}mv^2 = e{\cdot}\Delta V \quad \text{or} \quad v = \sqrt{\frac{2e{\cdot}\Delta V}{m}}$$

So,

$$v = \sqrt{\frac{2(1.60\times10^{-19}\ \text{C})(50.0\ \text{V})}{9.11\times10^{-31}\ \text{kg}}} = 4.19\times10^{6}\ \text{m/s}$$

Thus, the electron's de Broglie wavelength is

$$\lambda = \frac{h}{mv} = \frac{6.63\times10^{-34}\ \text{J}\cdot\text{s}}{(9.11\times10^{-31}\ \text{kg})(4.19\times10^{6}\ \text{m/s})} = 1.74\times10^{-10}\ \text{m}$$

**(b)** Since this result is the same order of magnitude as the distance between the crystal's atoms, diffraction could be observed. Thus, passing electrons through a crystal lattice should prove de Broglie's hypothesis.

**FOLLOW-UP EXERCISE.** In this Example, what would change if the particle were a proton? In other words, would the de Broglie wave of a proton be more or less likely to exhibit diffraction than an electron under the same conditions? Explain, and give a numerical answer.

---

## 28.1.1 Problem-Solving Hint

In Example 28.2, if the accelerating voltage were changed, the calculation of $v$ and $\lambda$ would have to be repeated. It is convenient to use the numerical values of $m$, $e$, and $h$ to derive a nonrelativistic expression for the de Broglie wavelength of an electron when it is accelerated through a potential difference $\Delta V$. From energy conservation (see Example 28.2a), the particle's speed is $v = \sqrt{((2e \cdot \Delta V)/m)}$. Thus, its de Broglie wavelength is $\lambda = (h/mv) = \left(h/\left(\sqrt{2me \cdot \Delta V}\right)\right) = \sqrt{(h^2/(2me \cdot \Delta V))}$. Inserting the values of $h$, $e$, and $m$ and rounding to three significant figures, yields

$$\lambda = \sqrt{\frac{1.50}{\Delta V(\text{in volts})}} \times 10^{-9} \text{ m}$$
$$= \sqrt{\frac{1.50}{\Delta V(\text{in volts})}} \text{ nm} \qquad (28.3)$$

(only for a nonrelativistic electron initially at rest)

For example, if $\Delta V = 50.0$ V, then $\lambda = \sqrt{(1.50/50.0)}$ nm = 0.173 nm, a result that differs in the last digit from that in Example 28.2 due to rounding. You may wish to derive a comparable expression for a proton, to use in solving similar problems. What would have to be changed?

The wavelike property of particles was first demonstrated in 1927 by two physicists in the United States, C. J. Davisson and L. H. Germer, in an experiment using a crystal to diffract a beam of electrons. A single crystal of nickel was cut to expose a spacing of $d = 0.215$ nm between the planes of atoms. When a beam of electrons, accelerated by 54.0 V, was directed normally onto the crystal face, a maximum in the intensity of scattered electrons was observed at an angle of 50° relative to the surface normal (▼ **Figure 28.2**).

Equation 28.3 can be used to determine the de Broglie wavelength of these electrons:

$$\lambda = \sqrt{\frac{1.50}{\Delta V}} \text{ nm} = \sqrt{\frac{1.50}{54.0}} \text{ nm} = 0.167 \text{ nm}$$

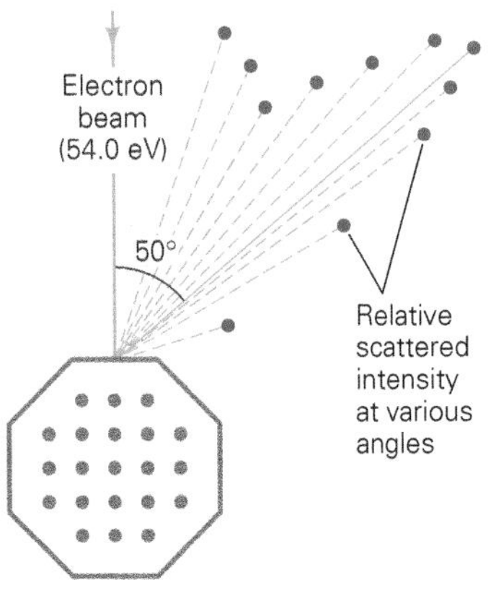

▲ **FIGURE 28.2 The Davisson–Germer experiment** When electrons with kinetic energy of 54.0 eV is incident on the face of a nickel crystal array, a maximum in the scattered intensity is observed at 50°.

From the condition of constructive interference, the first-order maximum (Equation 24.3) should be observed at an angle whose sine is given by $\sin\theta = (\lambda/d) = (0.167/0.215 \text{ nm}) = 0.775$. Thus $\theta = \sin^{-1}(0.775) = 50.8°$.

The agreement was well within the range of experimental uncertainty, and thus the Davisson–Germer experiment gave convincing proof of de Broglie's matter wave hypothesis.

The de Broglie hypothesis led to the development of many important practical applications – for example, the *electron microscope*. The first such microscope was built in Germany in 1931. In it, electron "waves" are focused using magnetic and electric fields to form images much like the light in a regular microscope. However, electrons have much shorter wavelengths, thus providing greater magnification and finer resolution than any light microscope. In fact, the resolving power of electron microscopes is on the order of a nanometers, typically just a bit larger than atomic sizes.

In another type of microscope, called a *transmission electron microscope* (TEM), an electron beam is directed onto a very thin specimen (typically about 100 atoms thick). Different numbers of electrons pass through different parts of the specimen, depending on its structure. The transmitted beam is then brought into focus by a magnetic objective coil revealing details of the specimen's atomic structure. The resolution of a typical TEM is about 500 000 times greater than that of a human eye.

The surfaces of thicker objects can be examined by the reflection of the electron beam from the surface using the *scanning electron microscope* (SEM). An electron beam spot is scanned across the specimen by means of deflecting magnetic coils. Surface irregularities cause directional variations in the intensity of the reflected electrons, which gives contrast to the image. With this technique, an SEM can give images with a remarkable three-dimensional quality.

# 28.2 The Schrödinger Wave Equation

De Broglie's hypothesis predicts that particles have associated waves that somehow govern their behavior. However, it does not tell us the *form* of these waves, only their wavelengths. To have a useful theory, an equation that will give the mathematical form of these matter waves is needed. In addition, the "rules" about how these waves govern particle motion are required. In 1926, Erwin Schrödinger (1887–1961), an Austrian physicist, presented a general equation that describes the de Broglie matter waves and their interpretation.

A traveling de Broglie wave varies with location and time in a way similar to everyday wave motion (Section 13.2). The de Broglie wave is denoted by $\psi$ (the Greek letter psi, pronounced "sigh") and is called the **wave function**. The wave function is associated with the particle's kinetic, potential, and total energy. Recall that for a conservative mechanical system (Section 5.5), the total mechanical energy $E$, the sum of the kinetic and potential energies, is constant; that is, $K + U = E$ = constant. Schrödinger proposed a similar

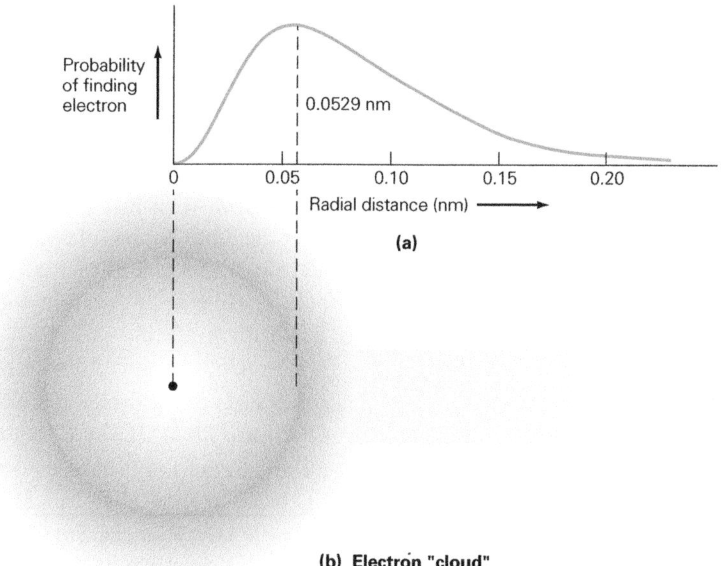


▲ FIGURE 28.3 **Electron probability for hydrogen atom orbits (a)** The square of the wave function is the probability of finding the hydrogen electron at a particular location. Here, it is in the ground state ($n = 1$), and this probability is plotted as a function of radial distance from the proton. The electron has the greatest probability of being at a distance of 0.0529 nm, which matches the radius of the first Bohr orbit. **(b)** The probability distribution gives rise to the idea of an electron "probability cloud" around the proton. The cloud's density reflects the probability density.

equation for the wave function. **Schrödinger's wave equation*** has the general form

$$(K+U)\psi = E\psi \quad \text{(Schrödinger wave equation)} \quad (28.4)$$

Equation 28.4 can be solved for $\psi$, but for us, the more important question is related to the physical significance of $\psi$. During the early development of quantum mechanics, it was not at all clear how $\psi$ should be interpreted. After much thought and investigation, Schrödinger and his colleagues hypothesized the following:

> The square of a particle's wave function $\psi$ is proportional to the probability of finding that particle at a given location.

The interpretation of $\psi^2$ as a *probability* altered the idea that the electron in a hydrogen atom could be found only in orbits at discrete distances from the proton, as in the Bohr theory. In fact, when the Schrödinger equation was solved for the hydrogen atom, the result was a nonzero probability of finding the electron at almost any distance from the proton. As an example of this, the relative probability of finding an electron [in the ground state ($n = 1$)] at a given distance from the proton, as determined by solving the Schrödinger wave equation, is shown in ▲ **Figure 28.3a**.

As can be seen, the maximum probability coincides with the Bohr radius of 0.0529 nm, but it is possible, with less probability, that the electron could be found closer to or further from the proton. Notice that, although the ground state wave function exists for distances well beyond 0.20 nm from the proton, there is little chance of finding an electron beyond this distance. The probability density distribution gives rise to the idea of an *electron cloud* around the proton (Figure 28.3b). This cloud is actually a probability density cloud, meaning that the electron can be found in different locations with varying probabilities.

An interesting quantum mechanical result that runs counter to our everyday experiences is *tunneling*. In classical physics, there are regions forbidden to particles by energy considerations. These regions are areas where a particle's potential energy would be greater than its total energy. Classically, the particle is not allowed in such regions because it would have a negative kinetic energy there ($K = E - U < 0$), which is impossible. In such situations, we say that the particle's location is limited by a *potential energy barrier*.

In certain instances, however, quantum mechanics predicts a small but finite probability of the particle's wave function penetrating the barrier and thus of the particle being found on the other side of the barrier. Thus, there is a certain probability of the particle "tunneling" through the barrier, especially on the atomic level, where the wave nature of particles is exhibited.

A practical result of quantum mechanical tunneling is the scanning tunneling microscope (STM) which was invented in the late 1970s and revolutionized the field of surface physics. The STM produces atomic-level images by positioning its sharp tip very close to a surface. A voltage is applied between tip and surface, enabling electron tunneling through the

* Although Equation 28.4 appears to be just a simple multiplication of $E = K + U$ by $\psi$, it is, in fact, much more complex – it is an example of a partial differential equation – one that requires calculus and more advanced mathematics to solve. This is clearly beyond the scope of this text.

vacuum gap. This tunneling current is extremely sensitive to this separation distance. A feedback circuit monitors this current and moves the probe vertically to keep the current constant. The separation distance is processed by computers for display. When the probe is passed over the surface in successive parallel movements, a three-dimensional "relief map" of the surface can be created, with resolution on the order of the size of individual atoms - allowing detailed surface images to be created such as that shown in the chapter opening photograph.

## 28.3 Atomic Quantum Numbers and the Periodic Table

### 28.3.1 The Hydrogen Atom

When the Schrödinger equation was solved for the hydrogen atom, the results predicted its energy levels to be the same as those given by the Bohr theory. Recall that the Bohr model energy values depended only on the **principal quantum number** $\boldsymbol{n}$. However, the solution to the Schrödinger equation yielded two other quantum numbers, designated as $\ell$ and $m_\ell$. The reasoning is that three quantum numbers are needed because the electron can move in three dimensions. Thus the Bohr theory, being essentially a two-dimensional (flat planar) model was deemed incomplete, requiring only one quantum number.

The quantum number $\ell$ is the **orbital quantum number**. It is associated with the orbital angular momentum of the electron. For each value of $n$, the $\ell$ quantum number has integer values from zero up to a maximum value of $n - 1$. For example, if $n = 3$, the three possible values of $\ell$ are 0, 1, and 2. Thus the number of different $\ell$ values for a given $n$ value is equal to $n$, and in the hydrogen atom, the energy depends only on $n$. ▼**Figure 28.4** shows three orbits with different angular momenta, but the same energy (i.e., the same $n$ value). Thus, orbits with the same $n$ value, but different $\ell$ values, have the same energy and are said to be **degenerate**.

The quantum numbers $m_\ell$ is the **magnetic quantum number**. The name originated from experiments in which an external magnetic field was applied to a sample. They showed that a particular energy level (with given values of $n$ and $\ell$) of a hydrogen atom actually consists of several orbits that differ slightly in energy only when in a magnetic field. Thus, in the absence of a magnetic field, there was additional energy degeneracy. Clearly, there must be more to the description of the orbit than just $n$ and $\ell$. The quantum number $m_\ell$ was introduced to enumerate the number of levels that existed for a given orbital quantum number $\ell$. When there is no magnetic field, the energy of the atom does not depend on either of these quantum numbers.

The magnetic quantum number $m_\ell$ is associated with the orientation of the orbital angular momentum vector $\vec{\mathbf{L}}$ in space (▼**Figure 28.5**). If there is no external magnetic field, then all orientations of $\vec{\mathbf{L}}$ have the same energy. For each value of $\ell$, $m_\ell$ is an integer that can range from zero to $\pm\ell$. That is, $m_\ell = 0, \pm1, \pm2, \ldots, \pm\ell$. For example, an orbit described by $n = 3$ and $\ell = 2$ can have $m_\ell$ values of $-2$, $-1$, 0, $+1$, and $+2$. In this case, the orbital angular momentum vector $\vec{\mathbf{L}}$ has five possible orientations, all with the same energy if no magnetic field is present. In general, for a given value of $\ell$, there are $2\ell + 1$ possible values of $m_\ell$. For example, with $\ell = 2$, there are five values of $m_\ell$, since $2\ell + 1 = (2 \times 2) + 1 = 5$.

However, this finding was not the end of the story. The use of high-resolution optical spectrometers showed that each emission line of hydrogen is, in fact, two very closely spaced

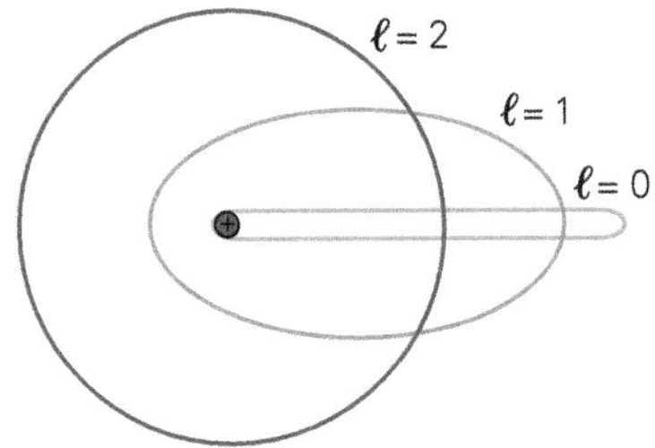


▲FIGURE 28.4 **The orbital quantum number $\ell$** The orbits of an electron are shown for the second excited state in hydrogen. For the principal quantum number $n = 3$, there are three different values of angular momentum (corresponding to the three differently shaped orbits and three different values of $\ell$), all of the same total energy. The circular orbit has the maximum angular momentum; the narrowest orbit classically has zero angular momentum.

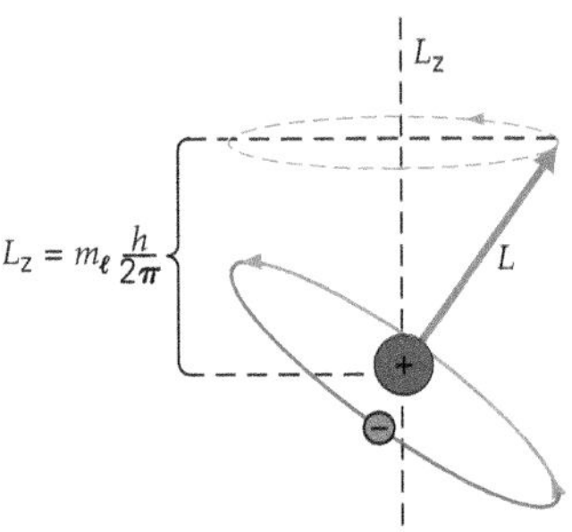


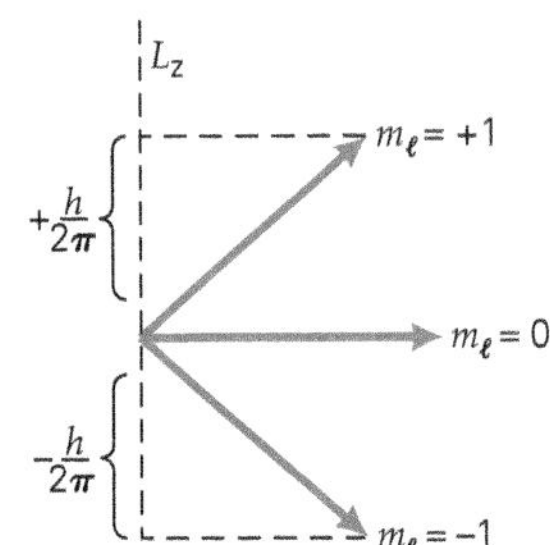


▲FIGURE 28.5 **The magnetic quantum number $m_\ell$** $\vec{\mathbf{L}}$ is the vector angular momentum associated with the orbiting electron. Because the energy of an orbit is independent of the orientation of its plane, orbits with the same $\ell$ that differ only in $m_\ell$ have the same energy. There are $2\ell + 1$ possible orientations for a given $\ell$. The value of $m_\ell$ tells the component of the angular momentum vector in a given direction, as shown in the lower drawing for $\ell = 1$.

lines. Thus, each emitted wavelength is actually two. This splitting is called *spectral fine structure.* Hence, a fourth quantum number was necessary in order to describe each atomic state completely. This number is called the **spin quantum number** $m_s$ of the electron. It is associated with an intrinsic angular momentum of the electron. This property, called *electron spin,* is sometimes described as analogous to the angular momentum associated with a spinning object (▼ **Figure 28.6**). Because each energy level is split into only two levels, the electron's intrinsic angular momentum (or, more simply, its "spin") can possess only two orientations, called "spin up" and "spin down."

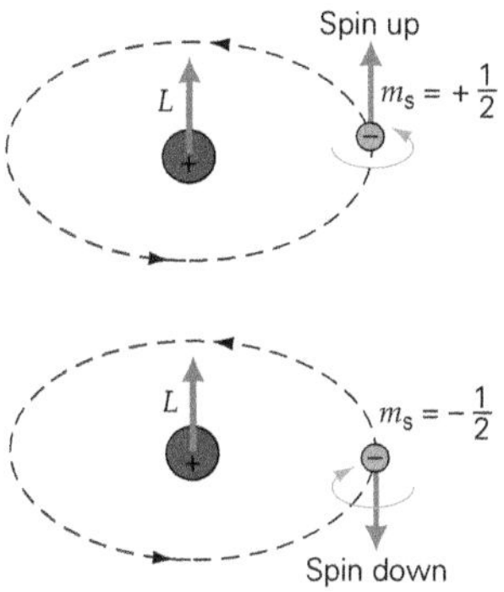


▲ **FIGURE 28.6 The electron spin quantum number $m_s$** The electron spin can be either up or down. Electron spin is strictly a quantum mechanical property and should not be identified with the physical spin of a macroscopic body, except when used in conceptual reasoning.

Thus, the fine structure of an atom's energy levels results from the electron spins having two orientations with respect to the atom's *internal* magnetic field, produced by the electron's orbital motion. Analogous to a magnetic moment (such as a compass) in a magnetic field, the atom possesses slightly less energy when its electron's spin is "lined up" with the field than when the spin is aligned opposite to the field. However, keep in mind that spin is fundamentally a purely quantum mechanical concept; it is not really analogous to a spinning top. This is because, as far as we know now, the electron possesses no size - that is, it is truly a point particle.

For example, in the excited state of hydrogen with $n = 3$ and $\ell = 2$, each value of $m_\ell$ also has two possible spin orientations. For example, when $m_\ell = +1$, there are two possible sets of the four quantum numbers: $n = 3$, $\ell = 2$, and $m_\ell = +1$, with $m_s = +1/2$; and $n = 3$, $\ell = 2$, and $m_\ell = +1$, with $m_s = -1/2$. Both sets have nearly the same energy, thus the energy is almost independent of the electron's spin direction. This condition results in another (approximate) energy degeneracy. In summary, the energy of the various states of the hydrogen atom are, to a very high degree, determined by the principal quantum number $n$.

The four quantum numbers for hydrogen are summarized in ▶ **Table 28.1.** Other particles, such as protons, neutrons, and composites of them called atomic nuclei, also possess spin. See Chapters 29 and 30 for details.

**TABLE 28.1** Quantum Numbers for the Hydrogen Atom

| Quantum Number | Symbol | Allowed Values | Number of Allowed Values |
|---|---|---|---|
| Principal | $n$ | 1, 2, 3, ... | no limit |
| Orbital | $\ell$ | 0, 1, 2, 3, ..., $(n-1)$ | $n$ (for each $n$) |
| Magnetic | $m_\ell$ | $0, \pm1, \pm2, \pm3, \ldots, \pm\ell$ | $2\ell + 1$ (for each $\ell$) |
| Spin | $m_s$ | $\pm1/2$ | 2 |

## 28.3.2 Multielectron Atoms

The Schrödinger equation cannot be solved exactly for atoms with more than one electron. However, a solution can be found, to a workable approximation, in which each electron occupies a state characterized by a set of quantum numbers similar to those for hydrogen. Because of the repulsive forces between the electrons, the description of a multielectron atom is much more complicated. For one, the energy depends not only on the principal quantum number $n$, but also on the orbital quantum number $\ell$. This condition gives rise to a subdivision (or "splitting") of the degeneracy seen in hydrogen atoms. In multielectron atoms, the energies of the atomic levels generally depend on all four quantum numbers.

It is common to refer to all the energy levels that share the same $n$ value as making up a **shell** and to all the levels that share the same $\ell$ values within that shell as **subshells**. Thus electrons with the same $n$ value are said to be "in the same shell." Similarly, electrons with the same $n$ and $\ell$ values are said to be "in the same subshell." The $\ell$ subshells can be designated by integers; however, it is common to use letters instead. The letters $s, p, d, f, g, \ldots$ correspond to the values of $\ell = 0, 1, 2, 3, 4, \ldots$, respectively. After $f$, the letters go alphabetically.

Because in multielectron atoms an electron's energy depends on both $n$ and $\ell$, both quantum numbers are used to label atomic energy levels (a shell and subshell, respectively). The labeling convention is as follows: $n$ is written as a number, followed by the letter that stands for the value of $\ell$. For example, 1*s* denotes an energy level with $n = 1$ and $\ell = 0$; 2*p* is for $n = 2$ and $\ell = 1$; 3*d* is for $n = 3$ and $\ell = 2$; and so on. Also, it is common to refer to the $m_\ell$ values as representing orbitals. For example, a 2*p* energy level has three orbitals, corresponding to the $m_\ell$ values of −1, 0, and +1 (because $\ell = 1$).

The hydrogen atom energy levels are not evenly spaced, but they do increase sequentially. In multielectron atoms, not only are the energy levels unevenly spaced, but their numerical sequence is also, in general, out of order. The shell–subshell (in $n$–$\ell$ notation) energy level sequence for a multielectron atom is shown in ▶ **Figure 28.7a.**

Notice, for example, that the energy of the 4*s* level is less than that of the 3*d* level. Such variations result in part from

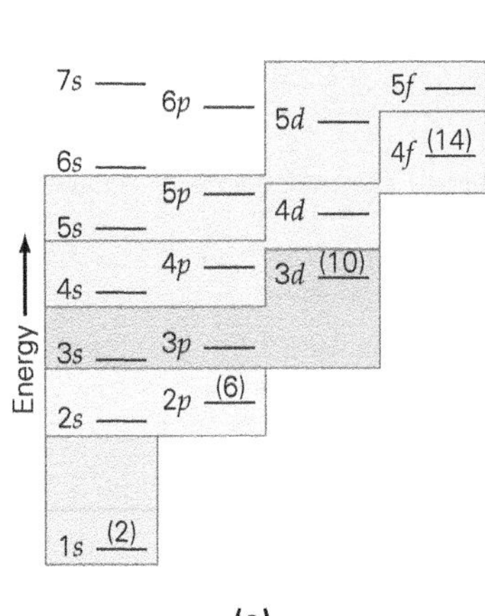


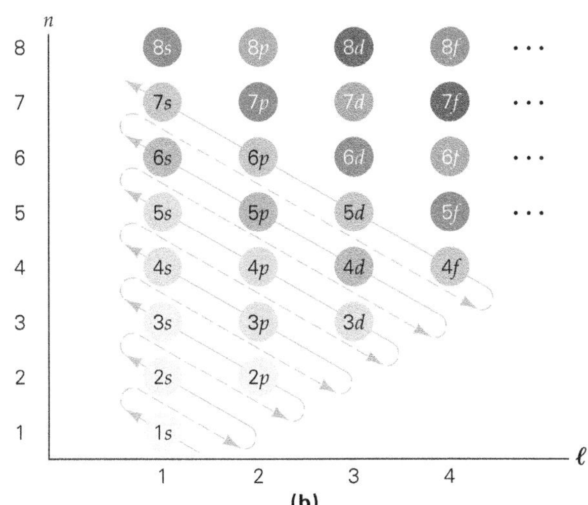


▲ **FIGURE 28.7 Energy levels of a multielectron atom (a)** The shell–subshell ($n$–$\ell$) sequence shows that the energy levels are not evenly spaced and that the sequence of energy levels has numbers that are out of numerical order. For example, the 4*s* level lies below the 3*d* level. The maximum number of electrons for a subshell, $2(2\ell + 1)$, is shown in parentheses on representative levels. (The vertical energy differences are not to scale.) **(b)** A convenient way to remember the energy level order for a multielectron atom is to list the $n$-versus-$\ell$ values as shown here. The diagonal lines then give the levels in ascending energy order.

electrical forces between the electrons. Furthermore, note that the electrons in the outer orbits are "shielded" from the attractive force of the nucleus by the electrons that are closer to the nucleus. For example, consider the highly elliptical orbit (Figure 28.4) of an electron in the 4*s* ($\ell = 0$) orbit. The electron clearly spends more time near the nucleus, and hence is more tightly bound than if it were in the more circular 3*d* ($\ell = 2$) orbit. A convenient way to remember the order of the levels is given in Figure 28.7b.

The ground state of a multielectron atom has some similarities to that of the hydrogen atom (with its single electron in the 1*s*, or lowest energy, level). In a multielectron atom, the electrons are still in the lowest possible energy levels. But to identify how electrons fill these levels, we must know how many electrons can occupy a particular energy level. For example, the lithium (Li) atom has three electrons. Can they all be in the 1*s* level? As will be seen, the answer, as given by the Pauli exclusion principle, is no.

### 28.3.3 The Pauli Exclusion Principle

Exactly how the electrons in a multielectron atom distribute themselves among the energy levels is governed by a principle set forth in 1928 by the Austrian physicist Wolfgang Pauli (1900–1958). The **Pauli exclusion principle** states the following:

> No two electrons in an atom can have the same set of quantum numbers ($n$, $\ell$, $m_\ell$, $m_s$). That is, no two electrons in an atom can be in the same quantum state.

This principle limits the number of electrons that can occupy a given energy level. For example, the 1*s* ($n = 1$ and $\ell = 0$) level can have only one $m_\ell$ value, $m_\ell = 0$, along with only two $m_s$ values, $m_s = \pm 1/2$. Thus, there are only two unique sets of quantum numbers ($n$, $\ell$, $m_\ell$, $m_s$) for the 1*s* level – (1, 0, 0, + ½) and (1, 0, 0, – ½) – so only two electrons can occupy the 1*s* level. If this situation is indeed the case, then it is said that such a shell is *full*; all other electrons are excluded from it by Pauli's principle. Thus, for a Li atom, with three electrons, the third electron must occupy the next higher level (2*s*) when the atom is in the ground state. This case is illustrated in ▶ **Figure 28.8**, along with the ground state energy levels for some other atoms.

---

**INTEGRATED EXAMPLE 28.3: THE QUANTUM SHELL GAME – HOW MANY STATES?**

(a) How does the number of possible electron states in the 2*p* subshell compare to that in the 4*p* subshell: (1) the number of states in the 2*p* subshell is greater than that in the 4*p* subshell; (2) the number of states in the 2*p* subshell is less than that in the 4*p* subshell; or (3) the number of states in the 2*p* subshell is the same as that in the 4*p* subshell? (b) Compare the number of possible electron states in the 3*p* subshell to the number in the 4*d* subshell.

**CONCEPTUAL REASONING.** The term *subshell* refers to the states with the same $\ell$ quantum number within a shell; that is, they all have the same principal quantum number $n$. All that matters is that their $\ell$ values are the same. Therefore, the number of states in the two subshells is the same, so the correct answer is (3).

**THINKING IT THROUGH.** This is a matter of following the quantum mechanical counting rules. The *p* level means $\ell = 1$, and the d level means $\ell = 2$. In each subshell, replace the letter designation by its number. Thus, the data given are as follows:

***Given:***

3*p* level means $n = 3$ and $\ell = 1$
4*d* level means $n = 4$ and $\ell = 2$

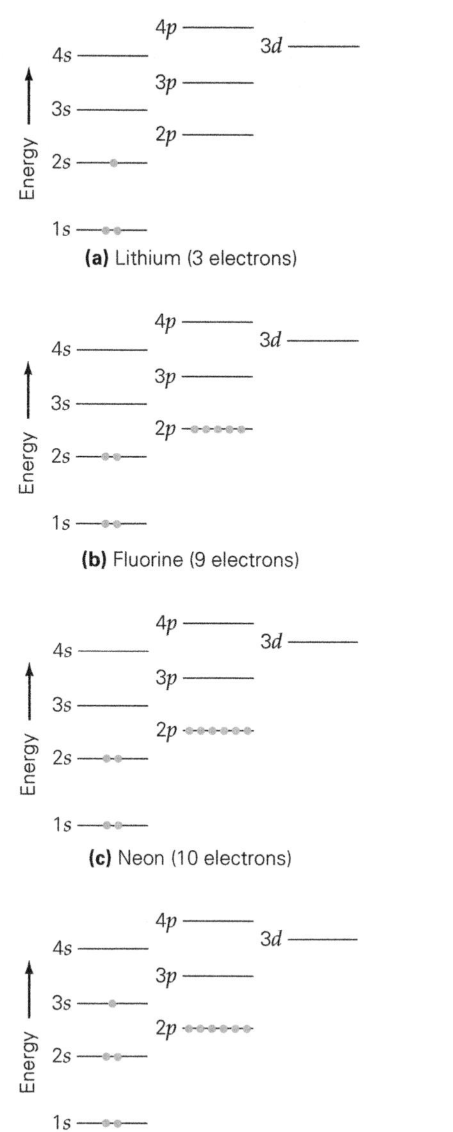


▲ **FIGURE 28.8 Filling subshells** The electron subshell distributions for several atoms in their ground state according to the Pauli exclusion principle. Because of spin, any *s* subshell can hold a maximum of two electrons, and any *p* subshell can hold a maximum of six electrons. What can you say about the *d* subshells?

***Find:*** the number of quantum states in the 3*p* subshell as compared with the number in the 4*d* subshell.

For a particular subshell, the $\ell$ value determines the number of states. Recall that there are $(2\ell+1)$ possible $m_\ell$ values for a given $\ell$. Thus, for $\ell = 1$, there are $[(2 \times 1) + 1] = 3$ values for $m_\ell$. (They are +1, 0, and −1.) Each of these values can have two $m_s$ values ($\pm$ ½), making six different combinations of $(n, \ell, m_\ell, m_s)$, or six states.

In general, the number of possible states for a given value of $\ell$ is $2(2\ell + 1)$, taking into account the two possible "spin states" for each orbital state:

Number of electron states for 3*p* ($\ell = 1$) is

$$2(2\ell + 1) = 2[(2 \times 1) + 1] = 6$$

Number of electron states for 4*d* ($\ell = 2$) is

$$2(2\ell + 1) = 2[(2 \times 2) + 1] = 10$$

Comparison shows that the *d* subshell has more possible electron states than does the *p* subshell, regardless of which shell they are in (i.e., their *n* value).

These results are summarized in ▼ **Table 28.2**. Notice the total number of states in a given shell (designated by *n*) is $2n^2$. For example, for the $n = 2$ shell, the total number of states in its combined *s* and *p* subshells ($\ell = 0, 1$) is $2n^2 = 2(2)^2 = 8$. Thus up to eight electrons can be accommodated in the $n = 2$ shell: two in the 2*s* subshell and six in the 2*p* subshell.

**FOLLOW-UP EXERCISE.** How many electrons could be accommodated in the 3*d* subshell if there were no spin quantum number?

### 28.3.4 Electron Configurations

The electron structure of the ground state of atoms can be determined by putting an increasing number of electrons in the lower energy subshells [hydrogen (H), one electron; helium (He), two electrons; lithium (Li), three electrons; and so on], as was done for four elements in Figure 28.8. However, rather than drawing diagrams, a shorthand notation called *electron configuration* is widely used.

In this notation, the subshells are written in order of increasing energy, and the number of electrons in each subshell is designated with a superscript. For example, $3p^5$ means that a 3*p* subshell is

**TABLE 28.2** Possible Sets of Quantum Numbers and States

| Electron Shell *n* | Subshell $\ell$ | Subshell Notation | Orbitals ($m_\ell$) | Number of Orbitals ($m_\ell$) in Subshell = $(2\ell + 1)$ | Number of States in Subshell = $2(2\ell + 1)$ | Total Electron States for Shell = $2n^2$ |
|---|---|---|---|---|---|---|
| 1 | 0 | 1s | 0 | 1 | 2 | 2 |
| 2 | 0 | 2s | 0 | 1 | 2 | 8 |
| | 1 | 2*p* | 1, 0, −1 | 3 | 6 | |
| 3 | 0 | 3*s* | 0 | 1 | 2 | 18 |
| | 1 | 3*p* | 1, 0, −1 | 3 | 6 | |
| | 2 | 3*d* | 2, 1, 0, −1, −2 | 5 | 10 | |
| 4 | 0 | 4*s* | 0 | 1 | 2 | 32 |
| | 1 | 4p | 1, 0, −1 | 3 | 6 | |
| | 2 | 4*d* | 2, 1, 0, −1, −2 | 5 | 10 | |
| | 3 | 4*f* | 3, 2, 1, 0, −1, −2, −3 | 7 | 14 | |

occupied by five electrons. The electron configurations for the atoms shown in Figure 28.8 can thus be written as follows:

| | | |
|---|---|---|
| Li | (3 electrons) | $1s^2 2s^1$ |
| F | (9 electrons) | $1s^2 2s^2 2p^5$ |
| Ne | (10 electrons) | $1s^2 2s^2 2p^6$ |
| Na | (11 electrons) | $1s^2 2s^2 2p^6 3s^1$ |

In writing an electron configuration, when one subshell is filled, you go to the next higher one. The total of the superscripts in any configuration add up to the number of electrons in the atom.

The energy spacing between adjacent subshells is not uniform, as Figures 28.7a and 28.8 show. In general, there are relatively large energy gaps between the $s$ subshells and the subshells immediately below them. (Compare the $4s$ subshell with the $3p$ one in Figure 28.7a.) The subshells just below the $s$ subshells are usually $p$ subshells, with the exception of the lowest subshell – the $1s$ subshell is below the $2s$ subshell. The gaps between other subshells – for example, between the $3s$ subshell and the $3p$ subshell above it, or between the $4d$ and $5p$ subshells – are considerably smaller.

This unevenness in energy differences gives rise to periodic large energy gaps, represented by vertical lines between certain subshells in the electron configuration:

$$\underset{(2)}{1s^2}\,|\,\underset{(8)}{2s^2 2p^6}\,|\,\underset{(8)}{3s^2 3p^6}\,|\,\underset{(18)}{4s^2 3d^{10} 4p^6}\,|\,\underset{(18)}{5s^2 4d^{10} 5p^6}\,|\,\underset{(32)}{6s^2 4f^{14} 5d^{10} 6p^6}\,|\ldots$$

(number of states)

The subshells between the lines have only slightly different energies. The grouping of subshells (e.g., $2s^2 2p^6$) that have about the same energy is referred to as an **electron period**.

Electron periods form the basis for the periodic table of elements. With the new knowledge of electron configurations, you are now in a position to understand the periodic table better than the person who originally developed it.

### 28.3.5 The Periodic Table of Elements

By 1860, more than sixty chemical elements had been discovered. Several attempts had been made to classify the elements into some orderly arrangement, but none were satisfactory. It had been noted in the early 1800s that the elements could be listed in such a way that similar chemical properties recurred periodically throughout the list. With this idea, in 1869, a Russian chemist, Dmitri Mendeleev (pronounced men-duh-*lay*-eff), created an arrangement of the elements, based on this periodic property. The modern version of his **periodic table of elements** is used today and can be seen on the walls of just about every science building (▶ **Figure 28.9**).

Mendeleev arranged the known elements in rows, called **periods**, in order of increasing atomic mass. When he came to an element that had chemical properties similar to those of one of the previous elements, he put this element below the previous similar one. In this manner, he formed both horizontal rows of elements and vertical columns called **groups**, or families, of elements with similar chemical properties. The table was later rearranged in order of increasing proton number (the number of protons in the nucleus of an atom is the number at the top of each of element box in Figure 28.9) to resolve some inconsistencies. Notice that if atomic masses were used, cobalt and nickel, atomic numbers 27 and 28, respectively, would fall in reversed columns.

Because only sixty-five elements were known at the time, there were vacant spaces in Mendeleev's table. The elements for these spaces were yet to be discovered. Because the missing elements were part of a sequence and had properties similar to those of other elements in a group, Mendeleev could predict their masses and chemical properties. Less than 20 years after Mendeleev devised his table, which showed chemists what to look for in order to find the undiscovered elements, three of the missing elements were discovered.

The periodic table puts the elements into seven horizontal rows, or periods. The first period has only two elements. Periods 2 and 3 each have eight elements, and periods 4 and 5 each have eighteen elements. Recall that the $s$, $p$, $d$, and $f$ subshells can contain a maximum of 2, 6, 10, and 14 electrons $[2(2\ell + 1)]$, respectively. You should begin to see a correlation between these numbers and the arrangements of elements in the periodic table.

The periodicity of the periodic table can be understood in terms of the electron configurations of the atoms. For $n = 1$, the electrons are in one of two $1s$ states; for $n = 2$, electrons can fill the $2s$ and $2p$ states, which gives a total of ten electrons; and so on. Thus, the period number for a given element is equal to the highest $n$ shell containing electrons in the atom. Notice the electron configurations for the elements in Figure 28.9. Also, compare the electron periods given earlier, as defined by energy gaps, and the periods in the periodic table (▶ **Figure 28.10**). There is a one-to-one correlation, so the periodicity comes from energy-level considerations in atoms.

Chemists refer to elements in which the last (least bound) electron enters an $s$ or $p$ subshell as *main group elements.* In *transition elements,* the last electron enters a $d$ subshell, and in *inner transition elements,* the last electron enters an $f$ subshell. So that the periodic table is not unmanageably wide, the $f$ subshell elements are usually placed in two rows at the bottom of the table. Each of the two rows is given a name – the *lanthanide series* and the *actinide series* – based on its position within the period.

Finally, it can also be understood why elements in vertical columns, or groups, have similar chemical properties. The chemical properties of an atom, such as its ability to react and form compounds, depend almost entirely on the atom's outermost electrons – that is, those electrons in the outermost *unfilled* shell. It is these electrons, called *valence electrons,* that form chemical bonds with other atoms. Because of the way in which the elements are arranged in the table, the outermost electron configurations of all the atoms in any one group are similar. The atoms in such a group would thus be expected to have similar chemical properties, and they do.

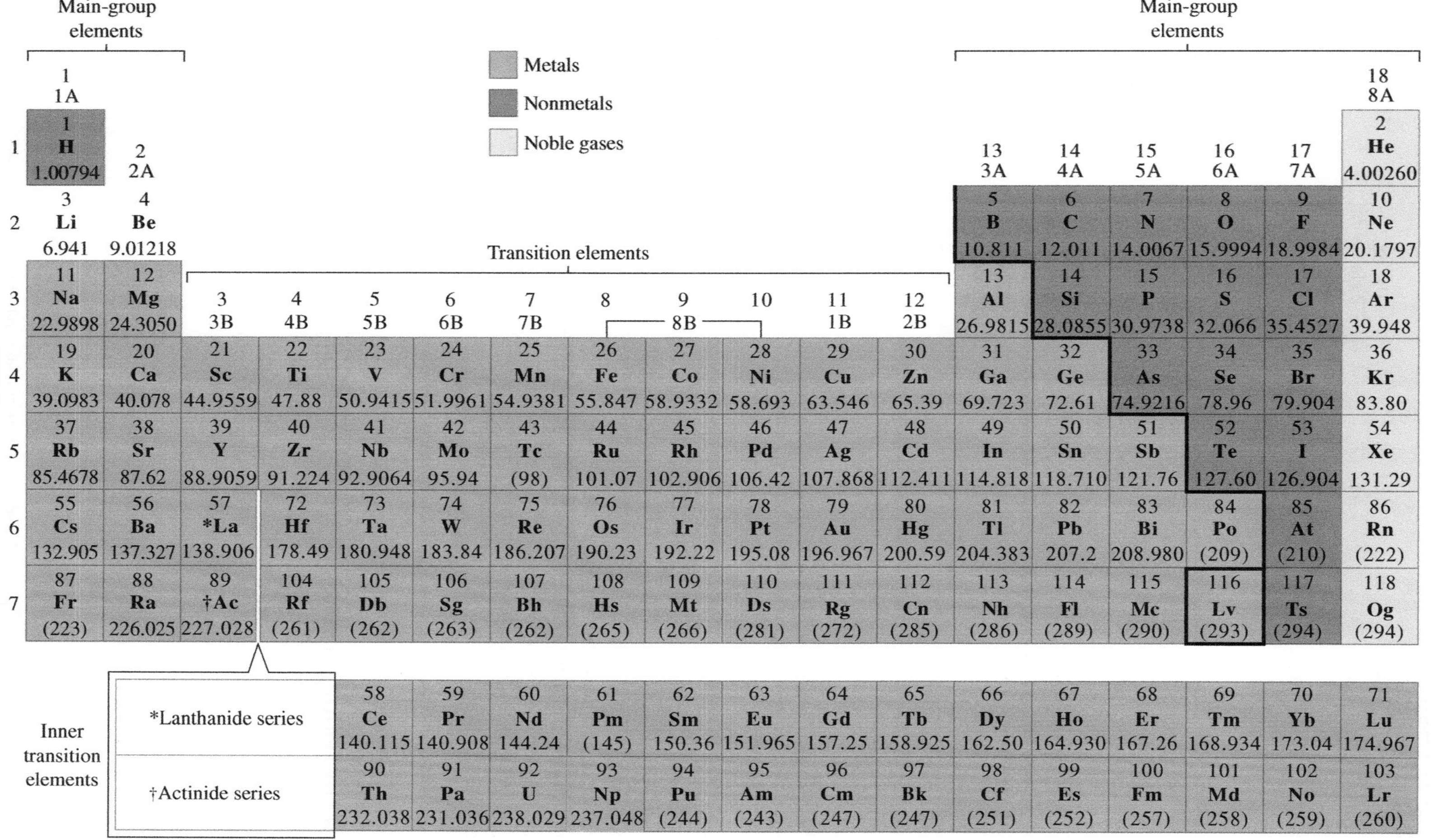

Main-group elements | Transition elements | Main-group elements

Metals | Nonmetals | Noble gases

| Period | 1<br>1A | 2<br>2A | 3<br>3B | 4<br>4B | 5<br>5B | 6<br>6B | 7<br>7B | 8<br>8B | 9<br>8B | 10<br>8B | 11<br>1B | 12<br>2B | 13<br>3A | 14<br>4A | 15<br>5A | 16<br>6A | 17<br>7A | 18<br>8A |
|---|---|---|---|---|---|---|---|---|---|---|---|---|---|---|---|---|---|---|
| 1 | 1 **H** 1.00794 | | | | | | | | | | | | | | | | | 2 **He** 4.00260 |
| 2 | 3 **Li** 6.941 | 4 **Be** 9.01218 | | | | | | | | | | | 5 **B** 10.811 | 6 **C** 12.011 | 7 **N** 14.0067 | 8 **O** 15.9994 | 9 **F** 18.9984 | 10 **Ne** 20.1797 |
| 3 | 11 **Na** 22.9898 | 12 **Mg** 24.3050 | | | | | | | | | | | 13 **Al** 26.9815 | 14 **Si** 28.0855 | 15 **P** 30.9738 | 16 **S** 32.066 | 17 **Cl** 35.4527 | 18 **Ar** 39.948 |
| 4 | 19 **K** 39.0983 | 20 **Ca** 40.078 | 21 **Sc** 44.9559 | 22 **Ti** 47.88 | 23 **V** 50.9415 | 24 **Cr** 51.9961 | 25 **Mn** 54.9381 | 26 **Fe** 55.847 | 27 **Co** 58.9332 | 28 **Ni** 58.693 | 29 **Cu** 63.546 | 30 **Zn** 65.39 | 31 **Ga** 69.723 | 32 **Ge** 72.61 | 33 **As** 74.9216 | 34 **Se** 78.96 | 35 **Br** 79.904 | 36 **Kr** 83.80 |
| 5 | 37 **Rb** 85.4678 | 38 **Sr** 87.62 | 39 **Y** 88.9059 | 40 **Zr** 91.224 | 41 **Nb** 92.9064 | 42 **Mo** 95.94 | 43 **Tc** (98) | 44 **Ru** 101.07 | 45 **Rh** 102.906 | 46 **Pd** 106.42 | 47 **Ag** 107.868 | 48 **Cd** 112.411 | 49 **In** 114.818 | 50 **Sn** 118.710 | 51 **Sb** 121.76 | 52 **Te** 127.60 | 53 **I** 126.904 | 54 **Xe** 131.29 |
| 6 | 55 **Cs** 132.905 | 56 **Ba** 137.327 | 57 ***La** 138.906 | 72 **Hf** 178.49 | 73 **Ta** 180.948 | 74 **W** 183.84 | 75 **Re** 186.207 | 76 **Os** 190.23 | 77 **Ir** 192.22 | 78 **Pt** 195.08 | 79 **Au** 196.967 | 80 **Hg** 200.59 | 81 **Tl** 204.383 | 82 **Pb** 207.2 | 83 **Bi** 208.980 | 84 **Po** (209) | 85 **At** (210) | 86 **Rn** (222) |
| 7 | 87 **Fr** (223) | 88 **Ra** 226.025 | 89 **†Ac** 227.028 | 104 **Rf** (261) | 105 **Db** (262) | 106 **Sg** (263) | 107 **Bh** (262) | 108 **Hs** (265) | 109 **Mt** (266) | 110 **Ds** (281) | 111 **Rg** (272) | 112 **Cn** (285) | 113 **Nh** (286) | 114 **Fl** (289) | 115 **Mc** (290) | 116 **Lv** (293) | 117 **Ts** (294) | 118 **Og** (294) |

| Inner transition elements | | | | | | | | | | | | | | | |
|---|---|---|---|---|---|---|---|---|---|---|---|---|---|---|---|
| *Lanthanide series | 58 **Ce** 140.115 | 59 **Pr** 140.908 | 60 **Nd** 144.24 | 61 **Pm** (145) | 62 **Sm** 150.36 | 63 **Eu** 151.965 | 64 **Gd** 157.25 | 65 **Tb** 158.925 | 66 **Dy** 162.50 | 67 **Ho** 164.930 | 68 **Er** 167.26 | 69 **Tm** 168.934 | 70 **Yb** 173.04 | 71 **Lu** 174.967 |
| †Actinide series | 90 **Th** 232.038 | 91 **Pa** 231.036 | 92 **U** 238.029 | 93 **Np** 237.048 | 94 **Pu** (244) | 95 **Am** (243) | 96 **Cm** (247) | 97 **Bk** (247) | 98 **Cf** (251) | 99 **Es** (252) | 100 **Fm** (257) | 101 **Md** (258) | 102 **No** (259) | 103 **Lr** (260) |

Notes: (1) Values in parentheses are the mass numbers of the most common or most stable isotopes of radioactive elements. (2) Some elements adjacent to the stair-step line between the metals and nonmetals have a metallic appearance but some nonmetallic properties. These elements are often called metalloids or semimetals. There is no general agreement on just which elements are so designated. Almost every list includes Si, Ge, As, Sb, and Te. Some also include B, At, and/or Po.

▲ **FIGURE 28.9 The periodic table of elements** The elements are arranged in order of increasing atomic, or proton, number. Horizontal rows are called *periods*, and vertical columns are called *groups*. The elements in a group have similar chemical properties. Each atomic mass represents an average of that element's isotopes, weighted to reflect their relative abundance in our immediate environment. The masses have been rounded to two decimal places; more precise values are given in Appendices III and IV. (A value in parentheses represents the mass number of the best-known or longest-lived isotope of an unstable element.) See Appendix III for an alphabetical listing of elements.

| Shell (last to be filled) | Subshells | Number of electrons in subshell, $2(2\ell + 1)$ | Corresponding period in periodic table |
|---|---|---|---|
| $n = 7$ | $7p$ | 6 | Period 7 (32 elements) |
| | $6d$ | 10 | |
| | $5f$ | 14 | |
| | $7s$ | 2 | |
| $n = 6$ | $6p$ | 6 | Period 6 (32 elements) |
| | $5d$ | 10 | |
| | $4f$ | 14 | |
| | $6s$ | 2 | |
| $n = 5$ | $5p$ | 6 | Period 5 (18 elements) |
| | $4d$ | 10 | |
| | $5s$ | 2 | |
| $n = 4$ | $4p$ | 6 | Period 4 (18 elements) |
| | $3d$ | 10 | |
| | $4s$ | 2 | |
| $n = 3$ | $3p$ | 6 | Period 3 (8 elements) |
| | $3s$ | 2 | |
| $n = 2$ | $2p$ | 6 | Period 2 (8 elements) |
| | $2s$ | 2 | |
| $n = 1$ | $1s$ | 2 | Period 1 (2 elements) |

Energy →

▲ **FIGURE 28.10** **Electron periods** The periods of the periodic table are related to electron configurations. The last $n$ shell to be filled is equal to the period number. The electron periods and the corresponding periods of the table are defined by relatively large energy gaps between successive subshells (such as that between $4s$ and $3p$) of the atoms.

For example, notice the first two groups at the left of the table. They have one and two outermost electrons in an $s$ subshell, respectively. These elements are all highly reactive metals that form compounds that have many similarities. The group at the far right, the noble gases, includes elements with completely filled subshells (and thus a full shell). These elements are at the ends of electron periods, or just before a large energy gap. These gases are nonreactive and can form compounds (by chemical bonding) only under very special conditions.

### CONCEPTUAL EXAMPLE 28.4: COMBINING ATOMS – PERFORMING CHEMISTRY WITH THE PERIODIC TABLE

Combinations of atoms, or molecules, can form if atoms come together and share outer electrons. This sharing process is called *covalent bonding*. In this "shared-custody" scheme, both atoms find it energetically beneficial (i.e., they lower their combined total energies) to have the equivalent of a filled outer shell of electrons, if only on a part-time basis. Using your knowledge of electron shells and the periodic table, determine which of the following atoms would most likely form a covalent arrangement with oxygen: (a) neon (Ne), (b) calcium (Ca), or (c) hydrogen (H).

**REASONING AND ANSWER.** Choice (a), neon, with a total of ten electrons, can be eliminated immediately, because it has a full outer shell of eight electrons and, as such, has nothing to be gained by losing or adding electrons. Looking at the periodic table, it can be seen that oxygen, with six outer-shell electrons, is two electrons shy of having a full complement of eight electrons. Choice (b), calcium, with its twenty electrons, is two electrons beyond the previous full shell of ten. You might think, therefore, that calcium is a possible covalent partner. However, you must remember that the covalent arrangement is a two-way street. In other words, the arrangement would also require calcium sometimes to have two more electrons than normal.

This situation would put the calcium atom in the awkward position of being four electrons beyond the full shell of ten and fourteen electrons away from the next complete shell. Hence, even though this attempt at covalent bonding might seem to work for oxygen, it certainly won't work well for calcium. Thus, answer (b), calcium, is not correct.*

The remaining candidate, (c) hydrogen, has one electron fewer than a full shell of two. Thus, if a hydrogen atom could add one electron, it would attain an electron configuration like that of the lightest inert gas, helium. Since each hydrogen needs to share only one electron, two of them can accomplish this by sharing with a single oxygen atom. Part of the time, the hydrogen atoms must share their electrons with the oxygen atom in order to create the latter's full outer shell of eight electrons. So, the correct answer is (c) – two hydrogen atoms covalently bound to a single oxygen atom. This combination has the molecular formula $H_2O$ – water.

* The two species can, however, combine to form calcium oxide (CaO) by a different method. The bonding that keeps calcium oxide together is based on the electrical attraction between a positive calcium ion $Ca^{+2}$, and a negative oxygen ion $O^{-2}$ after the two *permanently* exchange two electrons, making each a doubly charged ion of the opposite sign. This bond is called an *ionic bond*.

**FOLLOW-UP EXERCISE.** In this Example, (a) what would be the electron configuration of the oxygen in the water molecule at some instant when it has "custody" of the electrons from both hydrogen atoms? (b) What would be the net charge on the oxygen in this case?

## 28.4 The Heisenberg Uncertainty Principle

An important aspect of quantum mechanics has to do with measurement and accuracy. In classical mechanics, there is no limit to the accuracy of a measurement. For example, if both the position and the velocity of an object are known *exactly* at a particular time, you can determine *exactly* where the object will be in the future and where it was in the past (assuming the forces that act on it are known).

However, quantum theory predicts otherwise and sets limits on the accuracy of measurements. This idea was introduced in 1927 by the German physicist Werner Heisenberg (1901–1976), who developed another approach to quantum mechanics that complemented Schrödinger's wave theory. The **Heisenberg uncertainty principle**, as applied to position and momentum, states:

> It is impossible to know simultaneously an object's exact position and momentum.

This concept can be illustrated with a thought experiment. Suppose that you want to measure the position and momentum (i.e., velocity) of an electron. In order to "see," or locate, the electron, at least one photon must bounce off it and arrive at an eye (or, more likely, a detector), as in ▶ **Figure 28.11.** However, in the collision process, some of the photon's energy and momentum are transferred to that electron and it recoils. Thus, in the very process of locating the electron's position accurately, uncertainty is introduced into the electron's velocity (or momentum, because here $\Delta\vec{\mathbf{p}} = m\Delta\vec{\mathbf{v}}$). This effect isn't noticed in our everyday macroscopic world, because the recoil produced by viewing an object with light is negligible. That is, the force exerted by the light cannot appreciably alter the motion or position of an object of everyday mass.

According to wave optics, the position of an electron can be measured *at best* to an uncertainty $\Delta x$ that is on the order of the wavelength of $\lambda$ the light used – that is, $\Delta x \approx \lambda$. The photon used to do this location has a momentum of $p = h/\lambda$. Because the amount of momentum transferred during collision isn't determined, the final momentum of the electron would have an uncertainty on the order of the momentum of the photon $\Delta p \approx h/\lambda$. Notice that the product of these two uncertainties is *at least* as large as $h$, since

$$(\Delta p)(\Delta x) \approx \left(\frac{h}{\lambda}\right)(\lambda) = h$$

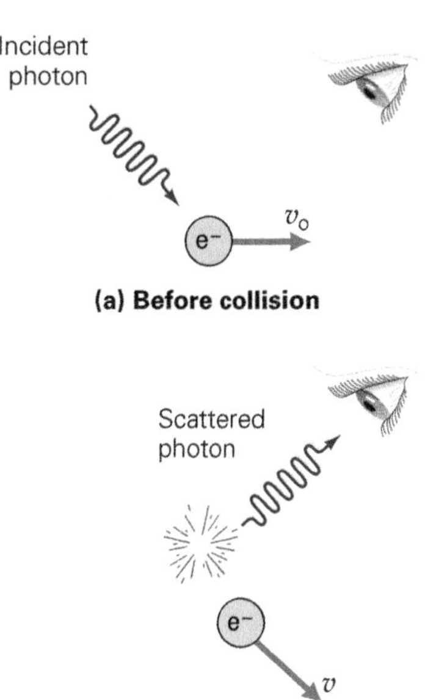


▲ FIGURE 28.11 **Measurement-induced uncertainty (a)** To measure the position and momentum (or velocity) of an electron, at least one photon must collide with the electron and be scattered toward the eye or detector. **(b)** In the collision process, energy and momentum are transferred to the electron, which induces uncertainty in its velocity.

This equation relates the minimum uncertainties, or maximum accuracies, of simultaneous measurements of the momentum and position. In actuality, the uncertainties could be worse, depending on the apparatus and technique used. Using more detailed considerations, Heisenberg found that the product of the two uncertainties would equal, *at a minimum*, $h/(2\pi)$. However, it could be higher. Hence,

$$(\Delta p)(\Delta x) \geq \frac{h}{2\pi} \tag{28.5}$$

(Heisenberg Uncertainly Principle)

That is, the product of the minimum uncertainties of simultaneous momentum and position measurements is on the order of Planck's constant ($h$) divided by $2\pi$, or about $10^{-34}$ J·s.

To locate the position of a particle accurately (i.e., make $\Delta x$ as small as possible), a photon with a very short wavelength must be used. However, this type of photon carries a lot of momentum, which results in an increased momentum uncertainty. To take the extreme case, if the location of a particle could be measured exactly (i.e., $\Delta x \to 0$), there would be no idea about the value of its momentum ($\Delta p \to \infty$). Thus, the measurement process itself limits the accuracy to which position and momentum can be measured simultaneously. In Heisenberg's words, "Since the measuring device has been constructed by the observer ... we have to remember that what we observe is not nature in itself but nature exposed to our method of questioning." To see how the Heisenberg uncertainty principle affects the microscopic and macroscopic worlds, consider Integrated Example 28.5.

**INTEGRATED EXAMPLE 28.5: AN ELECTRON VERSUS A BULLET – THE UNCERTAINTY PRINCIPLE**

Suppose an electron and a bullet have the same speed, measured to the same accuracy. (a) How would their minimum uncertainties in position compare: (1) the electron's location would be more uncertain than that of the bullet; (2) the bullet's location would be more uncertain than that of the electron; or (3) their location uncertainties would be the same? (b) If the bullet's mass is 20.0 g and both the bullet and the electron have a speed of 300 m/s with an uncertainty of $\pm 0.010\%$, determine the minimum uncertainty in the position of each.

(a) **CONCEPTUAL REASONING.** The minimum uncertainty in location is related to the minimum uncertainty in momentum. Since both have the same uncertainty in speed, the bullet's momentum is much more uncertain (since $\Delta p = m\Delta v$). Therefore, the bullet's location will be much less uncertain (i.e., more accurate) than that of the electron, so the answer is (1).

(b) **QUANTITATIVE REASONING AND SOLUTION.** The uncertainty principle (Equation 28.5) can be solved for $\Delta x$ in each case, because $\Delta p$ can be determined from the uncertainty in speed, $\Delta v$, and mass. Listing the quantities, including the known electron mass,

***Given:***

$m_e = 9.11 \times 10^{-31}$ kg
$m_b = 20$ g $= 0.020$ kg
$v_b = v_e = 300$ m/s $\pm$ 0.01%

***Find:*** $\Delta x_e$ and $\Delta x_b$ (minimum uncertainties in positions)

The uncertainty in speed is $\pm 0.01\%$ (or, as a decimal, $\pm 0.00010$) of the speed for both. Numerically this is $\pm(300 \text{ m/s})(0.00010) = \pm 0.030$ m/s. Thus for both, $v = 300 \pm 0.030$ m/s. The total uncertainty in speed is twice this amount, because the measurements can be off both above and below the measured values; hence for both: $\Delta v = 0.060$ m/s.

So for the electron, the minimum uncertainty in position is

$$\Delta x_e = \frac{h}{2\pi\Delta p} = \frac{h}{2\pi m_e \Delta v} = \frac{6.63\times10^{-34}\,\text{J·s}}{2\pi(9.11\times10^{-31}\,\text{kg})(0.060\,\text{m/s})}$$
$$= 0.0019\,\text{m} = 1.9\,\text{mm}$$

Similarly, the bullet's minimum uncertainty in position is

$$\Delta x_b = \frac{h}{2\pi m_b \Delta v} = \frac{6.63\times10^{-34}\,\text{J·s}}{2\pi(0.020\,\text{kg})(0.060\,\text{m/s})} = 8.8\times10^{-32}\,\text{m}$$

The uncertainty in the bullet's position is much, much smaller than the diameter of a nucleus! And its position uncertainty is many orders of magnitude less than that of the electron. The lesson here is that uncertainty in location for everyday objects traveling at ordinary speeds is negligible. However, for electrons, 1.9 mm is significant and measurable.

**FOLLOW-UP EXERCISE.** In this Example, what would the minimum uncertainty in the electron's speed have to be for the minimum uncertainty in its position to be on the order of atomic dimensions, or 0.10 nm?

An equivalent form of the uncertainty principle relates uncertainties in energy and time. As with position and momentum, a detailed analysis shows that, *at best*, the product of the uncertainties in energy and time is $h/(2\pi)$. Since it could be larger, it is written as

$$(\Delta E)(\Delta t) \geq \frac{h}{2\pi} \tag{28.6}$$

(alternative Heisbenberg Uncertainty Principle)

This form of the uncertainty principle shows that the energy of an object may be uncertain by an amount $\Delta E$, depending on the time taken to measure it, $\Delta t$. For longer times, the energy measurement becomes increasingly more accurate. Once again, these uncertainties are important only for very light objects. This version of the uncertainty principle is of particular importance in nuclear physics and elementary particle interactions (Chapters 29 and 30).

## 28.5 Particles and Antiparticles

In 1928, when the British physicist Paul Dirac (1902–1984), extended quantum mechanics to include relativistic considerations, something new and very different was predicted – a particle called the **positron**. The positron was predicted to have the same mass as the electron, but to carry a *positive* charge. The oppositely charged positron is the **antiparticle** of the electron. (Antiparticles will be discussed in more depth in Chapters 29 and 30.)

The positron was first observed experimentally in 1932 by the American physicist C. D. Anderson in cloud chamber experiments with cosmic rays. The curvature of the particle tracks in a magnetic field showed two types of particles, with opposite charge and the same mass. Anderson had, in fact, discovered the positron.

Because electric charge is conserved, a positron can be created only with the simultaneous creation of an electron (so that the net charge remains zero). This process is called **pair production**. In Anderson's experiment, positrons were observed to be emitted from a thin lead plate exposed to cosmic rays from outer space, which contain highly energetic X-rays. Pair production occurs when an X-ray photon nears a nucleus – the nuclei of the lead atoms in the Anderson experiment. In this process, the photon goes out of existence, and an *electron–positron pair* (an electron and a positron) is created, as illustrated in ▼**Figure 28.12**. This

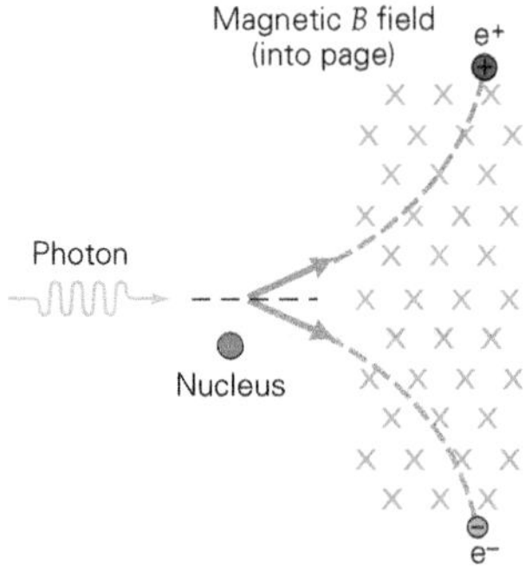


▲**FIGURE 28.12 Pair production** An electron ($e^-$) and a positron ($e^+$) can be created when an energetic photon passes near a heavy nucleus.

result represents a direct conversion of electromagnetic (photon) energy into mass. By the conservation of energy (the recoil energy of the massive nearby nucleus is usually negligible),

$$hf = 2m_ec^2 + K_{e^-} + K_{e^+}$$

where $hf$ is energy of the photon, $2m_ec^2$ is the total mass–energy equivalent of the electron–positron pair, and the $K$s represent the kinetic energies of the produced particles.

The minimum energy $E_{min}$ to produce such a pair occurs when they are produced at rest – when $K_{e^-}$ and $K_{e^+}$ are zero. Thus

$$E_{min} = hf = 2m_ec^2 = 1.022\,\text{MeV} \tag{28.7}$$

(minimum electron-positron production energy)

(Recall from Section 26.5 that the rest energy of an electron is 0.511 MeV.) This minimum energy is called the **threshold energy** for pair production.

But if they are created by cosmic rays, why aren't positrons commonly found in nature? This is because, almost immediately after their creation, positrons go out of existence by a process called **pair annihilation**. When an energetic positron appears, it loses kinetic energy by collision as it passes through matter. Finally, almost at rest, it combines with an electron and forms a hydrogen-like atom, called a *positronium atom*, in which they circle each other about their common center of mass. The positronium atom is unstable and quickly decays into two photons, each with an energy of 0.511 MeV (▼ **Figure 28.13**). Pair annihilation is then a direct conversion of mass energy back into electromagnetic energy – the inverse of pair production, so to speak. Pair annihilation is the basis for a medical diagnostic tool called a *positron emission tomography* (or PET) *scan*.

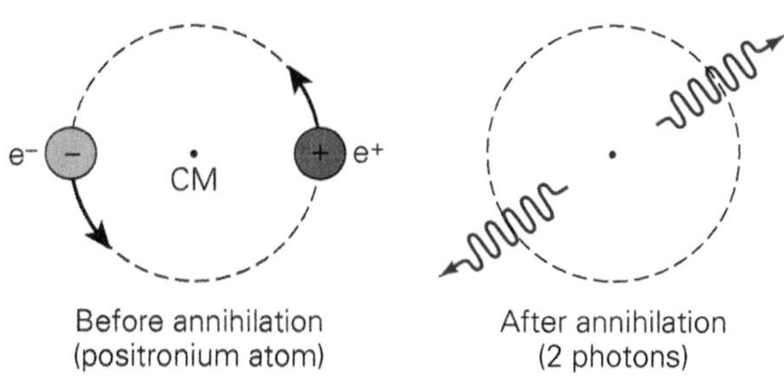


▲ **FIGURE 28.13 Pair annihilation** A positron and electron can form a positronium atom. The disappearance of a positronium atom is signaled by the appearance of two photons, each with an energy of 0.511 MeV. (Why would we not expect one photon with an energy of 1.022 MeV?) (In the left figure, CM stands for center of mass of the positronium.)

More generally, all particles have antiparticles. For example, there is an antiproton with the same mass as a proton, but with a negative charge. Even a neutral particle such as the neutron has an antiparticle – the antineutron. It is even conceivable that antiparticles predominate in some parts of the universe. If so, atoms made of the **antimatter** in these regions would consist of negatively charged nuclei composed of antiprotons and antineutrons, surrounded by orbiting positrons (antielectrons). It would be difficult to distinguish a region of antimatter from regular matter, since their physical behaviors would presumably be the same.

# Chapter 28 Review

- The momentum ($p$) of a photon carrying energy $E$ is inversely related to the wavelength of its associated electromagnetic wave by

$$p = \frac{E}{c} = \frac{hf}{c} = \frac{h}{\lambda} \tag{28.1}$$

- The **de Broglie hypothesis** assigns a wavelength to material particles. The **de Broglie wavelength** of a particle is

$$\lambda = \frac{h}{p} = \frac{h}{mv} \tag{28.2}$$

- A quantum mechanical **wave function** $\psi$ is a "probability wave" associated with a particle. The probability density, the square of the wave function, gives the relative probability of finding a particle at a particular location.
- Electron orbital energies are determined primarily by the **principal quantum number *n*.**
- The quantum number $\ell$ is the **orbital quantum number** and is associated with the orbital angular momentum of the orbit. For each value of $n$, the $\ell$ quantum number has one of $n$ possible integer values from zero up to a maximum value of $n - 1$.
- The quantum number $m_\ell$ is the **magnetic quantum number** and is associated with the $z$-component of the orbital angular momentum. For a given $\ell$, there are $2\ell + 1$ possible $m_\ell$ values (integers), given by $m_\ell = 0, \pm 1, 2, \ldots \pm \ell$.
- The quantum number that describes an electron's intrinsic angular momentum is the **spin quantum number $m_s$**, which, for electrons, protons, and neutrons, can have only two values: $m_s = \pm 1/2$. These values correspond to the two spin directions: up and down.
- Orbits with different quantum numbers that have the same energy are **degenerate**.
- Orbits that share a common principal quantum number $n$ are in the same **shell**.
- Within a shell, orbits that have a common orbital quantum number $\ell$ are in the same **subshell**.
- The **Pauli exclusion principle** states that in a given atom, no two electrons can have exactly the same set of four quantum numbers.

- The **Heisenberg uncertainty principle** states that you cannot simultaneously measure both the position and the momentum of a particle exactly. The same condition holds for a particle's energy and the time interval during which that energy is measured. The uncertainties are

$$(\Delta p)(\Delta x) \geq \frac{h}{2\pi} \quad (28.5)$$

and

$$(\Delta E)(\Delta t) \geq \frac{h}{2\pi} \quad (28.6)$$

- **Pair production** refers to the creation of a particle and its **antiparticle** by a photon. The reverse process is **pair annihilation**, in which a particle and its antiparticle annihilate and their mass–energy is converted into two photons.

# End of Chapter Questions and Exercises

## Multiple Choice Questions

### 28.1 Matter Waves: The de Broglie Hypothesis

1. Which color light consists of photons with the least momentum: (a) red, (b) green, (c) violet, or (d) yellow?
2. If the following were traveling at the same speed, which would have the longest de Broglie wavelength: (a) an electron, (b) a proton, (c) a carbon atom, or (d) a hockey puck?
3. An electron can travel at different speeds. Which of the following speeds (all in m/s) would be associated with the shortest de Broglie wavelength: (a) $10^3$, (b) $10^4$, (c) $10^2$, or (d) the wavelength is the same at all speeds?

### 28.2 The Schrödinger Wave Equation

4. Schrödinger's wave equation involves a particle's (a) kinetic energy, (b) potential energy, (c) total energy, (d) all of the preceding.
5. The square of a particle's wave function is related to (a) the energy of the particle, (b) the radius of the particle, (c) the probability of locating the particle, (d) the quantum number of its state.
6. An electron can tunnel through a "potential energy barrier." Classically if it were in the barrier region, its kinetic energy would be (a) negative, (b) zero, (c) positive, or (d) zero.

### 28.3 Atomic Quantum Numbers and the Periodic Table

7. The principle quantum number $n$ is associated with (a) the total energy of a state, (b) the orbital angular momentum of an orbit, (c) the $z$-component of the orbital angular momentum, (d) the electron's spin angular momentum.
8. The orbital quantum number $\ell$ determines (a) the total energy of a state, (b) the $z$-component of the orbital angular momentum, (c) the orbital angular momentum of an orbit, (d) the electron's spin angular momentum.
9. The magnetic quantum number $m_\ell$ tells (a) the $z$-component of an orbit's angular momentum, (b) an orbit's angular momentum, (c) the total energy of a state, (d) the electron's spin angular momentum.
10. For an electron, the number of values the spin quantum number $m_s$ can have is (a) one, (b) two, (c) three, (d) four.

### 28.4 The Heisenberg Uncertainty Principle

11. If the uncertainty in the position of a moving particle increases, (a) the particle may be located more exactly, (b) the uncertainty in its momentum decreases, (c) the uncertainty in its speed increases, (d) the time to measure the position increases.
12. If the speed of an electron could be exactly measured, that would mean that (a) the particle could be located more exactly, (b) the particle could not be located at all, (c) the uncertainty in its momentum increases, (d) the time to measure the position decreases.
13. According to the uncertainty principle, if a measurement time interval is done more quickly, the uncertainly in an energy measurement is (a) relatively large, (b) relatively small, (c) exactly the same, or (d) none of the preceding.

### 28.5 Particles and Antiparticles

14. Pair production involves (a) the production of two electrons, (b) the production of two positrons, (c) a hydrogen atom, (d) the production of a particle and its antiparticle.
15. Due to momentum considerations, pair annihilation could not result in the emission of how many photons: (a) one, (b) two, (c) three, or (d) four?
16. When a stationary particle and its stationary antiparticle annihilate, a total energy of 25 MeV is released. How do the masses of the particles compare to that of an electron: (a) they are less massive than an electron, (b) they are more massive than an electron, (c) they have the same mass as an electron?

## Conceptual Questions

### 28.1 Matter Waves: The de Broglie Hypothesis

1. The de Broglie hypothesis predicts that a wave is associated with any object that has momentum. Why isn't the wave associated with a moving car observed?
2. If a baseball and a bowling ball were traveling at the same speed, which one would have a shorter de Broglie wavelength? Why?
3. An electron is accelerated from rest through an electric potential difference. Will increasing the potential difference result in a longer or shorter de Broglie wavelength? Explain.

### 28.2 The Schrödinger Wave Equation

4. According to modern quantum theory and the Schrödinger equation, there is a probability that if you ran into a wall you could end up on the other side. (Don't try this!) Explain the idea behind such an event and discuss why it has never been observed to happen.

5. How would the radius for the maximum probability in Figure 28.3a change if the charge on the proton in the hydrogen atom were suddenly decreased? Explain your reasoning.

**28.3 Atomic Quantum Numbers and the Periodic Table**

6. According to the Pauli exclusion principle, can two electrons in an atom have the same spin? Explain.
7. What is the basis of the periodic table of elements in terms of quantum theory? What do the elements in a particular group have in common? How about in a particular period?

**28.4 The Heisenberg Uncertainty Principle**

8. Why is it impossible to simultaneously and accurately measure the position and speed of a particle?
9. A bowling ball has well-defined position and speed, whereas an electron does not. Why?
10. A laser requires a metastable state that is a relatively long-lived excited atomic state. What is the uncertainty in the energy of the metastable state compared to non-metastable excited states?

**28.5 Particles and Antiparticles**

11. Why is the energy threshold for electron–positron pair production actually higher than the sum of their two masses (1.022 MeV)? [*Hint*: Linear momentum must be conserved.]
12. Can the production of two electrons and two positrons be accomplished using a high-energy photon? Explain. Why can't two electrons and one positron result?
13. Explain why there are always two photons created in pair annihilation. Why can't the process create just one photon?

## Exercises*

*Integrated Exercises (IEs) are two-part exercises. The first part typically requires a conceptual answer choice based on physical thinking and basic principles. The following part is quantitative calculations associated with the conceptual choice made in the first part of the exercise.*

**28.1 Matter Waves: The de Broglie Hypothesis**

1. • What is the de Broglie wavelength associated with a 1000-kg car moving at 25 m/s?
2. • If the de Broglie wavelength associated with an electron is $7.50 \times 10^{-7}$ m, what is its speed?
3. IE • An electron and a proton are moving with the same speed. (a) Compared with the proton, will the electron have (1) a shorter, (2) an equal, or (3) a longer de Broglie wavelength? Why? (b) If the speed of the electron and proton is 100 m/s, what are their de Broglie wavelengths?
4. •• An electron is accelerated from rest through a potential difference of 100 V. What is the de Broglie wavelength of the electron?
5. •• An electron is accelerated from rest through a potential difference so that its de Broglie wavelength is 0.010 nm. What is the potential difference?
6. IE •• Electrons are accelerated from rest through an electric potential difference. (a) If this potential difference increases to four times the original value, the new de Broglie wavelength will be (1) four times, (2) twice, (3) one-fourth, (4) one-half that of the original. Why? (b) If the original potential is 250 kV and the new potential is 600 kV, what is the ratio of the new de Broglie wavelength to the original?
7. IE •• A proton traveling at a speed $4.5 \times 10^4$ m/s is then accelerated through a potential difference of 37 V. (a) Will its de Broglie wavelength (1) increase, (2) remain the same, or (3) decrease, due to the potential difference? Why? (b) By what percentage does the de Broglie wavelength of the proton change?
8. •• A proton and an electron are accelerated from rest through the same potential difference. What is the ratio of the de Broglie wavelength of the electron to that of the proton?
9. •• A charged particle is accelerated through a given potential difference. If the voltage is doubled, what is the ratio of the new de Broglie wavelength to the original value?
10. •• A scientist wants to use an electron microscope to observe details on the order of 0.25 nm. Through what potential difference must the electrons be accelerated from rest so that they have a de Broglie wavelength of this magnitude?
11. •• What is the energy of a beam of electrons that exhibits a first-order maximum at an angle of 50° when diffracted by a crystal grating with a lattice plane spacing of 0.215 nm?
12. ••• According to the Bohr theory of the hydrogen atom, the speed of the electron in the first Bohr orbit is $2.19 \times 10^6$ m/s. (a) What is the wavelength of the matter wave associated with the electron? (b) How does this wavelength compare with the circumference of the first Bohr orbit?
13. ••• (a) What is the de Broglie wavelength of the Earth in its orbit about the Sun? (b) Treating the Earth as a de Broglie wave in a large "gravitational" atom, what would be the principal quantum number, *n*, of its orbit? (c) If the principal quantum number increased by 1, how would the radius of the orbit change? (Assume a circular orbit.)

**28.2 The Schrödinger Wave Equation**

14. • If the absolute value of the wave function of a proton is twice as large at location A than at location B, how many times is it more likely to find it at A than at B?
15. • If you are twice as likely to find an electron at a distance of 0.0400 nm than 0.0500 nm from the

* The bullets denote the degree of difficulty of the exercises: •, simple; ••, medium; and •••, more difficult.

nucleus, what is the ratio of the absolute value of its wave function at 0.0400 nm to that at 0.0500 nm?

16. •• A particle in box is constrained to move in one dimension, like a bead on a wire, as illustrated in ▼**Figure 28.14**. Assume that no forces act on the particle in the interval $0 < x < L$ and that it hits a perfectly rigid wall. The particle will exist only in states of certain kinetic energies that can be determined by analogy to a standing wave on a string (Section 13.5). This means that an integral number $n$ of half-wavelengths "fit" into the box's length $L$, or $n(\lambda_n/2) = L$, where $n = 1, 2, 3, \ldots$. Using this relationship, show that the "allowed" kinetic energies of the particle are given by $K_n = n^2[h^2/(8mL^2)]$, where $n = 1, 2, 3, \ldots$ and $m$ is the particle's mass. [*Hint*: Recall that kinetic energy is related to momentum by $K = p^2/(2m)$ and the de Broglie wavelength of the particle is related to its momentum.]

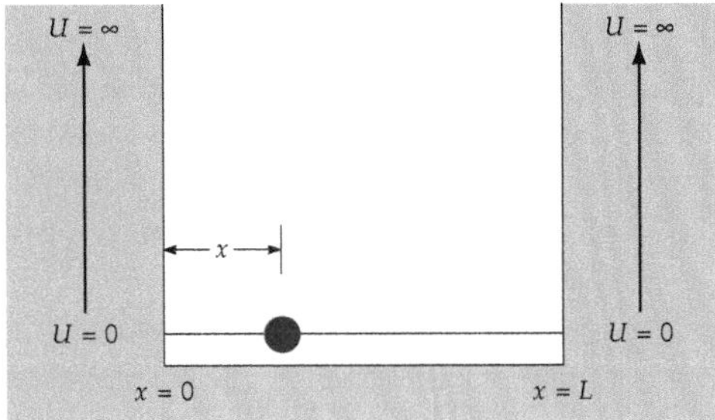


▲**FIGURE 28.14 Particle in a box** See Exercises 28.16 and 28.17.

17. •• Let's model a proton in a nucleus as a particle trapped in the one-dimensional box. Assume the particle is in a one-dimensional nucleus of length 7.11 fm (approximate diameter of a $^{208}$Pb nucleus). (a) Using the results of Exercise 28.16, find the energies of the proton in the ground state and first two excited states. (b) The nucleus is to absorb a photon of just the right energy to enable it to make an upward transition from the ground state to the second excited state. How much energy would this take and what type of photon would it be classified as? (Neglect recoil of the absorbing nucleus after the photon is absorbed.)

## 28.3 Atomic Quantum Numbers and the Periodic Table

18. • (a) How many possible sets of quantum numbers are there for the $n = 1$ and $n = 2$ shells? (b) Write the explicit values of all the quantum numbers ($n$, $\ell$, $m_\ell$, $m_s$) for these levels.

19. • How many possible sets of quantum numbers are there for the subshells with (a) $\ell = 2$ and (b) $\ell = 3$?

20. **IE** • (a) Which has more possible sets of quantum numbers associated with it, $n = 2$ or $\ell = 2$? (b) Prove your answer to part (a).

21. • An electron in an atom is in an orbit that has a magnetic quantum number of $m_\ell = 2$. What are the minimum values that (a) $\ell$ and (b) $n$ could be for that orbit?

22. •• Draw the ground state energy-level diagrams like those in Figure 28.8 for (a) nitrogen (N) and (b) potassium (K).

23. •• Draw schematic diagrams for the electrons in the subshells of (a) sodium (Na) and (b) argon (Ar) atoms in the ground state.

24. •• Identify the atoms of each of the following ground state electron configurations: (a) $1s^2\,2s^2$; (b) $1s^2\,2s^2\,2p^3$; (c) $1s^2\,2s^2\,2p^6$; (d) $1s^2\,2s^2\,2p^6\,3s^2\,3p^4$.

25. •• Write the ground state electron configurations for each of the following atoms: (a) boron (B), (b) calcium (Ca), (c) zinc (Zn), and (d) tin (Sn).

26. **IE** ••• (a) If there were no electron spin, the 1$s$ state would contain a maximum of (1) zero, (2) one, (3) two electrons. Why? (b) What would be the first two inert or noble gases if there were no electron spin?

27. ••• How would the electronic structure of lithium differ if electron spin were to have three possible orientations instead of just two?

## 28.4 The Heisenberg Uncertainty Principle

28. • A 1.0-kg ball has a position uncertainty of 0.20 m. What is its minimum momentum uncertainty?

29. **IE** • An electron and a proton each have a momentum of $3.28470 \times 10^{-30}$ kg·m/s $\pm$ $0.00025 \times 10^{-30}$ kg·m/s. (a) The minimum uncertainty in the position of the electron compared with that of the proton will be (1) larger, (2) the same, (3) smaller. Why? (b) Calculate the minimum uncertainty in the position for each.

30. •• What is the minimum uncertainty in the speed of an electron that is known to be somewhere between 0.050 nm and 0.10 nm on a horizontal axis?

31. •• What is the minimum uncertainty in the position of a 0.50-kg ball that is known to have a speed uncertainty of $3.0 \times 10^{-28}$ m/s?

32. •• The energy of a 2.00-keV electron is known to within ±3.00%. How accurately can its position be measured?

## 28.5 Particles and Antiparticles

33. • What is the energy of the photons produced in proton–antiproton pair annihilation, assuming that both particles are essentially at rest initially?

34. **IE** •• A muon, or $\mu$ meson, has the same charge as an electron, but is 207 times as massive. (a) Compared with electron–positron pair production, the pair production of a muon and an antimuon requires a photon of (1) more, (2) the same amount of, (3) less energy. Why? (b) What would be the minimum energy and frequency for such a photon?

35. **IE** •• (a) The minimum photon energy to create a proton–antiproton pair is (1) more than, (2) the same as, or (3) less than the minimum photon energy to create a neutron–antineutron pair. Explain. (b) Calculate minimum photon frequency to create the proton–antiproton pair and the neutron–antineutron pair (to four significant figures).

# 29

# The Nucleus



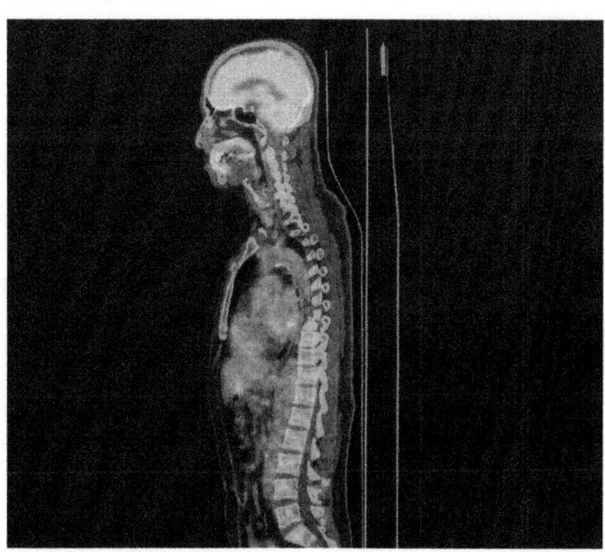

**This image was created by detecting radiation emitted from radioactive nuclei in the patient.**

The lateral view of the body of a person in the chapter-opening photograph was created by radiation from a radioactive source. *Radiation* and *radioactivity* are words that sometimes may sound menacing, but the beneficial uses of radiation are often overlooked. For instance, exposure to high-energy radiation can cause cancer – yet precisely the same sort of radiation, in relatively small doses, can be useful in the diagnosis and treatment of cancer.

The bone scan in this photograph was created by radiation released when unstable nuclei spontaneously decayed after being administered to and taken up by the body. This nuclear decay is more commonly known as radioactive decay. But what makes some nuclei stable, while others decay? What determines the rate at which they decay and the type of particle(s) they emit? These are some of the questions that will be explored in this chapter. The detection and measurement of radiation will also be discussed, as well as more about its dangers and uses.

The study of radioactivity and nuclear stability fundamentally concerns the nature of the nucleus, its structure, its energy, and how this energy can be released. Nuclear energy is one of our major energy sources and will be considered in Chapter 30. In this chapter the focus is on understanding the nucleus and its properties and characteristics.

## 29.1 Nuclear Structure and the Nuclear Force

Many experiments show that atoms contain electrons. Since an atom is typically electrically neutral, it must contain a positive charge equal in magnitude to the total charge of its electrons. Because the electron's mass is small compared with the mass of even the lightest of atoms, most of an atom's mass must be associated with that positive charge.

Based on these observations, J. J. Thomson (1856–1940), a British physicist who had proven the existence of the electron in 1897, proposed a model of the atom. In his model, the electrons are uniformly distributed within a continuous sphere of positive charge. It was called a "plum pudding" model, because the electrons in the positive charge are analogous to raisins in a plum pudding. The region of positive charge was assumed to have a radius on the order of $10^{-10}$ m, or 0.1 nm, roughly the diameter of an atom. As you probably know, the modern model of the atom is quite different. This model concentrates all the positive charge, and practically all of the mass, in a central *nucleus*, surrounded by orbiting electrons. This alternate atomic model was first proposed by British physicist Ernest Rutherford (1871–1937). Combining this idea with the Bohr theory of electron orbits (Section 27.4) led to the simplistic "solar system" model, or **Rutherford–Bohr model**, of the atom.

Rutherford's insight came from the results of alpha particle scattering experiments performed in his laboratory around 1911. An alpha ($\alpha$) particle is a doubly positively charged particle ($q_\alpha = +2e$) that is naturally emitted from some radioactive materials. (See Section 29.2.) A beam of these particles was directed at a thin gold foil "target," and the deflection angles and percentage of scattered particles were observed (▼ **Figure 29.1**).

An alpha particle is more than 7000 times as massive as an electron. Thus, the Thomson model predicts only tiny deflections for the alpha particles – the result of collisions with the light electrons as the alpha particle passes through such a model of a gold atom (▶ **Figure 29.2**). Surprisingly, however, Rutherford observed alpha particles scattered at appreciable angles. In about 1 in every 8000 scatterings, the alpha particles were actually

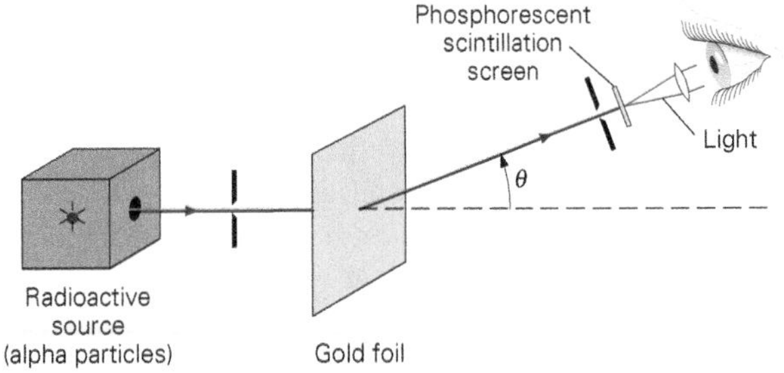


▲ FIGURE 29.1 **Rutherford's scattering experiment** A beam of alpha particles from a radioactive source was scattered by gold nuclei in a thin foil, and the scattering was observed as a function of the scattering angle $\theta$. The observer detects the light (viewed through a lens) given off by a phosphorescent scintillation screen.

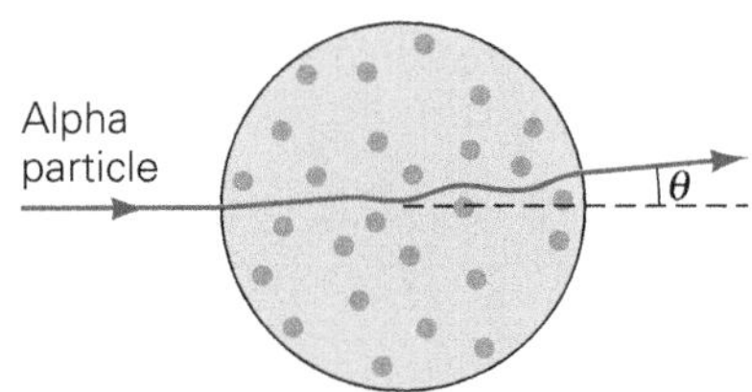


▲ FIGURE 29.2 **The plum pudding model** In Thomson's plum pudding model of the atom, massive alpha particles were expected to be only slightly deflected by collisions with the electrons (dots) in the atom. The experimental results were quite different.

*backscattered*; that is, they were scattered through angles greater than 90° (▶ **Figure 29.3**).

Calculations showed that the probability of backscattering in the Thomson model was minuscule – certainly much, much less than 1 in 8000. As Rutherford described the backscattering, "It was almost as incredible as if you had fired a 15-inch shell at a piece of tissue paper and it came back and hit you."

The experimental results led Rutherford to the concept of a nucleus: "On consideration, I realized that this scattering backward must be the result of a single collision, and when I made calculations I saw that it was impossible to get anything of that order of magnitude unless you took a system in which the greater part of the mass of the atom was concentrated in a minute nucleus. It was then that I had the idea of an atom with a minute massive center carrying a charge."

If all of the positive charge of a target atom were concentrated in a small region, then an alpha particle coming close to this region would experience a large deflecting electrical repulsion force. The mass of this positive "nucleus" would be larger than that of the alpha particle, and in this model backscattering is much more likely to occur than in the plum pudding model.

A simple estimate can give an idea of the approximate size of a nucleus. It is during a head-on collision that an alpha particle comes closest to the nucleus ($r_{\min}$ in Figure 29.3). That is, the alpha particle approaching the nucleus stops at $r_{\min}$ and is accelerated back along its original path. Assuming a spherical charge distribution, the electric potential energy of the alpha particle (symbol $\alpha$) and nucleus (symbol n) when separated by a center-to-center distance $r$ is $U_e = kq_\alpha q_n/r = k(2e)(Ze)/r$ (Equation 16.5). Here, $Z$ is the **atomic number**, or the number of protons in the nucleus. Therefore, the charge of the nucleus is given by $q_n = +Ze$. By conservation of energy, the kinetic energy of the incoming alpha particle is completely converted into electric potential energy at the turnaround point $r_{\min}$. Using $q_\alpha = +2e$, and equating the magnitudes of these two energies

$$\frac{1}{2}mv^2 = \frac{k(2e)Ze}{r_{\min}}$$

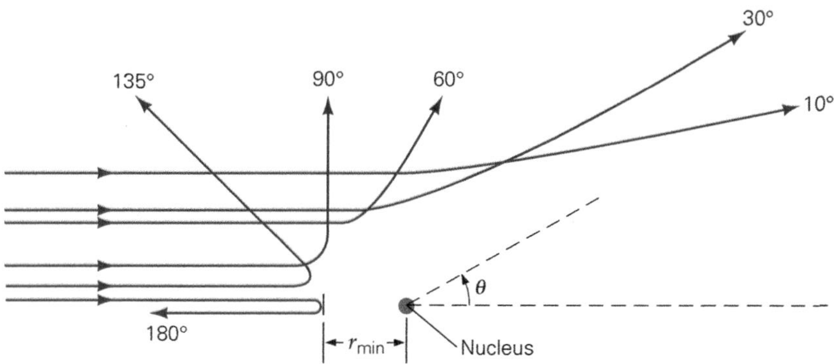


▲ FIGURE 29.3 **Rutherford scattering** A compact, dense atomic nucleus with a positive charge accounts for the observed scattering. An alpha particle in a head-on collision with the nucleus would be scattered directly backward ($\theta = 180°$) after coming within a distance $r_{min}$ of the nucleus. At this scale, the electron orbits (about the nucleus) are too far away to be seen.

or, solving for $r_{min}$

$$r_{min} = \frac{4kZe^2}{mv^2} \quad \text{(closest nuclear approach of } \alpha) \qquad (29.1)$$

In Rutherford's experiment, the kinetic energy of the alpha particles from his source had been measured, and Z was known to be 79 for gold. Using these values, along with the constants in Equation 29.1, Rutherford found $r_{min}$ to be on the order of $10^{-14}$ m, much, much smaller than the atomic radius, which is on the order of $10^{-10}$ m.

Although the nuclear model of the atom is useful, the nucleus is much more than a volume of positive charge. It is actually composed of two types of particles - protons and neutrons - collectively referred to as **nucleons**. The nucleus of the hydrogen atom is a single proton. Rutherford suggested that the hydrogen nucleus be named *proton* (from the Greek for "first") after he became convinced that no nucleus could be less massive than the hydrogen nucleus. A **neutron** is an electrically neutral particle with a mass slightly greater than that of a proton. The existence of the neutron was not experimentally verified until 1932.

## 29.1.1 The Nuclear Force

The forces in the nucleus certainly include the attractive gravitational force between nucleons. But in Chapter 15, this gravitational force was shown to be negligible compared with the repulsive electrical force between the positive protons. Taking only these repulsive forces into account, the nucleus should fly apart. Yet the nuclei of many atoms are stable. Therefore, there must be an attractive force between nucleons that overcomes the electrical repulsion and holds the nucleus together. This strong attractive force is called the **strong nuclear force**, or simply the *nuclear force*.

The exact expression for the nuclear force is complex. However, some general features of it are as follows:

- The nuclear force is strongly attractive and much larger in magnitude than both the electrostatic force and the gravitational force between nucleons.
- The nuclear force is very short-ranged; that is, a nucleon interacts only with its nearest neighbors, over distances on the order of $10^{-15}$ m.
- The nuclear force is independent of electric charge; that is, it results in the same force between *any* two nucleons - two protons, a proton and a neutron, or two neutrons.

Thus, nearby protons repel each other electrically but also attract each other and nearby neutrons by the strong force, with the latter force winning the battle. Having no electric charge, neutrons only attract nearby protons and neutrons.

## 29.1.2 Nuclear Notation

To describe the composition of nuclei, it is convenient to use the notation illustrated in ▼**Figure 29.4a**, which uses the chemical symbol of the element with subscripts and a superscript. The subscript on the left is called the *atomic number* (Z), which indicates the number of protons in the nucleus. A more descriptive

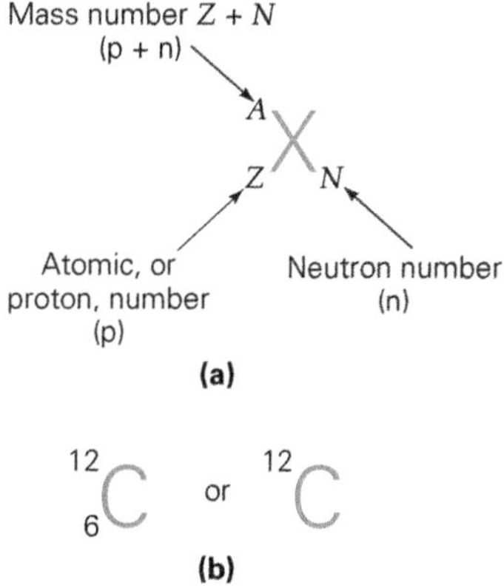


▲ FIGURE 29.4 **Nuclear notation (a)** The composition of a nucleus is shown by the chemical symbol of the element with the mass number *A* (sum of protons and neutrons) as a left superscript and the proton (atomic) number Z as a left subscript. The neutron number *N* may be shown as a right subscript, but both Z and *N* are routinely omitted, because the letter symbol tells you Z, and $N = A - Z$. **(b)** The two most common nuclear notations for a nucleus of one of the stable isotopes of carbon – carbon-12.

name is the **proton number Z**, which will be used in this book. For electrically neutral atoms, Z must be equal to the number of orbital electrons.

The number of protons in the nucleus of an atom determines the species of the atom – that is, the element to which the atom belongs. In Figure 29.4b, the proton number $Z = 6$ indicates that the nucleus belongs to a carbon atom. The proton number thus defines which chemical symbol is used. Electrons can be removed from (or added to) an atom to form an ion, but this does not change the atom's species. For example, a nitrogen atom with an electron removed, $N^+$, is still nitrogen – a nitrogen *ion*. It is the proton number, not the electron number, that determines the species of atom.

The superscript to the left of the chemical symbol is the **mass number (*A*)**, equal to the total number of protons and neutrons in the nucleus. Since protons and neutrons have roughly equal masses, the mass numbers of nuclei give a relative comparison of nuclear masses. For the carbon nucleus in Figure 29.4b, the mass number is $A = 12$, because there are six protons and six neutrons.

The number of neutrons, called the **neutron number (*N*)**, is sometimes indicated by a subscript on the right side of the chemical symbol. However, this subscript is usually omitted, because it can be determined from *A* and *Z*; that is, $N = A - Z$. Similarly, the proton number is routinely omitted, because the chemical symbol uniquely specifies the value of *Z*.

Even though the atoms of an element have the same number of protons in their nuclei, they may have different numbers of neutrons. For example, stable nuclei of different carbon atoms ($Z = 6$) may contain six or seven neutrons. In nuclear notation, these nuclei would be written as ${}^{12}_{6}C_6$ and ${}^{13}_{6}C_7$. Atoms whose nuclei have the same number of protons but different numbers of neutrons are called **isotopes**. The two listed here are the only two stable isotopes of carbon.

Isotopes are like members of a family. They all have the same *Z* number and the same surname (element name), but they are distinguishable by the number of neutrons in their nuclei, and therefore by their mass. Isotopes are commonly referred to by their mass numbers; for example, these isotopes of carbon are called *carbon-12* and *carbon-13*, and respectively. There are other isotopes of carbon that are unstable; for example, ${}^{14}C$. A particular nuclear species or isotope of any element is also called a **nuclide**. Generally, only a few isotopes of a given species are stable. But the number of stable isotopes can vary from none to several or more. For example, in the case of carbon, ${}^{14}C$ is unstable (although long-lived) and ${}^{12}C$ and ${}^{13}C$ are the only truly stable isotopes.

An important family of isotopes is that for hydrogen, which has three: ${}^{1}H$, ${}^{2}H$, and ${}^{3}H$. These isotopes are given special names. ${}^{1}H$ is called *ordinary hydrogen*, or simply *hydrogen*; ${}^{2}H$ is called *deuterium*. Deuterium, which is stable, is sometimes known as *heavy hydrogen*. It can combine with oxygen to form heavy water (written $D_2O$). The third isotope of hydrogen, ${}^{3}H$, called *tritium*, is unstable.

## 29.2 Radioactivity

Most elements have at least one stable isotope. Atoms with stable nuclides are the ones that are most familiar in our environment. However, some nuclei are unstable and disintegrate spontaneously (or decay), emitting particles and photons. Unstable isotopes are said to be *radioactive* or to exhibit **radioactivity**. For example, tritium (${}^{3}_{1}H$) has a radioactive nucleus. Of all the unstable nuclides, only a small number "live" long enough to still exist naturally. Many others can be produced artificially (Chapter 30).

Radioactivity is unaffected by normal physical or chemical processes, such as heat, pressure, and chemical reactions. Processes such as these do not affect the source of the radioactivity – the nucleus. Nor can nuclear instability be explained by a simple imbalance of attractive and repulsive forces within the nucleus. This is because, experimentally, nuclear disintegrations (of a given isotope) occur at a fixed rate. That is, the nuclei in a given sample do not all decay at the same time. According to classical theories, identical nuclei should decay at the same time. Therefore, radioactive decay suggests that the probability effects of quantum mechanics might be in play.

The discovery of radioactivity in 1896 is credited to the French scientist Henri Becquerel. While studying the fluorescence of a uranium compound, he discovered that a photographic plate near a sample had been darkened, even though the compound had not been activated by exposure to light and was not fluorescing. Apparently, this darkening was caused by some new type of radiation emitted from the compound. In 1898, Pierre and Marie Curie* announced the discovery of two radioactive elements, radium and polonium, which they had isolated from uranium pitchblende ore.

Experiment shows that there are three different kinds of radiation emitted by radioactive isotopes. When a radioactive isotope is placed in a chamber so that the emitted radiation passes through a magnetic field to a photographic plate, the various types of radiation expose the plate, producing characteristic spots by which the types of radiation may be identified (▶ **Figure 29.5**). The positions of the spots show that some isotopes emit radiation that is deflected to the left, some emit radiation that is deflected to the right, and some emit radiation that is undeflected. These spots are characteristic of what came to be known as *alpha*, *beta*, and *gamma* radiations.

From the opposite deflections of two of the types of radiation in the magnetic field, it is evident that positively charged particles are associated with alpha decay and that negatively charged particles are emitted during beta decay. Because of their much

* Marie Sklodowska Curie (1867–1934) was born in Poland and studied in France, where she met and married physicist Pierre Curie (1859–1906). In 1903, Madame Curie (as she is commonly known) and Pierre shared the Nobel Prize in physics with Henri Becquerel (1852–1908) for their work on radioactivity. She was also awarded the Nobel Prize in chemistry in 1911 for the discovery of radium.

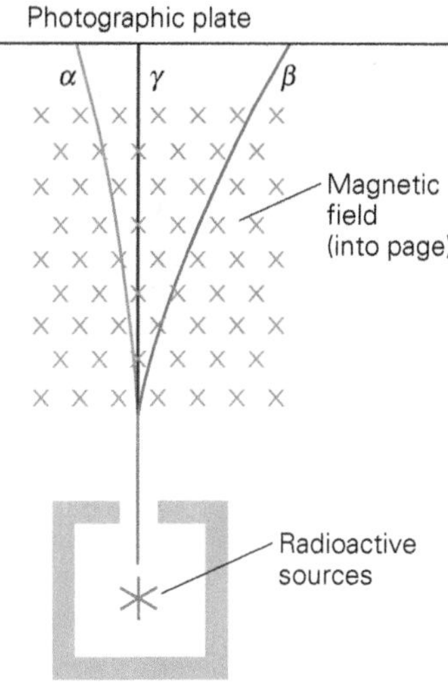


▲ **FIGURE 29.5 Nuclear radiation** Different types of radiation from radioactive sources can be distinguished by passing them through a magnetic field. Alpha and beta particles are deflected. From the magnetic force rule, alpha particles are positively charged and beta particles are negatively charged. The radii of curvature (not drawn to scale) allow the particles to be distinguished by mass. Gamma rays are not deflected and thus are uncharged.

smaller deflection, alpha particles must be considerably more massive than beta particles. The undeflected gamma radiation must be electrically neutral.

Detailed investigations of the three different radiation types revealed the following:

- **Alpha particles** are actually doubly charged ($q = +2e$) particles that contain two protons and two neutrons. They are identical to the nucleus of the helium atom ($^4_2$He).
- **Beta particles** are electrons (positive electrons, or *positrons*, were discovered later).
- **Gamma rays** are high-energy quanta of electromagnetic energy (photons).

For a few radioactive elements, two spots are found on the plate, indicating that the elements decay by two different modes. Let's look at some details of each of these decay modes.

### 29.2.1 Alpha Decay

When an alpha particle is emitted by a nucleus, that nucleus loses two protons and two neutrons, so the mass number *(A)* is decreased by four ($\Delta A = -4$) and the proton number (*Z*) is also decreased by two ($\Delta Z = -2$). Because the *parent nucleus* (the original nucleus) loses two protons, the *daughter nucleus* (the resulting nucleus) is the nucleus of a different element, defined by the new proton number. Thus, the alpha-decay process is one of nuclear transmutation, or conversion, in which the nuclei of one element change into the nuclei of a lighter element.

An example of an isotope, or nuclide, that undergoes alpha decay is polonium-214. The decay process is represented as a nuclear equation (usually written without neutron numbers):

$$\underset{\text{polonium}}{^{214}_{84}\text{Po}} \rightarrow \underset{\text{lead}}{^{210}_{82}\text{Pb}} + \underset{\text{alpha particle (helium nucleus)}}{^{4}_{2}\text{He}}$$

Notice that both the mass number and proton number totals are equal on each side of the equation: ($214 = 210 + 4$) and ($84 = 82 + 2$), respectively. This condition reflects the experimental facts that *two conservation laws apply to all nuclear processes*. The first is the **conservation of nucleons**:

> The total number of nucleons (*A*) remains constant in any nuclear process.

The second is the familiar **conservation of charge**:

> The total charge remains constant in any nuclear process.

These conservation laws allow us to predict the daughter nucleus, as in Example 29.1.

---

**EXAMPLE 29.1: PLUTONIUM'S DAUGHTER – ALPHA DECAY AND A FAMILIAR NAME**

A $^{239}$Pu nucleus undergoes alpha decay. What is the resulting daughter nucleus?

**THINKING IT THROUGH.** Nucleon conservation allows the prediction of the daughter's proton number if plutonium's proton number is known. This can be found in the periodic table (Figure 28.9). From that, the daughter element's name can be determined, also from the periodic table.

**SOLUTION**

From the periodic chart, the proton number for plutonium is 94. Since $\Delta Z = -2$ for alpha decay, the parent plutonium-239 $\left(^{239}_{94}\text{Pu}\right)$ nucleus loses two protons, and the daughter nucleus has a proton number $Z = 94 - 2 = 92$, which is uranium's proton number (see the periodic table). The equation for this decay is therefore

$$^{239}_{94}\text{Pu} \rightarrow {}^{235}_{92}\text{U} + {}^{4}_{2}\text{He} \quad \text{or} \quad ^{239}_{94}\text{Pu} \rightarrow {}^{235}_{92}\text{U} + {}^{4}_{2}\alpha$$

Note the helium nucleus is commonly written as an alpha particle $^4_2\alpha$ (or just $\alpha$).

**FOLLOW-UP EXERCISE.** Using high-energy accelerators, it is possible to *add* an alpha particle to a nucleus – essentially the reverse of the reaction in this Example. Write the equation for this nuclear reaction and predict the identity of the resulting nucleus if an alpha particle is added to a $^{12}$C nucleus.

---

From experiments, it has been found that the kinetic energies of alpha particles from radioactive sources are typically a few million electron-volts (MeV). (See Section 16.2.) For example, the energy of the alpha particle from the decay of $^{214}$Po is about 7.7 MeV, and that from $^{238}$U decay is about 4.14 MeV. Alpha particles from such sources were used in the scattering experiments that led to the Rutherford nuclear model.

Outside the nucleus, the repulsive electric force increases as an alpha particle approaches the nucleus. Inside the nucleus, however, the strongly attractive nuclear force dominates. These conditions are depicted in ▼**Figure 29.6**, which shows a graph of the potential energy $U$ of the alpha nucleus system as a function of $\boldsymbol{r}$, the distance from the center of the nucleus. Consider alpha particles (with kinetic energy of 7.7 MeV) from a $^{214}$Po source incident on $^{238}$U (Figure 29.6). The alpha particles don't have enough kinetic energy to overcome the electric potential energy "*barrier*," whose maximum exceeds 7.7 MeV. Thus, Rutherford scattering occurs. On the other hand, it is known that the $^{238}$U nucleus *does* undergo alpha decay, emitting an alpha particle with an energy of 4.4 MeV, which is *below* the height of the barrier. How can these lower-energy alpha particles cross a barrier from the inside to the outside, when higher-energy alpha particles cannot cross from outside to the inside? According to classical theory, this is impossible, since it violates the conservation of energy. However, quantum mechanics offers an explanation.

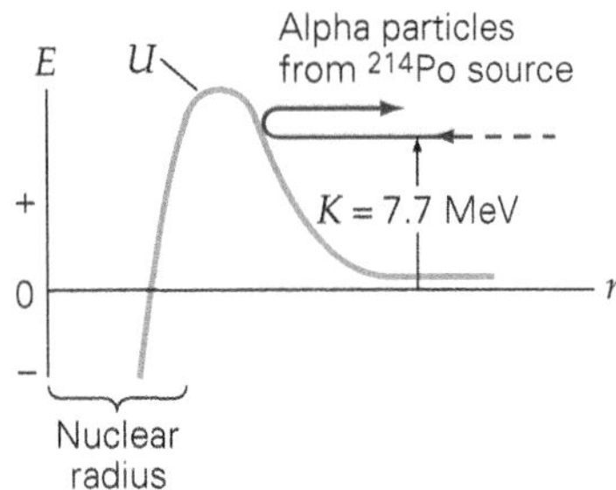


▲ FIGURE 29.6 **Potential energy barrier for alpha particles** Alpha particles from radioactive polonium with energies of 7.7 MeV do not have enough energy to overcome the electrostatic potential energy barrier of the $^{238}$U nucleus and are scattered.

Quantum mechanics predicts a nonzero probability of finding an alpha particle, initially inside the nucleus, to be outside the nucleus (▶ **Figure 29.7**). This phenomenon is called **tunneling**, or **barrier penetration**, since the alpha particle, with its probability wave function, has a chance of tunneling through the barrier and appearing outside the nucleus.

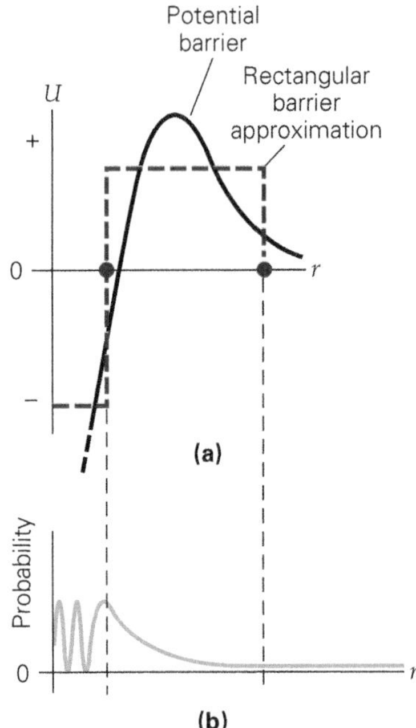


▲ FIGURE 29.7 **Tunneling or barrier penetration (a)** The potential energy barrier presented by a nucleus to an alpha particle can be approximated by a rectangular barrier. **(b)** The probability of finding the alpha particle at a given location, according to quantum mechanical calculations, is shown. If the particle is initially inside the nucleus, it has a likelihood of "tunneling" through the barrier and appearing outside the nucleus. Typically, this event has a very small, but nonzero, probability of occurring for elements above lead on the periodic table.

## 29.2.2 Beta Decay

The emission of an electron (a beta particle) in a nuclear decay process might seem contradictory to the proton–neutron model of the nucleus. Note, however, that the electron emitted in **beta decay** is not part of the original nucleus. *The electron is created during the decay.* There are several types of beta decay. When a negative electron is emitted, the process is called **$\beta^-$ decay**. An example of this type of beta decay is that of $^{14}$C:

$$\underset{\text{carbon}}{{}^{14}_{6}\text{C}} \rightarrow \underset{\text{nitrogen}}{{}^{14}_{7}\text{N}} + \underset{\substack{\text{beta particle}\\ \text{(electron)}}}{{}^{0}_{-1}\text{e}}$$

The parent nucleus (carbon-14) has six protons and eight neutrons, whereas the daughter nucleus (nitrogen) has seven protons and seven neutrons. Notice that the electron symbol has a nucleon number of zero (the electron is not a nucleon) and a charge number of –1. Thus, both nucleon number (14) and electric charge (+6) are conserved.

In this type of beta decay, the neutron number of the parent nucleus decreases by one, and the proton number of the daughter nucleus increases by one. Thus, the nucleon number remains unchanged. In essence, a neutron within such an unstable nucleus decays into a proton and an electron which is then emitted:

$$\underset{\text{neutron}}{{}^{1}_{0}\text{n}} \rightarrow \underset{\text{proton}}{{}^{1}_{1}\text{p}} + \underset{\text{electron}}{{}^{0}_{-1}\text{e}} \qquad \text{(basic } \beta^- \text{ decay)}$$

Negative beta decay generally happens when a nucleus is unstable because of having too many neutrons compared to protons. (See Section 29.5 on nuclear stability.) The most massive stable isotope of carbon is $^{13}C$, with seven neutrons. But $^{14}C$ has too many neutrons (eight) for a nucleus with six protons and is unstable. Since negative beta decay simultaneously decreases the neutron number and increases the proton number, the product is more stable. In this case, the product nucleus is $^{14}N$, which is actually stable. For completeness and correctness, it should be noted that another elementary particle, called a *neutrino*, is emitted during beta decay. For simplicity, it will not be shown in the nuclear decay equations here. Its important role in beta decay will be discussed more fully in Chapter 30.

There are actually two modes of beta decay, $\beta^-$ and $\beta^+$, as well as a third process called *electron capture* (EC). Whereas $\beta^-$ decay involves the emission of an electron, **$\beta^+$ decay**, or *positron decay*, involves the emission of a positron. The positron is a positive electron – the antiparticle of the electron (see Section 28.5). A positron is symbolized as $_{+1}^{0}e$. Nuclei that undergo $\beta^+$ decay have too many protons relative to the number of neutrons. The net effect of $\beta^+$ decay is to convert a proton into a neutron. As in $\beta^-$ decay, this process serves to create a more stable daughter nucleus. An example of $\beta^+$ decay is:

$$\underset{\text{oxygen}}{{}^{15}_{8}O_7} \rightarrow \underset{\text{nitrogen}}{{}^{15}_{7}N_8} + \underset{\text{positron}}{{}^{0}_{+1}e}$$

Positron emission is also accompanied by a neutrino (but a different type from that associated with $\beta^-$ decay). This will also be discussed in more detail in Chapter 30.

A process that competes with $\beta^+$ decay is called **electron capture** (abbreviated as **EC**). This process involves the capture of *orbital* electrons (most likely the innermost orbital electrons) by a nucleus. The net result is the same daughter nucleus that would have been produced by positron decay – hence the use of the descriptive word "competing." That is, there is usually a certain probability that *both* processes can happen. A specific example of electron capture is:

$$\underset{\text{orbital electron}}{{}^{0}_{-1}e} + \underset{\text{beryllium}}{{}^{7}_{4}Be} \rightarrow \underset{\text{lithium}}{{}^{7}_{3}Li}$$

As in $\beta^+$ decay, a proton changes into a neutron, but note that no beta particle is emitted in EC.

### 29.2.3 Gamma Decay

In **gamma decay**, the nucleus emits a gamma ($\gamma$) ray, a high-energy photon of electromagnetic energy. The emission of a gamma ray by a nucleus in an excited state is analogous to the emission of a photon by an excited atom. It is common that the nucleus emitting the gamma ray is a daughter nucleus left in an excited state following alpha decay, beta decay, or EC.

Nuclei possess energy levels analogous to those of atoms. However, nuclear energy levels are much farther apart and more complicated than those of an atom. The nuclear energy levels are typically separated by kiloelectron-volts (keV) and megaelectron-volts (MeV), rather than the few electron-volts (eV) that separate atomic energy levels. As a result, gamma rays are very energetic, with energies larger than those of X-rays. Thus gamma rays have extremely short wavelengths. It is common to indicate a nucleus in an excited state with a superscript asterisk. For example, the decay of $^{61}Ni$ from an excited nuclear state (indicated by the asterisk) to one of lesser energy (in the same nucleus) is written as follows:

$$\underset{\substack{\text{nickel}\\\text{(excited)}}}{{}^{61}_{28}Ni^*} \rightarrow \underset{\text{nickel}}{{}^{61}_{28}Ni} + \underset{\text{gamma ray}}{\gamma}$$

*Note that in gamma decay, the mass and proton numbers do not change.* The daughter nucleus is simply the parent nucleus with less energy. As an example of gamma emission following beta decay, consider Integrated Example 29.2.

---

INTEGRATED EXAMPLE 29.2: TWO FOR ONE – BETA DECAY AND GAMMA DECAY

Naturally occurring cesium has only one stable isotope, $^{133}_{55}Cs$. However, the unstable isotope $^{137}_{55}Cs$ is a common nucleus found in spent nuclear fuel rods at nuclear power plants. (See Chapter 30.) When $^{137}_{55}Cs$ decays, its daughter nucleus is sometimes left in an excited state. After the initial decay, the daughter emits a gamma ray to produce a final stable nucleus. (a) Does $^{137}_{55}Cs$ first decay by (1) $\beta^+$ decay, (2) $\beta^-$ decay, or (3) electron capture? Explain. (b) Find the final daughter product by writing the chain of decay equations. Show all the steps leading to the final stable nucleus.

**(a) CONCEPTUAL REASONING.** $^{137}_{55}Cs$ has too many neutrons to be stable, as $^{133}_{55}Cs$, with four fewer neutrons, is stable. Both (1) and (3) increase the number of neutrons. Lowering the number of neutrons calls for $\beta^-$ decay, so the correct choice is (2), $\beta^-$ decay.

**(b) QUANTITATIVE REASONING AND SOLUTION.** Since $^{133}_{55}Cs$ must decay by emitting a $\beta^-$ particle, its daughter (in an excited state) can be determined from charge and nucleon conservation. The final state of the daughter results after a gamma-ray photon is emitted.

The data is as follows:

***Given:*** Initial nucleus of $^{137}_{55}Cs$

***Find:*** Decay schemes leading to the stable nucleus

During $\beta^-$ decay, the proton number increases by one; thus, the daughter will be barium ($Z = 56$). (See the periodic table.) The decay equation should indicate that barium is left in an excited state, ready to decay via gamma emission. (As usual in this chapter, the neutrino is omitted.) Thus, the decay equation is

$$\underset{\text{cesium}}{{}^{137}_{55}Cs} \rightarrow \underset{\substack{\text{barium}\\\text{(excited)}}}{{}^{137}_{56}Ba^*} + \underset{\text{electron}}{{}^{0}_{-1}e}$$

This process is then followed by the emission of a gamma ray from the excited barium nucleus:

$$\underset{\text{barium (excited)}}{{}^{137}_{56}\text{Ba}^*} \rightarrow \underset{\text{barium}}{{}^{137}_{56}\text{Ba}} + \underset{\text{gamma ray}}{\gamma}$$

Sometimes this process is written as a combined equation to show the sequential behavior:

$$\begin{array}{ccccc} {}^{137}_{55}\text{Cs} & \rightarrow & {}^{137}_{56}\text{Ba}^* & + & {}^{0}_{-1}\text{e} \\ \text{cesium} & & \text{barium (excited)} & & \text{electron} \\ & & \downarrow & & \\ & & {}^{137}_{56}\text{Ba} & + & \gamma \\ & & \text{barium} & & \text{gamma ray} \end{array}$$

**FOLLOW-UP EXERCISE.** An unstable isotope of sodium, $^{22}$Na, can be produced in nuclear reactors. The only stable isotope of sodium is $^{23}$Na. $^{22}$Na is known to decay by beta decay. (a) Which type of beta decay is it? Explain. (b) Write down the beta decay scheme and predict the daughter nucleus.

### 29.2.4 Radiation Penetration and Decay Series

The absorption, or penetration, of nuclear radiation is an important consideration in many modern applications. A familiar use of radiation is the radioisotope treatment of cancer. Radiation penetration is also important, for example, in determining the shielding needed around nuclear facilities. In the food industry, gamma radiation is used to penetrate some foods to kill bacteria and thus sterilize the food.

The three types of radiation (alpha, beta, and gamma) are absorbed quite differently. As they move along their paths, the electrically charged alpha and beta particles interact with the electrons of the atoms of a material and may ionize some of them. The charge and speed of the particle determine the rate at which it loses energy along its path (since ionizing an atom takes energy) and thus the degree of penetration. The amount of penetration also depends on properties of the material, such as its density. In general, what happens when these particles enter a material is as follows:

- Alpha particles are doubly charged, have a relatively large mass, and move relatively slowly. Thus a few centimeters of air or a sheet of paper will usually completely stop them.
- Beta particles are much less massive and are singly charged. They can travel a few meters in air or a few millimeters in materials such as aluminum before being stopped.
- Gamma rays are uncharged and are therefore more penetrating than alpha and beta particles. A significant portion of a beam of high-energy gamma rays can penetrate a centimeter or more of a dense material, such as lead, which is a commonly used shielding material. Photons can lose energy or be removed from a beam by a combination of mechanisms: Compton scattering, the photoelectric effect, and pair production (Chapter 27).

Radiation passing through matter can do considerable damage. Structural materials can become brittle and lose their strength when exposed to strong radiation, as can happen in nuclear reactors and to space vehicles exposed to cosmic radiation. In biological tissue, radiation damage is chiefly due to ionizations in living cells (Section 29.5). We are continually exposed to normal background radiation from radioisotopes in our environment and cosmic radiation from outer space. The damage inflicted to cells from exposure to everyday levels of such radiation is usually too low to be harmful.

However, concern has been expressed about the radiation exposure of people employed, for example, in jobs where radiation levels may be considerably higher. Thus workers at nuclear plants are constantly monitored for absorbed radiation and subject to strict rules that govern the amount of time for which they can work in a given period. Also, airplane crews who spend many hours aboard high-flying aircraft receive significant exposure to radiation from cosmic rays.

Of the many unstable nuclides, only a small number occur naturally. Most of the radioactive nuclides in nature are products of the decay series of heavy nuclei. There is continual radioactive decay progressing in a series creating successively lighter elements. For example, the $^{238}$U decay series (or "chain") is shown in ▼**Figure 29.8**. It stops when the stable isotope of

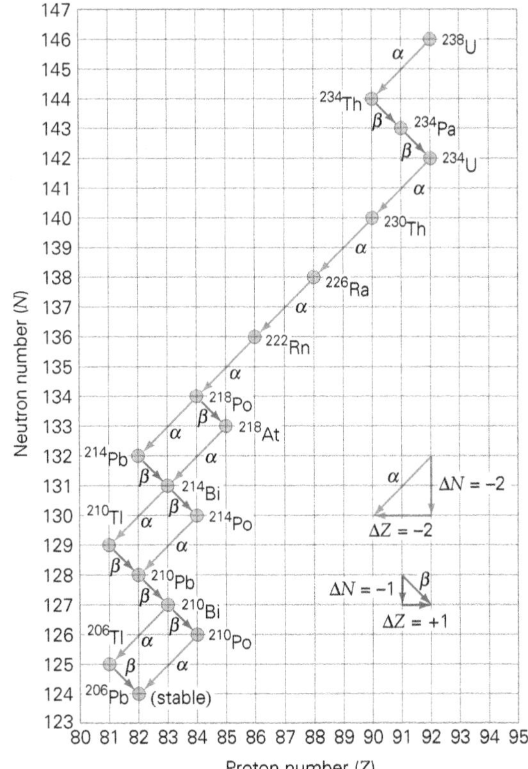


▲**FIGURE 29.8 Decay series of uranium-238** On this plot of $N$ versus Z, a diagonal transition from right to left is an alpha decay process, and a diagonal transition from left to right is a $\beta^-$ decay process. (How can you tell?) The decay series continues until the stable nucleus $^{206}$Pb is reached.

lead, $^{206}Pb$, is reached. Note that some nuclides in the series decay by two modes and that radon ($^{222}Rn$) is part of this decay series. This radioactive gas has received a great deal of attention by health officials because it can accumulate in poorly ventilated buildings.

## 29.3 Decay Rate and Half-Life

The nuclei in a sample of radioactive material do not decay all at once, but rather do so randomly at a rate characteristic of the particular nucleus and unaffected by external influences. It is impossible to tell exactly when a particular unstable nucleus will decay. What can be determined, however, is how many nuclei in a sample will decay during a given period of time.

The **activity ($R$)** of a sample of radioactive nuclide is defined as the number of nuclear disintegrations, or decays, per second. For a given amount of material, activity decreases with time, as fewer and fewer radioactive nuclei remain. Each nuclide has its own characteristic rate of decrease. The rate at which the number of parent nuclei ($N$) decreases is proportional to the number present, or expressed mathematically: $\Delta N/\Delta t \propto N$. This can be rewritten in equation form (with a constant of proportionality $\lambda$) as:

$$\frac{\Delta N}{\Delta t} = -\lambda N$$

The constant $\lambda$ is the **decay constant**. It has SI units of $s^{-1}$ (you should check it) and depends only on the particular nuclide. The larger the decay constant, the greater the rate of decay. The minus sign in the previous equation indicates that $N$ is decreasing, thus $\Delta N/\Delta t$ must be negative. The activity ($R$) of a radioactive sample is the magnitude of $\Delta N/\Delta t$, or the *decay rate*, expressed in decays per second, but without the minus sign (see Example 29.3):

$$R = \text{activity} = \left|\frac{\Delta N}{\Delta t}\right| = \lambda N \quad (29.2)$$

Using calculus, Equation 29.2 (with the minus sign) can be solved for the number of the remaining (i.e., undecayed) parent nuclei ($N$) at any time $t$ compared with the number $N_o$ present at $t = 0$. Without proof, the result is:

$$N = N_o e^{-\lambda t} \quad \text{(radioactive decay law)} \quad (29.3)$$

Thus, the number of undecayed (parent) nuclei decreases *exponentially* with time as illustrated in ▶ **Figure 29.9**. This graph is essentially a plot of the exponentially decaying function $e^{-\lambda t}$. (Remember $e \approx 2.718$ as the base of natural logarithms; it should be available on your calculator. See Appendix I for a primer on exponential functions and natural logarithms.)

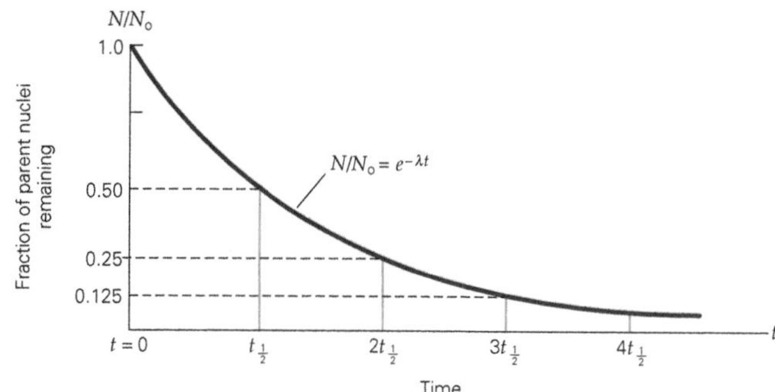


▲ **FIGURE 29.9 Radioactive decay** The fraction of the remaining parent nuclei ($N/N_o$) in a radioactive sample plotted as a function of time follows exponential decay. The curve's shape and steepness depend on the decay constant $\lambda$ or, equivalently, the half-life $t_{1/2}$.

The decay rate of a nuclide is commonly expressed in terms of its *half-life*. The **half-life** (symbolized by $t_{1/2}$) is the time it takes for half of the nuclei in a sample to decay. Thus this time is when $N/N_o = 1/2$ in Figure 29.9. In that time, activity (decays/sec) is also cut in half, since the activity is proportional to the number of undecayed nuclei remaining. Because of this proportionality, the decay rate, not the number of undecayed nuclei, is usually used to measure half-lives. Therefore what is usually measured experimentally is the time it takes for the decay rate to drop by one-half.

For example, by plotting decay rates, ▼ **Figure 29.10** illustrates the half-life of strontium-90 ($^{90}Sr$) to be 28 years. An alternative view of the half-life is to consider the mass of parent material. For example, if there were initially 100 $\mu g$ of $^{90}Sr$ present, then only 50 $\mu g$ would remain after 28 years. The other 50 $\mu g$ would have decayed by beta decay:

$$\underset{\text{strontium}}{^{90}_{38}\text{Sr}} \rightarrow \underset{\text{yttrium}}{^{90}_{39}\text{Y}} + \underset{\text{electron}}{^{0}_{-1}\text{e}}$$

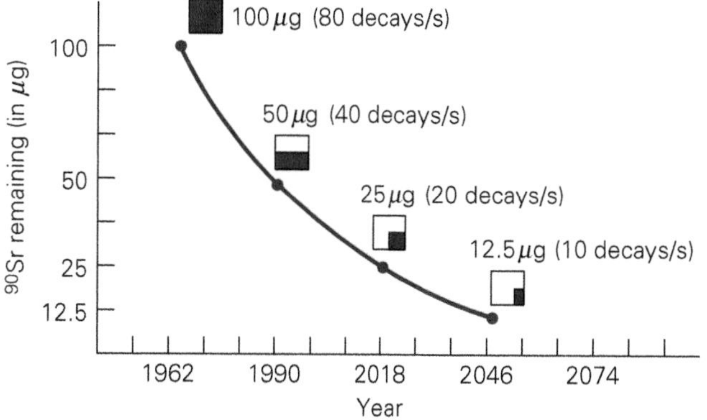


▲ **FIGURE 29.10 Radioactive decay and half-life** As shown here for $^{90}Sr$, after each half-life $t_{1/2} = 28y$, half of the amount of $^{90}Sr$ present at the start of that period of time remains, with the other half having decayed to $^{90}Y$ via beta decay. Similarly, the sample's activity (decays per second) will also decrease by half after 28 years.

After another 28 years, half of these $^{90}$Sr nuclei would decay, leaving only 25 μg, and so on.

The half-lives of radioactive nuclides vary greatly, as ▼**Table 29.1** shows. Nuclides with very short half-lives are generally created in nuclear reactions (Chapter 30). If these nuclides had existed before the Earth was formed (about 4.8 billion years ago), they would have long since decayed. In fact, this is the case for technetium (Tc) and promethium (Pm, not shown in Table 29.1). These elements do *not* exist naturally, as they have no stable configurations and their half-lives are short compared to 4.8 billion years. However, they can be produced in laboratories. At the other end of the half-life spectrum, that of naturally occurring $^{238}$U isotope is about 4.5 billion years. This means that about half of the original $^{238}$U present when the Earth was formed exists today.

The longer the half-life of a nuclide, the more slowly it decays and the smaller the decay constant λ. Thus the half-life and the decay constant have an inverse relationship, or $t_{1/2} \propto 1/\lambda$. To show this explicitly, consider Equation 29.3. When $t = t_{1/2}$, we have $N/N_o = 1/2$ and therefore, $N/N_o = 1/2 = e^{-\lambda t_{1/2}}$. But since $e^{-0.693} \approx 1/2$ (check this on your calculator), the exponents can be equated to give the relationship between half-life and decay constant:

$$t_{1/2} = \frac{0.693}{\lambda} \quad \text{(half-life from decay constant)} \qquad (29.4)$$

The concept of half-life is important in medical applications, as is shown in Example 29.3.

**TABLE 29.1** The Half-Lives of Some Radioactive Nuclides (in Order of Increasing Half-Life)

| Nuclide | Primary Decay Mode | Half-Life of Decay Mode |
|---|---|---|
| Beryllium-8 $\left({}^{8}_{4}\text{Be}\right)$ | α | $1 \times 10^{-16}$ s |
| Polonium-213 $\left({}^{213}_{84}\text{Po}\right)$ | α | $4 \times 10^{-16}$ s |
| Oxygen-19 $\left({}^{19}_{8}\text{O}\right)$ | $\beta^-$ | 27 s |
| Fluorine-17 $\left({}^{17}_{9}\text{F}\right)$ | $\beta^+$, EC | 66 s |
| Polonium-218 $\left({}^{218}_{28}\text{Po}\right)$ | α, $\beta^-$ | 3.05 min |
| Technetium-104 $\left({}^{104}_{43}\text{Tc}\right)$ | $\beta^-$ | 18 min |
| Iodine-123 $\left({}^{123}_{53}\text{I}\right)$ | EC | 13.3 h |
| Krypton-76 $\left({}^{76}_{36}\text{Kr}\right)$ | EC | 14.8 h |
| Magnesium-28 $\left({}^{28}_{12}\text{Mg}\right)$ | $\beta^-$ | 21 h |
| Radon-222 $\left({}^{222}_{86}\text{Rn}\right)$ | α | 3.82 days |
| Iodine-131 $\left({}^{131}_{53}\text{I}\right)$ | $\beta^-$ | 8.0 days |
| Cobalt-60 $\left({}^{60}_{27}\text{Co}\right)$ | $\beta^-$ | 5.3 y |
| Strontium-90 $\left({}^{90}_{38}\text{Sr}\right)$ | $\beta^-$ | 28 y |
| Radium-226 $\left({}^{226}_{88}\text{Ra}\right)$ | α | 1600 y |
| Carbon-14 $\left({}^{14}_{6}\text{C}\right)$ | $\beta^-$ | 5730 y |
| Plutonium-239 $\left({}^{239}_{94}\text{Pu}\right)$ | α | $2.4 \times 10^4$ y |
| Uranium-238 $\left({}^{238}_{92}\text{U}\right)$ | α | $4.5 \times 10^9$ y |
| Rubidium-87 $\left({}^{87}_{37}\text{Rb}\right)$ | $\beta^-$ | $4.7 \times 10^{10}$ y |

### EXAMPLE 29.3: AN "ACTIVE" THYROID – HALF-LIFE AND ACTIVITY

The half-life of iodine-131 ($^{131}$I), used in thyroid treatments, is 8.0 days. At a certain time, about $4.0 \times 10^{14}$ $^{131}$I nuclei are present in a hospital patient's thyroid gland. (a) What is the $^{131}$I activity in the thyroid at that time? (b) How many $^{131}$I nuclei remain after 1.0 day?

**THINKING IT THROUGH.** (a) Equation 29.4 enables the determination of the decay constant λ from the half-life, and then Equation 29.2 can be used to find the initial activity. (b) To get *N*, Equation 29.3 can be used with the aid of the $e^x$-button on a calculator.

**SOLUTION**

Listing the data and converting the half-life into seconds:

***Given:***

$t_{1/2} = 8.0 \text{ days} = 6.9 \times 10^5$ s
$N_o = 4.0 \times 10^{14}$ nuclei (initially)
$t = 1.0$ day

***Find:***

**(a)** $R_o$ (activity at $t = 0$)
**(b)** $N$ (number of undecayed nuclei after 1.0 day)

**(a)** The decay constant is determined from its relationship to the half-life (Equation 29.4):

$$\lambda = \frac{0.693}{t_{1/2}} = \frac{0.693}{6.9 \times 10^5 \text{ s}} = 1.0 \times 10^{-6} \text{ s}^{-1}$$

Using the initial number of undecayed nuclei, $N_o$, the initial activity $R_o$ is

$$R_o = \frac{\Delta N}{\Delta t} = \lambda N_o = (1.0 \times 10^{-6} \text{ s}^{-1})(4.0 \times 10^{14}) = 4.0 \times 10^8 \text{ decays/s}$$

**(b)** With $t = 1.0$ day and $\lambda = 0.693/t_{1/2} = 0.693/8.0\text{days} = 0.087\text{day}^{-1}$:

$$\begin{aligned} N = N_o e^{-\lambda t} &= (4.0 \times 10^{14} \text{ nuclei}) e^{-(0.087 \text{ day}^{-1})(1.0 \text{ day})} \\ &= (4.0 \times 10^{14} \text{ nuclei}) e^{-0.087} \\ &= (4.0 \times 10^{14} \text{ nuclei})(0.917) = 3.7 \times 10^{14} \text{ nuclei} \end{aligned}$$

*Note:* The $e^x$ calculator button is sometimes labeled as the inverse of the ln *x*. You should become familiar with it. Here it was used to find that $e^{-0.087} \approx 0.917$ to three significant figures.

**FOLLOW-UP EXERCISE.** In this Example, suppose that the attending physician will not allow the patient to go home until the activity is reduced to 1/64 of its original level. (a) How long would the patient have to remain? (b) In practice, the amount of time is much shorter than your answer to part (a). Can you think of a possible biological reason(s) for this?

The "strength" of a radioactive sample usually refers to its activity *R*. A common unit of radioactivity is named in honor of Pierre and Marie Curie. One **curie** (Ci) is defined as

$$1\,\text{Ci} = 3.70\times10^{10}\ \text{decays/s}$$

This definition is historical and is based on the known activity of 1.00 g of pure radium. However, the modern SI unit is the Becquerel (Bq), defined as 1 Bq = 1 decay/s. Therefore,

$$1\,\text{Ci} = 3.70\times10^{10}\,\text{Bq}$$

Even with today's emphasis on SI units, the "strengths" of radioactive sources are still commonly specified in curies. The curie is a relatively large unit, however, so the *millicurie* (mCi), the *microcurie* ($\mu$Ci), and even smaller multiples such as the *nanocurie* (nCi) and *picocurie* (pCi) are used. Teaching laboratories, for example, typically use samples with activities of 1 $\mu$Ci or less. The strength of a source is calculated in Example 29.4.

---

EXAMPLE 29.4: DECLINING SOURCE STRENGTH - GET A HALF-LIFE!

A $^{90}$Sr beta source has an initial activity of 10.0 mCi. How many decays per second will be taking place after 84.0 years?

**THINKING IT THROUGH.** Table 29.1 contains the half-life for this source. In this Example, we can use the fact that in each successive half-life, the activity decreases by half from what it was at the start of that interval. Thus, Equation 29.3 need not be used, because the elapsed time is exactly three half-lives. (This approach is advisable only when the elapsed time is an integral multiple of the half-life, as it is here.)

**SOLUTION**

***Given:***

initial activity = 10.0 mCi
$t = 84.0$ y
$t_{1/2} = 28.0$ y (from Table 29.1)

***Find:*** *R* (activity after 84.0 years)

Since 84 years is exactly three half-lives, the activity after that amount of time has elapsed will be one-eighth as great since (1/2) × (1/2) × (1/2) = 1/8, and the strength of the source will then be

$$R = 10.0\,\text{mCi}\times\frac{1}{8} = 1.25\,\text{mCi} = 1.25\times10^{-3}\,\text{Ci}$$

Expressed in becquerels, this is

$$R = (1.25\times10^{-3}\,\text{Ci})\left(3.70\times10^{10}\,\frac{\text{decay/s}}{\text{Ci}}\right)$$
$$= 4.63\times10^{7}\ \text{decay/s} = 4.63\times10^{7}\,\text{Bq}$$

**FOLLOW-UP EXERCISE.** For the material in this Example, suppose a radiation safety officer tells you that this sample can go into the low-level waste disposal only when its activity drops to one-millionth of its initial activity. Estimate, to two significant figures, how long the sample must be kept before it can be disposed.

---

## 29.3.1 Radioactive Dating

Because their decay rates are constant, radioactive nuclides can be used as nuclear clocks. In Example 29.4, the half-life of a radioactive nuclide was used to determine how much of the sample will exist in the future. Similarly, by using the half-life to calculate backward in time, scientists can determine the age of objects that contain known radioactive nuclides. As you might surmise, some idea of the initial amount of the nuclide present must be known.

Let's look at how this dating is done with $^{14}$C, a very common method used in archeology. **Carbon-14 dating** is used on materials that were once part of living things and on the remnants of objects made from or containing such materials (such as wood, bone, leather, or parchment). The process depends on the fact that living things (including yourself) contain a known amount of radioactive $^{14}$C. The concentration is very small - about one $^{14}$C atom for every $7.2\times10^{11}$ atoms of ordinary $^{12}$C. Even so, the $^{14}$C present in our bodies *cannot* be due to $^{14}$C that was present when the Earth was formed, because the half-life of $^{14}$C is 5730 years, which is very short in comparison with the age of the Earth.

The $^{14}$C nuclei that exist in living things are present because that isotope is continuously produced in the atmosphere by cosmic rays. *Cosmic rays* are high-speed charged particles that reach us from various sources such as the Sun and nearby exploding stars called supernovae. These "rays" are primarily protons. When they enter our upper atmosphere, they can cause nuclear reactions that produce neutrons (▶ **Figure 29.11**). These neutrons are then absorbed by the nuclei of the nitrogen in the air, which, in turn, decay by emitting a proton (written as p or $^1_1$H) to produce $^{14}$C by the reaction $^{14}_{7}\text{N} + ^{1}_{0}\text{n} \rightarrow ^{14}_{6}\text{C} + ^{1}_{1}\text{H}$. $^{14}$C then eventually $\beta^-$ decays $\left(^{14}_{6}\text{C} \rightarrow ^{14}_{7}\text{N} + ^{0}_{-1}\text{e}\right)$ since it is neutron rich. The concentration of $^{14}$C in the atmosphere is relatively constant, because of atmospheric mixing and the fixed decay rate.

The $^{14}$C atoms are oxidized into carbon dioxide ($CO_2$), so a small fraction of the $CO_2$ molecules in the air is radioactive. Plants take in this radioactive $CO_2$ by photosynthesis, and animals ingest the plant material. As a result, the concentration of $^{14}$C in living organic matter is the same as the concentration in the atmosphere, one part in $7.2\times10^{11}$. However, once an organism dies, the $^{14}$C in that organism is no longer replenished, and thus the $^{14}$C concentration decreases. Thus, the concentration of $^{14}$C in dead matter relative to that in living things can be used to establish when the organism died. Thus to do this, the $^{14}$C activity in organisms now alive must be found. Example 29.5 shows how this is done.

---

EXAMPLE 29.5: LIVING ORGANISMS - NATURAL CARBON-14 ACTIVITY

For $^{14}$C, determine the activity *R* (decays per minute per gram of natural carbon), found in living organisms, assuming the concentration of $^{14}$C is the same as in the atmosphere.

**THINKING IT THROUGH.** From the previous discussion, the amount of $^{14}$C relative to $^{12}$C is known. To calculate the $^{14}$C activity, the decay constant ($\lambda$) is needed. This can be computed from the half-life of $^{14}$C ($t_{1/2} = 5730$ years) and the number of $^{14}$C atoms (*N*)

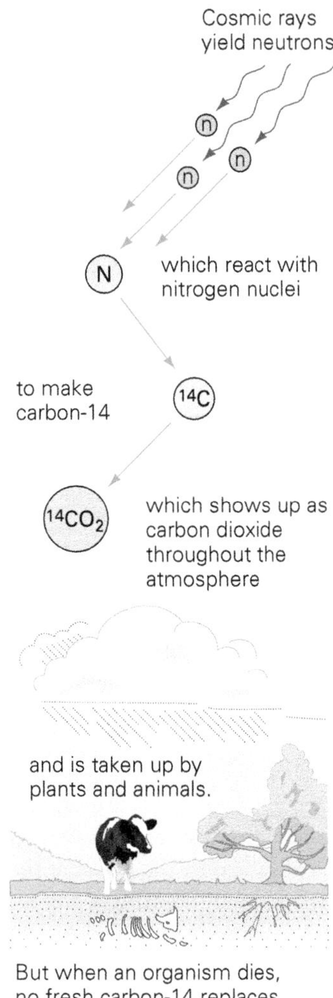
Cosmic rays
yield neutrons
n
n
n
N
which react with
nitrogen nuclei
to make
carbon-14
14C
14CO2
which shows up as
carbon dioxide
throughout the
atmosphere
and is taken up by
plants and animals.
But when an organism dies,
no fresh carbon-14 replaces
the carbon-14 decaying in
its tissues, and the carbon-14
radioactivity decreases by
half every 5730 years.

▲ **FIGURE 29.11 Carbon-14 radioactive dating** The formation of carbon-14 in the atmosphere and its entry into the biosphere.

per gram. Carbon has an atomic mass of 12.0, so $N$ can be found from Avogadro's number (recall that $N_A = 6.02 \times 10^{23}$ atoms/mole) and the number of moles, $n = N/N_A$ (see Section 10.3).

**SOLUTION**

Using the known isotopic ratio and the half-life (converting it into minutes), we have:

***Given:***

$$\frac{^{14}\mathrm{C}}{^{12}\mathrm{C}} = \frac{1}{7.2\times10^{11}} = 1.4\times10^{-12}$$

$$t_{1/2} = (5730\text{ years})\,(5.26\times10^{5}\text{ min/year}) = 3.01\times10^{9}\text{ min}$$

Carbon has 12.0 g per mole

***Find:*** Average activity $R$ per gram

From the half-life, the decay constant is

$$\lambda = \frac{0.693}{t_{1/2}} = \frac{0.693}{3.01\times10^{9}\text{ min}} = 2.30\times10^{-10}\text{ min}^{-1}$$

For 1.0 g of carbon, the number of moles is $n = 1.0$ g/(12 g/mol) = 1/12 mol so the number of atoms ($N$) is

$$N = nN_A = \left(\frac{1}{12}\text{mol}\right)(6.02\times10^{23}\text{C nuclei/mol}) = 5.0\times10^{22}\ \text{C nuclei (per gram)}$$

The number of $^{14}$C nuclei per gram is given by the isotopic ratio:

$$N\left(\frac{^{14}\mathrm{C}}{^{12}\mathrm{C}}\right) = (5.0\times10^{22}\text{ C nuclei/g})\left(1.4\times10^{-12}\frac{^{14}\text{C nuclei}}{\text{C nuclei}}\right) = 7.0\times10^{10}\ {}^{14}\text{C nuclei/g}$$

Thus the activity in decays per gram of carbon per minute (to two significant figures) is

$$\left|\frac{\Delta N}{\Delta t}\right| = \lambda N = (2.30\times10^{-10}\text{ min}^{-1})\,(7.0\times10^{10\,14}\text{ C/g}) = 16\,\frac{^{14}\text{C decays}}{\text{g·min}}$$

For example, if an artifact such as a bone has a current activity of 8.0 decays per gram of carbon per minute, then the original living organism would have died about one half-life, or about 5700 years, ago, thus putting the date when the bone's organism died to be about 3700 B.C.

**FOLLOW-UP EXERCISE.** Suppose that your instruments could only measure $^{14}$C activity down to 1.0 decay/min. To how far back (two significant figures) could you estimate the ages of dead organisms?

Consider how the activity calculated in Example 29.5 can be used to date ancient organic finds.

**EXAMPLE 29.6: OLD BONES – CARBON-14 DATING**

A bone is unearthed in an archeological dig and analysis determines there are 20 beta emissions per minute from 10 g of carbon in the bone. What is the approximate age of the bone?

**THINKING IT THROUGH.** Since the initial activity of a living sample is known (Example 29.5), we can work backward to determine the amount of time elapsed.

**SOLUTION**

For comparison purposes, the activity *per gram* is the relevant number.

***Given:*** 20 decays/min (in 10 g of carbon) = 2.0 decay/s(g·min)

***Find:*** approximate age of the bone

When the organism died the $^{14}C$ activity would have been 16 decays/g·min (Example 29.5). Afterward, the decay rate decreases by half for each half-life:

$$16 \xrightarrow{t_{1/2}} 8 \xrightarrow{t_{1/2}} 4 \xrightarrow{t_{1/2}} 2 \text{ decays/(g·min)}$$

So, with the observed activity it can be concluded that the $^{14}C$ in the bone has gone through approximately three half-lives. Thus

$$\text{Age} \approx 3.0t_{1/2} = (3.0)(5730\,\text{y}) = 1.7 \times 10^4\,\text{y}$$
$$\approx 17\,000\,\text{y}$$

**FOLLOW-UP EXERCISE.** Studies indicate that on Earth the stable isotope $^{39}K$ represents about 93.2% of all the potassium. A long lived (but unstable) nuclide, $^{40}K$, represents only about 0.010%. $^{40}K$ has a half-life of $1.28 \times 10^9$ y. (a) The remainder of the existing potassium (6.8%) consists of just one other stable isotope of potassium. What isotope is this most likely to be? (b) How many times more abundant would $^{40}K$ have been on the Earth when it first formed ($4.7 \times 10^9$ years ago) than it is now?

The limit of radioactive carbon dating depends on the ability to measure the very low activity in old samples. Current techniques give an age-dating limit of about 40 000 – 50 000 years, depending on the sample size. After about ten half-lives, the radioactivity is barely measurable (less than two decays per gram per hour).

Another radioactive dating process uses lead-206 ($^{206}Pb$) and uranium-238 ($^{238}U$). This dating method is used extensively in geology, because of the long half-life of $^{238}U$. Lead-206 is the stable end isotope of the $^{238}U$ decay series. (See Figure 29.8.) If a rock sample contains both of these isotopes, the lead is assumed to be a decay product of the uranium that was there when the rock first formed. Thus, the ratio of $^{206}Pb/^{238}U$ can be used to determine the age of the rock.

## 29.4 Nuclear Stability and Binding Energy

Now that some of the properties of unstable isotopes have been considered, let's turn to the stable ones. Stable isotopes exist naturally for all elements having proton numbers from 1 to 83, except those with Z=43 (technetium) and Z=61 (promethium). The nuclear interactions (forces) that determine nuclear stability are extremely complicated. However, by looking at some of the properties of stable nuclei, it is possible to obtain general criteria for nuclear stability.

### 29.4.1 Nucleon Populations

One of the first considerations is the relative number of protons and neutrons in stable nuclei. Nuclear stability depends on the dominance of the attractive nuclear force between nucleons over the repulsive Coulomb force between protons. This force dominance depends on the ratio of protons to neutrons.

For stable nuclei of low mass numbers (about $A < 40$), the ratio of neutrons to protons ($N/Z$) is approximately 1; that is, the number of protons and neutrons are equal or nearly equal. Examples of this are $^4_2He$, $^{12}_6C$, $^{23}_{11}Na$, and $^{27}_{13}Al$. For stable nuclei of higher mass numbers ($A > 40$), the number of neutrons exceeds the number of protons ($N/Z > 1$). The heavier the nuclei, the larger this ratio becomes. In other words, the $N/Z$ ratio increases with $A$.

This trend is illustrated in ▼ **Figure 29.12**, which is a plot of neutron number ($N$) versus proton number ($Z$) for stable nuclei. The heavier stable nuclei lie above the $N = Z$ line (where $N > Z$). Examples of heavy stable nuclei include $^{62}_{28}Ni$, $^{114}_{50}Sn$, and $^{208}_{83}Pb$.

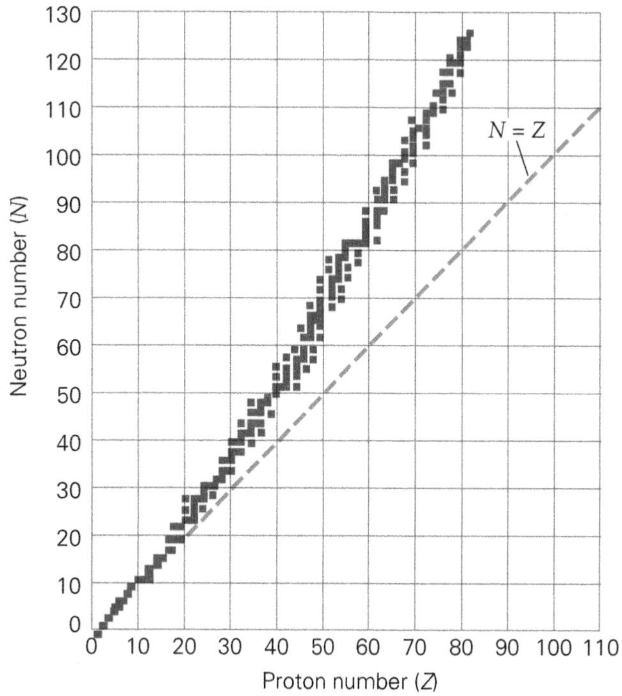


▲**FIGURE 29.12 A plot of *N* versus *Z* for stable nuclei** For nuclei with mass numbers $A < 40$ ($Z < 20$ and $N < 20$), the number of protons and the number of neutrons are equal or nearly equal. For nuclei with $A > 40$, the number of neutrons exceeds the number of protons, so these nuclei lie above the $N = Z$ line.

Radioactive decay "adjusts" the proton and neutron numbers of an unstable nuclide until a stable nuclide is produced – that is, until the product nucleus lands on the stability curve in Figure 29.12. Since alpha decay decreases the numbers of protons and neutrons by equal amounts, alpha decay alone would give nuclei with neutron populations that are larger than those of the stable nuclides on the curve. However, $\beta^-$ decay following alpha decay can lead to a stable combination, since the effect of $\beta^-$ decay is to convert a neutron into a proton. Thus, very heavy unstable nuclei undergo a chain, or sequence, of alpha and beta decays until a stable nucleus is reached (recall Figure 29.8 for $^{238}U$).

### 29.4.2 Pairing

Many stable nuclei have even numbers of both protons and neutrons, and very few have odd numbers of both protons and neutrons. In fact, a survey of the stable isotopes (▼**Table 29.2**) shows that only four contain odd numbers of both. These four are isotopes of the elements with the four lowest odd proton numbers: ${}^{2}_{1}H$, ${}^{6}_{3}Li$, ${}^{10}_{5}B$, and ${}^{14}_{7}N$.

TABLE 29.2 Pairing Effect of Stable Nuclei

| Proton Number | Neutron Number | Number of Stable Nuclei |
|---|---|---|
| Even | Even | 168 |
| Even | Odd | 57 |
| Odd | Even | 50 |
| Odd | Odd | 4 |

The dominance of even–even combinations indicates that the protons and neutrons in nuclei tend to "pair up." That is, two protons pair up and, separately, two neutrons pair up. Aside from the four nuclei mentioned, all odd–odd nuclei are unstable. Also, odd–even and even–odd nuclei tend to be less stable than the even–even variety.

This **pairing effect** provides a qualitative criterion for stability. For example, you might expect the aluminum isotope ${}^{27}_{13}Al$ to be stable (even–odd), but not ${}^{26}_{13}Al$ (odd–odd). This is, in fact, the case.

The *general criteria for nuclear stability* can be summarized as follows:

1. All isotopes with a proton number greater than 83 ($Z > 83$) are unstable.
2. (a) Most even–even nuclei are stable.
   (b) Many odd–even and even–odd nuclei are stable.
   (c) Only four odd–odd nuclei are stable $\left({}^{2}_{1}H, {}^{6}_{3}Li, {}^{10}_{5}B, \text{and } {}^{14}_{7}N\right)$.
3. (a) Stable nuclei with mass numbers $A < 40$ have approximately the same number of protons and neutrons.
   (b) Stable nuclei with mass numbers $A > 40$ have more neutrons than protons.

---

CONCEPTUAL EXAMPLE 29.7: RUNNING DOWN THE CHECKLIST - NUCLEAR STABILITY

Is the sulfur isotope ${}^{38}_{16}S$ likely to be stable?

**REASONING AND ANSWER.** The general criteria for nuclear stability enable the analysis.

1. *Satisfied.* Isotopes with $Z > 83$ are unstable. With $Z = 16$, this criterion is satisfied.
2. *Satisfied.* The isotope ${}^{38}_{16}S_{22}$ has an even–even nucleus, so it could be stable.
3. *Not satisfied.* Here, $A < 40$, but $Z = 16$ and $N = 22$ are not approximately equal.

Thus ${}^{38}S$ is not likely to be stable. (It actually decays by $\beta^-$ emission, since it is neutron rich.)

**FOLLOW-UP EXERCISE.** (a) List likely isotopes of copper ($Z = 29$). (b) Apply the criteria for nuclear stability to see which on the list are likely to be stable. Use Appendix IV to check your conclusions.

---

### 29.4.3 Binding Energy

An important quantitative aspect of nuclear stability is the binding energy of the nucleons. Binding energy can be calculated by considering the mass–energy equivalence along with known nuclear masses. Since nuclear masses are so small in relation to the kilogram, another unit, the **atomic mass unit (u)**, is used to quantify them. The conversion factor (to six significant figures) between the atomic mass unit and the kilogram is $1\text{ u} = 1.660\ 54 \times 10^{-27}$ kg.

The masses of the various particles can be expressed either in atomic mass units (▼**Table 29.3**) or their energy equivalents. The energy equivalents come from Einstein's mass–energy equivalence relationship $E = mc^2$. The following conversion shows that 1 u of mass has an energy equivalent 931.5 MeV:

$$mc^2 = (1.66054 \times 10^{-27}\text{ kg})(2.9977 \times 10^{8}\text{ m/s})^2 = 1.4922 \times 10^{-10}\text{ J}$$
$$= \frac{1.4922 \times 10^{-10}\text{ J}}{1.602 \times 10^{-13}\text{ J/Mev}} = 931.5\text{MeV}$$

as is shown in the first entry of Table 29.3. Our text will use 931.5 MeV/u (four significant figures) as a handy conversion factor when needed to avoid having to multiply by $c^2$.

TABLE 29.3 The Atomic Mass Unit (u), Particle Masses, and Their Energy Equivalents

| Particle | Mass (u) | Mass (kg) | Equivalent Energy (MeV) |
|---|---|---|---|
| 1 u energy equiv. | 1 (exact) | $1.660\ 54 \times 10^{-27}$ | 931.5 |
| Electron | $5.485\ 78 \times 10^{-4}$ | $9.109\ 35 \times 10^{-31}$ | 0.511 |
| Proton | 1.007 276 | $1.672\ 62 \times 10^{-27}$ | 938.27 |
| Hydrogen atom | 1.007 825 | $1.673\ 56 \times 10^{-27}$ | 938.79 |
| Neutron | 1.008 665 | $1.675\ 00 \times 10^{-27}$ | 939.57 |

Note in Table 29.3 that the proton and hydrogen atom have different masses due to the mass of the atomic electron. Typically, laboratories measure the masses of *atoms* (nucleons plus *Z* electrons) rather than those of their *nuclei.* Keep this factor in mind. Since nuclear energy calculations usually involve small differences in mass, the electron mass can be significant.

Nuclear stability can be thought of in terms of energy. For example, if the mass of a helium-4 nucleus is compared with the total mass of nucleons that compose it, a significant inequity emerges: A neutral helium atom (including its *two* electrons) has a mass of 4.002 603 u. Atomic masses of various atoms are given in Appendix IV.) The total mass of two hydrogen atoms (${}^{1}H$) including two electrons and two neutrons is

$$2m({}^{1}H) = 2.015\,650\text{u}$$
$$2m_n = 2.017\,330\text{u}$$
$$\text{Total} = 4.032\,980\text{u}$$

This total is greater than the mass of the helium atom (4.002 603 u). In other words, the helium nucleus is less massive than the sum of its parts by an amount

$$\Delta m = [2m(^1\text{H}) + 2m_n] - m(^4\text{He})$$
$$= 4.032980\,\text{u} - 4.002603\,\text{u} = 0.030377\,\text{u}$$

(Note the two electron masses of helium subtract out, since the mass of two hydrogen atoms includes two electrons.) This difference in mass, the **mass defect**, has an energy equivalent of (0.030 377 u) (931.5 MeV/u) = 28.30 MeV. This energy is the **total binding energy** ($E_b$) of the $^4$He nucleus. In general, for any nucleus, $E_b$ is related to the mass defect $\Delta m$ by

$$E_b = (\Delta m)c^2 \quad \text{(total binding energy)} \qquad (29.5)$$

An alternative interpretation of binding energy is that it represents the minimum energy required to separate the constituent nucleons completely into free particles. This concept is illustrated in ▼ **Figure 29.13** for the helium nucleus, for which a minimum of 28.30 MeV is required to separate it into four nucleons.

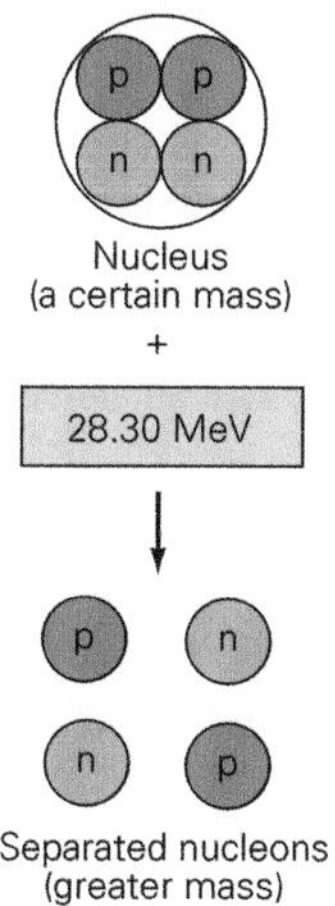


▲ FIGURE 29.13 **Binding energy** 28.30 MeV is required to separate a helium nucleus into its constituent protons and neutrons. Conversely, if two protons and two neutrons combine to form a helium nucleus, 28.30 MeV of energy would be released.

An insight into the nature of the nuclear force can be gained by considering the *average binding energy per nucleon* for stable nuclei. This quantity is the total binding energy of a nucleus, divided by the total number of nucleons, or $E_b/A$, where $A$ is the mass number. For example, the helium nucleus ($^4$He) in Figure 29.13 has an average binding energy per nucleon of

$$\frac{E_b}{A} = \frac{28.30\,\text{MeV}}{4} = 7.075\,\text{MeV/nucleon}$$

Compared with typical binding energies of atomic electrons (e.g., 13.6 eV for a hydrogen electron in the ground state), nuclear binding energies are millions of times larger, indicative of a very strong binding force.

---

**EXAMPLE 29.8: THE STABLEST OF THE STABLE–BINDING ENERGY PER NUCLEON**

Compute the average binding energy per nucleon of the iron-56 nucleus ($^{56}_{26}$Fe).

**THINKING IT THROUGH.** The atomic mass of iron-56 is found in Appendix IV, and the other needed masses are in Table 29.3. The mass defect $\Delta m$ can then be determined and from that the binding energy $E_b$. Last, the average binding energy per nucleon, $E_b/A$, can be found.

**SOLUTION**

***Given:***

$m_{\text{Fe}} = 55.934\,939$ u
$m_{\text{H}} = 1.007\,825$ u
$m_n = 1.008\,665$ u

***Find:*** $E_b/A$ (average binding energy per nucleon)

(Note the use of the iron *atom* mass and the hydrogen *atom* mass rather than nuclear masses.)
The mass defect is the difference between the iron atom mass and that of its separated constituents. The total mass of the constituents (26 hydrogen atoms and 30 neutrons) is:

$$26m_{\text{H}} = 26(1.007\,825\,\text{u}) = 26.203\,450\,\text{u}$$
$$30m_n = 30(1.008\,665\,\text{u}) = 30.259\,950\,\text{u}$$
$$\text{Total} = 56.463\,400\,\text{u}$$

Thus, the mass defect is

$$\Delta m = 26m_{\text{H}} + 30m_n - m_{\text{Fe}}$$
$$= 56.463\,400\,\text{u} - 55.934\,939\,\text{u} = 0.528\,461\,\text{u}$$

The total binding energy is calculated using the energy equivalence of 1 u:

$$E_b = (\Delta m)c^2 = (0.528\,461\,\text{u})(931.5\,\text{MeV/u}) = 492.3\,\text{MeV}$$

This iron nuclide has 56 nucleons, so the average binding energy per nucleon is

$$\frac{E_b}{A} = \frac{492.3\,\text{MeV}}{56} = 8.791\,\text{MeV/nucleon}$$

**FOLLOW-UP EXERCISE.** (a) To illustrate the pairing effect, compare the average binding energy per nucleon for $^4$He (calculated previously to be 7.075 MeV) with that of $^3$He. (Find the atomic masses in Appendix IV.) (b) Which one is more tightly bound, on average, and how does your answer reflect pairing?

---

If $E_b/A$ is plotted versus mass number, it is found that the values lie along a curve as shown in ▼**Figure 29.14**. The value of $E_b/A$ rises rapidly with increasing $A$ for light nuclei and starts to level off (around $A=15$) at about 8.0 MeV/nucleon, with a maximum value of about 8.8 MeV/nucleon in the vicinity of iron, which has the most stable nucleus. (See Example 29.8.) For $A>60$, $E_b/A$ values decrease slowly, indicating that the nucleons are, on average, less tightly bound.

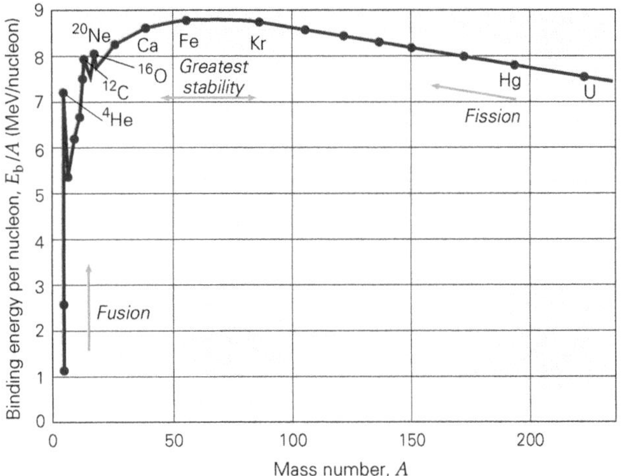


▲**FIGURE 29.14 A plot of binding energy per nucleon versus mass number** If the binding energy per nucleon ($E_b/A$) is plotted versus mass number ($A$), the curve has a maximum near iron (Fe). The nuclei in this region are, on average, the most tightly bound and have the greatest stability. Extremely heavy nuclei can release energy by splitting (*fission*). Extremely light nuclei can combine to release energy (*fusion*).

The importance of the maximum in the curve cannot be understated. Consider what could happen on either side of it. If a massive nucleus were to split, in a process called **fission**, into two lighter nuclei, the resulting nuclei would be more tightly bound, and energy would be released. On the low-mass side of the maximum, if two nuclei could be fused, in a process called **fusion**, a more tightly bound nucleus would be created, and energy would also be released. The details and application of these processes will be discussed in detail in Chapter 30.

From Figure 29.14 it can be seen that the average binding energy per nucleon does not vary much from $E_b/A \approx 8$ MeV/nucleon. Thus to a good approximation, $E_b \propto A$. In other words, the total binding energy is (approximately) proportional to the total number of nucleons. This proportionality points to a characteristic of the nuclear force that is quite different from the electrical force, for example. Suppose that the attractive nuclear force acted between all the pairs of nucleons in a nucleus. Each pair of nucleons would then contribute to the total binding energy. Considering all combinations, statistics tell us that in a nucleus containing $A$ nucleons, there are $A(A-1)/2$ pairs. Thus there would be $A(A-1)/2$ contributions to the total binding energy. For nuclei with $A \gg 1$ (heavy nuclei), $A(A-1) \approx A^2$, and the binding energy would be expected to be proportional to the square of $A$, or $E_b \propto A^2$. This would be the expected result if the nucleon–nucleon force acted over long range. But, as has been seen experimentally, $E_b \propto A$. This result indicates that any nucleon in a nucleus is *not* bound to all the other nucleons. This phenomenon, called *saturation*, means that the nuclear force acts over a short range and that any particular nucleon interacts only with its nearest neighbors.

### 29.4.4 Magic Numbers

The concept of filled shells in atoms was investigated in Chapter 28. There is an analogous effect in the nucleus. Although the concept of individual nucleon "orbits" inside the nucleus is hard to visualize, experimental evidence does indicate the existence of "closed nuclear shells" when the number of protons or neutrons is 2, 8, 20, 28, 50, 82, or 126. Major work on the nuclear shell model was done by Nobel Prize winner Maria Goeppert-Mayer (1906–1972), a German-born physicist.

The number of stable isotopes of various elements provides solid evidence of the existence of such **magic numbers**. If an element has a magic number of protons, it has an unusually high number of stable isotopes. On the other hand, an element whose proton number is far away from a magic number may have only one or two (or even no) stable isotopes. Aluminum, for example, with 13 protons, has just one stable isotope, $^{27}$Al. But tin, with $Z=50$ (a magic number), has ten stable isotopes, ranging from $N=62$ to $N=74$. Neighboring indium, with $Z=49$, has only two stable isotopes, and antimony, with $Z=51$, also has only two.

Another piece of experimental evidence for magic numbers is related to binding energies. High-energy gamma-ray photons can knock out single nucleons from a nucleus (a phenomenon called the *photonuclear effect*) in a manner analogous to the photoelectric effect in metals. Experimentally, a nuclide with a proton magic number, such as tin, requires about 2 MeV *more* photon energy to eject a proton from its nucleus than does a nuclide that does not have a magic number of protons. Thus, magic numbers are associated with extra-large binding energies, another sign of higher-than-average stability.

## 29.5 Radiation Detection, Dosage, and Applications

### 29.5.1 Detecting Radiation

In general, our senses cannot detect radioactive decay directly. Such detection must be accomplished by indirect means. For example, people who work near nuclear reactors wear film badges to indicate exposure to radiation darkening of the film. If more immediate ("real-time") quantitative methods are needed, other instruments are available.

These instruments, as a group, are known as **radiation detectors**. Fundamentally, they are based on the ionization or excitation of atoms, a phenomenon caused by the passage of energetic particles through matter. The electrically charged alpha and beta particles transfer energy to atoms by electrical interactions, removing electrons and creating ions. Gamma-ray photons can produce ionization by the photoelectric effect and Compton scattering.

They may also produce electrons and positrons by pair production, if their energy is large enough. Regardless of the source, the ionization produced by these interactions, not the actual particles, are what is "detected" by a radiation detector.

One of the most common radiation detectors is the *Geiger counter*, developed by Hans Geiger (1882–1945), a student and then colleague of Ernest Rutherford. The principle of the Geiger counter is illustrated in ▼**Figure 29.15**. A voltage is applied between a wire electrode and the outer electrode (a metallic tube) of the Geiger tube, which contains a gas (such as argon) at low pressure. When an ionizing particle enters the tube, the particle ionizes some gas atoms. The freed electrons are accelerated toward the positive anode. On their way, they strike and ionize other atoms. This process snowballs, and the resulting "avalanche" produces a current pulse. The pulse is amplified and sent to an electronic counter that counts the pulses. The pulses are sometimes used to drive a loudspeaker so that particle detection is heard as a click.

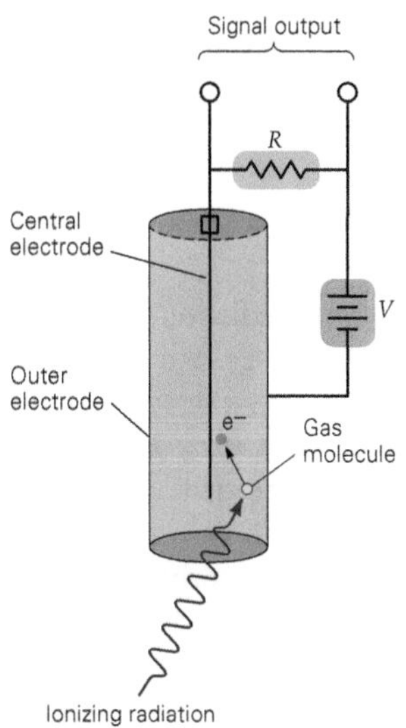


▲ **FIGURE 29.15 The Geiger counter** Incident radiation ionizes a gas atom, freeing an electron that is, in turn, accelerated toward the central (positive) electrode. On the way, this electron produces additional electrons through ionization, resulting in a current pulse that is detected as a voltage across the external resistor *R*.

Another method of detection is the *scintillation counter* (▼**Figure 29.16**). Here, the atoms of a phosphor material [e.g., sodium iodide (NaI)] are excited by an incident particle. A visible-light pulse is emitted when the atoms return to their ground state. This is converted to an electrical pulse by a photoelectric material. This electrical pulse is then amplified in a *photomultiplier tube*, which consists of a series of electrodes of successively higher potential. The photoelectrons are accelerated toward the first electrode and acquire sufficient energy to cause several secondary electrons from ionization to be emitted when they strike the electrode. This process continues, and relatively weak scintillations are converted into sizable electrical pulses, which are then counted electronically.

The previous detectors determine the number of particles that interact in their material. Other methods allow the actual trajectory, or "tracks," of charged particles to be seen and/or recorded. For example, the *cloud chamber* was developed in the early 1900s by C. T. R. Wilson, a British atmospheric physicist. Here supercooled vapor condenses into droplets on the sites of ionized molecules, thus showing the path of an energetic particle. The *bubble chamber*, invented by the American physicist D. A. Glazer in 1952, uses a similar principle. A reduction in pressure causes a liquid to be superheated and able to boil. Ions produced along the path of an energetic particle become sites for bubble formation, and a trail of bubbles is created.

In a *spark chamber*, the path of a charged particle is registered by a series of sparks. The charged particle passes between a pair of electrodes that have a high difference in potential and are immersed in an inert (noble) gas. The charged particle causes the ionization of gas molecules, giving rise to a visible spark (flash of light) between the electrodes as the released electrons travel to the positive electrode. A spark chamber is an array of such electrodes in the form of parallel plates. A series of sparks then marks the particle's path.

Regardless of the detector, once a particle's trajectory is determined, its energy can be found. Typically, a magnetic field is applied across the chamber, charged particles are deflected, and their energy is found from the radius of curvature of its

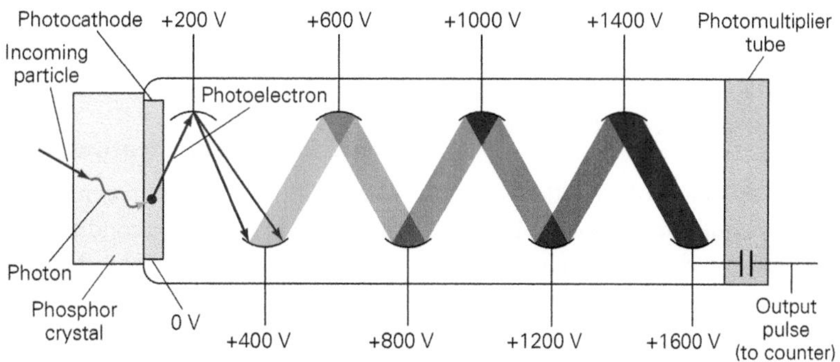


▲ **FIGURE 29.16 The scintillation counter** A photon emitted by a phosphor atom excited by an incoming particle causes the emission of a photoelectron from the photocathode. Accelerated through a difference in potential in a photomultiplier tube, the photoelectrons free secondary electrons when they collide with successive electrodes at higher potentials. After several steps, a relatively weak scintillation is converted into a measurable electric pulse.

path. Gamma rays will not leave tracks in any of these detectors. However, their presence can be detected indirectly because they can produce electrons by the photoelectric effect and/or pair production. The gamma-ray energy can then be found from the measured energy of these electrons.

### 29.5.2 Biological Effects and Medical Applications of Radiation

In medicine, nuclear radiation can be beneficial in both the diagnosis and treatment of some diseases, but it also is potentially harmful if not administered properly. Nuclear radiation and X-rays can penetrate human tissue without pain. However, early investigators quickly learned that large doses or repeated small doses lead to reddened skin, lesions, and other conditions. It is now known that certain types of cancers can be caused by excessive exposure to radiation.

The chief hazard of radiation is damage to living cells, due primarily to ionization. Ions produced by radiation may be highly reactive. Such ions can interfere with the normal chemical operations of the cell. If enough cells are damaged or killed, cell reproduction might not be fast enough, and the irradiated tissue could eventually die. In other instances, genetic damage, or mutation, may occur in a chromosome in the cell nucleus. To continue this topic, let us investigate how the radiation "dose" is quantified.

#### 29.5.2.1 Radiation Dosage

An important consideration in radiation therapy and radiation safety is the amount, or *dose*, of radiation energy absorbed. Several quantities are used to describe this amount in terms of *exposure*, *absorbed dose*, or *equivalent dose*. The earliest unit of dosage, the **roentgen (R)**, is based on exposure and defined in terms of ionization produced in air. One roentgen is the quantity of X-rays or gamma rays required to produce an ionization charge of $2.58 \times 10^{-4}$ C/kg in air.

The **rad** (**r**adiation **a**bsorbed **d**ose) is an *absorbed dose* unit. One rad is an absorbed dose of radiation energy of $10^{-2}$ J/kg of absorbing material. Note that the rad is based on energy absorbed from radiation rather than simply ionization caused by radiation in air (as the roentgen is). As such, it is more directly related to the biological damage. Because of this, the rad has largely replaced the roentgen. Note that the rad is not an SI unit. The SI unit for absorbed dose is the **gray (Gy)**, defined as 1 Gy = 1 J/kg = 100 rad.

An even more meaningful assessment of radiation effects involves measuring the *biological damage* produced, because it is well known that equal doses (in rads) of different types of radiation produce different effects. For example, a relatively massive alpha particle with a charge of $+2e$ moves through the tissue rather slowly, with a great deal of electrical interaction. The ionization sites thus are more localized along a short penetration path. Because of this high density of biological damage, alpha particles can (potentially) do more damage than electrons or gamma rays.

To take into account the difference in biological damage created by different types of radiation, the concept of *effective dose* is used. This dose is measured in terms of the **rem** (**r**ad **e**quivalent **m**an). The various degrees of effectiveness (creating cellular damage) of particles are characterized by the **relative biological effectiveness (RBE)**, or *quality factor* (QF), which is tabulated for various particles in ▼**Table 29.4**. (Note that X-rays and gamma rays have, by definition, an RBE of 1.)

TABLE 29.4 Typical Relative Biological Effectiveness (RBE) Values of Various Types of Radiation

| Type | RBE (or QF) |
|---|---|
| X-rays and gamma rays | 1 |
| Beta particles | 1.2 |
| Slow neutrons | 4 |
| Fast neutrons and protons | 10 |
| Alpha particles | 20 |

The effective dose is given by the product of the dose in rads and the appropriate RBE:

$$\text{effective dose (in rems)} = \text{dose (in rads)} \times \text{RBE} \qquad (29.6a)$$

Thus, 1 rem of *any* type of radiation does approximately the same amount of biological damage. For example, a 20-rem effective dose of alpha particles does the same amount of damage as a 20-rem effective dose of X-rays. However, to administer this dose, 20 rad of X-rays is needed, compared with only 1 rad of alpha particles.

Recall that the SI unit of absorbed dose is the gray. The SI unit of effective dose is the **sievert (Sv)**:

$$\text{effective dose (in sieverts)} = \text{dose (in grays)} \times \text{RBE} \qquad (29.6b)$$

Since (1 Sv = 100 rem), it follows that 1 Sv = 100 rem.

It is difficult to set a maximum permissible radiation dosage, but the general standard for humans is an average dose of 5 rem/y after age 18, with no more than 3 rem in any three-month period. In the United States, the average annual dose per capita is about 200 mrem. About 125 mrem of this comes from the background of cosmic rays and naturally occurring radioactive isotopes in soil, building materials, etc. The remainder is chiefly due to medical applications.

#### 29.5.2.2 Medical Treatment Using Radiation

Some radioactive isotopes can be used for medical treatment, typically for cancerous conditions. Since a radioactive isotope, or a *radioisotope*, behaves chemically like a stable isotope of the element, it can participate in chemical reactions associated with normal bodily functions. One such radioisotope, used to treat thyroid cancer, is $^{131}$I. Under usual conditions, the thyroid gland absorbs normal iodine. However, if $^{131}$I is absorbed in a large enough dose, it can kill cancer cells. To see how a dose of radiation to the thyroid from $^{131}$I can be estimated, consider Example 29.9.

### EXAMPLE 29.9: RADIATION DOSAGE–IODINE-131 AND THYROID CANCER

One method of treating a cancerous thyroid is to administer a hefty amount of the radioactive isotope $^{131}$I. The thyroid absorbs this iodine, and the iodine's gamma rays kill cells in the thyroid. (For data on $^{131}$I, see Example 29.3.) (a) Write the decay scheme of $^{131}$I, and predict the daughter nucleus, which, in this case, is stable after emitting a gamma ray. (b) The charged particle [part (a) tells the type] has an average kinetic energy of 200 keV. Assume that the patient was given 0.0500 mCi of $^{131}$I and the thyroid absorbs only 25%. Further assume that only 40% of that 25% actually decays in the thyroid. If all of the energy carried by the charged particles is deposited in the thyroid, estimate the dose received by the thyroid (mass of 50.0 g) due to the ionization created by the charged particles only. (Do not include gamma rays.)

**THINKING IT THROUGH.** (a) The decay type must be $\beta^-$, since the initial nucleus has too many neutrons (78 compared with 74 for the stable iodine nucleus). The daughter nucleus can be determined from its proton number. The daughter is left in an excited state and emits a gamma ray to become stable. (b) The dose depends on the energy deposited per kilogram of thyroid. Hence, what is needed is the number of $\beta^-$ particles emitted (and therefore absorbed). This is determined from the initial number of $^{131}$I nuclei that are present in the thyroid. The effective dose will depend on the RBE for $\beta^-$ particles, found in Table 29.4.

**SOLUTION**

***Given:***

0.0500 mCi of $^{131}$I ingested
25% of the $^{131}$I makes it to the thyroid
40% of that $^{131}$I that makes it to the thyroid decays there
$\bar{K}_\beta = 200\,\text{keV}$
$m = 50.0\text{ g} = 0.0500\text{ kg}$
RBE (from Table 29.4)

***Find:***

**(a)** The decay scheme for $^{131}$I
**(b)** The dose (in rem) to the thyroid

**(a)** The element with $Z=54$ is xenon (*Xe*). The decay scheme is therefore

$$\begin{array}{ccccc} {}^{131}_{53}\text{I} & \rightarrow & {}^{131}_{54}\text{Xe}^* & + & \beta^- \\ \text{iodine} & & \text{xenon (excited)} & & \text{beta} \\ & & \downarrow & & \\ & & {}^{131}_{54}\text{Xe} & + & \gamma \\ & & \text{xenon} & & \text{gamma ray} \end{array}$$

**(b)** Of the 0.0500 mCi, only 0.0125 mCi makes it to the thyroid. Of that, only 40%, or 0.00500 mCi ($5.00 \times 10^{-6}$ Ci), decays in the thyroid. From this and Equation 29.2, the number of $^{131}$I nuclei that decay in the thyroid can be found. From Example 29.3, the decay constant for $^{131}$I is $\lambda = 1.0 \times 10^{-6}\text{ s}^{-1}$.

Thus, the number of $^{131}$I nuclei, *N*, that decay in the thyroid is

$$N = \frac{R}{\lambda} = \frac{5.00\times10^{-6}\text{Ci}(3.7\times10^{10}(\text{nuclei/s/Ci}))}{1.0\times10^{-6}\,\text{s}^{-1}}$$
$$= 1.85\times10^{11}\ {}^{131}\text{I nuclei}$$

Each $^{131}$I nucleus releases one $\beta^-$ particle, with an average kinetic energy of 200 keV. Remembering the conversion factor $1.60 \times 10^{-19}$ J/eV or $1.60\times10^{-16}$ J/keV, the energy, *E*, deposited in the thyroid is

$$E = (1.85\times10^{11}\ {}^{131}\text{I nuclei})\left(200\ \frac{\text{keV}}{{}^{131}\text{I nuclei}}\right)\left(1.60\times10^{-16}\ \frac{\text{J}}{\text{keV}}\right)$$
$$= 5.92\times10^{-3}\,\text{J}$$

The absorbed dose is

$$\text{absorbed dose} = \frac{5.92\times10^{-3}\,\text{J}}{0.0500\,\text{kg}} = 0.118\,\text{J/kg}$$
$$= 0.118\,\text{Gy or } 11.8\,\text{rad}$$

The effective dose (in sieverts and rems) is this result multiplied by the RBE for betas (1.2):

$$\text{effective dose} = (0.118\,\text{Gy})(1.2) = 0.142\,\text{Sv} = 14.2\,\text{rem}$$

**FOLLOW-UP EXERCISE.** In this Example, determine the absorbed and effective dose from the gamma rays, assuming that 10% of the gamma rays are absorbed in the thyroid and that their energy is 364 keV. Compare these results with the dose from the $\beta^-$ radiation.

Radiation has always been a double-edged sword. Its potentially harmful side is well known, yet radiation can also provide solutions to problems. For example, gamma radiation exposure is an approved method of food sterilization in the United States. External radiation sources such as $^{60}$Co, an emitter of high-energy gamma rays with a relatively long half-life, are routinely used to treat cancer.

#### 29.5.2.3 Medical Diagnosis Applications That Use Radiation

Besides therapeutic use, such as previously described for $^{131}$I, radioactive isotopes are used for diagnosis. Since the radioisotope behaves chemically like a stable isotope, attaching radioisotopes to molecules enables those molecules to act as tracers as they travel to different organs. For example, the activity of the thyroid gland can be determined by monitoring its iodine uptake with small amounts of radioactive $^{123}$I. This isotope emits gamma rays and has a half-life of 13.3 h. The uptake of radioactive iodine by a thyroid can be monitored by a gamma detector and compared with the function of a normal thyroid to check for abnormalities.

Another commonly used diagnostic tracer is technetium-99 ($^{99}$Tc). It has a convenient half-life of 6 h, emits gamma rays, and combines with a large variety of compounds. When injected into the bloodstream, $^{99}$Tc will not be absorbed by the brain, because of the blood–brain barrier. However, tumors do not have this barrier, and brain tumors readily absorb $^{99}$Tc. These tumors will show up as gamma-ray emitting sites as measured by external detectors. Similarly, other areas of the body can be scanned, and unusual activities noted and measured.

***Positron emission tomography*** (PET) uses tracers that are positron emitters, such as $^{11}$C and $^{15}$O. When a positron is emitted, it is quickly annihilated, and two gamma rays are produced that travel in opposite directions. These are recorded in coincidence by a ring of detectors surrounding the patient, thus pinpointing the radioisotope's location. In a common application, PET is used to detect fast-growing cancer cells. The positron emitter $^{18}$F is chemically attached to glucose molecules and administered to the patient. Normal cells absorb glucose, but *very* active cancer cells absorb considerably more. By comparing the emissions coming from a patient to those from a healthy person, these "overactive" cancer cells can be detected.

### 29.5.3 Domestic and Industrial Applications of Radiation

A common application of radioactivity in the home is the smoke detector. In this detector, a weak radioactive source ionizes air molecules. The freed electrons and the positive ions are collected using the voltage of a battery, thus setting up a small current in the detector circuit. If smoke enters the detector, the ions there become attached to the smoke particles, causing a reduction in the current. This drop is sensed electronically, which triggers an alarm (▼**Figure 29.17**).

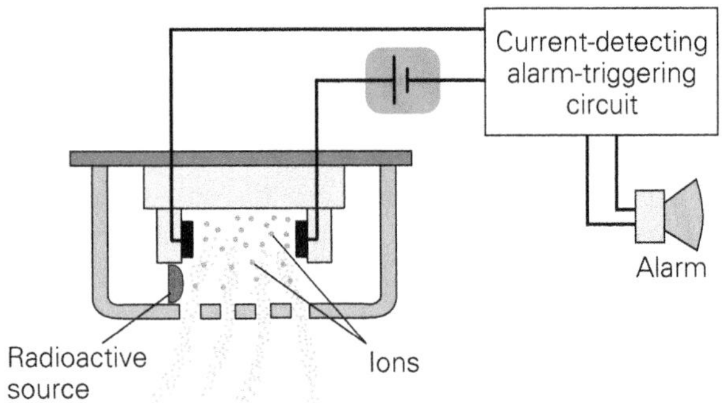


▲FIGURE 29.17 **Sketch of a smoke detector** A weak radioactive source ionizes the air and sets up a small current. Smoke particles that enter the detector attach to some of the ionized electrons, thereby reducing the current, causing an alarm to sound.

Industry also makes good use of radioactive isotopes. Radioactive tracers can be used to determine flow rates in pipes and to detect leaks. Also, it is possible to *radioactivate* certain compounds using a technique called **neutron activation analysis**. This technique provides an important method of identifying elements in a sample. A big advantage of neutron activation analysis over chemical and optical methods is that the former requires much smaller samples and can detect very minute amounts of an element.

As an example, consider starting with californium-252, an unstable neutron emitter that can be produced artificially. Its decay scheme is $^{252}_{98}\text{Cf} \rightarrow ^{251}_{98}\text{Cf} + ^{1}_{0}\text{n}$. These neutrons can then be used to bombard a sample and create gamma rays characteristic of the nuclei in that sample. A common target is nitrogen, consisting mostly of $^{14}$N. When $^{14}$N absorbs a neutron, the $^{15}$N nucleus is usually created in an excited state. The excited $^{15}$N nucleus then decays with the emission of a gamma ray with a distinctive energy. This sequential reaction is

$$^{1}_{0}\text{n} + ^{14}_{7}\text{N} \rightarrow \underset{\text{(excited nucleus)}}{^{15}_{7}\text{N}} \rightarrow ^{15}_{7}\text{N} + \underset{\text{(gamma ray)}}{\gamma}$$

An energy-sensitive gamma-ray detector is placed near the sample, and detection of gamma rays of this distinctive energy indicates the presence of nitrogen. The detection of nitrogen via this activation technique is now used as an important anti-terrorism tool. This is because virtually all explosives contain nitrogen. Thus, neutron activation checks packages, luggage, etc. for explosive devices without opening or dismantling the object of interest. (Other non-explosive materials may also contain nitrogen, so manual checks are made to confirm any findings.)

## Chapter 29 Review

- The nucleus of an atom contains protons and neutrons, collectively called **nucleons**.
- The nucleus is characterized by its **proton number *Z*** and its **neutron number *N***. Its **mass number, *A***, is the total number of nucleons, so $A = N + Z$.
- **Isotopes** of a given element differ only in the number of neutrons in their nucleus.
- Nuclei may undergo **radioactive decay** by the emission of an **alpha particle** (a helium nucleus) ($\alpha$); a **beta particle**, which can be either an electron ($\beta^-$) or a positron ($\beta^+$); or a **gamma ray** ($\gamma$), a high-energy photon of electromagnetic radiation. Some nuclei become more stable by capturing an orbital electron (electron capture, or **EC**).
- In any nuclear process, two conservation rules pertain: **conservation of nuclear number** and **conservation of charge**.
- The **half-life** of a nuclide is the time required for the number of undecayed nuclei in a sample to fall to half of its initial value. The number of undecayed nuclei remaining after a time *t* is given by

$$N = N_o e^{-\lambda t} \tag{29.3}$$

where the decay constant ($\lambda$) is inversely related to the half-life by

$$t_{1/2} = \frac{0.693}{\lambda} \tag{29.4}$$

- The **activity (*R*)** of a radioactive sample is the rate at which its nuclei decay. It is proportional to the number of undecayed nuclei and is given by

$$R = \text{activity} = \left|\frac{\Delta N}{\Delta t}\right| = \lambda N \qquad (29.2)$$

Activity is measured in units of the curie (Ci) or the becquerel (Bq). $1\text{ Ci} = 3.70 \times 10^{10}$ decays/s, and $1\text{ Bq} \equiv 1$ decay/s; thus, $1\text{ Ci} = 3.70 \times 10^{10}$ Bq.

- The **atomic mass unit (u)** is related to the kilogram by $1\text{ u} = 1.660\ 54 \times 10^{-27}$ kg. 1 u of mass is equivalent to 931.5 MeV.
- The **total binding** energy (***E***$_b$) of a nucleus is the minimum amount of energy needed to separate a nucleus into its constituent nucleons:

$$E_b = (\Delta m)c^2 \qquad (29.5)$$

where Δm is the mass defect. The mass defect is the difference between the sum of the masses of the constituent nucleons and the mass of the nucleus.

- The **effective dose** of radiation [**rems** or **sieverts (Sv)**] is determined by the energy deposited per kilogram of material [**rads** or **grays (Gy)**] and the type of particle depositing that energy (as expressed by the **relative biological effectiveness**, or **RBE**):

$$\text{effective dose (in rems)} = \text{dose (in rads)} \times \text{RBE} \qquad (29.6c)$$

$$\text{effective dose (in sieverts)} = \text{dose (in grays)} \times \text{RBE} \qquad (29.6d)$$

# End of Chapter Questions and Exercises

## Multiple Choice Questions

**29.1 Nuclear Structure and the Nuclear Force**

1. In the Rutherford scattering experiment, which target nucleus would alpha particles of a given kinetic energy approach more closely: (a) carbon, (b) iron, or (c) lead?
2. The nuclei of oxygen-16 and oxygen-17 have the same number of (a) nucleons, (b) neutrons, (c) protons, or (d) none of the preceding.
3. At a given distance, between which pair of particles is the nuclear force largest: (a) neutron–proton, (b) neutron–neutron, (c) proton–proton, or (d) the force is the same for all pairs?
4. How many neutrons are there in the nuclide $^{25}$Mg: (a) 25, (b) 12, or (c) 13?
5. The element X has a nucleus given by $^{7}X_{4}$. This element is (a) nitrogen, (b) lithium, (c) beryllium, (d) none of these.

**29.2 Radioactivity**

6. The conservation of nucleons and the conservation of charge apply to (a) only alpha decay, (b) only beta decay, (c) only gamma decay, (d) all nuclear decay processes.
7. $\beta^-$ decay can occur only in nuclei with what *Z* values: (a) $Z > 82$, (b) $Z \leq 82$, or (c) it can occur regardless of the *Z* value?
8. Aluminum has only one stable isotope, $^{27}$Al. $^{26}$Al would be expected to decay by which decay mode: (a) $\beta^+$, (b) $\beta^-$, or (c) neither type of beta decay is an option for $^{26}$Al?
9. The alpha decay of $^{237}Np_{144}$ results in a nucleus containing how many protons: (a) 233, (b) 91, or (c) 142?
10. The gamma decay of $^{89}Y^*$ results in a nucleus containing how many neutrons: (a) 89, (b) 39, or (c) 50?

**29.3 Decay Rate and Half-Life**

11. After one half-life, a sample of a radioactive material (a) is half as massive, (b) has its half-life reduced by half, (c) is no longer radioactive, (d) has its activity reduced by half.
12. In two half-lives, the activity of a radioactive sample will have decreased from its original value to what percent: (a) 25%, (b) 50%, (c) 75%, or (d) 87.5%?
13. A sample containing an alpha emitter gives off alpha particles of kinetic energy of 4.4 MeV. After three half-lives, what is the energy of the alpha particles given off: (a) 2.2 MeV, (b) 1.1 MeV, (c) 8.8 MeV, or (d) 4.4 MeV?
14. Radioactive element A has a half-life of 3.5 days and B's half-life is 7.0 days. How do their decay constants compare: (a) $\lambda_A = \lambda_B$, (b) $\lambda_A < \lambda_B$, or (c) $\lambda_A > \lambda_B$?
15. Radioactive element A has a decay constant three times greater than that of B. How do their half-lives compare: (a) *A*'s = 3 × *B*'s, (b) *B*'s = 3 × *A*'s, or (c) their relative half-lives can't be determined from the data?

**29.4 Nuclear Stability and Binding Energy**

16. For nuclei with $A > 40$, which of the following is correct: (a) the number of protons is approximately equal to the number of neutrons; (b) the number of protons exceeds the number of neutrons; (c) all such nuclei are stable up to $Z = 92$; or (d) none of these?
17. The average binding energy per nucleon of the daughter nucleus in a decay process is (a) greater than, (b) less than, or (c) equal to that of the parent nucleus.
18. From which nucleus is it easier to remove a neutron: (a) $^{26}$Mg, (b) $^{25}$Mg, or (c) they require the same amount of energy?
19. Isotopes of which of these elements would require more energy per nucleon to completely break them apart: (a) Fe, (b) Cd, or (c) Au?
20. During which process is the average binding energy of the product nuclei less than the average binding energy of the initial nuclei: (a) fission, (b) fusion, (c) neither of these processes, or (d) both of these processes?

**29.5 Radiation Detection, Dosage, and Applications**

21. Which type of detector records the trajectory of charged particles: (a) Geiger counter, (b) scintillation counter, or (c) spark chamber?
22. A bubble chamber (in a magnetic field) shows two tracks of equal curvature but in opposite directions, emanating from a point in space with no apparent incoming particle. It is highly likely that this event is (a) alpha decay, (b) beta decay, (c) pair production.
23. The same effective radiation dose (in rems) is given by a source of slow neutrons and an X-ray machine. What is the ratio of the dose (in rads) from the neutrons to that from the X-rays: (a) 1 : 2, (b) 1:4, (c) 2:1, or (d) 4:1?
24. The same radiation dose (in rads) is given by a source of fast neutrons and an alpha particle source. What is the ratio of the effective dose (in rems) from the neutrons to that from the alpha source: (a) 1:2, (b) 1:4, (c) 2:1, or (d) 4:1?

## Conceptual Questions

**29.1 Nuclear Structure and the Nuclear Force**

1. In the Rutherford scattering experiment, the minimum distance of approach for the alpha particle is given by Equation 29.1. Explain why this distance does not necessarily represent the nuclear radius. Is it larger or smaller than the nuclear radius?
2. A Rutherford scattering experiment uses a beam of alpha particles of a known kinetic energy. Qualitatively compare the distance of closest approach ($r_{min}$) to a target consisting of lead atoms to that to a target of uranium atoms. Give a numerical answer as a ratio of the closest approach for lead to uranium.
3. A Rutherford scattering experiment is performed on a gold foil target with alpha beams of two different kinetic energies, one of 3.00 MeV, the second of 6.00 MeV. Qualitatively compare the distance of closest approach ($r_{min}$) for the two different energy alpha beams. Give a numerical answer as a ratio of the closest approach for the more energetic alphas to the less energetic ones.
4. Nuclei with the same number of neutrons are called *isotones*. What nitrogen nucleus is an isotone of carbon-13?
5. Nuclei with the same number of nucleons are called *isobars*. What nuclide of nitrogen is an isobar of carbon-13?
6. Nuclei with the same number of protons are called *isotopes*. List the stable nuclides that are isotopes of the most common form of nitrogen.

**29.2 Radioactivity**

7. Neither neutron nor proton number is conserved in either type of beta decay. Is this a violation of the conservation of nucleons? Explain.
8. $^{19}F$ is the only stable isotope of fluorine. What two possible decay modes would you expect $^{18}F$ to decay by? What would be the resulting nucleus in each case?
9. When an excited nucleus decays to a lower energy level by gamma-ray emission, the actual energy of the gamma-ray photon is a bit less than the difference in energy between the two levels involved in the transition. Explain why this is true. [*Hint*: Think about conservation of linear momentum in a two-body "explosion" and see Chapter 6.]
10. Suppose, in an alpha decay, the emitted alpha particle has a kinetic energy of 5.25 MeV. The total kinetic energy released (from mass energy) is actually 5.30 MeV. Explain where the difference went. [*Hint*: See the hint in Conceptual Question 9.]
11. During a particular decay sequence in a decay chain, a nucleus first decays by alpha emission, followed by a $\beta^-$ emission, and last a gamma-ray emission. Compare the number of neutrons, protons, and nucleons in the final nucleus to that in the initial nucleus. How would your answer differ if the $\beta^-$ decay had occurred before the alpha decay, with the gamma emission still coming at the end? Explain.
12. A basic assumption of radiocarbon dating is that the cosmic-ray intensity has been generally constant for the last 40 000 years or so. Suppose it were found that the intensity 100 000 years ago was much less than it is today. How would this finding affect the results (ages of samples) of carbon-14 dating?

**29.3 Decay Rate and Half-Life**

13. How do physical or chemical properties affect the decay rate, or half-life, of a radioactive isotope?
14. Nuclide A has a decay constant that is half that of nuclide B. Samples of both types start with the same number of undecayed nuclei, *N*. In terms of *N*, after two of A's half-lives have elapsed, (a) how many of A have decayed? (b) How many of B have decayed?
15. What are the (a) half-life and (b) decay constant for a stable isotope?
16. Nuclide A has a half-life that is one-third that of nuclide B. The type A sample starts with twice the number of undecayed nuclei as in the sample of type B. Compare the initial sample activities expressed as a ratio of the activity of sample A to that of B.
17. What is the expression for the number of nuclei that have decayed, *N*, in a given time *t* in terms of the initial number of undecayed nuclei $N_o$, the time elapsed *t*, and the decay constant $\lambda$? [*Hint:* See Equation 29.3.]

**29.4 Nuclear Stability and Binding Energy**

18. Using the general guidelines for nuclear stability, explain why aluminum has only one stable isotope ($^{27}Al$). Why aren't other isotopes, such as $^{28}Al$ or $^{26}Al$, stable?
19. Explain why, of the two main uranium isotopes, $^{238}U$ is more abundant than $^{235}U$. [*Hint*: Although they both are unstable, $^{238}U$ is closer to stability; why?]

20. Compared to $^3$He, the probability of absorbing a neutron is much less likely for $^4$He. Explain this using what you know about odd–even proton and neutron numbers.
21. Explain how fusing of light nuclei and splitting of heavy nuclei can *both* release energy.
22. If one nucleus has a larger average binding energy than another, does this mean that it takes more total energy to break the former into its constituent nucleons? Explain.

**29.5 Radiation Detection, Dosage, and Applications**

23. If X-rays and alpha particles give the same dose, how will their effective doses compare?
24. If X-rays and slow neutrons give the same effective dose, how will their doses compare?
25. PET scans require extremely fast computers coupled with gamma-ray detectors capable of accurate energy measurements. Explain why both energy accuracy and comparison of arrival times are crucial to the success of a PET scan. [*Hint:* In a PET scan, two opposing detectors pick up pair annihilation gamma-ray photons.]
26. A Geiger counter is not 100% efficient. That is, the number of events it records is smaller than the number of particles that enter it. Explain why the efficiency is not 100%.
27. Theoretically, the tumor-killing efficiency of particle beams is much greater than that of a beam of X-rays or gamma rays. Explain why this is true.

## Exercises*

*Integrated Exercises (IEs) are two-part exercises. The first part typically requires a conceptual answer choice based on physical thinking and basic principles. The following part is quantitative calculations associated with the conceptual choice made in the first part of the exercise.*

**29.1 Nuclear Structure and the Nuclear Force**

1. • Determine the number of protons, neutrons, and electrons in a neutral atom with the following nuclei: (a) $^{90}$Zr and (b) $^{208}$Pb.
2. • Magnesium has three stable isotopes. Write these isotopes in nuclear notation including nucleon, proton, and neutron number on the elemental symbol.
3. • An isotope of potassium has the same number of neutrons as argon-40. Write this potassium isotope in nuclear notation.
4. •$^{35}$Cl and $^{37}$Cl are two isotopes of chlorine. What are the numbers of protons, neutrons, and electrons in each if (a) the atom is electrically neutral, (b) the ion has a –2 charge, and (c) the ion has a +1 charge?
5. • One isotope of uranium has a mass number of 235, and another has a mass number of 238. What are the numbers of protons, neutrons, and electrons in a neutral atom of each?
6. IE • (a) Isotopes of an element must have the same (1) atomic number, (2) neutron number, (3) mass number. (b) Write two possible isotopes for gold-197.
7. •• An experimental expression for the radius ($R$) of a nucleus is $R = R_o A^{1/3}$, where $R_o = 1.2 \times 10^{-15}$ m and $A$ is the mass number of the nucleus. (a) Find the nuclear radii of the noble gases: He, Ne, Ar, Kr, Xe, and Rn. (b) Determine the density of the nuclei associated with each of these species and compare them. Does your answer surprise you?
8. IE ••• Assume Rutherford used alpha particles with a kinetic energy of 5.25 MeV. (a) To which of the following nuclei would the alpha particle come closest in a head-on collision: (1) aluminum, (2) iron, or (3) lead? (b) Determine the distance of closest approach for the three nuclei in part (a) and compare them to the nuclear radii given in Exercise 29.7. Are any of these distances comparable to the radius of the target nucleus?

**29.2 Radioactivity**

9. IE • Tritium is radioactive. (a) Would you expect it to (1) $\beta^+$, (2) $\beta^-$, or (3) alpha decay? Why? (b) Write the nuclear equation for the correct decay and identify the daughter nucleus. Is it stable?
10. • Write the nuclear equations for (a) beta decay of $^{60}$Co and (b) alpha decay of $^{222}$Rn.
11. • Write the nuclear equations for (a) alpha decay of $^{237}$Np, (b) $\beta^-$ decay of $^{32}$P, (c) $\beta^+$ decay of $^{56}$Co, (d) electron capture in $^{56}$Co, and (e) $\gamma$ decay of $^{42}$K from an excited nuclear state to the ground state.
12. IE • Polonium-214 can decay by alpha decay. (a) The product of its decay has how many fewer protons than polonium-214: (1) zero, (2) one, (3) two, or (4) four? (b) Write the nuclear equation for this decay and determine the daughter nucleus.
13. • A lead-209 nucleus results from both alpha-beta sequential decays and beta–alpha sequential decays. What was the grandparent nucleus? Show this result for both decay routes by writing the nuclear equations for both sequential decay processes.
14. •• Complete the following nuclear decay equations by filling in the blanks:
    a. $^{8}_{4}\text{Be} \rightarrow \_\_\_ + ^{4}_{2}\text{He}$
    b. $^{240}_{94}\text{Pu} \rightarrow \_\_\_ + ^{139}_{56}\_\_\_ + 2\left(^{1}_{0}\text{n}\right)$
    c. $\_\_\_ \rightarrow ^{47}_{21}\text{Sc} + \gamma$
    d. $^{22}_{11}\text{Na} \rightarrow \_\_\_ + ^{22}_{10}\_\_\_$
15. •• Complete the following nuclear-decay equations by filling in the blanks:
    a. $^{238}_{92}\_\_\_ \rightarrow ^{234}_{90}\_\_\_ + \_\_\_$
    b. $^{40}_{19}\text{K} \rightarrow ^{40}_{20}\text{Ca} + \_\_\_$
    c. $^{236}_{92}\text{U} \rightarrow ^{131}_{53}\text{I} + \_\_\_\left(^{1}_{0}\text{n}\right) + ^{102}_{39}\_\_\_$
    d. $^{23}_{11}\text{Na}^* \rightarrow \gamma + \_\_\_$
    e. $^{11}\text{C} + \_\_\_ \rightarrow \_\_\_\text{B}$

* The bullets denote the degree of difficulty of the exercises: •, simple; ••, medium; and •••, more difficult.

16. •• Actinium-227 decays by alpha decay or beta decay and is part of a long decay sequence, shown in Figure 29.8. Write all the possible nuclear decays for the decay series from $^{227}$Ac to $^{215}$Po. Identify the daughter nucleus at the end of each decay.
17. ••• The decay series for neptunium-237 is shown in ▼**Figure 29.18**. (a) What is the decay mode of each of the decays? (b) Determine the daughter nucleus at the end of each decay.

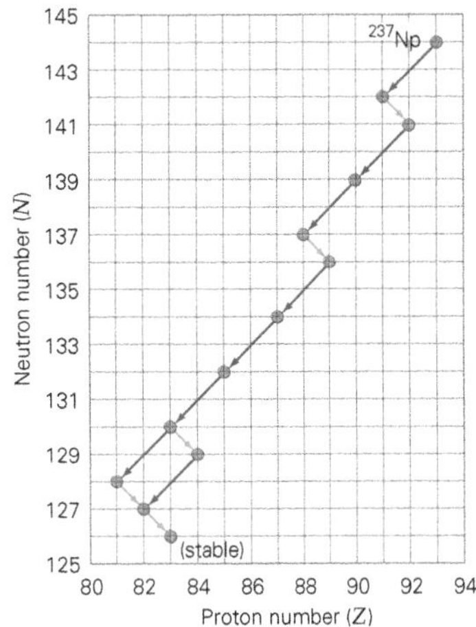


▲**FIGURE 29.18 Neptunium-237 decay series** See Exercise 29.17.

### 29.3 Decay Rate and Half-Life

18. • A particular radioactive sample undergoes $2.50 \times 10^6$ decays/s. What is the activity of the sample in (a) curies and (b) becquerels?
19. • At present, a radioactive beta source with a long half-life has an activity of 20 mCi. (a) What is the present decay rate in decays per second? (b) Assuming that one beta particle is emitted per decay, how many are emitted per minute?
20. IE • The half-life of a radioactive isotope is known to be exactly 1 h. (a) What fraction of a sample would be left after exactly 3 h: (1) one-third, (2) one-eighth, or (3) one-ninth? (b) What fraction of a sample would be left after exactly 1 day?
21. • A 1.25-$\mu$Ci alpha source gives off alpha particles each with a kinetic energy of 2.78 MeV. At what rate (in watts) is kinetic energy being produced?
22. •• A sample of technetium-104, which has a half-life of 18.0 min, has an initial activity of 10.0 mCi. Determine the activity of the sample after exactly 1 h has elapsed.
23. •• Calculate the time required for a sample of radioactive tritium to lose 80.0% of its activity. (Tritium has a half-life of 12.3 years.)
24. •• $^{131}$I is given to a patient for use in a diagnostic procedure on her thyroid. What percentage of the $^{131}$I sample remains after exactly one day, assuming that all of the $^{131}$I is retained in the patient's thyroid gland? (Answer to three significant figures.)
25. IE •• Carbon-14 dating is used to determine the age of some unearthed bones. (a) If the activity (in decays/min per gram of carbon) of bone A is higher than that of B, then A is (1) older than, (2) younger than, (3) the same age as B. Explain your reasoning. (b) A sample of A is found to have 4.0 beta decays/(min·g) of carbon, while B's activity is only 1.0 beta decay/(min·g) of carbon. What is the age difference between the bones?
26. •• Prove that the number $N$ of radioactive nuclei *remaining* in a sample after an integer number ($n$) of half-lives has elapsed is given by $N = (N_o/2^n) = (1/2)^n N_o$, where $N_o$ is the initial number of nuclei present.
27. •• Suppose that some ancient writings on parchment are found sealed in a jar in a cave. If carbon-14 dating shows the parchment to be 28 650 years old, what percentage of the original carbon-14 atoms still remains in the sample?
28. •• (a) What is the decay constant of fluorine-17 if its half-life is known to be 66.0 s? (b) How long will it take for the activity of a sample of $^{17}$F to decrease to 80% of its initial value? (c) Repeat part (b), but instead determine the time to decrease to an additional 20% to 60% of its initial value. Does it take twice as long to decay to 60% compared to 80% of its initial activity? Explain.
29. •• Francium-223 $\left(^{223}_{87}\text{Fr}\right)$ has a half-life of 21.8 min. (a) How many nuclei are initially present in a 25.0-mg sample of $^{223}_{87}$Fr? (b) What is its initial activity? (c) How many nuclei will be present 1 h and 49 min later? (d) What will be the sample's activity at this later time?
30. •• A basement room containing radon gas ($t_{1/2} = 3.82$ days) is sealed to be airtight. (a) If $7.50 \times 10^{10}$ radon atoms are trapped in the room, estimate how many radon atoms remain in the room after one week. (b) Radon undergoes alpha decay. After 30 days, is the number of its daughter nuclei equal to, or less than, the number of radon parents that have decayed? Explain your reasoning.
31. •• In 1898, Pierre and Marie Curie isolated about 10 mg of radium-226 from eight tons of uranium ore. If this sample had been placed in a museum, (a) how much of the radium would remain in the year 2109? (b) How many radium nuclei would have decayed during this time?
32. •• An ancient artifact is found to contain 250 g of carbon and has an activity of 475 decays/min. (a) What is the approximate age of the artifact, to the nearest thousand years? (b) What would its initial activity have been?
33. ••• The recoverable U.S. reserves of high-grade uranium-238 ore (high-grade ore contains about 10 kg of $^{238}U_3O_8$ per ton) are estimated to be about 500 000 tons. Neglecting any geological changes, what mass of $^{238}$U existed in this high-grade ore when the Earth was formed, about 4.8 billion years ago?
34. ••• Nitrogen-13, with a half-life of 10.0 min, decays by beta emission. (a) Write down the decay equation to determine the daughter product and whether the beta particle is a positron or electron. (b) If a sample of pure

$^{13}N$ has a mass of 1.50 g at a certain time, what is the activity min later? (c) What percentage of the sample is $^{13}N$ at this time? (d) What alternative process could have happened to the nitrogen-13? Write its decay equation and determine the daughter product for this process.

## 29.4 Nuclear Stability and Binding Energy

35. • Which one of each of the following pairs of nuclei should it be easier to remove a neutron from? (a) $^{16}_{8}O$ or $^{17}_{8}O$, (b) $^{40}_{20}Ca$ or $^{42}_{20}Ca$, (c) $^{10}_{5}B$ or $^{11}_{5}B$, (d) $^{208}_{82}Pb$ or $^{209}_{83}Bi$? State your reasoning in each case.
36. • Only two isotopes of Sb (antimony, $Z=51$) are stable. Pick the two most likely stable isotopes from the following list and explain your rationale. (a) $^{120}Sb$, (b) $^{121}Sb$, (c) $^{122}Sb$, (d) $^{123}Sb$, (e) $^{124}Sb$.
37. • The total binding energy of $^{2}_{1}H$ is 2.224 MeV. Use this information to compute the mass (in u) of a $^{2}H$ nucleus from the known mass of the proton and the neutron.
38. • Use Avogadro's number to show that $1\ u=1.66\times10^{-27}$ kg. [*Hint*: Recall that a $^{12}C$ atom has a mass of exactly 12 u.]
39. •• (a) What is the total binding energy of the $^{12}C$ nucleus? (b) What is its average binding energy per nucleon?
40. •• The mass of the nuclide $^{16}_{8}O$ is 15.994 915 u. (a) What is the total binding energy for this nucleus? (b) Determine its average binding energy per nucleon.
41. •• Which isotope of hydrogen has (a) the highest total binding energy and (b) the lowest average binding energy per nucleon, deuterium or tritium? Justify your answer mathematically.
42. •• Near high-neutron areas, such as a nuclear reactor, neutrons will be absorbed by protons (the hydrogen nucleus in water molecules) and will give off a gamma ray of a characteristic energy in the process. (a) Write the equation for the neutron absorption process and the subsequent gamma decay of the product nucleus. (b) What is the energy of the gamma ray (to three significant figures)? (Neglect nuclear recoil effects.)
43. •• (a) How much energy (to four significant figures) would be required to completely separate all the nucleons of a nitrogen-14 nucleus, the atom of which has a mass of 14.003 074 u? (b) Compute the average binding energy per nucleon of this nuclide.
44. •• Calculate the binding energy of the last neutron in the $^{40}_{19}K$ nucleus.
45. •• Suppose an alpha particle could be removed intact from an aluminum-27 nucleus ($m=26.981\ 541$ u). (a) Write the equation that represents this process and determine the daughter nuclide. (b) If the daughter nuclide has mass of 22.989 770 u, how much energy would be required to perform this operation?
46. •• On average, determine whether the nucleons are more tightly bound in a $^{27}Al$ nucleus or in a $^{23}Na$ nucleus.
47. •• The atomic mass of $^{236}_{92}U$ is 235.043 925 u. Find the average binding energy per nucleon for this isotope.
48. ••• The mass of $^{8}_{4}Be$ is 8.005 305 u. (a) Which is less, the total mass of two alpha particles or the mass of the $^{8}Be$ nucleus? (b) Which is greater, the total binding energy of the $^{8}Be$ nucleus or the total binding energy of two alpha particles? (c) On the basis of your answers to parts (a) and (b) alone, do you expect the $^{8}Be$ nucleus to decay spontaneously into two alpha particles?

## 29.5 Radiation Detection, Dosage, and Applications

49. • In a diagnostic procedure, a patient in a hospital ingests 80 mCi of gold-198 ($t_{1/2}=2.7$ days). What is the activity at the end of one month, assuming none of the gold is eliminated from the body by biological functions?
50. • A technician working at a nuclear reactor facility is exposed to a slow neutron radiation and receives a dose of 1.25 rad. (a) How much energy is absorbed by 200 g of the worker's tissue? (b) Was the maximum permissible radiation dosage exceeded?
51. • A person working with several nuclear isotope separation processes for a two-month period receives a 0.50-rad dose from a gamma source, a 0.30-rad dose from a slow-neutron source, and a 0.10-rad dose from an alpha source. Was the maximum permissible radiation dosage exceeded?
52. •• *Neutron activation analysis* was performed on small pieces of hair that had been taken from the exiled Napoleon after he died on the island of St. Helena in 1821. This procedure involves exposing the samples to a source of neutrons. Some (stable) arsenic nuclei, if present in the sample, will absorb a neutron. In Napoleon's case the samples contained abnormally high levels of arsenic, which supported the theory that his death was not due to natural causes. (a) These results came from studying beta emissions of the resulting $^{76}As$, nucleus. Write the nuclear equation for the neutron absorption and use it to determine the arsenic isotope initially present in the hair. (b) Write the nuclear equation for the subsequent beta decay of $^{76}As$. Use it to determine the nucleus after this decay.
53. ••• A cancer treatment called the *gamma knife* uses focused $^{60}Co$ sources to treat tumors. Each $^{60}Co$ nucleus emits two gamma rays, of energy 1.33 and 1.17 MeV, in quick succession. Assume that 50.0% of the total gamma-ray energy is absorbed by a tumor. Further assume that the total activity of the $^{60}Co$ sources is 1.00 mCi, the tumor's mass is 0.100 kg, and the patient is exposed to the gamma radiation for an hour. Determine the effective radiation dose received by the tumor. (Since the $^{60}Co$ half-life is 5.3 years, changes in its activity during treatment are negligible.)

# 30

# Nuclear Reactions and Elementary Particles

**This neutrino detection laboratory is helping to unravel the details of the history and structure of our universe.**

This chapter-opening photo shows the surface structure of the IceCube Neutrino Observatory at the South Pole. It is designed to detect elementary particles called neutrinos from deep space by their interactions with the ice pack. With this instrument and others such as the Large Hadron Collider (LHC), scientists hope to examine details of the Big Bang and better understand the nature of our universe and the laws that govern the building blocks of matter – elementary particles, including these neutrinos.

This chapter begins by considering nuclear reactions involving the age-old alchemist dream – the transmutation, or conversion, of one element into another. This will lead us to practical applications of energy generation, as well as an understanding of the mechanism that generates the tremendous energy streaming from stars like our Sun. Last, current theories of elementary particles and their forces are examined.

## 30.1 Nuclear Reactions

Ordinary chemical reactions between atoms and molecules involve only orbital electrons. The nuclei of these atoms do not participate, so the atoms retain their identity. Conversely, in **nuclear reactions**, the original nuclei are converted into nuclei of other elements. Scientists first became aware of this type of reaction during studies involving bombardment of nuclei with energetic particles.

The first artificially induced nuclear reaction was produced by Ernest Rutherford in 1919. Nitrogen was bombarded with alpha particles from a natural bismuth source ($^{214}$Bi). The particles produced by the reactions were identified as protons. Rutherford reasoned that an alpha particle colliding with a nitrogen nucleus must sometimes induce a reaction that produces a proton. In this case, it is said that the nitrogen nucleus is artificially transmuted, or converted, into an oxygen nucleus by the following reaction:

$$\underset{\substack{\text{nitrogen}\\(14.003074\,\text{u})}}{{}^{14}_{7}\text{N}} + \underset{\substack{\text{alpha particle}\\(4.002603\,\text{u})}}{{}^{4}_{2}\text{He}} \rightarrow \underset{\substack{\text{oxygen}\\(16.99133\,\text{u})}}{{}^{17}_{8}\text{O}} + \underset{\substack{\text{proton}\\(1.007825\,\text{u})}}{{}^{1}_{1}\text{H}}$$

(The masses are given for later use.)

In this particular reaction and many others like it, a short-lived (intermediate or temporary) *compound nucleus* is formed in an excited state. Thus the preceding reaction can be written more correctly (showing the compound nucleus explicitly) as

$${}^{14}_{7}\text{N} + {}^{4}_{2}\text{He} \rightarrow \left({}^{18}_{9}\text{F}^{*}\right) \rightarrow {}^{17}_{8}\text{O} + {}^{1}_{1}\text{H}$$

The intermediate nucleus here is an isotope of fluorine, ${}^{18}_{9}\text{F}^{*}$, formed in an excited state as indicated by the asterisk. A compound nucleus typically loses its excess energy by ejecting a particle (or particles) – here, a proton. Since a compound nucleus lasts only a short time, it is commonly omitted from nuclear reaction equations.

There are many reactions that do not involve a compound nucleus. In this type, individual nucleons are transferred between the target and incident particle on a short time scale, with no formation of a temporary nucleus. This type of reaction is called a *direct* or *transfer reaction*. It is distinguished from a compound nucleus reaction by the scattering pattern of the outgoing particles. A specific of this type is a *pickup reaction*, in which a nucleon is taken directly from the target nucleus with little disturbance to other nucleons. An example of this type of reaction is ${}^{12}_{6}\text{C} + {}^{3}_{2}\text{He} \rightarrow {}^{11}_{6}\text{C} + {}^{4}_{2}\text{He}$. Can you tell which type of nucleon was "picked up" from the ${}^{12}_{6}\text{C}$ target?

Regardless of the reaction details, what Rutherford had discovered was a way to change one element into another. This was the age-old dream of alchemists, although their main goal was to change common metals, such as mercury and lead, into valuable ones such as gold. This seemingly profitable metamorphosis and many other transmutations can be initiated today with particle accelerators, machines that accelerate charged particles to high speeds. When these particles strike target nuclei, they can initiate nuclear reactions. One reaction that can occur when a proton strikes a nucleus of mercury is

$$\underset{\substack{\text{mercury}\\(199.968321\text{u})}}{{}^{200}_{80}\text{Hg}} + \underset{\substack{\text{proton}\\(1.007825\text{u})}}{{}^{1}_{1}\text{H}} \rightarrow \underset{\substack{\text{gold}\\(196.96656\text{u})}}{{}^{197}_{79}\text{Au}} + \underset{\substack{\text{alpha particle}\\(4.002603\text{u})}}{{}^{4}_{2}\text{He}}$$

Here mercury is converted into gold, so it would seem that modern physics has fulfilled the dream. However, making such tiny amounts of gold costs far more than the gold is worth.

Reactions such as the foregoing ones (involving only two initial and two final nuclei) are of the general form

$$\text{A} + \text{a} \rightarrow \text{B} + \text{b}$$

where the uppercase letters represent nuclei and lowercase letters represent the incident and outgoing particles. Such reactions are often written in a shorthand notation:

$$\text{A(a, b)B}$$

For example, in this form, two of the previous reactions can be rewritten more compactly as

$${}^{14}\text{N}(\alpha,\text{p}){}^{17}\text{O} \quad \text{and} \quad {}^{200}\text{Hg}(\text{p},\alpha){}^{197}\text{Au}$$

The periodic table (see Figure 28.9 and inside the back cover of this text) contains more than 100 elements, but only 90 stable elements occur naturally. There are elements with unstable nuclei that exist in our environment, due to the decay chains that continually create them from heavier elements. Those with proton numbers greater than uranium, such as plutonium-239, as well as technetium (Z = 43) and promethium (Z = 61), can be created artificially by nuclear reactions. (Being radioactive, if technetium and promethium had been present when the Earth was formed, they would have long since decayed away.) The name *technetium* comes from the Greek word *technetos*, meaning "artificial"; technetium was the first unknown element to be created by artificial means. Elements with Z values up to Z = 118 have been created artificially.

### 30.1.1 Conservation of Mass–Energy and the *Q* Value

In every nuclear reaction, total relativistic energy $E$ $(= K + mc^2)$ must be conserved. Consider the previous Rutherford reaction: $^{14}\text{N}(\alpha, p)^{17}\text{O}$. By the conservation of total relativistic energy,

$$(K_{\text{N}} + m_{\text{N}}c^2) + (K_{\alpha} + m_{\alpha}c^2) = (K_{\text{O}} + m_{\text{O}}c^2) + (K_{\text{p}} + m_{\text{p}}c^2)$$

where the subscripts refer to the particular particle or nucleus. Rearranging this equation,

$$K_O + K_p - (K_N + K_\alpha) = (m_N + m_\alpha - m_O - m_p)c^2.$$

The **Q value** of the reaction is defined as the change in total kinetic energy, and

$$\begin{aligned} Q &= \Delta K = K_f - K_i \\ &= (K_O + K_p) - (K_N + K_\alpha) \quad (Q \text{ value defined}) \end{aligned} \tag{30.1}$$

Note that if $Q > 0$, the total kinetic energy increases after the reaction. If it is negative, there is a loss of kinetic energy. Thus, $Q$ is a measure of the kinetic energy that is released or "lost" in a reaction. Equation 30.1 can be expressed in terms of the masses:

$$Q = (m_N + m_\alpha - m_O - m_p)c^2 \tag{30.2}$$

More generally for a reaction of the form $A + a \rightarrow B + b$,

$$Q = \Delta K = (m_A + m_a - m_B - m_b)c^2 = (\Delta m)c^2 \quad (Q \text{ value in terms of masses}) \tag{30.3}$$

From Equation 30.3, an alternate interpretation of $Q$ is that it represents the difference in the mass (or rest) energies of the reactants (initial) and the products (final) of a reaction. This reflects the fact that mass, since it is a form of energy, can be converted into kinetic energy and vice versa. The mass difference $\Delta m$ can be positive or negative. (*Note:* Care must be taken here when working with $\Delta m = m_i - m_f$ since it is the opposite of our usual "difference" convention. This convention is to guarantee that the sign of $Q$ is correctly related to that of $\Delta K$.) Thus, if the total mass of the system increases during the reaction then the kinetic energy *decreases*, $Q$ must be *negative*. This type of reaction is labeled as *endoergic*. On the other hand, if the total mass decreases and the kinetic energy increases, then $Q$ must be positive and this is reaction type is *exoergic*.

If $Q$ is negative, the reaction will not happen on its own. In this case, the reaction requires a minimum kinetic energy before it can happen. To see this, let us look at Rutherford's original reaction. Using the masses given under the $^{14}N(\alpha, p)^{17}O$ reaction equation:

$$\begin{aligned} Q &= (m_N + m_\alpha - m_O - m_p)c^2 \\ &= \left[(14.003\,074\,u + 4.002\,603\,u - 16.999\,133\,u + 1.007\,825\,u)\right]c^2 \\ &= (-0.001\,281\,u)c^2 \end{aligned}$$

or (using the mass–energy equivalence factor from Section 29.4) the value in MeV is

$$Q = (-0.001\,281\,u)(931.5\text{ MeV/u}) = -1.193\text{ MeV}$$

With a negative $Q$ value this reaction is endoergic, hence the kinetic energy of the reacting nuclei is partially converted into mass. As an example of a reaction with a positive $Q$ value, consider the following Example.

---

**EXAMPLE 30.1: A POSSIBLE ENERGY SOURCE – Q VALUE OF A REACTION**

Determine whether the following reaction is endoergic or exoergic, and calculate its $Q$ value.

| $^2_1H$ | + | $^2_1H$ | $\rightarrow$ | $^3_2He$ | + | $^1_0n$ |
|---|---|---|---|---|---|---|
| deuteron | | deuteron | | helium | | neutron |
| (2.014 102 u) | | (2.014 102 u) | | (3.016 029 u) | | (1.008 665 u) |

**THINKING IT THROUGH.** The reaction is endoergic if $Q < 0$, and exoergic if $Q > 0$. The mass difference ($\Delta m$) is needed to determine $Q$ from Equation 30.3.

**SOLUTION**

$\Delta m$ is found by subtracting the final masses from the initial masses. Therefore,

$$\begin{aligned} \Delta m &= 2m_D - m_{He} - m_n \\ &= 2(2.014\,102\,u) - 3.016\,029\,u - 1.008\,665\,u = +0.003\,51\,u \end{aligned}$$

So total mass has been reduced, the total kinetic energy has increased, and the reaction is exoergic with a $Q$ value of

$$Q = (+0.003\,51\,u)(931.5\text{ MeV/u}) = +3.27\text{ MeV}$$

**FOLLOW-UP EXERCISE.** Determine whether the following reaction is endoergic or exoergic, and calculate its $Q$ value:

| $^{12}_6C$ | + | $^4_2He$ | $\rightarrow$ | $^{13}_6C$ | + | $^3_2He$ |
|---|---|---|---|---|---|---|
| carbon | | helium | | carbon | | helium |
| (12.000 000 u) | | (4.002 603 u) | | (13.003 355 u) | | (3.016 029 u) |

---

Note that radioactive decay (Chapter 29) is a special type of reaction involving one initial unstable nucleus and two (or more) product particles. The $Q$ value for radioactive decay must be positive, since there is a gain in total kinetic energy. For decay reactions, $Q$ is called the *disintegration energy*.

When a reaction's $Q$ value is negative, you might think that it could occur only if the incident particle had a kinetic energy at least equal to $Q$ – that is, $K_{min} \geq |Q|$.* However, if all the kinetic energy were converted to mass, the particles would be at rest afterward, which would violate conservation of

* Kinetic energy is written in terms of $|Q|$ because kinetic energy cannot be negative. The sign of $Q$ arises from a mass difference and indicates the gain or loss of mass. Thus, even if $Q$ is negative, $K_{min}$ must be positive requiring the absolute value.

linear momentum. Thus, for an endoergic reaction, the kinetic energy of the incident particle must be greater than $|Q|$. The minimum kinetic energy to initiate an endoergic reaction is the **threshold energy** ($K_{min}$). For nonrelativistic energies, the threshold energy is

$$K_{min} = \left(1 + \frac{m_a}{M_A}\right)|Q| \quad \text{(threshold energy)} \tag{30.4}$$

Here, $m_a$ and $M_A$ are the masses of the incident particle and target nucleus, respectively. A typical calculation for a threshold energy is illustrated in Example 30.2.

---

EXAMPLE 30.2: NITROGEN INTO OXYGEN - THRESHOLD ENERGY

What is the threshold energy for the reaction $^{14}N(\alpha, p)^{17}O$?

**THINKING IT THROUGH.** The $Q$ value for this reaction was calculated previously in the text. To get the threshold energy, Equation 30.4 should be used.

**SOLUTION**

The following data are taken from the text:

***Given:***

$m_a = m_\alpha = 4.002\ 603$ u
$M_A = m_N = 14.003\ 074$ u
$Q = -1.193$ MeV

***Find:*** $K_{min}$ (threshold energy)

From Equation 30.4,

$$K_{min} = \left(1 + \frac{m_\alpha}{M_N}\right)|Q| = \left(1 + \frac{4.002\,603\,\text{u}}{14.003\,074\,\text{u}}\right)|-1.193\,\text{MeV}| = 1.534\,\text{MeV}$$

**FOLLOW-UP EXERCISE.** In this Example, how much of the kinetic energy goes into increasing the system mass, and how much shows up as kinetic energy in the final state? Explain.

---

### 30.1.2 Reaction Cross-Sections

In an endoergic reaction, when the incident particle has more than the threshold energies for several competing reactions, any of them may occur, with differing probabilities. A measure of the probability that a particular reaction will occur is called the *cross-section* for that reaction and depends on many factors. Usually, it depends on the kinetic energy of the initiating particle, sometimes very dramatically. For positively charged incident particles, the presence of the (repulsive) Coulomb barrier means that the probability of a given reaction occurring generally increases with the kinetic energy of the incident particle.

Being electrically neutral, neutrons are unaffected by the Coulomb barrier. As a result, the cross-section for a given reaction involving neutrons can be quite large, even for low-energy neutrons. Reactions involving neutrons such as $^{27}Al(n, \gamma)^{28}$ Al are called *neutron capture reactions*. As the energy of the neutron increases, the cross-section can vary a great deal, as ▼ **Figure 30.1** shows. The peaks in the curve, called *resonances*, are associated with nuclear energy levels in the nucleus being formed. If the neutron's energy is "just right" to create the final nucleus in one of its energy levels, there is a relatively high probability that neutron absorption will occur, leaving the product nucleus in an excited state. Usually this nucleus de-excites by emitting a gamma ray, as shown in the shorthand notation previously mentioned.

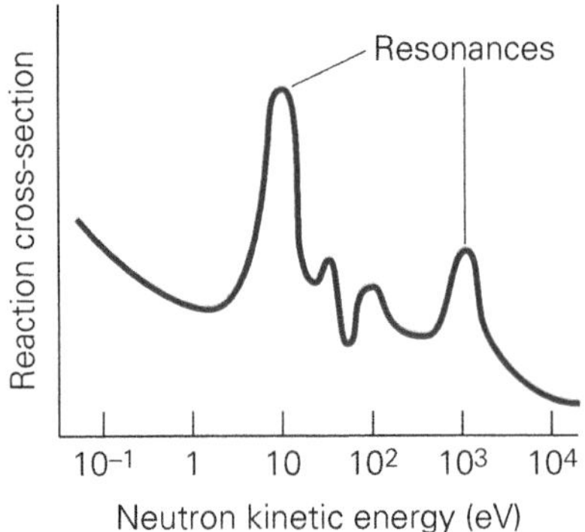


▲ FIGURE 30.1 **Neutron capture cross-section** A typical graph of a neutron capture cross-section versus energy. The peaks where the probabilities of capture are greatest are called *resonances*. They correspond to energy levels in the compound nucleus formed when the neutron is captured.

## 30.2 Nuclear Fission

In early attempts to create heavier elements artificially, uranium, the heaviest element known at the time, was bombarded with neutrons. An unexpected result was that the uranium nuclei sometimes split into fragments. These fragments were identified as the nuclei of lighter elements. The process was dubbed *nuclear fission*, after the biological fission process of cell division.

In a **fission reaction**, a heavy nucleus divides into two lighter nuclei with the emission of neutrons. Some of the initial mass is converted into kinetic energy of the neutron and fragments. Some heavy nuclei undergo *spontaneous fission*, but at very slow rates. However, fission can be *induced*, and this is the important process in nuclear energy production. For example, when a $^{235}U$ nucleus absorbs a neutron, one possibility is that it can fission into xenon and strontium by the (compound nucleus) reaction

$$^{235}_{92}U + ^{1}_{0}n \rightarrow \left(^{236}_{92}U^*\right) \rightarrow ^{140}_{54}Xe + ^{94}_{38}Sr + 2\left(^{1}_{0}n\right)$$

According to the *liquid drop model*, due to the absorbed energy, this intermediate nucleus (here $^{236}U$) undergoes

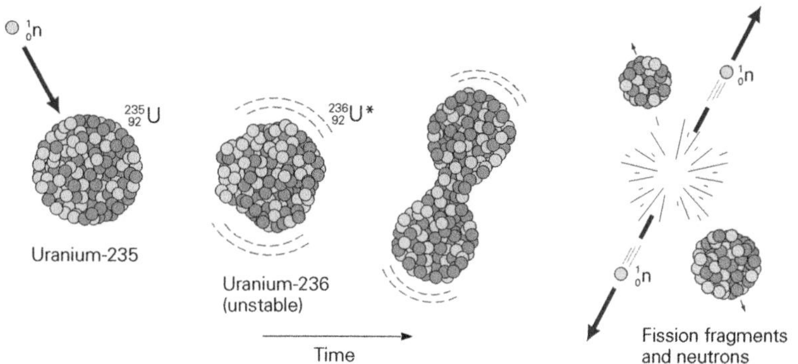


▲ FIGURE 30.2 **Liquid drop model of fission** When an incident neutron is absorbed by a fissionable nucleus, such as $^{235}$U, the unstable compound nucleus ($^{236}$U) undergoes oscillations and breaks apart like a liquid drop, typically emitting two or more neutrons and yielding two radioactive fragments.

oscillations much like a liquid drop (▲ **Figure 30.2**). If parts of the nucleus oscillate far enough away from one another, thus weakening the attractive nuclear force, the repulsive electrical force between the parts of this "nuclear drop" can cause it to split, or fission.

Note that the preceding reaction involving $^{235}$U is not unique. There are many other possible outcomes, including, for example, the following (the compound nuclei are omitted):

$$^1_0\text{n} + ^{235}_{92}\text{U} \rightarrow ^{141}_{56}\text{Ba} + ^{92}_{36}\text{Kr} + 3\left(^1_0\text{n}\right)$$

and

$$^1_0\text{n} + ^{235}_{92}\text{U} \rightarrow ^{150}_{60}\text{Nd} + ^{81}_{32}\text{Ge} + 5\left(^1_0\text{n}\right)$$

Only certain nuclei undergo fission. For them, the probability of induced fission depends on the energy of the incident neutrons. For example, the largest probabilities for fission of $^{235}$U and $^{239}$Pu occur for "slow" neutrons; that is, neutrons with kinetic energies less than 1 eV. This value is on the order of the average kinetic energy in a gas sample at room temperature; thus such neutrons are called *thermal neutrons*. On the other hand, for nuclei such as $^{232}$Th, "fast" neutrons with energies of 1 MeV (or greater) are more likely to trigger fission.

An estimate of the kinetic energy released in a fission reaction can be obtained by considering the $E_b/A$ curve for stable nuclei (Figure 29.14). When a nucleus with a high mass number ($A$), such as uranium, splits into two nuclei, it is, in effect, moving inward and upward along the curve toward more stable nuclei. As a result, the average binding energy per nucleon increases from about 7.8 MeV to about 8.8 MeV. Thus energy liberated is on the order of 1 MeV per nucleon in the fission products. In the first fission reaction, $140 + 94 = 234$ nucleons are bound in the products. Thus, the kinetic energy release is approximately (1 MeV/nucleon) $\times$ 234 nucleons $\approx$ 234 MeV. At first glance, this might not seem like much energy since 234 MeV is only about $3.7 \times 10^{-11}$ J, which pales in comparison to everyday energies. In fact, 234 MeV is only about 0.1% of the energy equivalent of the mass of the $^{235}$U nucleus. Nevertheless, on a percentage basis, it is many times larger than the energy released from the burning of oil or coal.

Practical amounts of energy from fission can, however, be obtained when huge numbers of these fissions occur per second. One way of accomplishing this is by a *chain reaction*. For example, suppose a $^{235}$U nucleus fissions (either on its own or triggered by an external neutron) with the release of two neutrons (▶ **Figure 30.3**).

Ideally, the released neutrons can then initiate two more fission reactions, a process that, in turn, releases four neutrons. These neutrons may initiate more reactions, and so on. Thus, the process can multiply, with the number of neutrons doubling with each generation. When this occurs, the neutron production rate (and the energy released) grows exponentially. In this case, the chain reaction is said to be uncontrolled, such as what occurs in "atomic" bombs (A-bombs) like the ones that were dropped to end World War II.

To maintain a sustained chain reaction, controlled or not, there must be an adequate quantity of fissionable material. The minimum mass required to produce a sustained chain reaction is called the **critical mass**. When critical mass is attained, there is enough fissionable material so that at least one neutron from each fission, on average, goes on to fission another nucleus.

Natural uranium consists of the isotopes $^{238}$U and $^{235}$U. The natural concentration of $^{235}$U is only about 0.7% of the total. The remaining 99.3% is $^{238}$U, which can absorb neutrons without fissioning, thereby inhibiting the chain reaction fissioning of $^{235}$U. To have more fissionable $^{235}$U in a sample and thus reduce the critical mass, the $^{235}$U is concentrated. This enrichment varies from 3% to 5% $^{235}$U for nuclear reactor–grade material to more than 99% for weapons-grade material.

Uncontrolled chain reactions take place almost instantly, and the quick and huge release of energy can cause an explosion as in the atomic bomb mentioned previously. (A more descriptive and physically correct name is the *fission bomb*.) In such a bomb,

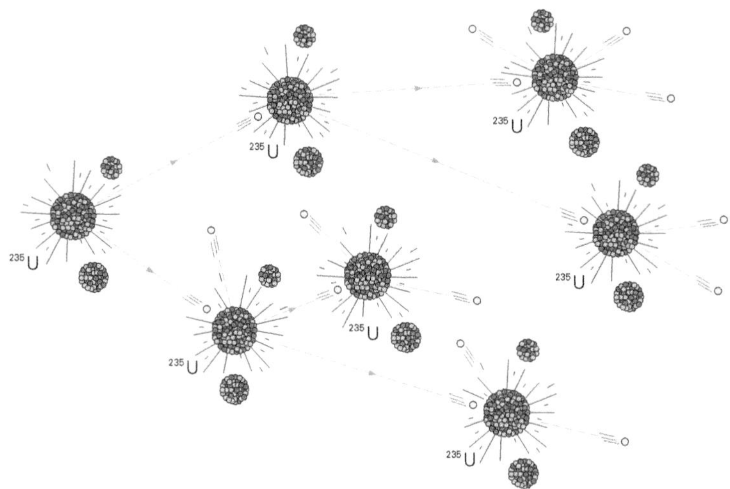


▲ **FIGURE 30.3 Fission chain reaction** The neutrons that result from one fission event can initiate other fission reactions, which, in turn, initiate further fission reactions, and so on. When enough fissionable material is present, the sequence of reactions can be adjusted to be self-sustaining (a chain reaction).

several subcritical pieces of fuel are quickly imploded to form a critical mass. The resulting chain reaction is then out of control, releasing an enormous amount of energy in a short period of time. For the steady production of energy from the fission process, the chain reaction process must be controlled. How this is accomplished in nuclear reactors is the topic of the following section.

## 30.2.1 Nuclear Reactors

### 30.2.1.1 The Power Reactor

A typical design for a nuclear reactor is shown in ▶ **Figure 30.4**. There are five key elements to a reactor: fuel rods, core, coolant, control rods, and moderator. Tubes packed with uranium oxide form **fuel rods**, located in the central portion of the reactor called the **core**. A typical commercial reactor contains fuel rods bundled into assemblies of approximately 200 rods each. Coolant flows around the rods to remove the heat generated during the chain reaction. Reactors in the United States are light-water reactors, which means that ordinary water is used as a **coolant** to remove heat. However, the hydrogen nuclei of ordinary water can capture neutrons to form deuterium, thus removing neutrons from the chain reaction. Hence, to offset this loss, fuel rods containing enriched uranium (3%–5% $^{235}U$) must be used.

The chain reaction rate, and therefore the energy output of a reactor, are controlled by boron or cadmium **control rods**, which can be inserted into or withdrawn from the reactor core. Cadmium and boron have a very high probability for absorbing neutrons. Thus when these rods are inserted between the fuel rod assemblies, some neutrons are removed from the chain reaction. The control rods are adjusted to create a sustainable fission chain reaction in which the average fission produces only one more fission.

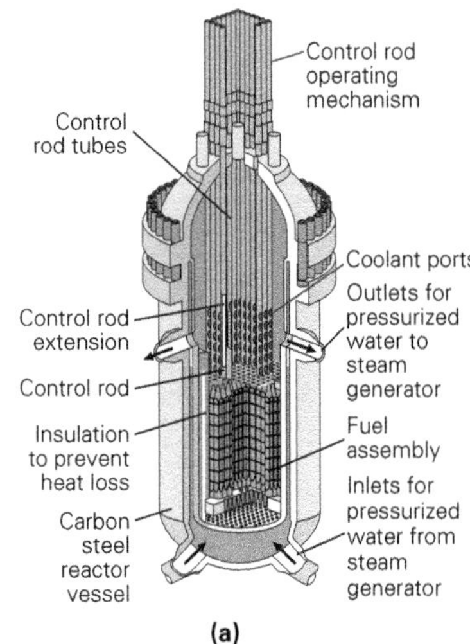


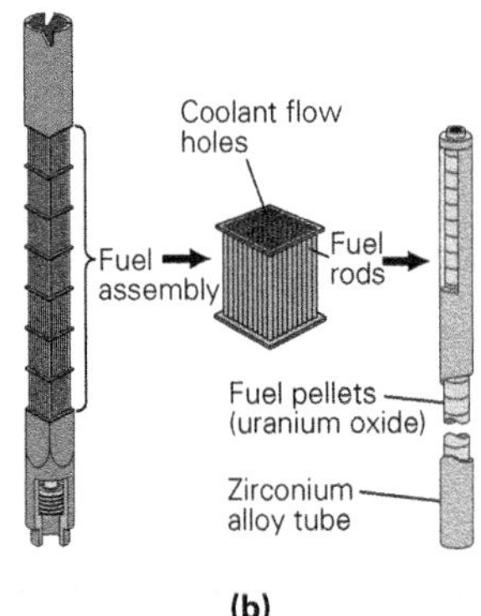


▲ **FIGURE 30.4 Nuclear reactor (a)** A schematic diagram of a reactor vessel. **(b)** A fuel rod and its assembly.

For refueling, or in an emergency, the control rods can be fully inserted, and enough neutrons are then removed to curtail the chain reaction and shut down the reactor. However, water must continue to circulate to prevent heat buildup due to the continuing decay of long-lived radioactive fission products in the rods. If not, damage to the rods can result. Melting and cracking of fuel rods due to inadequate cooling was the cause of the accident at Three Mile Island in 1979, and the (sealed) core of that reactor is still highly radioactive.

The water flowing through the fuel rod assemblies also acts as a neutron **moderator**. Recall that the fission cross-section for $^{235}$U is largest for thermal neutrons. However, neutrons emitted from fission are fast neutrons. Their speed is reduced, or moderated, by collisions with the water molecules, thus reducing fast neutrons to the needed thermal energies.

### 30.2.1.2 The Breeder Reactor

In a commercial power reactor, the $^{238}$U is unlikely to fission and goes along for the ride, so to speak. However, $^{238}$U can be involved in reactions caused by the unmoderated fast neutrons. For example, conversion of $^{238}$U to $^{239}$Pu via successive beta decays after fast neutron absorption can happen as follows:

$$\begin{array}{ccccccc} {}^{1}_{0}\text{n} & + \; {}^{238}_{92}\text{U} & \rightarrow & \left({}^{239}_{92}\text{U}^{*}\right) & \rightarrow & {}^{239}_{93}\text{Np} & + \; \beta^{-} \\ (\textit{fast}) & & & & & \downarrow & \\ & & & & & {}^{239}_{94}\text{Pu} & + \; \beta^{-} \end{array}$$

$^{239}$Pu, with a half-life of 24 000 years, is fissionable. Since it is possible to actively promote the conversion of $^{238}$U to $^{239}$Pu in a reactor by reducing the degree of moderation, the same amount of fissionable fuel (or more) can be produced ($^{239}$Pu) as is consumed ($^{235}$U). This is the principle behind the **breeder reactor.** Notice that this isn't a case of getting something for nothing. Rather, the reactor is converting the unfissionable $^{238}$U percentage of the fuel to fissionable $^{239}$Pu, while continuing to produce energy via $^{235}$U fission.

Developmental work on the breeder reactor in the United States was essentially stopped in the 1970s. However, countries such as France went on to develop operational breeder reactors that provide nuclear fuel ($^{239}$Pu) for their power reactors. Note that some $^{239}$Pu is always produced, even in non-breeder (power) reactors. This has caused worldwide concern over the spread of nuclear armaments. In principle, any power plant can be used to create $^{239}$Pu (as a byproduct), extract it from the spent fuel rods, and create fission bombs with plutonium instead of uranium.

### 30.2.1.3 Electricity Generation

The components of a typical pressurized water reactor used in the United States are shown in ▼ **Figure 30.5**. The heat generated by the (controlled) chain reaction is carried away by the water passing through the rods in the fuel assembly. The water is pressurized to several hundred atmospheres so that it can reach temperatures over 300°C for more efficient heat removal. The hot water is then pumped to a heat exchanger, where the heat is transferred to the water of a steam generator. Notice that reactor coolant and exchanger water are in separate and distinct closed systems. (Why?) Next, high-pressure steam turns a turbine that operates an electrical energy generator, just as in any nonnuclear power plant. The "exhausted" steam is then cooled and condensed. This last loop of coolant water transfers the (wasted) thermal energy to a nearby body of water or into the air.

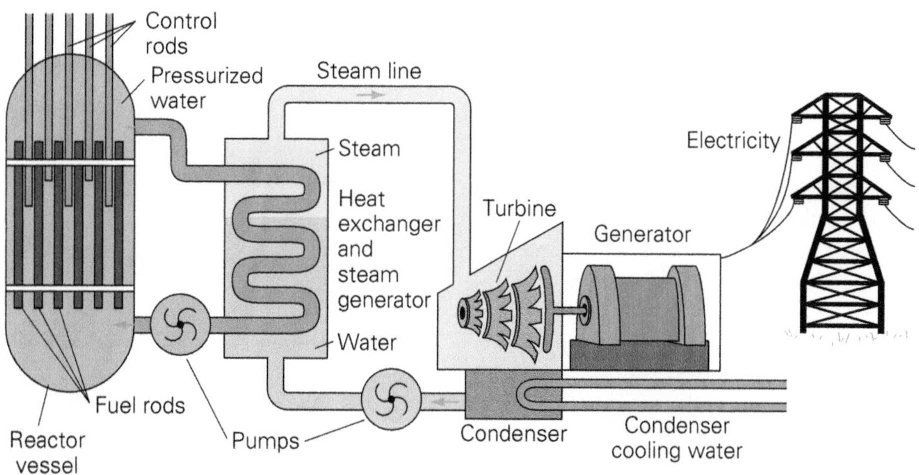


▲ **FIGURE 30.5** **Pressurized water reactor** The components of a pressurized water reactor. The heat energy from the reactor core is carried away by the circulating water. The water in the reactor is pressurized so that it can be heated to high temperatures for more efficient heat removal. The energy is used to generate steam, which drives the turbine that turns the generator to produce electrical energy.

## 30.3 Nuclear Fusion

In a nuclear **fusion reaction**, light nuclei fuse to form a more massive nucleus, releasing energy in the process. For example, the fusion of two deuterium nuclei $\left({}^{2}_{1}\mathrm{H}\right)$, called a D–D reaction, was examined in Example 30.1, where it was shown that this reaction releases 3.27 MeV of energy per fusion. Another example is the fusion of deuterium and tritium (a D–T reaction):

$$\underset{(2.014\,102\,\mathrm{u})}{{}^{2}_{1}\mathrm{H}} \quad + \quad \underset{(3.016\,049\,\mathrm{u})}{{}^{3}_{1}\mathrm{H}} \quad \rightarrow \quad \underset{(4.002\,603\,\mathrm{u})}{{}^{4}_{2}\mathrm{He}} \quad + \quad \underset{(1.008\,665\,\mathrm{u})}{{}^{1}_{0}\mathrm{n}}$$

Using the listed masses, you should be able to show that this reaction releases 17.6 MeV per fusion.

A fusion reaction releases much less energy in comparison with the more than 200 MeV released from a typical single fission. However, in equal masses of hydrogen and uranium, there are many, many more hydrogen nuclei than uranium nuclei. As a result, *per kilogram*, the fusion of hydrogen yields almost three times the energy released from uranium fission.

In a sense, our lives depend crucially on nuclear fusion because it is the source of the energy output for most stars, including our Sun. The fusion sequence responsible for the Sun's energy output starts with a proton–proton fusion that creates a deuteron (a bound proton + neutron):

$${}^{1}_{1}\mathrm{H} + {}^{1}_{1}\mathrm{H} \rightarrow {}^{2}_{1}\mathrm{H} + \beta^{+} + \nu$$

(here $\nu$ represents a particle called a neutrino, to be discussed in the next section). Next, another proton fuses with the previously created deuteron creating an isotope of helium and releasing a gamma ray:

$${}^{1}_{1}\mathrm{H} + {}^{2}_{1}\mathrm{H} \rightarrow {}^{3}_{2}\mathrm{He} + \gamma$$

Finally, two of these $^{3}\mathrm{He}$ nuclei themselves then fuse into $^{4}\mathrm{He}$:

$${}^{3}_{2}\mathrm{He} + {}^{3}_{2}\mathrm{He} \rightarrow {}^{4}_{2}\mathrm{He} + {}^{1}_{1}\mathrm{H} + {}^{1}_{1}\mathrm{H}$$

The net result of this sequence, called the *proton–proton cycle*, is that four protons $\left({}^{1}_{1}\mathrm{H}\right)$ combine to form one helium nucleus $\left({}^{4}_{2}\mathrm{He}\right)$ plus two positrons ($\beta^{+}$), two gamma rays ($\gamma$) , and two neutrinos ($\nu$) with a release of about 25 MeV of energy:

$$4\left({}^{1}_{1}\mathrm{H}\right) \rightarrow {}^{4}_{2}\mathrm{He} + 2\beta^{+} + 2\gamma + 2\nu + Q \quad (Q = +24.7\,\mathrm{MeV})$$

In a star such as our Sun, the gamma-ray photons then scatter off nuclei randomly on their way to the surface. Each scattering results in a reduction in energy (Compton scattering; see Section 27.3), until each photon is reduced to having only a few electron-volts of energy. Thus upon reaching the Sun's surface, they have become mostly visible-light photons. In our Sun, fusion occurs only in the central 10% of the Sun's mass. This process has been going on for about 5 billion years and should continue for approximately another 5 billion. As Example 30.3 shows, an enormous number of fusion reactions per second is required to power the Sun.

---

**EXAMPLE 30.3: STILL GOING – THE FUSION POWER OF THE SUN**

Incoming sunlight energy falls on the Earth at the rate of $1.40 \times 10^{3}\ \mathrm{W/m^{2}}$. Using this energy and the proton–proton cycle, calculate the mass lost by the Sun per second.

**THINKING IT THROUGH.** To find the mass loss rate, the Sun's total power output is needed. Imagine the power flow through a sphere centered on the Sun, with a radius equal to the distance between the Earth and Sun ($R_{\mathrm{E-S}}$), and then calculate that sphere's area. The total power can be found from the power per square meter and the sphere's area. The Sun's mass loss rate can then be determined from this total power, based on mass–energy equivalence (Table 29.3).

**SOLUTION**

The data are as follows (using data in Appendix II):

***Given:***

$R_{\mathrm{E-S}} = 1.50 \times 10^{8}\ \mathrm{km} = 1.50 \times 10^{11}\ \mathrm{m}$
$M_{S} = 2.00 \times 10^{30}\ \mathrm{kg}$
$P_{S}/A = 1.40 \times 10^{3}\ \mathrm{W/m^{2}}$ (power per unit area)

***Find:*** $\Delta m/\Delta t$ (mass loss rate)

The surface area of the imaginary sphere that intercepts all the Sun's energy is

$$A = 4\pi R_{\mathrm{E-S}}^{2} = 4\pi(1.50 \times 10^{11}\ \mathrm{m})^{2} = 2.83 \times 10^{23}\ \mathrm{m^{2}}$$

Thus, the total power output of the Sun is

$$\begin{aligned} P_{S} &= (1.40 \times 10^{3}\ \mathrm{W/m^{2}})(2.83 \times 10^{23}\ \mathrm{m^{2}}) \\ &= 3.96 \times 10^{26}\ \mathrm{W} = 3.96 \times 10^{26}\ \mathrm{J/s} \end{aligned}$$

To find the mass loss rate, convert this to MeV per second:

$$\frac{3.96 \times 10^{26}\ \mathrm{J/s}}{1.60 \times 10^{-13}\ \mathrm{J/MeV}} = 2.48 \times 10^{39}\ \mathrm{MeV/s}$$

Then, using the mass–energy equivalence of 931.5 MeV/u

$$\begin{aligned} \frac{\Delta m}{\Delta t} &= \frac{(2.48 \times 10^{39}\ \mathrm{MeV/s})(1.66 \times 10^{-27}\ \mathrm{kg/u})}{931.5\,\mathrm{MeV/u}} \\ &= 4.42 \times 10^{9}\ \mathrm{kg/s} \end{aligned}$$

**FOLLOW-UP EXERCISE.** In this Example, how many proton–proton cycles happen per second? [*Hint*: Each cycle releases 24.7 MeV.]

---

### 30.3.1 Fusion as a Source of Energy Produced on the Earth

In several ways, fusion appears to be an ideal energy source for the future. Enough deuterium exists in the oceans, in the form of heavy water ($D_2O$), to supply our needs for centuries. In addition, fusion, unlike fission, does not result in radioactive spent fuel rods. The unstable elements fusion does produce tend to have relatively short half-lives. For example, tritium, produced in some fusion reactions, has a half-life of only 12.3 years, compared with hundreds or thousands of years for some fission products.

However, unresolved technical problems must be overcome before controlled fusion can be used to produce commercial energy. A primary problem is that very high temperatures (on the order of millions of degrees) are needed to initiate fusion, because of the electrical repulsion between nuclei. Confining nuclei at these temperatures and at the required densities long enough to achieve commercial energy output so far has not been achieved. However, uncontrolled fusion has been demonstrated in the form of the *hydrogen (H) bomb*. In this case, the fusion reaction is initiated by an implosion created by a small atomic (fission) bomb. This implosion provides the necessary density and temperatures to begin the fusion process resulting in the release of a huge amount of energy in a very short time – an explosion.

## 30.4 Beta Decay and the Neutrino

At first glance, beta decay *appears* to be a two-body (emitted particle and daughter nucleus) process with the emission of an electron $\left({}_{-1}^{0}\mathrm{e}\right)$ or a positron $\left({}_{+1}^{0}\mathrm{e}\right)$ such as:

$$\begin{array}{ccccc} {}^{14}_{6}\mathrm{C} & \rightarrow & {}^{14}_{7}\mathrm{N} & + & {}_{-1}^{0}\mathrm{e}\quad \beta^- \text{ decay} \\ (14.003\,242\,\mathrm{u}) & & (14.003\,074\,\mathrm{u}) & & \end{array}$$

$$\begin{array}{ccccc} {}^{13}_{7}\mathrm{N} & \rightarrow & {}^{13}_{6}\mathrm{C} & + & {}_{+1}^{0}\mathrm{e}\quad \beta^+ \text{ decay} \\ (13.005\,739\,\mathrm{u}) & & (13.003\,355\,\mathrm{u}) & & \end{array}$$

However, when analyzed in detail, these appear to violate the conservation of energy and linear momentum, as well as other conservation laws.

The energy released (or disintegration energy) in the foregoing $\beta^-$ process, as calculated from the mass defect, using the given masses, is $Q = 0.156$ MeV.* (You should show be able to show this.) Therefore, if the decay involves only two particles (electron and daughter), the electron, being much lighter than the daughter, should always have a kinetic energy of just slightly less than 0.156 MeV (the more massive daughter would recoil and have a small amount). However, this is not what happens. When the electron's kinetic energies are measured, a *continuous spectrum* of energies is observed up to $K_{max} \approx Q$ (▶ **Figure 30.6**). (See Conceptual Example 30.4.) In other words, the released energy $Q$ is not accounted for by just the two particles, nor is this the only difficulty. The emitted $\beta^-$ and the daughter nucleus hardly ever leave the disintegration site in opposite directions. Thus, linear momentum conservation appears to be violated as well.†

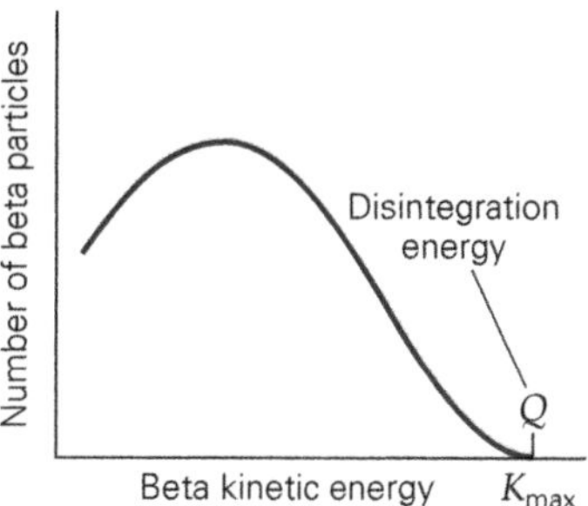


▲ FIGURE 30.6 **Beta ray spectrum** For a typical beta decay process, all beta particles are emitted with $K < Q$, leaving unaccounted-for energy.

What, then, is the problem? It would be hoped that our long-held conservation laws would not be proven invalid. An alternative explanation is that these apparent violations are telling us something about nature that had not yet been recognized. These apparent difficulties can be resolved if it is assumed that an *undetected* particle is also emitted during the decay. The existence of such a particle was first proposed in 1930 by Wolfgang Pauli. Fermi christened this particle the **neutrino** ("little neutral one" in Italian). For charge to be conserved, the neutrino had to be electrically neutral. Because the neutrino had been virtually impossible to detect, it must interact very weakly with matter. In fact, scientists eventually discovered that the neutrino interacts with matter through a second nuclear force, much, much weaker than the strong force, called the *weak interaction*, or the *weak nuclear force*. (See Section 30.5.)

Initial experimental results suggested that the neutrino had exactly zero mass and therefore traveled at the speed of light.‡ Recall that for massless particles (i.e., the photon), their linear momentum $p$ is related to their total energy $E$ by $E = pc$. Furthermore, the neutrino had to have a spin quantum number of ½. It took until 1956, when a particle with these properties was finally detected, that the neutrino's existence became firmly established experimentally.

Thus the two previous beta decay equations can now be written correctly and completely as

$${}^{14}_{6}\mathrm{C} \rightarrow {}^{14}_{7}\mathrm{N} + \beta^- + \bar{\nu}_e$$

and

$${}^{13}_{7}\mathrm{N} \rightarrow {}^{13}_{6}\mathrm{C} + \beta^+ + \nu_e$$

* The use of the atomic mass of the daughter $^{14}$N (seven electrons) is necessary to take into account the emitted electron, since the $^{14}$N resulting from beta decay would have only the six electrons that orbit the parent $^{14}$C nucleus.

† This process also violates the conservation of angular momentum. Careful analysis of the nuclear spin (angular momentum) before the decay shows that it does not match the combined spin (angular momentum) of the daughter plus electron.

‡ Whereas it was originally thought that neutrinos were massless, more sensitive experiments have determined that they do, in fact, have a very small amount of mass, on the order of a millionth of the electron's mass.

where $\nu$ (Greek letter nu) symbolizes the neutrino. The symbol with a bar over it represents an *antineutrino*. The overbar notation is a common way of indicating an antiparticle. Thus two *different* neutrinos are associated with beta decay. By definition, a neutrino is emitted in $\beta^+$ decay ($\nu_e$) and an antineutrino is emitted in $\beta^-$ decay ($\bar{\nu}_e$). The subscript *e* identifies the neutrinos as associated with electron–positron beta decay. As will be seen in Section 30.5, other types of neutrinos are associated with similar decays also related to the weak interaction.

**CONCEPTUAL EXAMPLE 30.4: HAVING IT ALL? MAXIMUM KINETIC ENERGY IN BETA DECAY**

Consider the decay of $^{14}C$ initially at rest: ${}^{14}_{6}C \rightarrow {}^{14}_{7}N + \beta^- + \bar{\nu}_e$. What can be said about the maximum possible kinetic energy ($K_{max}$) of the beta particle: (a) $K_{max} > Q$, (b) $K_{max} = Q$, or (c) $K_{max} < Q$? Explain your reasoning clearly.

**REASONING AND ANSWER.** Answer (a) cannot be correct, as it violates energy conservation. No one particle can have more than the total energy available. So, can answer (b) be correct? Suppose that the beta particle did get all the released energy. That would mean that neither the neutrino nor the daughter nucleus had any energy, and therefore, neither would have any momentum. But to conserve total momentum, which is zero (why?), at least one of these particles must move off in a direction opposite to that of the beta particle. Hence, at the very least the neutrino and/or daughter must have some energy after the decay. Thus the beta particle can't have it all, and answer (c) must be correct: $K_{max} < Q$.

**FOLLOW-UP EXERCISE.** A typical energy spectrum of emitted beta particles is in Figure 30.6. This indicates that there is a small probability of the beta particle having almost no kinetic energy. In this case, what would the daughter nucleus and neutrino trajectories look like?

## 30.5 Fundamental Forces and Exchange Particles

The forces involved in everyday activities are complicated because of the large numbers of atoms that make up ordinary objects. Contact forces (such as the normal force) between two objects, for example, are fundamentally due to the repulsive electromagnetic forces between the atomic electrons in the atoms that make up their surfaces. Looking at *fundamental* interactions between particles makes things simpler. On this level, there are only four known **fundamental forces**: the *gravitational force*, the *electromagnetic force*, the *strong nuclear force*, and the *weak nuclear force*.*

* It is now known that electric and magnetic forces are components of one force – called the electromagnetic force.

The most familiar of these forces are the gravitational and electromagnetic forces. Gravity acts between all particles, while the electromagnetic force is restricted to charged particles. Both forces decrease with increasing particle separation distance and have large ranges.

Recall that the field concept is routinely employed to describe these forces classically. Modern physics, however, provides an alternative, more fundamental, description of how forces are transmitted. *The idea is that the force transmittal process is described by an exchange of particles.* As a classical analog, suppose you and another person interacted by tossing a ball back and forth. As you throw the ball and the other person catches it, each of you feels a backward force. Hence an observer who couldn't see the ball could conclude that there is a repulsive force existing between the two of you.

The creation of such force-carrying particles would seem to violate the law of energy conservation. However, because of the uncertainty principle (Section 28.4), a particle can be created for a *short time* with no outside energy input without violating energy conservation, as long as it disappears later on. This means that, over long time intervals (on the nuclear scale, long can mean $10^{-15}$ s), energy is still conserved. For extremely short time intervals, the uncertainty principle permits a large uncertainty in energy ($\Delta E \propto 1/\Delta t$), creation of a particle (i.e., mass energy) is allowed, and energy conservation can be briefly violated but never detected. In other words, since the created particle is absorbed before detected, non-conservation of energy is never detected experimentally. Any particle created and then absorbed in such a manner is called a **virtual particle**. In this sense, *virtual* means "undetected."

In the modern view, the fundamental forces are transmitted by virtual **exchange particles**. Note that exchange particles for the four forces *must* differ in mass because of the widely different ranges for these four forces. For example, the greater the particle mass, the greater the energy $\Delta E$ required to create it, and thus the time $\Delta t$ that exists is shorter. It then follows that the distance it can travel must be smaller. That is, in summary:

> The range of a force associated with its exchange particle is inversely proportional to the mass of that particle.

### 30.5.1 The Electromagnetic Force and the Photon

The exchange particle of the electromagnetic force is a **virtual photon**. As a "massless" particle, the photon provides a force of infinite range, as is known to be true for the electromagnetic force. (Remember that the electric force between two charged particles is inversely related to the square of the distance between them and thus is never zero.)

A particle exchange can be visualized graphically by employing a *Feynman diagram*, as in ▶**Figure 30.7**, which illustrates how the exchange idea can explain the electrical repulsion between electrons. Such *space–time* diagrams are named after American scientist and Nobel Prize winner

Richard Feynman (1918–1988), who used them to analyze electromagnetic interactions in his *quantum electrodynamics* theory. The important points are the *vertices* (intersections) of the diagram. One electron creates a virtual photon at point A, and the other electron absorbs it at point B. Each of the electrons undergoes a change in energy and in momentum (including direction) by virtue of the photon exchange. To an experimenter, this process appears as the well-known repulsive electric force.

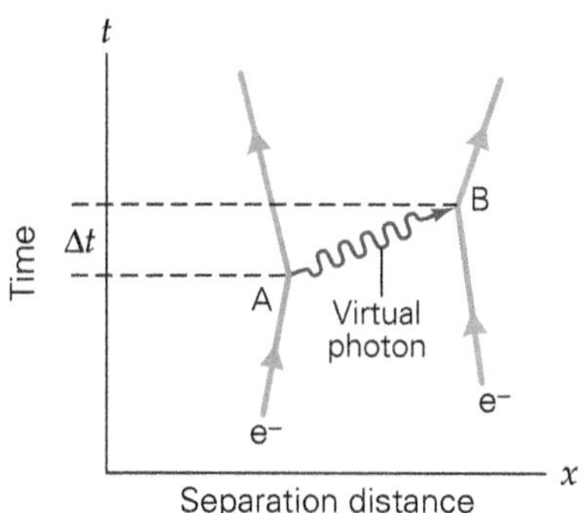


▲FIGURE 30.7 **Feynman diagram of an electron-electron interaction** The interacting electrons each undergo a change in energy and momentum due to the exchange of a virtual photon, created at A and absorbed at B in a time interval ($\Delta t$) consistent with the uncertainty principle.

## 30.5.2 The Strong Nuclear Force and Mesons

In 1935, Japanese physicist Hideki Yukawa (1907–1981) postulated that the short-range strong nuclear force between two nucleons is associated with an exchange particle he called the **meson**. An estimate of the meson's mass can be made from the uncertainty principle. If a nucleon were to create a virtual meson, conservation of energy would be (undetectably) violated by an amount of energy at least equal to the energy equivalent of the meson's mass, or $\Delta E=(\Delta m)c^2=m_{\mathrm{m}}c^2$ where $m_{\mathrm{m}}$ is the mass of the meson.

By the uncertainty principle, the meson would have to be absorbed in the exchange process in a time on the order of

$$\Delta t \approx \frac{h}{2\pi\Delta E}=\frac{h}{2\pi m_{\mathrm{m}}c^2}$$

In that time, the meson would travel a distance $R$, the range of the nuclear force.

Taking $R$ to be the experimental range for the nuclear force between two nucleons ($R\approx 1.4\times 10^{-15}$ m) and solving for $m_{\mathrm{m}}$ gives $m_{\mathrm{m}}\approx 270m_{\mathrm{e}}$, where $m_{\mathrm{e}}$ is the electron's mass. Thus if Yukawa's meson existed, it would have a mass about 270 times that of an electron.

Virtual mesons in an exchange process cannot be observed; however, physicists reasoned that if sufficient energy were involved in the collision of nucleons, *free* and detectable mesons might be created from the energy available in the collision. At the time of Yukawa's prediction, there were no known particles with masses between that of the electron ($m_{\mathrm{e}}$) and the proton ($m_{\mathrm{p}}=1836m_{\mathrm{e}}$).

In 1936, Yukawa's prediction seemed to come true when a new particle with a mass of about 200 $m_{\mathrm{e}}$ was discovered in cosmic rays. Originally called the $\mu$ (Greek mu) meson, and now just the **muon**, it was shown to have two charge varieties, $\pm$e, each with a mass of $m_{\mu^\pm}=207m_{\mathrm{e}}$. But further investigation showed that the muon did *not* behave like the meson of Yukawa's theory.

This situation was a source of controversy and confusion for years. But, in 1947, more particles in this mass range were discovered in cosmic radiation. These particles (one positive, one negative, and one with no charge) were called $\pi$ (pi) mesons (for *primary* mesons) and now commonly called **pions**. Measurement showed their masses to be $m_{\pi^\pm}=273m_{\mathrm{e}}$ and $m_{\pi^0}=264m_{\mathrm{e}}$. Moreover, pions were found to interact strongly with matter. Thus these pions fulfilled the requirements of Yukawa's theory and were generally accepted as the particles primarily responsible for the strong nuclear force. The Feynman diagram for a neutron and proton interacting via the exchange of a negative pion (i.e., via the strong force) is shown in ▼**Figure 30.8**.

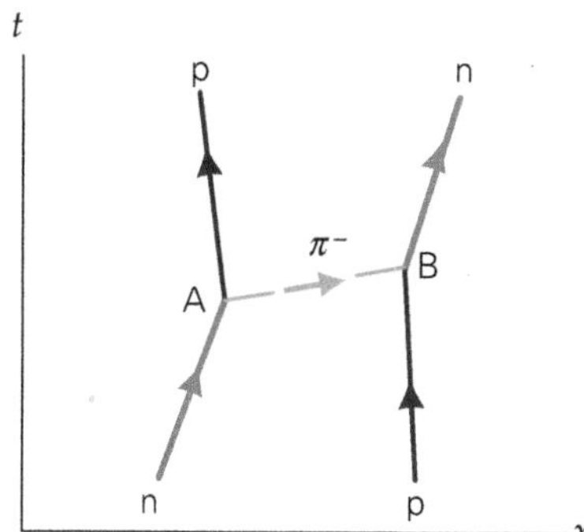


▲FIGURE 30.8 **A Feynman diagram for nucleon-nucleon interactions via pion exchange** Nucleon-nucleon interactions via the strong nuclear force occur through the exchange of virtual pions, here a negative one. This diagram shows one of many possible n-p interactions. It is called a *charge exchange* reaction; can you tell why?

Free pions and muons are unstable. For example, the $\pi^+$ particle decays in about $10^{-8}$ s into a muon and another type of neutrino: $\pi^+\rightarrow\mu^+ + \nu_\mu$. The $\nu_\mu$ is called a *muon neutrino* since it differs from the electron neutrino ($\nu_e$) found in beta decay. Muons can also decay into positrons and electrons with the emission of both types of neutrinos.

**TABLE 30.1** Fundamental Forces

| Force | Relative Strength | Action Distance | Exchange Particle | Particles That Experience the Interaction |
|---|---|---|---|---|
| Strong nuclear | 1 | Short range ($\approx 10^{-15}$ m) | Pion ($\pi$ meson) | Hadrons[a] |
| Electromagnetic | $10^{-3}$ | Inverse square (infinite) | Photon | Electrically charged |
| Weak nuclear | $10^{-8}$ | Extremely short range ($\approx 10^{-17}$ m) | W particle[b] | All |
| Gravitational | $10^{-45}$ | Inverse square (infinite) | Graviton | All |

[a] Hadrons are discussed in Section 30.6.
[b] Three particles are involved, as described in Section 30.8.

### 30.5.3 The Weak Nuclear Force and the W Particle

The discrepancies in beta decay discussed in the preceding section led to another discovery. Electrons and neutrinos are emitted from unstable nuclei, but they do not exist inside the nucleus before the decay takes place. Enrico Fermi proposed that these particles are actually created at the time the nucleus decays. For $\beta^-$ decay, this means that a neutron is in some way changed, or *transmuted*, into three particles: $n \rightarrow p^+ + \beta^- + \bar{\nu}_e$.

Experiments confirmed that free neutrons do, in fact, disintegrate by this scheme, with a half-life of about 10.4 min. But which force could cause a neutron to disintegrate in this manner? Since this neutron is outside the nucleus (free) when it decays, none of the known forces, including the strong nuclear force, seemed applicable. Thus, some other fundamental force must be acting in beta decay. Decay rate measurements indicated that the force was extraordinarily weak – weaker than the electromagnetic force, but still much stronger than the gravitational force. This force was dubbed the **weak nuclear force**.

Originally, it was thought that this *weak interaction* was extremely localized, without any measurable range. We now know that the weak force has a range of about $10^{-17}$ m. While this range is much smaller than that of the strong force, it isn't zero. This means that the exchange particles associated with the weak force (the virtual weak force carriers) must be much more massive than the pions of the strong force. The virtual exchange particles associated with the weak force were named **W particles**.* W (*w*eak) particles have masses about 100 times that of a proton, a fact that correlates with the extremely short range of the weak force. The existence of the W particle was confirmed in the 1980s when accelerators were built with enough energy to create the first real (non-virtual) W particles.

The weak force is the only force that acts on neutrinos, which explains why these particles are so difficult to detect. Research has shown that the weak force is involved in the transmutation of other subatomic particles as well. In general, the weak force is related to transmuting the identities of particles. It mainly manifests its existence through the emitted neutrinos. For example, the Sun's fusion reactions create neutrinos that continuously pass through the Earth. One highly noticeable, but infrequent, announcement of the weak force at work occurs during the explosion of a stellar *supernova*. In a supernova, the collapse of the core of an aging star gives rise to a huge energy release, accompanied by a great number of neutrinos. In a relatively "nearby" supernova ($1.5 \times 10^{18}$ km away) observed in 1987, a burst of neutrinos was detected. Today, neutrino research is a growing field that will play an important role in basic physics for years to come with detectors such as the IceCube Laboratory shown in the chapter-opening photo.

* The weak force is actually carried by *three* exchange particles: $W^+$, $W^-$, and $Z^\circ$ (neutral).

### 30.5.4 The Gravitational Force and Gravitons

The exchange particles thought to be associated with the gravitational force are called **gravitons**. There is still no firm evidence for the existence of this massless particle. (Why must gravitons be massless?) Ongoing experiments to detect the graviton have, as yet, proven unsuccessful, due to the relative weakness of its interaction. A comparison of the relative strengths of the fundamental forces is given in ▲**Table 30.1**.

## 30.6 Elementary Particles

The fundamental building blocks of matter are referred to as **elementary particles**. Simplicity reigned when it was thought that an atom was an indivisible particle and therefore was *the* elementary particle. Early in the twentieth century, the proton, neutron, and electron were discovered to be constituents of atoms. It was hoped that these three particles were nature's elementary particles. However, physicists have since discovered a huge variety of subatomic particles and are now working to simplify and reduce this list to a smaller set of truly elementary particle building blocks for all the other "composite" particles – if this task is indeed possible.

Several systems classify elementary particles on the basis of their various properties. One classification uses the distinction of nuclear force interactions. Particles that interact via the weak nuclear force, but not the strong force, are called leptons ("light ones"). The lepton family includes the electron, the muon, and their neutrinos. Other particles, called **hadrons**, are the only particles to interact by the strong nuclear force. These particles include the proton, neutron, and pion. Let's look briefly at the lepton and hadron families.

### 30.6.1 Leptons

The most familiar **lepton** is the electron. It is the only lepton that exists naturally in atoms. There is no evidence that it has any internal structure, at least down to distances on the order of $10^{-17}$ m. Thus, at present, the electron is considered to be a point particle. Its antiparticle (the positron) is also a lepton.

Muons also appear not to have any internal structure and are therefore sometimes referred to as *heavy electrons.* Muons occur as both positive and negative particles, and decay in about $2\times10^{-6}$ s in modes such as $\mu^- \rightarrow \beta^- + \bar{\nu}_e + \nu_\mu$.

A third charged lepton is the tau ($\tau^\pm$) particle, or **tauon**. It has a mass about twice that of a proton and comes in both charge states.

The remaining leptons are the neutrinos, which are present in cosmic rays, emitted by the Sun, and appear in some radioactive decays. Neutrinos have a very small mass and travel close to, but slightly less than, the speed of light. They interact via neither the electromagnetic force nor the strong force and pass through matter, only very rarely interacting via the weak force.

There are three types of neutrinos, each associated with a different lepton ($e^\pm$, $\mu^\pm$, and $\tau^\pm$). They are named, not surprisingly, *electron neutrino* ($\nu_e$), *muon neutrino* ($\nu_\mu$), and *tau neutrino* ($\nu_\tau$). There is an antineutrino (indicated by an overbar) for each, for a total of six neutrinos. This completes the list of leptons. Current theories predict there should be no others than the twelve listed here.

### 30.6.2 Hadrons

Another family of elementary particles is the **hadrons.** All hadrons interact by the strong force, the weak force, and gravity. The electrically charged members can also interact by the electromagnetic force.

The hadrons are subdivided into *baryons* and *mesons.* **Baryons** include the familiar nucleons – the proton and neutron. They are distinguished from mesons in that they possess half-integer intrinsic spin values $\frac{1}{2}, \frac{3}{2}, \ldots$. Except for the stable proton, baryons decay into products that eventually include a proton. For example, recall the beta decay of a free neutron into a proton.

**Mesons**, which include pions, have integer spin values (0, 1, 2 …) and eventually decay into leptons and photons. For example, the neutral pion decays into two gamma rays. (Why must there be at least two?)

The large number of hadrons suggests that they may be composites of other truly elementary particles. Some help in sorting out the hadron "zoo" came in 1963 when Murray Gell-Mann and George Zweig of the California Institute of Technology put forth the *quark theory*, which is discussed in the next section.

Leptons and some hadrons, as well as their properties, are summarized in ▶**Table 30.2**.

TABLE 30.2 Some Elementary Particles and Their Properties

| Family Name | Particle Type | Particle Symbol | Antiparticle symbol | Rest Energy (MeV) | Lifetime[a] (s) |
|---|---|---|---|---|---|
| **Lepton** | | | | | |
| | Electron | $e^-$ | $e^+$ | 0.511 | Stable |
| | Muon | $\mu^-$ | $\mu^+$ | 105.7 | $2.2\times10^{-6}$ |
| | Tauon | $\tau^-$ | $\tau^+$ | 1784 | $\approx 3\times 0^{-13}$ |
| | Electron neutrino | $\nu_e$ | $\bar{\nu}_e$ | 0[b] | Stable |
| | Muon neutrino | $\nu_\mu$ | $\bar{\nu}_\mu$ | 0[b] | Stable |
| | Tauon neutrino | $\nu_\tau$ | $\bar{\nu}_\tau$ | 0[b] | Stable |
| **Hadron** | | | | | |
| *Mesons* | Pion | $\pi^+$ | $\pi^-$ | 139.6 | $2.6\times10^{-8}$ |
| | | $\pi^0$ | Same | 135.0 | $8.4\times10^{-17}$ |
| | Kaon | $K^+$ | $K^-$ | 493.7 | $1.2\times10^{-8}$ |
| | | $K^0$ | $\overline{K^0}$ | 497.7 | $8.9\times10^{-11}$ |
| *Baryons* | Proton | $P$ | $\bar{p}$ | 938.3 | Stable (?)[c] |
| | Neutron | $n$ | $\bar{n}$ | 939.6 | $\approx 9\times10^{2}$ |
| | Lambda | $\Lambda^0$ | $\overline{\Lambda^0}$ | 1116 | $2.6\times10^{-10}$ |
| | Sigma | $\Sigma^+$ | $\overline{\Sigma^-}$ | 1189 | $8.0\times10^{-10}$ |
| | | $\Sigma^0$ | $\overline{\Sigma^0}$ | 1192 | $0.6\times10^{-20}$ |
| | | $\Sigma^-$ | $\overline{\Sigma^+}$ | 1197 | $1.5\times10^{-10}$ |
| | Xi | $\Xi^0$ | $\overline{\Xi^0}$ | 1315 | $2.9\times10^{-10}$ |
| | | $\Xi^-$ | $\overline{\Xi^+}$ | 1321 | $1.6\times10^{-10}$ |
| | Omega | $\Omega^-$ | $\Omega^+$ | 1672 | $8.2\times10^{-11}$ |

[a] Lifetimes are to two significant figures or fewer.
[b] Neutrinos are now known to possess a small amount of mass. Experiments yield upper limits and the mass (in energy units) of the electron neutrino is known to be less than $7\times10^{-6}$ MeV, but not zero.
[c] Electroweak theory predicts that the proton is unstable, with a half-life of 1000 trillion times the age of the universe.

## 30.7 The Quark Model

Gell-Mann and Zweig proposed that, in fact, hadrons are not fundamental building blocks. Hadrons, they theorized, are composites made of the truly elementary (fundamental) particles. They named these particles **quarks** (taken from James Joyce's novel *Finnegan's Wake**). However, Gell-Mann and Zweig noted that because leptons and photons are essentially point particles, they are likely to be truly elementary particles with no internal structure.

Noting that some hadrons are electrically charged, Gell-Mann and Zweig reasoned that quarks must also possess charge. Their initial **quark model** consisted of three different quarks (with fractional charges) and their antiparticles (called *antiquarks*). ▶**Table 30.3** shows that, to account for hadrons that were

* A line in the novel exclaims, "Three quarks for Muster Mark!" The "three quarks" denote the three children of a character in the novel, Mister (Muster) Mark, also known as Mr. Finn.

TABLE 30.3 Types of Quarks[a]

| Name | Symbol | Charge |
|---|---|---|
| Up | $u$ | $+\frac{2}{3}e$ |
| Down | $d$ | $-\frac{1}{3}e$ |
| Strange | $s$ | $-\frac{1}{3}e$ |
| Charm | $c$ | $+\frac{2}{3}e$ |
| Top (Truth) | $t$ | $+\frac{2}{3}e$ |
| Bottom (Beauty) | $b$ | $-\frac{1}{3}e$ |

[a] Antiquarks are designated by an overbar and have opposite charge.

undiscovered at the time, the list of quarks eventually had to be expanded to include six types. The original idea required only three quarks (plus three antiquarks). These were named the *up* quark ($u$), the *down* quark ($d$), and the *strange* quark ($s$). By using various combinations of these three types, the relatively heavy hadrons, the baryons (whose name means "heavy ones") could be built. In addition, quark–antiquark pairs could account for the lighter hadrons, the mesons. Thus, Gell-Mann and Zweig had proposed a radical new idea:

Quarks are the fundamental particles of the hadron family.

Since several quarks have to combine to give the charge on the hadron, the quarks must have fractions of an electron charge $e$. The theory proposed that $u$, $d$, and $s$ quarks have charges of $+\frac{2}{3}e, \pm\frac{1}{3}e$, and $-\frac{1}{3}e$, respectively. The antiquarks, designated by overbars, such as $\bar{u}$, have opposite charges. Thus, three quark combinations could produce any baryon. For instance, the quark composition of the proton and neutron would be $uud$ and $udd$, respectively. Mesons could also be constructed from quark/antiquark pairs, such as $u\bar{d}$ for the positive pion, $\pi^+$, see ▼**Figure 30.9**. For other quark considerations, consider the next Example.

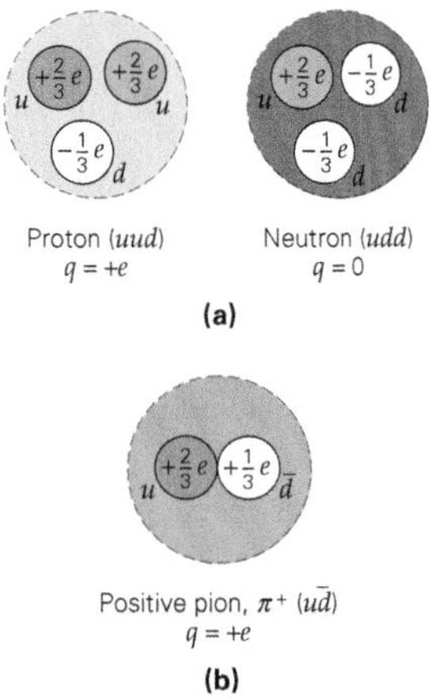


▲ FIGURE 30.9 **Hadronic quark structure (a)** Three combinations of quarks can be used to construct all baryons, such as the proton and neutron. **(b)** Quark–antiquark combinations can be used to construct all mesons, such as the positive pion.

### CONCEPTUAL EXAMPLE 30.5: BUILDING MESONS – QUARK ENGINEERING, INC.

Using the data in Tables 30.2 and 30.3, explain why it is not possible to build a meson from two quarks (i.e., without using antiquarks).

**REASONING AND ANSWER.** From Table 30.3, mesons are either neutral or charged, and if charged, by an integral multiple of $e$. Note that all the positively charged quarks have a charge of $+\frac{2}{3}e$. Thus, any two of them would add up to a meson charge of $+\frac{4}{3}e$, which is contrary to observation. Similar reasoning holds if two negatively charged quarks are used. Finally, consider combining one positively charged quark and one negatively charged quark. The net charge of this combination is $+\frac{1}{3}e$, again in disagreement with observation. Thus, no combination of two quarks (or two antiquarks) can produce a meson. The combinations must include at least one antiquark and one quark.

**FOLLOW-UP EXERCISE.** The antiparticle of the positive pion ($\pi^+$) is the negative pion ($\pi^-$). What is the quark structure of the negative pion? Show that it is composed of the antiquarks of the quarks that make up the positive pion.

The discovery of new subatomic particles in the 1970s led to the addition of the last three quark types: *charm (c)*; *top*, or *truth (t)*; and *bottom*, or *beauty (b)*. Today, there is firm experimental evidence of the existence of all six quarks and their six antiquarks. How many more such particles will be needed to keep up with the growing "zoo" of elementary particles? The hope is, of course, that this list will not need to be expanded. In summary, the present picture includes the following truly elementary particles: leptons (and antileptons), quarks (and antiquarks), and exchange particles such as the photon.

## 30.7.1 Quark Confinement, Color Charge, and Gluons

Thus far in the discussion of quarks, one thing has been missing – direct experimental observation of a quark. Unfortunately, even in the most energetic particle collisions, a *free quark has never been observed*. Physicists now believe that quarks are permanently confined within their particles by a nonlinear springlike force; that is, a force between quarks that grows as they separate from one another. This force grows very rapidly with distance and prevents the ejection of a quark. This idea is called **quark confinement**.

To explain the force between quarks and to clear up problems with the apparent violation of the Pauli exclusion principle (see Section 28.3), quarks were endowed with another characteristic called **color charge**, or simply *color*. There are three types of color charge: red, green, and blue. (These have nothing to do with visual color.) In analogy to electric charge, the quark confinement force exists because different color charges attract each other.

Recall from Section 30.5 that the electromagnetic force is due to virtual photon exchanges between charged particles. Similarly, the force between quarks of different color is due to exchanges of virtual particles called **gluons** (▶**Figure 30.10**). This force is sometimes called the **color force**. This theory is

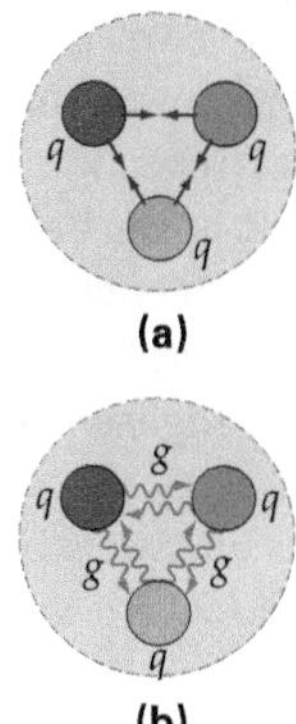


▲FIGURE 30.10 **Quarks, color charge, confinement, and gluons** **(a)** Quarks of different color charge (shown here as different shades of red) attract each other via the color force, which keeps them confined (here, inside a baryon – how do you know it is a baryon?). **(b)** Gluons (wiggly arrows) are exchanged between quarks of different color charge, creating the color force – analogous to virtual photons as related to the electric force.

named quantum chromodynamics (QCD); chromo for "color," in analogy to Feynman's quantum electrodynamics (QED), which successfully explains the electromagnetic force.

The concept of the force between quarks can be extended to explain the force between hadrons – the strong force. Consider the previous explanation of the strong force between a neutron and proton (Figure 30.8) as the result of the exchange of a virtual pion. More fundamentally, this force can be depicted in terms of an exchange of quarks (▶**Figure 30.11**).

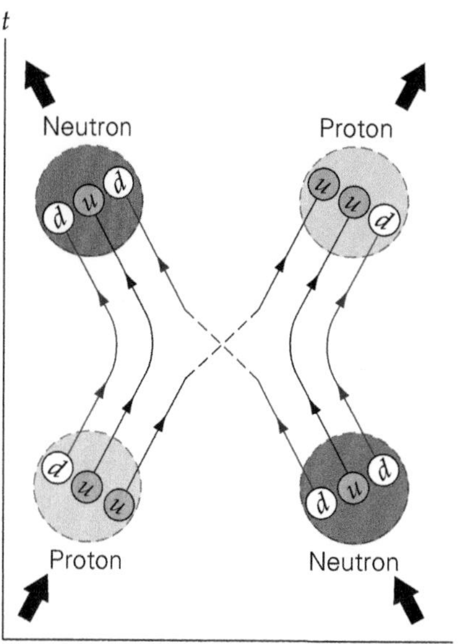


▲FIGURE 30.11 **Quark depiction of the strong nuclear force** Instead of envisioning an n–p interaction as an exchange of a virtual pion (Figure 30.8), a quark exchange model can be used. Here a pair of quarks (equivalent to a pion) is transferred. A proton becomes a neutron, and vice versa.

## 30.8 Force Unification Theories, the Standard Model, and the Early Universe

### 30.8.1 Early Unification Success

Einstein was one of the first to conjecture that it might be possible to unify the fundamental forces of nature. The idea was that perhaps these four, apparently very different, forces might really be different manifestations of one force. Each manifestation would appear under a different set of conditions. Thus, even if an electrically neutral particle could not interact with other particles via electromagnetism, it would still interact with them gravitationally, and perhaps via other nonelectromagnetic interactions as well. Since then, it has been the dream of physicists to understand the "fundamental interactions" in the universe using just one force. Attempts to unify the various forces are called **unification theories**. Maxwell took a first step toward unification in the nineteenth century when he combined the electric and magnetic forces into a single electromagnetic force. Einstein later showed that the electric and magnetic forces are simply different aspects of that single electromagnetic force, depending on the frame from which they are measured (a result of special relativity).

### 30.8.2 Modern Unification Success

The next major step occurred in the 1960s when Sheldon Glashow, Abdus Salam, and Steven Weinberg successfully combined the electromagnetic force with the weak force into a single **electroweak force**. For their efforts, they were awarded the Nobel Prize in 1979.

How can such apparently very different forces be unified? After all, their exchange particles are so different – recall the wide difference between the massless photon ($\gamma$) for the electromagnetic force and the massive W particle for the weak force. It was theorized that, at extremely high energies, the mass-energy of the W particle would become negligible compared with its total energy, in effect making it massless, like the photon. To understand this reasoning, recall (see Chapter 26) that the total energy of a particle is the sum of its kinetic and rest energies ($E = K + mc^2$). If $K \gg mc^2$ (i.e., high energies), the particle's rest energy is negligible compared with its kinetic energy. Perhaps, Glashow, Salam, and Weinberg reasoned, this would mean that the W particle could act more "photon-like" than it seems at first thought. Such high energy conditions were, for example, probably present at the origin of the Big Bang.

In the electroweak theory, weakly interacting particles such as electrons and neutrinos carry a *weak charge*. This weak charge is responsible for the exchange of particles, thus creating the combined electroweak force. The electroweak unification theory predicted the existence of three electroweak exchange particles: $W^+$ and $W^-$ when there is charge exchange, and the neutral $Z^o$ when there is no charge transfer (▼**Figure 30.12**). The eventual discovery of these particles led to the Nobel Prize for Carlo Rubbia and Simon van der Meer in 1984.

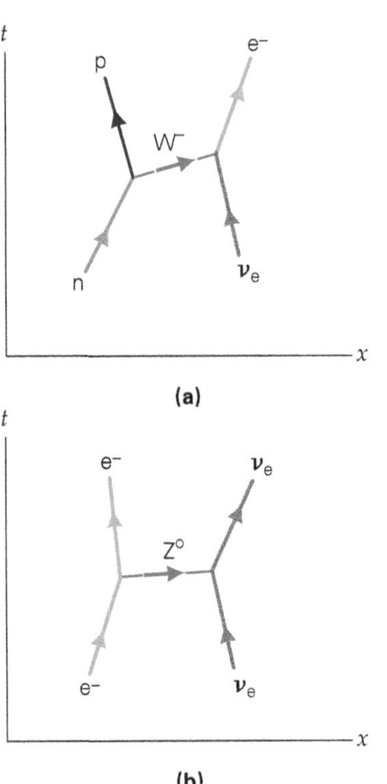


▲ **FIGURE 30.12 Weak force interactions** The Feynman diagrams for **(a)** the neutrino-induced conversion of a neutron to a proton and an electron via the exchange of a $W^-$ particle and **(b)** the scattering of a neutrino by an electron through the exchange of a neutral $Z^\circ$ particle.

## 30.8.3 Future Unification: The Standard Model and Beyond

A goal now is to unify the electroweak force with the strong nuclear force. If successful, the QCD model of the strong force and the already unified electroweak force would be combined into a **grand unified theory (GUT)** describing the *electrostrong force*. Such unification would thus reduce the number of fundamental forces to two. Most viable GUT theories require many exchange particles. In addition, leptons and quarks are combined into one family and can change into each other through the exchange of these particles.

Taken together, the electroweak theory and the QCD model for the strong interaction are referred to as the **standard model**. In this model, the gluons carry the strong force. This force keeps quarks together to form composite particles such as protons and pions. Leptons do not experience the strong force and participate in only the gravitational and electroweak interactions. Presumably, the gravitational interaction is carried by the yet undiscovered graviton. The electromagnetic part of the electroweak interaction is carried by the photon, and the W and Z bosons carry the weak portion of the electroweak force.

One apparent inconsistency in the standard model is the large mass difference among the exchange particles. In the 1960s, a British physicist named Peter Higgs proposed a massive particle, which is now known as the Higgs boson.* If it did exist, then its electroweak interactions would explain the masses of the W and Z bosons and it could be responsible for the masses of the other particles as well. The Higgs boson (mass 125 GeV, symbol $H^\circ$) was finally identified in 2012 as one of the reaction products of high-energy collisions at the Large Hadron Collider (LHC).

The LHC is at the French–Swiss border under the auspices of CERN (a nuclear research consortium of many European countries). Protons whirl around the 27-km circumference at close to *c*, while superconducting magnets provide the magnetic (centripetal) force. The available collisional energy (14 TeV) of the colliding proton beams was enough to create a free Higgs particle. In addition to the Higgs, some theories predict that these high-energy collisions may yield other forms of matter, such as dark matter, micro black holes, and magnetic monopoles. Whatever happens, the results are sure to be interesting, so stay tuned!

Finally, scientists are optimistic that experimental verification of one of the GUT theories might occur in this century. The ultimate unification would be to include the gravitational force into the GUT, creating a single **superforce**. How this would be done, or even *if* it can be done, is not clear. One problem is that, although the three (strong, weak, electromagnetic) components of the GUT can be represented as fields in space and time, in our current view, gravity *is* space and time!

## 30.8.4 Evolution of the Universe and Force Unification

An interesting connection now exists between elementary particle physicists and astrophysicists interested in the evolution of the universe – in particular, its very early evolution. This connection occurs because, according to present ideas, the universe began about 13.7 billion years ago with the *Big Bang*. It is theorized that temperatures during the first $10^{-45}$ s of the universe were on the order of $10^{32}$ K. This corresponds to particle kinetic energies of about $10^{22}$ MeV, high enough that the rest energy of even the most massive elementary particles would be negligible. In effect, the particles would be massless. This would perhaps, astrophysicists conjecture (taking a page from particle physics), place the four forces on an equal footing.

As the universe expanded and cooled, these early elementary particles condensed into what we see today. First, protons, neutrons, and electrons formed; in turn, they combined into atoms and, eventually, molecules. Over the billions of years since, the average temperature of the universe has cooled off to its present value of about 3 K. In the process, scientists think that the superforce symmetry (the equal footing of all four forces) has been lost, leaving us with four very different-looking forces that are really components of a superforce.

* Higgs and his Belgian collaborator Francois Englert received the Nobel Prize for their theoretical prediction of the boson.

It is hoped that future experiments might tell us more about the early moments of the universe and thus about the superforce. The ultimate goal of physics might even be within reach - to understand the basic interactions that govern the universe. While great strides are being made, it is likely to take well into this century, if not longer, to achieve this goal.

## Chapter 30 Review

- A nuclear reaction (including decay) involves the interactions of nuclei and particles, usually resulting in different product nuclei or particles.
- The Q value of a reaction (or decay) is the energy released or absorbed in the process. For a two-body reaction of the form $A + a \rightarrow B + b$,

$$Q = (m_A + m_a - m_B - m_b)c^2 = (\Delta m)c^2 \qquad (30.3a)$$

If $Q > 0$, the reaction is **exoergic** and energy is released.
If $Q < 0$, the reaction is **endoergic** and energy is absorbed.

- In **fission**, an unstable heavy nucleus decays by splitting into two fragments and several neutrons, which together have less total mass than that of the original nucleus; kinetic energy is thus released.
- A **chain reaction** occurs when the neutrons released from one fission trigger other fissions, which trigger further fissions, and so on.
- In a **nuclear power reactor**, the fission chain reaction is kept from going out of control by a set of control rods. The heat energy released is usually used to create steam, which eventually turns generator turbine blades to create electricity.
- In **fusion**, two light nuclei fuse, producing a nucleus with less total mass than that of the original nuclei; energy is thus released.
- The **beta decay** of an unstable nucleus produces a daughter nucleus, an electron or positron, and an antineutrino or neutrino.
- **Exchange particles** are virtual particles associated with various forces. The **pion** (a meson) is the exchange particle primarily responsible for the strong nuclear force.
- The **weak nuclear force**, transmitted by the **W particle**, is primarily responsible for beta decay and the instability of the neutron.
- **Leptons** are the family of elementary particles that interact through the weak force, electromagnetism, and gravity, but not the strong force. Electrons, muons, tauons, and neutrinos make up the lepton family.
- **Hadrons** are the family of elementary particles that interact by the strong force, the weak force, and gravity. If electrically charged, hadrons also interact by the electromagnetic force. The hadrons are subdivided into baryons and mesons. Baryons include the familiar nucleons the proton and neutron. Except for the stable proton, baryons decay into products that eventually include a proton. Mesons, which include pions, eventually decay into leptons and photons.
- **Quarks** are elementary, fractionally charged particles that make up hadrons.
- **Charge color** describes the color force between quarks. Color force explains why a free quark is never likely to be seen, a phenomenon called quark confinement. Quark exchange is the fundamental explanation for the strong nuclear force.
- The **electroweak force** is the name given to the unified electromagnetic and weak forces.
- The **grand unified theory (GUT)** is an attempt to unify the electroweak force with the strong nuclear force.
- The **superforce** is the single force that will result if the long-pursued unification of the "fundamental" forces ever materializes.
- In the **standard model**, gluons carry the strong force, which keeps quarks together in composite particles such as protons. Leptons participate in only the gravitational and electroweak interactions. The former is carried by the graviton. The electroweak interaction is carried by the photon and the W and Z bosons.

## End of Chapter Questions and Exercises

### Multiple Choice Questions

**30.1 Nuclear Reactions**

1. The absorption of a slow neutron by $^{235}$U temporarily results in the compound nucleus (a) $^{235}$U, (b) $^{235}$U*, (c) $^{236}$U*, (d) $^{236}$Np*.
2. If an alpha particle is incident on a carbon-12 nucleus and the outgoing particle is a proton, what is the product nucleus: (a) $^{16}$O, (b) $^{16}$N, (c) $^{15}$N, or (d) $^{16}$C?
3. A $^{27}$Mg nucleus and a free proton are the products of a nuclear reaction involving a target nucleus of $^{27}$Al. What was the incident particle that triggered this: (a) a neutron, (b) a proton, or (c) an alpha particle?

**30.2 Nuclear Fission**

4. Nuclear fission (a) is endoergic, (b) occurs only for $^{235}$U, (c) releases about 500 MeV per fission, (d) requires a critical mass for a sustained reaction.
5. A nuclear reactor (a) can operate on natural (unenriched) uranium, (b) has its chain reaction controlled by neutron-absorbing materials, (c) can be partially controlled by the amount of moderator, (d) all of the preceding.
6. A standard fission breeder reactor creates fissionable material by converting (a) $^{238}$U into $^{235}$U, (b) $^{238}$U into $^{239}$Pu, (c) $^{235}$U into $^{238}$U, (d) $^{239}$Pu into $^{235}$U.

**30.3 Nuclear Fusion**

7. A nuclear fusion reaction (a) has a positive Q value, (b) may occur in laboratories spontaneously, (c) is an example of "splitting" the atom.

8. The net effect of the fusion cycle presently occurring in our Sun is to convert (a) hydrogen into helium, (b) helium into carbon, (c) helium into hydrogen.

**30.4 Beta Decay and the Neutrino**

9. In $\beta^-$ decay, if the daughter nucleus is stationary, how do the moment a (magnitude) of the $\beta^-$ and the $\bar{\nu}_e$ compare: (a) the $\beta^-$ has more momentum, (b) the $\bar{\nu}_e$ has more momentum, or (c) they have the same momentum?
10. $\beta^+$ decay is associated with what type of neutrino: (a) an antineutrino, (b) a neutrino, or (c) no neutrino?
11. A beta decay of the unstable nuclide $^{28}$Al would produce what type of neutrino: (a) an antineutrino, (b) a neutrino.

**30.5 Fundamental Forces and Exchange Particles**

12. Virtual particles (a) form virtual images, (b) exist only in an amount of time permitted by the uncertainty principle, (c) make up positrons.
13. The exchange particle associated with the strong nuclear force is the (a) pion, (b) W particle, (c) muon, (d) positron.
14. The exchange particle associated with the weak nuclear force is the (a) pion, (b) W particle, (c) muon, (d) positron.

**30.6 Elementary Particles**

15. Particles that interact by the strong nuclear force are called (a) muons, (b) hadrons, (c) W particles, (d) leptons.
16. Which one of the following is *not* a hadron: (a) proton, (b) neutron, (c) pion, (d) electron, or (e) all are hadrons?
17. Which one of the following leptons has the least mass: (a) neutrino, (b) electron, (c) muon, or (d) tauon?

**30.7 The Quark Model**

18. Quarks are thought to make up which of the following particles: (a) hadrons, (b) muons, (c) electrons, or (d) neutrinos?
19. Which of the following cannot be the charge on a quark: (a) $(1/3)e$, (b) $-(2/3)e$, (c) $(2/3)e$, or (d) zero?
20. The virtual particle responsible for the color force between quarks is (a) the pion, (b) the muon, (c) the tauon, (d) the gluon.
21. Individual free quarks may never be observed due to a phenomenon known as (a) quark invisibility, (b) quark cloaking, (c) quark confinement, (d) quark life sentences.

**30.8 Force Unification Theories, the Standard Model, and the Early Universe**

22. The *grand unified theory* would reduce the number of fundamental forces to (a) one, (b) two, (c) three, (d) four.
23. The magnetic force is part of the (a) electroweak force, (b) weak force, (c) strong force, (d) superforce.
24. A prediction of the grand unified theory is that (a) protons are unstable, (b) neutrons are unstable, (c) neutrinos are unstable.

## Conceptual Questions

**30.1 Nuclear Reactions**

1. Using conservation of linear momentum, explain how a colliding beam reaction (in which both particles move toward a head-on collision) requires less incident kinetic energy than a reaction in which one particle is at rest.
2. For a given $Q$ value and incident particle (labeled a), how does the threshold energy required to initiate a specific nuclear reaction vary as the mass of the target (labeled A) changes? Sketch a graph of threshold energy versus target mass ranging from target masses on the order of the incident mass to target masses much greater than the incident mass. What is the threshold energy if $M_A \gg m_a$? Explain what this means physically.
3. To capture a slow neutron, which nucleus would have a larger capture cross-section: $^{24}$Mg or $^{25}$Mg? Explain.
4. A $^3$He particle initiates a reaction when incident on a target nucleus. (a) Find a reaction scenario that would yield a product nucleus that is of the same species as the target nucleus. (b) Find a reaction scenario that would yield a product nucleus that has the same number of nucleons as the target nucleus but one less neutron.

**30.2 Nuclear Fission**

5. Explain clearly why the cooling water in a U.S. nuclear reactor helps keep the chain reaction from multiplying in addition to serving as a neutron moderator.
6. Used fuel rods from most U.S. reactors are currently stored on site in pools of water that contain a salt of boron (borated water). Why is the boron needed?
7. Countries entering the nuclear age by constructing nuclear power plants are of concern to the world community because these plants could supply material for nuclear weapons (fission bombs). Explain how this can happen. [*Hint:* Power reactors use $^{235}$U as the fissionable material, but it is only a fraction of the uranium in the rods.]
8. As world oil prices generally rise there has been a call to move electric generation from fossil fuel plants to nuclear fission plants. Discuss some of the pros and cons involved in replacing fossil fuel plants with nuclear versions. Military, terrorism, global warming, and environmental concerns should be on your list.

**30.3 Nuclear Fusion**

9. Explain why there are both density and temperature requirements to sustain a viable fusion energy plant. Why are these requirements at odds with one another?
10. Does it take a higher temperature to fuse two protons or to fuse two $^{12}$C nuclei? Explain.

**30.4 Beta Decay and the Neutrino**

11. A $\bar{\nu}_e$ can interact with protons (very weakly but relatively strongly compared to other nuclei), usually

the nucleus of the hydrogen atom in a water molecule in a target of water. The reaction is essentially the inverse of $\beta^-$ decay; that is, the proton is converted into a neutron. Neutrino experimenters can detect this reaction, and thus the neutrinos, by using $\gamma$-detectors. The neutrino "signature" occurs when the detector simultaneously records a $\gamma$-ray with energy of 2.22 MeV and two others each of 0.511 MeV. Write the reaction and explain where the three $\gamma$-rays come from. [*Hint:* 2.22 MeV is the binding energy of the deuteron and 0.511 MeV is the electron rest energy.]

12. It is known that neutrinos have a very small mass. If they were massless (as previously assumed), neutrinos of all energies would travel at the same speed, *c*. Since they have mass, their speed must be less than *c*. Is it still true that they would travel at the same speed regardless of their energy? Explain.
13. In beta decay, explain why the beta particle kinetic energy must be less than the *Q* value of the decay.
14. (a) In beta decay, if the daughter nucleus is stationary, what can you say about the directions of the beta particle and the neutrino? Explain. (b) In a $\beta$ decay with a *Q* value of 1.5 MeV, suppose the $\beta$ is emitted with very little kinetic energy. How would the momenta of the daughter nucleus and neutrino compare? Would they equally share the 1.5 MeV, or would one end up with most of it? Explain.

**30.5 Fundamental Forces and Exchange Particles**

15. If virtual exchange particles are unobservable by themselves, how is their existence verified?
16. When a proton interacts with another proton, which of the four fundamental forces would be involved? How about when an electron interacts with another electron?
17. Compare the effective range of the strong nuclear force with that of the weak nuclear force. What does that tell you about the relative masses of the exchange particle(s) associated with these two forces? What should be the approximate (numerical) ratio of their masses (weak exchange particle to strong exchange particle)? [*Hint*: see Table 30.1.]

**30.6 Elementary Particles**

18. Explain how you can distinguish between a hadron and a lepton based on their interactions and exchange particles.
19. Which of the leptons are stable? Which of the hadrons are stable?

**30.7 The Quark Model**

20. What is meant by quark flavor and color? Can these attributes be changed? Explain.
21. With so many types of hadrons, why aren't fractional electronic charges observed?
22. What are some of the important differences and distinctions between baryons and mesons?

**30.8 Force Unification Theories, the Standard Model, and the Early Universe**

23. Theoretical physicists often assert that at the extremely high energies thought to have existed in the first moments of the universe, the W particle (known to have a mass several hundred times that of the proton) would behave, mass-wise, similarly to the (massless) photon. Explain this apparently contradictory statement. [*Hint:* Check Chapter 26 and the relativistic relationships between rest energy, kinetic energy, and total energy.]
24. Successful unification of the electromagnetic force with the weak force led physicists to what surprising conclusion about the proton? Describe how you might experimentally detect and verify this prediction.

## Exercises*

*Integrated Exercises (IEs) are two-part exercises. The first part typically requires a conceptual answer choice based on physical thinking and basic principles. The following part is quantitative calculations associated with the conceptual choice made in the first part of the exercise.*

**30.1 Nuclear Reactions**

1. • Complete the following nuclear reactions:
   (a) ${}^{1}_{0}\text{n} + {}^{40}_{18}\text{Ar} \rightarrow \_\_\_\_ + \alpha$
   (b) ${}^{1}_{0}\text{n} + {}^{235}_{92}\text{U} \rightarrow {}^{98}_{40}\text{Zr} + {}^{135}_{\_\_}\_\_ + \_\_\left({}^{1}_{0}\text{n}\right)$
   (c) ${}^{14}_{7}\text{N}(\alpha, p)\_\_\_$
   (d) ${}^{13}\text{C}(\_\_\_,\alpha){}^{10}\text{B}$
2. IE • (a) Consider the reaction ${}^{13}_{6}\text{C} + {}^{1}_{1}\text{H} \rightarrow {}^{4}_{2}\text{He} + {}^{10}_{5}\text{B}$. Determine whether it is endoergic or exoergic. (b) If it is exoergic, find the amount of energy released; if it is endoergic, find the threshold energy.
3. • Determine what the daughter nuclei are in each of the following decays. Can any of these decays occur spontaneously? Explain your reasoning in each case.
   (a) ${}^{22}_{10}\text{Ne} \rightarrow \_\_ + {}^{0}_{-1}\text{e}$.
   (b) ${}^{226}_{88}\text{Ra} \rightarrow \_\_ + {}^{4}_{2}\text{He}$
   (c) ${}^{16}_{8}\text{O} \rightarrow \_\_ + {}^{4}_{2}\text{He}$
4. • Find the *Q* value for ${}^{1}_{1}\text{H} + {}^{2}_{1}\text{H} \rightarrow {}^{3}_{2}\text{He} + \gamma$.
5. IE • ${}^{238}\text{U}$ undergoes alpha decay as follows:

$$\begin{array}{ccccc} {}^{238}_{92}\text{U} & \rightarrow & {}^{234}_{90}\text{Th} & + & {}^{4}_{2}\text{He} \\ (238.050\,786\,\text{u}) & & (234.043\,583\,\text{u}) & & (4.002\,603\,\text{u}) \end{array}$$

   (a) Do you expect the *Q* value to be (1) positive, (2) negative, or (3) zero? Why? (b) Find the *Q* value.

* The bullets denote the degree of difficulty of the exercises: •, simple; ••, medium; and •••, more difficult.

6. •• Find the threshold energy for the following reaction:

$$\begin{array}{ccccccc} {}^{16}_{8}\text{O} & + & {}^{1}_{0}\text{n} & \rightarrow & {}^{13}_{6}\text{C} & + & {}^{4}_{2}\text{He} \\ (15.994\,915\,\text{u}) & & (1.008\,665\,\text{u}) & & (13.003\,355\,\text{u}) & & (4.002\,603\,\text{u}) \end{array}$$

7. •• Find the threshold energy for the following reaction:

$$\begin{array}{ccccccc} {}^{3}_{2}\text{He} & + & {}^{1}_{0}\text{n} & \rightarrow & {}^{2}_{1}\text{H} & + & {}^{2}_{1}\text{H} \\ (3.016\,029\,\text{u}) & & (1.008\,665\,\text{u}) & & (2.014\,102\,\text{u}) & & (2.014\,102\,\text{u}) \end{array}$$

8. •• Is this reaction endoergic or exoergic? Prove your answer by determining the $Q$ value.

$$\begin{array}{ccccccc} {}^{7}_{3}\text{Li} & + & {}^{1}_{1}\text{H} & \rightarrow & {}^{4}_{2}\text{He} & + & {}^{4}_{2}\text{He} \\ (7.016\,005\,\text{u}) & & (1.007\,825\,\text{u}) & & (4.002\,603\,\text{u}) & & (4.002\,603\,\text{u}) \end{array}$$

9. •• Is this reaction endoergic or exoergic? Prove your answer by finding the $Q$ value.

$$\begin{array}{ccccccc} {}^{200}_{80}\text{Hg} & + & {}^{1}_{1}\text{H} & \rightarrow & {}^{197}_{79}\text{Au} & + & {}^{4}_{2}\text{He} \\ \text{mercury} & & \text{proton} & & \text{gold} & & \text{alpha particle} \\ (199.968\,321\,\text{u}) & & (1.007\,825\,\text{u}) & & (196.966\,56\,\text{u}) & & (4.002\,603\,\text{u}) \end{array}$$

10. •• Determine the $Q$ value of the following reaction:

$$\begin{array}{ccccccc} {}^{9}_{4}\text{Be} & + & {}^{4}_{2}\text{He} & \rightarrow & {}^{12}_{6}\text{C} & + & {}^{1}_{0}\text{n} \\ (9.012\,183\,\text{u}) & & (4.002\,603\,\text{u}) & & (12.000\,000\,\text{u}) & & (1.008\,665\,\text{u}) \end{array}$$

11. •• What is the minimum kinetic energy a proton must have in order to initiate the reaction ${}^{3}_{1}\text{H}(p, d)^{2}_{1}\text{H}$? ("d" stands for a deuterium nucleus – the deuteron.)
12. •• $^{226}$Ra decays and emits a 4.706-MeV alpha particle. Find the kinetic energy of the recoiling daughter nucleus if the $^{226}$Ra was originally stationary.
13. **IE** ••• The same type of incident particle is used for two endoergic reactions. In one reaction, the mass of the target nucleus is 15 times that of the incident particle, and in the other reaction, it is 20 times the incident particle's mass. The $Q$ value of the first reaction is known to be three times that of the second. (a) Compared with the second reaction, the first reaction has (1) greater, (2) the same amount of, (3) less minimum threshold energy. (b) Prove your answer to part (a) by calculating the ratio of the minimum threshold energy for the first reaction to that for the second reaction.

### 30.2 Nuclear Fission

14. • Find the approximate energy released in the following fission reactions: (a) ${}^{235}_{92}\text{U} + {}^{1}_{0}\text{n} \rightarrow$ fission products plus five neutrons, (b) ${}^{235}_{94}\text{Pu} + {}^{1}_{0}\text{n} \rightarrow$ fission products plus three neutrons.
15. •• (a) In power reactors, using water as a moderator ("neutron slower") works well, because the proton and neutron have nearly the same mass. Explain why this is true. [*Hint:* Consider an elastic head-on collision of objects of equal mass.] (b) From your reasoning in part (a), it follows that in a head-on elastic collision, we might expect a neutron to lose all of its kinetic energy in one collision, whereas for an "almost miss," it might be expected to lose essentially none. Let's therefore assume that, on average, the neutron loses 50% of its kinetic energy during each proton collision. Estimate how many collisions are needed to reduce a 2.0-MeV neutron to a neutron with a kinetic energy of only 0.02 eV (which qualifies as a "thermal" neutron).

### 30.3 Nuclear Fusion

16. •• Fill in the blanks: (a) ${}^{1}_{1}\text{H} + \_\_ \rightarrow {}^{2}_{1}\text{H} + \gamma$, (b)$\_\_\_ + {}^{3}_{2}\text{He} \rightarrow {}^{4}_{2}\text{He} + 2({}^{1}_{1}\text{H})$. (c) Find the energy released in each.
17. •• Fill in the blanks: (a) ${}^{2}_{1}\text{H} + {}^{2}_{1}\text{H} \rightarrow {}^{3}_{2}\text{He} + \_\_$, (b) ${}^{2}_{1}\text{H} + {}^{3}_{1}\text{H} \rightarrow \_\_ + {}^{1}_{0}\text{n}$. (c) Find the energy released in each.

### 30.4 Beta Decay and the Neutrino

18. • In a decay process the beta particle has a kinetic energy of 0.65 MeV, and the neutrino energy is 0.25 MeV. Neglecting daughter recoil, find the disintegration energy.
19. **IE** • A neutrino created in a beta decay process has an energy of 2.65 MeV. What is the maximum possible kinetic energy of the beta particle if the disintegration energy is 5.35 MeV: (a) (1) zero, (2) less than 5.35 MeV but not zero, (3) 5.35 MeV, or (4) greater than 5.35 MeV? (b) Under these conditions, determine the kinetic energy of the beta particle. (c) What is the direction, relative to the neutrino direction, of the momentum of the beta particle after the decay? Explain your reasoning.
20. •• Show that the disintegration energy for $\beta^-$ decay is $Q = (m_P - m_D - m_e)c^2 = (M_P - M_D)c^2$ where the *m*s represent the masses of the parent and daughter nuclei and the *M*s represent the masses of the neutral atoms. [*Hint:* The number of electrons before and after are the same. Why?]
21. •• What is the maximum kinetic energy of the electron emitted when a $^{12}$B nucleus beta decays into a $^{12}$C nucleus? (See Exercise 20.)
22. •• The kinetic energy of an electron emitted from a $^{32}$P nucleus that beta decays into a $^{32}$S nucleus is 1.00 MeV. What is the energy of the accompanying neutrino? Neglect the recoil energy of the daughter nucleus. (See Exercise 20.)
23. ••• Show that the disintegration energy for $\beta^+$ decay is $Q = (m_P - m_D - m_e)c^2 = (M_P - M_D - 2m_e)c^2$, where the *m*s represent the masses of the parent and daughter nuclei and the *M*s represent the masses of the neutral atoms. [*Hint:* Count the number of electrons both before and after the decay. They are not the same.]
24. ••• The kinetic energy of a positron emitted from the $\beta^+$ decay of a $^{13}$N nucleus into a $^{13}$C nucleus is 1.190 MeV. What is the energy of the accompanying neutrino? Neglect the recoil energy of the daughter nucleus. (See Exercise 23.)

25. ••• The expressions for the *Q* values associated with both $\beta^-$ and $\beta^+$ decay are in Exercises 30.20 and 30.23. Assume that the daughter's atomic mass $M_D$ is the same in both processes. (a) Use the $\beta^-$ expression to show that the requirement for $\beta^-$ decay to occur is simply that the mass of the parent atom be greater than the mass of the daughter atom; that is, $M_p > M_D$. (b) Use the $\beta^+$ expression to show that for $\beta^+$ decay to occur, the mass of the parent atom must be larger than the mass of the daughter atom by two electron masses; that is, $M_P > M_D + 2m_e$.

**30.5 Fundamental Forces and Exchange Particles**

26. • (a) Assuming the range of the nuclear force to be $1.0 \times 10^{-15}$ m, predict the mass (in kilograms) of the exchange particle related to this force. (b) Convert the answer into the particle's rest energy (in MeV) and identify it.
27. IE •• (a) In an interaction using the virtual particle model, the range of the interaction (1) increases, (2) remains the same, (3) decreases as energy of the exchange particle increases. Why? (b) If the W particle has a rest energy on the order of 1.0 GeV, what is the approximate range (to two significant figures) for the interaction involving it?
28. •• By what minimum amount is energy conservation "violated" during a $\pi^o$ exchange process?
29. •• How long is the conservation of energy "violated" in a $\pi^o$ exchange process?

**30.6 Elementary Particles**

30. •• (a) What is the mass difference between the charged pion and its neutral version? Express your answer in both kilograms and rest energy (MeV). (b) What is the kinetic energy of a neutral pion traveling at 0.75*c*?
31. •• If the (electron) neutrino mass were $6.0 \times 10^{-6}$ MeV, what is its speed if its total energy is 0.50 MeV? [*Hint:* See Section 26.4 and binomial expansion usage since $E \gg mc^2$.]
32. •• (a) Using Table 30.2, calculate the mass of the $\Omega^-$ particle in kilograms. (b) Determine its total energy if it is moving at a speed of 0.800*c*.
33. •• (a) Using Table 30.2, estimate the average distance a $\tau^-$ particle would travel in the laboratory if it were moving at 0.95*c*. (b) What would its kinetic energy be?

**30.7 The Quark Model**

34. IE • (a) The quark combination for an antiproton is (1) $uud$, (2) $udd$, (3) $\overline{uud}$, (4) $u\overline{dd}$. (b) Show that your answer to (a) gives the correct electric charge for the antiproton.
35. IE • (a) The quark combination for a antineutron is (1) $\overline{udd}$, (2) $uud$, (3) $\overline{uud}$, (4) $ddd$. (b) Show that your answer to (a) gives the correct electric charge for the antineutron.
36. • (a) Show that the neutral pion cannot be composed solely of any pair of quarks in which one is an up quark (or an anti-up quark) and one is a down quark (or an anti-down quark). (b) According to quantum theory, the $\pi^o$ can be thought of as a sum of two different pairs of quarks/antiquarks. Each pair would be either both up or both down quarks. What are these two pairs?

**30.8 Force Unification Theories, the Standard Model, and the Early Universe**

37. • Suppose the grand unified theory (GUT) was correct and the half-life of a proton was $1.2 \times 10^{35}$ y. Estimate the decay rate for the protons in a liter of water both in decays/second and curies. How does this compare to a small (by laboratory standards) radioactive source of one microcurie?
38. •• (a) Referring to Exercise 37, what would be the proton decay constant $\lambda$? (b) Suppose your experiment required detection of at least one decay per week. What would be the minimum length of one side of a cube-shaped sample of water?
39. •• Referring to Exercise 37, estimate the activity of the Earth's oceans (in curies) due to proton decay. Assume the oceans are 3 km deep covering 75% of the Earth's surface.

# Appendices



## Appendix I* Mathematical Review for College Physics

### A Symbols, Arithmetic Operations, Exponents, and Scientific Notation

#### Commonly Used Symbols in Relationships

$=$ means two quantities are equal, such as $2x = y$.

$\equiv$ means "defined as," such as the definition of pi:

$$\pi \equiv \frac{\text{circumstance of a circle}}{\text{the diameter of that circle}}$$

$\approx$ means approximately equal, as in 30 m/s $\approx$ 60 m/h.

$\neq$ means inequality, such as $\pi \neq 22/7$.

$\geq$ means that one quantity is greater than or equal to another. For example, if the age of the universe $\geq$ 10 billion years, its minimum age is 10 billion years.

$\leq$ means that one quantity is less than or equal to another. For example, if a lecture room holds $\leq$ 45 students, the maximum number of students is 45.

$>$ means that one quantity is greater than another, such as 14 eggs $>$ 1 dozen eggs.

$\gg$ means that one quantity is *much* greater than another. For example, the number of people on Earth $\gg$ 1 million.

$<$ means that one quantity is less than another, such as $3 \times 10^{22} < 10^{24}$.

$\ll$ means that one quantity is much less than another, such as $10 \ll 10^{11}$.

$\propto$ means proportional to. That is, if $y = 2x$ then $y \propto x$.

This means that if $x$ is increased by a certain multiplicative factor, $y$ is also increased the same way. For example, if $y = 3x$, if $x$ is changed by a factor of $n$ (i.e., if $x$ becomes $nx$), then so is $y$, because $y' = 3x' = (3nx) = n(3x) = ny$.

$\Delta Q$ means "change in the quantity $Q$." This means "final minus initial." For example, if the value $V$ of an investor's stock portfolio in the morning is $V_i$ = \$10 100 and at the close of trading it is $V_f$ = \$10 050, then $\Delta V = V_f - V_i$ = \$10 050 − \$10 100 = −\$50.

The Greek letter capital sigma ($\Sigma$) indicates the sum of a series of values for the quantity $Q_i$ where $i = 1, 2, 3, \ldots, N$, that is,

$$\sum_{i=1}^{N} Q_i = Q_1 + Q_2 + Q_3 + \cdots Q_N$$

* This appendix does not include a discussion of significant figures, since a thorough discussion is presented in Section 1.6.

$|Q|$ denotes the absolute value of a quantity $Q$ without a sign. If $Q$ is positive, then $|Q| = Q$; if $Q$ is negative then $|-Q| = Q$. Thus $|-3| = 3$.

### Arithmetic Operations and Their Order of Usage

Basic arithmetic operations are addition (+) subtraction (–), multiplication (× or ·), and division (÷ or /). Another common operation, exponentiation ($x^n$), involves raising a quantity ($x$) to a power ($n$). If several of these operations are included in one equation, they are performed in this order: (a) parentheses, (b) exponentiation, (c) multiplication, (d) division, (e) addition, and (f) subtraction.

A handy mnemonic used to remember this order is: "**P**lease **E**xcuse **M**y **D**ear **A**unt **S**ally," where the capital letters stand for the various operations in order: **P**arentheses, **E**xponents, **M**ultiplication, **D**ivision, **A**ddition, **S**ubtraction. Note that operations within parentheses are always first, so to be on the safe side, appropriate use of parentheses is encouraged. For example, $24^2/8{\cdot}4 + 12$ could be evaluated several ways. However, according to the agreed-on order, it has a unique value: $24^2/8{\cdot}4 + 12 = 576/8{\cdot}4 + 12 = 576/32 + 12 = 18 + 12 = 30$. To avoid possible confusion, the quantity could be written using two sets of parentheses as follows: $[24^2/(8{\cdot}4)] + 12 = [576/(32)] + 12 = 18 + 12 = 30$.

### Exponents and Exponential Notation

Exponents and exponential notation are very important when employing scientific notation (see the next section). You should be familiar with power and exponential notation (both positive and negative, fractional and integral) such as the following:

$$x^0 = 1$$

$$x^1 = x \qquad x^{-1} = \frac{1}{x}$$

$$x^2 = x \cdot x \qquad x^{-2} = \frac{1}{x_2} \qquad x^{1/2} = \sqrt{x}$$

$$x^3 = x \cdot x \cdot x \qquad x^{-3} = \frac{1}{x^3} \qquad x^{1/3} = \sqrt[3]{x} \quad \text{etc.}$$

Exponents combine according to the following rules:

$$x^a \cdot x^b = x^{(a+b)} \qquad x^a/x^b = x^{(a-b)} \qquad (x^a)^b = x^{ab}$$

### Scientific Notation (Powers-of-10 Notation)

In physics, many quantities have values that are very large or very small. To express them, **scientific notation** is frequently used. This notation is sometimes referred to as *powers-of-10* notation for obvious reasons. (See the previous section for a discussion of exponents.) When the number 10 is squared or cubed, we can write it as $10^2 = 10 \times 10 = 100$ or $10^3 = 10 \times 10 \times 10 = 1000$. You can see that the number of zeros is equal to the power of 10. Thus $10^{23}$ is a compact way of expressing the number 1 followed by 23 zeros.

A number can be represented in many different ways. For example, the distance from the Earth to the Sun is 93 million miles. This value can be written as 93 000 000 miles. Expressed in a more compact scientific notation, there are many correct forms, such as $93 \times 10^6$ miles, $9.3 \times 10^7$ miles, or $0.93 \times 10^8$ miles. Any of these is correct, although the second is preferred, because when using powers of 10 notation, it is customary to leave only one digit to the left of the decimal point, in this case 9.3. (This is called customary or standard form.) Note that the exponent, or power of 10, changes when the decimal point of the prefix number is shifted.

Negative powers of 10 also can be used. For example, $10^{-2} = (1/10^2) = (1/100) = 0.01$. So, if a power of 10 has a negative exponent, the decimal point may be shifted to the left once for each power of 10. For example, $5.0 \times 10^{-2}$ is equal to 0.050 (two shifts to the left).

The decimal point of a quantity expressed in powers-of-10 notation may be shifted to the right or left irrespective of whether the power of 10 is positive or negative. General rules for shifting the decimal point are as follows:

1. The exponent, or power of 10, is *increased* by 1 for every place the decimal point is shifted to the left.
2. The exponent, or power of 10, is *decreased* by 1 for every place the decimal point is shifted to the right.

This is simply a way of saying that as the coefficient (prefix number) gets smaller, the exponent gets correspondingly larger, and vice versa. Overall, the number is the same.

When multiplying using this notation, the exponents are added. Thus, $10^2 \times 10^4 = (100)(10\,000) = 1\,000\,000 = 10^6 = 10^{2+4}$. Division follows similar rules using negative exponents, for example, $(10^5/10^2) = (100\,000/100) = 1000 = 10^3 = 10^{5+(-2)}$.

Care should be taken when adding and subtracting numbers written in scientific notation. Before doing so, all numbers must be converted to the same power of 10. For example,

$$\begin{aligned} 1.75 \times 10^3 - 5.0 \times 10^2 &= 1.75 \times 10^3 - 0.50 \times 10^3 \\ &= (1.75 - 0.50) \times 10^3 = 1.25 \times 10^3 \end{aligned}$$

## B Algebra and Common Algebraic Relationships

### General

The basic rule of algebra, used for solving equations, is that if you perform any legitimate operation on both sides of an equation, it remains an equation, or equality. (An example of an illegal operation is dividing by zero; why?) Thus such operations as adding a number to both sides, cubing both sides, and dividing both sides by the same number all maintain the equality.

For example, suppose you want to solve $\frac{x^2+6}{2} = 11$ for $x$.

To do this, first multiply both sides by 2, giving $\left(\frac{x^2+6}{\cancel{2}}\right) \times \cancel{2} = 11 \times 2 = 22$ or $x^2 + 6 = 22$. Then subtract 6 from both sides to obtain $x^2 + 6 - 6 = 22 - 6 = 16$ or $x^2 = 16$. Finally, taking the square root of both sides, the solutions are $x = \pm 4$.

### Some Useful Results

Many times, *the square of the sum or difference of two numbers* is required. For any numbers $a$ and $b$:

$$(a \pm b)^2 = a^2 \pm 2ab + b^2$$

Similarly, *the difference of two squares* can be factored:

$$(a^2 - b^2) = (a + b)(a - b)$$

A quadratic equation is one that can be expressed in the form $ax^2 + bx + c = 0$. In this form it can always be solved (usually for two different roots) using the *quadratic formula*: $x = \dfrac{-b \pm \sqrt{b^2 - 4ac}}{2a}$. In kinematics, this result can be especially useful as it is common to have equations of this form to solve; for example: $4.9t^2 - 10t - 20 = 0$. Just insert the coefficients (making sure to include the sign) and solve for $t$ (here $t$ represents the time for a ball to reach the ground when thrown straight upward from the edge of a cliff; see Chapter 2). The result is

$$t = \frac{10 \pm \sqrt{10^2 - 4(4.9)(-20)}}{2(4.9)} = \frac{10 \pm 22.2}{9.8}$$
$$= +3.3\text{ s} \quad \text{or} \quad -1.2\text{ s}$$

In all such problems, time is "stopwatch" time and starts at zero; hence the negative answer can be ignored as physically unreasonable although it is a solution to the equation.

### Solving Simultaneous Equations

Occasionally solving a problem might require solving two or more equations simultaneously. In general, if there are $N$ unknowns in a problem, exactly $N$ independent equations will be needed. If there are less than $N$ equations, there are not enough for a complete solution. If there are more than $N$ equations, then some are redundant, and a solution is usually still possible, although more complicated. In general in this textbook, such concerns will usually be with two simultaneous equations, and both will be linear. Linear equations are of the form $y = mx + b$. Recall that when plotted on an $x$–$y$ Cartesian coordinate system, the result is a straight line with a slope of $m$ ($=\Delta y/\Delta x$) and a $y$-intercept of $b$, as shown for the red line here.

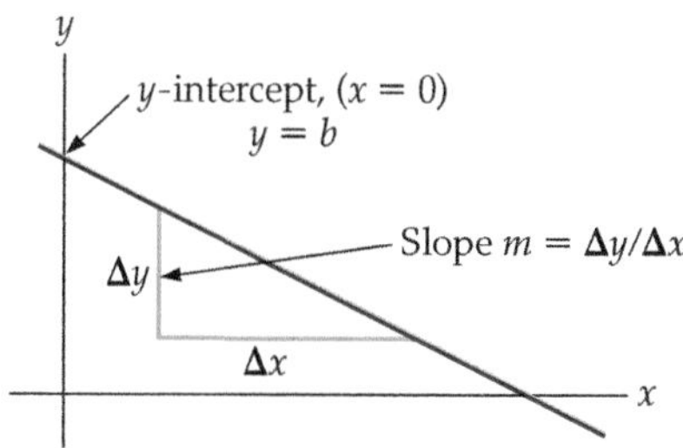


To solve two linear equations simultaneously graphically, simply plot them on the axes and evaluate the coordinates at their intersection point. While this can always be done in principle, it is only an approximate answer and usually takes quite a bit of time.

The most common (and exact) method of solving simultaneous equations involves the use of algebra. Essentially, you solve one equation for an unknown and substitute the result into the other equation, ending up with one equation and one unknown. Suppose you have two equations and two unknown quantities ($x$ and $y$), but in general, any two unknown quantities:

$$3y + 4x = 4 \quad \text{and} \quad 2x - y = 2$$

Solving the second equation for $y$ yields $y = 2x - 2$. Substituting this value for $y$ into the first equation, $3(2x - 2) + 4x = 4$. Thus, $10x = 10$ and $x = 1$. Putting this value into the second of the original two equations, the result is $2(1) - y = 2$ and therefore $y = 0$. (Of course, at this point a good double-check is to substitute the answers and see if they solve both equations.)

## C Geometric Relationships

In physics and many other areas of science, it is important to know how to find circumferences, areas, and volumes of some common shapes. Here are some equations for such shapes.

### Circumference ($c$), Area ($A$), and Volume ($V$)

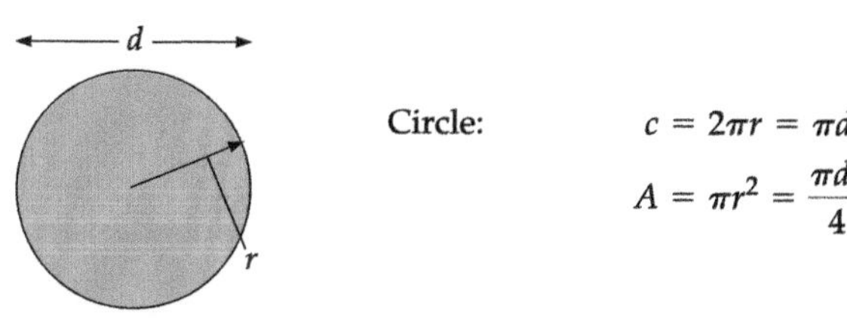


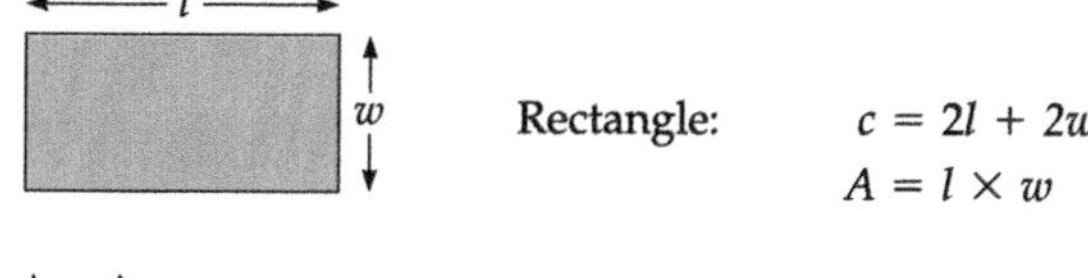


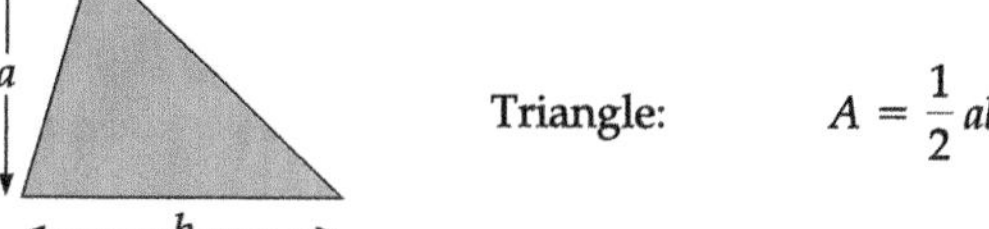


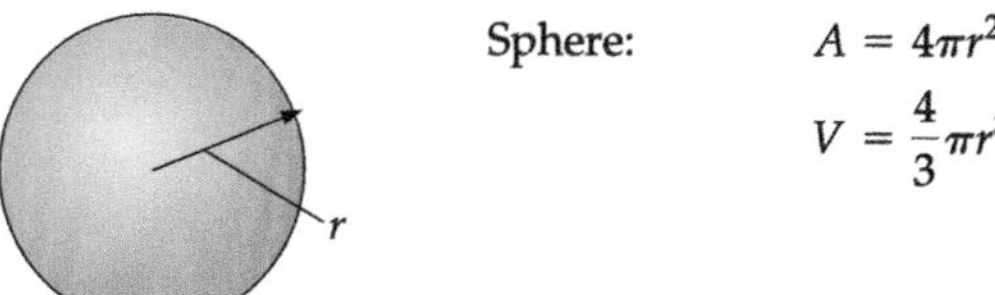


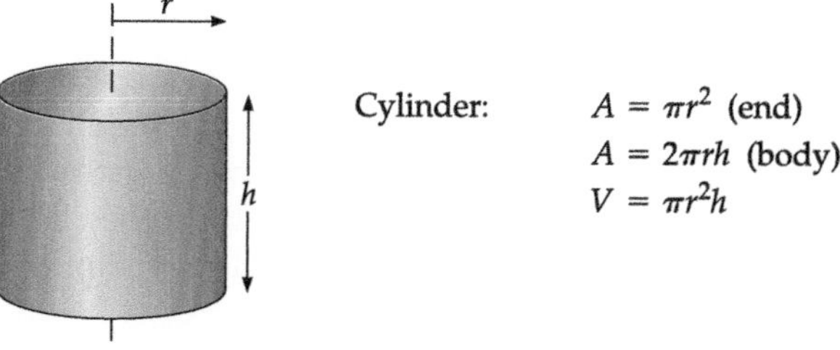

## D Trigonometric Relationships

Understanding elementary trigonometry is crucial in physics, especially since many of the quantities are vectors. Here is a summary of definitions of the three most common trig functions, which you should commit to memory.

### Definitions of Trigonometric Functions

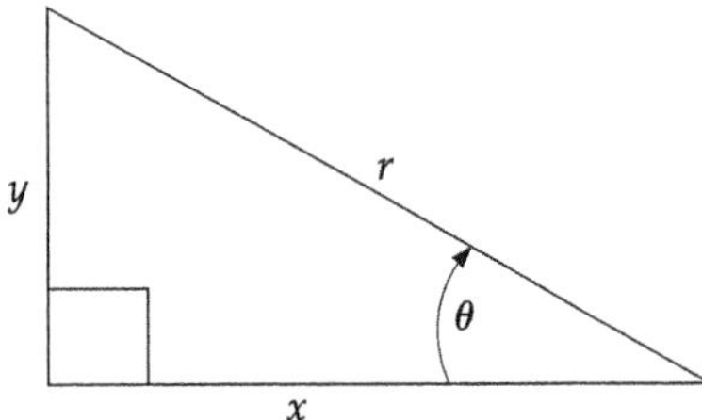


$$\sin\theta = \frac{y}{r} \quad \cos\theta = \frac{x}{r} \quad \tan\theta = \frac{\sin\theta}{\cos\theta} = \frac{y}{x}$$

| $\theta°$ (rad) | $\sin\theta$ | $\cos\theta$ | $\tan\theta$ |
|---|---|---|---|
| 0° (0) | | 1 | 0 |
| 30° ($\pi/6$) | 0.500 | 0.866 | 0.577 |
| 45° ($\pi/4$) | 0.707 | 0.707 | 1.00 |
| 60° ($\pi/3$) | 0.866 | 0.500 | 1.73 |
| 90° ($\pi/2$) | 1 | 0 | $\to\infty$ |

For very small angles,

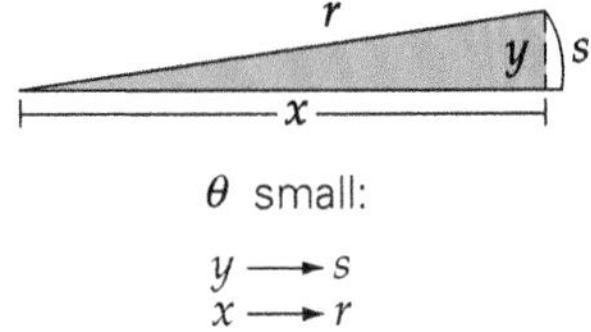


$$\theta\,(\text{in rad}) = \frac{s}{r} \approx \frac{y}{r} \approx \frac{y}{x}$$

$$\theta\,(\text{in rad}) \approx \sin\theta \approx \tan\theta$$

$$\cos\theta \approx 1 \quad \sin\theta \approx \theta \quad (\text{radians})$$

$$\tan\theta = \frac{\sin\theta}{\cos\theta} \approx \theta \quad (\text{radians})$$

The sign of a trigonometric function depends on the quadrant, or the signs of $x$ and $y$. For example, in the second quadrant $x$ is negative and $y$ is positive; therefore, $\cos\theta = x/r$ is negative and $\sin\theta = y/r$ is positive. [Note that $r$ (shown as the dashed lines) is always taken as positive.] In the figure, the dark red lines are positive and the light red lines negative.

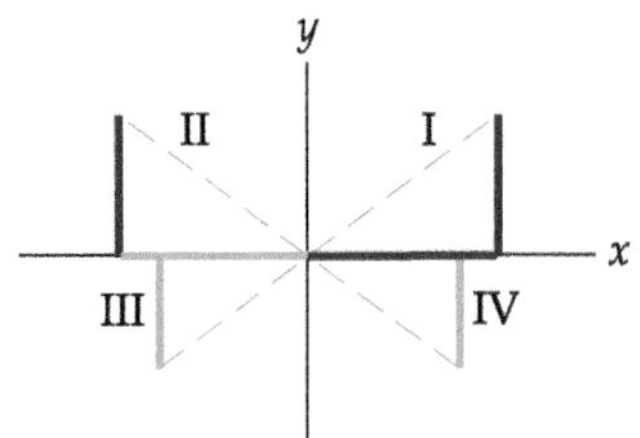


### Some Useful Trigonometric Identities

$$1 = \sin^2\theta + \cos^2\theta$$

$$\sin 2\theta = 2\sin\theta\cos\theta$$

$$\cos 2\theta = \cos^2\theta - \sin^2\theta = 2\cos^2\theta - 1 = 1 - 2\sin^2\theta$$

$$\sin^2\theta = \frac{1}{2}(1 - \cos 2\theta)$$

$$\cos^2\theta = \frac{1}{2}(1 + \cos 2\theta)$$

For half-angle ($\theta/2$) identities, simply replace $\theta$ with $\theta/2$; for example,

$$\sin^2(\theta/2) = \frac{1}{2}(1 - \cos\theta)$$

$$\cos^2(\theta/2) = \frac{1}{2}(1 + \cos\theta)$$

Trigonometric values of sums and differences of angles are sometimes of interest. Here are several basic relationships.

$$\sin(\alpha \pm \beta) = \sin\alpha\cos\beta \pm \cos\alpha\sin\beta$$

$$\cos(\alpha \pm \beta) = \cos\alpha\cos\beta \mp \sin\alpha\sin\beta$$

$$\tan(\alpha \pm \beta) = \frac{\tan\alpha \pm \tan\beta}{1 \mp \tan\alpha\tan\beta}$$

### Law of Cosines

For a triangle with angles $A$, $B$, and $C$ with opposite sides $a$, $b$, and $c$, respectively:

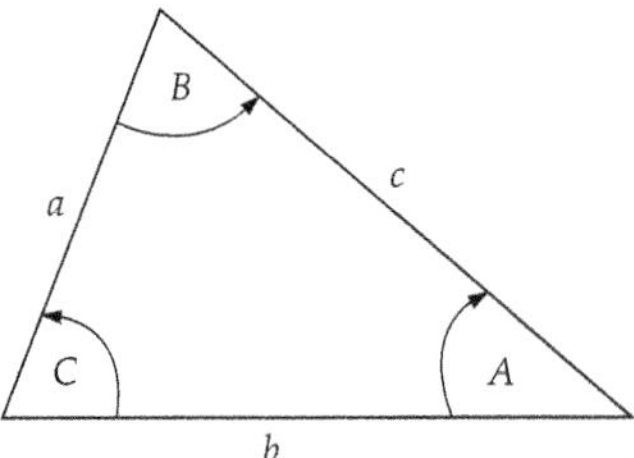


$a^2 = b^2 + c^2 - 2bc\cos A$ (with similar results for $b^2 = \ldots$ and $c^2 = \ldots$)

If $A = 90°$ then $\cos A = 0$ and this reduces to the Pythagorean theorem as it should: $a^2 = b^2 + c^2$.

### Law of Sines

For a triangle with angles *A*, *B*, and *C* with opposite sides *a*, *b*, and *c*, respectively:

$$\frac{a}{\sin A} = \frac{b}{\sin B} = \frac{c}{\sin C}$$

## E Logarithms

Presented here are some of the fundamental definitions and relationships for logarithms. Logarithms are commonly used in science, so it is important that you know what they are and how to use them. Logarithms are useful because, among other things, they allow you to more easily multiply and divide very large and very small numbers.

### Definition of a Logarithm

If a number *x* is written as another number *a* to some power *n*, as $x = a^n$, then *n* is defined to be the *logarithm of the number x to the base a*. This is written compactly as

$$n \equiv \log_a x$$

### Common Logarithms

If the base *a* is 10, the logarithms are called *common logarithms*. When the abbreviation *log* is used, without a base specified, base 10 is assumed. If another base is being used, it will be specifically shown. For example, $1000 = 10^3$; therefore, $3 = \log_{10} 1000$, or simply $3 = \log 1000$. This is read "3 is the log of 1000."

### Identities for Common Logarithms

For any two numbers *x* and *y*:

$$\log(10^x) = x$$
$$\log(xy) = \log x + \log y$$
$$\log\left(\frac{x}{y}\right) = \log x - \log y$$
$$\log(x^y) = y \log x$$

### Natural Logarithms

The natural logarithm uses as its base the irrational number *e*. To six significant figures, its value is $e \approx 2.71828\ldots$ Fortunately, most calculators have this number (along with other irrational numbers, such as pi) in their memories. (You should be able to find both *e* and $\pi$ on yours.) The natural logarithm received its name because it occurs naturally when describing a quantity that grows or decays at a constant percentage (rate). The natural logarithm is abbreviated *ln* to distinguish it from the common logarithm, *log*. That is, $\log_e x \equiv \ln x$, and if $n = \ln x$, then $x = e^n$. Similarly to the common logarithm, we have the following relationships for any two numbers *x* and *y*:

$$\ln(e^x) = x$$
$$\ln(xy) = \ln x + \ln y$$
$$\ln\left(\frac{x}{y}\right) = \ln x - \ln y$$
$$\ln(x^y) = y \ln x$$

Occasionally you must convert between the two types of logarithms. For that, the following relationships can be handy:

$$\log x = 0.43429 \ln x$$
$$\ln x = 2.3026 \log x$$

# Appendix II Planetary Data

| Name | Equatorial Radius (km) | Mass (compared with Earth's)[a] | Mean Density ($\times 10^3$ kg/m$^3$) | Surface Gravity (compared with Earth's) | Semimajor Axis | | Orbital Period | | Eccentricity | Inclination to Ecliptic |
|---|---|---|---|---|---|---|---|---|---|---|
| | | | | | $\times 10^6$ km | AU[b] | Years | Days | | |
| Mercury | 2439 | 0.0553 | 5.43 | 0.378 | 57.9 | 0.3871 | 0.24084 | 87.96 | 0.2056 | 7°00′26″ |
| Venus | 6052 | 0.8150 | 5.24 | 0.894 | 108.2 | 0.7233 | 0.615 15 | 224.68 | 0.0068 | 3°23′40″ |
| Earth | 6378.140 | 1 | 5.515 | 1 | 149.6 | 1 | 1.000 04 | 365.25 | 0.0167 | 0°00′14″ |
| Mars | 3397.2 | 0.1074 | 3.93 | 0.379 | 227.9 | 1.5237 | 1.8808 | 686.95 | 0.0934 | 1°51′09″ |
| Jupiter | 71 398 | 317.89 | 1.36 | 2.54 | 778.3 | 5.2028 | 11.862 | 4337 | 0.0483 | 1°18′29″ |
| Saturn | 60 000 | 95.17 | 0.71 | 1.07 | 1427.0 | 9.5388 | 29.456 | 10 760 | 0.0560 | 2°29′17″ |
| Uranus | 26 145 | 14.56 | 1.30 | 0.8 | 2871.0 | 19.1914 | 84.07 | 30 700 | 0.0461 | 0°48′26″ |
| Neptune | 24 300 | 17.24 | 1.8 | 1.2 | 4497.1 | 30.0611 | 164.81 | 60 200 | 0.0100 | 1°46′27″ |
| Pluto[c] | 1500–1800 | 0.02 | 0.5–0.8 | ~0.03 | 5913.5 | 39.5294 | 248.53 | 90 780 | 0.2484 | 17°09′03″ |

[a] Planet's mass/Earth's mass, where $M_E = 6.0 \times 10^{24}$ kg.
[b] Astronomical unit: 1 AU $= 1.5 \times 10^8$ km, the average distance between the Earth and the Sun.
[c] Pluto is now classified as a "dwarf" planet.

# Appendix III Alphabetical Listing of the Chemical Elements through Atomic Number 109 (The periodic table is provided inside the back cover.)

| Element | Symbol | Atomic Number (Proton Number) | Atomic Mass | Element | Symbol | Atomic Number (Proton Number) | Atomic Mass | Element | Symbol | Atomic Number (Proton Number) | Atomic Mass |
|---|---|---|---|---|---|---|---|---|---|---|---|
| Actinium | Ac | 89 | 227.0278 | Hafnium | Hf | 72 | 178.49 | Praseodymium | Pr | 59 | 140.9077 |
| Aluminum | Al | 13 | 26.981 54 | Hahnium | Ha | 105 | (262) | Promethium | Pm | 61 | (145) |
| Americium | Am | 95 | (243) | Hassium | Hs | 108 | (265) | Protactinium | Pa | 91 | 231.0359 |
| Antimony | Sb | 51 | 121.757 | Helium | He | 2 | 4.002 60 | Radium | Ra | 88 | 226.0254 |
| Argon | Ar | 18 | 39.948 | Holmium | Ho | 67 | 164.9304 | Radon | Rn | 86 | (222) |
| Arsenic | As | 33 | 74.9216 | Hydrogen | H | 1 | 1.007 94 | Rhenium | Re | 75 | 186.207 |
| Astatine | At | 85 | (210) | Indium | In | 49 | 114.82 | Rhodium | Rh | 45 | 102.9055 |
| Barium | Ba | 56 | 137.33 | Iodine | I | 53 | 126.9045 | Rubidium | Rb | 37 | 85.4678 |
| Berkelium | Bk | 97 | (247) | Iridium | Ir | 77 | 192.22 | Ruthenium | Ru | 44 | 101.07 |
| Beryllium | Be | 4 | 9.01218 | Iron | Fe | 26 | 55.847 | Rutherfordium | Rf | 104 | (261) |
| Bismuth | Bi | 83 | 208.9804 | Krypton | Kr | 36 | 83.80 | Samarium | Sm | 62 | 150.36 |
| Bohrium | Bh | 107 | (264) | Lanthanum | La | 57 | 138.9055 | Scandium | Sc | 21 | 44.9559 |
| Boron | B | 5 | 10.81 | Lawrencium | Lr | 103 | (260) | Seaborgium | Sg | 106 | (263) |
| Bromine | Br | 35 | 79.904 | Lead | Pb | 82 | 207.2 | Selenium | Se | 34 | 78.96 |
| Cadmium | Cd | 48 | 112.41 | Lithium | Li | 3 | 6.941 | Silicon | Si | 14 | 28.0855 |
| Calcium | Ca | 20 | 40.078 | Lutetium | Lu | 71 | 174.967 | Silver | Ag | 47 | 107.8682 |
| Californium | Cf | 98 | (251) | Magnesium | Mg | 12 | 24.305 | Sodium | Na | 11 | 22.989 77 |
| Carbon | C | 6 | 12.011 | Manganese | Mn | 25 | 54.9380 | Strontium | Sr | 38 | 87.62 |
| Cerium | Ce | 58 | 140.12 | Meitnerium | Mt | 109 | (268) | Sulfur | S | 16 | 32.066 |
| Cesium | Cs | 55 | 132.9054 | Mendelevium | Md | 101 | (258) | Tantalum | Ta | 73 | 180.9479 |
| Chlorine | Cl | 17 | 35.453 | Mercury | Hg | 80 | 200.59 | Technetium | Tc | 43 | (98) |
| Chromium | Cr | 24 | 51.996 | Molybdenum | Mo | 42 | 95.94 | Tellurium | Te | 52 | 127.60 |
| Cobalt | Co | 27 | 58.9332 | Neodymium | Nd | 60 | 144.24 | Terbium | Tb | 65 | 158.9254 |
| Copper | Cu | 29 | 63.546 | Neon | Ne | 10 | 20.1797 | Thallium | Tl | 81 | 204.383 |
| Curium | Cm | 96 | (247) | Neptunium | Np | 93 | 237.048 | Thorium | Th | 90 | 232.0381 |
| Dubnium | Db | 105 | (262) | Nickel | Ni | 28 | 58.69 | Thulium | Tm | 69 | 168.9342 |
| Dysprosium | Dy | 66 | 162.50 | Niobium | Nb | 41 | 92.9064 | Tin | Sn | 50 | 118.710 |
| Einsteinium | Es | 99 | (252) | Nitrogen | N | 7 | 14.0067 | Titanium | Ti | 22 | 47.88 |
| Erbium | Er | 68 | 167.26 | Nobelium | No | 102 | (259) | Tungsten | W | 74 | 183.85 |
| Europium | Eu | 63 | 151.96 | Osmium | Os | 76 | 190.2 | Uranium | U | 92 | 238.0289 |
| Fermium | Fm | 100 | (257) | Oxygen | O | 8 | 15.9994 | Vanadium | V | 23 | 50.9415 |
| Fluorine | F | 9 | 18.998 403 | Palladium | Pd | 46 | 106.42 | Xenon | Xe | 54 | 131.29 |
| Francium | Fr | 87 | (223) | Phosphorus | P | 15 | 30.973 76 | Ytterbium | Yb | 70 | 173.04 |
| Gadolinium | Gd | 64 | 157.25 | Platinum | Pt | 78 | 195.08 | Yttrium | Y | 39 | 88.9059 |
| Gallium | Ga | 31 | 69.72 | Plutonium | Pu | 94 | (244) | Zinc | Zn | 30 | 65.39 |
| Germanium | Ge | 32 | 72.561 | Polonium | Po | 84 | (209) | Zirconium | Zr | 40 | 91.22 |
| Gold | Au | 79 | 196.9665 | Potassium | K | 19 | 39.0983 | | | | |

# Appendix IV Properties of Selected Isotopes

| Atomic Number (Z) | Element | Symbol | Mass Number (A) | Atomic Mass[a] | Abundance (%) or Decay Mode[b] (If Radioactive) | Half-Life (If Radioactive) |
|---|---|---|---|---|---|---|
| 0 | (Neutron) | n | 1 | 1.008 665 | $\beta^-$ | 10.6 min |
| 1 | Hydrogen | H | 1 | 1.007 825 | 99.985 | |
| | Deuterium | D | 2 | 2.014 102 | 0.015 | |
| | Tritium | T | 3 | 3.016 049 | $\beta^-$ | 12.33 y |
| 2 | Helium | He | 3 | 3.016 029 | 0.00014 | |
| | | | 4 | 4.002 603 | ≈100 | |
| 3 | Lithium | Li | 6 | 6.015 123 | 7.5 | |
| | | | 7 | 7.016 005 | 92.5 | |
| 4 | Beryllium | Be | 7 | 7.016 930 | EC, $\gamma$ | 53.3 d |
| | | | 8 | 8.005 305 | $2\alpha$ | $6.7 \times 10^{-17}$ s |
| | | | 9 | 9.012 183 | 100 | |
| 5 | Boron | B | 10 | 10.012 938 | 19.8 | |
| | | | 11 | 11.009 305 | 80.2 | |
| | | | 12 | 12.014 353 | $\beta^-$ | 20.4 ms |
| 6 | Carbon | C | 11 | 11.011 433 | $\beta^-$, EC | 20.4 ms |
| | | | 12 | 12.000 000 | 98.89 | |
| | | | 13 | 13.003 355 | 1.11 | |
| | | | 14 | 14.003 242 | $\beta^-$ | 5730 y |
| 7 | Nitrogen | N | 13 | 13.005 739 | $\beta^-$ | 9.96 min |
| | | | 14 | 14.003 074 | 99.63 | |
| | | | 15 | 15.000 109 | 0.37 | |
| 8 | Oxygen | O | 15 | 15.003 065 | $\beta^+$, EC | 122 s |
| | | | 16 | 15.994 915 | 99.76 | |
| | | | 18 | 17.999 159 | 0.204 | |
| 9 | Fluorine | F | 19 | 18.998 403 | 100 | |
| 10 | Neon | Ne | 20 | 19.992 439 | 90.51 | |
| | | | 22 | 21.991 384 | 9.22 | |
| 11 | Sodium | Na | 22 | 21.994 435 | $\beta^+$, EC, $\gamma$ | 2.602 y |
| | | | 23 | 22.989 770 | 100 | |
| | | | 24 | 23.990 964 | $\beta^-$, $\gamma$ | 15.0 h |
| 12 | Magnesium | Mg | 24 | 23.985 045 | 78.99 | |
| 13 | Aluminum | Al | 27 | 26.981 541 | 100 | |
| 14 | Silicon | Si | 28 | 27.976 928 | 92.23 | |
| | | | 31 | 30.975 364 | $\beta^-$, $\gamma$ | 2.62 h |
| 15 | Phosphorus | P | 31 | 30.973 763 | 100 | |
| | | | 32 | 31.973 908 | $\beta^-$ | 14.28 d |
| 16 | Sulfur | S | 32 | 31.972 072 | 95.0 | |
| | | | 35 | 34.969 033 | $\beta^-$ | 87.4 d |
| 17 | Chlorine | Cl | 35 | 34.968 853 | 75.77 | |
| | | | 37 | 36.965 903 | 24.23 | |
| 18 | Argon | Ar | 40 | 39.962 383 | 99.60 | |
| 19 | Potassium | K | 39 | 38.963 708 | 93.26 | |
| | | | 40 | 39.964 000 | $\beta^-$, EC, $\gamma$, $\beta^+$ | $1.28 \times 10^9$ y |
| 20 | Calcium | Ca | 40 | 39.962 591 | 96.94 | |
| 24 | Chromium | Cr | 52 | 51.940 510 | 83.79 | |
| 25 | Manganese | Mn | 55 | 54.938 046 | 100 | |
| 26 | Iron | Fe | 56 | 55.934 939 | 91.8 | |
| 27 | Cobalt | Co | 59 | 58.933 198 | 100 | |
| | | | 60 | 59.933 820 | $\beta^-$, $\gamma$ | 5.271 y |

*(Continued)*

| Atomic Number (Z) | Element | Symbol | Mass Number (A) | Atomic Mass[a] | Abundance (%) or Decay Mode[b] (If Radioactive) | Half-Life (If Radioactive) |
|---|---|---|---|---|---|---|
| 28 | Nickel | Ni | 58 | 57.935 347 | 68.3 | |
| | | | 60 | 59.930 789 | 26.1 | |
| | | | 64 | 63.927 968 | 0.91 | |
| 29 | Copper | Cu | 63 | 62.929 599 | 69.2 | |
| | | | 64 | 63.929 766 | $\beta^-, \beta^+$ | 12.7 h |
| | | | 65 | 64.927 792 | 30.8 | |
| 30 | Zinc | Zn | 64 | 63.929 145 | 48.6 | |
| | | | 66 | 65.926 035 | 27.9 | |
| 33 | Arsenic | As | 75 | 74.921 596 | 100 | |
| 35 | Bromine | Br | 79 | 78.918 336 | 50.69 | |
| 36 | Krypton | Kr | 84 | 83.911 506 | 57.0 | |
| | | | 89 | 88.917 563 | $\beta^-$ | 3.2 min |
| 38 | Strontium | Sr | 86 | 85.909 273 | 9.8 | |
| | | | 88 | 87.905 625 | 82.6 | |
| | | | 90 | 89.907 746 | $\beta^-$ | 28.8 y |
| 39 | Yttrium | Y | 89 | 89.905 856 | 100 | |
| 43 | Technetium | Tc | 98 | 97.907 210 | $\beta^-, \gamma$ | $4.2 \times 10^6$ y |
| 47 | Silver | Ag | 107 | 106.905 095 | 51.83 | |
| | | | 109 | 108.904 754 | 48.17 | |
| 48 | Cadmium | Cd | 114 | 113.903 361 | 28.7 | |
| 49 | Indium | In | 115 | 114.903 88 | 95.7; $\beta^-$ | $5.1 \times 10^{14}$ y |
| 50 | Tin | Sn | 120 | 119.902 199 | 32.4 | |
| 53 | Iodine | I | 127 | 126.904 477 | 100 | |
| | | | 131 | 130.906 118 | $\beta^-, \gamma$ | 8.04 d |
| 54 | Xenon | Xe | 132 | 131.904 15 | 26.9 | |
| | | | 136 | 135.907 22 | 8.9 | |
| 55 | Cesium | Cs | 133 | 132.905 43 | 100 | |
| 56 | Barium | Ba | 137 | 136.905 82 | 11.2 | |
| | | | 138 | 137.905 24 | 71.7 | |
| | | | 144 | 143.922 73 | $\beta^-$ | 11.9 s |
| 61 | Promethium | Pm | 145 | 144.912 75 | EC, $\alpha, \gamma$ | 17.7 y |
| 74 | Tungsten | W | 184 | 183.950 95 | 30.7 | |
| 76 | Osmium | Os | 191 | 190.960 94 | $\beta^-, \gamma$ | 15.4 d |
| | | | 192 | 191.961 49 | 41.0 | |
| 78 | Platinum | Pt | 195 | 194.964 79 | 33.8 | |
| 79 | Gold | Au | 197 | 196.966 56 | 100 | |
| 80 | Mercury | Hg | 202 | 201.970 63 | 29.8 | |
| 81 | Thallium | Tl | 205 | 204.974 41 | 70.5 | |
| | | | 210 | 209.990 069 | $\beta^-$ | 1.3 min |
| 82 | Lead | Pb | 204 | 203.973 044 | $\beta^-$, 1.48 | $1.4 \times 10^{17}$ y |
| | | | 206 | 205.974 46 | 24.1 | |
| | | | 207 | 206.975 89 | 22.1 | |
| | | | 208 | 207.976 64 | 52.3 | |
| | | | 210 | 209.984 18 | $\alpha, \beta^-, \gamma$ | 22.3 y |
| | | | 211 | 210.988 74 | $\beta^-, \gamma$ | 36.1 min |
| | | | 212 | 211.991 88 | $\beta^-, \gamma$ | 10.64 h |
| | | | 214 | 213.999 80 | $\beta^-, \gamma$ | 26.8 min |
| 83 | Bismuth | Bi | 209 | 208.980 39 | 100 | |
| | | | 211 | 210.987 26 | $\alpha, \beta^-, \gamma$ | 2.15 min |
| 84 | Polonium | Po | 210 | 209.982 86 | $\alpha, \gamma$ | 138.38 d |
| | | | 214 | 213.995 19 | $\alpha, \gamma$ | 164 ms |
| 86 | Radon | Rn | 222 | 222.017 574 | $\alpha, \beta$ | 3.8235 d |

(*Continued*)

| Atomic Number (Z) | Element | Symbol | Mass Number (A) | Atomic Mass[a] | Abundance (%) or Decay Mode[b] (If Radioactive) | Half-Life (If Radioactive) |
|---|---|---|---|---|---|---|
| 87 | Francium | Fr | 223 | 223.019 734 | $\alpha, \beta^-, \gamma$ | 21.8 min |
| 88 | Radium | Ra | 226 | 226.025 406 | $\alpha, \gamma$ | $1.60 \times 10^3$ y |
| | | | 228 | 228.031 069 | $\beta^-$ | 5.76 y |
| 89 | Actinium | Ac | 227 | 227.027 751 | $\alpha, \beta^-, \gamma$ | 21.773 y |
| 90 | Thorium | Th | 228 | 228.028 73 | $\alpha, \gamma$ | 1.9131 y |
| | | | 232 | 232.038 054 | 100; $\alpha, \gamma$ | $1.41 \times 10^{10}$ y |
| 92 | Uranium | U | 232 | 232.037 14 | $\alpha, \gamma$ | 72 y |
| | | | 233 | 233.039 629 | $\alpha, \gamma$ | $1.592 \times 10^5$ y |
| | | | 235 | 235.043 925 | 0.72; $\alpha, \gamma$ | $7.038 \times 10^8$ y |
| | | | 236 | 236.045 563 | $\alpha, \gamma$ | $2.342 \times 10^7$ y |
| | | | 238 | 238.050 786 | 99.275; $\alpha, \gamma$ | $4.468 \times 10^9$ y |
| | | | 239 | 239.054 291 | $\beta^-, \gamma$ | 23.5 min |
| 93 | Neptunium | Np | 239 | 239.052 932 | $\beta^-, \gamma$ | 2.35 d |
| 94 | Plutonium | Pu | 239 | 239.052 158 | $\alpha, \gamma$ | $2.41 \times 10^4$ y |
| 95 | Americium | Am | 243 | 243.061 374 | $\alpha, \gamma$ | $7.37 \times 10^3$ y |
| 96 | Curium | Cm | 245 | 245.065 487 | $\alpha, \gamma$ | $8.5 \times 10^3$ y |
| 97 | Berkelium | Bk | 247 | 247.070 03 | $\alpha, \gamma$ | $1.4 \times 10^3$ y |
| 98 | Californium | Cf | 249 | 249.074 849 | $\alpha, \gamma$ | 351 y |
| 99 | Einsteinium | Es | 252 | 254.088 02 | $\alpha, \gamma, \beta^-$ | 276 d |
| 100 | Fermium | Fm | 257 | 253.085 18 | EC, $\alpha, \gamma$ | 3.0 d |

[a] The masses given throughout this table are those for the neutral atom, including the Z electrons.
[b] "EC" stands for electron capture.

# Appendix V Answers to Follow-Up Exercises

## Chapter 15

**15.1** $1.52 \times 10^{-20}$ %

**15.2** No. If the comb were positive, it would polarize the paper in the reverse way and still attract it.

**15.3** $\vec{F}_1$ has a magnitude of $3.8 \times 10^{-7}$ N at 57° above the $+x$-axis or $\vec{F}_1 = (-0.22\,\mu\text{N})\hat{\mathbf{x}} + (0.32\,\mu\text{N})\hat{\mathbf{y}}$.

**15.4** 0.12 m or 12 cm

**15.5** $F_e/F_g = 4.2 \times 10^{42}$ or $F_e = 4.2 \times 10^{42}\,F_g$. The magnitude of the electrical force is the same as that between a proton and electron because they all have the same (magnitude) charge. However, the gravitational force is reduced because the masses are now two very low-mass electrons rather than an electron and a much more massive proton.

**15.6** The field is zero at 0.60 m (or 60 cm) to the left of $q_1$.

**15.7** $\vec{E} = (-797\,\text{N/C})\hat{\mathbf{x}} + (359\,\text{N/C})\hat{\mathbf{y}}$ or $E = 874$ N/C at 24.2° above the -$x$-axis.

**15.8** The field line pattern would, instead, point inward and be spaced twice as far apart.

**15.9** (a) The larger of the two fields is due to the closer positive end and points upward. The smaller field due to the negative end points downward, thus the field line is straight and vertically upward away from the positive end. (b) The larger of the two fields is due to the closer negative end and points upward. The smaller field due to the positive end points downward, thus the field line is straight and vertically upward toward the negative end. (c) Both fields point downward, thus the field line is straight and downward, away from the positive end and toward the negative end.

**15.10** (a) The electric field is upward from ground to cloud. (b) $2.3 \times 10^3$ C.

**15.11** Positive charge would reside completely on the outside surface, thus only the electroscope attached to the outside surface would show deflection.

## Chapter 16

**16.1** (a) $\Delta U_e$ doubles to $+7.20 \times 10^{-18}$ J because the particle's charge is doubled. (b) $\Delta V$ is unchanged since it is unrelated to the amount of charge on the particle. (c) $v = 4.65 \times 10^4$ m/s.

**16.2** $6.63 \times 10^7$ m/s

**16.3** (a) $\Delta V = +6.79$ V-(+27.2 V) $= -20.4$ V
(b) $\Delta U_e = +\,3.27 \times 10^{-18}$ J

**16.4** $6.60 \times 10^{-20}$ C

**16.5** (a) Surface 1 is at a higher electric potential than surface 2 because it is closer to the positively charged surface. (b) At large distances, the charged object would "look like" a point charge, thus the equipotential surfaces gradually become spherical as the distance from the object increases.

**16.6** $d = 8.9 \times 10^{-16}$ m, which is much smaller than the size of an atom (or a nucleus for that matter), and thus this design is unfeasible.

**16.7** $7.90 \times 10^3$ V

**16.8** The capacitance decreases as the spacing increases. Since the voltage across the capacitor remains constant, the charge on the capacitor would have to decrease. Thus charge would flow off of the capacitor. $\Delta Q = -3.30 \times 10^{-12}$ C.

**16.9** $U_{\text{parallel}} = 1.20 \times 10^{-4}$ J and $U_{\text{series}} = 5.40 \times 10^{-4}$ J, so the parallel arrangement stores more energy.

**16.10** (a) 0.50 $\mu$F (b) The energy stored in capacitor 3 is six times that stored in # 1.

## Chapter 17

**17.1** The result is the same: $\Delta V_{AB} = \Delta V$.

**17.2** About 32 years

**17.3** 100 V

**17.4** Since $R = \rho L/A$, if resistivity is doubled and length halved, the numerator stays the same. If the diameter is halved, the area *decreases* by a factor of 4. Thus the resistance increases by a factor of 4, to $3.0 \times 10^3$ Ω. The current in the second fish is $I = \Delta V/R = 0.133$ A $= 133$ mA.

**17.5** 0.67 Ω. The material with the largest temperature coefficient of resistivity makes a more sensitive thermometer because it produces a larger (and thus easier to measure and more accurate) change in resistance for a given temperature change.

**17.6** (a) $R_1 = (\Delta V)^2/P_1 = 11.0$ Ω and $R_2 = 0.900R_1 = 9.92$ Ω
(b) $I_1 = \Delta V/R_1 = 10.5$ A and $I_2 = 1.11I_1 = 11.6$ A

**17.7** (a) $Q = mc\Delta T = 1.67 \times 10^5$ J;
$P = Q/t = 1.67 \times 10^5$ J/180 s $= 930$ W $= 0.930$ kW.
So $R = (\Delta V)^2/P = (120\text{ V})^2/(1.67 \times 10^5\text{ J}) = 15.5$ Ω.
(b) Two cups a day for 30 days (60 cups) means a total operation time of 3.00 min × 60 = 180 min = 3.00 h. The energy usage is (0.930 kW)(3.00 h) = 2.79 kWh, which is a monthly cost of 2.79 kWh × ($0.15/kWh) ≈ 42 cents.

**17.8** The power difference is 150 W = 0.150 kW so the difference in energy usage per year is (0.150 kW)(5 h/d)(250 d/y) = 188 kWh.
At 0.15$/kWh this is a cost difference (savings) of about $28 per year for more efficient TV. To recoup the $900 initial expenditure (replacing the plasma TV) would take $900/($28/y) or about 32 years.

## Chapter 18

**18.1** (a) Series: $P_1 = 4.0$ W, $P_2 = 8.0$ W, $P_3 = 12$ W; parallel: $P_1 = 1.4 \times 10^2$ W, $P_2 = 72$ W, $P_3 = 48$ W.
(b) In series, the most power is dissipated in the largest resistance. In parallel, the most power is dissipated in the least resistance. (c) Series: Total resistor power is 24 W, and $P_b = I_b(\Delta V_b) = (2.0\text{ A})(12\text{ V}) = 24$ W, so

yes, as required by energy conservation. Parallel: Total power is $P_{tot} = 2.6 \times 10^2$ W and $P_b = I_b(\Delta V_b) = (22\ \text{A})(12\ \text{V}) = 2.6 \times 10^2$ W (to two significant figures), so yes, as required by energy conservation.

**18.2** $P_2 = I_2^2 R_2 = 9.0$ W, $P_3 = I_3^2 R_3 = 0.87$ W, $P_4 = I_4^2 R_4 = 2.55$ W, and $P_5 = I_5^2 R_5 = 5.63$ W. The sum is 72.1 W (three significant figures). The power output of the battery is $P_b = I_b(\Delta V_b) = (3.00\ \text{A})(24.0\ \text{V}) = 72.0$ W. This equality (difference due to rounding) is conservation of energy on a per unit time (power) basis.

**18.3** At the junction, we still have $I_1 = I_2 + I_3$ (Equation 1). Using the loop theorem around #3 in the clockwise direction (all numbers are volts, deleted for convenience): $6 - 6I_1 - 9I_2 = 0$ (Equation 2). For loop 1: $6 - 6I_1 - 12 - 2I_3 = 0$ (Equation 3). Solve Equation 1 for $I_2$ and substitute into Equation 2. Then solve Equations 2 and 3 simultaneously for $I_1$ and $I_3$. All answers are the same as in the Example, as they should be.

**18.4** (a) The maximum energy storage at 9.00 V is 4.05 J. At 7.20 V, the capacitor stores only 2.59 J or 64% of the maximum. This is because the energy storage varies as the square of the voltage across the capacitor and $0.8^2 = 0.64$. (b) 8.64 V, because the voltage does not rise linearly, but levels off in an exponential fashion.

**18.5** (a) Place the two voltmeter leads on opposite ends of the resistance. (b) $\Delta V_2 = (2.00\ \text{A})(4.00\ \Omega) = 8.00$ V. (c) The parallel connection of the voltmeter and resistance produces an equivalent resistance less than 4.00 Ω; hence the overall circuit resistance drops and the total current rises. This produces an increase in voltage across $R_1$ and thus a decrease in voltage across voltmeter/$R_2$ combination. We therefore expect the voltmeter to read less. (d) The resistance of the voltmeter/$R_2$ combination is $\frac{R_2 R_{VM}}{(R_2 + R_{VM})} = 3.70\ \Omega$. Thus the overall circuit resistance would be 5.70 Ω resulting in an increased current of 12.0 V/5.70 Ω = 2.11 A. The voltage drop across $R_1$ would then be (2.11 A)(2.00 Ω) = 4.22 V and the voltage drop across the voltmeter/$R_2$ combination (i.e., voltmeter reading) would be 12.0–4.22 = 7.78 V.

**18.6** (a) In parallel all the voltages are the same so you can place the voltmeter leads on opposite sides of the battery or either of the two resistors. (b) We now know the battery voltage is 12.0 V. The battery is the only element in the circuit carrying the total current so the ammeter must be placed either just before or just after the battery. (c) Since the 6.00 Ω resistor has 12.0 V across it, it must have a 2.00 A current. Since the total current is 4.00 A the unknown resistor must also have 2.00 A (i.e., the current splits equally) and thus also has a resistance of 6.00 Ω.

## Chapter 19

**19.1** East, since reversing both the velocity direction and the sign of the charge leaves the direction the same.

**19.2** (a) Using the force right-hand rule, the proton would initially deflect in the $-x$ direction. (b) 0.21 T.

**19.3** 0.500 V

**19.4** (a) At the poles, the magnetic field is perpendicular to the ground. Since the current is parallel to the ground, according to the force right-hand rule the force on the wire would be in a plane parallel to the ground. Thus it would not be able to cancel the downward force of gravity. (b) The wire's mass is 0.041 g, which is unrealistically low for a wire 1 meter long.

**19.5** (a) At 45°, the torque is 0.27 m·N, or about 71% of the maximum torque (b) 30°

**19.6** (a) South (b) 37.5 A

**19.7** 1500 turns

**19.8** (a) The force becomes repulsive. You should be able to show this using the right-hand source and force rules. (b) The force is proportional to the product of the currents, which will increase by a factor of 9. To offset this, the distance between the wires must increase by a factor of 9 to 27 mm.

**19.9** The permeability would be 40% of that in the Example, $\mu \geq 480\ \mu_o = 6.0 \times 10^{-4}$ T·m/A.

## Chapter 20

**20.1** (a) Clockwise (b) 0.335 mA

**20.2** Any way that will increase the flux, such as increasing the loop area or the number of loops. Changing to a lower resistance would also help.

**20.3** $7.36 \times 10^{-4}$ T

**20.4** 1.5 m/s

**20.5** 0.28 m

**20.6** (a) $6.1 \times 10^3$ J (b) $5.0 \times 10^2$ J, so about 12 times more energy is used during startup

**20.7** (a) She would use a step-up transformer because European appliances are designed to work at 240 V, which is twice the U.S. voltage of 120 V. (b) The output current would be 1500 W/240 V or 6.25 A. Thus the input current would be 12.5 A. (Voltage would be stepped up by a factor of 2, and thus the input current is twice as large as the output current.)

**20.8** (a) Higher voltages allow for lower current usage. This in turn reduces joule heat losses in the delivery wires and in the motor windings, making more energy available for doing mechanical work and therefore a higher efficiency. (b) Since the voltage is doubled, the current is halved. The heat loss in the wire is proportional to the *square* of the current. Thus, losses will be cut by a factor of 4 or be reduced to 25% of their value at 120 V.

**20.9** 0.38 cm/s

## Chapter 21

**21.1** (a) 0.25 A (b) 0.35 A (c) $9.6 \times 10^2\ \Omega$, larger than the 240 Ω required for a bulb of the same power in the United States. The voltage in Great Britain is larger than that in the United States. To keep the power constant, the current must be reduced by using a larger resistance.

**21.2** If the resistance of the appliance is constant, the power will quadruple, since $P \propto (\Delta V)^2$. Even if the resistance increased, the power would probably be much more than the appliance was designed for and it would likely burn out, or at least blow a fuse.

**21.3** (a) $\sqrt{2}(120\,\text{V}) = 170\,\text{V}$ (b) 120 Hz

**21.4** (a) $\sqrt{2}(2.55\,\text{A}) = 3.61\,\text{A}$ (b) 180 Hz

**21.5** (a) The current would increase to 0.896 A. (b) With a frequency increase, the capacitive reactance $X_c$ decreases. Resistance is independent of frequency, so it remains constant. Thus the total circuit impedance decreases because of the capacitor.

**21.6** (a) In an RLC circuit, the phase angle $\phi$ depends on the difference $X_L$-$X_C$. If you increase the frequency, $X_L$ would increase and $X_c$ would decrease, thus their difference would increase and so would $\phi$. (b) $\phi = 84.0°$, an increase as expected.

**21.7** 6.98 W

**21.8** (a) If you have a receiver tuned to a frequency between the two station frequencies, you would not receive the maximum strength signal from either station, but there might be enough power from each to hear them simultaneously. (b) 651 kHz.

## Chapter 22

**22.1** Unless interrupted, light rays travel in straight lines and their travel path is reversible. If you can see someone in a mirror, that person can see you. Conversely, if you can't see the trucker's mirror, then they can't see your image in their mirror.

**22.2** $n = 1.25$ and $\lambda_m = 400$ nm

**22.3** By Snell's law, $n_2 = 1.24$ so $v = c/n_2 = 2.42 \times 10^8$ m/s.

**22.4** With a larger index of refraction, $\theta_2$ is smaller so the refracted light inside the glass is toward the lower left. Therefore, the lateral displacement is larger. Answer: 0.72 cm.

**22.5** (a) The light frequency is the same in the different media, so the emerging light has the same frequency as the source. (b) In general, the index of refraction of glass is greater than water and air has the lowest value. Therefore, $\lambda_{air} > \lambda_{water} > \lambda_{glass}$.

**22.6** Because of total internal reflections, the diver could not see anything above water. Instead, he would see the reflection of objects on the sides and/or bottom of the pool.

**22.7** $n = 1.4574$. Green light will be refracted more than red as green has a shorter wavelength, and thus a larger index of refraction than red. By Snell's law, green will have a smaller angle of refraction, in other words, it is refracted more.

## Chapter 23

**23.1** No effect. Note that the solution to the Example does not depend on the distance from the mirror. The geometry of the situation is the same regardless of the distance.

**23.2** $d_i \approx 60$ cm; real, inverted, magnified.

**23.3** $d_i = d_o$ and $M = -1$; real, inverted, same size.

**23.4** The virtual image is also always upright and reduced.

**23.5** $d_i = -20$ cm (in front of the lens); virtual, upright, and magnified.

**23.6** $d_o = 2f = 24$ cm

**23.7** Blocking half of the lens would result in half the amount of light focused at the image, so the resulting image would be less bright but still full size.

**23.8** The virtual image is also always upright and reduced.

**23.9** 3 cm behind $L_2$; real, inverted, reduced ($M_{total} = -0.75$).

**23.10** (a) If the lens is immersed in water, Equation 23.8 should be modified to $\frac{1}{f} = \left(\left[\frac{n}{n_m}\right] - 1\right)\left(\frac{1}{R_1} + \frac{1}{R_2}\right)$, where $n_m = 1.33$ (water). Since $n = 1.52 > n_m = 1.33$, the lens is still converging. (b) $P = 0.238$ D; $f = 4.20$ m.

## Chapter 24

**24.1** $\Delta y = y_r - y_b = 1.2 \times 10^{-2}$ m $= 1.2$ cm

**24.2** Twice as thick, $t = 199$ nm.

**24.3** In brass instruments, the sound comes from a relatively large, flared opening. There is little diffraction, thus most of the energy travels in the forward direction. In woodwind instruments, much of the sound comes from tone holes along the column of the instrument. These holes are small compared to the wavelength of the sound, so there is appreciable diffraction. As a result, the sound is radiated in nearly all directions, even backward.

**24.4** The width would increase by a factor of 700/550 = 1.27.

**24.5** $\Delta\theta_2 = \theta_2(700\text{ nm}) - \theta_2(400\text{ nm}) = 47.8° - 25.0° = 22.8°$

**24.6** 45°

**24.7** $\theta_2 = 41.2°$

**24.8** 589 nm; yellow

## Chapter 25

**25.1** It wouldn't work; a real image would form on person's side of lens ($d_i = +0.76$ m).

**25.2** For an object at $d_o = 25$ cm, the image for the left eye would be formed at 1.0 m; this is beyond the near point for that eye, so the object could be seen clearly. The image for the right eye would be formed at 0.75 m; this is inside the near point for that eye, so the object would not be seen clearly.

**25.3** Lens for near-point viewing, 2.0 cm longer.

**25.4** Length doubles to 40 cm.

**25.5** $f_i = 8.0$ cm

**25.6** The erecting lens (focal length $f_i$) should be placed between the objective and the eyepiece at a distance of

$2f_i$ from the image formed by the objective, which acts as an object. The erecting lens produces an inverted image of the same size at $2f_i$ on the opposite side of the lens, which acts as an object for the eyepiece. The use of the erecting lens lengthens the telescope by $4f_i = 16$ cm. The total length is slightly less than 25 cm + 16 cm or 41 cm.

**25.7** 3.4 cm

## Chapter 26

**26.1** Light waves from two simultaneous events on the $y$-axis meet at some midpoint receptor on the $y$-axis. Since there is no relative motion along that axis, a simultaneous recording of the two events will also be recorded along the $y'$-axis. Hence the two observers agree on simultaneity for this situation.

**26.2** $v = 0.9995c$; no, not twice as fast. Travel is limited to less than $c$, so this is only about a 0.15% increase.

**26.3** (a) 0.667 $\mu$s (b) 0.580 $\mu$s. The observer watching the ship measures the proper time interval. To the person on the ship, that time interval is lengthened.

**26.4** $v = 0.991c$

**26.5** (a) 1.17 MeV (b) 0.207 MeV

**26.6** (a) 0.319 $mc^2$ (b) 3.33 $mc^2$

**26.7** $1.95 \times 10^7$ more massive

## Chapter 27

**27.1** (a) 3000 K (b) 10 000 K

**27.2** The new wavelength is 496 nm and it is shorter than the 550 nm in the Example meaning a higher photon energy. Thus the maximum kinetic energy of the photoelectrons is higher, requiring a higher stopping voltage.

**27.3** 2.50 V

**27.4** (a) $\Delta\lambda/\lambda_o = 324$, meaning a wavelength increase of $3.24 \times 10^4$ %. (b) Percentagewise, this is much larger than in the Example because the wavelength of the incoming light ($\lambda_o$) is much smaller for gamma rays.

**27.5** (a) $7.29 \times 10^5$ m/s (b) 1.51 eV $= 2.42 \times 10^{-19}$ J

**27.6** 365 nm (UV)

**27.7** The least energetic photon in the Lyman series results in a transition from $n = 2$ (first excited state) to the ground state $n = 1$. This results in a photon of energy 10.2 eV. The wavelength of this light is 122 nm, which is UV.

**27.8** (a) There are six possible transitions from the $n = 4$ state to the ground state, and thus the emitted light has six different possible wavelengths. (b) If the atom is excited from the ground state to the first excited state ($n = 1$ to $n = 2$), then it has no choice but to emit a single photon during the de-excitation state ($n = 2$ to $n = 1$) because there are no intermediate states.

## Chapter 28

**28.1** (a) $8.8 \times 10^{-33}$ m/s (b) $2.3 \times 10^{33}$ s, or $7.2 \times 10^{25}$ y. This is about $4.8 \times 10^{15}$ times longer than the age of the universe. This movement would definitely not be noticeable.

**28.2** The proton's de Broglie wavelength would be $4.1 \times 10^{-12}$ m, or about twenty times smaller than atomic spacing distances. With a wavelength much smaller than the atomic spacing, these protons would not be expected to exhibit significant diffraction effects.

**28.3** Only five electrons could be accommodated in the $3d$ subshell if there were no spin (with spin there can be up to ten).

**28.4** (a) $1s^2 2s^2 2p^6$ (b) $-2e$ or $-3.2 \times 10^{-19}$ C

**28.5** $1.2 \times 10^6$ m/s

## Chapter 29

**29.1** ${}^{12}_{6}C + {}^{4}_{2}He \rightarrow {}^{16}_{8}O$; thus the resulting nucleus is oxygen-16.

**29.2** (a) Since ${}^{23}_{11}Na$ is the stable isotope with 11 protons and 12 neutrons, ${}^{22}_{11}Na$ is 1 neutron shy of being stable. In other words, it is proton-rich or neutron-poor. Thus, the expected decay mode is $\beta^+$ (positron decay). (b) Neglecting the emitted neutrino, the decay is ${}^{22}_{11}Na \rightarrow {}^{22}_{10}Ne + {}^{0}_{+1}e$. The daughter nucleus is neon-22.

**29.3** (a) 48 d because reducing activity by a factor of 64 requires six half-lives ($1/2^6 = 1/64$). (b) The process of excretion from the body can also remove $^{131}I$.

**29.4** The closest integer is 20, since $2^{20} \approx 1.05 \times 10^6$. So it takes about 20 half-lives, or 560 y.

**29.5** The measurement can be made as far back as four $^{14}C$ half-lives, or $2.3 \times 10^4$ y (23 000 y).

**29.6** (a) $^{40}K$ is an odd–odd nucleus (19 protons, 21 neutrons) and thus is unstable. $^{41}K$, an odd–even potassium isotope (19 protons, 22 neutrons), is the likely candidate for the remainder of the stable potassium. $^{43}K$ has too many neutrons (24) compared to protons (19) for this region of the periodic chart. (b) Using $N = N_o e^{-\lambda t}$, it follows that $N_o/N = e^{\lambda t} = 12.8$. Thus there was about 13 times more $^{40}K$ than now at the Earth's formation.

**29.7** (a) Starting with 29 protons and 29 neutrons, we have the following candidates: ${}^{58}_{29}Cu, {}^{59}_{23}Cu, {}^{60}_{29}Cu, {}^{61}_{29}Cu, {}^{62}_{29}Cu, {}^{63}_{29}Cu, {}^{64}_{29}Cu$, etc. Now delete the odd–odd isotopes to get the most likely (stable) isotopes: ${}^{59}_{29}Cu, {}^{61}_{29}Cu, {}^{63}_{29}Cu, {}^{65}_{29}Cu$, etc. (b) Further trimming of the list can be done by deleting those with $N \approx Z$ (why?) and those with $N$ significantly larger than $Z$ (why?). Since $Z$ should be just a bit smaller than $N$ in this mass region, we expect neutron numbers in the mid-30s. Hence good guesses would be ${}^{63}_{29}Cu$ and ${}^{65}_{29}Cu$. These are, in fact, the only two stable isotopes of copper.

**29.8** (a) The result for $^3He$ is 2.573 MeV/nucleon, which is considerably smaller than the 7.075 MeV/nucleon value for $^4He$. (b) $^4He$ is the more tightly bound of the two.

Unlike $^3$He, all protons and neutrons in $^4$He are paired, resulting in a more tightly bound nucleus.

**29.9** The absorbed dose is 0.0215 Gy or 2.15 rad. Since the RBE for gamma rays is 1, the effective dose is 0.0215 Sv or 2.15 rem or about one-seventh the dose from the beta radiation.

## Chapter 30

**30.1** $Q = -15.63$ MeV, so it is endoergic (takes energy to happen).

**30.2** The increase in mass has an energy equivalent of 1.193 MeV. The rest of the incident kinetic energy (1.534 MeV − 1.193 MeV, or 0.341 MeV) must be distributed between the kinetic energies of the proton and the oxygen-17.

**30.3** There are about $1.00 \times 10^{38}$ proton-proton cycles per second.

**30.4** Because the beta particle has little energy and therefore little momentum, the neutrino and the daughter nucleus would have to recoil in (almost) opposite directions to conserve linear momentum. This assumes the original nucleus had zero linear momentum.

**30.5** Since the negative pion has a charge of $-e$, its quark structure must be $\bar{u}d$. The structure of the positive pion is $u\bar{d}$. (See Figure 30.9.) Thus the composition of the antiparticle of the $\pi^+$ must be $\bar{u}\bar{\bar{d}}$. However, the antiquark of an antiquark is the original quark ($\bar{\bar{u}} = u$). So the structure of the antiparticle of $\pi^+$ is $\bar{u}\bar{\bar{d}} = \bar{u}d$, which is the $\pi^-$ quark structure, as expected.

# Appendix VI Answers to Odd-Numbered Questions and Exercises

## Chapter 15 Multiple Choice

1. (c)
3. (c)
5. (d)
7. (a)
9. (c)
11. (c)
13. (a)
15. (d)
17. (b)

## Chapter 15 Conceptual Questions

1. There would be no effect since it is simply a sign convention.
3. If an object is positively charged, its mass decreases because it loses electrons. If an object is negatively charged, its mass increases because it gains electrons.
5. No. The charges simply change location. There is no gain or loss of electrons.
7. The water would bend the same way as with the balloon. In this case, the positive charges in the rod would attract the negative ends of the water molecules, causing the stream to bend toward the balloon.
9. A positive charge can be placed midway between the two electrons so that each will experience a repulsive force from the other and an attractive force from the positive charge. With the proper amount of positive charge, the two forces on each electron will cancel.
11. (a) The object must be positively charged because the downward repulsion of the nearby positive charge of the dipole is greater than the upward attraction of the more distant negative charge of the dipole. (b) The dipole would accelerate downward because the downward attraction on the negative end is greater than the upward repulsion on its positive end.
13. It is determined by the relative density or spacing of the field lines. The closer the lines are together, the greater the field magnitude.
15. If a positive charge is at center of the spherical shell, the electric field is *not* zero inside. The field lines run radially outward to the inside surface of the shell where they stop at the induced negative charges on this surface. The field lines reappear on the outside shell surface (which is positively charged) and continue radially outward as if emanating from the point charge at the center. If the charge were negative, the field lines would point in the reverse directions.
17. (a) Yes, this is possible when the electric fields are equal in magnitude and opposite in direction at some location. For example, at the midway point in between and along a line joining two charges of the same type and magnitude, the electric field is zero. (b) No, this is not possible. Between the charges, the fields are in the same direction, so they could not cancel. At other points, the fields could not cancel because they would either have different magnitudes (for points along the line connecting the two charges) or would not have opposite directions (for all other points).
19. At very large distances, the object looks very small – like a point. So the electric field pattern will closely resemble that of a point charge.
21. The surface must be spherical.
23. Note that electric field is zero inside the metallic slab. Since charges are mobile, the negative charges are attracted toward the upper portion of the slab, while the positive charges move toward the lower portion. The amount of charge induced on each side of the slab is the same in magnitude as that on each of the plates.

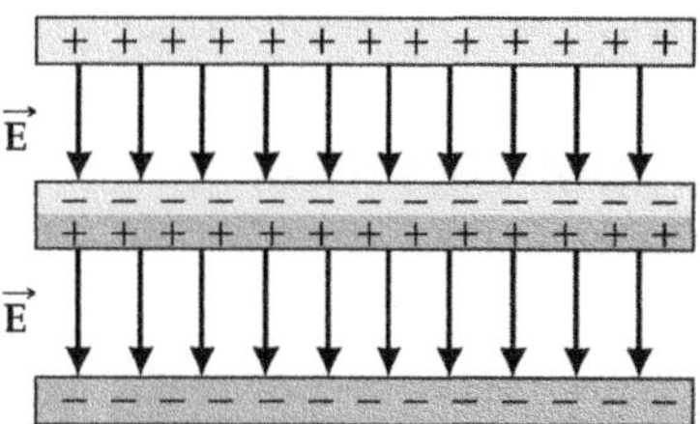


## Chapter 15 Exercises

1. $-1.6 \times 10^{-13}$ C
3. (a) $6.4 \times 10^{-19}$ C (b) two electrons
5. (a) (1) positive (b) $+4.8 \times 10^{-9}$ C; $2.7 \times 10^{-20}$ kg (c) $2.7 \times 10^{-20}$ kg
7. $5.15 \times 10^{-11}$ C = 51.5 pC
9. (a) the same (b) ¼ (c) ½
11. (a) $5.8 \times 10^{-11}$ N (b) zero
13. 2.24 m
15. (a) $x = 0.25$ m (b) no location (c) $x = -0.294$ m for $\pm q_3$
17. (a) At 5.0 cm: $2.7 \times 10^{-18}$ N; at 10.0 cm: $2.7 \times 10^{-19}$ N; at 15.0 cm: $5.8 \times 10^{-19}$ N; at 20.0 cm: $8.1 \times 10^{-19}$ N; at 25.0 cm: $2.7 \times 10^{-18}$ N (b) The least force is midway between the two charges. See graph

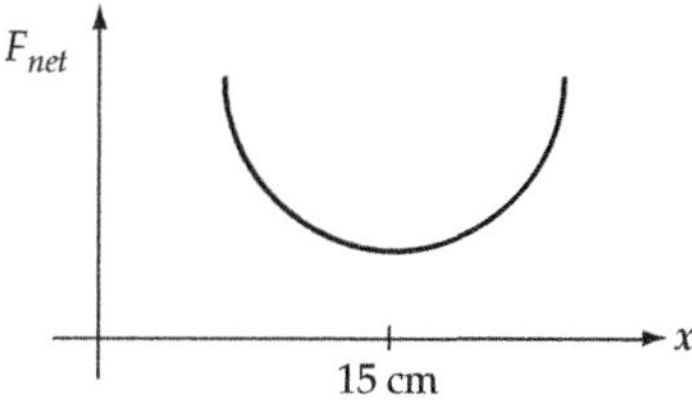


19. (a) 96 N, 39° below the positive $x$-axis (b) 61 N, 84° above the negative $x$-axis
21. (a) (2) decreased (b) $2.5 \times 10^{-5}$ N/C
23. $2.9 \times 10^{5}$ N/C
25. $1.2 \times 10^{-7}$ m away from the charge
27. Proton: $1.0 \times 10^{-5}$ N/C upward; electron: $5.6 \times 10^{-11}$ N/C downward
29. $(2.2\times10^{5}\ \text{N/C})\hat{\mathbf{x}} + (-4.1\times10^{5}\ \text{N/C})\hat{\mathbf{y}}$
31. $5.4 \times 10^{6}$ N/C toward the $-4.0\ \mu$C charge
33. $3.8 \times 10^{7}$ N/C in the $+y$ direction

35. (a) $1.5 \times 10^{-5}$ C/m$^2$ = 15 $\mu$C/m$^2$ (b) $3.4 \times 10^{-7}$ C = 0.34 $\mu$C
37. $(-4.4\times10^6\ \text{N/C})\hat{\mathbf{x}}+(7.3\times10^7\ \text{N/C})\hat{\mathbf{y}}$
39. (a) (1) Negative (b) zero (c) $+Q$ (d) $-Q$ (e) $+Q$
41. (a) Zero (b) $kQ/r^2$ (c) zero (d) $kQ/r^2$
43.

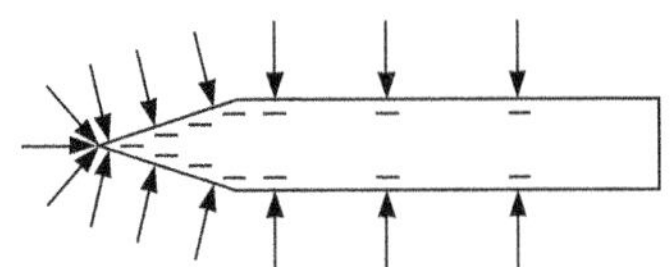

45. (a) Up (b) The bottom surface of the top slab is negative and the lower slab has the same charge distribution as the top slab. (c) 500 N/C up (d) $Q = 1.1 \times 10^{-11}$ C
47. (a) To the right (b) $8.55 \times 10^3$ N/C
49. (a) Positive on right plate, negative on left (b) right to left (c) $1.13 \times 10^{-13}$ C

## Chapter 16 Multiple Choice

1. (d)
3. (b)
5. (b)
7. (a)
9. (b)
11. (b)
13. (c)
15. (a)
17. (b)
19. (d)
21. (c)
23. (b)
25. (b)

## Chapter 16 Conceptual Questions

1. (a) Electrical potential is the electrostatic potential energy *per unit charge*. (b) No difference.
3. Approaching a negative charge means moving toward a region of lower electric potential. Positive charges tend to move toward regions of lower potential, thus losing potential energy and gaining kinetic energy (speeding up).
5. It does not accelerate. It is in a region where potential is constant, so $E = 0$ and it feels no force.
7. The ball would accelerate in the direction from the beach to the ocean (from higher potential energy to lower potential

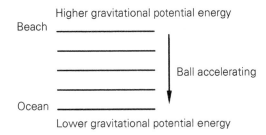

Higher gravitational potential energy
Beach
Ball accelerating
Ocean
Lower gravitational potential energy

energy). If it were initially rolled up the beach, then it would slow down as it goes to higher gravitational potential.
9. It takes zero work because there is no change in kinetic or potential energy.
11. The surfaces are spherical in shape due to symmetry. The electric field is radial, so the surfaces must be spheres.
13. (a) $K = q\Delta V = (1.60 \times 10^{-19}\ \text{C})\ (10^6\ \text{V}) = 1.60 \times 10^{-13}$ J
(b) The kinetic energy is directly proportional to the particle's charge, so it the charge were doubled, the kinetic energy would double.
15. The energy stored in a capacitor is $U_C = Q^2/2C$ and the capacitance of a parallel plate capacitor is $C = \varepsilon_0 A/d$. If the plates are moved apart, $d$ increases, so $C$ decreases. Since $Q$ remains constant for an isolated capacitor, decreasing $C$ will increase the energy $U_C$.
17. (a) $Q' = Q/3$ (b) $U_c' = U_c/3$ (c) $E' = E/3$
19. (a) increased by a factor of $\kappa$ (b) reduced by a factor of $\kappa$ (c) reduced by a factor of $\kappa$
21. In series they would have to have the same capacitance. In parallel the voltage across them is the same regardless.
23. (a) Parallel (b) Series

## Chapter 16 Exercises

1. 1.0 cm
3. (a) 2.7 $\mu$C (b) negative to positive
5. (a) $5.9 \times 10^5$ m/s down (b) loses potential energy
7. (a) (2) 3 (b) 0.90 m (c) −6.7 kV
9. (a) The electron gains $6.2 \times 10^{-19}$ J of potential energy.
(b) The electron loses $6.2 \times 10^{-19}$ J of potential energy.
(c) The electron gains $4.8 \times 10^{-19}$ J of potential energy.
11. 1.1 J
13. (a) +0.27 J (b) no, since electric force is conservative
15. −0.72 J
17. (a) $3.1 \times 10^5$ V (b) $2.1 \times 10^5$ V
19. (a) (3) a lower potential (b) $4.2 \times 10^7$ m/s (c) $6.0 \times 10^{-9}$ s
21. 70 cm
23. 8.3 mm
25. (a) (1) the smaller sphere (b) +297 eV
27. (a) $\Delta V = 3.5$ V, $v = 1.1 \times 10^6$ m/s (b) $\Delta V = 4.1$ kV, $v = 3.8 \times 10^7$ m/s (c) $\Delta V = 5.0$ kV, $v = 4.2 \times 10^7$ m/s
29. $2.4 \times 10^{-5}$ C
31. 0.418 mm
33. (a) 4.54 nC (b) $2.72 \times 10^{-8}$ J (c) 2290 V/m
35. 2.2 V
37. (a) $2.2 \times 10^4$ V/m (b) $1.1 \times 10^{-5}$ C (c) $5.7 \times 10^{-4}$ J (d) $\Delta Q$ = zero; $-1.7 \times 10^{-3}$ J
39. (a) $3.1 \times 10^{-9}$ C (b) $3.7 \times 10^{-8}$ J
41. (a) 2.4 (b) (2) decreased (c) −63 $\mu$J
43. (a) 0.24 $\mu$F (b) 1.0 $\mu$F
45. (a) (1) more energy (b) In series, 15 $\mu$J. In parallel, 76 $\mu$J.
47. (a) (3) $Q/3$ (b) 3.0 $\mu$C (c) 9.0 $\mu$C
49. For $C_1$: 2.4 $\mu$C and 7.2 $\mu$J. $C_2$ is the same as $C_1$, since it has the same capacitance as $C_1$ and they are in parallel. For $C_3$: 1.2 $\mu$C and 3.6 $\mu$J. For $C_4$: 3.6 $\mu$C and 11 $\mu$J.

## Chapter 17 Multiple Choice

1. (b)
3. (c)
5. (c)
7. (a)
9. (a)
11. (c)
13. (a)
15. (a)
17. (b)
19. (d)
21. (d)
23. (b)

## Chapter 17 Conceptual Questions

1. Although electrode A is negative, it is *less negative* than electrode B and is therefore at a higher potential than B. We often say that electrode A is at a positive potential relative to B.
3. No. When current is flowing through the battery, the terminal voltage will be *less than* 12 V due to the voltage drop across the internal resistance of the battery.
5. (a) The electron flow in the resistor is upward. (b) The current in the resistor is downward. (c) The current in the battery is upward.
7. Electrons flow from A to B inside the battery, but in the wire the flow is from B to A, thereby completing a closed loop and ensuring continuous current.
9. Write the relationship between voltage and current as: $\Delta V = (R)I$. If resistance is constant, then the plot will be that of a straight line of the form $y = mx$ where $m$ is the slope. Thus the slope is the resistance. Therefore the smaller resistance has a less steep (smaller) slope.
11. (a) The resistance is $R = \rho L/A$, so the ratio of the two resistances is $\frac{R_2}{R_1} = \frac{L_2}{L_1}\frac{A_1}{A_2} = (2)\left(\frac{1}{2}\right) = 1$. Since $I = \Delta V/R$, the current is the same. (b) Since $A = \pi d^2/4$, half the diameter means (1/4) the area, so $\frac{A_2}{A_1} = \frac{1}{4}$. Therefore $R_2/R_1 = (1)(4) = 4$. Thus, $I_2/I_1 = R_1/R_2 = (¼)$, so we have one-quarter of the current.
13. Since $R = \rho\frac{L}{A} = \rho\frac{L}{\pi(D/2)^2}$, if $L \rightarrow L/2$, then the denominator must also be reduced by one-half to keep the resistance the same. Therefore $(D/2)^2 \rightarrow 1/2\ (D/2)^2$, which means that $D \rightarrow D/\sqrt{2}$.
15. Since $P = (\Delta V)^2/R$ the bulb of higher power has a smaller resistance, which means it has a *thicker* wire if the length is the same. So the wire in the 60-W bulb would be thicker.
17. The current is also determined by the resistance, so $P = I^2R = \left(\frac{\Delta V}{R}\right)^2 R = \frac{(\Delta V)^2}{R}$. Thus a high-wattage bulb has *less* resistance than a low-wattage bulb.

## Chapter 17 Exercises

1. (a) 4.5 V (b) 1.5 V
3. (a) 24 V (b) The two 6.0-V in series, together in parallel with the 12-V.
5. (a) (2) The same (b) A: Series gives 3.0 V for each group. Parallel gives the same 3.0 V for the total. B: Parallel gives the same 1.5 V for each group. Series gives 3.0 V for the total.
7. 0.25 A
9. (a) 0.30 C (b) 0.90 J
11. 56 s
13. (a) (2) To the left (b) 1.8 A to the left (c) 1.5 A to the left (d) 3.3 A to the left
15. (a) 11.4 V (b) 0.32 Ω
17. 1.0 V
19. (a) (1) A greater diameter (b) 1.29
21. (a) 4 (b) 4
23. (a) 0.13 Ω (b) 0.038 Ω
25. (a) 4.6 mΩ (b) 8.5 mA
27. (a) 5.4 cm (b) 0.11 mΩ
29. (a) (1) greater than (b) 1.6
31. 144 Ω
33. $2.0 \times 10^3$ W
35. 1.2 Ω
37. (a) (4) 1/4 times (b) 5.63 W
39. (a) 4.3 kW (b) 13 Ω (c) It is not ohmic
41. (a) 58 Ω (b) 86 Ω
43. (a) 0.600 kWh (b) 9 cents (c) 417 days
45. (a) 0.15 A (b) $1.4 \times 10^{-4}$ Ω·m (c) 2.3 W
47. (a) 110 J (b) 6.8 J
49. (a) 22 Ω (b) 5.6 A
51. 4.3
53. (a) Central air: $130; blender: $0 (to nearest dollar); dishwasher: $1; microwave oven: $1; refrigerator: $6; stove (oven and burners): $13; color TV: $1. (b) Central air: 86%; blender: 0%; dishwasher: 0.7%; microwave oven: 0.7%; refrigerator: 4%; stove: 8.6%; color TV: 0.7%. (c) Central air: $I = 41.7$ A, $R = 2.88$ Ω; blender: $I = 6.7$ A, $R = 18$ Ω; dishwasher: $I = 10.0$ A, $R = 12$ Ω; microwave oven: $I = 5.2$ A, $R = 33$ Ω; refrigerator: $I = 4.2$ A, $R = 28$ Ω; stove (oven): $I = 37.5$ A, $R = 3.2$ Ω; stove (top burners): $I = 50.0$ A, $R = 2.4$ Ω; color TV: $I = 0.83$ A, $R = 150$ Ω.

## Chapter 18 Multiple Choice

1. (b)
3. (b)
5. (a)
7. (c)
9. (a)
11. (d)
13. (b)
15. (c)
17. (b)
19. (b)
21. (a) and (b)
23. (b) and (d)

## Chapter 18 Conceptual Questions

1. No, not generally. However, if all resistors are equal, the voltages across them are the same.
3. No, not generally. However, if all resistors are equal, the currents in them are the same.
5. If they are in series, the effective resistance will be closer in value to that of the large resistance because $R_s = R_1 + R_2$. If $R_1 \gg R_2$, then $R_s \approx R_1$. If they are in parallel, the effective resistance will be closer in value to that of the small resistance because $R_p = R_1R_2/(R_1 + R_2)$. If $R_1 \gg R_2$, then $R_p \approx R_1R_2/R_1 = R_2$.
7. (a) The third resistor has the largest current, because the total current through the two other resistors is equal to the current through the third resistor. (b) The third resistor also has the largest voltage, because the current through it is the largest and all the resistors have the same resistance value (since $\Delta V = IR$). (c) The third resistor also has the largest power output because it has the largest current and largest voltage ($P = I\Delta V$).
9. Not necessarily. If two batteries of unequal emfs are connected with opposite polarity in series with a resistor, the larger battery will force current to enter the positive terminal of the smaller battery.
11. The 60-W bulb has a higher resistance than the 100-W bulb. When these are in series, they have the same current. Therefore, the 60-W bulb will have a higher voltage. Thus, the 60-W bulb has more power because $P = I\Delta V$.
13. In series, the current is the same in all resistors. Since each resistor's voltage drop is related to its resistance by $\Delta V_i = IR_i$, it is clear that the greater the resistance the larger the voltage drop.
15. During the charging of a the capacitor, its charge as a function of time is $Q(t) = Q_o(1 - e^{-t/\tau})$. After one time constant, $\tau$, has elapsed, the charge is $Q(\tau) = Q_o(1 - e^{-1}) \approx 0.632Q_o$, so the time it takes to charge up to $0.25Q_o$ is less than one time constant. When discharging, the capacitor's charge is $Q(t) = Q_oe^{-t/\tau}$. After one time constant, the charge is $Q(\tau) = Q_oe^{-1} \approx 0.368Q_o$, so the time to discharge to $0.25\ Q_o$ is longer than one time constant.
17. (a) An ammeter has very low resistance, so if it were connected in parallel in a circuit, the circuit current would be very high and its galvanometer could burn out. (b) A voltmeter has very high resistance, so if it were connected in series in a circuit, it would read the voltage of the source because it has the highest resistance (most probably) and therefore the most voltage drop among the circuit elements. The circuit current would drop close to zero.
19. An ammeter is used to measure current when it is in series with a circuit element. If it has very small resistance, there will be very little voltage across it, so it will not affect the voltage across the circuit element, not its current.
21. (a) The three resistors are in parallel and the voltmeter is across any one of them.

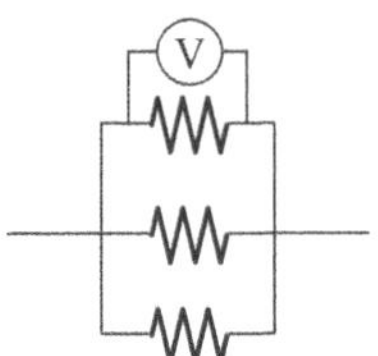


(b) Same as (a).
(c) The three resistors are in series and the voltmeter is connected in parallel with this series combination.

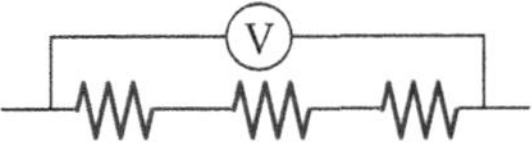


(d) The three resistors are in series and the voltmeter is connected in parallel with just one of them (the one across which you want to measure the potential difference across).

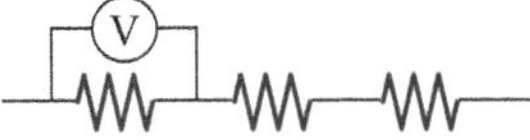


23. No, because high voltage can produce high or low current depending on the resistance of the path.
25. It is safer to jump. If you step off the car one foot at a time, there will be a high voltage between your feet. If you jump, the voltage (i.e., potential *difference*) between your feet is zero because your feet will be at the same potential all the time.

## Chapter 18 Exercises

1. (a) In series for maximum resistance: 60 Ω. (b) In parallel for minimum resistance: 5.5 Ω.
3. 30 Ω
5. (a) 30 Ω (b) 0.30 A (c) 1.4 W
7. (a) (1) $R/4$ (b) 3.0 $\mu\Omega$
9. 1.0 A for all. So $\Delta V_{8.0} = 8.0$ V and $V_{4.0} = 4.0$ V.
11. 2.7 Ω
13. (a) $I_1 = 1.0$ A, $I_2 = I_3 = 0.50$ A (b) $\Delta V_1 = 20$ V, $\Delta V_2 = \Delta V_3 = 10$ V (c) 30 W
15. (a) The currents in the upper and middle arms will not change. The current in the lower arm will be half of what it was. (b) $I_1 = I_2 = 3.0$ A, $I_3 = I_5 = 30$ A, $I_4 = 6.0$ A.
17. (a) 0.085 A (b) $P_{15} = 7.0$ W, $P_{40} = 2.6$ W, $P_{60} = 0.24$ W, $P_{100} = 0.41$ W
19. (a) $I_1 = I_2 = 0.67$ A, $I_3 = 1.0$ A, $I_4 = I_5 = 0.40$ A (b) $\Delta V_1 = 6.7$ V, $\Delta V_2 = 3.3$ V, $\Delta V_3 = 10$ V, $\Delta V_4 = 2.0$ V, $\Delta V_5 = 8.0$ V (c) $P_1 = 4.4$ W, $P_2 = 2.2$ W, $P_3 = 10$ W, $P_4 = 0.80$ W, $P_5 = 3.2$ W (d) $P = 21$ W
21. $I_1 = 1.0$ A down, $I_2 = 0.5$ A right, $I_3 = 0.5$ A down
23. $I_1 = 0.33$ A left, $I_2 = 0.33$ A right
25. $I_1 = 3.75$ A up, $I_2 = 1.25$ A left, $I_3 = 1.25$ A right

27. $I_1 = 0.664$ A left, $I_2 = 0.786$ A right, $I_3 = 1.450$ A up, $I_4 = 0.770$ A down, $I_5 = 0.016$ A down, $I_6 = 0.664$ A right
29. 0.693 time constants
31. (a) 0.86 mJ (b) 17 V (c) 480 kΩ (d) 35 $\mu$A
33. (a) $9.4 \times 10^{-4}$ C (b) After a long time, $\Delta V_C = 24$ V and $\Delta V_R = 0$ V
35. (a) $2.0 \times 10^{-6}$ A (b) 0.86% (c) $1.7 \times 10^{-6}$ C. Charge is a maximum as $t \to \infty$ or $t \gg \tau$ (d) 99% (e) $2.0 \times 10^{-6}$ J
37. (a) 0.60000 A (b) 6.0000 V. (The current in the resistor and voltage across the resistor are not affected by the voltmeter but the total current delivered by the battery is.)
39. (a) $3.3 \times 10^{-4}$ (b) 1200 A

## Chapter 19 Multiple Choice

1. (a)
3. (c)
5. (d)
7. (a)
9. (d)
11. (b)
13. (d)
15. (b)
17. (c)
19. (b)
21. (b)
23. (d)
25. (a)

## Chapter 19 Conceptual Questions

1. The magnet would attract the unmagnetized iron bar when a pole end is placed at the center of its long side. If the end of the unmagnetized bar were placed at the center of the long side of the magnet it would not be attracted.
3. (a) The spacing of the iron filings gets farther apart. Thus field strength decreases with distance from the middle. (b) The magnetic field points up and down (parallel to the line along which the magnets are aligned). But we cannot tell if it points up or down from looking at the filing pattern alone.
5. Not necessarily, because there still could be a field. If the magnetic field and the velocity of the charged particle make an angle of either 0° or 180°, there is no magnetic force on the particle.
7. (a) The fields should be uniform and of equal magnitude but should point in opposite directions. The lower field should point into the paper and the upper field should point out of the paper. (b) The emerging kinetic energy is the same as the initial kinetic energy. Since the magnetic force is perpendicular to the velocity, it changes only the direction of the velocity, not its magnitude, so the kinetic energy does not change.
9. The magnetic force can be used to "separate" particles of various charge states and/or momenta because in the same field, curvature of their trajectory depends on both of these quantities.
11. The electric force is $qE$ and the magnetic force is $qvB$. Since both forces depend linearly on the charge, the selected speed, which we get by equating their magnitudes, is independent of the charge.
13. (a) If the electric field is reduced, the magnetic force will be greater than the electric force. Therefore the positive charges in the velocity selector will be deflected upward. Hence they will not enter the region of $B_2$. (b) If $B_1$ is reduced, the electric force will be greater than the magnetic force. Thus the positive charges will be deflected downward and will not enter the region of $B_2$.
15. The spring should shorten because the coils of the spring attract each other due to the magnetic fields created in the coils. (Parallel wires with current in same direction will attract each other.)
17. (a) $(A{\cdot}m^2)[N/(A{\cdot}m)] = m{\cdot}N$
(b) Curl your right-hand fingers counterclockwise and your thumb points upward, toward you, which is the direction of the magnetic moment.
19. The compass points downward so the magnetic field is downward at the center of the loop. The current direction is clockwise according to the right-hand source rule.
21. Not necessarily. The magnetic field in a solenoid depends on the current in it and the number of turns per unit length (the density of turns), not just the number of turns. For example, if the 200 turns is over 0.20 m and the 100 turns is over 0.10 m, then they will have the same turns per unit length and the same magnetic field.
23. The direction of the current should be counterclockwise to cancel the magnetic field of the outer loop. Its current should be smaller than 10 A, because the field created by a loop is *inversely* related to the radius of that loop. With a smaller radius for the inner loop, its current must be less than 10 A.
25. The purpose of the iron core is to increase the magnetic permeability and magnetic field, because the magnetic field is proportional to the magnetic permeability of the material.
27. Hawaii is slightly north of the Equator so the magnetic field there points northward and mostly parallel to the surface of the Earth but slightly downward. This would be the direction of the remnant magnetism.
29. It will be the north magnetic pole. Right now, the pole near geographical north is a south magnetic pole.

## Chapter 19 Exercises

1. (a) The sketch should show that the second magnet is directly below the first one, oriented vertically, with its north pole directly above its south pole. The two north poles are 2.5 cm apart. (b) Since the magnets are identical and the north poles of #2 and #3 are both 2.5 cm from the north pole of #1, they will exert forces of equal magnitude on the north pole of #1. The force due to #2 is upward and the force due to #3 is to the left, so the net force is 2.1 mN at 45° above the horizontal toward the left.

3. (a) $B_1$ is in the $-x$-direction and $B_2$ is in the $-y$-direction, so the net field at the origin will lie in the third quadrant at an angle of 27° below the $-x$-axis. (b) The field now is in the fourth quadrant at 27° below the $+x$-axis.
5. $3.5 \times 10^3$ m/s
7. $2.0 \times 10^{-14}$ T and to the left looking in the direction of the velocity
9. (a) $3.8 \times 10^{-18}$ N (b) $2.7 \times 10^{-18}$ N (c) zero (d) zero
11. (a) $8.6 \times 10^{12}$ m/s² southward (b) Same magnitude but northward (c) Same magnitude but opposite the direction of the force on the proton (d) 1830
13. (a) $1.8 \times 10^3$ V (b) Same voltage independent of particle charge
15. $5.3 \times 10^{-4}$ T
17. (a) $4.8 \times 10^{-26}$ kg (b) $2.4 \times 10^{-18}$ J (c) No increase since the magnetic force does no work
19. (a) (i) To the right (ii) upward (iii) into the page (iv) to the left (v) into or out of the page (b) In each case except (v), the force is $8.3 \times 10^{-4}$ N. In (v) the force is zero.
21. 1.2 N.
23. (a) Zero (b) 4.0 N/m in the $+z$ direction (c) 4.0 N/m in the $-y$ direction (d) 4.0 N/m in the $-z$ direction (e) 4.0 N/m in the +y direction (f) 2.8 N/m in the $+z$ direction
25. (a) 0.400 N/m in the $+z$ direction (b) same as (a) (c) 0.500 N/m in the $-z$ direction.
27. 7.5 N upward in the plane of the paper.
29. (a) 0.013 m·N (b) Doubling the field doubles the torque (c) Doubling the current doubles the torque (d) Doubling the area doubles the torque (e) It can't be done by increasing the angle
31. One-fourth of its initial value
33. 11 A
35. 0.25 m
37. (a) $2.0 \times 10^{-5}$ T (b) 9.6 cm from wire 1
39. At the right-hand point, $2.9 \times 10^{-6}$ T and into the page. At the left-hand point, the magnitude is the same but the field points out of the page.
41. $1.0 \times 10^{-4}$ T and away from the observer
43. 4.0 A
45. (a) $8.8 \times 10^{-2}$ T (b) to the right
47. (a) (1) attractive (b) 170 A (c) zero
49. (a) $3.74 \times 10^{-3}$ T·m/A (b) $3.0 \times 10^3$
51. (a) 10° north of east (b) $9.6 \times 10^6$ m/s²

## Chapter 20 Multiple Choice

1. (a) and (b)
3. (d)
5. (d)
7. (a)
9. (a)
11. (c)
13. (b)
15. (a)
17. (d)
19. (f)

## Chapter 20 Conceptual Questions

1. (a) When the bar magnet enters the coil, the needle deflects to one side, and when it leaves the coil, the needle reverses direction. (b) No, because of induced currents. According to Lenz's law, it is repelled as it moves toward the loop and attracted as it leaves the loop.
3. Move coil with the same velocity as magnet, so there is no change in flux through the coil.
5. Faraday's law is $\mathscr{E} = -N\dfrac{\Delta\Phi}{\Delta t}$.
   The units of $\mathscr{E}$ are J/C = N·m/C. The units of $N\dfrac{\Delta\Phi}{\Delta t}$ are $\text{T·m}^2/\text{s} = \left[\dfrac{\text{N}}{\text{C·m/s}}\right]\text{m}^2/\text{s} = \dfrac{\text{N·s·m}^2}{\text{C·m·s}} = \text{N·m/C}$.
7. To prevent induced current, the magnetic flux must remain constant. Since the magnitude of the magnetic field has increased, the area must decrease, so the diameter must also decrease.
9. (a) The magnetic flux through the coil is proportional to $\cos \omega t$ while the induced emf is proportional to $\sin \omega t$. The max emf occurs when $\sin \omega t = 1$, which is when $\omega t = \pi/2$. But at this time, $\cos \omega t = \cos \pi/2 = 0$. (b) As in (a), the flux is maximum when $\cos \omega t = 1$, thus $\omega t = 0$. But then $\sin \omega t = \sin 0 = 0$, so the emf is zero. The basic answer to (a) and (b) is that flux and emf are 90° out of phase. When one is a maximum, the other is zero.
11. If the armature is jammed or turns very slowly, there is no back emf and there is a large current.
13. Power is transmitted at low current to reduce joule heating since $P = I^2R$.
15. (a) The induced current is clockwise since the induced field points away from you. (b) The induced current is clockwise since the induced field points away from you. (c) The induced current is counterclockwise since the induced field points toward you. (d) The induced current is zero since the magnetic flux does not change.
17. UV radiation causes sunburn and much of UV can penetrate the clouds. You feel cool because infrared radiation, which is partially responsible for the sensation of "feeling hot," is absorbed by water molecules in the clouds.
19. (a) The car is like a moving observer. Due to its oncoming motion, it strikes the waves at a higher rate than if it were at rest. Therefore the reflected waves have a higher frequency than the outgoing waves. (b) The frequency is higher (see above). The wavelength is shorter since $\lambda = c/f$.

## Chapter 20 Exercises

1. 32 cm
3. 42° or 138°
5. (a) $1.3 \times 10^{-6}$ Wb (b) 3.0 A
7. (a) 1.6 V (b) 0.40 V
9. (a) 0.30 s (b) 0.60 s
11. (a) (1) At the equator (b) 0 mV at the equator and 0.20 mV at the North Pole
13. (a) 0.60 V (b) zero (c) 4.0 A

15. (a) 2.6 V (b) Zero (c) The induced emf is a maximum when the flux is changing at its maximum rate, which is when the magnetic field is zero. This occurs twice per cycle. The minimum emf is zero when the magnetic field is at its maximum value, which occurs twice per cycle.
17. (a) 0.057 V (b) 10 loops
19. (a) 100 V (b) zero (c) (100 V) $\sin(120\pi t)$ (d) 1/120 s (e) 200 V
21. (a) 16 Hz (b) 24 V
23. (a) (3) lower than 44 A (b) 4.0 A
25. (a) 216 V (b) 160 A (c) 8.1 Ω
27. (a) 16 turns (b) 500 A
29. (a) 24 (b) primary: 83 mA; secondary: 2.0 A
31. (a) 17.5 A (b) 15.7 V
33. (a) (2) Nonideal (the power in the secondary is lower than that in the primary) (b) 45%
35. (a) 128 kWh (b) $1840
37. (a) 53 W (b) 200
39. 326 m and 234 m, respectively
41. 2.6 s
43. AM: 67 m; FM: 0.77 m

## Chapter 21 Multiple Choice

1. (c)
3. (a)
5. (b)
7. (b)
9. (b)
11. (d)
13. (d)

## Chapter 21 Conceptual Questions

1. The average current is zero due to directional change, so it can be either positive or negative. However, power is delivered regardless of current direction. Therefore power does not average out to zero.
3. The current is cut in half because the voltage is halved. Power is proportional to the *square* of the current, the power drops to 25% of the designed power.
5. The time averaged power is inversely proportional to the resistance, so the power is cut in half and the rms voltage does not change since it depends only on the ac power source. The rms current is cut in half because it is inversely proportional to the resistance.
7. For a capacitor, the lower the frequency, the longer the charging time in each cycle. If the frequency is very low (dc), then the charging time is very long, so it acts as an ac open circuit. For an inductor, the lower the frequency, the more slowly the current changes in the inductor. The slower the current changes, the less back emf is induced in the inductor, resulting in less impedance to current.
9. At $t = 0$, $I = 120$ A, or at maximum. The voltage is then zero, because current leads voltage by 90° in a capacitor. When current is maximum, voltage is one quarter of a period behind, or at zero. They are out of phase.
11. Capacitive reactance is proportional to the product of the frequency and the capacitance. To keep it constant, the frequency should be halved if the capacitance is doubled.
13. (a) The series resonance frequency is $f_o = \dfrac{1}{2\pi\sqrt{LC}}$. If $C$ is quadrupled, $\sqrt{C}$ increases by 2×, so $f_o$ is halved. (b) If $L$ is increased by 9 times, $\sqrt{L}$ increases by 3×, so $f_o$ is reduced to 1/3 of its initial value. (c) Changing the resistance has no effect on the resonance frequency. (d) Doubling both $L$ and $C$ increases the denominator by $\sqrt{2}\sqrt{2} = 2$, and reduces $f_o$ to 1/2 its original value.
15. We know the circuit is not at resonance because the inductive reactance and capacitance are not equal. Since the inductive reactance (250 Ω) is greater than the capacitive reactance (150 Ω), we know that the driving frequency must be greater than the resonance frequency.

## Chapter 21 Exercises

1. For a 120-V line: 170 V; for a 240-V line: 339 V
3. 1.2 A
5. (a) 10.0 A (b) 14.1 A (c) 12.0 Ω
7. (a) $I_{rms} = 4.47$ A; $I_{max} = 6.32$ A (b) $\Delta V_{rms} = 112$ V; $\Delta V_{max} = 158$ V
9. (170 V) $\sin(119\pi t)$
11. $I_{rms} = 0.333$ A; $I_{max} = 0.471$ A; $R = 360$ Ω
13. (a) 7.5 Ω (b) 20 Hz; $T = 0.050$ s (c) 240 W
15. (a) 60 Hz (b) 1.4 A (c) 120 W (d) $\Delta V = (120\text{ V}) \sin(380t)$ (e) $P = I\,\Delta V = (240\text{ W}) \sin^2(380t)$ (f) $P = 120$ W, the same as in (c)
17. 1.3 kΩ
19. 2.3 A
21. (a) −38% (b) 60%
23. (a) 250 Hz (b) 990 Hz
25. (a) 90 V (b) Voltage leads by 90°
27. (a) 4.42 μF (b) 0.10 A
29. (a) 170 Ω (b) 200 Ω
31. (a) $X_L = 38$ Ω; $Z = 110$ Ω (b) 1.1 A
33. (a) (3) negative (b) −27°
35. (a) (3) in resonance (b) 72 Ω
37. (a) 50 W (b) 115 W
39. (a) 53 pF (b) 31 pF
41. (a) 1.2 kW; $Z = 9.0$ Ω (b) 13 A
43. $(\Delta V_{rms})_R = 12$ V; $(\Delta V_{rms})_L = 270$ V; $(\Delta V_{rms})_C = 270$ V
45. (a) (2) equal to 25 Ω (b) 362 Ω
47. (a) 55% (b) 30%

## Chapter 22 Multiple Choice

1. (b)
3. (d)
5. (b)
7. Both (a) and (c) must be satisfied.
9. (c)

## Chapter 22 Conceptual Questions

1. The angle of reflection is always equal to the angle of incidence.
3. Specular reflection. After rain, water fills the crevices and turns the road into a smooth surface. Diffuse reflection becomes specular.
5. Laser beam. The fish does not appear at its true location due to refraction. The laser beam refracts and retraces the light the hunter sees from the fish. The arrow travels a near-straight line and passes above the fish.
7. The severed look occurs because the refraction angle is different for the air-glass interface than for the water-glass interface. The top portion refracts from air to glass, and the bottom portion refracts from water to glass. This is different from Figure 22.15b. There we see the top portion in air directly and the bottom portion in water through refraction from water to air. The angle of refraction made the pencil appear to be bent.
9. No, total internal reflection could not occur because medium 2 is more optically dense than medium 1. We know this because the light is bent toward the normal in medium 2.
11. In a prism, there are two refractions and two dispersions, therefore doubling the effect on dispersion.
13. No, because the light will be further dispersed by the second prism.
15. A glass pane is a few mm thick, so the separation of colors is too small for our detection.

## Chapter 22 Exercises

1. Both the incident and reflected rays make an angle of 30° with the normal, so the angle between the rays is 60°.
3. (a) (2) $90°-\alpha$ (b) 57°
5. (a) (3) $\tan^{-1}(w/d)$ (b) 27°
7. When the mirror rotates through an angle of $\theta$, the normal will rotate through an angle of $\theta$ and the angle of incidence is $35° + \theta$. The angle of reflection is also $35° + \theta$. Since the original angle of reflection is 35°, the reflected ray will rotate through an angle of $2\theta$. If the mirror rotates in the opposite direction, the angle of reflection will be $35° - \theta$. However, the normal will also rotate through an angle of $\theta$ but in the opposite direction. The reflected ray still rotates through $2\theta$.

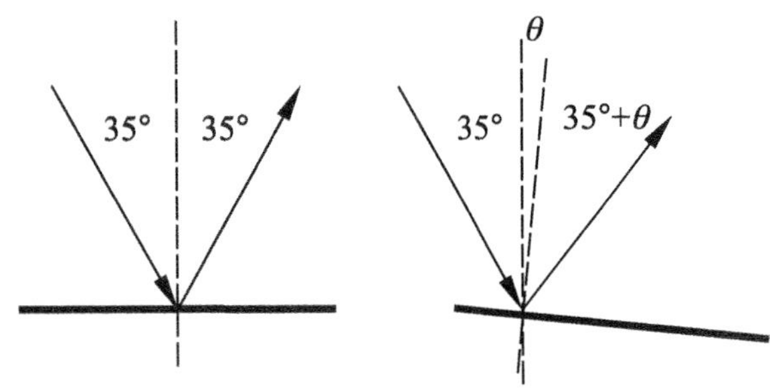


9. If $\alpha = 90°$, $\theta_{i_1} + \theta_{r_2} = 90°$, and the normals are perpendicular. Reflected and incident rays are parallel for any $\theta_1$.
11. 1.41
13. 1.34
15. (a) (1) Greater than (b) 17°
17. (a) (2) From water to air (b) 48.8°
19. 47°
21. $\lambda_m = 140$ nm; $f = f_m = 1.55 \times 10^{15}$ Hz
23. (a) (3) Less than (b) 15/16
25. (a) This is caused by refraction at the water-air interface. The angle of refraction in air is greater than the angle of incidence in water, so the object immersed in water appears closer to the surface. (b) The distance $a$ is common to $d$ and $d'$, so $\tan\theta_1 = \dfrac{a}{d'}$ and $\tan\theta_2 = \dfrac{a}{d}$. Combining yields
$$d' = \frac{\tan\theta_2}{\tan\theta_1}d. \quad \text{If } \theta < 15°, \tan\theta \approx \sin\theta.$$
$$\text{So } \frac{d'}{d} = \frac{\tan\theta_2}{\tan\theta_1} \approx \frac{\sin\theta_2}{\sin\theta_1} = \frac{1}{n}(\text{Snell}). \text{ Hence } d' \approx \frac{d}{n}$$
27. 66.7%
29. (a) (3) Less than 45° (b) 20°
31. See for 40° but not 50°
33. (a) (3) the indices of refraction of both (b) air: 1.41; water: 1.88.
35. 43°
37. (a) 25° (b) $1.97 \times 10^8$ m/s (c) 362 nm
39. $n_R = 1.362$; $n_B = 1.371$
41. 1.4980.

## Chapter 23 Multiple Choice

1. (c)
3. (b)
5. (a)
7. (a)
9. (d)
11. (c)
13. (b)
15. (c)
17. (b)

## Chapter 23 Conceptual Questions

1. No, virtual images cannot be seen on a screen because no rays intersect at the image.
3. During the day, the reflection is mainly from the silvered back surface. During the night, when the switch is flipped, the reflection comes from the front side. There is a reduction of intensity and glare because the front side reflects only about 5% of the light, which is more than enough to see due to the dark background.
5. When viewed in a rearview mirror, right-left image reversal by a plane mirror reads "FIRE."
7. No, convex mirrors produce only reduced images. See Example 23.8 and Figure 23.21.

9. The image is smaller than the object, and it is possible to "see your full body in 10 cm" in a diverging mirror.
11. The object distance should be between $f$ and $2f$, where $f$ is the focal length.
13. For a magnified upright image, the object should be inside the focal point of the lens. We could not do this with a diverging lens because the image it produces is always reduced in size from the object.
15. +, +; +, ∞; +, −; −, −; ∞, −; +, −
17. No. The more general lens maker's equation is $\frac{1}{f}=[(n/n_m)-1]\left(\frac{1}{R_1}+\frac{1}{R_2}\right)$, where $n_m$ is the index of refraction of the surrounding material. If $n_m > n$, $f$ is negative, meaning the lens is diverging.
19. Spherical aberration is caused by a spherical lens surface. Rays that pass through the outer parts of the lens are not focused to the same place as rays that pass through the center of the lens. This causes a fuzzy image.

## Chapter 23 Exercises

1. 5.0 m
3. (a) 80 cm (b) 50 cm (c) +1.0
5. (a) The dog's image is 3.0 m behind the mirror. (b) The dog moves at a relative speed of 2.0 m/s.
7. (a) (3) You see multiple images caused by reflections from two mirrors. (b) The first image by the north mirror is 3.0 m behind the north mirror. The second image by the south mirror is 11 m behind the south mirror. The first image by the south mirror is 5.0 m behind the south mirror. This image is 13 m from the north mirror. So the second image by the north mirror is 13 m behind the north mirror.
9. The two triangles (with $d_o$ and $d_i$ as base, respectively) are similar to each other, because all three angles of one triangle are the same as those of the other triangle due to the law of reflection. Furthermore, the two triangles share the same height, the common vertical side. Therefore the two triangles are identical. Hence $d_o = d_i$.
11. (a) (1) Real, (2) inverted, and (3) reduced. (b) 66.7 cm (in front of the mirror), the lateral magnification is −0.667.
13. $f = -33.3$ cm and $M = 0.400$
15. $d_i = -30$ cm and the image is 9.0 cm high, virtual, upright and magnified.
17. For $d_o = 2f$, $1/d_i = 1/f - 1/(2f) = 1/(2f)$, so $d_i = 2f$. $M = -d_i/d_o = -(2f)/(2f) = -1$. The image is real, inverted, and the same size as the object.
19. $d_i = 37.5$ cm, the image is 3.0 cm high, real and inverted.
21. (a) (1) Convex (b) −10 cm
23. −25 cm (behind the mirror)
25. (a) (2) Concave (b) 24 cm
27. (a) (1) Virtual and upright (b) 1.5 m
29. 2.3 cm
31. (a)

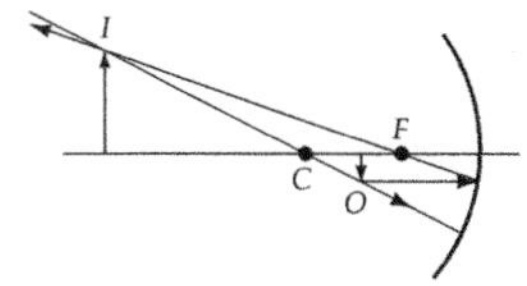


(b) $d_i = 60$ cm; $M = -3.0$, real and inverted
33. (a) (2) Concave (b) 36 cm
35. (a) $d_i = \frac{d_o f}{d_o - f} = \frac{f}{1 - f/d_o}$

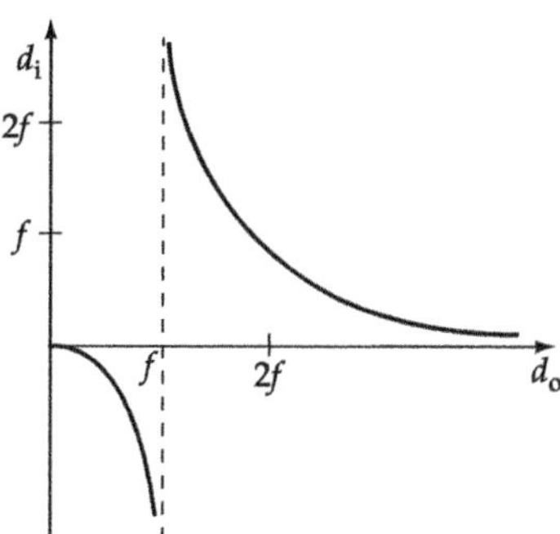


$$|M| = \frac{d_i}{d_o} = \frac{f}{d_o - f}$$

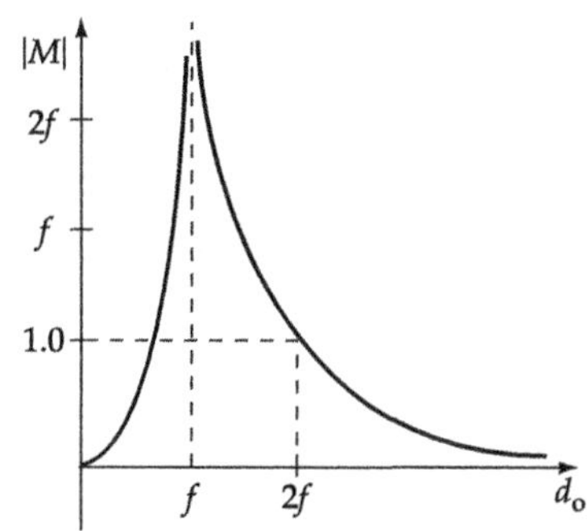


(b) $d_i = \frac{d_o(-f)}{d_o + f} = \frac{-f}{1 + f/d_o}$; $|M| = \frac{d_i}{d_o} = \frac{-f}{d_o + f}$

37. $d_i = 12.5$ cm; $M = -0.250$
39. $d_o = 22$ cm; $M = -9.0$
41. (a) (1) Virtual, upright and magnified (b) $d_i = -47$ cm; $M = +3.1$
43. (a) $d_i = -6.4$ cm; $M = +0.64$ virtual and upright (b) $d_i = -10.5$ cm; $M = +0.42$ virtual and upright
45. (a) $d_o = 18$ cm (b) $d_o = 6.0$ cm
47. Since the mirror and lens equations are the same and the definition of the lateral magnification are also the same mathematically, the graphs are exactly the same as those in Exercise 23.35.
49. 0.5 mm and inverted
51. (a) 40 cm from the object (b) −1.0
53. (a) From similar triangles, $\frac{d_i - f}{f} = -\frac{y_i}{y_o}$, where the negative is introduced, because the image is inverted. Also $\frac{y_i}{y_o} = \frac{d_i}{d_o}$. So $\frac{d_i - f}{f} = \frac{d_i}{d_o}$, and rearranging gives $\frac{1}{d_o} + \frac{1}{d_i} = \frac{1}{f}$. (b) $M = \frac{y_i}{y_o} = -\frac{d_i}{d_o}$ from the similar triangles in (a).
55. −37 cm
57. (a) (1) Convex to produce a magnified image (b) 6.25 cm

59. $M_1 = -\frac{h_{i1}}{h_{o1}}, M_2 = -\frac{h_{i2}}{h_{o2}}$, and $M = -\frac{h_{i2}}{h_{o1}}$. Since $h_{o2} = h_{i1}$ (the image formed by the first lens is the object for the second lens), we have $M_1 M_2 = \frac{h_{i1}}{h_{o1}}\frac{h_{i2}}{h_{o2}} = \frac{h_{i2}}{h_{o1}} = M_{total}$.
61. −25 cm
63. (a) (2) +, − (b) 27.2 cm
65. −40 cm and concave

## Chapter 24 Multiple Choice

1. (b)
3. (a)
5. (a)
7. (b)
9. (a)
11. (d)
13. (b)
15. (c)

## Chapter 24 Conceptual Questions

1. The interference maxima occur at angles given by $\sin\theta = n\lambda/d$. As $d$ decreases, $\sin\theta$ and $\theta$ both increase, which causes the pattern to spread out.
3. No, this is not a violation of the conservation of energy. Energy is redistributed (moved from the minima to the maxima). Total energy is still conserved.
5. It is always dark because of destructive interference due to the 180° phase shift. If there had not been the 180° phase shift, zero thickness would have corresponded to constructive interference.
7. If slit distance is comparable to width, a second diffraction pattern perpendicular to the first will also be observed.
9. Since $d \sin\theta = n\lambda$, the advantage is a wider diffraction pattern, as $d$ is smaller.
11. (a) Twice (b) four times (c) none (d) six times.
13. The display appears and disappears as the sunglasses rotate because the light from the numbers on a calculator is polarized.
15. There is no air on the Moon, so an astronaut would see a black sky; no sunlight can scatter.

## Chapter 24 Exercises

1. 0.75 m = 0.50 m + 0.25 m = 1.5(0.50 m) = 1.5$\lambda$. So the waves will interfere destructively.
   1.0 m = 0.50 m + 0.50 m = 2(0.50 m) = 2$\lambda$. So the waves will interfere constructively.
3. 0.37°
5. 489 nm
7. (a) (2) also decrease (b) 600 nm (c) 0.41 cm < 0.45 cm
9. (a) 402 nm (b) 3.45 cm
11. (a) (1) increase (b) 0.63 cm (c) 0.94 cm > 0.63 cm
13. $4.2 \times 10^{-5}$ m
15. 450 nm
17. (a) Yes (b) 113 nm (c) 196 nm
19. (a) 158.2 nm (b) 316.4 nm
21. $2.0 \times 10^{-4}$ m
23. (a) 5.4 cm (b) 2.7 cm
25. (a) 4.3 mm (b) $f = 6.9 \times 10^{10}$ Hz, microwave
27. (a) (1) Increase (b) 3.6 mm (c) 6.2 mm
29. $7.1 \times 10^{-10}$ m
31. $\theta_1 = \pm 22.31°$ $\theta_2 = \pm 49.41°$. For $n \geq 3$, $\sin\theta > 1$, so there are no more maxima. Thus the total number of side maxima is 4: two on each side.
33. $\theta_o = 0°$; $\theta_1 = \pm 23°$; $\theta_2 = \pm 52°$
35. (a) (1) Closer to (b) Red: 34.1°, Blue: 18.7°
37. 39.2°
39. (a) (1) Also increase (b) 58° and 61°
41. 31.7°
43. 40.5°
45. (a) (2) Less than (b) air: 58.9°; water: 51.3°
47. No, because the maximum angle of incidence at water-glass interface is less than polarizing (Brewster) angle (48.75° < 48.81°).
49. (a) (1) Longer than (b) 822 nm (IR)

## Chapter 25 Multiple Choice

1. (c)
3. (a)
5. (d)
7. (b)
9. (b)
11. (c)
13. (d)

## Chapter 25 Conceptual Questions

1. The iris, crystalline lens, and retina of the eye correspond to the aperture, lens, and film/sensor, respectively, of a camera.
3. Yes. The image is smaller for nearsightedness and larger for farsightedness.
5. Weaker prescription. To correct nearsightedness, a diverging lens is needed to form an image at the far point of an object at infinity. Thus $f = d_i$, where $d_i$ is the distance to the far point. When replacing ordinary glasses with contacts, the distance from the lens to the image is now a few centimeters longer, which makes the focal length slightly longer. Since $P = 1/f$, an increase in $f$ means a decrease in $P$, thus the contacts are weaker than the regular lenses.
7. A short focal length lens has a small radius. The aberration (angle approximation is no longer valid if the object is large compared with the size of the lens) will get more important as the focal length of the lens gets smaller. This limits the magnification to about 3× to 4×.
9. A telescope is supposed to magnify objects, so its angular magnification should be greater than 1. Since the magnification is $m = -f_o/f_e$, the eyepiece should have a

shorter focal length than the objective lens. Thus, use the lens with the shorter focal length as the eyepiece and the other lens as the objective.

11. Chromatic aberration occurs because different wavelengths of light (i.e., different colors) are refracted to different focal points. This behavior does not occur for reflection. Spherical aberration is eliminated in a reflecting telescope by grinding a parabolic, rather than a spherical, surface for the mirror.
13. A smaller minimum angle of resolution corresponds to a higher resolution because a smaller angle of resolution means more details can be resolved.
15. From a resolution point of view, the smaller lens has lower resolution. The smaller the lens, the greater the minimum angle of resolution, and the lower the resolving power.

## Chapter 25 Exercises

1. (a) +5.0 D (b) −2.0 D
3. (a) (3) diverging (b) −1.1 D
5. −200 cm and thus a diverging lens
7. (a) 2.8 D (b) leave the glasses on
9. (a) (2) farsightedness (b) (1) converging lens (c) +3.3 D
11. (a) −0.505 D (b) −0.500 D
13. 83 cm
15. Top half of the lens: 85.0 cm from his eyes; Bottom half of the lens: 50.0 cm from his eyes.
17. 3.1×
19. 2.5×
21. (a) (2) A short focal length (b) 1.9× (with the 28-cm lens); 1.6× (with the 40-cm lens)
23. 6.0 D
25. 77 D
27. (a) −340× (b) 3900%
29. (a) (2) The one with the shorter focal length (b) 0.45 cm as objective: −280×; 0.35-cm as objective: −360×
31. 25×
33. (a) −4.0× (b) 75 cm
35. $f_o = 1.0$ m and $f_e = 2.0$ cm
37. (a) 13 cm (b) 7.0×
39. (a) Maximum magnification: 60.0 cm and 0.80 cm; minimum magnification: 40.0 cm and 0.90 cm (b) $m_1 = -7.5\times$; $m_2 = -44\times$
41. 650 nm
43. $\theta_{min} = 1.32 \times 10^{-7}$ rad; Hale is 1.6 times larger.
45. (a) (3) Blue (b) $\theta_{min} = 0.0040°$ (for 400-nm light); $\theta_{min} = 0.0070°$ (for 700-nm light).
47. 17 km
49. (a) $\theta_{min} = 0.00126°$ (b) $1.8 \times 10^{-5}$ mm .

## Chapter 26 Multiple Choice

1. (d)
3. (b)
5. (b)
7. (a)
9. (a)
11. (c)
13. (a)
15. (b)
17. (b)
19. (a)
21. (a)
23. (d)

## Chapter 26 Conceptual Questions

1. No, she cannot. Newton's laws apply only in *inertial* reference frames. The carousel is a noninertial reference frame because it is accelerated due to its spin.
3. They are exactly the same because an elevator moving at constant velocity is an inertial reference frame. Thus in both frames, the acceleration is zero.
5. Yes, there is such a reference frame. It would have to move along the $x$-axis in the direction of A toward B.
7. In $O$, the bullet takes 1 s to hit the target. Light takes $10^{-6}$ s to get to the target (the time for light at $300 \times 10^8$ m/s to travel 300 m). $O'$ would have to travel to the right. The light flashes from the gun reach the target in $10^{-6}$ s The observer in $O'$ would have to cover the 300 m in less than $10^{-6}$ s to intercept the signals at the same time, which means $v > c$. Since $v > c$ is not possible, all observers agree that the gun fires before the bullet hits the target.
9. Greater speed means more length contraction. Rocket B's length has contracted by 21 cm while A's has contracted by only 11 cm, so B is moving faster than A. The faster the speed, the slower clocks run, so B has a slower-running clock than A.
11. (a) You are measuring the proper time, because you and the clock are in the same frame (no relative motion). (b) Your professor measures the proper length of the spacecraft, because your professor and the spacecraft are in the same frame of reference (no relative motion).
13. No, the acceleration cannot be constant. Since $v$ must be less than $c$, it cannot accelerate at a constant rate. If it did, eventually, $v$ would become more than $c$.
15. The rest energy of the proton is 938 MeV. Since its kinetic energy (2 MeV) is much less than this, classical physics is adequate. On the other hand, if its kinetic energy is 2000 MeV, which is large in comparison to its rest energy, then relativistic physics is required.
17. Drop the cup with the pole vertical. By the principle of equivalence, the weight of the ball appears to be zero in the downward accelerating reference frame. In that frame, the ball is subject only to the tension force of the stretched rubber band and is pulled inside the cup.
19. Light is bent by the gravity of the black hole. At a certain distance, this light would be bent enough to go into orbit around the black hole. Therefore light from the back of your head could go into orbit and come around to your eyes.

## Chapter 26 Exercises

1. (a) 0.10 s less time (b) 0.10 s more time
3. (a) 45 m/s (b) 4.0 s
5. (a) B could have caused A (b) B could not have caused A.
7. 23 min
9. (a) 14.1 m (b) $c/3$ relative to you
11. $0.998c$
13. (a) 4.8 y. (b) The traveler's time is the proper time, so 2.1 y. (c) The distance as measured on Earth is the proper length of 4.3 ly. (d) Since the distance measured from Earth is the proper length, the distance the traveler measures is 1.9 ly.
15. (a) 5.3 m (b) 27 ns (c) 36 ns because they do not agree on the travel distance
17. (a) $0.952c = 2.86 \times 10^8$ m/s (b) 16.4 min
19. (a) $2.14 \times 10^{-14}$ m (b) $6.00 \times 10^5$ m/s
21. (a) $0.985c$ (b) 2.50 MeV (c) $1.56 \times 10^{-21}$ kg·m/s
23. 0.45 kg
25. $1.6 \times 10^{24}$ J $= 4.6 \times 10^{17}$ kWh
27. (a) 79 keV (b) 79 keV (c) $1.6 \times 10^{-22}$ kg·m/s
29. (a) $2.9 \times 10^8$ m/s (b) $3.8 \times 10^{-10}$ J (c) $1.7 \times 10^{-18}$ kg·m/s
31. (a) (1) More (b) $3.7 \times 10^{-12}$ kg. No, this is not detectable on the everyday scale
33. (a) $E = m\gamma c^2$ and $p = m\gamma v$.
    Thus $p^2c^2 = (m\gamma v)^2c^2 = m^2\gamma^2 v^2c^2$ and $(mc^2)^2 = m^2c^4$.
    Adding these two and simplifying gives
    $p^2c^2 + (mc^2)^2 = m^2\gamma^2v^2c^2 + m^2c^4 = m^2c^2(\gamma^2v^2 + c^2)$.
    Last, using $v = c\sqrt{1-\left(\frac{1}{\gamma^2}\right)}$ gives:
    $p^2c^2 + (mc^2)^2 = m^2c^2[\gamma^2c^2(1 - (1/\gamma^2)) + c^2] = (m\gamma c^2)^2 = E^2$.
    (b) $p = 1700$ MeV/$c = 9.0 \times 10^{-19}$ kg·m/s.
35. As a black hole, the Sun's density would be $1.8 \times 10^{19}$ kg/m$^3$. The Sun's present density is $1.4 \times 10^3$ kg/m$^3$. Thus as a black hole, our Sun would be about $10^{16}$ times denser than it is now.
37. At two Schwarzschild radii, the escape velocity is $2.1 \times 10^8$ m/s $= 0.71c$. At twice the Sun's present radius, the escape velocity is $4.4 \times 10^5$ m/s $= 0.0015c$.
39. $2R_S$

## Chapter 27 Multiple Choice

1. (b)
3. (c)
5. (b)
7. (a)
9. (d)
11. (d)
13. (d)
15. (d)

## Chapter 27 Conceptual Questions

1. According to Wien's displacement law, the temperature of a black body is inversely proportional to the wavelength at which the maximum amount of energy is radiated. Since a red star radiates most of its energy at longer wavelengths than a blue star. The latter must be hotter.
3. From Wien's displacement law

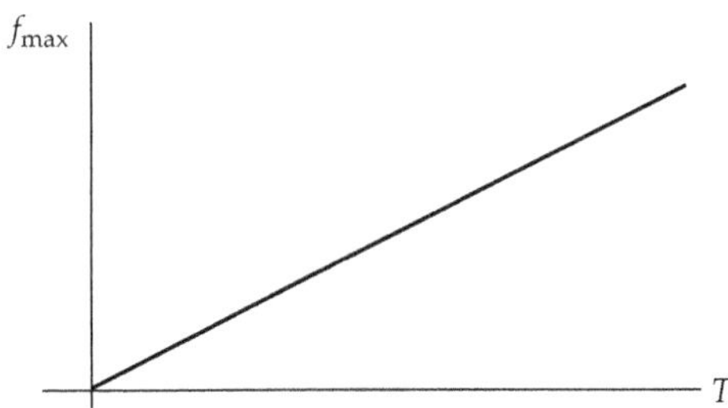


   $\lambda_{max} T = 2.9 \times 10^{-3}$ m · K. Hence
   $f_{max} = (c/(2.9 \times 10^{-3}$ m · K)) $T$, and so
   $f_{max}$ is proportional to $T$.
   The graph is a straight line. Thus if $T$ is tripled, $f_{max}$ becomes 3 times as large.
5. It is not possible. The frequency of IR radiation is less than the frequency of UV radiation. Since the energy of a photon is $E = hf$, the IR photon has less energy than the UV photon.
7. The greater the work function, the more energy it takes to dislodge photoelectrons and thus the less kinetic energy these electrons have. A smaller kinetic energy means that a low stopping potential is needed to stop them. Therefore a larger work function results in a lower stopping potential.
9. For each scattering, the wavelength shift is on the order of $\Delta\lambda = \lambda_C = 0.00243$ nm. The wavelength change from x-ray to visible light is about 550 nm − 0.01 nm = 550 nm. So it would take about 200 000 scatterings, for an x-ray photon to become a visible light photon.
11. Calling the $+x$-axis the original direction of the incident photon and the $+y$-axis the direction in which the scattered photon goes, the electron would have momentum components in the $+x$-direction and the $-y$-direction (due to momentum conservation). Hence it would move off at an angle below the $+x$-axis in the fourth quadrant.
13. It takes less energy to ionize the electron that is in an excited state than in the ground state. The excited state already has more energy.
15. During an optical pumping process, light is used to "pump" electrons from lower energy levels to higher ones. This results in a deviation of population of quantized energy states from its thermal equilibrium distribution.
17. In a *spontaneous* emission, electrons jump from a higher-energy state to a lower-energy state without any external stimulation, and a photon is released in the process. *Stimulated* emission, on the other hand, is an induced process. The electron in the higher-energy orbit can jump to a lower-energy orbit when a photon of energy equal to the difference of the energy between the two orbits is introduced. Once atoms are prepared with enough electrons in the higher-energy state, stimulating photons trigger them to jump down to the lower-energy state. The emitted photons trigger the rest of the electrons and eventually all the electrons will be in the lower-energy state.

## Chapter 27 Exercises

1. 9670 nm
3. $\lambda_{max} = 1.06 \times 10^{-5}$ m; $f = 2.83 \times 10^{13}$ Hz
5. 690°C
7. 4800°C
9. $3.8 \times 10^{19}/\text{m}^2$ per second
11. 306 nm, UV
13. (a) $1.32 \times 10^{-18}$ J. (b) 8.27 eV
15. $\phi_o = 4.0 \times 10^{-19}$ J $= 2.5$ eV
17. 354 nm
19. 254 nm
21. (a) $6.7 \times 10^{-34}$ J·s (b) $2.9 \times 10^{-19}$ J
23. 4.5 eV
25. (a) 0.625 V (b) 2.33 eV (c) zero since there is no emission
27. 0.44 eV
29. 180° or backscattering
31. 54°
33. (a) (2) Less than 5.0 keV but not zero (b) 20 eV
35. (a) 66° (b) $3.19 \times 10^6$ m/s
37. (a) $E_2 = -3.40$ eV (b) $E_3 = -1.51$ eV
39. 310
41. (a) 10.2 eV (b) 1.89 eV (c) the first is UV, the second is visible
43. (a) (1) Increase but more than double (b) $K_1 = 15.4$ eV and $K_2 = 44.4$ eV
45. (a) (1) 5→3 Transition produces the smallest energy and the longest wavelength (b) $\lambda_{5\to3} = 1280$ nm; $\lambda_{6\to2} = 411$ nm; $\lambda_{2\to1} = 122$ nm
47. (a) 2.55 eV (b) $\Delta E = (-13.6\text{ eV})\left(\frac{1}{n_f^2} - \frac{1}{n_i^2}\right) = 2.55$ eV. Using trial and error $n_i = 2$ and $n_f = 4$ is the only transition with an energy difference of 2.55 eV.
49. (a) (1) One (b) 2 to 3 (c) $\Delta E = 1.89$ eV and $\lambda = 656$ nm (red)
51. $v_n = \frac{nh}{2\pi m r_n}$ but $r_n$ (in nm) $= (0.0529)n^2$. Inserting this into the previous expression and converting from nm to m: $$v_n = \frac{n(6.63\times10^{-34}\text{ J}\cdot\text{S})}{2\pi(9.11\times10^{-31}\text{kg})(0.0529\times10^{-9}\text{m})n^2} = \frac{2.2\times10^6}{n}\text{ m/s}$$
53. (a) For the 2.0-eV state, 620 nm; for the 4.0-eV state, 310 nm (b) 2.0 eV is in the visible (red-orange)

## Chapter 28 Multiple Choice

1. (a)
3. (b)
5. (a)
7. (a)
9. (a)
11. (b)
13. (a)
15. (a)

## Chapter 28 Conceptual Questions

1. Its wavelength is too short compared to everyday dimensions, so we won't observe its wave nature.
3. The wavelength will be shorter, as a higher potential difference yields more linear momentum and the de Broglie wavelength is inversely proportional to momentum.
5. If the proton's charge were decreased, it would attract the electron less strongly than now, so the electron would not be held as close to the proton. Therefore the electron would be less likely to be found as close to the proton as it now is. The size of the probability cloud would increase.
7. The atoms in a group all have similar outer shells with similar numbers of valence electrons and similar chemical properties. The atoms in a given period all have the same maximum principal quantum number *n*.
9. According to the uncertainty principle, the product of uncertainty in position and the uncertainty in momentum is in the order of Plank's constant. A bowling ball's large diameter and momentum (mostly due to its large mass) make the uncertainty in them (determined by the extremely small value of Planck's constant) negligible. However, for an electron, with its very small mass, the uncertainty in both its location and momentum cannot be ignored.
11. Linear momentum needs to be conserved so the particles will be moving afterward because the initial photon has momentum. Therefore the electron–positron pair must have some kinetic energy. The input energy is converted into not only their rest energy but also their kinetic energy.
13. Momentum must be conserved in any process. If the annihilating particles are at rest, their total momentum is zero. Therefore the momentum of the resulting photons must also be zero. This momentum could not be zero if there were only one photon; there must be two photons of equal energy traveling in opposite directions. Even if the annihilating particles are not at rest, two photons are still produced because there is always some reference frame in which they have zero initial momentum.

## Chapter 28 Exercises

1. $2.7 \times 10^{-38}$ m
3. (a) (3) a longer (b) $\lambda_{electron} = 7.28 \times 10^{-6}$ m; $\lambda_{proton} = 3.97 \times 10^{-9}$ m
5. $1.5 \times 10^4$ V
7. (a) (3) Decrease (b) a 53% decrease
9. 0.71
11. $8.89 \times 10^{-18}$ J $= 55.6$ eV
13. (a) $3.71 \times 10^{-63}$ m (b) $n = 1.18 \times 10^{72}$ (c) increase but undetectable
15. 1.41
17. (a) $K_1 = 4.07$ MeV; $K_2 = 16.3$ MeV; $K_3 = 36.6$ MeV. (b) $\Delta E = 32.5$ MeV (a gamma ray)

19. (a)

$$\ell = 2, m_l = 0, m_s = \pm\tfrac{1}{2} \rightarrow 2 \text{ states}$$

$$\ell = 2, m_l = +2, m_s = \pm\tfrac{1}{2} \rightarrow 2 \text{ states}$$

$$\ell = 2, m_l = +1, m_s = \pm\tfrac{1}{2} \rightarrow 2 \text{ states}$$

$$\ell = 2, m_l = -2, m_s = \pm\tfrac{1}{2} \rightarrow 2 \text{ states}$$

$$\ell = 2, m_l = -1, m_s = \pm\tfrac{1}{2} \rightarrow 2 \text{ states}$$

There are 10 possible sets for each value of n $\geq 3$.
(b) Since $\ell \leq n - 1$, if $\ell = 3$, $n \geq 4$, $m_l = 0, \pm1, \pm2, \pm3$, and $m_s = \pm\frac{1}{2}$. The possibilities are:

$$\ell = 3, m_l = 0, m_s = \pm\tfrac{1}{2} \rightarrow 2 \text{ states}$$

$$\ell = 3, m_l = +3, m_s = \pm\tfrac{1}{2} \rightarrow 2 \text{ states}$$

$$\ell = 3, m_l = +2, m_s = \pm\tfrac{1}{2} \rightarrow 2 \text{ states}$$

$$\ell = 3, m_l = +1, m_s = \pm\tfrac{1}{2} \rightarrow 2 \text{ states}$$

$$\ell = 3, m_l = -3, m_s = \pm\tfrac{1}{2} \rightarrow 2 \text{ states}$$

$$\ell = 3, m_l = -2, m_s = \pm\tfrac{1}{2} \rightarrow 2 \text{ states}$$

$$\ell = 3, m_l = -1, m_s = \pm\tfrac{1}{2} \rightarrow 2 \text{ states}$$

There are 14 possible sets for each value of $n \geq 4$.

21. (a) $\ell = 2$ (b) $n = 3$
23. (a) Na has 11 electrons

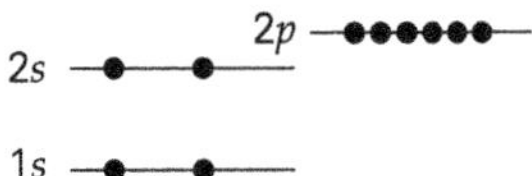


(b) Ar has 18 electrons

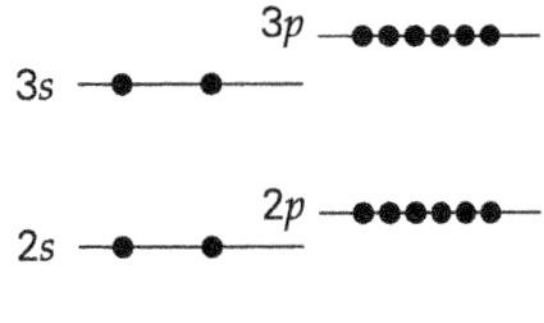


25. (a) B: $1s^2 2s^2 2p^1$
(b) Ca: $1s^2 2s^2 2p^6 3s^2 3p^6 4s^2$
(c) Zn: $1s^2 2s^2 2p^6 3s^2 3p^6 3d^{10} 4s^2$
(d) Sn: $1s^2 2s^2 2p^6 3s^2 3p^6 3d^{10} 4s^2 4p^6 4d^{10} 5s^2 5p^2$
27. $1s^3$
29. (a) (2) The same (b) both 0.21 m
31. $7.0 \times 10^{-7}$ m
33. 938.27 MeV
35. (a) (3) Less than (b) $f_p = 4.541 \times 10^{23}$ Hz; $f_n = 4.547 \times 10^{23}$ Hz

## Chapter 29 Multiple Choice

1. (a)
3. (d)
5. (b)
7. (c)
9. (b)
11. (d)
13. (d)
15. (b)
17. (a)
19. (a)
21. (c)
23. (b)

## Chapter 29 Conceptual Questions

1. The minimum distance of approach is larger than the nuclear radius. If the alpha particles got closer than the nuclear radius, they would feel the strong nuclear force and the scattering would have a different pattern since it would *not* be just due to the Coulomb force.
3. The higher energy alpha particles will get closer because they have more kinetic energy to be converted to electrical potential energy.
5. Carbon-13 has 13 nucleons. The nitrogen isobar of carbon-13 has 13 nucleons, so it is nitrogen-13.
7. Nucleon number (the sum of the proton number and the neutron number) is still conserved in the process $n \rightarrow p + e^- + \bar{\nu}_e$ even though neutron and proton numbers are not. So this is not a violation of nucleon conservation.
9. Due to momentum conservation, the decay particles must go in opposite directions to carry some of kinetic energy.
11. The decay processes are:

$$^{A}_{Z}X_{N} \xrightarrow{\alpha} {}^{A-4}_{Z-2}Y_{N-2} \xrightarrow{\beta-} {}^{A-4}_{Z-1}Z^{*}_{N-3} \xrightarrow{\gamma} {}^{A-4}_{Z-1}Z_{N-3}$$

The final nucleus has mass number A–4, atomic number Z–1, and neutron number N–3. The final product remains the same regardless of whether alpha or beta decay occurs first; however, the intermediate products will vary.
13. They do not affect the decay rate (or half-life); it is independent of temperature, environment, and chemistry.
15. (a) A stable isotope does not decay, so its half-life is infinite. (b) Since the decay constant is inversely proportional to the half-life, the decay constant is zero.
17. After time $t$, the number of undecayed nuclei is $N_o e^{-\lambda t}$, so the number of *decayed* nuclei $N$ is $N = N_o - N_o e^{-\lambda t} = N_o(1 - e^{-\lambda t})$.
19. $^{238}_{92}U$ is even-even so it tends to be more stable than $^{235}_{92}U$ which is even–odd. Even so, both are unstable because $Z > 82$ but $^{238}_{92}U$ has a longer half-life; that is, it is closer to stability.

21. In both the fusion of very light nuclei and the fission of very heavy nuclei, the average binding energy per nucleon increases (see the graph in Figure 29.14). The binding energy is the energy released during each of these processes.
23. The RBE for x-rays is 1 and 20 for alphas from Table 29.4. Equal absorbed doses mean 20 times more effective dose for alphas, thus the effective dose of an alpha particle would be 20 times that of an x-ray.
25. The detectors need to be accurate to measure the gamma ray energy of 0.511 MeV. Also the two photons must arrive in coincidence. The detectors need to identify both photons as originating from the same annihilation event.
27. X-rays and gamma rays are absorbed more or less continuously as they pass through tissue and therefore release their energy gradually over a fairly long range. Particle beams, on the other hand, are charged, and the interaction of these charged particles with the tissue causes them to be stopped suddenly, and hence to release their energy very quickly and thus over a short distance. This energy is therefore all deposited to a small portion of tissue (the tumor).

## Chapter 29 Exercises

1. (a) $^{90}$Zr has 40 protons, 40 electrons and 50 neutrons. (b) $^{208}$Pb has 82 protons, 82 electrons and 126 neutrons.
3. $^{21}_{19}\mathrm{K}$
5. U has 92 protons. U-235: 143 neutrons, 92 electrons; U-238: 145 neutrons, 92 electrons.
7. (a) $R_{He} = 1.9 \times 10^{-15}$ m; $R_{Ne} = 3.3 \times 10^{-15}$ m;
   $R_{Ar} = 4.1 \times 10^{-15}$ m; $R_{Kr} = 5.3 \times 10^{-15}$ m;
   $R_{Xe} = 6.1 \times 10^{-15}$ m; $R_{Rn} = 7.3 \times 10^{-15}$ m.
   (b) $M_{He} = 6.68 \times 10^{-27}$ kg; $M_{Ne} = 3.34 \times 10^{-26}$ kg;
   $M_{Ar} = 6.68 \times 10^{-26}$ kg; $M_{Kr} = 1.40 \times 10^{-25}$ kg;
   $M_{Xe} = 2.20 \times 10^{-25}$ kg; $M_{Rn} = 3.71 \times 10^{-25}$ kg.
   (c) $2.3 \times 10^{17}$ kg/m$^3$ is the same for all nuclei. Yes, because the density is huge.
9. (a) (2) beta minus decay (b) He-3, stable.
11. (a) $^{237}_{93}\mathrm{Np} \rightarrow {}^{233}_{91}\mathrm{Pa} + \alpha$
   (b) $^{32}_{15}\mathrm{P} \rightarrow {}^{32}_{16}\mathrm{S} + \beta^-$
   (c) $^{56}_{27}\mathrm{Co} \rightarrow {}^{56}_{26}\mathrm{Co} + \beta^-$
   (d) $^{56}_{27}\mathrm{Co} + \beta^- \rightarrow {}^{56}_{26}\mathrm{Fe}$
   (e) $^{42}_{19}\mathrm{K}^* \rightarrow {}^{42}_{19}\mathrm{K} + \gamma$
13. $\alpha$–$\beta$: $^{209}_{82}\mathrm{Pb} + \beta^- \rightarrow {}^{209}_{81}\mathrm{Tl}$, $^{209}_{81}\mathrm{Tl} + \alpha \rightarrow {}^{213}_{83}\mathrm{Bi}$
   $\beta$–$\alpha$: $^{209}_{82}\mathrm{Pb} + \alpha \rightarrow {}^{213}_{84}\mathrm{Po}$, $^{213}_{84}\mathrm{Po} + \beta^- \rightarrow {}^{213}_{83}\mathrm{Bi}$
15. (a) The first blank is U. The second blank is Th. The third blank is an alpha particle. (b) The particle is an electron ($\beta^-$). (c) The first blank is 3. The isotope in the second blank is Y. (d) The missing isotope is $^{23}_{11}\mathrm{Na}$. (e) The first blank: $_{-1}^{0}\mathrm{e}$ and second blank: $^{11}_{5}\mathrm{B}$.
17. (a) Alpha: $^{237}_{93}\mathrm{Np} \rightarrow {}^{233}_{91}\mathrm{Pa} + {}^{4}_{2}\mathrm{He}$
   Beta: $^{233}_{91}\mathrm{Pa} \rightarrow {}^{233}_{92}\mathrm{U} + {}_{-1}^{0}\mathrm{e}$
   Alpha: $^{233}_{92}\mathrm{U} \rightarrow {}^{229}_{90}\mathrm{Th} + {}^{4}_{2}\mathrm{He}$
   Alpha: $^{229}_{90}\mathrm{Th} \rightarrow {}^{225}_{88}\mathrm{Ra} + {}^{4}_{2}\mathrm{He}$
   Beta: $^{225}_{88}\mathrm{Ra} \rightarrow {}^{225}_{89}\mathrm{Ac} + {}_{-1}^{0}\mathrm{e}$
   Alpha: $^{225}_{89}\mathrm{Ac} \rightarrow {}^{221}_{87}\mathrm{Fr} + {}^{4}_{2}\mathrm{He}$
   Alpha: $^{221}_{87}\mathrm{Fr} \rightarrow {}^{217}_{85}\mathrm{At} + {}^{4}_{2}\mathrm{He}$
   Alpha: $^{217}_{85}\mathrm{At} \rightarrow {}^{213}_{83}\mathrm{Bi} + {}^{4}_{2}\mathrm{He}$
   There are two possible decay modes for $^{213}$Bi: alpha: $^{213}_{83}\mathrm{Bi} \rightarrow {}^{209}_{81}\mathrm{Tl} + {}^{4}_{2}\mathrm{He}$ and beta: $^{213}_{83}\mathrm{Bi} \rightarrow {}^{213}_{84}\mathrm{Po} + {}_{-1}^{0}\mathrm{e}$.
   These two daughter nuclei decay as follows:
   Beta: $^{209}_{81}\mathrm{Tl} \rightarrow {}^{209}_{82}\mathrm{Pb} + {}_{-1}^{0}\mathrm{e}$ and alpha: $^{213}_{84}\mathrm{Po} \rightarrow {}^{209}_{82}\mathrm{Pb} + {}^{4}_{2}\mathrm{He}$
   Beta: $^{209}_{82}\mathrm{Pb} \rightarrow {}^{209}_{83}\mathrm{Bi} + {}_{-1}^{0}\mathrm{e}$ ($^{209}$Bi is stable).
   (b) The daughter nuclei are shown in each reaction above.
19. (a) $7.4 \times 10^{8}$ decays/s (b) $4.4 \times 10^{10}$ betas/min
21. $2.06 \times 10^{-8}$ W
23. 28.6 y
25. (a) (2) Younger than (b) $1.1 \times 10^{4}$ y
27. 3.1%
29. (a) $6.75 \times 10^{19}$ nuclei
   (b) $3.58 \times 10^{16}$ decays/s $= 3.58 \times 10^{16}$ Bq
   (c) $2.11 \times 10^{18}$ nuclei
   (d) Since $R \propto N$, $R$ will decrease by the same factor
31. (a) 9.1 mg (b) $2.4 \times 10^{18}$ nuclei
33. $7.6 \times 10^{6}$ kg
35. (a) $^{17}_{8}\mathrm{O}$ (b) $^{42}_{20}\mathrm{Ca}$ (c) $^{10}_{5}\mathrm{B}$ (d) approximately the same
37. 2.013 553 u
39. (a) 92.2 MeV (b) 7.68 MeV/nucleon
41. (a) For deuterium the binding energy is 2.224 MeV. For tritium the binding energy is 8.482 MeV. Therefore tritium has the greater total binding energy. (b) For deuterium, the binding energy per nucleon is 1.112 MeV/nucleon and for tritium it is 2.827 MeV/nucleon. Therefore deuterium has the lower binding energy per nucleon.
43. (a) 104.7 MeV (b) 7.476 MeV/nucleon
45. (a) $^{27}_{13}\mathrm{Al} \rightarrow {}^{23}_{11}\mathrm{Na} + {}^{4}_{2}\mathrm{He}$ (b) 10.09 MeV
47. 7.59 MeV/nucleon
49. 36 $\mu$Ci
51. Dose = 3.7 rem in two months. So yes, the maximum permissible radiation dosage is exceeded.
53. 26.6 rem or 0.266 Sv

## Chapter 30 Multiple Choice

1. (c)
3. (a)
5. (d)
7. (a)
9. (c)
11. (a)
13. (a)
15. (b)
17. (a)
19. (d)
21. (c)
23. (a)

## Chapter 30 Conceptual Questions

1. When both particles are moving toward each other, the magnitude of the total momentum is only the difference between the magnitudes of the two momenta of the two particles. According to the conservation of momentum, the total momentum after the collision will be less than if one would be at rest. In turn, the particles have less kinetic energies, so the reaction requires less incident kinetic energy.
3. $^{25}$Mg would have a larger capture cross-section because it has an odd neutron number and thus the likelihood of neutron capture (so as to create a neutron pair in the nucleus) is higher.
5. The hydrogen nuclei in water can capture neutrons to form deuterium and remove neutrons from the chain reaction. Since the proton and neutron have nearly the same mass the collisions result in reduction (moderation) of the neutrons to thermal energies.
7. Most of the uranium in a reactor is $^{238}$U. Although this isotope is not fissionable, it can be converted into $^{239}$Pu by fast neutron bombardment. $^{239}$Pu is fissionable and can be used to construct a bomb.
9. For a sustained viable fusion plant, a large number of fusion reactions and thus a high plasma density are required. To have the nuclei get close enough to fuse, high kinetic energies and hence very high temperatures are also needed. These requirements are "at odds" because very high temperatures tend to make the plasma expand and hence be less dense. Alternatively, if the higher density is achieved, it would be at a temperature that is not high enough for fusion.
11. $p+\bar{\nu}_e \rightarrow n+\beta^+$ and $p+n\rightarrow {}^2_1H+\gamma$. The 2.22 MeV gamma ray is from the formation of deuterium. The two 0.511 MeV gamma rays are the result of the pair annihilation when the positron meets an electron.
13. The $Q$ value is the energy released in the decay. Some of this energy goes to the beta particle, some to the recoil of the nucleus, and some to the neutrino. Therefore the electron does not get all of the released energy.
15. The forces created by the virtual exchange particles can be predicted and these predictions agree with experiment.
17. The range of the strong nuclear force is about 100 times as great as the range of the weak nuclear force. As shown in this chapter, as a result of the uncertainty principle, the range of the force is given approximately by $R \approx h/(2\pi mc)$, where $m$ is the mass of the exchange particle. Solving for $m$ gives $m \approx h/(2\pi Rc)$, so the mass of the short-range weak force exchange particle should be greater than the mass of the relatively long-range strong force exchange particle. Therefore the ratio of the masses of the exchange particles, using Table 30.2, is 100.
19. The stable leptons are the electron and all three neutrinos ($\nu_e$, $\nu_\mu$, $\nu_\tau$). Of the hadrons, only the proton is truly stable (and there is even some question about that). The neutron is *much* more stable than the other hadrons, but outside the nucleus it does decay in about 90 s.
21. All hadrons are thought to be composed of quarks and/or antiquarks. Quarks are not believed to be able to exist freely outside the nucleus, so no fractional charges are observed.
23. The total energy $E$ of a particle is the sum of its rest energy $E_0$ and its kinetic energy $K$. At *very* high energies, the kinetic energy is so much greater than the rest energy that the particle behaves as though it has no rest energy: $E = E_0 + K \approx K$. This is just like a photon, which has no rest energy and is truly massless.

## Chapter 30 Exercises

1. (a) proton number: 16 (S); nucleon number: 37 thus it is ${}^{37}_{16}$S. (b) The blanks are ${}^{135}_{52}$Te and $3\left({}^1_0 n\right)$ (3 neutrons). (c) The species is ${}^{17}_8$O. (d) The species is ${}^1_1$H.
3. (a) ${}^{22}_{11}$Na and $Q < 0$, so no (b) ${}^{222}_{86}$Rn and $Q > 0$, so yes (c) ${}^{12}_6$C and $Q < 0$, so no
5. (a) (1) positive (b) + 4.28 MeV
7. 4.36 MeV
9. Exoergic; $Q = +6.50$ MeV
11. 5.38 MeV
13. (a) (1) greater (b) 3.05
15. (a) The proton and neutron have nearly the same mass, hence during collisions the neutrons will lose the maximum amount of energy by colliding with protons, thus "moderating" or slowing them at the quickest rate. (b) 36 collisions
17. (a) neutron (b) $^4$He (c) Part (a): 3.270 MeV; Part (b): 17.59 MeV
19. (a) (2) less than 5.35 MeV but not zero. (b) 2.70 MeV. (c) If there is no appreciable momentum given to the daughter nucleus, then they will move apart in the opposite direction.
21. 13.37 MeV
23. $Q = (m_p - m_D - m_e)c^2$
 $= \{(m_p + Zm_e) - [m_D + (Z-1)m_e] - 2m_e\}c^2$
 $= (M_p - M_D - 2m_e)c^2$
25. (a) From Exercise 20, $Q = (M_P - M_D)c^2$. For $\beta^-$ decay to occur, $Q$ must be positive. Thus $(M_P - M_D)c^2 > 0$, which can only happen if $M_P > M_D$. (b) By the same reasoning as in (a), $Q > 0$. From Exercise 30.23: $Q = (M_P - M_D - 2m_e)c^2 > 0$, or $M_P - M_D - 2m_e > 0$, so $M_P > M_D + 2m_e$
27. (a) (3) decreases (b) $1.98 \times 10^{-16}$ m
29. $4.71 \times 10^{-24}$ s
31. $0.999\,999\,999\,928c$
33. (a) 0.27 mm (b) 3930 MeV
35. (a) (1) $\bar{u}\bar{d}\bar{d}$. (b) Adding the charges on the three antiquarks gives zero, the charge on the neutron
37. $6.1 \times 10^{-17}$ decays/s or $1.7 \times 10^{-27}$ Ci. This is much much less that 1.0 Ci.
39. 2.0 $\mu$Ci

# Photo Credits

| | |
|---|---|
| **Figure 15.6** | Jerry D. Wilson |
| **Figure 17.13** | Anthony J. Buffa |
| **Figure 22.10** | fir0002 [Creative Commons] |
| **Figure 22.19** | US Navy [Public domain] |
| **Figure 22.21a** | Sai2020 [Public domain] |
| **Figure 23.4a** | USA Library of Congress [Public Domain] |
| **Figure 23.26** | Benjah-bmm27 [Public Domain] |
| **Figure 23.27** | Bo Lou |
| **Figure 23.28b** | Bo Lou |
| **Figure 24.9a** | Jeroen den Otter [Free photo] |
| **Figure 24.11b** | Ulfbastel [Public Domain] |
| **Figure 24.20b** | Jeff Dahl [Creative Commons] |
| **Figure 24.23c** | Bo Lou |
| **Figure 24.27b** | Bo Lou |
| **Figure 24.29** | Bo Lou |
| **Figure 25.14a** | NASA [Public Domain] |
| **Figure 25.14b** | GMTO corporation [Creative Commons] |
| **Figure 25.15** | NASA [Public Domain] |

# Index

*Note*: Page numbers followed by "*fn*" indicate footnotes.

## F



## G



## H



## I

## Physical Data*

| Quantity | Symbol | Approximate Value |
|---|---|---|
| Universal gravitational constant | $G$ | $6.67 \times 10^{-11}\ \text{N}\cdot\text{m}^2/\text{kg}^2$ |
| Acceleration due to gravity (generally accepted value on surface of Earth) | $g$ | $9.80\ \text{m/s}^2 = 980\ \text{cm/s}^2 = 32.2\ \text{ft/s}^2$ |
| Speed of light | $c$ | $3.00 \times 10^8\ \text{m/s} = 3.00 \times 10^{10}\ \text{cm/s} = 1.86 \times 10^5\ \text{mi/s}$ |
| Boltzmann's constant | $k_B$ | $1.38 \times 10^{-23}\ \text{J/K}$ |
| Avogadro's number | $N_A$ | $6.02 \times 10^{23}\ \text{mol}^{-1}$ |
| Gas constant | $R = N_A k_B$ | $8.31\ \text{J/(mol}\cdot\text{K)} = 1.99\ \text{cal/(mol}\cdot\text{K)}$ |
| Coulomb's law constant | $k = 1/4\pi\epsilon_o$ | $9.00 \times 10^9\ \text{N}\cdot\text{m}^2/\text{C}^2$ |
| Electron charge | $e$ | $1.60 \times 10^{-19}\ \text{C}$ |
| Permittivity of free space | $\epsilon_o$ | $8.85 \times 10^{-12}\ \text{C}^2/(\text{N}\cdot\text{m}^2)$ |
| Permeability of free space | $\mu_o$ | $4\pi \times 10^{-7}\ \text{T}\cdot\text{m/A} = 1.26 \times 10^{-6}\ \text{T}\cdot\text{m/A}$ |
| Atomic mass unit | u | $1.66 \times 10^{-27}\ \text{kg} \leftrightarrow 931\ \text{MeV}$ |
| Planck's constant | $h$ | $6.63 \times 10^{-34}\ \text{J}\cdot\text{s}$ |
| | $\hbar = h/2\pi$ | $1.05 \times 10^{-34}\ \text{J}\cdot\text{s}$ |
| Electron mass | $m_e$ | $9.11 \times 10^{-31}\ \text{kg} = 5.49 \times 10^{-4}\ \text{u} \leftrightarrow 0.511\ \text{MeV}$ |
| Proton mass | $m_p$ | $1.672\ 62 \times 10^{-27}\ \text{kg} = 1.007\ 276\ \text{u} \leftrightarrow 938.27\ \text{MeV}$ |
| Neutron mass | $m_n$ | $1.674\ 93 \times 10^{-27}\ \text{kg} \times 1.008\ 665\ \text{u} \leftrightarrow 939.57\ \text{MeV}$ |
| Bohr radius of hydrogen atom | $r_1$ | 0.053 nm |

*Values from NIST Reference on Constants, Units, and Uncertainty.

## Solar System Data*

| | |
|---|---|
| Equatorial radius of the Earth | $6.378 \times 10^3\ \text{km} = 3963\ \text{mi}$ |
| Polar radius of the Earth | $6.357 \times 10^3\ \text{km} = 3950\ \text{mi}$ |
| | Average: $6.4 \times 10^3$ km (for general calculations) |
| Mass of the Earth | $5.98 \times 10^{24}\ \text{kg}$ |
| Diameter of Moon | $3500\ \text{km} \approx 2160\ \text{mi}$ |
| Mass of Moon | $7.4 \times 10^{22}\ \text{kg} \approx \frac{1}{81}$ mass of Earth |
| Average distance of Moon from the Earth | $3.8 \times 10^5\ \text{km} = 2.4 \times 10^5\ \text{mi}$ |
| Diameter of Sun | $1.4 \times 10^6\ \text{km} \approx 864\ 000\ \text{mi}$ |
| Mass of Sun | $2.0 \times 10^{30}\ \text{kg}$ |
| Average distance of the Earth from Sun | $1.5 \times 10^8\ \text{km} = 93 \times 10^6\ \text{mi}$ |

*See Appendix III for additional planetary data.

## Mathematical Symbols

| | |
|---|---|
| $=$ | is equal to |
| $\neq$ | is not equal to |
| $\approx$ | is approximately equal to |
| $\sim$ | about |
| $\propto$ | is proportional to |
| $>$ | is greater than |
| $\geq$ | is greater than or equal to |
| $\gg$ | is much greater than |
| $<$ | is less than |
| $\leq$ | is less than or equal to |
| $\ll$ | is much less than |
| $\pm$ | plus or minus |
| $\mp$ | minus or plus |
| $\bar{x}$ | average value of $x$ |
| $\Delta x$ | change in $x$ |
| $\lvert x \rvert$ | absolute value of $x$ |
| $\Sigma$ | sum of |
| $\infty$ | infinity |

## The Greek Alphabet

| | | | | | |
|---|---|---|---|---|---|
| Alpha | A | $\alpha$ | Nu | N | $\nu$ |
| Beta | B | $\beta$ | Xi | $\Xi$ | $\xi$ |
| Gamma | $\Gamma$ | $\gamma$ | Omicron | O | $o$ |
| Delta | $\Delta$ | $\delta$ | Pi | $\Pi$ | $\pi$ |
| Epsilon | E | $\varepsilon$ | Rho | P | $\rho$ |
| Zeta | Z | $\zeta$ | Sigma | $\Sigma$ | $\sigma$ |
| Eta | H | $\eta$ | Tau | T | $\tau$ |
| Theta | $\Theta$ | $\theta$ | Upsilon | $\Upsilon$ | $\upsilon$ |
| Iota | I | $\iota$ | Phi | $\Phi$ | $\phi$ |
| Kappa | K | $\kappa$ | Chi | X | $\chi$ |
| Lambda | $\Lambda$ | $\lambda$ | Psi | $\Psi$ | $\psi$ |
| Mu | M | $\mu$ | Omega | $\Omega$ | $\omega$ |

Printed and bound by CPI Group (UK) Ltd, Croydon, CR0 4YY
17/10/2024
01775663-0015